MICROBIOLOGY LABORATORY

Fundamentals and Applications

George A. Wistreich

East Los Angeles College

Prentice Hall
Upper Saddle River, New Jersey 07458

Executive Editor: David Kendric Brake
Acquisitions Editor: Linda Schreiber
Assistant Vice President and Director of Production: David W. Riccardi
Special Projects Manager: Barbara A. Murray
Total Concept Coordinator: Kimberly P. Karpovich
Cover Design: Joseph Sengotta
Cover Photos: Background: Live gram stain
  Insets (top to bottom)
    Penicillium with antibiotic droplets
    Spore stain
    Microscope with Koch: photograph contributed by Molecular Probes, Inc.,
    courtesy of Paul Millard and Bruce Roth.

Learning Resources
Centre

12082279

Printed in the United States of America
10 9 8 7 6 5 4 3 2 1

ISBN   0-02-428980-9

Prentice-Hall International (UK) Limited, *London*
Prentice-Hall of Australia Pry. Limited, *Sydney*
Prentice-Hall Canada Inc., *Toronto*
Prentice-Hall Hispanoamericana, S.A., *Mexico*
Prentice-Hall of India Private Limited, *New Delhi*
Prentice-Hall of Japan, Inc., *Tokyo*
Simon & Schuster Asia Pte. Ltd., *Singapore*
Editora Prentice-Hall do Brasil. Ltda., *Rio de Janeriro*

To Reneé, my wife, and to my sons Eddie and Phillip, whose love and support have been such an important part of my personal and professional life, and to my colleagues, former instructors, and students who have provided the many opportunities to apply Microbiology Fundamentals through the years.

Contents

Preface

The purpose of *Microbiology Laboratory Fundamentals and Applications* is to provide students beginning their study of microbiology with functional and effective experiences that will enable them to work with, understand, and appreciate the importance of microorganisms. The exercises have been chosen, designed, and arranged to help students learn to observe and recognize fundamental similarities and differences among microorganisms. The elements of critical thinking are also interwoven with the various exercises and experiments to build the scientific perspective and abilities of students.

Learning Aids

To make this manual an effective tool for students, we have developed and incorporated a number of learning aids. These include:

Pronunciation Guide to Organisms. The scientific names of each organism studied are presented in a way to aid the student in acquiring a greater ease with which to discuss, describe, and use microorganisms. A phonetic pronunciation guide is listed on the front and back inside covers of this manual. These lists contain all organisms discussed and/or used. In addition, the phonetic pronunciation of an organism's *genus* and *species* is given the first time the organism appears in the manual.

Section Overviews. Each major section of the laboratory manual is preceded by a description of the range of topics and experiments to be covered.

Learning Objectives. Each exercise begins with a list of learning objectives to emphasize the purpose of the procedures and what the student is to accomplish.

Illustrations. More than 500 photographs and diagrams have been incorporated into the manual to provide a clear and functional understanding of techniques, properties of microorganisms, and specific test results.

Procedure Diagrams. Exercises that introduce basic techniques contain step-by-step illustrations of the procedures to show the proper handling of equipment and cultures. An alphabetical listing of *Procedure Diagrams* can be found on pages xxiii-xxiv.

Referral Boxes. In exercises requiring the use of specific procedures and/or techniques described in other parts of the laboratory manual, special *Referral Boxes* have been inserted to help students to quickly locate the needed procedure.

Laboratory Review and Photograph Quizzes. *Laboratory Reviews* accompany all exercises to help students determine whether they understood the major features of the exercise and related topics. Several exercises also contain photograph quizzes. Here, the student is expected to answer a question pertaining to specific results of the exercise. Answers to the photograph quizzes can be found in Appendix 1.

Key Terms. Each exercise contains a specific section of definitions together with the phonetic pronunciations of selected terms.

Experimental Exercises. A separate section containing a number of specially developed exercises, techniques, and applications can be found at the end of the manual. These exercises lend themselves to experimentation and additional applications of the principles of microbiology and molecular biology.

Index. An index has been included in this manual because of the large number of tests and procedures it contains. Students should have no difficulty in locating a particular test or procedure.

Organization of the Manual

The 66 instructional exercises, 4 experimental exercises, and related items in this manual are arranged into 15 sections, as follows:

Microbiology Laboratory Fundamentals and Applications begin with a short visual survey of equipment generally used in a microbiology laboratory. Familiarity with these and the other equipment items presented in specific exercises will contribute to a smoother operating laboratory.

Section I teaches certain fundamental procedures used in microscopy and in the study of microorganisms. These procedures include oil-immersion objective usage, smear preparations, simple staining, and hanging-drop and temporary wet-mount techniques.

Section II stresses methods employed in the isolation and study of both aerobic and anaerobic microbial cultures. Dilution techniques and the use of pipettes together with an application of the pour technique to determine colony-forming units of cultures also are included.

Section III compares the representative microbial groups, which include bacteria, algae, protozoa, and fungi, as to their structure, morphological arrangement, and selected activities. Exercises in the section also deal with the distribution of microorganisms in the environment, and spontaneous generation. Microscopic measurements and an examination of Leeuwenhoek's peppercorn water also are important aspects of this section.

Section IV demonstrates specific staining techniques, such as the Gram Stain and acid-fast procedures, which are used in the identification and classification of bacteria. Both the hot and cold forms of acid-fast staining techniques are presented.

Section V familiarizes students with the structure and function of bacterial cells. Techniques used to observe flagella and bacterial motility by means of motility agar are included.

Section VI demonstrates the extracellular and intracellular metabolic activities of bacteria. The section begins with a comparison of various selective and differential plating media, and then proceeds to the use of a number of tube and plate media to show the enzymatic activities of various bacterial species. The use of rapid and miniaturized testing procedures to identify microorganisms is also covered. The section ends by introducing the use of a biochemical flowchart (key) in the identification of an unknown enteric bacterial species.

Section VII stresses the principles of sterilization and disinfection, and the effectiveness of chemical and physical agents in the control of microorganisms. Particular attention is given to the thermal resistance of microorganisms, the use of ultraviolet light in sterilization and as a mutagenic agent, and the inhibitory actions of dyes, different pH levels, and heavy metals. Antibiotic sensitivity testing and the detection of antibiotic resistance also are demonstrated.

Section VIII emphasizes certain properties of viruses. Both bacterial and animal viruses are studied and techniques for their detection and cultivation are included.

Section IX includes exercises involving gene transfer processes found in bacteria. Experiments contain safe procedures designed to present the basic features of bacterial transformation, conjugation, and transduction. Rapid colony transformation with plasmid DNA and the application of the Ames test for the detection of mutations and/or cancer-causing chemicals also are contained in this section.

Section X demonstrates some of the industrial applications of microorganisms. Exercises covering the distribution of microbes in food and their involvement in the production of sauerkraut, cheese, yogurt, and wine are included.

Section XI contains exercises designed to increase students' understanding of basic immunological principles and their application. Modifications of standard procedures and concepts and a survey of commercially available tests are presented. Such tests include latex agglutination and radial immunodiffusion. The section begins with the process of phagocytosis and concludes with an enzyme immunoassay.

Section XII provides a number of exercises dealing with epidemiology. Particular topics covered include a demonstration of Koch's Postulates, the effectiveness of universal precautions, and the proper handling of specimens and their transport. Consideration also is given to aerosols and their importance to laboratory and hospital sepsis.

Section XIII includes exercises relating to several aspects of basic medical microbiology. Experiments dealing with the normal flora *(microbiota)* of the body demonstrate the fact that the presence of microbes on or in the body is not necessarily indicative of disease. An experiment showing the antimicrobial activity of tears is also included. Procedures for handling and identifying various unknown laboratory specimens are considered. In addition, representative disease-producing microorganisms and procedures used for their identification and study are stressed in specific exercises.

Section XIV provides an introduction to the fundamental areas of medical helminthology. The distinguishing features of helminths and arthropods associated with specific infectious diseases are considered.

Section XV presents a number of experimental exercises that reinforce critical thinking and the scientific method. Several of the topics considered lend themselves to class demonstrations and to special student projects. The exercises include topics such as determining the effects of temperature on the bacterial growth curve, immunoelectrophoresis, the polymerase chain reaction (PCR), and DNA restriction analyses.

Instructor's Manual

An updated *Instructor's Manual* to accompany the *Laboratory Manual* is available and contains specific directions for the preparation of laboratory materials, alternate procedures, the use of supplementary aids, suggested sources for audiovisual aids, laboratory equipment and supplies, questions for laboratory examinations and quizzes, and answers to exercise questions.

George A. Wistreich

Acknowledgments

The author would like to thank Millipore Corporation; Whittaker, M. A. Bioproducts; Difco Laboratories, Becton Dickinson and Company; Becton Dickinson Microbiology Systems; Terumo Medical Corporation; Marion Scientific; Diatech Diagnostics Inc; Biomed Diagnostics; and Evergreen Scientific, for providing samples of their products for use in developing this manual.

I am deeply grateful to the members of the editorial and production staff of Prentice-Hall Publishing Company for their untiring and imaginative talents, and especially to Linda Schreiber and David Brake for their continued support, interest, and efforts in the publication of this manual. Kim Karpovich and Barbara Murray deserve special mention for their constant vigilance in working with this complex project. I would also like to thank Sylvia Fernandez for the effort and care she exercised in typing portions of the manuscript.

George A. Wistreich

Notes to the Student

Materials to Be Provided by the Student

1. One clean laboratory coat or apron
2. Textbook and laboratory manual
3. One 4H drawing pencil
4. Colored pencils (red, blue, yellow, and green)
5. One package of glass slides and cover slips
6. One package of lens paper
7. Two packages of matches
8. One lint-free cloth
9. Slide box (optional)
10. One wax marking pencil or a washable-ink felt pen
11. Surgical or rubber gloves for protection against stains (optional)

Laboratory Operating Procedures

1. Each laboratory class will begin promptly at the beginning of the period. At each meeting, the instructor will give a lecture demonstration that will include general directions designed to make your work easier and your methods more efficient.
2. You are expected to read the assigned laboratory exercise and carry out the suggested reading assignments before each class. The instructor will designate the parts of an exercise scheduled to be undertaken in class.
3. You will be assigned a microscope and laboratory drawer for the duration of the course. It will be your responsibility to keep all equipment issued to you in good working condition.
4. Store your personal belongings, such as textbooks and coats, in the area specified by your instructor. Never let them clutter the operating space on the laboratory bench, where they might become contaminated through exposure to microbial cultures.
5. As you perform your experiments, record data in ink and make sketches and labels in pencil. Always perform separate steps of an exercise in the sequence given in the manual. To help you in this regard, each **Materials** and **Procedure** section includes check-off boxes (❑) in front of each item. Once the item in the **Materials** section has been obtained or a specific step of a **Procedure** has been completed, **MARK THE BOX.** Following this practice will help you to know exactly where you are in an exercise.
6. Unless your instructor specifies otherwise, each experiment is to be completed within one week. Your laboratory notebook is meant to serve as a record of your work. Keep it up to date, using procedures outlined by your instructor.
7. Leave all laboratory facilities and equipment in good order at the close of each period. Place waste paper and contaminated glassware in the receptacles provided. Inform the instructor of any defects in equipment.
8. Conserve water and gas. Turn off water faucets and gas burners when they are not in use.

A Short Illustrated Survey of Microbiology Laboratory Equipment

A number and variety of equipment items are used in a microbiology laboratory. Becoming familiar with them and their applications is an important aspect of the smooth operation of a laboratory. Here are a few of the items all students generally will use. Later exercises will introduce other instruments that are used in the application of microbiology concepts and fundamentals

General Items

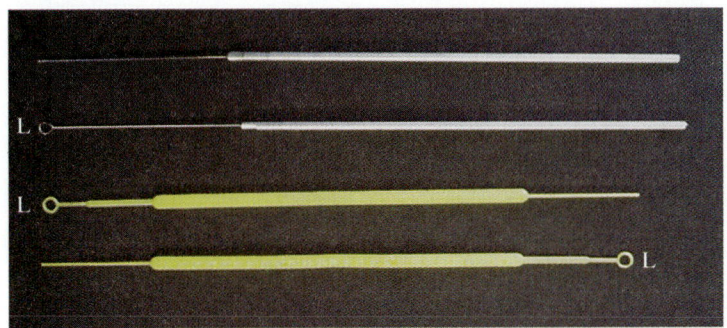

Figure i–1
Inoculating instruments are among the major tools used in a microbiology laboratory. The inoculating loop (L) generally is used for broth (liquid) cultures, while the inoculating needle is used with microorganisms growing on solid (agar) surfaces. The wire portion of these devices can be made of stainless steel or platinum. Disposable sterile plastic inoculating instruments also are in use.

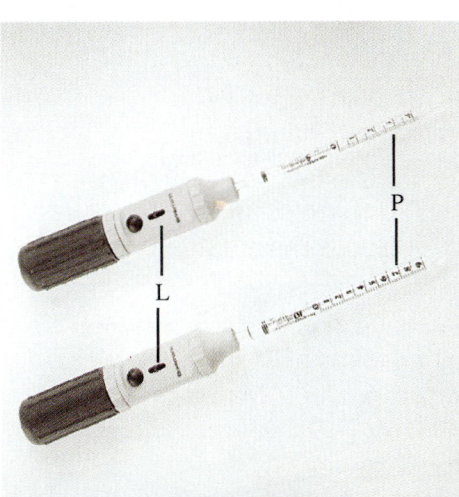

Figure i–2
Devices such as the one shown are used for the safe, rapid, and accurate delivery of volumes ranging from 0.1 to 100 milliliters (mL). Removal and dispensing of fluids are easily controlled by a single lever (L). An appropriate sterile pipette (P) is inserted in the device as shown.
(Photo courtesy of Brinkmann Instruments, Inc. Westbury, N.Y.)

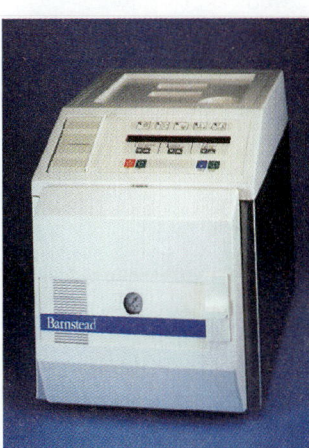

Figure i–3
Autoclaves, which come in several different sizes, are used for the sterilization of cultures, media, infectious wastes, and related items. The proper disposal of contaminated materials in the laboratories requires the use of such important instruments.
(Courtesy Barnstead Thermolyne Corporation, Dubuque, Iowa.)

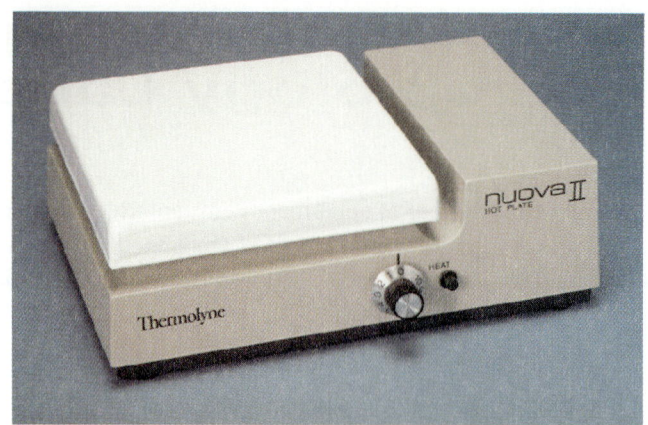

Figure i–4

Hot plates are used in a variety of laboratory situations, including the preparation of laboratory media and solutions, determining the effects of temperature on cultures and selected procedures. (Courtesy Barnstead Thermolyne Corporation, Dubuque, Iowa.)

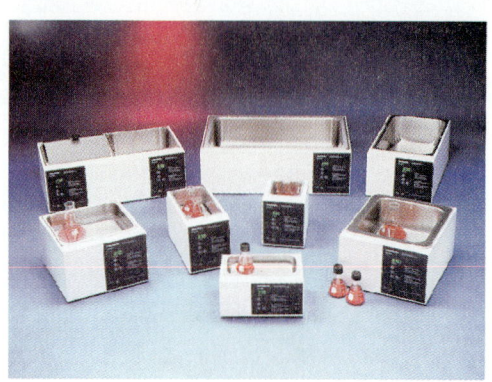

Figure i–5

Water baths are used for the cultivation of various types of microorganisms. Modern-day water baths are equipped with an internal circulating pump to move water without disturbing cultures and microprocessor controls to maintain required temperatures. (Courtesy of Fisher Scientific.)

Figure i–6

An example of a typical incubator found in a microbiology laboratory. Incubators maintain specific temperatures for the cultivation of microbial cultures. (Courtesy of Fisher Scientific.)

Safety Materials

Figure i–7
Laboratory safety is of major importance in microbiology. Accidents can happen. Eye wash stations are ready for immediate use to rinse away chemical contamination. Wash bottles contain a sterile, buffered, eye wash solution.
(Courtesy of Scienceware® from Bel-Art Products, Pequannock, N.J.)

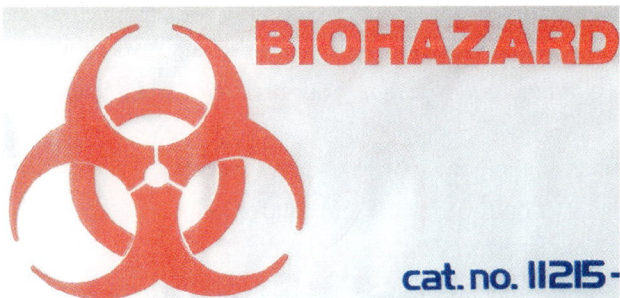

Figure i–8
Biohazard bags are commonly used for the disposal of microbial cultures. The red or orange color and/or a large **biohazard** symbol makes these bags quite visible in a laboratory. Laboratory precautions are generally imprinted in four languages on these bags.

Laboratory Safety

The smooth and safe operation of a microbiology laboratory depends on observing specific safety rules and regulations, and carefully applying standard culture-handling procedures. Read the following sections. If you have any questions or do not understand any of the items, ask your instructor for a clarification. Verify your reading and understanding of this *Laboratory Safety* section by signing and dating the student contract that follows on page xix.

General Rules and Regulations

1. Never eat, store food, or smoke in the laboratory. **Do not bring food to the laboratory.**
2. To prevent contamination of your street clothes, always wear a laboratory coat or apron during class. Remove this protective apparel before leaving the laboratory.
3. Report fires or personal injuries such as cuts or burns to your instructor immediately. Your instructor will show you the location and demonstrate the use of emergency equipment such as fire extinguishers, fire blankets, and first-aid kits. Know their locations and how to use them.
4. Always double-check the label of any chemical or microbial culture you use.
5. Keep any and all flammable liquids away from an open flame (Bunsen burner).
6. Do not inhale the fumes of any substance directly. When required to note the odor of a culture, for example, take a deep breath of fresh air and exhale normally. With your hand waft the vapors toward your nose and sniff slowly.
7. Do not pipette cultures or chemicals by mouth. Always use a *pipette aid* or *pump.* (Your instructor and a later exercise will demonstrate the proper way to pipette.)
8. Do not use your fingers as a stopper for a tube or a bottle when shaking the tube to mix its contents.
9. If you get a chemical on your skin, or in your eyes, immediately flood with water and call for help.
10. Never remove equipment, media, or microbial cultures from the laboratory.
11. Dispose of chemicals, biological materials, and paper as indicated by your instructor. Do not put paper or related materials in the sink.
12. Wash your hands thoroughly with detergent and water before leaving the laboratory, even for coffee and health breaks.
13. Whenever you are in doubt about a personal health problem or laboratory procedure, ask your instructor before you attempt to do anything.

Culture-Handling Procedures

1. Before and after each class period, wipe table tops with the disinfectant provided for this purpose. Your instructor will show you how and where to discard contaminated items.
2. Carry and/or store microbial cultures and inoculated media in racks, baskets, or other designated containers. Avoid spillage.
3. Always flame inoculating loops and/or needles before and after use.
4. Place contaminated materials, old cultures, and the results of exercises into the containers designated by the instructor.
5. Never pipette microbial cultures or chemicals by mouth.
6. In the event of a laboratory accident, such as spilling or dropping a live culture, remain calm and do the following:
   a. Report the accident to your instructor as soon as possible.
   b. Place paper towels over the spilled material.
   c. Pour disinfectant liberally over the towels.
   d. After 15 minutes, remove and dispose of the towels in the receptacle used for the disposal of contaminated materials.

Precautions for the Handling of Blood or Other Body Fluids

The *Centers for Disease Control and Prevention* have developed a series of guidelines for health care workers in order to prevent the transmission of human immunodeficiency viruses (HIVs) as well as other blood-related infectious diseases such as hepatitis B virus infection. The guidelines in the form of laboratory precautions and procedures are clearly specified in the appropriate exercises of this manual. They are generally contained in a *Caution Box.* Following the specific steps will enable you to work with body fluids safely and effectively and without any danger not only to your safety and well-being, but to others in the laboratory.

Laboratory Safety Rules and Regulations Contract

I have read, understand, and agree to follow the safety procedures described on page xvi as well as any other written or verbal instruction provided by my instructor.

_____

Student Signature

_____

Date

How to Use the Laboratory Manual

This laboratory manual has been designed to help you to observe, learn, remember, and apply fundamental concepts, principles, and techniques of microbiology. Think of it as an extension of your textbook. An organized use of the manual will provide considerable benefits. Undoubtedly, you will develop your own method of using the manual; however, here are some suggestions that may help you to gain the most from your efforts and time.

1. Before starting a specific exercise, read the introduction to the section. It will provide important background information and give you a general overview of the exercises in the section and their relationship to one another.

2. Next, read the introduction to the exercise to be performed. Use the stated objectives as indications of what you should know and the skills you should acquire upon completion of the exercise.

3. Examine the items listed in the Materials section, and refer to the *"Pronunciation Guide to Organisms Studied in This Manual"* inside the front and back covers. This procedure will acquaint you with the microorganisms and items to be used in carrying out specific procedures. Check the box in front of each item as it is obtained. Refer to this section if you are ever in doubt about which microorganisms or materials are to be used.

4. Next, read the Procedure section and look at any figures specified. A number of exercises contain color photographs. Use them during and upon the completion of an exercise to confirm your results. Underlining steps of a procedure and all important information can be helpful in locating specific steps of a procedure and can save time. Place a check in the box before each step after it is completed. This practice will help you to quickly locate your place in the procedure. Certain exercises contain step-by-step illustrations of procedures to show the proper handling of equipment and cultures.

5. Then read the questions and examine the tables to be completed in the Results and Observations and in the Questions sections. This practice will help you to determine the types of information you will need to obtain from the exercise.

6. Now you are ready to perform the procedure. If you have gone through the earlier steps of this section, you should not find the exercise difficult. It is always easier and less frustrating to carry out an exercise if you know what is expected and what is to be done before starting.

7. Following the procedure, examine your preparations and complete the Results and Observations and Questions sections. Use the color photographs of the manual and your textbook as needed.

8. Finally, review the main portions of each exercise to reinforce your understanding of the concepts, principles, terminology, and techniques presented.

Pronunciation Guide

In the exercises of this manual, you will find *phonetic pronunciations* for several terms and specific microorganisms. Taking time to sound out new terms and to say them aloud once or twice will help you to master one of the tasks in a microbiology course—*learning its specialized vocabulary.* The following key explains the system used for the pronunciations.

1. The strongest accented syllable appears in capital letters, e.g., *microbe* (MĪ-krōb) and *microscope* (MĪ-Krō-skōp).

2. Vowels pronounced with long sounds are indicated by a line above the vowel and are pronounced as in the following common words.

   ā as in *māke*
   ē as in *bē*
   ī as in *īvy*
   ō as in *pōle*

3. Vowels not marked for long sounds are pronounced with the short sound, as in the following words.

   e as in *bet*
   i as in *sip*
   o as in *not*
   u as in *bud*

4. Other phonetic symbols are used to indicate the following sounds.

   a as in *above*
   soo as in *sue*
   kyoo as in *cute*
   oy as in *oil*

Alphabetical Listing of Procedure Diagrams

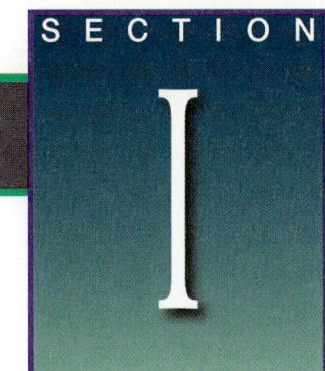

S E C T I O N

I

An Introduction to Microscopy and Specimen Preparation

In passing, it is worth noting that the word amateur . . . means "one who loves." Almost any intelligent and industrious person can become a true scientist, if he [or she] is an amateur in the original sense of the word and really loves what he [or she] does.

René Dubos

One of the tools essential for studying microorganisms is the bright-field microscope. Two types of this instrument are in general use: the simple microscope and the compound microscope.

Anton van Leeuwenhoek—considered to be largely responsible for establishing the foundations of microscopy, initiating the use of the microscope as a research tool, and developing the technique of specimen viewing by transmitted light—used a simple single lens microscope effectively to describe bacteria and protozoa for the first time in 1674 (Figure I–1).

Unfortunately, the effectiveness of the simple microscope (or ordinary magnifying glass) is restricted, because, among other disadvantages, its power to magnify is limited. On the other hand, the compound microscope, which contains a series of lenses and is much more powerful than the simple microscope, has had widespread application in science laboratories and a variety of industries.

The issue of who invented and developed the compound microscope is still disputed. Hans and Zacharias Janssen are generally credited with basic contributions towards its development. The introduction of the Abbé substage condenser and superior oil-immersion lenses in 1866 led to the refinement of the compound microscope. Over the years, the shape and size of both the simple and the complex microscopes have changed (Figure I–2).

The primary function of a microscope is to enable the observer to distinguish structures and points separated by short distances. This function is called **resolution.** Resolution is more important than magnification, because obtaining the

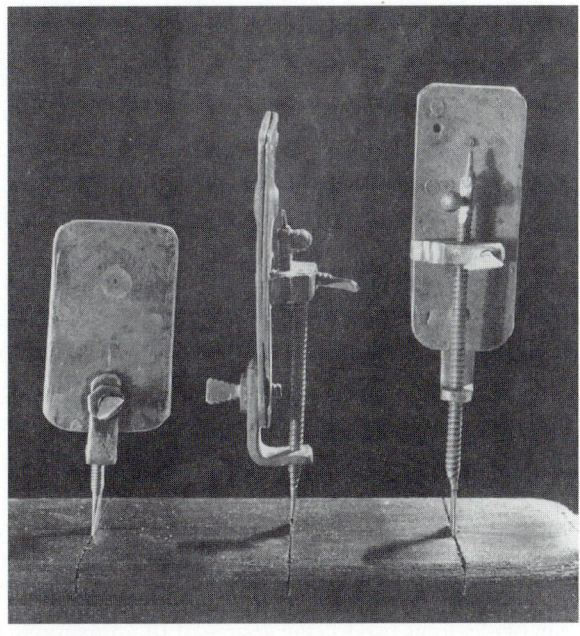

Figure I–1
The Anton van Leeuwenhoek simple (one lens) microscope. The first views of microorganisms such as bacteria and other types of cells were made with this instrument. Specimens were placed on the tip of an adjustable point (s) and viewed through a small lens (L). Focusing was possible with the aid of specific focusing (F) screws.
(Courtesy of Ruksmuseum voor De Geschiedenis Der Natuurweienschappen, Holland.)

1

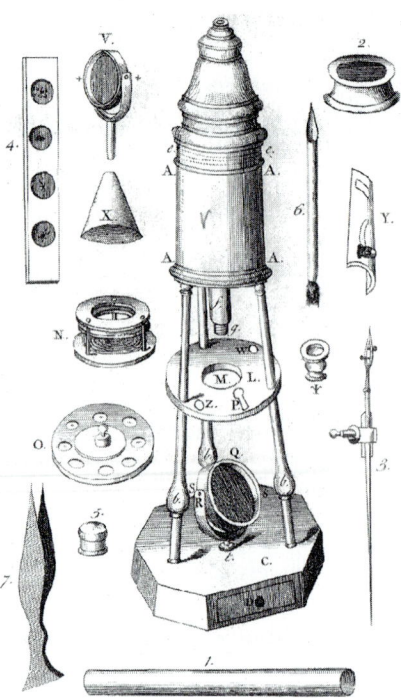

Figure I–2

Scarlett's light microscope of approximately 1760. Note the large number of microscope parts. Do any of them look like the parts of a modern-day instrument? (Refer to Figure 1–2.)

largest image possible is not always necessary but obtaining sharp detail is. Resolution is dependent upon many factors, including illumination, the nature of the specimens, and the investigator.

The two general types of microscopes used by microbiologists and biologists are the **light-field** and **electron microscopes.** The light-field instrument incorporates optical lenses and visible light as the source of illumination. The electron microscope contains electromagnetic lenses and utilizes a beam of electrons as the source of illumination. Electron microscopes are used to obtain higher magnifications and greater detail of biological materials (Figures I–3 and I–5).

Certain changes of the bright-field system have been developed to increase the effectiveness of investigations of microorganisms and their parts in a natural state. Microscopes so modified included phase-contrast and dark-field instruments. In phase-contrast microscopy, the contrast between a transparent specimen and the surrounding background is increased while the transparency of the specimen remains unchanged. In dark-field microscopy, the image is formed from diffracted, reflected, or refracted light, and specimen contrast against a dark background is thereby increased. The result is a brilliant image against a dark background in which minute structures usually too small to be observed directly are visible (Figure I–4a).

The resolving power of a microscope is inversely related to the wavelength of light used; that is, the shorter the wavelength, the greater the resolving power. (A **wavelength** is the distance between two corresponding points on a wave, such as the distance between two successive *peaks* or *crests* [Figure I–5].) In an early attempt to improve microscopy, instruments were developed that employed ultraviolet light, which has a shorter wavelength than ordinary light. Thus, with ultraviolet light as the source of illumination, finer details could be seen than with visible light. Because the human eye cannot see the short wavelengths of ultraviolet light, photographic plates were used to record specimen images. Because the use of ultraviolet light increases resolving power, these instruments are being used with greater frequency in microbiology. In addition, they have achieved considerable importance in the examination of fluorescent materials, which absorb the invisible wavelength and emit a visible image (Figure I–6).

In electron microscopy, the beam of electrons used as the source of illumination is focused by electromagnetic lenses instead of the optical lenses characteristic of other microscopes. This electron beam has a considerably shorter wavelength than ordinary light. For this reason, the resolving power and the useful magnification obtainable with electron microscopes far exceed those in the optical systems described above. Two general types of electron microscopes are currently in use: the *transmission* and the *scanning* instruments.

(a)

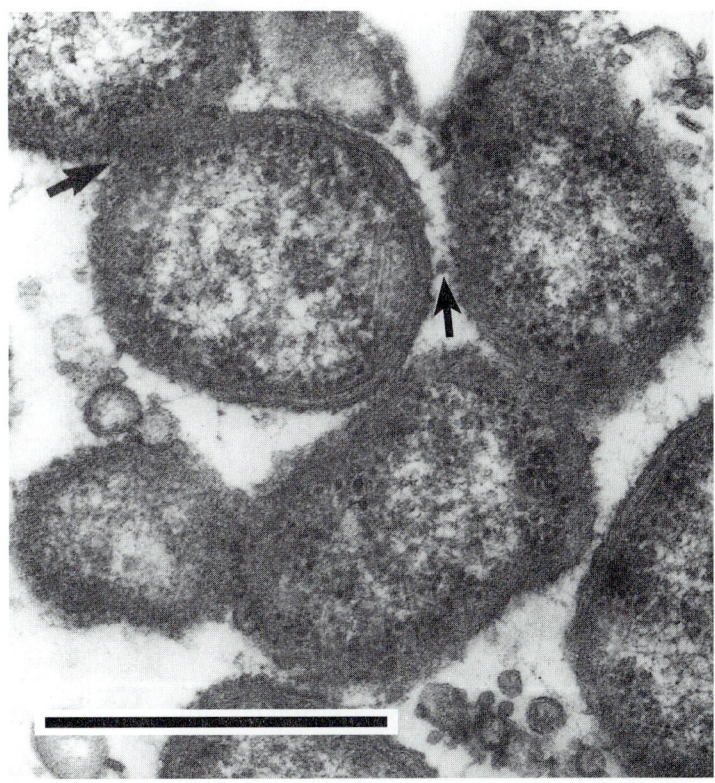

(b)

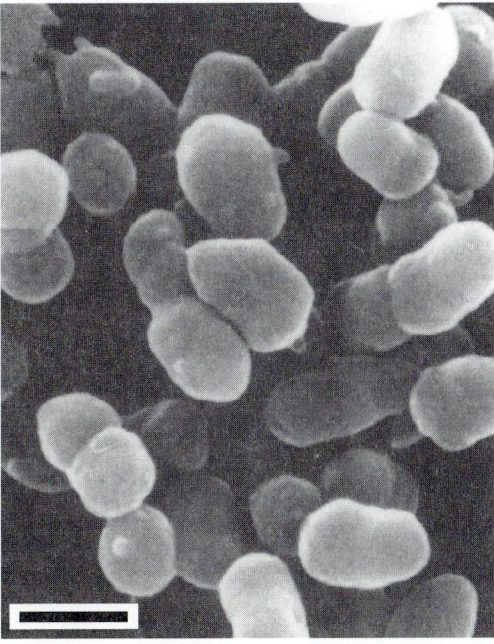

Figure I–3

Electron micrographs (a) An example of a transmission electron micrograph showing the *internal* appearance of a bacterial rod. The arrows are pointing to small blister-like elevations (vesicles) on cells. (b) A scanning electron micrograph showing only the external features of similar cells. Arrows here also are pointing to vesicles. The bar markers measure 1 μm and 0.5 μm respectively.
(From Kinder, S.A. and S.C. Holt, *J. Bacteriolo,* 175: 840–850, [1993].)

(a)

(b)

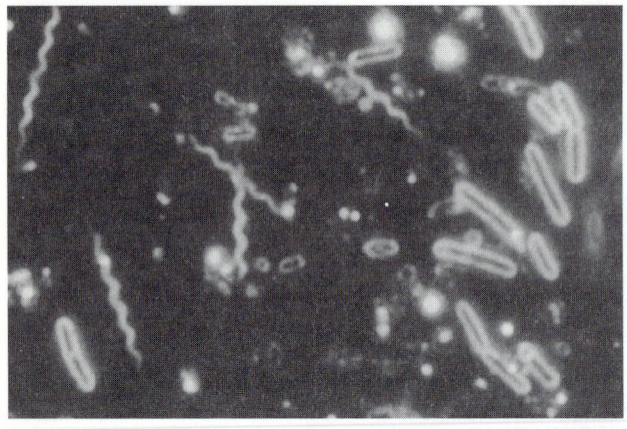
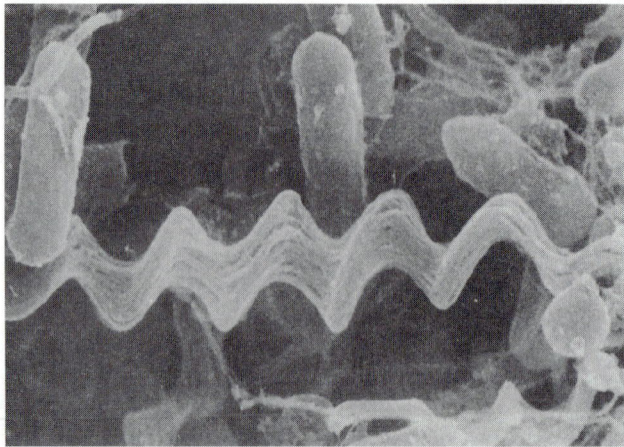

Figure I–4

Micrographs taken with the aid of microscopes used in the visual identification of bacteria and other microbial forms. Microscopes magnify specimens and are used in distinguishing arrangements, microscopic structures, and points separated by short distances. These micrographs show similar bacterial samples at different magnifications and with variations in the amount of observable detail. (a) A dark-field preparation showing bacterial rods (B) and spirochetes (A). (b) A scanning micrograph clearly showing distinct cells and their arrangements.

Several elaborate techniques are used in electron microscopy in preparing specimens and providing specimen contrast. These techniques include ultrathin sectioning (thin slicing) and staining, depositing of heavy metals in a vacuum (shadow casting), and the use of electron stains such as phosphotungstic acid and uranyl acetate. The instrument used is a deciding factor in the selection of the specimen preparation technique. The scanning electron microscope (an important innovation in electron microscopy) is particularly advantageous, because it provides two-dimensional images of the three dimensions features of specimens without requiring extensive preparation techniques (Figures I–3b and I–4b). Another instrument, the *scanning transmission electron microscope,* which is similar in operation to a scanning electron microscope, can be used to examine transmission electron microscope preparations such as negatively stained specimens and ultrathin sections. Electron microscopes are employed in various research fields to reveal details of biological and physical systems that would otherwise be impossible to examine. Table I–1 summarizes the general properties of selected microscopes.

A considerable amount of information can be obtained from examining microorganisms microscopically. Living organisms can be studied to determine cell size, cellular arrangement, motility, and responses to environmental factors. But observing cells in their natural or unstained state is sometimes difficult because of the semitransparency of the cells. Therefore, stained preparations of killed microorganisms are more frequently employed. Staining solutions are applied to thin films of microorganisms (called **smears**) that have been passed through a flame three or four times (a process

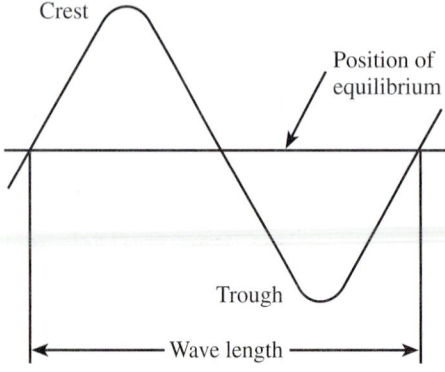

Figure I–5

The anatomy of a wavelength.

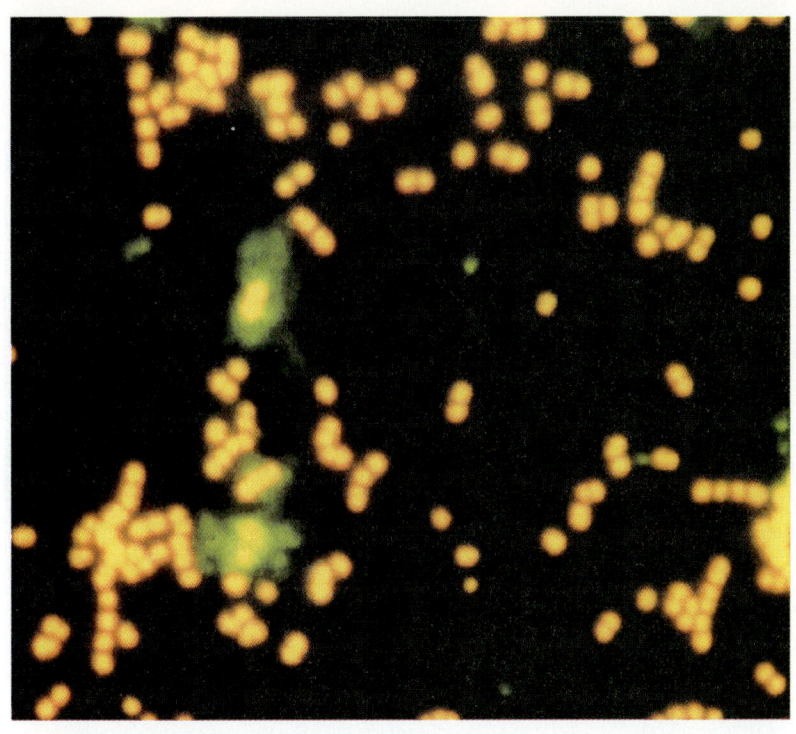

Figure I–6
Bacterial cocci (spherical cells) shown by fluorescence microscopy. (Courtesy Difco Laboratories, Detroit, MI.)

Table I–1

Comparison of Microscope Types

Magnification Type	Range	Source of Illumination	Resolution (in μm)	Applications Include
Compound light (bright) field	1–2,000	Visible light	300	Used extensively for the examination of various cell types, either stained or unstained
Dark-field	1,000	Visible light	100–200	Used to observe unstained living or hard to stain cells and to detect motility
Phase contrast	10–1,500	Visible light	200	Used to examine and study cellular structures; staining of specimens is not required
Fluorescence	10–3,000	Ultraviolet	200	Useful in many medical diagnostic procedures for identifying cells, including cancer and micro-organisms and/or their parts; fluorescent dyes are used for staining
Transmission electron (TEM)	200–1,000,000	Electrons	1	Used to examine and study the ultra-structure of cells and viruses and certain biochemicals
Scanning electron (SEM)	200–500,000 and greater	Electrons	5.0–10.0	Used to examine and study surface features of cells and nonbiological materials; images are presented in three-dimensional form
Scanning transmission electron (STEM)	Combination of TEM and SEM	Electrons	Combination of TEM and SEM	Used for applications similar to those listed for TEM and SEM; also used for analysis of chemical elements in specimens; equipped with computer for measurements and calculations
Scanning-probe (tunneling)	500,000 and greater	Electrons	0.2	Used for the examination of surfaces both biological and nonbiological

called **heat fixation**). These stained preparations are examined for the presence of internal and external structures, as well as for the ability of the organism to retain certain dyes. Both types of information can aid in the successful identification of microorganisms. It should be noted that each method of specimen preparation possesses certain specific advantages.

Size and shape are important features in the identification of microorganisms and other cells. In microbiology as well as other branches of the biological sciences, several common metric units are used for measuring microorganisms or their parts: the *millimeter* (mm), the *micrometer* (μm), and the *nanometer* (nm). The relative sizes of various cell types are shown in Figure I–7.

A short, straight line, called a **bar marker,** is frequently placed at one corner of a photograph as a size reference. The length of the marker represents the length of a metric unit at a particular magnification. Typical units include 1 μm, 1 nm, and fractions of multiples of these units. (See Figure I–3.)

The exercises in this section introduce a number of fundamental techniques and equipment used by microbiologists and investigators in allied scientific areas.

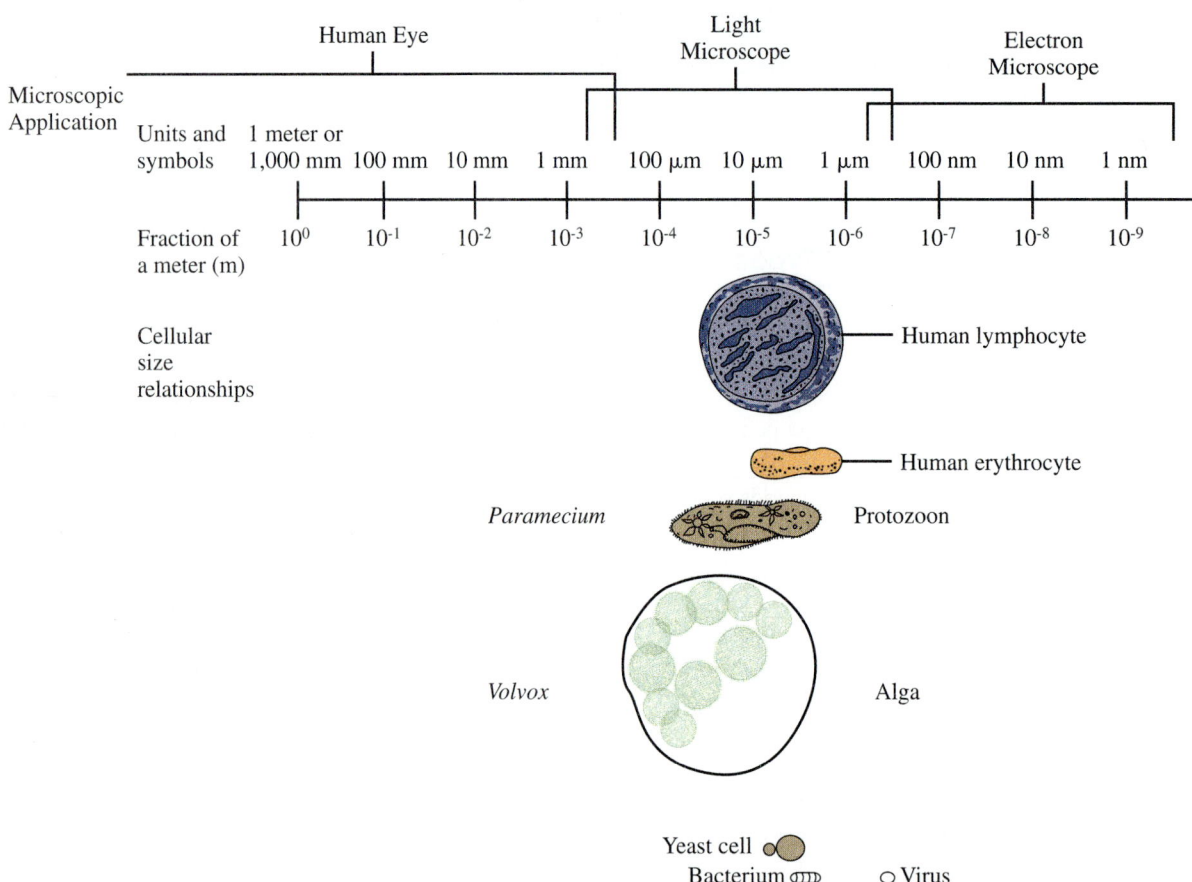

Figure I–7

Metric units of length commonly in the biological sciences. 1 meter = 1000 millimeters (mm); 1 mm = 1000 micrometers (μm); 1 μm = 1000 nanometers (nm). Comparisons of different cell types including microorganisms, and the respective magnification ranges of selected microscopes. Later exercises will show the size relationships of a number of microorganisms.

After completing this exercise, you should be able to:

1. Identify the parts of the microscope, and understand their functions.
2. Demonstrate the proper method of focusing, changing objectives, carrying the microscope, and cleaning the microscope.
3. Calculate the total magnification of any ocular and objective combination.
4. Prepare a temporary wet mount for the examination of a specimen.
5. Use the microscope, especially the oil-immersion lens, effectively.
6. Identify several types of human blood cells and their parts.
7. Define and understand the following terms and concepts: *resolving power, parfocal, working distance, depth of field,* and *magnification.*

The compound microscope is a precision instrument. It is essentially a system of accurately ground lenses arranged to give sharp, clear, magnified images of minute objects. Magnification that may be obtained with different lens systems ranges from approximately 100 to 2,000 times ($\times$) the diameter of the specimen being observed.

While magnification is an important microscope property, the limiting factor that determines the usefulness of a microscope is its **resolving power.** The resolving power of an optical system is its ability to distinguish fine detail by producing separate images of small parts of a specimen that are only a short distance apart. In practice, the limit of resolution obtainable with a compound light microscope is about 0.2 μm, or roughly 1,000 times the resolving power of the unaided human eye.

The quality of the image obtained depends on the degree of *contrast* between the specimen and its surroundings and at times between parts of a specimen. Such contrast is due in part to differences in light absorption and various techniques that affect absorption used to improve contrast. For example, the use of chemical stains results in color differences (Figure 1–1a). Reducing the intensity of the light illuminating a specimen makes it easier to distinguish cells that have a *refractive index* similar to how the light illuminating a specimen makes it easier to distinguish cells that have a refractive index similar to the fluid in which they are suspended (Figure 1–1b). In this exercise, the wet-mount technique and the stained blood smear are used to emphasize certain properties of the light microscope and to introduce the features of living and nonliving cells, respectively.

A. Use and Maintenance of the Microscope

Microscopes vary greatly from laboratory to laboratory. Your instructor will demonstrate and explain the general construction and manipulation of the particular instrument you will use. There is, however, a set of basic rules that should always be observed for all instruments to ensure maximum efficiency and proper maintenance.

1. Always carefully remove the microscope from its place of storage by holding it firmly by the arm with one hand and supporting it at the base with the other. Keep the instrument in an upright position (Figure 1–2).
2. Place the microscope at least 6 inches (about 15 cm) from the edge of your laboratory table.
3. Do not tamper with any parts of the instrument. If the microscope does not seem to be functioning, notify the instructor immediately. Since students in other laboratory sections may use the same instrument, any part previously damaged and not reported could be charged to you.
4. Do not touch the lenses with your fingers. Perspiration contains fatty acids and other substances that can damage the lenses. Always use lens paper for cleaning the optical system. Never use paper towels or cloths. These materials will scratch the delicate lens surfaces.

(a)

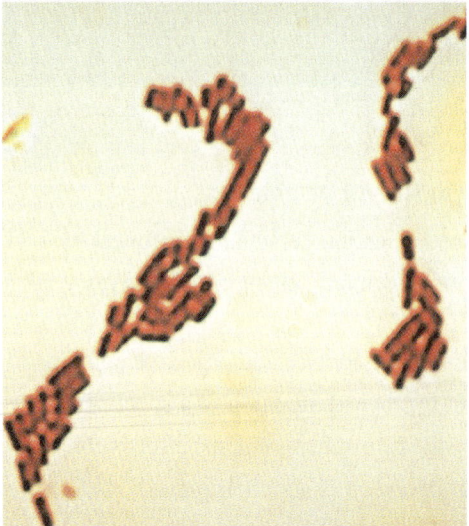

(b)

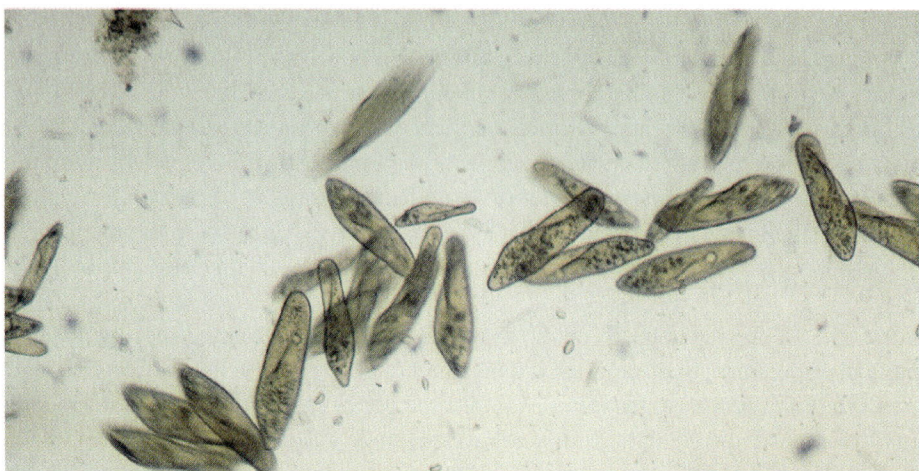

Figure 1–1

Microscopic observations. (a) The results of a stained bacterial preparation. (b) A temporary wet-mount preparation showing the well known protozoon, *Paramecium*.

5.   Wipe the lenses of the microscope before and after use.

6.   Use a clean, soft, linen cloth to remove dirt, dust, or other materials from parts other than the lenses.

7.   Do not allow liquids, particularly acids and alcohol, to come in contact with any part of the microscope.

8.   Always use a cover slip when examining objects of organisms mounted in water or other fluids.

9.   Always raise the body tube sufficiently before either placing a slide on, or removing a slide from, the microscope stage.

10.   Remove immersion oil from the microscope with lens paper only.

11.   Before putting the microscope away, always rotate the low-power objective into working position (i.e., in line with the body tube), open the diaphragm, and raise the condenser to its higher fixed position.

12.   When disconnecting the light source, do not pull on the wire; instead, grasp the plug firmly to remove it from the socket.

13.   If a cover is provided for the microscope, be certain that it covers the entire instrument.

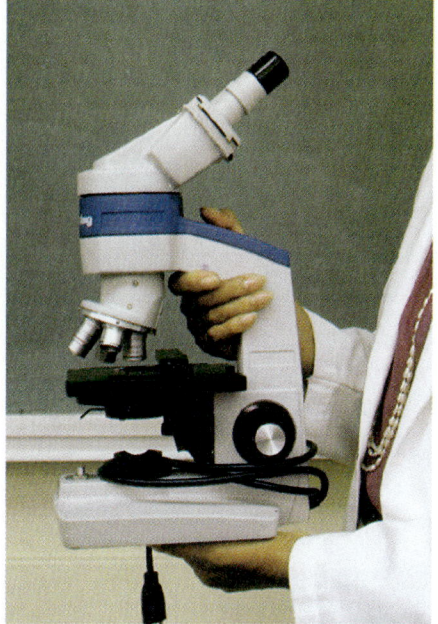
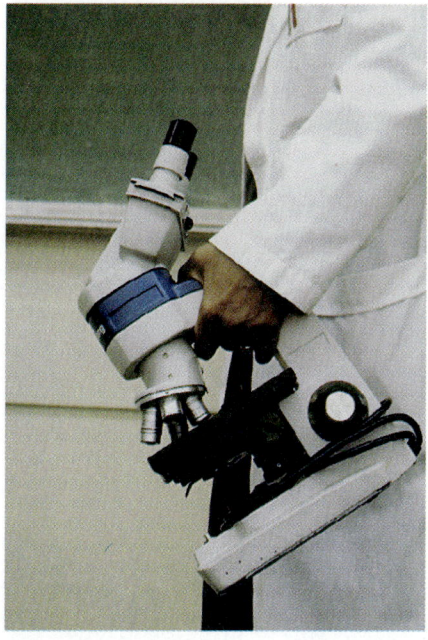

(a)   (b)

Figure 1–2

The appropriate way to carry a microscope. The microscope should always be carried in an upright position as shown in (a). Holding the microscope in a position such as the one shown in (b) can cause eyepieces to fall out of the instrument.

B. Microscope Parts and Principles of Microscopy

The ordinary compound microscope consists of a series of optical lenses, mechanical adjustment parts, and supportive structures for various components. The optical lenses include the *ocular,* or *eyepiece;* usually three *objectives* with different magnifying powers; and the *substage condenser.* The *coarse-* , *fine-* , and *condenser-adjustment knobs* together with the *iris-diaphragm lever* are the major mechanical parts concerned with the operation of the instrument. The various components of the scope are held in position by, or contained within, such supportive structures as the *base, arm, pillar, body tube* (barrel), and *revolving nosepiece* (Figure 1–3).

To serve its purpose, a microscope must provide adequate magnifying power so that the finest details of a specimen are separated enough to be visible to the eye. The magnifying power of microscope lenses usually is indicated by markings on objectives and eyepieces (e.g., 4×, 10×, 45×, 100×). The total magnifications produced by the lenses individually.

While magnification is an important microscope property, the limiting factor that determines the usefulness of the light microscope is its resolving power. The resolving power of an optical system can be defined as the ability to show clearly two points that are close together as separate entities. This feature is largely determined by the wavelength of the light source and the angular aperture (opening) of the lens system being used. Resolution is also affected by the refractive index of the medium through which light passes before entering the microscope objective. (The path of light rays are bent or refracted from a straight path as they pass through different media or substances. The refractive index refers to the ratio of the speed of light in air to its speed in another medium.) The relationship of these factors is expressed in the following combined formula:

$$\text{Numerical aperture (NA)} = \eta \sin \theta$$

where η represents the refractive index of the medium through which light passes before entering the objective lens and $\sin \theta$ is the trigonometric sine of one-half the angle formed by light rays (in the shape of a cone)[1] coming from the condenser and passing through the specimen. Figure 1–4*a* depicts the explanation of numerical aperture. Values for NA are engraved on the barrel of objectives and specify the maximum resolution obtainable with the proper use of the appropriate accessories.

[1]Light in this form is frequently referred to as "a pencil of rays."

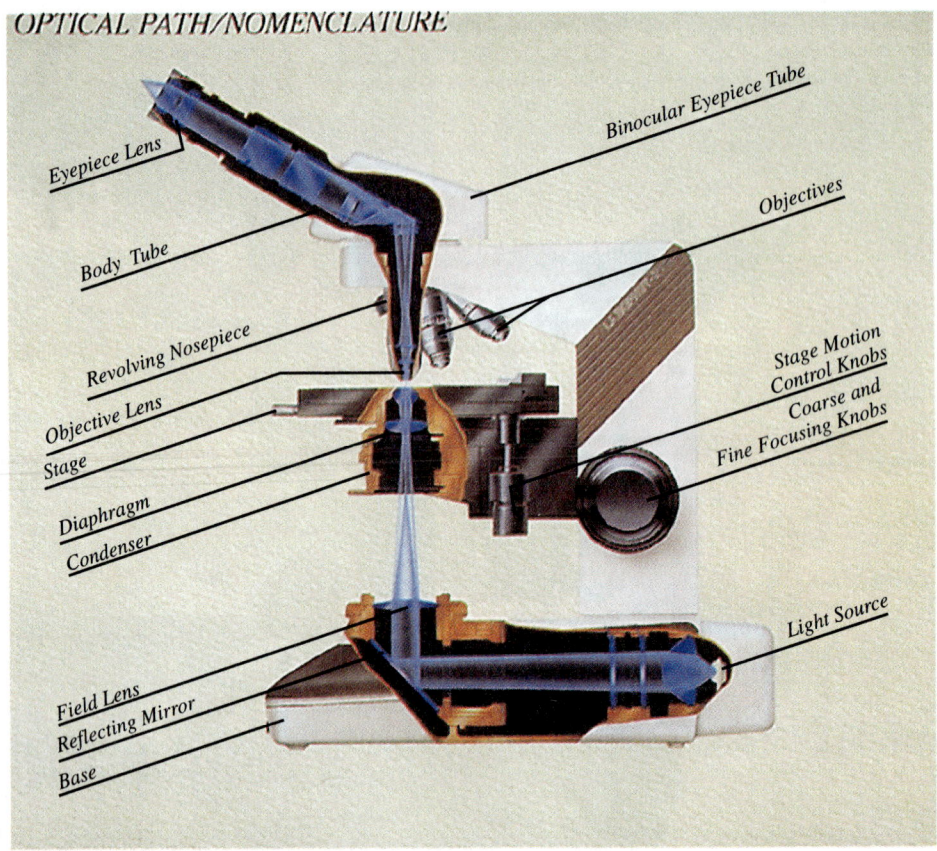

OPTICAL PATH/NOMENCLATURE

Eyepiece Lens

Binocular Eyepiece Tube

Objectives

Body Tube

Revolving Nosepiece

Stage Motion Control Knobs

Objective Lens

Coarse and Fine Focusing Knobs

Stage

Diaphragm

Condenser

Light Source

Field Lens

Reflecting Mirror

Base

Figure 1–3

A cut-away diagram from a compound microscope showing major components. Many instrument components are combined. Note that the fine-adjustment and coarse-adjustment knobs are combined, and that the condenser adjustment is permanently fixed in position.
(Courtesy of Carl Zeiss Inc., Thornwood, N.Y.)

The objective lenses serve not only to magnify but also to focus the light rays from a specimen in order to form a real image within the body tube. This resulting image is further magnified by the ocular lens system of the microscope. The total magnification obtained is determined by multiplying the magnification of the objective used by the magnifying power of the ocular lens.

The individual objectives of a microscope consist of a combination of convex and concave lenses. These objectives are contained in a revolving nosepiece that is used to rotate the desired objective into place. The distance from the specimen and a point roughly midway between the component lenses of the objective is known as the **focal length.** It is important in determining the distance required for focusing. The distance between an objective and a specimen is the **working distance.** This distance decreases as the magnification is increased; more light is needed for better and more accurate viewing of a specimen. Figure 1–4b shows the relationship between the working distances of objectives and the iris-diaphragm adjustments needed to provide functional illumination.

Although compound microscopes from different manufacturers vary in appearance or design, they consist of the same basic components. The parts of a typical microscope and their functions are described in Table 1–1. Use the information in Table 1–1 with Figure 1–3 as you examine and locate specific parts of the microscope.

Materials

❏ 1. A compound microscope and an appropriate light source (1 each per student)

❏ 2. A large cutaway diagram of a representative microscope

Table 1–1

Microscope Components and Their Functions

Microscope Component	Location or Description	Functions
Ocular (eyepiece)	Uppermost series of lenses	Magnification
Body tube (barrel)	Main cylindrical part	Holds oculars, conducts light rays from specimen
Nosepiece	Movable, usually circular plate at bottom of body tube	Holds objectives
Low-power objective	Shortest objective, usually magnifies 10 ×	Magnification
High-power (dry) objective	Intermediate-sized objective, usually magnifies 43× to 45×	Magnification
Oil-immersion objective	Longest objective; marked with either etched, red, or black circles; usually magnifies 100×	Magnification
Condenser	Lens system located below central stage opening	Concentrates and directs light beam through specimen
Stage	Platform on which specimens for examination are placed	Specimen support
Mechanical stage	Control devices that permit movement of slides from left to right and forward and backward on stage	Used to move specimen
Iris diaphragm	Located beneath stage in association with condenser unit, controlled with a lever	Regulates brightness or intensity of light passing through lenses
Coarse-adjustment knob[a]	Generally located below stage	Used for preliminary and coarse focusing by raising or lowering body tube
Fine-adjustment knob[a]	Generally located below stage	Used for final or fine focusing by raising or lowering body tube
Condenser-adjustment knob	Control knob located below stage	Used to obtain full illumination by raising or lowering condenser
Base	Heavy, bottom portion on which instrument rests	Microscope support
Arm	Somewhat curved portion of microscope, used in carrying instrument	Microscope support
Mirror[b]	Double-faced mirror for reflecting light through specimen	Reflects light from source through lens systems

[a]On several instruments, the coarse and fine adjustments are combined in one knob device.
[b]Found on older instruments.

Procedure

This procedure is to be performed by students individually.

❑ 1. Remove the microscope assigned to you from the cabinet. Hold the arm with one hand and support the instrument at the base with the other (Figure 1–2a).

❑ 2. Place the microscope approximately 6 inches (15 cm) from the edge of the laboratory table with the microscope arm facing you.

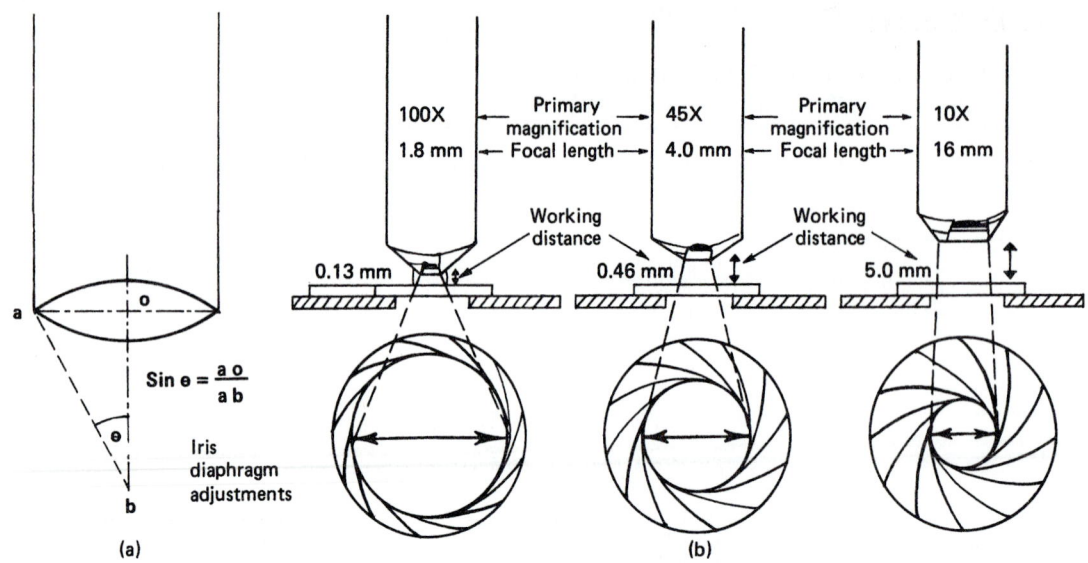

Figure 1–4

(a) A diagrammatic representation of numerical aperture. (b) The relationships between the working distances of the low, high, and oil-immersion objectives, and the iris diaphragm adjustments needed to provide adequate illumination of specimens. The primary magnification, focal lengths, and working distance are indicated for each objective.

❑ 3. Examine the microscope and, with the aid of the diagrams provided, locate the components listed below that are applicable to your instrument:

❑ a. Ocular
❑ b. Body tube (or barrel)
❑ c. Revolving nosepiece
❑ d. Low-power objective
❑ e. High-dry objective
❑ f. Oil-immersion lens
❑ g. Base

❑ h. Condenser
❑ i. Iris-diaphragm lever
❑ j. Coarse-adjustment knob
❑ k. Fine-adjustment knob
❑ l. Condenser-adjustment knob
❑ m. Mechanical stage

C. Specimen Preparation: Temporary Wet-Mount Technique

Direct examination of living specimens can be extremely useful in determining the size and shape relationships, motility, and reactions to various chemicals. The temporary wet-mount method described here is adequate for both living and non-living materials to be examined for a short time. (*Note:* The specimens prepared in the following two procedures will be used for viewing in Part D of this exercise.)

Materials

❑ 1. Glass slides and cover slips
❑ 2. Lens paper and xylene
❑ 3. Eye dropper (1 per 4 students)
❑ 4. Newspaper

❑ 5. Scissors and inoculating needle (2 pairs per 4 students)
❑ 6. Live culture of *Euglena* species (an alga)

Procedure 1: Letter *e*

This procedure is to be performed by students individually.

❏ 1. Carefully clean a glass slide and a cover slip as indicated by the instructor.

❏ 2. Cut a series of letters from the newspaper provided. Be certain that at least one of the letters is a lowercase *e*.

❏ 3. Perform the three steps shown in Procedure Diagram 1.

❏ 4. Save this preparation for viewing in the next part of the exercise.

Procedure Diagram 1
Temporary Wet-Mount Technique

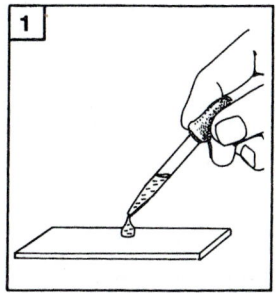 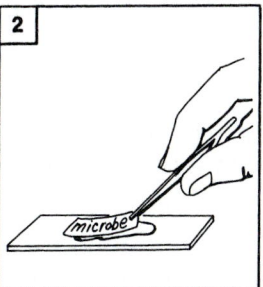 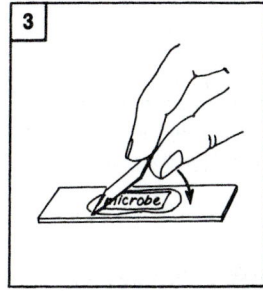

1. Place 1 drop of water on the center of a clean glass slide.

2. Place the string of letters in the drop of water and make certain that the liquid covers them.

3. Place the edge of the cover slip on the drop of water on the slide so that the fluid completely wets the edge. Then slowly lower it into place over the letters.

Procedure 2: *Euglena* (an Alga)

This procedure is to be performed by students individually.

❏ 1. Place 2 drops of the *Euglena* culture on the center of a clean glass slide.

❏ 2. Apply the cover slip as shown in Procedure Diagram 1.

❏ 3. Save this preparation for viewing in the next part of the exercise.

D. Use of the Microscope

This portion of the exercise is designed to familiarize you with certain properties of microscope components. It is to be performed by students individually.

Materials

The following materials should be provided for class use:

❏ 1. Temporary wet mounts of the letter *e* and of *Euglena* culture prepared earlier

❏ 2. Lens paper

Procedure 1: Microscope Operation

❑   1. Wipe all lenses with lens paper.

❑   2. Move the low-power objective into position under the body tube. You will feel a click when it is correctly in place.

❑   3. Connect your light source and turn it on.

❑   4. Place the letter *e* slide on the mechanical stage and secure it with the slide holder. The finger control lever for the slide holder is generally located at the rear left-hand corner of the mechanical stage. Forward pressure on the lever will cause the slide holder to open, and the slide may be inserted into the holding area. The bottom of the slide should rest on the microscope stage and not on the mechanical stage apparatus. Release the finger control lever to allow the slide holder to close and to hold the slide within the mechanical stage.

❑   5. Using the control knobs on the mechanical stage, move the slide until the specimen on the slide is over the center of the condenser opening in the stage. One control knob moves the slide left to right and the other moves it front to back.

❑   6. Use the coarse-adjustment knob to lower the low-power objective to approximately one-quarter inch above the cover slip.

❑   7. Use the condenser-adjustment knob to raise the condenser as far as it will go.

❑   8. While looking into the ocular, move the iris-diaphragm lever back and forth. Note the change in illumination. The intensity or brightness of illumination is controlled by adjusting the iris diaphragm.

❑   9. Locate your specimen. Focus *upward,* that is, move the objectives upward with the fine-adjustment knob until the specimen comes into clear view. In a microscope with a fixed body tube, the objectives remain in a fixed position. Focus is controlled by moving the stage *downward* until the specimen comes into view. If you have difficulty in locating your specimen, look for a region with a color or intensity different from the other portions of the slide.

❑ 10. In order to change objectives (switch from low power to high power), do not raise the body tube; simply turn the nosepiece to bring the desired objective into place. The specimen will generally stay in focus with any objective, but a slight turn of the fine-adjustment knob may be necessary. Most microscopes are **parfocal,** that is, once the specimen is in focus, it will remain so when the objective lens is changed.

❑ 11. When looking through the microscope, keep both eyes open at all times. Keeping both eyes open prevents eye strain and enables you to observe and draw microscopic views of specimens without moving your head.

Procedure 2: Examination of Temporary Wet Mounts

This procedure is to be performed by students individually.

Letter *e*

❑ 1. Examine the letter *e* as it appears on the slide to the unaided eye in a normal reading position. Make a sketch of the letter in the Results and Observations section.

❑ 2. Focus and examine the preparation under the low- and high-power objectives following the steps listed in Procedure 1 in this section (D).

❑ 3. Slowly move the slide first side to side and then forward (up) and backward (down) with the mechanical stage control knobs. Compare this view of the orientation of the *e* with the view with the unaided eye.

❑ 4. Sketch representative microscopic fields in the Results and Observations section.

❑ 5. On your sketch of the *e* as observed under low power, draw a circle around the portion of the letter seen under high power.

❑ 6. Refer to the Questions section and answer the questions pertaining to this portion of the exercise.

Euglena

❑ 1. Focus and examine the preparation under the low- and high-power objectives following the steps listed in Procedure 1 of this section.

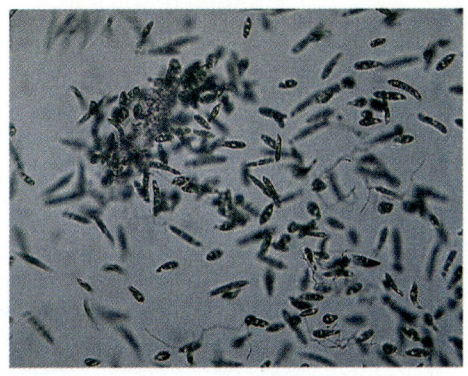

Figure 1–5
A view of the alga, *Euglena*

❏ 2. As you move the slide from side to side, move the iris-diaphragm lever to reduce the brightness of the light. This adjustment will cause the *Euglena* cells to stand out from their background.

❏ 3. Compare your view with Figure 1–5. Identify the orange, photosensitive eyespot, chloroplasts, and flagellum in your specimen.

❏ 4. Note the rapid movement and the direction of movement exhibited by *Euglena*.

❏ 5. Refer to the Results and Observations section and answer the questions pertaining to this portion of the exercise.

Results and Observations

1. Sketch your observations here. _____

Letter *e*	Letter *e*	Letter *e*
(natural size)	(low power)	(high power)

2. Were all *Euglena* moving in the same direction? _____

3. Under which objective was the movement of *Euglena* easier to follow? _____

4. Could you see specific structures of the *Euglena* studied? If so, which ones? _____

_____

E. The Oil-Immersion Lens

The oil-immersion lens is the highest magnification objective used in general microbiology courses. The best possible results are obtained when the objective is immersed in a medium that has approximately the same index of refraction as glass. One such medium is cedarwood oil. The oil has the advantage of not evaporating when exposed to air for long periods of time, and allows the lens to form a clearer and more detailed image. This image is brought into focus, with contact maintained between the oil and the front lens of the objective.

In using the oil-immersion objective, a drop of oil is placed on the portion of the slide where the specimen is located, and the bottom lens of the objective is lowered *(immersed)* into the oil. The oil acts as an additional lens in the system and prevents the loss of necessary light rays (Figure 1–6).

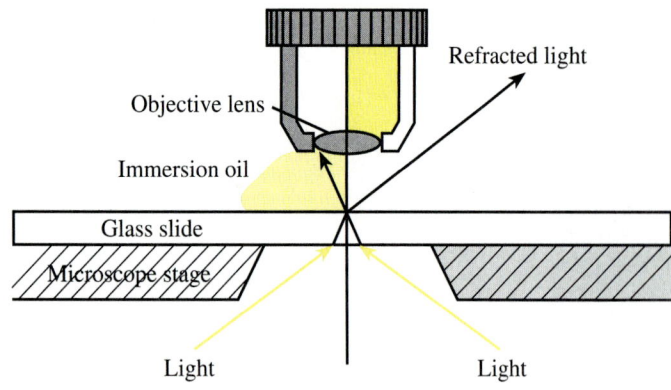

Figure 1–6
A diagrammatic comparison of the effects of immersion in oil and in air on the path of light leaving a microscope glass slide.

Materials

❏ 1. Prepared human blood smears

❏ 2. Charts showing blood cell types

❏ 3. Immersion oil

❏ 4. Lens paper and xylene

Procedure

This procedure is to be performed on an individual basis. The blood cells of mammals most readily visible in smears include the erythrocytes (red cells) and the leukocytes (white blood cells). The thrombocytes (platelets) can also be observed in some preparations. Figure 1–7 shows the general appearance and relative sizes and shapes of most blood cells. Representative human blood cells are shown in color in Figures 1–7 and 1–8. The appearance and numbers of these cells are important to detecting and diagnosing various diseases.

A typical erythrocyte has the appearance of a circular, bioconcave disk and does not possess a nucleus in its mature state. Erythrocytes are by far the most numerous of all types of cells found in the peripheral blood (bloodstream). The red cell usually measures approximately 7 μm in diameter and 2 μm in thickness.

The various kinds of leukocytes are classified as either *granulocytes* or *agranulocytes*. Granulocytes contain distinct cytoplasmic granules that upon staining (as in the case of bloodsmear preparation), react with dyes to yield characteristic colors. Based upon such chemical reactions, three types of granulocytes may be distinguished: (1) eosinophils (in which the granules react with acid dyes and become red); (2) basophils (in which the granules react with basic dyes and become blue); and (3) neutrophils (in which the granules react with neutral dyes or acid and basic dye mixtures and become neutral- or orange-colored). In addition to the characteristic appearance of the granules, granulocytes also possess irregular and multilobed (lobose) nuclei.

Agranulocytes differ from granulocytes in that they do not possess granules and their nuclei are rounded rather than lobose. The two general cell types in this group are the lymphocyte and the monocyte. In lymphocytes, individual rounded nuclei occupy the major portion of the cell. Monocytes are larger than lymphocytes, and their nuclei are generally kidney shaped.

The oil-immersion lens must be used correctly for dependable results. The immersion oil employed usually has the same optical characteristic (refractive index) as glass. Consequently, with this system light rays pass directly from the glass slide into the objective lens. If oil were not used with the objective, the lightrays would be refracted (bent) by the air present between the glass slide and the objective, and less light would enter the microscope. If you carefully follow the steps listed below, good illumination should result.

❏ 1. Place the blood smear in position on the microscope stage.

❏ 2. Raise the body tube and rotate the high-dry objective into position.

❏ 3. Lower the objective until it reaches a point just short of contact with the specimen.

❏ 4. Focus upward with the fine-adjustment knob until the specimen comes into view.

   Note: (a) If the body tube of the microscope is in a fixed position, refer to step 9 in the preceding section on microscope operation. (b) In order to locate your specimen quickly, always look for a region that differs in color or intensity from other portions of the slide.

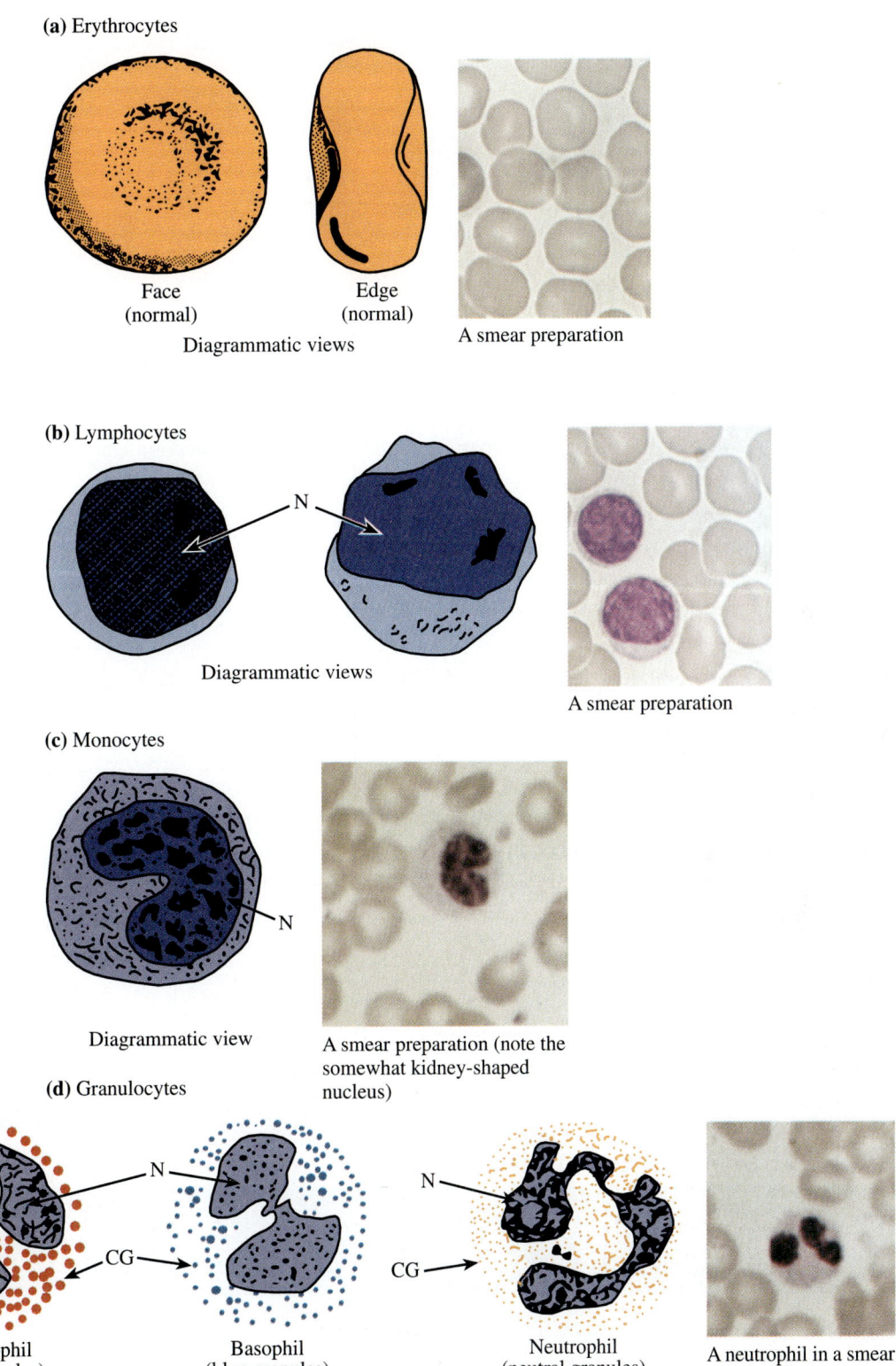

(a) Erythrocytes

Face
(normal)

Edge
(normal)

Diagrammatic views

A smear preparation

(b) Lymphocytes

N

Diagrammatic views

A smear preparation

(c) Monocytes

N

Diagrammatic view

A smear preparation (note the somewhat kidney-shaped nucleus)

(d) Granulocytes

N

CG

Eosinophil
(red granules)

N

Basophil
(blue granules)

N

CG

Neutrophil
(neutral granules)

A neutrophil in a smear preparation

Figure 1–7

Selected normal human blood cells. CG = cytoplasmic granules; N = nucleus

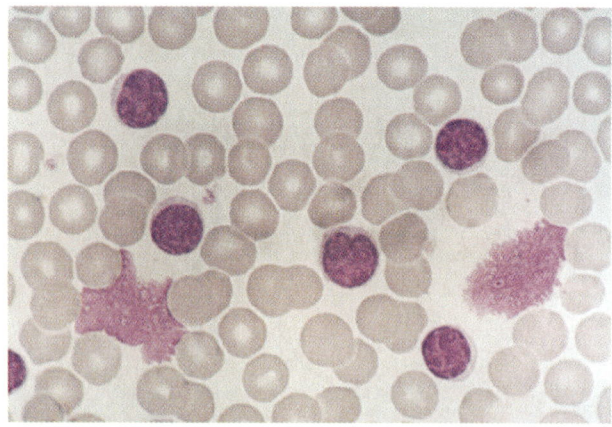

Figure 1–8
A representative human blood smear preparation.

❑ 5. Make light adjustments with the iris-diaphragm lever, if necessary.

❑ 6. Swing the high-dry objective aside.

❑ 7. Place a drop of immersion oil on the preparation and move the oil-immersion lens into position. The objective should touch the oil.

❑ 8. Look through the ocular and proceed as indicated in step 4.

   Note: In working with bacterial smears, the general tendency is to bypass steps 2 through 4 and to employ the oil-immersion lens directly. In this case, place oil on the slide and lower the lens into the oil just short of touching the slide. Focus as in steps 4 and 5. Locate the bacteria on slides by looking for areas of different color or intensity.

❑ 9. Locate and identify both red and white blood cells. Label the representative field of the blood smear shown in the Results and Observations section. Refer to Figure 1–7.

Results and Observations

Identify the cells indicated in Figure 1–8.

a. _____   d. _____

b. _____   e. _____

c. _____   f. _____

Laboratory Review 1   The Use and Care of the Microscope

1. Identify the labeled microscope parts in Figure 1–9.

    a. _____   e. _____   i. _____

    b. _____   f. _____   j. _____

    c. _____   g. _____   k. _____

    d. _____   h. _____

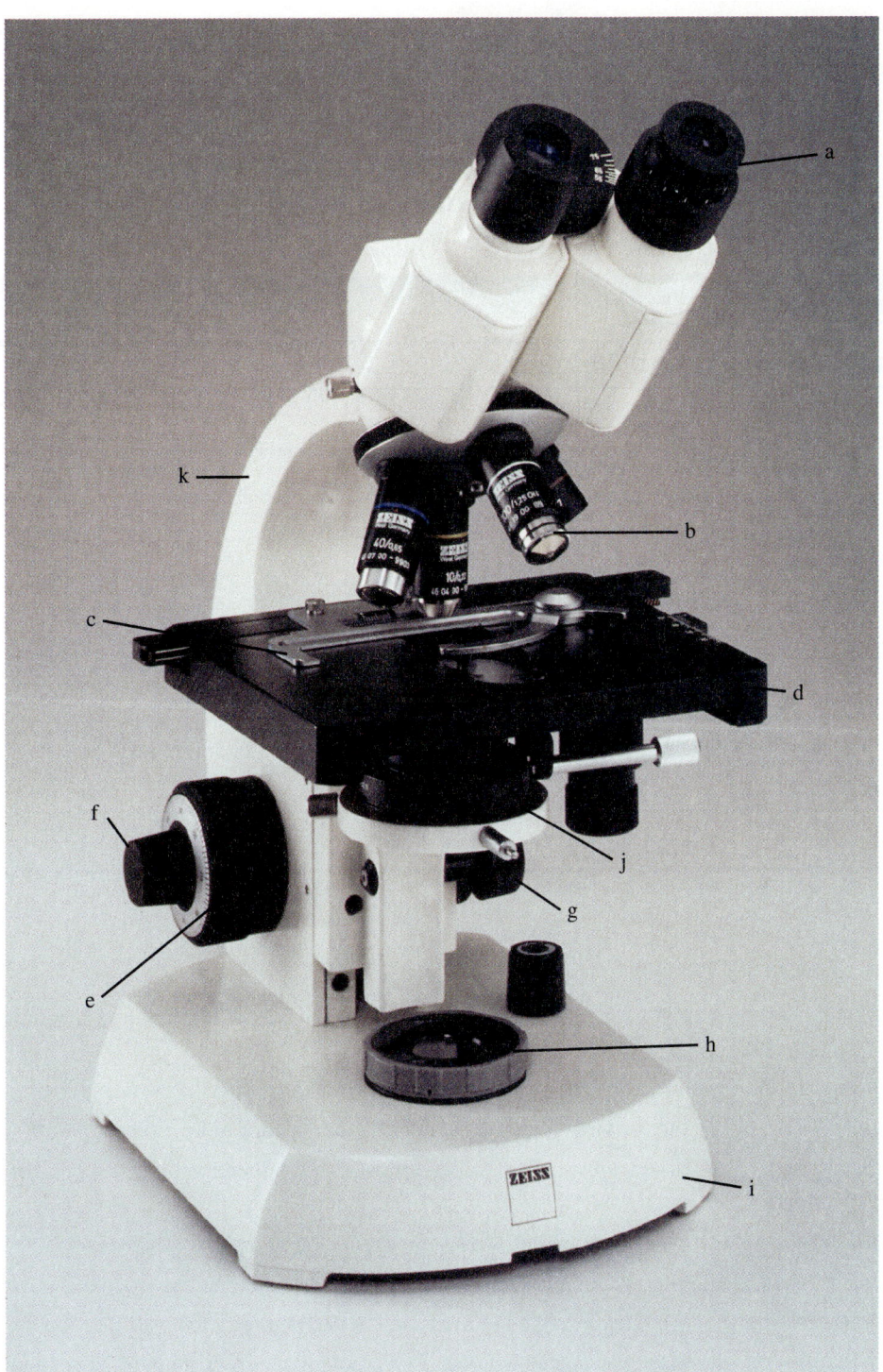

Figure 1–9
A compound microscope.
(Courtesy of Carl Zeiss, Inc.,
Thornwood, N.Y.)

2. What is the magnification of the following objectives without an ocular?

   a. Low power _____   b. High power _____   c. Oil immersion _____

3. Which of the objectives comes closest to a specimen and is most likely to break a slide if proper precautions are

   not taken? _____

4. a. What is working distance? _____

   _____

   b. Does the working distance increase or decrease as the magnification is increased? _____

5. As the magnification used is increased (e.g., going from low power to oil immersion), is more or less illumination needed? _____

6. List 3 factors that affect resolving power.

    a. _____   b. _____   c. _____

7. Complete the following table by giving the function(s) of each part listed.

Table 1–2

Microscope Parts and Functions

Microscope Part	Function(s)
a. Ocular	
b. Objective	
c. Condenser	
d. Iris-diaphragm lever	
e. Fine-adjustment knob	
f. Coarse-adjustment knob	
g. Condenser-adjustment knob	
h. Mechanical stage	
i. Microscope base	

Key Terms

condenser: lens system used to concentrate and direct the source of illumination through a specimen being viewed

focal length: the distance from a specimen and a point midway between the component lenses of the objective

image (IM-ij): the picture or view of an object such as that produced by a lens

iris diaphragm (Ī-ris DĪ-a-fram): microscope part used to regulate the intensity of light passing through a specimen

magnification: the ratio of the apparent size to the actual size of an object when observed through a microscope

objective: the magnifying lens of a microscope closest to the object being viewed

ocular (OK-ū-lar): uppermost lens having the function of magnification

resolving power: the shortest distance between two points that allows them to be seen as two separate points

working distance: the distance from the front of the objective lens of a compound microscope to the top of a specimen

EXERCISE 2

The Preparation of Bacterial and Oral Smears, and the Use of Simple Stains

After completing this exercise, you should be able to:

1. Effectively perform the appropriate aseptic techniques required in the handling of bacterial cultures.
2. Prepare and stain bacterial and oral smears.
3. Locate, examine, and interpret stained bacterial and blood smears.
4. Distinguish among basic bacterial shapes.
5. Develop a perspective on size relationships among bacteria, protozoa, and blood cells.

The shape or morphology of bacteria can be observed with microorganisms in either the living or killed (fixed) state. The microscopic study of live cells is limited, however, in that usually only the outline and structural arrangement of cells are revealed with bright-field microscopy. Stained preparations of fixed cells permit (1) a more detailed examination of cells, (2) the observation of internal cellular components, and (3) the preliminary differentiation of microorganisms.

The first step in the preparation of smears of microorganisms involves the removal of a small amount of microbial growth from a culture, which is then spread on the surface of a clean glass slide. This is done with the aid of an inoculating instrument (Figure 2–1). The wire portion of the instrument, which may be in the form of a loop or needle, is sterilized by holding it in a flame until it is entirely heated to redness. Such flaming incinerates any microorganisms on the wire.

After allowing a few seconds for cooling, the needle or loop can be used to lift bacteria from surfaces or remove them from broth cultures, and transfer them to a slide or to other media. The inoculating instrument then is flamed (resterilized) once more to kill any remaining bacteria. Flaming the wire portion from where it is connected to the handle to its tip is the accepted practice. This use of the inoculating loop or needle is one of several that belong to the group referred

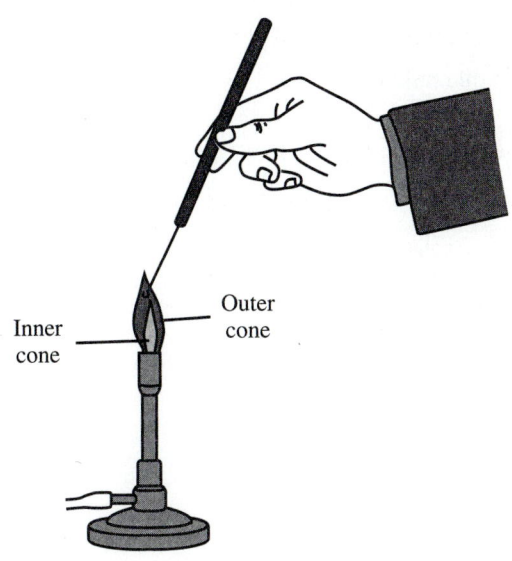

Inner cone

Outer cone

Figure 2–1
The correct way to use an inoculating tool. Grasp the inoculating loop or needle with your thumb and index finger as if it were a pencil or pen. To sterilize an inoculating tool insert it at a 60-degree angle into the outer or upper cone (hot portion) of the Bunsen burner flame.

to as **aseptic (sterile) techniques.** Aseptic techniques are procedures used to prevent contamination of cultures and to protect individuals and the environment from microbial exposure. Other techniques will be presented in later exercises.

After the small amount of bacterial growth is spread on a glass slide, the resulting thin film is allowed to air-dry, and is then usually passed through the flame of a Bunsen burner. The latter step not only kills but also coagulates the proteinaceous substances of cells, thereby fastening, or fixing, the organisms to the slide. The resulting **heat-fixed smear** is then ready for the application of staining solutions.

The chemical substances commonly used to stain bacteria are called *dyes.* Such dyes are in solutions called stains and are either acidic or basic. The acidic dyes, such as acid fuchsin and eosin, stain the cytoplasmic components of cells that are alkaline in nature. The basic dyes, such as crystal violet, methylene blue, and safranin, combine with those cellular elements that are acidic in nature. In the preparation of simple staining solutions, a particular dye is dissolved in either distilled water or alcohol. The stain is applied to smears and left standing for approximately 30 to 60 seconds. The smear is then washed, dried, and examined under the microscope. This procedure usually enables the microbiologists to see the general shapes and arrangements of cells. Structures, such as cell walls, spores, and capsules, are usually not stained. Exercises in Sections IV and V will present the procedures used to stain these bacterial structures. In this exercise, smear preparation and the simple-staining technique will be applied to bacterial cultures and to material obtained from the mouth.

The examination of a patient's blood may be extremely important in the diagnosis of disease. One integral part of hematological examination is a properly prepared blood smear, or spread. The clinician makes such a smear by obtaining a drop of blood from the patient, placing it on a slide, and then spreading it, with the aid of a second slide, into a thin layer. After the preparation dries, any one of several staining procedures—for example, Wright's and Giemsa—can be applied. Results obtained from the examination of the smear may be diagnostic in themselves or may serve as a basis for additional tests. This exercise also includes observing stained blood smears prepared with specimens from infected individuals. Examination of these smears will serve to develop a perspective on size relationships among bacteria, protozoa, and blood cells.

A. Bacterial Smear Preparation and Simple Staining

Materials

❏ 1. Forty-eight-hour cultures of each of the following microorganisms grown in or on the media indicated (1 culture per 4 students):
   ❏ a. *Bacillus subtilis* (ba-SIL-us, SA-til-us) on a nutrient agar slant
   ❏ b. *Escherichia coli* (esh-er-IK-ē-a, KOH-lī) in nutrient broth
   ❏ c. *Micrococcus luteus* (mī-krō-KOK-us, lū-TĒ-us) in nutrient broth

❏ 2. Staining solutions (1 bottle per 4 students)
   ❏ a. Crystal violet
   ❏ b. Safranin

❏ 3. Clean, lint-free glass slides

❏ 4. Immersion oil

❏ 5. Xylene

❏ 6. Lens paper

❏ 7. Inoculating loops (1 per student)

Procedure

This procedure is to be performed by students individually.

❏ 1. Before carrying out any staining technique, clean slides and pass them through the hot portion of the flame to remove any grease that may be present. *Note: For smears made with material from solid media, first place 1 loopful of water on the slide* (Figure 2–2). Continue with the steps shown in Procedure Diagrams 2 and 3.

(a) (b)

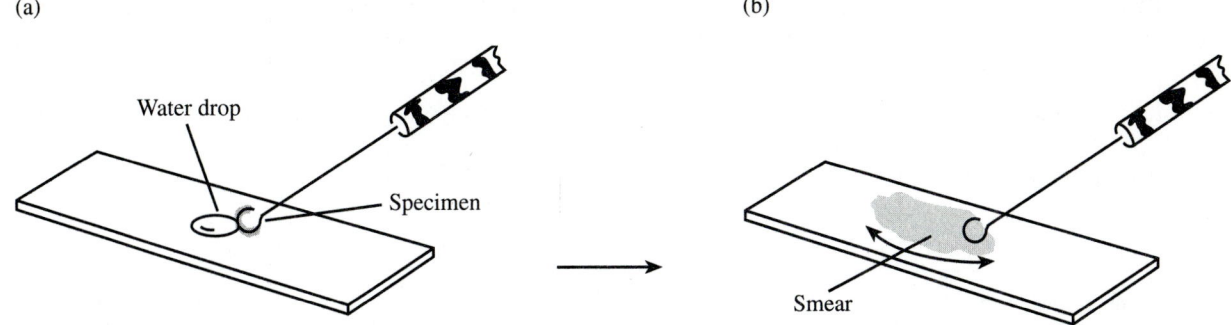

Figure 2–2

With material from agar culture or simply non-liquid sources, first place a loopful of water on clean slide (a). Then mix the specimen with the water drop, and spread over a small area (b).

Procedure Diagram 2
Bacterial (Heat-Fixed) Smear Preparation

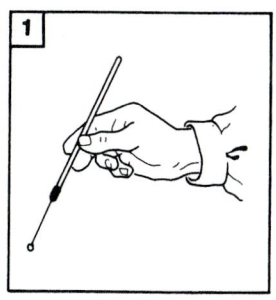

1. Grasp the inoculating instrument with your thumb and index finger as if it were a pencil or pen. An inoculating loop is used for liquid cultures, and a needle is used for material from solid media.

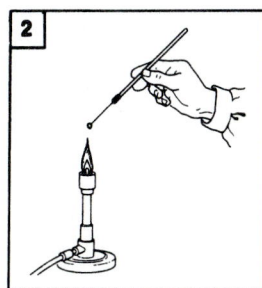

2. Insert the inoculating instrument at a 60-degree angle into the upper cone (hot portion) of the Bunsen burner flame. Heat the entire wire portion to redness.

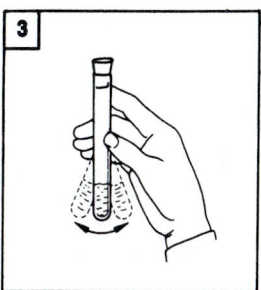

3. Pick up a tube of culture with your free hand. With broth (liquid) cultures, shake from side to side to bring the organisms into suspension.

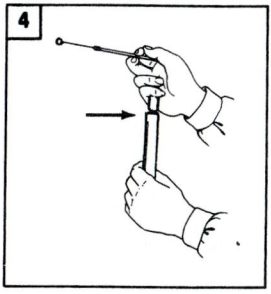

4. Remove the plug or cap of the tube with free fingers of the hand holding the inoculating loop.

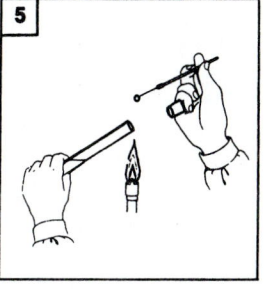

5. Flame the tip of the tube by passing it quickly through the upper cone of the flame.

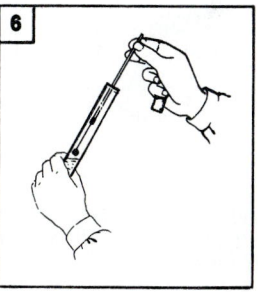

6. Insert the sterilized inoculating loop and remove a small amount of culture. (*Note:* Do not be disturbed by a sizzling noise.)

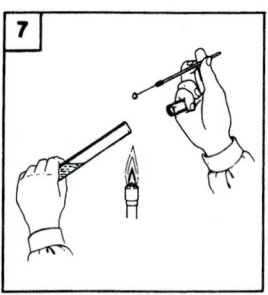

7. Flame the tube again and replace the plug or cap. Put the tube into a rack.

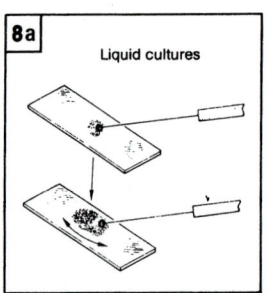

Liquid cultures

8a. With liquid cultures, place a loopful of broth on a slide, spread the broth over a small area, and allow the smear to dry.

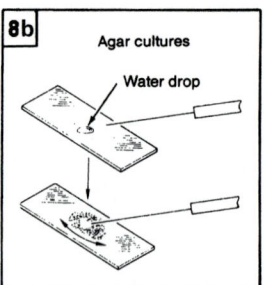

Agar cultures

Water drop

8b. With agar cultures, place a small amount of growth in a loopful of water already on the slide, mix, spread over a small area, and allow to air dry.

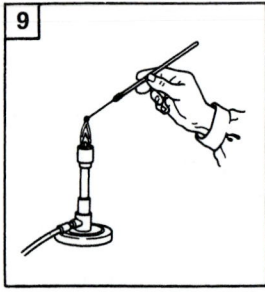

9. Flame the inoculating instrument and put it on a metal surface.

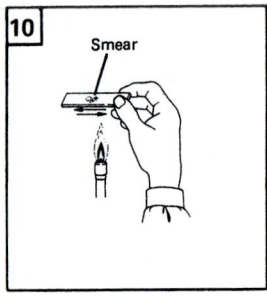

Smear

10. Pass the smear through the hot portion of the flame several times. *Do not overheat.*

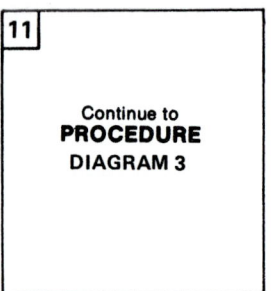

Continue to
PROCEDURE
DIAGRAM 3

Procedure Diagram 3
Simple Staining

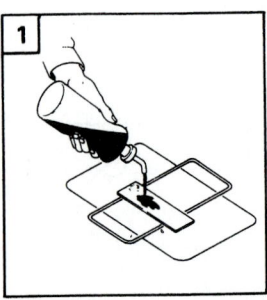

1. Apply only enough stain to cover the smear, and allow it to remain for 30 sec.

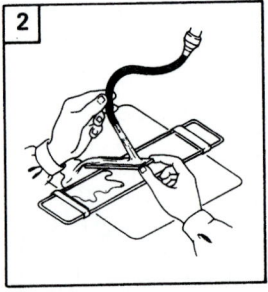

2. Pour off the stain. Rinse with slowly running water.

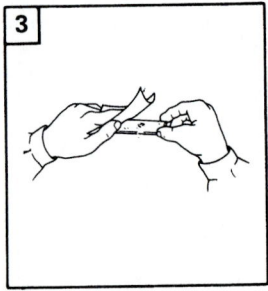

3. Dry the slide by placing it between pieces of blotting paper. *Do not rub.*

(a)

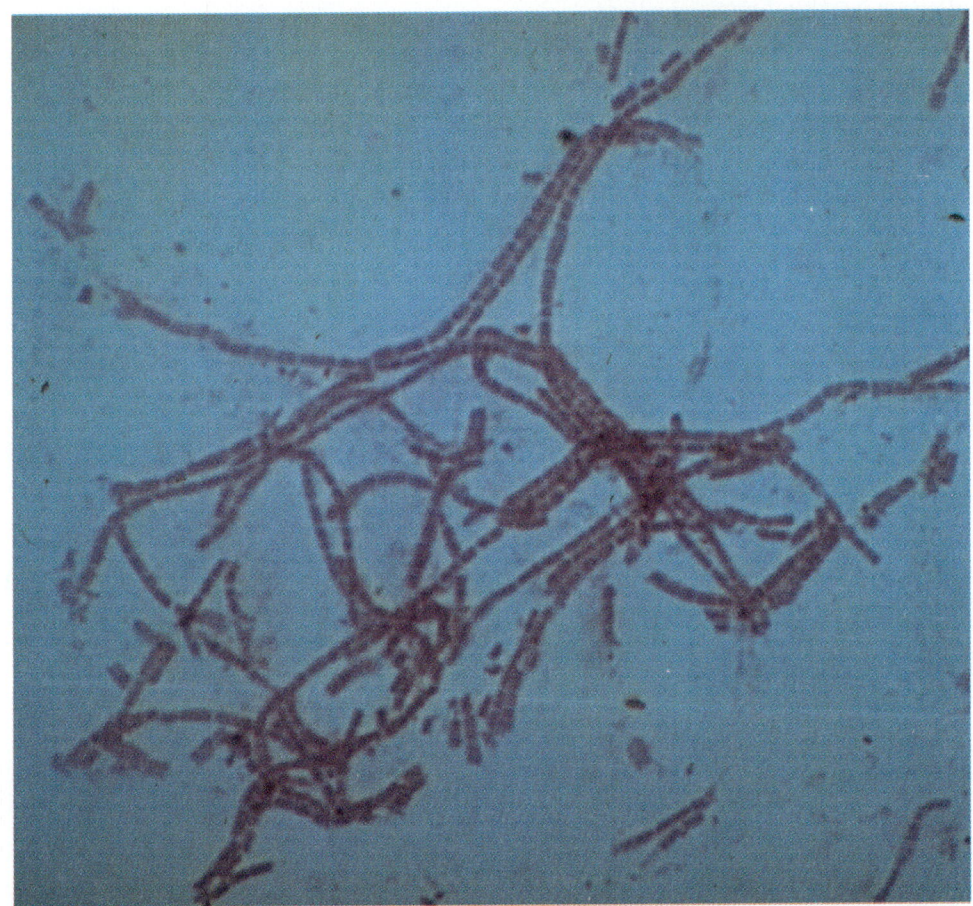

(b)

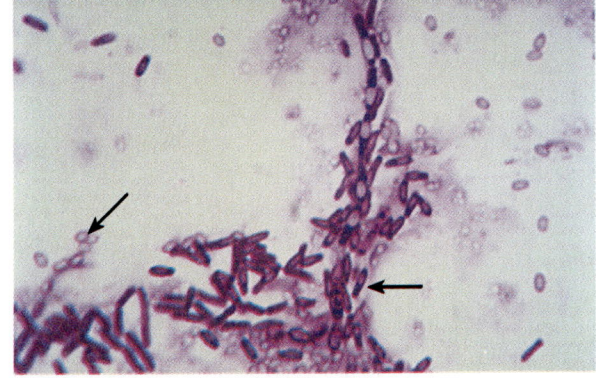

(c)

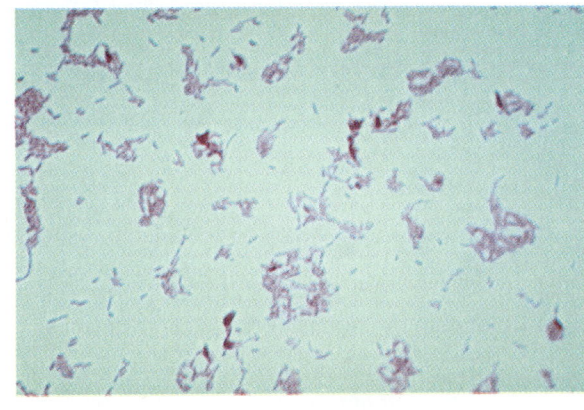

(d)

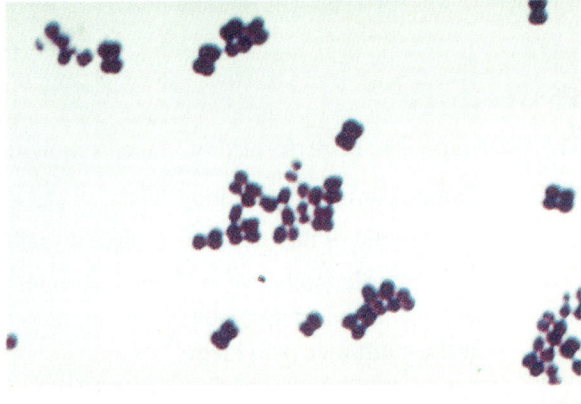

Figure 2–3

Examples of simple stained bacterial smears observed under oil immersion. (a) The rod *Bacillus subtilis* stained with safranin. (b) *B. subtilis* stained with crystal violet. This bacterium can form heat-resistant structures known as spores. In a simple stained smear such as the one shown, spores do not stain but appear as clear zones (arrows) within the bacterial cells in which they were formed or as free structures. (c) The small rod, *Escherichia coli,* stained with safranin. This microorganism is quite small and frequently gives the appearance of being a coccus. In such cases, it is not uncommon to find the term coccoid or coccobacillus being used (d) A young culture of the coccus *Micrococcus luteus* stained with crystal violet. Note the appearance of cocci in pairs, and other arrangements.

❏ 2. After performing the techniques shown in Procedure Diagrams 2 (10 steps) and 3 (3 steps), examine all preparations under the oil-immersion lens.

❏ 3. Record your findings by indicating the stain used in the numbered space under each circle in the Results and Observations section, and make a sketch of your findings in the appropriate portion of each circle. Refer to the representative morphological features of the several bacterial species shown by simple staining in Figure 2–3. (Pay particular attention to the arrangements, shapes, and relative sizes of the cells.)

❏ 4. Repeat the procedure for each of the cultures and staining solutions provided.

Results and Observations

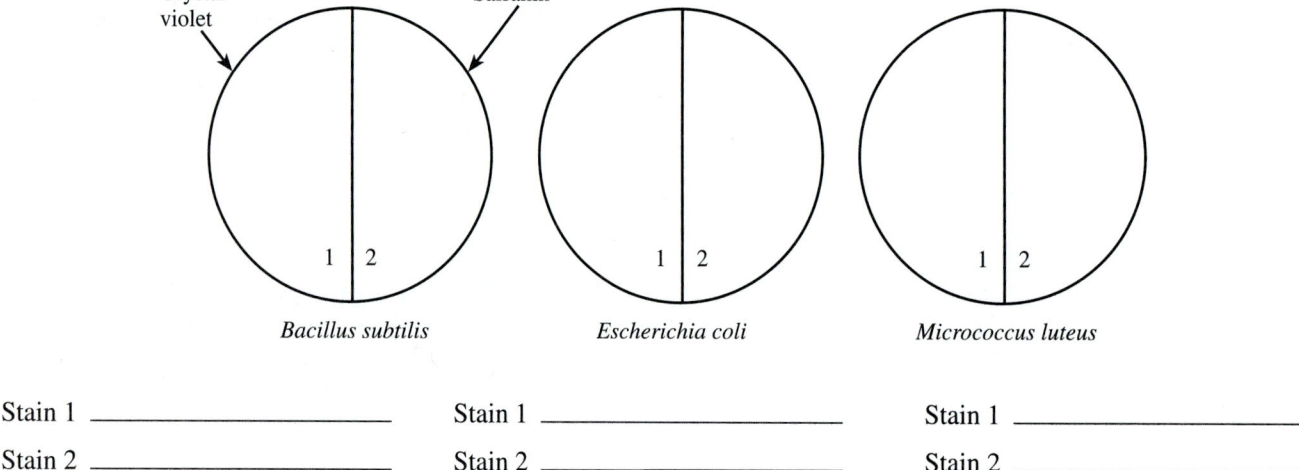

Stain 1 _____

Stain 2 _____

Stain 1 _____

Stain 2 _____

Stain 1 _____

Stain 2 _____

B. Oral Smear Preparation

Materials

The following items should be provided for general class use:

❏ 1. Sterile toothpicks

❏ 2. Staining solutions (1 bottle per 4 students)

    ❏ a. Crystal violet

    ❏ b. Safranin

❏ 3. Clean, lint-free glass slides

❏ 4. Immersion oil

❏ 5. Xylene

❏ 6. Lens paper

❏ 7. Separate containers with disinfectant for used toothpicks

Procedure

This procedure is to be performed by students individually.

❏ 1. Flame your inoculating loop.

❏ 2. With the aid of this sterile loop place a small drop of tap water on the center of a clean glass slide.

❏ 3. Take a sterile toothpick and remove a small amount of material from the area between your teeth and mix it with the tap water. Spread this mixture into a smear about the size of a dime. **Dispose of the used toothpick in the container provided.**

❏ 4. Allow the smear to air-dry, and then heat-fix.

❏ 5. Prepare a simple stained smear using either crystal violet or safranin. (Refer to Procedure Diagram 3.)

❏ 6. Examine the stained smear under oil immersion. Sketch a representative view of the microscopic field in the appropriate area of the Results and Observations section. Compare your observations with that shown in Figure 2–4.

❏ 7. Based on your findings, answer the questions in the Results and Observations section.

Results and Observations

1. Sketch your results here.

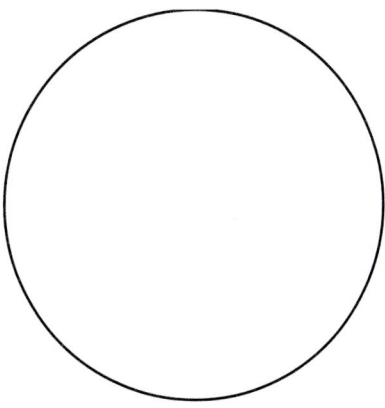

Simple stain preparation

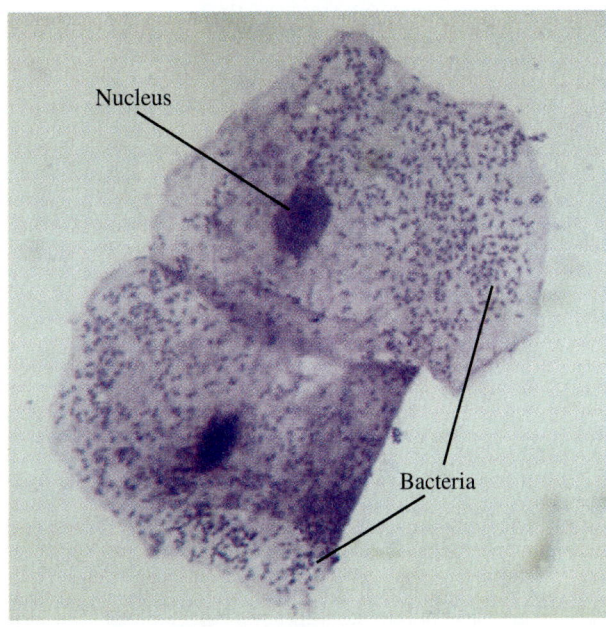

Figure 2–4

A simple stained oral smear. Note the presence of the large squamous epithelial cell. The nucleus of the cell is quite obvious.

2.   What types of cells were observed in the oral smear? _____

_____

3.   What was the major bacterial cell type in your smear? _____

_____

4.   Can bacteria be easily distinguished from other cells in a stained oral smear? _____

_____

C. Blood Smear Examination

Materials

The following items should be provided 1 per 4 students:

❏ 1. Demonstration slides of blood smears with:

    ❏ a. *Borrelia recurrentis* (bor-RE-lē-a, re-kur-EN-tis)
    ❏ b. *Treponema pallidum* (trep-ō-NĒ-ma, PAL-e-dum)
    ❏ c. *Trypanosoma gambiense* (trī-pan-ō-SŌ-ma, gam-BĒ-en-zē)

Procedure

❏ 1. Examine and compare the demonstration slides with Figure 2–5. (Refer to Exercise 1, Part E, for a discussion of blood cell types.)

❏ 2. Answer the questions in the Results and Observations section and in the Laboratory Review section.

(a)

(b)

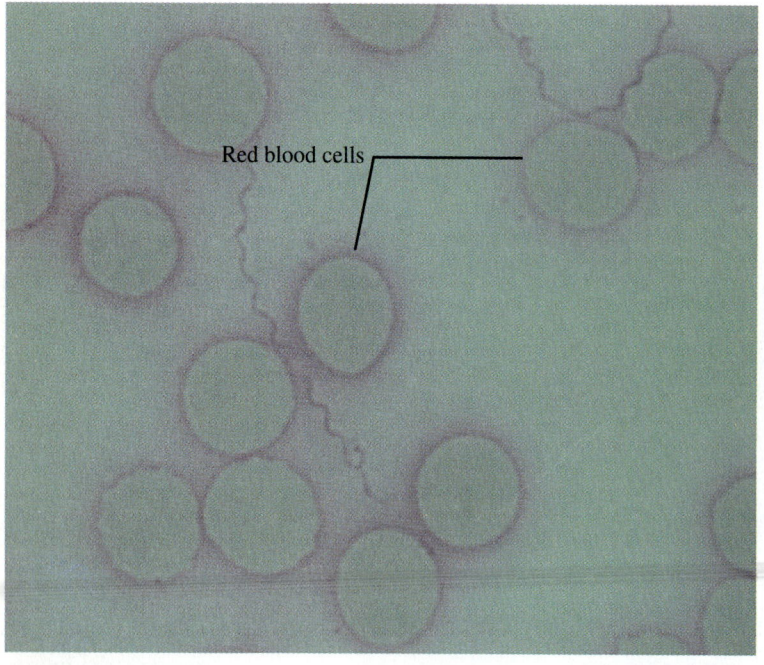
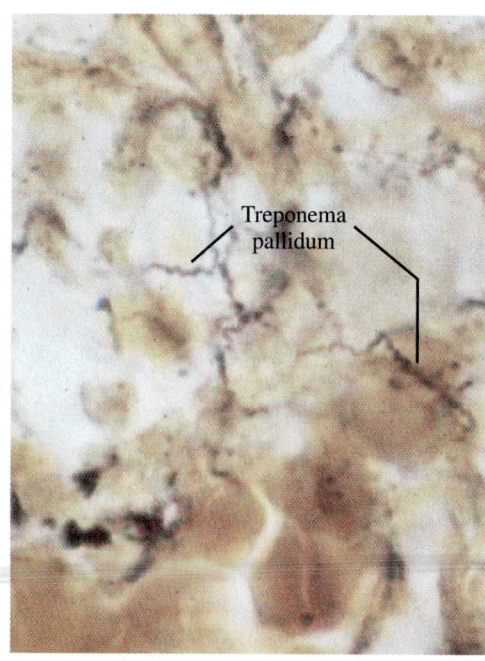

Red blood cells

Treponema pallidum

Figure 2–5

Microorganisms in blood and/or tissue smears. (a) Photomicrograph of a blood smear with the spirochete *Borrelia recurrentis,* the causative agent of relapsing fever. (b) Photomicrograph of testicular tissue containing the spirochete *Treponema pallidum,* the causative agent of syphilis. (continued)

(c)

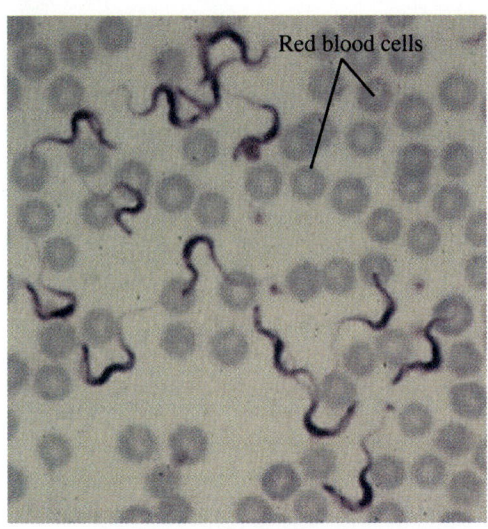

Red blood cells

Figure 2–5 (continued)
(c) A blood smear containing the causative protozoon of African sleeping sickness, *Trypanosoma gambiense*. This pathogen is spread by *Glossina* spp. (tsetse flies).

Results and Observations

1. a. Observe and describe any abnormalities in your blood smear. Pay particular attention to such details as crenation, broken cells, and uneven staining. _____

    _____

    _____

    b. Explain any and all abnormalities. _____

    _____

    _____

2. Are there differences in the location of microorganisms in relation to erythrocytes on the demonstration slides? __

    _____

Laboratory Review **2**   The Preparation of Bacterial and Oral Smears, and the Use of Simple Stains

1. Define or explain:

    a. flaming _____

    _____

    b. morphology and morphological arrangement _____

    _____

    c. dye _____

    _____

Key Terms

aseptic (ā-SEP-tik): being free of all life

aseptic technique: precautionary measures used in microbiological methods, and in dental and medical practice to prevent contamination of cultures, media, and/or individuals by environmental sources of microorganisms

dye: a chemical substance used to stain cells

micrometer (mī-KROM-e-ter): one-millionth of a meter, or 10^{-6} m (abbreviated μm)

smear: a thin film of cells

Photograph Quiz 1

1. What morphological types of bacteria are apparent in Figure 2–6. _____

_____

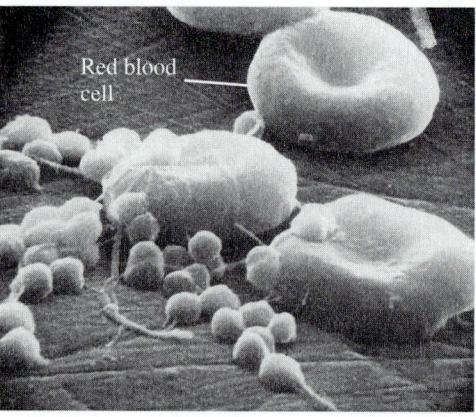

Figure 2–6

A scanning electron micrograph showing human red blood cells and bacteria.

2. If the average diameter of the red blood cells shown in this scanning micrograph is 7 μm, what would you

estimate the diameters of the bacterial cocci to be? _____

After completing this exercise, you should be able to:

1. Prepare and examine microscopically hanging-drop specimens.

2. Use the temporary wet-mount technique to detect motility.

3. Distinguish size differences among microbial types (e.g., algae, bacteria, yeast, and protozoa).

4. Distinguish between motility (vital movement) and Brownian movement.

The direct examination of live microorganisms can be extremely useful in determining size and shape relationships, motility, and reactions to various chemicals or immune sera. Two methods are in general use for studies of this type: *hanging-drop* and *wet-mount* techniques. Both methods maintain the natural shape of organisms and reduce the distorted effects that can occur when specimens are dried and fixed. Because the majority of microorganisms are not too different in either color or refractive index from the fluid in which they are suspended, a light source reduced in intensity is advisable for viewing purposes. Examination of the respective microbial cultures in this exercise will serve to emphasize size differences and other features of selected microorganisms. Compare the size differences and appearances of the organisms shown in Figure 3–1.

Live microorganisms suspended in a liquid environment always appear to be moving. The extent of movement is determined by their being **motile** or **nonmotile.** A motile organism moves slowly or rapidly in one direction. Some organisms, such as the protozoon *Paramecium,* exhibit a turning or twisting movement. Others, such as the well-known *Amoeba,* exhibit an oozing motion, caused by an extension of itself known as a **pseudopod.** Several structures can be found among protozoa that enable them to move. These are **pseudopodia, flagella,** and **cilia.**

(a)

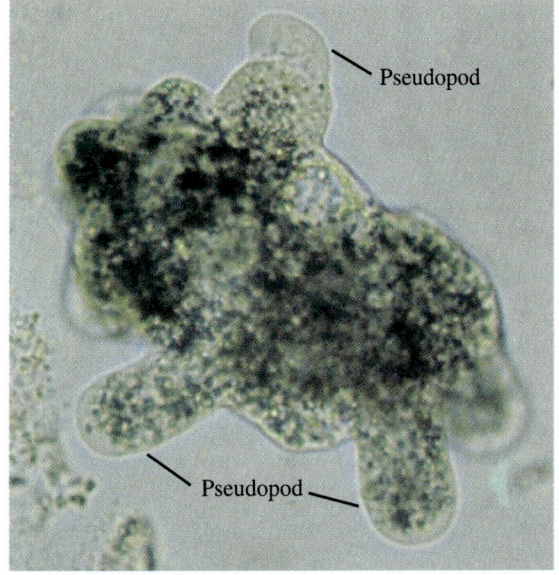

(b)

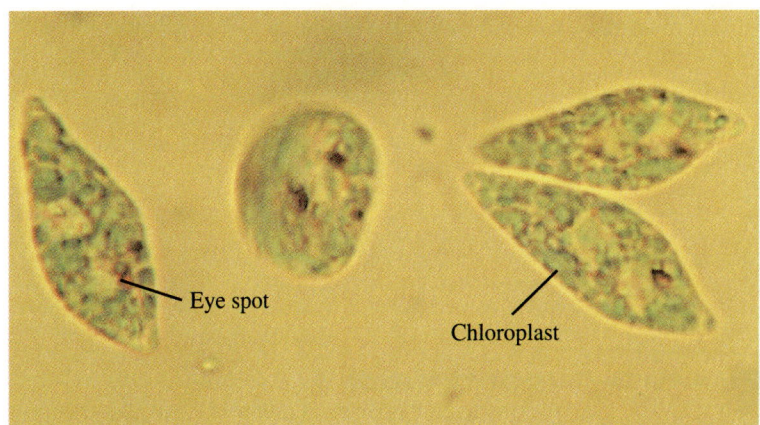

Figure 3–1

Wet-mount preparations of selected living organisms. (a) The protozoon, *Amoeba proteus* showing pseudopods. (b) The alga *Euglena* species. (continued)

(c)

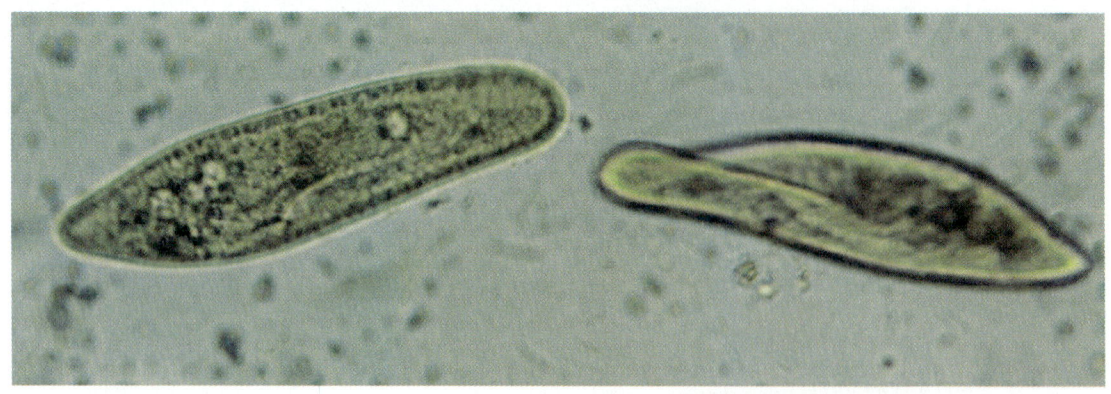

(d)

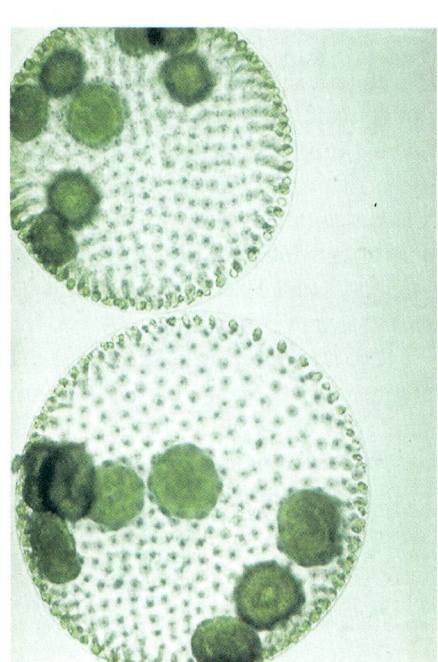

Bacillus

Saccharomyces

(e)

(f)

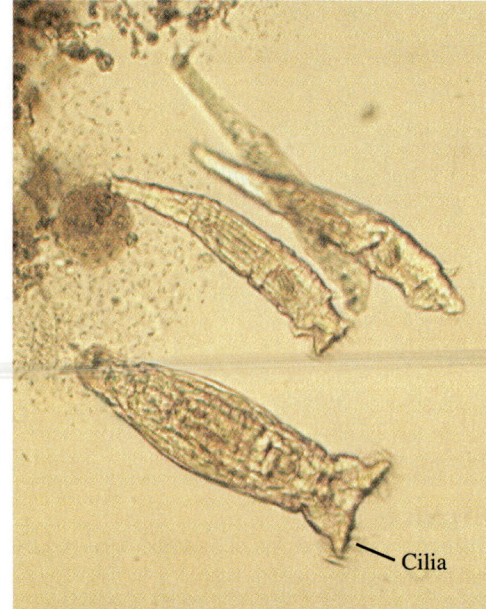

Figure 3–1 (continued)

(c) Another protozoon, *Paramecium* species. A group of these cells collected near the edge of a cover slip shown under low-power. (d) A combined preparation of the yeast *Saccharomyces cerevisiae* and the bacterium *Bacillus subtilis.* This view is under high-power. (e) Another alga, *Volvox.* (f) A low-power view of highly active worms known as rotifers. These microscopic worms are commonly found in aquatic environments.

Cilia

Motile bacteria move themselves through liquids by rotating one or more corkscrew-shaped flagella. Nonmotile organisms exhibit a dancing or jiggling motion that does not result in traveling any distance. This motion is known as **Brownian movement** and is caused by the bombardment of cells by molecules in the suspending fluid.

A number of organisms used in this exercise are rapid movers. This type of activity makes studying them difficult. One way to slow the movement of organisms is to add a thick fluid to the specimen. A number of such preparations are commercially available.

Materials

The following items should be provided per 4 students:

❑ 1. Cultures with medicine droppers of the following:

    ❑ a. *Amoeba proteus* (a-MĒ-ba, PRŌ-tē-us)

    ❑ b. *Euglena* species (ū-GLĒ-na)

    ❑ c. *Paramecium* species (par-a-MĒ-sē-um)

    ❑ d. Rotifers (RŌ-te-fers)

    ❑ e. *Volvox* (VOL-vox)

❑ 2. One 24-hour nutrient broth culture of the following:

    ❑ a. *Bacillus subtilis* (ba-SIL-us, SA-til-us)

    ❑ b. *Saccharomyces cerevisiae* (sak-a-rō-MĪ-sēz, ser-e-VIS-ī)

❑ 3. Four hollow-ground depression slides and cover slips

❑ 4. One container of each of the following:

    ❑ a. Vaseline

    ❑ b. Xylene

    ❑ c. A 2% solution of carboxymethylcellulose dissolved in 0.2 molar sucrose

❑ 5. Eight applicator sticks or toothpicks

❑ 6. Four inoculating loops

❑ 7. One wax marking pencil or felt-tip marking pen

❑ 8. Seven glass slides and cover slips

Additional Technique Required in This Exercise:

Temporary Wet-Mount Technique, Procedure Diagram 1, Exercise 1

A. Hanging-Drop Preparation

Procedure

This procedure is to be performed by students individually.

❑ 1. Perform the steps shown in Procedure Diagram 4 with each of the bacterial and yeast cultures provided.

❑ 2. Shake all cultures gently before use.

❑ 3. Observe all preparations first under low power and then under high power. **Do not use the oil-immersion objective.** *Note:* Reducing the intensity of the light source will help greatly to eliminate difficulties in locating specimens.

❑ 4. Record your findings in the Results and Observations section.

❑ 5. At the conclusion of the experiment, use xylene to remove the Vaseline from the depression slides and cover slips.

Procedure Diagram 4
Hanging-Drop Preparation

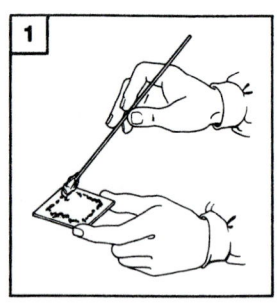

1. Apply vaseline to the edges on one surface of a cover slip and lay it down with treated surface facing upward.

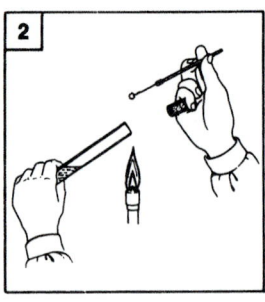

2. Remove the plug from one of the cultures provided and flame the lip of the tube.

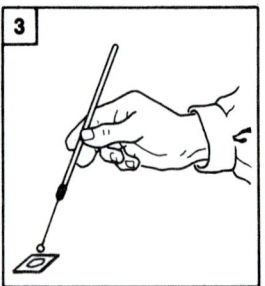

3. Remove 1 loopful of culture and place it in the center of the cover slip.

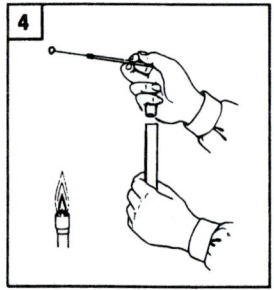

4. Flame the lip of the tube and replace the plug.

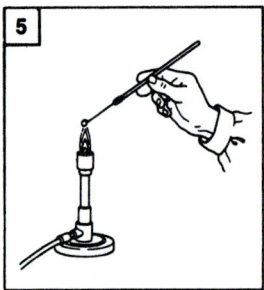

5. Flame the inoculating loop before putting it down.

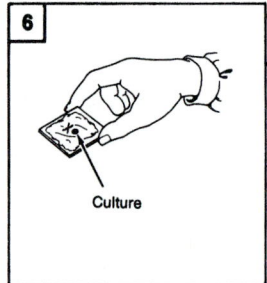

6. Make a small mark with a wax marking pencil next to the culture's edge.

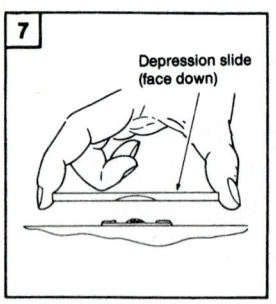

7. Lower a depression slide onto the prepared cover slip so that the depression covers the suspension. Press on the slide gently.

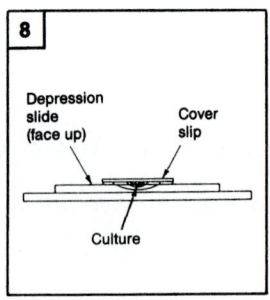

8. Lift the preparation and turn it right side up.

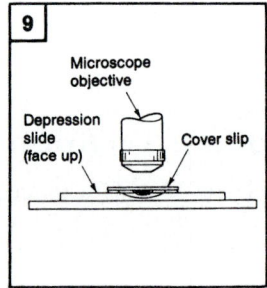

9. Locate the wax pencil mark to find the hanging drop, and examine the preparation.

B. Temporary Wet-Mount Technique

Procedure

This procedure is to be performed by students individually.

❑ 1. Perform the general steps shown in Procedure Diagram 1 (Exercise 1) with each of the protozoan, algal, and rotifer cultures provided.

❑ 2. Take specimens from the bottom of the respective containers. (The organisms are concentrated in this area.)

❑ 3. Place 1 drop of a culture on a clean slide.

❑ 4. Place 1 edge of a cover slip in contact with the culture and lower it gently until it comes into contact with the slide.

❑ 5. Repeat steps 2 through 4 and mix 1 drop of a 2% carboxymethylcellulose solution with the drop of each culture. Mix the solutions with an applicator stick before placing a cover slip on the preparation. Note any changes in the mobility of the cultures.

❑ 6. Examine all preparations first under low power and then under high power. *Note:* Reducing the intensity of the light source will help greatly to eliminate difficulties in locating specimens.

❑ 7. Record your findings in the Results and Observations section.

Results and Observations

1. Which of the cultures exhibited vital or true motility? _____

_____

2. Which cultures demonstrated Brownian movement? _____

_____

3. How did the size of the protozoa in this exercise compare with that of the other organisms observed? (Be certain to note the total magnifications used.) _____

_____

_____

Laboratory Review 3 Techniques for the Observation of Live Organisms: The Hanging-Drop and Temporary Wet Mount

1. Does the hanging-drop technique have any practical uses? List 3. _____

_____

_____

2. a. Are there any disadvantages to the hanging-drop technique? Explain. _____

_____

_____

b. Are there any disadvantages to the wet-mount technique? _____

_____

_____

3. Differentiate between vital and Brownian movements. _____

_____

_____

4. Can the possession of flagella by bacteria be utilized for classification purposes? Explain. _____

_____

_____

Key Terms

Brownian movement: a vibrating movement resulting from a bombardment of surrounding molecules

motility: the property of a cell to move a definite distance by its own power

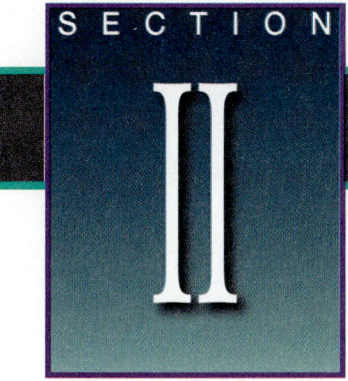

SECTION II

Cultivation Techniques

Accordingly, I took (with the help of a magnifying mirror) the stuff off and from between my teeth further back in my mouth where the heat of the coffee couldn't get at it. This stuff I mixed with a little spit out of my mouth (in which there were no air bubbles) . . . then I saw with as great a wonderment as ever before, an inconceivable great number of little animalcules . . . the whole stuff seemed to be alive and moving . . . .

A. van Leeuwenhoek, 1692

Several procedures and types of equipment are employed in the microbiology laboratory for the isolation, cultivation, and identification of microorganisms. Because pure cultures are required for determination of biological and biochemical activities of microorganisms, and because of the potential pathogenic nature of some microorganisms, the methods for handling microbes are designed to maintain culture purity and prevent contamination. Before any laboratory handling of cultures can be undertaken, you must be able to manipulate the inoculating loop and needle and to use proper transfer techniques. Exercise 4 presents the fundamental techniques used in the handling and transfer of microbial cultures. Such techniques are used throughout this manual.

Various techniques are used to (1) isolate microorganisms from different environments, (2) separate microorganisms in mixed cultures from one another, and (3) obtain pure microbial cultures. These methods include broth and agar dilution procedures and the use of enriched, selective, and differential media. For any bacterial species to be cultured for any purpose, it is necessary to provide the appropriate biochemical nutrients and biophysical environment. Several of the biophysical factors affecting bacterial growth are controlled mainly by the ingredients of the culture medium used. Such factors include osmotic pressure, pH, and water activity. Other biophysical factors are controlled by the external environment and include oxygen and temperature. Still another factor of major importance in the culturing of anaerobic bacteria is the oxidation-reduction potentials. It is important to note that no one set of conditions is satisfactory for the laboratory cultivation of all bacterial species. Bacteria have extremely diverse nutritional and physical requirements for growth. It is for this reason that nutritional media, incubation conditions, and related factors need to be adjusted precisely for each type of bacterium to insure bacterial growth. In this section, several techniques, incubation temperatures and environments, and media will be used to obtain pure bacterial cultures.

Once satisfactory conditions for cultivation have been provided, it is possible to observe, identify, and study specific cultural characteristics of particular bacterial species. It is also possible to determine the effects of various media and environmental conditions on bacterial growth and reproduction. Several exercises in this section demonstrate selected cultural properties of bacteria and emphasize the significance of these characteristics in species identification.

In the study of microorganisms, there are situations when it is necessary to determine cell numbers or activity in culture, and/or to follow the growth of a specific species. Since most microbial cultures contain thousands and even hundreds of millions of cells per milliliter (**mL**), dilutions of such preparations must be made before plating them. Exercise 6 introduces dilution techniques and their application to estimate and to determine the number of bacteria in a culture.

Obtaining a pure bacterial species involves special microbiological techniques. To insure success, students should take care during the performance of all procedures to avoid air currents created by open windows, heating systems, random movements of hands, or even talking. Laboratory workers involved in isolating or transferring microorganisms must constantly guard against the potential hazards of contaminating cultures and of acquiring a laboratory infection.

Transfer and Colony Selection Techniques

After completing this exercise, you should be able to:

1. Perform basic bacteriological transfer techniques using broth and agar cultures.
2. Handle bacteriological cultures aseptically.
3. Recognize selected properties of bacterial broth and agar slant cultures.

The inoculation of laboratory media for the cultivation and identification of microorganisms is a fundamental and important procedure. Extreme care must be exercised in applying this technique to prevent contamination of cultures. The precautionary steps used to avoid contamination constitute the procedure known as *aseptic technique.*

This exercise is designed to introduce the student to the fundamental techniques used to isolate and grow pure microbial cultures.

Most laboratory culture media are prepared in either a liquid (broth) or solid form. Meticulous care is taken to provide the proper concentration of nutrients, pH, and other factors needed for the cultivation of microorganisms. Since the ingredients and glassware employed in the preparation of media can contain unwanted microorganisms, freshly prepared media must be sterilized immediately. Test tubes containing media are stoppered with cotton plugs or loose-fitting autoclavable plastic or metal caps to prevent the entrance of foreign microorganisms. The tubes with their contents are then sterilized.

Other essential glassware items used in the microbiological laboratory are glass Petri plates. These are sterilized in the autoclave. Media are then poured into them aseptically. Many laboratories are now using presterilized plastic test tubes and Petri dishes in place of the glass varieties.

A wide variety of nutrient preparations or culture media (singular, *medium*) are in use for the cultivation of bacteria (Figure 4–1). Two examples of general-purpose media are nutrient broth and nutrient agar. The broth preparation contains beef extract, and peptone, dissolved in distilled water. Nutrient agar contains all of the ingredients found in nutrient broth plus the solidifying agent, agar, generally in a 1.5 to 2.0% concentration. The preparation of beef extract involves boiling beef in water until the water is totally evaporated. The related mixture is a sticky brown mass that contains all of the water-soluble nutrients normally found in beef tissue. The *peptone* in the medium is an enzymatically digested protein obtained from either animal or vegetable sources. Peptones contain a mixture of amino acids, carbohydrates, mineral salts, and polypeptides. Because of its composition, nutrient broth is considered to be a **complex medium.** Such media consist of water-soluble extracts of partially digested animal or plant tissues. The exact type and quantities of nutrients in such preparations are unknown. Other types of media are described in later exercises.

After the inoculation of broth media, bacteria may exhibit a particular form of growth. These include clouding of the medium *(turbidity),* accumulations of cells at the tube bottom *(sediment),* and the formation of a thin surface film *(pellicle).* Agar media may be used in the form of plates or slants (tube preparations). On agar plates, bacteria form visible accumulations of cells known as *colonies.* The appearance of colonies as to pigmentation, shape, elevation, and the patterns of growth on agar plates and the patterns of growth on agar slants are important characteristics used in the identification of bacterial species. Figures 4–2 and 4–3 show selected properties of broth and agar slant cultures, respectively. Exercises 7 and 8 provide additional information about such bacterial growth properties.

Several special precautions are used in isolation and culturing procedures. These precautions include flaming the lips of tubes and heating the entire inoculating loop or needle to redness. It is important to note that an inoculating loop is used to transfer microorganisms from a broth medium, and an inoculating needle (straight wire) is used to transfer microorganisms from a solid medium. Several isolation and culturing techniques are presented in this and later exercises. The instructor first will demonstrate these procedures, and each student in turn will perform them.

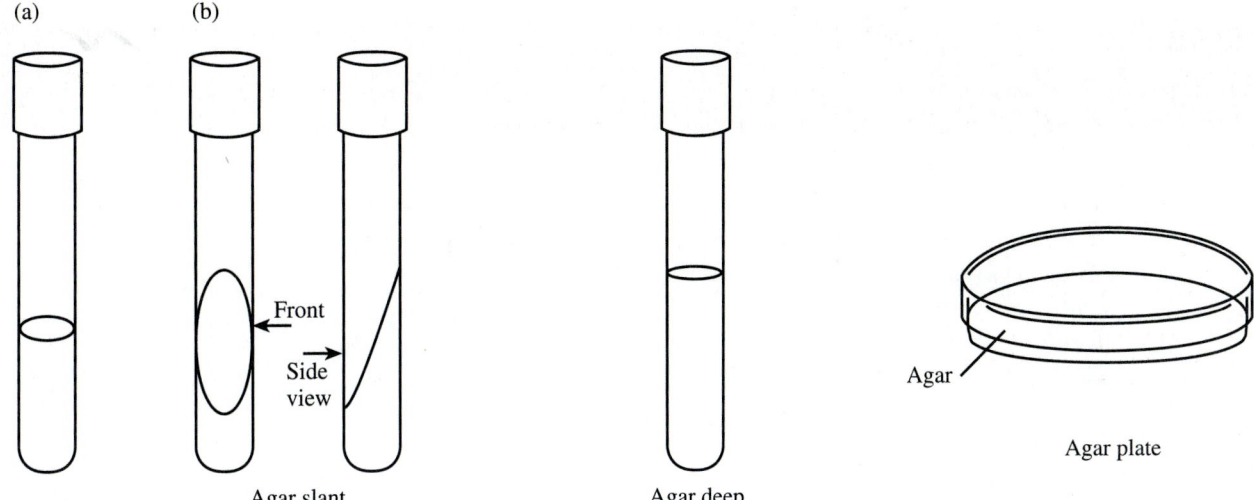

Figure 4–1
The general forms of media found in a microbiology laboratory. (a) Broth. (b) Agar preparations: agar slant, agar deep, and agar plate.

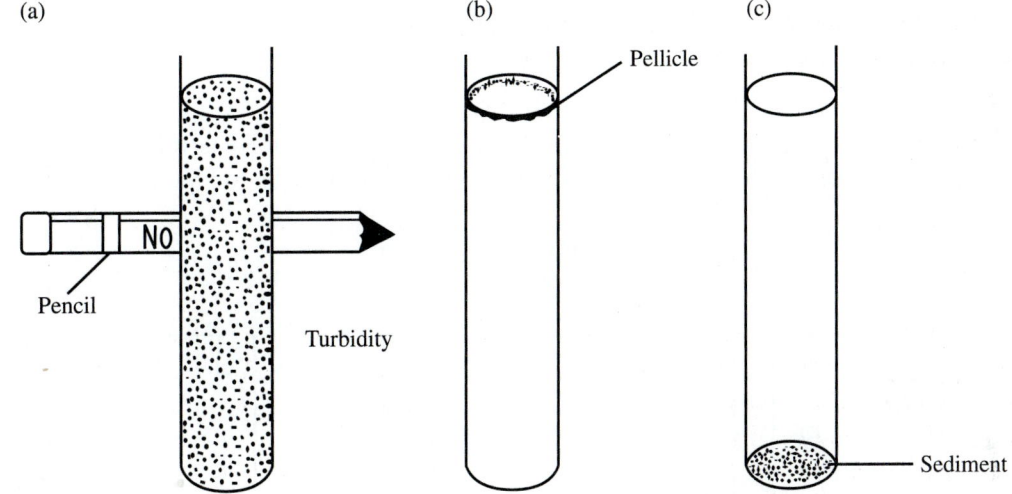

Figure 4–2
Selected broth culture properties. (a) Turbidity or cloudiness. (b) Pellicle formation. (c) Sediment formation. Note that the growth is located at the bottom of the tube. The rest of the broth is clear.

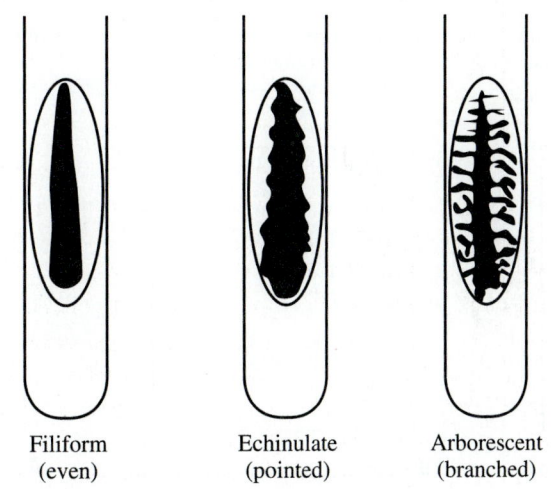

Filiform
(even)

Echinulate
(pointed)

Arborescent
(branched)

Figure 4–3
Examples of agar slant patterns. (Refer to Exercise 7, Figure 7–4 for additional patterns.)

Materials

❑ 1. Twenty-four-hour cultures of each of the following microorganisms grown in or on the media indicated (1 per 4 students):

❑ a. *Serratia marcescens* (ser-Ā-shē-a, mar-SES-enz) in nutrient broth

❑ b. *Micrococcus luteus* on a nutrient agar slant

❑ c. *Bacillus subtilis* on a nutrient agar streak plate

❑ 2. The following media should be provided per student:

❑ a. Three tubes of nutrient broth

❑ b. One nutrient agar slant

❑ 3. Additional materials:

❑ a. Inoculating loops and needles (1 per student)

❑ b. Wax marking pencil (1 per 4 students)

A. Broth Transfer

Procedure

This procedure is to be performed by students individually.

❑ 1. The transfer technique will be demonstrated first by the instructor. Observe all steps carefully. Then carry out the inoculations as shown in the 9 steps of Procedure Diagram 5.

❑ 2. Place the Bunsen burner in front of you and put all other equipment items where you will be able to reach them without difficulty and without burning yourself.

❑ 3. Incubate the inoculated cultures at room temperature for 48 hours or until the next laboratory period, and record your findings in the Results and Observations section, Part A. Refer to Figure 4–2.

Procedure Diagram 5
Broth Transfer

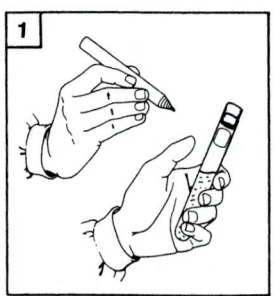

1. Label the tube to be used with the microorganism used, the date, and your name or initials.

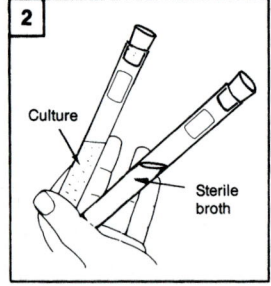

2. Take in one hand the *Serratia* broth culture and one labeled tube of sterile broth.

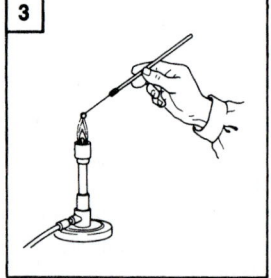

3. Take the inoculating loop with your other hand and flame the entire wire portion to redness.

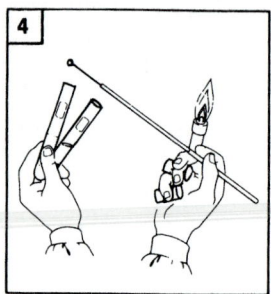

4. Remove the plugs or caps from the tubes by grasping them between the fingers of the hand holding the inoculating loop.

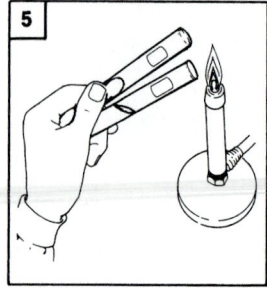

5. Flame the mouths (lips) of both tubes.

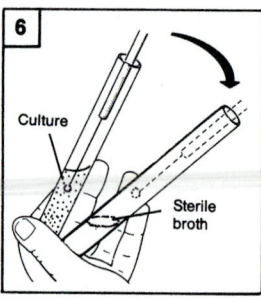

6. Insert the loop into the broth culture and obtain a loopful of culture.

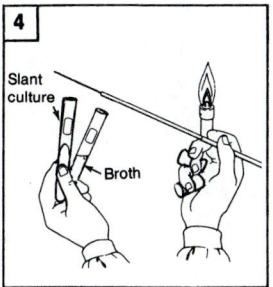

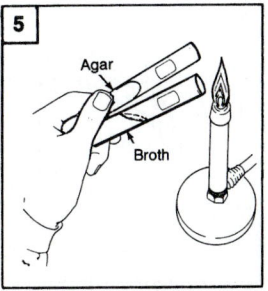

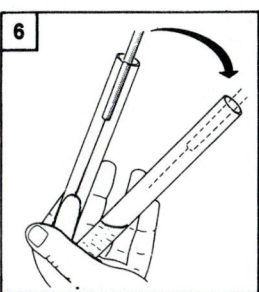

7. Introduce the inoculum by immersing the loop into the sterile broth. (Before removing the loop, touch it to the inner tube surface to eliminate any excess broth.)

8. Flame the lips of both tubes again, and replace the plugs or caps in their respective tubes.

9. Flame the inoculating loop again, and put it down on a metal surface. Incubate tubes as directed.

B. Agar Slants as Sources of Inoculum

Procedure

This procedure is to be performed by students individually.

❏ 1. The technique will be demonstrated by the instructor. Observe all 9 steps carefully. Then carry out the inoculation as shown in Procedure Diagram 6.

❏ 2. Incubate the inoculated culture at room temperature for 48 hours or until the next laboratory period, and enter your findings in the Results and Observations section, Part B. Refer to Figure 4–2.

Procedure Diagram 6
Agar Slants as Sources of Inoculum

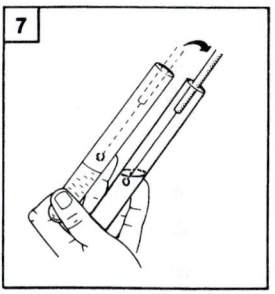

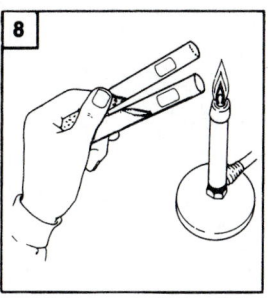

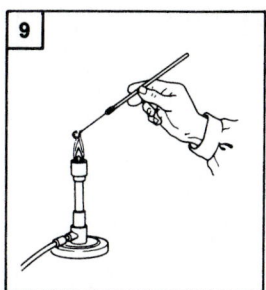

1. Label the tube with the microorganism to be used, the date, and your name or initials.

2. Take in one hand the agar slant *Micrococcus* culture and 1 tube of sterile broth. Hold the agar slant so that it faces you.

3. Take the inoculating needle with your other hand and flame the wire portion to redness.

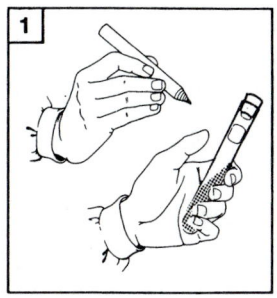

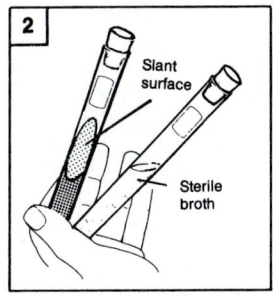

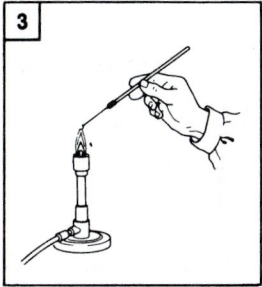

4. Remove the plugs or caps from the tubes by grasping them between the fingers holding the inoculating needle.

5. Flame the lips of both tubes.

6. Obtain an inoculum by removing a small portion of the *Micrococcus* surface growth. *Do not dig into the agar.*

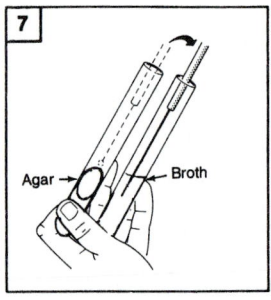

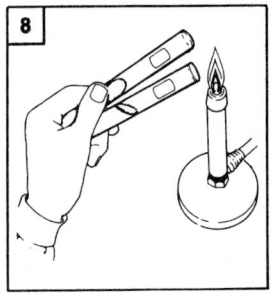

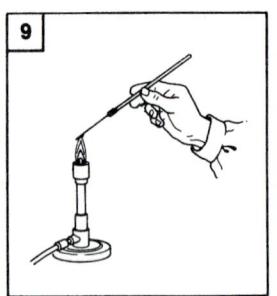

7. Immerse the inoculum in the broth and gently shake the needle to free the microorganisms sticking to it.

8. Flame the lips of both tubes again and replace the plugs or caps in their respective tubes.

9. Flame the inoculating needle and put it down in a suitable place. Incubate as directed.

C. Inoculation of an Agar Slant

Procedure

❏ 1. This technique will be demonstrated by the instructor. Following the demonstration, carry out the inoculation as shown in the 9 steps of Procedure Diagram 7.

❏ 2. Incubate the inoculated tube at room temperature at 48 hours or until the next laboratory period, and record your findings in the Results and Observations section, Part D. Refer to Figure 4–3.

Procedure Diagram 7
Inoculation of an Agar Slant

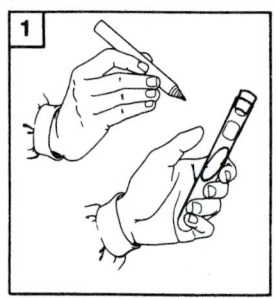

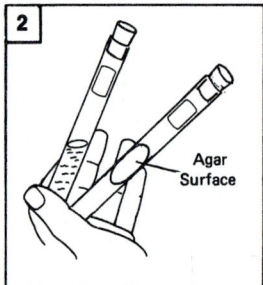

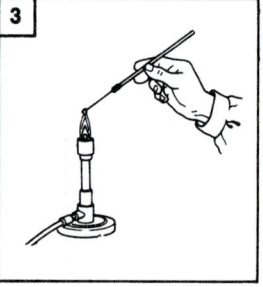

1. Label the agar slant with the organism to be used, the date, and your name or initals.

2. Take in one hand the *Serratia* broth culture and 1 sterile nutrient agar slant. Hold the slant surface so that it faces upward and toward you.

3. Take the inoculating loop with your other hand and flame the entire wire portion to redness.

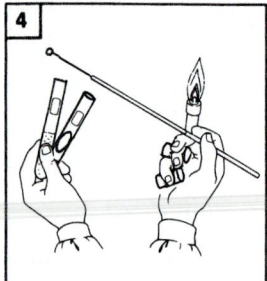

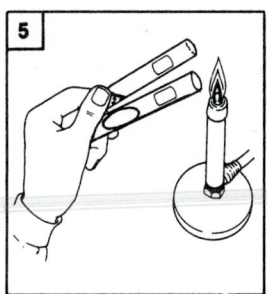

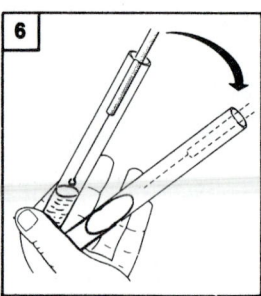

4. Remove the plugs or caps from the tubes by grasping them between the fingers of the hand holding the inoculating loop.

5. Flame the lips of both tubes.

6. Obtain an inoculum by removing a loopful of the broth culture.

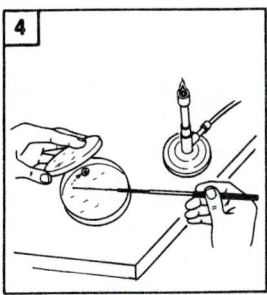

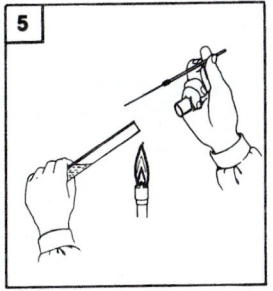

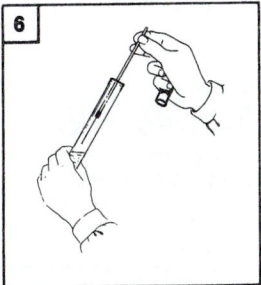

7. Place the loop into the sterile agar slant's surface at its bottom. Move the loop from side to side as you pull it upward out of the tube. *Do not dig into the agar.*

8. Flame the lips of the tubes and replace the plugs or caps.

9. Flame the inoculating loop and place it on an appropriate surface. Incubate as directed.

D. Colony Selection

Procedure

❑ 1. This technique will be demonstrated first by the instructor. Observe all steps carefully. Then carry out the inoculation as shown in the 8 steps of Procedure Diagram 8.

❑ 2. Incubate the inoculated tube at room temperature for 48 hours or until the next laboratory period, and record your findings in the Results and Observations section, Part C. Refer to Figure 4–2.

**Procedure Diagram 8
Colony Selection**

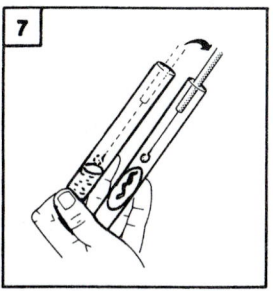

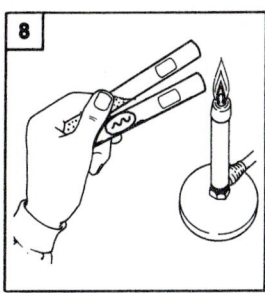

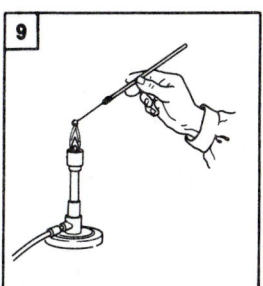

1. Label the tube with the microorganism to be used, the date and your name or initials.

2. Select an isolated *Bacillus* colony and circle it with a wax pencil on the Petri plate's bottom.

3. Flame an inoculating needle to redness.

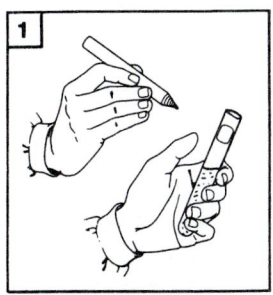

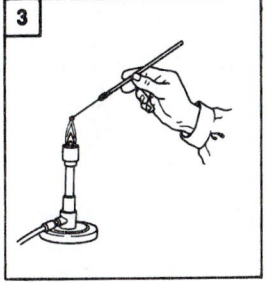

4. Lift the Petri plate top, cool the needle by touching a clear agar surface, and remove a small amount of growth from the circle colony. Close the lid.

5. Remove the plug or cap from one tube of sterile broth and flame the lip of the tube.

6. Immerse the inoculum in the broth and shake the needle to free the microbial growth sticking to it.

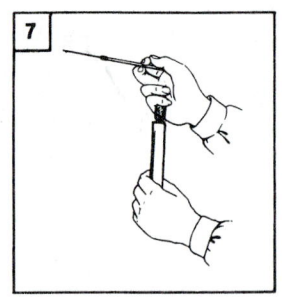

7. Flame the lip of the tube and replace the plug or cap.

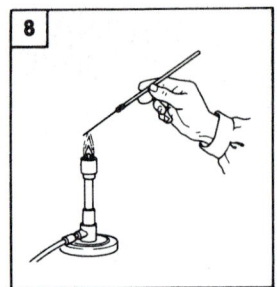

8. Flame the inoculating needle and put it down on an appropriate surface. Incubate as directed.

Results and Observations (A through D)

1. Description and comparison of broth cultures. Refer to Figure 4–2 and Exercise 7 for descriptions of broth culture properties. Enter your findings in Table 4–1.

Table 4–1

Inoculation Results

Properties	Part A[a]	Part B[b]	Part C[c]
Pellicle			
Turbidity			
Sediment			
Pigment			
Odor			

[a]Broth transfer
[b]Agar slants as sources of inoculum
[c]Inoculation of an agar slant

2. Description of agar slant culture. Refer to Figure 4–3 and Exercise 7 for descriptions of agar slant culture properties. Enter your findings in Table 4–2.

Table 4–2

Agar Slant Properties

Properties	Part D
Amount of growth	
Pigment	
Agar slant pattern	
Other	

Laboratory Review **4**

Transfer and Colony Selection Techniques

1. List 4 specific characteristics of broth cultures.

   a. _____   c. _____

   b. _____   d. _____

2. What is a bacterial colony?

   _____

3. List 4 general forms of media used in a microbiology laboratory.

   a. _____   c. _____

   b. _____   d. _____

4. List 4 properties of bacterial colonies visible on a nutrient agar plate or related medium.

   a. _____   c. _____

   b. _____   d. _____

5. a. What is the composition of nutrient broth?

   _____

   _____

   b. How does nutrient agar differ from nutrient broth?

   _____

   _____

6. List 2 ways to sterilize laboratory media.

   a. _____   b. _____

7. Define:

   a. protein _____

   b. carbohydrate _____

   c. peptone _____

   d. beef extract _____

Key Terms

agar (Ā-gar): a dried polysaccharide extract obtained from species of red algae, and used as the solidifying agent of various types of microbiological media; agar generally melts at 99–100°C, is solid at 42°C, and is attacked enzymatically by very few microbial species

aseptic (a-SEP-tik) technique: in this exercise, aseptic technique refers to precautionary measures taken to prevent the contamination of cultures or sterile media

bacterial colony: a visible accumulation of bacteria on a solid culture medium

broth medium: a liquid preparation used to grow microorganisms such as bacteria

contamination (kon-tam-i-NĀY-shun): the introduction of unwanted microorganisms into or onto normally sterile materials

sterile (STER-il): free from any living form of microbial life

Pour Plate and Streak Plate Techniques for Isolating Pure Cultures

After completing this exercise, you should be able to:

1. Isolate individual colonies from mixed cultures by means of the streak plate and pour plate techniques.
2. Recognize the advantages and disadvantages of the streak plate and pour plate techniques.
3. Recognize selected properties of bacterial colonies on agar plates.

Certain procedures have become indispensable to bacteriologists. Among them are the standard pour plate and streak plate techniques. These methods can be effective in both the detection and enumeration of different microorganisms present in typical specimens used for study.

The forerunner of the present pour plate method was developed in the laboratory of the famous bacteriologist Robert Koch. Today this technique consists of: (1) cooling a melted agar-containing medium (1.5% agar) to approximately 43° to 45°C and (2) inoculating the medium with a specimen just prior to pouring it into a sterile Petri plate. Thus, bacteria are distributed throughout the agar and trapped in position as the medium hardens. Although the solidified medium restricts bacterial movement from one area to another, it is of a soft enough consistency to permit growth. Growth occurs both on the surface and within the inoculated medium. Unfortunately, there are several disadvantages in this technique, including the following: (1) colonies of several species may present a similar appearance in the agar environment; (2) certain species of bacteria may not grow in this environment; and (3) difficulty may be encountered in removing (picking) colonies for further study. Figure 5–2 shows the features of the pour plate.

The streak plate procedure is another example of a dilution technique. It was originally developed by two bacteriologists, Loeffler and Gaffky, in the laboratory of Robert Koch. The modern method for the preparation of a streak plate involves the spreading of a single loopful of material containing microorganisms over the surface of an agar medium that has been allowed to solidify. Figure 5–1 shows streaking methods for spreading the initial inoculum over the agar surface. In practice, the streaks are closer together and greater in number than those shown here. (Variations of these techniques are in general use. The type of streaking shown in Figure 5–1*d*, for example, is the basis for several different streaking methods.) Figure 5–3 shows the features of a typical streak plate.

Before the inoculation of any medium is performed, the plate bottom should be labeled to identify the type of culture and other important information. The plate bottom is used since it cannot be separated from the medium. Plate tops can accidentally be interchanged.

After inoculation by any method, plates are incubated at desired temperatures. Water of condensation forms in most Petri plates as a consequence of the high concentration of water in agar. To prevent water from forming on media surfaces and causing bacterial colonies to run together, incubate Petri plates in an inverted position.

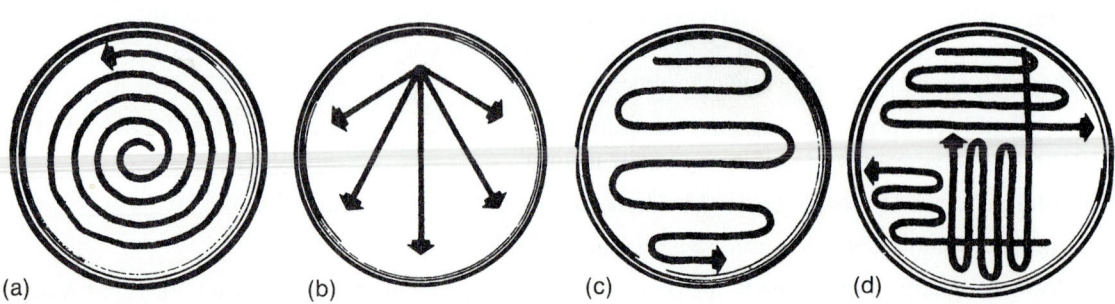

(a)   (b)   (c)   (d)

Figure 5–1
Representative streaking methods.

Materials

❏ 1. Twenty-four-hour nutrient broth mixed culture containing *Micrococcus luteus* and *Serratia marcescens* (1 per 4 students)

❏ 2. The following materials should be provided per student:
   ❏ a. One nutrient agar deep
   ❏ b. One sterile Petri plate
   ❏ c. One nutrient agar plate

❏ 3. Water bath facility to melt agar deeps

❏ 4. A wax marking pencil or other marking device (1 per 4 students)

A. Pour Plate Technique

Procedure

This procedure is to be performed by students individually.

❏ 1. This technique for obtaining a pure culture from a mixture will be demonstrated by your instructor. Observe the steps and perform the procedure as shown in Procedure Diagram 9.

❏ 2. Before beginning the procedure, label the bottom of a sterile Petri plate with your name, the date, the organism used, and the temperature of incubation.

❏ 3. Incubate the pour plate in an inverted position in your desk drawer or as directed until the next laboratory period.

❏ 4. Repeat the procedure if indicated by your instructor.

❏ 5. In the appropriate space of the Results and Observations section, sketch the appearance of your plate after incubation. Refer to Figure 5–2.

(a)

(b)

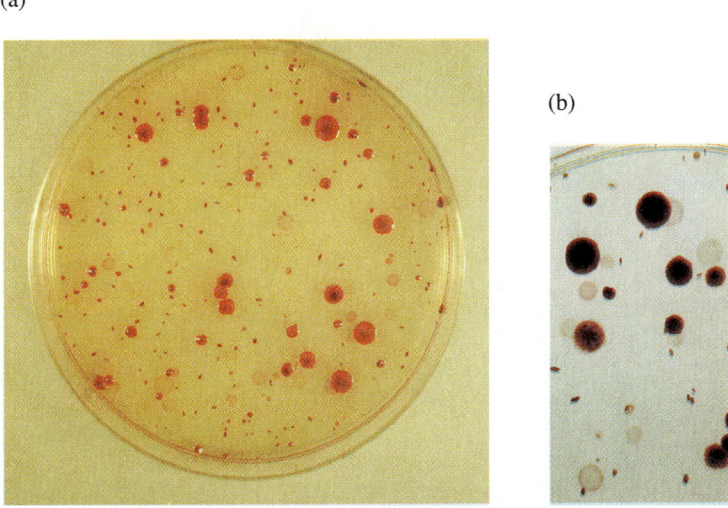

Figure 5–2

Pour plate preparation of *Micrococcus luteus* and *Serratia marcescens.* Note the distribution of different colonies. A higher magnification of this preparation shows the various shapes of bacterial colonies. Compare the colonial forms with the appropriate Figure in Exercise 7. Would you expect some difficulty in removing a colony embedded in the agar medium?

Procedure Diagram 9
Pour Plate Technique

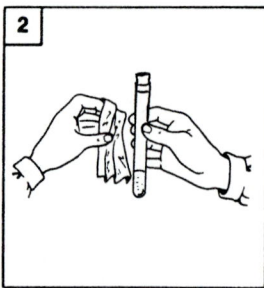

1. Place nutrient agar deeps into the boiling water bath for melting.

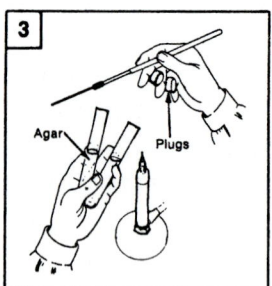

2. Remove a melted deep, cool to 45°C (this temperature can be tolerated on the inner surface of the forearm), and wipe the tube to remove any moisture.

3. Take the tubes of cool melted agar and the mixed culture in one hand. Flame the inoculating loop and the lips of the tubes after removing their plugs or caps.

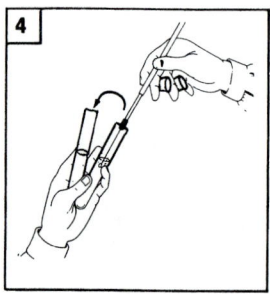

4. Remove 1 loopful of the culture and inoculate the melted deep.

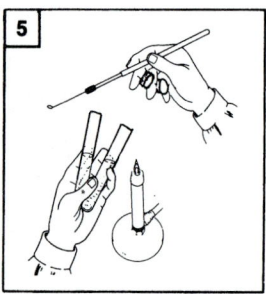

5. Flame the lips of both tubes, replace the plugs or caps correctly, and put the tubes down. Flame the loop and put it down.

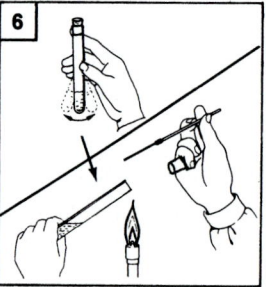

6. Pick up and gently shake the freshly inoculated deep. Remove the plug or cap and flame the lip of the tube.

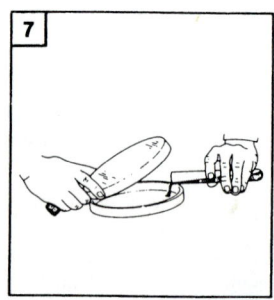

7. Pour the contents of the tube into the bottom of a labeled Petri dish by slightly raising the dish's top.

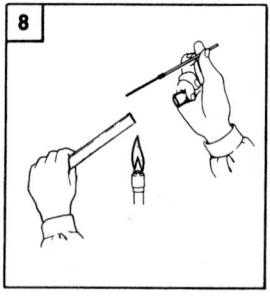

8. Flame the lip of the tube, replace the plug, and dispose of the contaminated tube as directed.

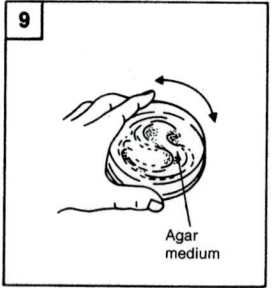

9. Rotate the plate gently to distribute the agar evenly, and allow the agar to harden (about 10 min).

B. Streak Plate Technique

Procedure

This procedure is to be performed by students individually.

❑ 1. This second technique for obtaining a pure culture from a mixture will be demonstrated by your instructor. Observe the steps and perform the technique as shown in Procedure Diagram 1.

❑ 2. Label the bottom of a sterile Petri plate with your name, the date, the organisms used, and the temperature of incubation.

❑ 3. Incubate your streak plate in an inverted position in your desk drawer or as directed until the next laboratory period.

❑ 4. Repeat the procedure as directed by your instructor.

❑ 5. Sketch the appearance of your plate after incubation in the appropriate space of the Results and Observations section. Refer to Figures 5–3 and 5–4.

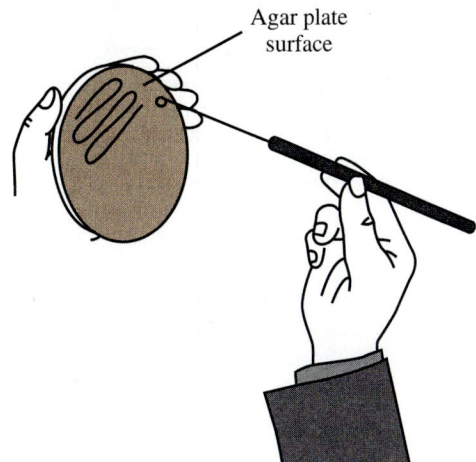

Agar plate surface

Figure 5–3
Starting the streaking of an agar plate. Note the way the inoculating instrument is held. This is a fundamental technique.

(b)

(a)

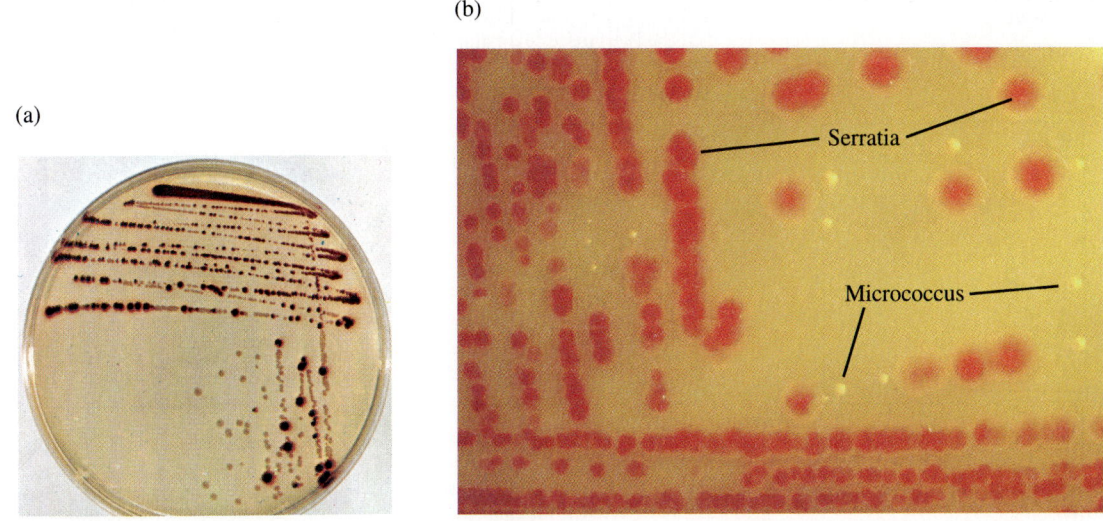

Serratia

Micrococcus

Figure 5–4
(a) Streak plate preparation with a mixed culture containing *Micrococcus luteus* and *Serratia marcescens*. (b) A close-up of the agar surface. Note the well-separated colonies.

**Procedure Diagram 10
Streak Plate Technique**

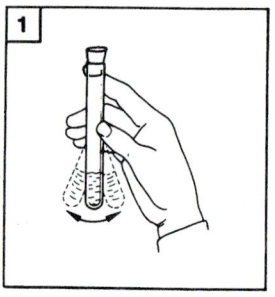

1. Shake the mixed culture tube gently.

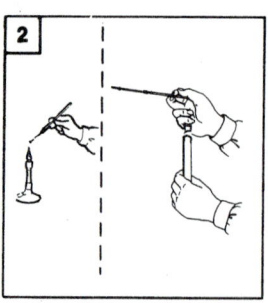

2. Flame the inoculating loop to redness, remove the plug or cap of the tube with the free fingers of the hand holding the inoculating loop, and flame the lip of the tube.

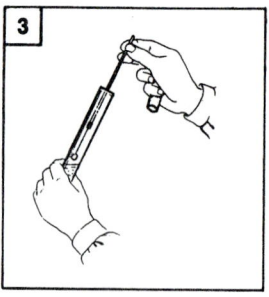

3. Remove a loopful of the mixed culture after the loop has cooled for at least 5 sec.

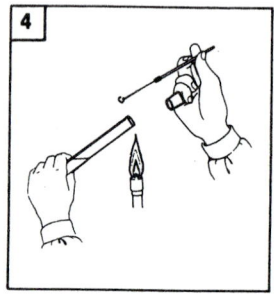

4. Flame the tube once again, replace the plug or cap, and put the tube in a rack or another appropriate place.

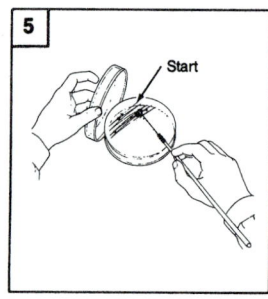

5. Hold the Petri plate so that the bottom rests on the palm of the left hand if you are right-handed or the right if you are left-hand. Lift the Petri plate cover and place the inoculum at the edge of the agar surface farthest from you.

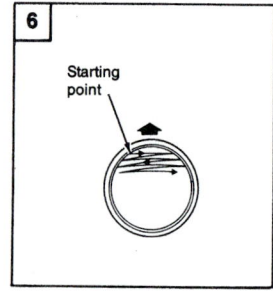

6. Streak the inoculum on the agar surface from side-to-side in parallel lines covering approximately one quarter of the plate and in the pattern shown. Do not dig into the agar.

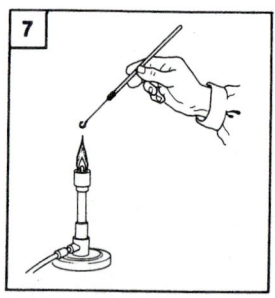

7. Lower the top of the plate and flame the inoculating loop. Rotate the Petri plate one quarter of a full turn.

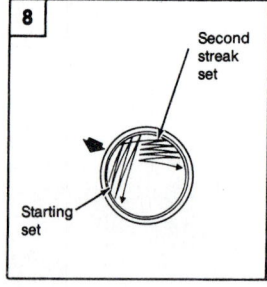

8. Lift the dish top and cool the inoculating loop by touching the agar surface away from the set of streaks. Skim the inoculating loop over the first set of streaks, once, and make a second set of streaks. Lower the plate top

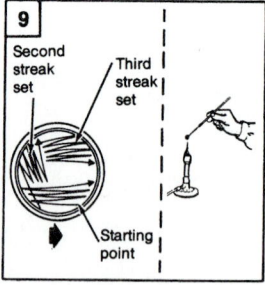

9. Repeat steps 7 and 8. The final result should be a set of distinct streaked regions as shown. Flame the inoculating loop before putting it down. Incubate the plate in an inverted position.

Results and Observations

1. Sketch your results here.

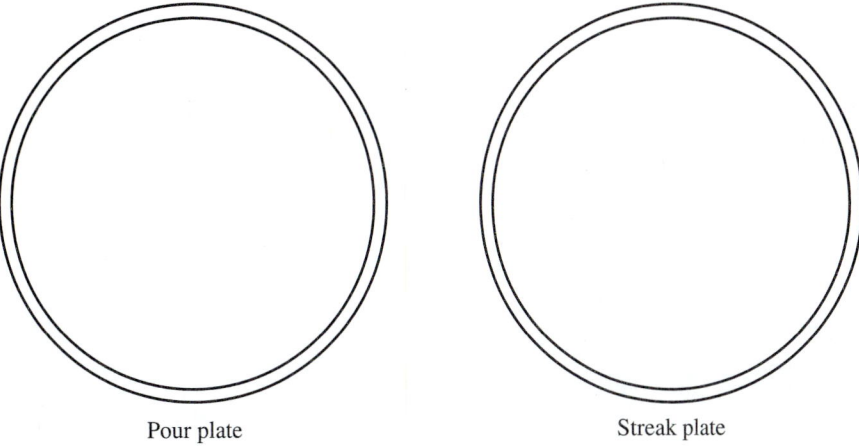

Pour plate   Streak plate

2. Complete Table 5–1 by indicating the colonial properties of *Serratia marcescens* and *Micrococcus luteus.* (Refer to Figure 5–5.)

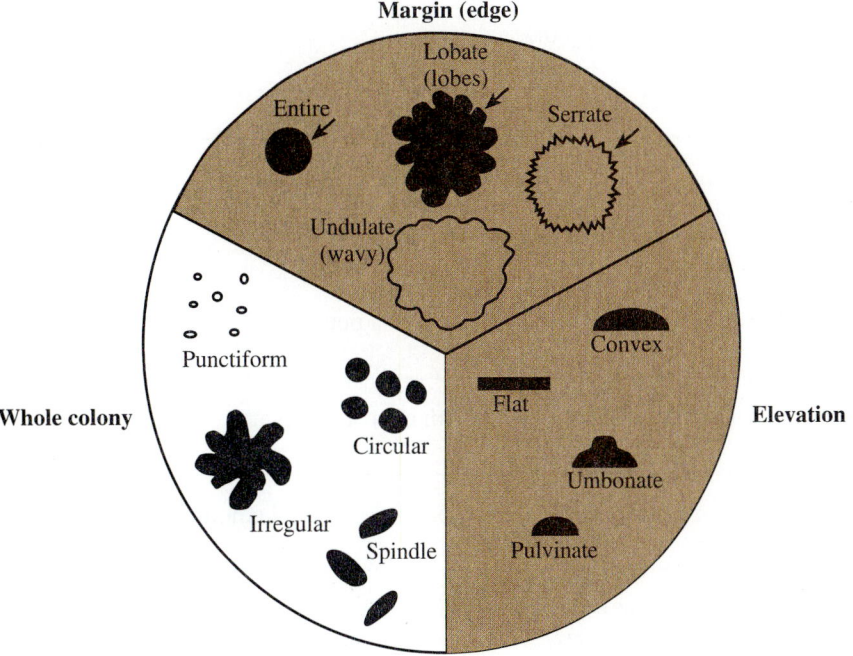

Figure 5–5
Selected bacterial colony properties on agar plate media.

Table 5-1

Colonial Properties

Colonial Characteristics	Microbial Cultures			
	Micrococcus luteus		Serratia marcescens	
	SP[a]	PP	SP	PP
Pigmentation				
Margin				
Whole Colony				
Elevation				

[a]Explanation of symbols: PP = pour plate; SP = streak plate.

Laboratory Review 5 — Pour Plate and Streak Plate Techniques for Isolating Pure Cultures

1. What is the purpose of inverting inoculated plates during incubation? _____
   _____

2. Where should colonies appear in the case of:
   a. streak plates _____
   b. pour plates _____

3. Indicate the temperature ranges for the following microbial categories (refer to your textbook):
   a. psychrophiles _____ b. mesophiles _____
   c. thermophiles _____

4. What factor(s) could account for an absence of growth on a pour plate? _____
   _____

5. What factor(s) could account for an absence of growth on a streak plate? _____
   _____

6. What explanation could be given for the failure of obtaining isolated colonies on a streak plate?_____
   _____

7. Compare gelatin with agar-agar with respect to the properties listed in Table 5–2.

Table 5–2

Gelatin and Agar Comparison

Properties	Gelatin	Agar
Chemical composition		
Temperature required for melting		
Temperature required to solidify		
Possibility of enzymatic attack by bacteria (yes or no)?		

Key Terms

agar (A-gar): a dried polysaccharide extract obtained from species of red algae, and used as the solidifying agent of various types of microbiological media; agar generally melts at 99–100°C, is solid at 42°C, and is attacked enzymatically by very few microbial species

bacterial colony: a visible accumulation of bacteria on a solid culture medium

dilution (di-LOO-shun): a process by which the contents of a preparation are thinned or spread out

pour (PŌR) plate: a basic technique used in the culturing of and/or isolation of bacteria in which a melted, yet sufficiently cooled, medium is inoculated with a bacterial culture, introduced into a sterile Petri dish, and allowed to harden; the individual bacteria trapped within the medium grow and eventually form colonies

streak (STRĒK) plate: another basic technique used in the culturing of and/or isolation of bacteria in which an inoculum is spread over the surface of a medium by means of an inoculating loop or needle; the isolated bacteria in the inoculum grow and form colonies on the surface of the medium

After completing this exercise, you should be able to:

1. Use and manipulate various pipettes and pipettors to safely transfer liquids and/or bacterial cultures.
2. Calculate and perform dilutions.
3. Use metric units of measurement for liquids.
4. Perform the standard plate count technique.
5. Determine the number of viable bacteria/milliliter (mL) by means of a spread-plate technique.

In various types of microbiological studies such as the examination or inspection of certain foods, beverages, water, and even cosmetics, it is important to know how many live bacteria are present. Performing such determinations demonstrates that if bacteria are alive or viable, they are generally capable of forming colonies in or on a suitable solid medium. Therefore, the number of cells in a specific volume of material capable of dividing on or in a solid agar medium is referred to as the **viable number.** For practical reasons, however, the viable number is expressed in terms of **colony-forming units (CFUs).**

Determining viable numbers in a culture or other types of specimens generally requires a procedure to *dilute* or sufficiently separate bacteria so that they can form distinct individual colonies. Because such cultures and specimens usually contain several thousand or even hundreds of millions of cells per milliliter, dilutions must be performed before the plating of culture samples takes place. A **serial dilution procedure,** which is a method of sequentially diluting a culture sample through a series of sterile dilution blanks, is shown in Figure 6–1. The fluid in the *dilution blanks* may be distilled water, saline, or other suitable liquid in which microorganisms can be suspended.

This exercise will use serial dilutions and will emphasize measurements used for liquids. Such measurements are based on the metric system unit for volume, namely the **liter** (L). In the laboratory, the two most useful subunits of the liter (which is approximately equal to a quart) are the **milliliter** (ml) and the **microliter** (μL). The relationship of these volumes are as follows:

$$1{,}000 \text{ mL} = 1 \text{ L}$$
$$1 \text{ ml} = 0.001 \ (10^{-3}) \text{ L}$$
$$1 \text{ mL} = 1{,}000 \ \mu\text{L}$$
$$1 \ \mu\text{L} = 0.000001 \ (10^{-6}) \text{ L}$$

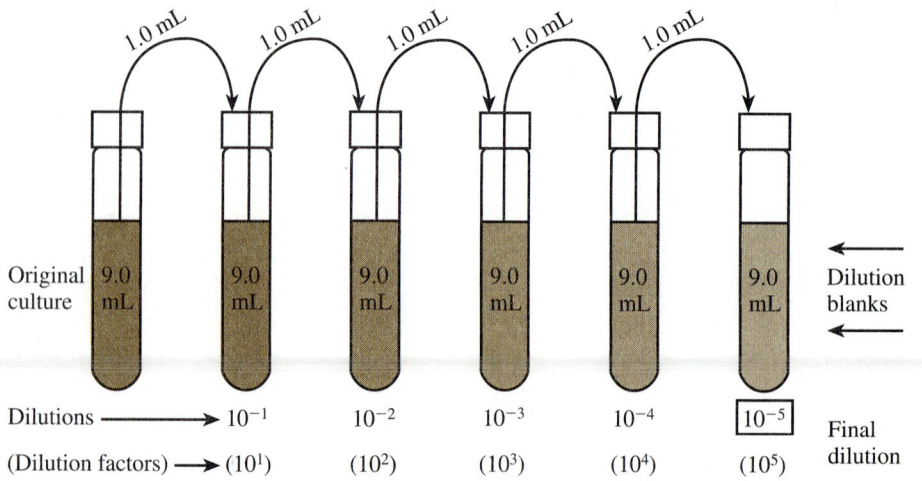

Figure 6–1

General steps in a dilution procedure. The *dilutions* and *dilution factors* are shown.

With this information in mind, examine Figure 6–1 to see how dilutions are made and how to calculate the final dilution of a sample or suspension. By transferring 1.0 mL from the original culture to 9 mL of a dilution blank *(diluent),* a 1:10 (one-to-ten) dilution is obtained. The dilution also can be expressed as 10^{-1} or 1/10. Taking 1.0 mL from this 1:10 dilution and placing it into another 9 mL of a dilution blank results in a 1:100 or 10^{-2} dilution. The dilution in this tube is calculated by multiplying the two *dilution factors,* $10^{-1} \times 10^{-1}$ (or 1/10 × 1/10) $= 10^{-2}$ or 1:100. The expression of this last dilution (10^{-2}) is referred to as the *dilution factor.*

The delivery or transfer of cultures and/or fluid volumes is done by means of pipettes or pipets (Figure 6–2) and one of several types of pipetting aids such as a *Pi-Pump* (Figure 6–3a) or a *Brinkmann Pipette Helper* (Figure 6–3b). In microbiology laboratories pipetting by mouth is unsafe since microorganisms may accidentally be sucked up through a pipette and then ingested. To avoid laboratory accidents pipette helpers are used.

Figure 6–2
Examples of commercially available, prepackaged, sterile, disposable pipettes. (Reprinted with permission from Corning Incorporated.)

Two types of pipettes are used in most microbiology laboratories. These are the *serological pipette,* which delivers a measured volume and is calibrated to deliver definite volume when completely empty, and the *measuring pipette,* which delivers a measured volume. Most manufacturers of serological pipettes print the type of pipettes on the device. Figure 6–4 shows general features of a serological pipette. Both types of pipettes are usually calibrated *to deliver* (**TD**) the volume of a nonviscous fluid indicated on the upper part of the pipette at 20°C. Pipettes generally have the *total volume* that can be held and the *smallest graduated division* imprinted near the top of the individual pipettes. In addition, some pipettes have a small upper chamber in which a cotton plug is inserted to prevent the accidental passage of cultures or other fluids into a mechanical pipetting device.

Pipetting is not difficult. However, to avoid pipetting errors, the eyes of the user should be horizontal with the top of the fluid column in the pipette (Figure 6–5). The term *meniscus* is used for the curved air-liquid interface at the top of the fluid.

The viable numbers or colony-forming units (CFU) of a bacterial suspension can be determined by use of the *pour plate technique* (Exercise 5) or the *spread plate technique* (Figure 6–6). The spread plate technique involves spreading a small, known volume of a cell suspension on the surface of an agar plate medium. By carefully spreading the cell suspension evenly over the agar surface, the colonies formed after incubation should be evenly distributed, which makes it easier to obtain a more accurate count. The spread plate technique is generally preferred for obligate aerobes, while the pour plate method is better for facultative anaerobes or microaerophilic cultures.

Serial dilution and plating procedures may present certain problems in determining the viable number of a bacterial or related suspension. Among the most common are *sampling* and *technical errors.* Sampling errors are generally caused by unequal distribution of cells in samples or in a dilution blank, while technical errors are associated with inaccuracies in preparing dilution blanks, with unequal volumes, or with making dilutions (pipetting errors).

(b)

(a)

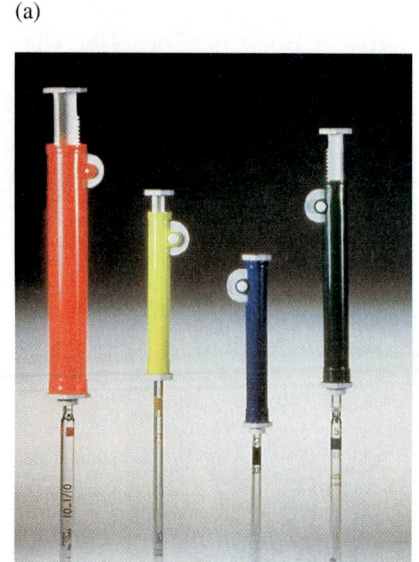
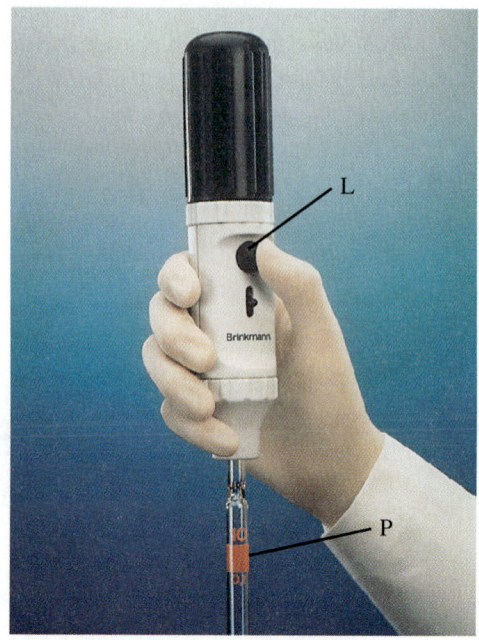

Figure 6–3

Pipetting helpers. In a microbiology laboratory, it is generally unsafe to pipette by mouth since microorganisms may accidentally be ingested. Two types of pipette aids are shown. The procedures for their use are also described in this exercise (a) The pipette pump is a fast-release device that provides safe, accurate, and trouble-free pipetting with a simple, one-handed operation. Pipettes fit smoothly into its flexible tapered chuck. The thumbwheel on the side is rotated for precision filling or dispensing and the plunger may be pressed for quick emptying.
(Courtesy of Scienceware® from Bel-Art Products.)

(b) The Brinkmann Pipet Helper. This is a lightweight device that is used to deliver fluids when attached to appropriate pipettes (P). A single lever (L) controls the uptake and/or delivery of liquids.
(Photo courtesy of Brinkmann Instruments, Inc. Westbury, N.Y.)

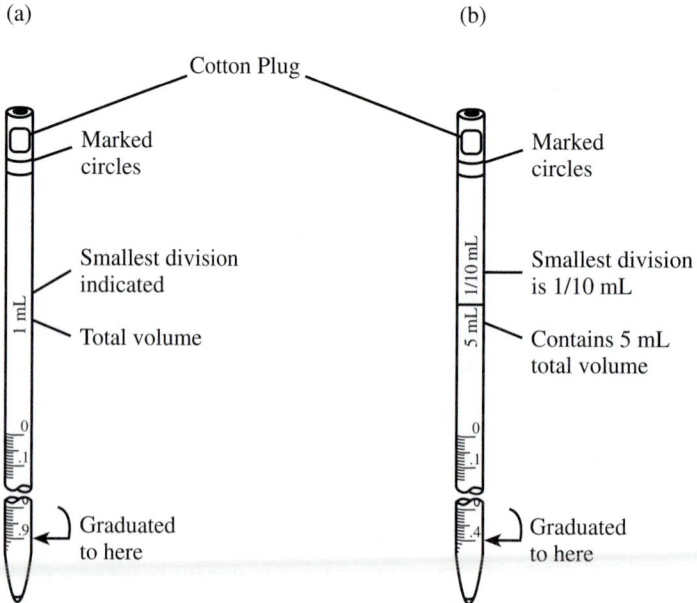

Figure 6—4

Pipettes. (a) The general features of a serological pipette. (b) A 5 mL serological pipette showing its calibrations. Each division is 0.1 mL. Serological pipettes are calibrated to deliver a measured volume of fluid.

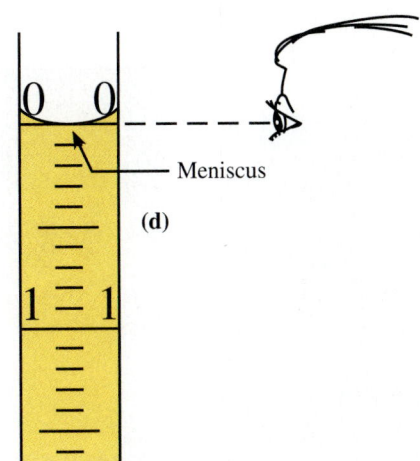

Figure 6–5

The location of the *meniscus* (men-IS-kus) the curved upper surfaces of a liquid in a container such as a pipette. The correct position of the eye for reading the volume or level of a solution also is shown.

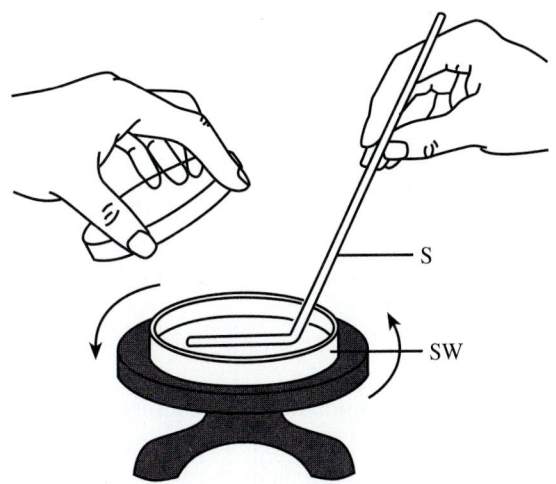

Figure 6–6

The spread-plate technique. After a known volume is applied to the surface of an agar plate medium, it is distributed over the surface by means of a sterile spreader (S). During the spreading the plate is turning on a spreading wheel (SW). The spreader is sterilized both before and after the procedure with ethanol and flaming.

The spreading technique or variations of it can be used to determine the number of living (viable) organisms in various substances such as water, milk, and foods, and during the various stages of bacterial growth curves. The technique is relatively simple to perform and produces excellent reproducible results. As indicated earlier, it is based on the assumption that each viable cell will form a colony. Thus, the number of resulting colonies on a plate is an indication of the number of viable cells in the original sample under study. Counting colonies on plates can be done with a *Quebec Colony Counter* or similar equipment (Figure 6–7). Such devices provide an indirect (oblique) source of light, a magnifying glass, and grid markings to help in the systematic examination of plates. If a colony counter is used, all plates should be placed upside down on the counter.

If colonies are large enough to count easily with the unaided eye, plates can be held up so that it is between the eyes of the person counting and a source of light. The colonies are counted by looking through the bottom of the plate.

For purposes of accuracy and reliability, after incubation only those plates with 30 to 300 colonies are used. Plates with higher counts are reported as "too numerous to count" **(TNTC)**. This **30 to 300 rule** is statistically valid. The number of viable organisms in the original sample per milliliter is determined by multiplying the number of colonies per plate

(a)

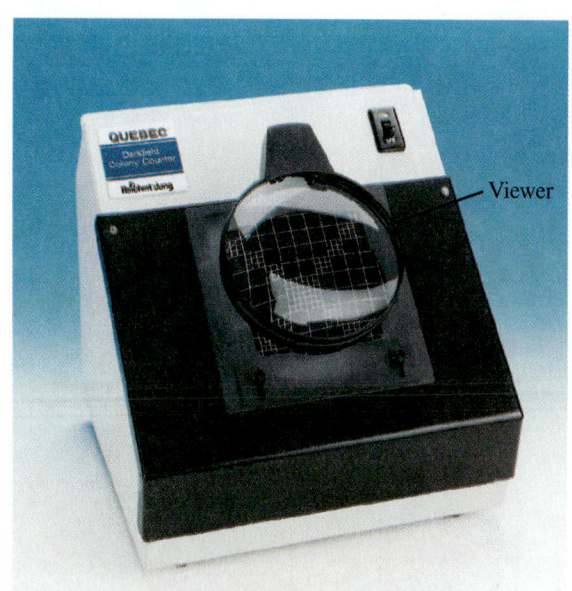

(b)

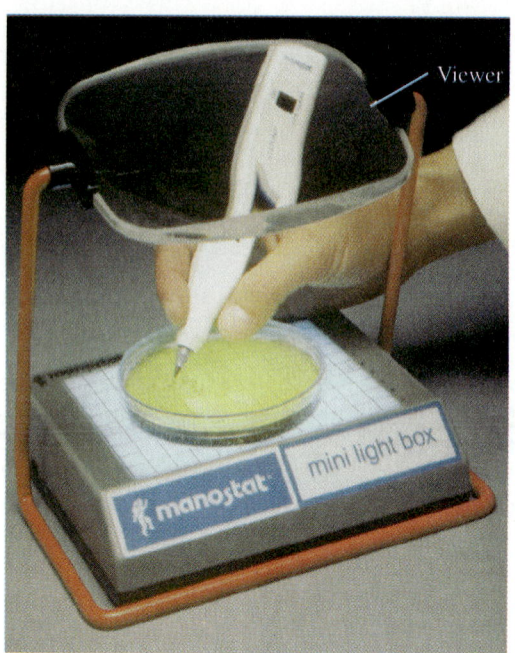

Figure 6–7

Bacterial counting systems. (a) A Quebec Colony Counter.
(Courtesy of Leica Inc)

(b) A modern colony counting system. This system is designed to count bacterial colonies, viral formed lesions in cultures known as plaques (PLAKS), or any repetitive counting task. The hand counter marks and even beeps to acknowledge that each colony has been counted.
(Photo courtesy of Lux Scientific.)

by the dilution factor for the particular plate. (The dilution factor here refers to the total number of times the original sample was diluted.) Here is an example of a calculation. If 213 colonies were counted on the 10^{-4} (1:10,000) dilution plate, the number of organisms in the original sample would be calculated as follows:

$$\frac{\text{Colony-forming units (CFUs)}}{\text{mL (original sample)}} = \frac{\text{Average number of colonies}}{\text{Plate}} \times \text{Dilution factor*}$$

*Remember that the dilution factor refers to the total number of times a culture has been diluted.

$$\text{CFUs/mL} = 213 \times 10{,}000 \text{ (or } 213 \times 10^4 \text{ or } 2.13 \times 10^6\text{)}$$
$$\text{CFUs/mL} = 2{,}130{,}000$$

The various parts of this exercise are designed to demonstrate the fundamental aspects of dilution techniques and their applications.

A. Pipetting with Pipettors

(*Note:* Pipetting fluids by mouth, especially microbial cultures, is no longer practiced in laboratories.)

Materials

The following items should be provided for class demonstration:

❑ 1. An assortment of individually wrapped and unwrapped pipettes

❑ 2. Pipetting aids

❑ 3. Quebec Colony Counter and/or other similar devices

The following items should be provided for general class use:

❑ 1. Individually wrapped sterile 1-mL, 5-mL, and 10-mL pipettes

❑ 2. Nonsterile 1-mL, 5-mL, and 10-mL pipettes

❑ 3. Pi-Pumps for 1-mL and 5-mL pipettes

❑ 4. A Brinkmann Pipette Helper

❑ 5. Rubber bulbs of various sizes

The following items should be provided per 4 students:

❑ 1. Twenty individually wrapped, sterile 1-mL pipettes

❑ 2. One Pi-Pump (green)

❑ 3. Sixteen water blanks each consisting of 9 mL in tubes

❑ 4. Four mL of a 2% methylene blue solution

❑ 5. Four test tube racks

❑ 6. Container for pipette disposal

❑ 7. Marking pen

Procedure 1: Introduction to Equipment

This procedure is to be performed individually.

❑ 1. Examine the different pipettes. Note the calibrations. (Refer to Figure 6–2.)

❑ 2. Examine the different pipetting devices and colony-counting devices. (Refer to Figures 6–3 and 6–7.)

❑ 3. Answer the questions in the Results and Observations and Laboratory Review sections.

Procedure 2: The Basic Dilution Technique

This procedure is to be performed by students individually.

❑ 1. Obtain four 9-mL water blanks, five 1-mL nonsterile pipettes, one tube containing 1-mL methylene blue, and an appropriate Pi-Pump (blue).

❑ 2. Label the water blank tubes 1, 2, 3, and 4 respectively.

❑ 3. Take a 1-mL pipette and note the numbers and markings inscribed on it.

❑ 4. Carefully attach the appropriate Pi-Pump to the pipette as shown in Procedure Diagram 11.

❑ 5. Place the pipette into the tube containing the methylene blue dye and carefully draw up 1 mL. (*Note:* The pipette holds 1 mL when the meniscus is aligned on the *O* line of the pipette when it is held vertically [see Figure 6–5].)

❑ 6. Pick up tube #1 and expel the 1 mL of the dye into it as shown in Procedure Diagram 11. Draw the fluid up and down into the pipette several times to insure a good mix.

❑ 7. Remove the pipette and place it in the disposal container or dispose of it as indicated by the instructor.

❑ 8. Attach a new pipette to the Pi-Pump and draw up 1 mL of the fluid from tube #1 and transfer it to tube #2 as described in step 6.

❑   9. Remove the pipette and dispose of it as indicated by your instructor.

❑ 10. Repeat steps 8 and 9 with the remaining tubes and starting with tube #2 follow the steps shown in Figure 6–1.

❑ 11. Note the color changes in the respective tubes.

❑ 12. Calculate the dilutions for each tube and determine the dilution factors. Enter all findings in the Results and Observations section and answer the questions in the Laboratory Review.

Procedure Diagram 11
The Use of a Pipette Pump

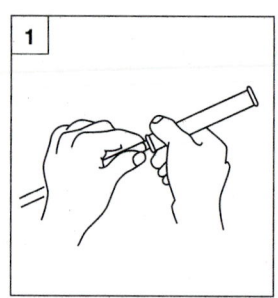

1. Hold the pipette at its blunt end and attach it to the Pi-Pump. Press the pipette into place firmly, but gently. DO NOT FORCE

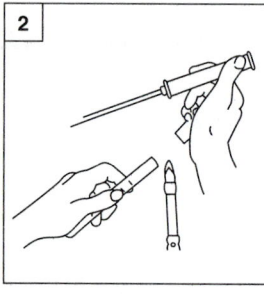

2. Take the tube from which a fluid volume will be taken. Using aseptic technique remove the plug or cap from the tube and quickly flame the lip of the tube.

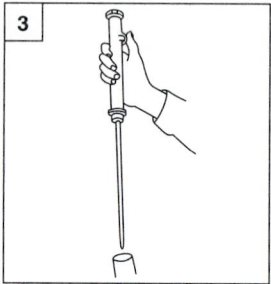

3. Hold the Pi-Pump as shown and insert the pipette into the fluid. Rapidly draw 1 mL of fluid into the pipette by rotating the Pi-Pump knob with your thumb. (The exact volume of the pipette is reached when the fluid, the meniscus (Figure 6 –5), is aligned with the 0 (zero) line of the pipette.

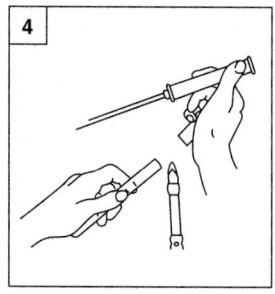

4. Flame the lip of the tube and return it to a rack or beaker.

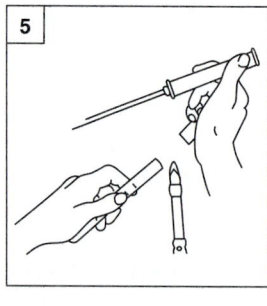

5. Pick up a tube containing a dilution blank with your free hand. Remove the plug or cap and flame the tube lip quickly.

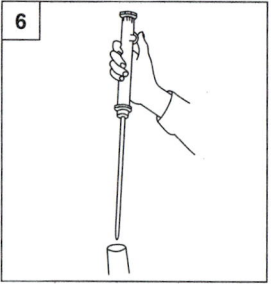

6. Rotate the Pi-Pump knob and deliver the volume of the fluid into the tube.

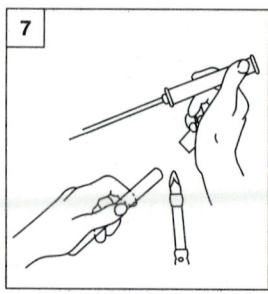

7. Flame the lip of the tube quickly and replace the plug or cap.

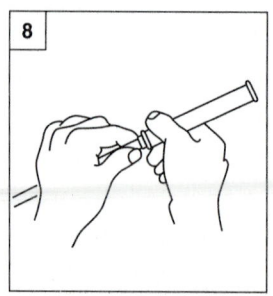

8. Disconnect the Pi-Pump from the pipette, and dispose of the pipette as indicated by the instructor.

B. The Spread Plate Technique and Determining CFUs

Materials

The following items should be provided per 4 students:

❏ 1. Five mL of a 24-hour trypticase soy or nutrient broth culture of *Escherichia coli*

❏ 2. Thirty-two sterile and individually wrapped 1-mL pipettes

❏ 3. Twenty 9-mL sterile distilled water dilution blanks in screw-capped tubes

❏ 4. Twenty-four trypticase soy or nutrient agar plates

❏ 5. One spreader wheel

❏ 6. One glass spreader rod

❏ 7. Two Pi-Pumps for 1-mL pipettes

❏ 8. One 250-mL glass beaker with 95% alcohol for glass rod sterilizing

❏ 9. One container with disinfectant for pipette disposal

The following items should be provided for general class use:

❏ 1. One or more Quebec or other colony-counting devices

❏ 2. Test tube racks

❏ 3. Wax pencils or marking pens

Procedure 1: Serial Dilutions with a Bacterial Culture

This procedure is to be performed by students individually.

❏ 1. Obtain the necessary equipment items for one set of dilution blanks and plates to perform the spread plate method shown in Figure 6–10.

❏ 2. Label one set of dilution blanks for each of the following dilutions: 10^{-1}, 10^{-2}, 10^{-3}, 10^{-4}, and 10^{-5}.

❏ 3. Carefully unwrap a sterile 1-mL pipette. (Refer to Figure 6–8.) Be careful to open the wrapper end at the opposite of the pipette tip.

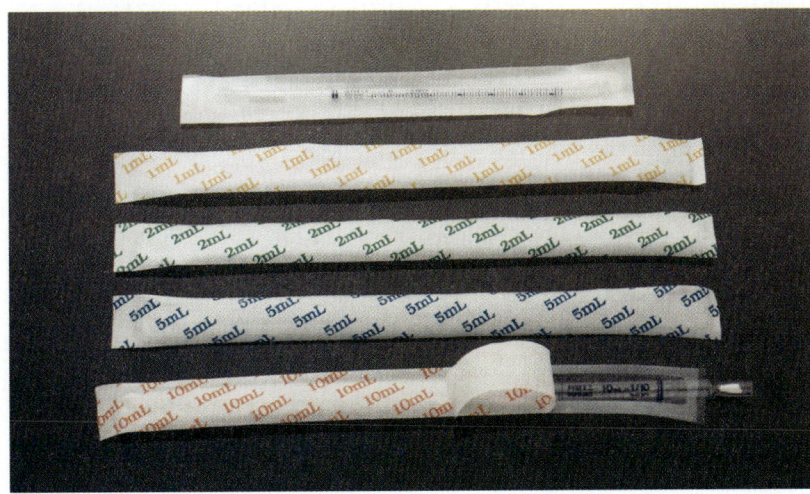

Figure 6–8
The general way to unwrap a prepackaged, disposable pipette.
(Reprinted with permission from Corning Incorporated.)

❑ 4. Unwrap only enough of the pipette to attach to the Pi-Pump. (Refer to Procedure Diagram 11.)

❑ 5. Aseptically transfer 1 mL of the *E. coli* culture to the first 9 mL dilution blank (10^{-1}).

❑ 6. Move the rotary wheel of the Pi-Pump up and down to mix the contents of the dilution blank.

❑ 7. Carefully disconnect the pipette and place it into the container with disinfectant.

❑ 8. Use a second sterile 1-mL pipette and the Pi-Pump to aseptically transfer 1 mL of the 10^{-1} dilution to the second 9 mL dilution blank (10^{-2}).

❑ 9. Mix the contents of the dilution blank by moving the rotary wheel up and down several times.

❑ 10. Carefully disconnect the pipette and place it into the container with disinfectant.

❑ 11. Use a third sterile 1-mL pipette and the Pi-Pump to aseptically transfer 1 mL of the 10^{-2} dilution to the third 9 mL dilution blank (10^{-3}).

❑ 12. Mix the contents of the dilution blank by moving the rotary wheel up and down several times.

❑ 13. Carefully disconnect the pipette and place it into the container with disinfectant.

❑ 14. Use a fourth sterile 1-mL pipette and the Pi-Pump to aseptically transfer 1 mL of the 10^{-3} dilution to the fourth 9 mL dilution blank (10^{-4}).

❑ 15. Mix the contents of the dilution blank by moving the rotary wheel up and down several times.

❑ 16. Carefully disconnect the pipette and place it into the container with disinfectant.

❑ 17. Use a fifth sterile 1-mL pipette and the Pi-Pump to aseptically transfer 1 mL of the 10^{-4} dilution to the fifth 9 mL dilution blank (10^{-5}).

❑ 18. Mix the contents of the dilution blank by moving the rotary wheel up and down several times.

❑ 19. Carefully disconnect the pipette and place it into the container with disinfectant.

❑ 20. Save the respective dilutions for the next procedure.

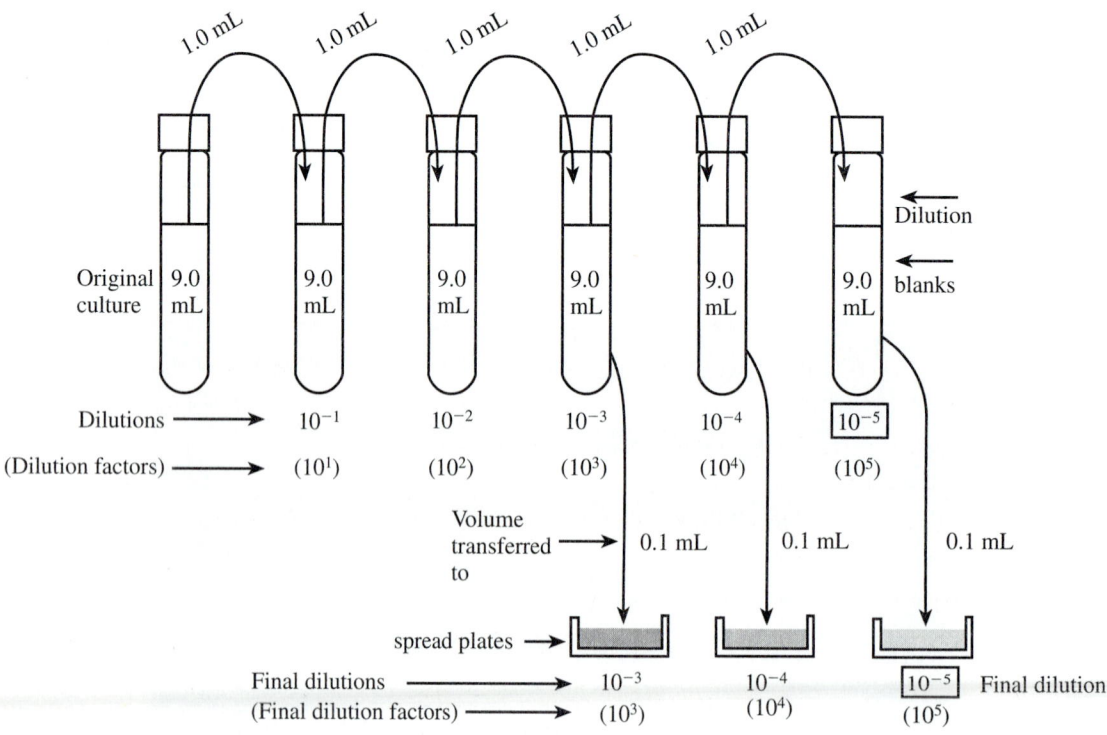

Figure 6–9
Serial dilutions and the spread plate technique

Procedure 2: The Spread Plate Technique

This procedure is to be performed by students individually.

❑ 1. Label the bottoms of two separate agar plates for each of the following dilutions: 10^{-3}, 10^{-4}, and 10^{-5}. **(Include your name or initials, the date, and the microorganism used in the exercise. Write small, but legibly, around the edges of the plate bottoms.)**

❑ 2. Carefully unwrap a sterile 1-mL pipette exposing only enough of the pipette to attach to the Pi-Pump.

❑ 3. Aseptically draw up 0.2 mL of the 10^{-3} dilution and place 0.1 mL onto the center of each of the correspondingly labeled agar plates.

❑ 4. Disconnect the pipette and place it into the container with disinfectant.

❑ 5. Place and center one of the plates onto the spreader wheel.

❑ 6. Dip the spreader rod into the beaker of ethanol. Remove the rod and touch it to the edge of the beaker to drain off excess fluid.

❑ 7. Pass the rod through a Bunsen burner flame to ignite the ethanol. **(Keep the rod in a downward position for a few seconds to allow the flame to go out.** Refer to Figure 6–10.) DO NOT PUT THE ROD DOWN.

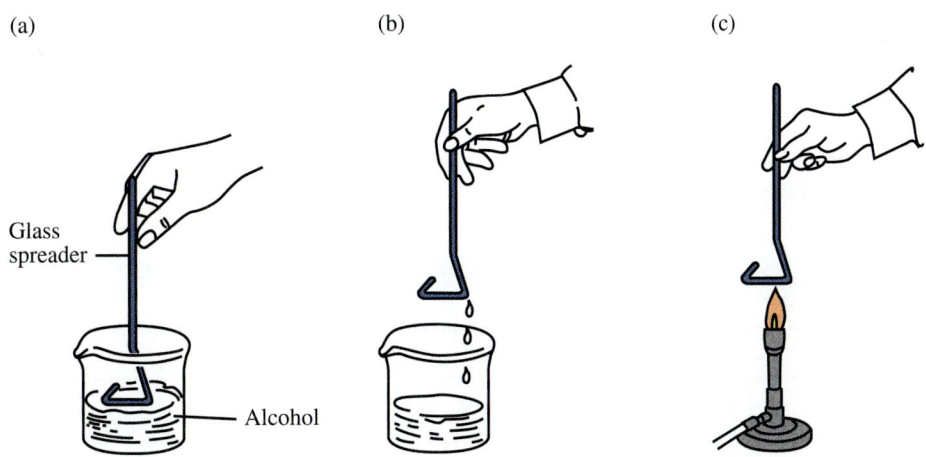

(a)　　　　　　　　　　　(b)　　　　　　　　　　　(c)

Glass spreader —

— Alcohol

Figure 6–10
Steps in sterilizing a bent glass rod spreader. (a) Inserting the spreader portion of the rod into alcohol. (b) Draining. (c) Flaming and sterilizing.

❑ 8. With the other hand, start the spreader wheel turning at a moderate speed, and then remove the Petri dish top. Keep the Petri dish top in the hand.

❑ 9. Touch the spreader rod to the inside surface of the Petric dish to cool it.

❑ 10. Gently place the sterilized spreader rod onto the center of the agar plate. The placement of the rod and the rotating plate should spread the 0.1 mL inoculum over the plate surface. **(This step should only take a few seconds.)** Refer to Figure 6–6.

❑ 11. Replace the Petri dish top.

❑ 12. Return the spreader rod to the beaker with ethanol.

❑ 13. Repeat steps 5 through 12 with the second agar plate containing the same dilution.

❑ 14. Next, repeat steps 3 through 13 with each of the following dilutions: 10^{-4} and 10^{-5}.

❑ 15. Incubate all plates at room temperature for 24 hours, or as indicated by your instructor.

❑ 16. If the plates will not be examined and the colonies counted after incubation, they should be refrigerated until the appropriate time.

Procedure 3: Colony Counts and CFU/mL Determinations

This procedure is to be performed by students individually.

- ❑ 1. Colony counters or simply holding plates up to a ceiling light or a somewhat strong light source can be used for colony counts after incubation.
- ❑ 2. If a Quebec Colony Counter or similar device (Figure 6–7) is used, place individual plates upside down on the counter so that marks can be made on the *bottoms* of the respective plates.
- ❑ 3. Use a marking pen or wax pencil to mark the area of each colony seen on the plate.
- ❑ 4. When all the colonies have been identified (marked), count all the markings and record your findings for the specific plate and dilution in the Results and Observations section.
- ❑ 5. Repeat steps 2 through 4 for each plate and dilution.
- ❑ 6. Calculate the CFU/mL for each dilution using the formula given in the Results and Observations sections.
- ❑ 7. Answer the questions in the Results and Observations section and the Laboratory Review.

Results and Observations

Basic Dilutions

1. Enter your findings in Table 6–1.

Table 6–1

Pipetting Exercise

Tube Number	Dilution	Dilution Factor	Description of Color
1			
2			
3			
4			

2. Were the volumes (the individual meniscus levels) at the same height in all tubes? If not, why? _____

_____

3. Were there color intensity changes observed with any of the tubes? Describe and explain the changes. _____

_____

Colony Counts and CFU/mL Calculations

1. Enter your findings and the final values for your CFU/mL calculations in Table 6–2. Use the following formula and space provided to calculate the CFU/mL values.

$$\text{CFU/mL} = \frac{\text{Average number of colonies}}{\text{Plate}} \times \text{Dilution factor}$$

Table 6–2

CFU Determinations

Dilution Counted	Total Number of Colonies		Average Count	CFU/mL
	Plate #1	Plate #2		
10^{-3}				
10^{-4}				
10^{-5}				

2. Were there large differences in the colony counts of plates with the same dilution? If so, explain the differences.

_____

3. Did any of the dilutions used result in a TNTC situation? If so, why did this situation occur? _____

_____

4. Did you find any particular problem or difficulty with performing the steps involved with dilutions and/or the spread plate technique? If you did, what were the difficulties, and indicate how they could be avoided. _____

_____

_____

Laboratory Review 6   Dilution Techniques and Colony Counting

1. Which plating method should be used to determine the CFUs of an obligate aerobe? _____

2. Which plating technique should be used to determine the viable numbers of facultative bacterial culture? _____

_____

3. How many $\mu L = 1$ L? _____

4. What would the final dilution be in the last tube shown in the following figure?

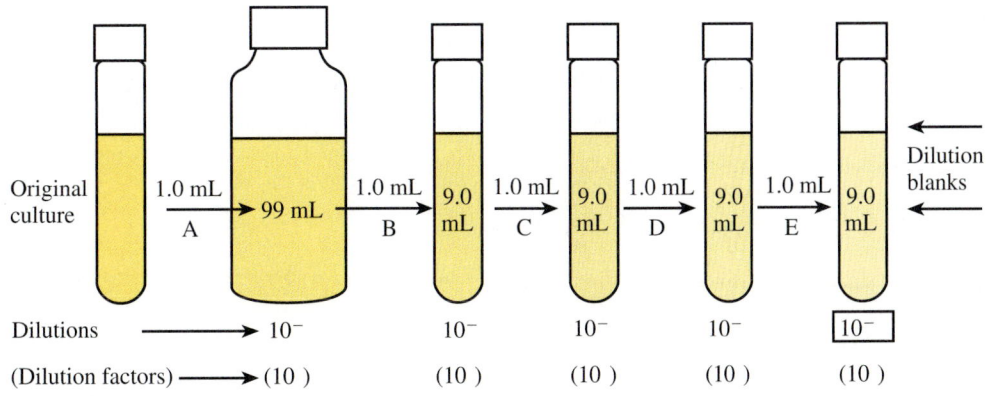

Figure 6–11
Serial dilution determination.

5. List two errors associated with serial dilutions.

   a. _____   b. _____

6. Why are the plates containing between 30 and 300 colonies the only ones used for CFU determinations: _____

   _____

7. Why are different pipettes used between dilutions? _____

   _____

8. Give the CFU/mL of the original culture in the following situations (Table 6–3):

Table 6–3

CFU/mL Determinations

Plate Count	Final Dilution	CFU/mL
a. 40	1:100	
b. 98	1:1000	
c. 157	10^{-6}	

Key Terms

calibration (kal-i-BRĀ-shun): refers to determining an instrument's accuracy by comparing the measurement provided with that of a known standard

facultative anaerobe: an organism capable of metabolizing in environments with or without oxygen

meniscus (men-IS-kus): the curved upper surface of a liquid in a container such as a pipette

microaerophilic bacterium: an organism capable of metabolizing in environments containing small amounts of oxygen

TD : to deliver the volume of a nonviscous fluid indicated on the upper part of a pipette at 20°C

viable number of bacteria: refers to the number of cells in a suspension or culture capable of division on a solid medium; also called the colony-forming unit

A Demonstration of Bacterial Culture Characteristics

After completing this exercise, you should be able to:

1. Distinguish basic features of bacterial colonies, broth cultures, and agar slant growths.
2. Recognize the advantages and limitations of culture characteristics in the identification of bacterial species.
3. Determine the influence of temperature on pigment production.

The cultivation of microorganisms requires the use of nutrient preparations called *culture media* (singular, *medium*). Natural media such as milk, vegetable slices, and certain meat infusions contain soluble organic and inorganic substances that are the necessary factors for growth. However, the exact chemical compositions of natural media are unknown and quite variable. On the other hand, in the case of *chemically defined* or *synthetic media,* the kinds and exact amounts of all ingredients are known. Media of this type can be duplicated to specification and used to study the effects of specific compounds on microbial growth. Bacterial cultures used in this exercise have been grown in or on trypticase soy media. A trypticase soy medium is another example of a complex medium as described in Exercise 4. It is an excellent rich medium and contains dipotassium phosphate, glucose, phytone peptone, sodium chloride, and trypticase peptone. Phytone is prepared from enzymatically digested soybean meal, while trypticase peptone is obtained from enzymatically digested milk protein (casein).

There are three general forms of culture media: *solid, semisolid,* and *broth* (liquid). Each type of medium has particular properties that make it more or less suitable for certain growth situations. Most solid and semisolid media contain agar as a solidifying agent. It is a complex polysaccharide extracted commercially from certain species of red marine algae such as *Gelidium, Gracilaria, and Rhodophyta.*

Microorganisms such as bacteria grow as visible accumulations of identical cells, forming *colonies* on the surfaces of agar media contained in Petri dishes. These dishes are constructed to allow air but not dust to pass through. Bacteria may be spread over the surface of an agar medium or inoculated into a melted agar medium and then poured into a sterile Petri dish. Exercise 5 describes these procedures more fully.

The pigmentation, size, shape, and overall general appearance of bacterial colonies when growth in agar plates (Figure 7–1) can serve as identifying features when viewed under both reflected and transmitted light (see Figure 7–6). Pigmentation and several other cultural characteristics are influenced by incubation temperature. Colonies may appear to be transparent, opaque, or translucent. Some species produce a distinguishing fluorescent pigment that is evident when cultures are examined by ultraviolet light. Other colonies exhibit specific surface properties and may appear as dry (powdery), contoured, rough, smooth (glistening), rugose (wrinkled), or with concentric rings or ridges (Figure 7–1). Some organisms also produce characteristic odors. Such odors may be sweet, foul, soil-like, or musty.

A large number of media have been developed to aid in the isolation, differentiation, and identification of microorganisms. Two examples of media used for demonstration and identification of bacterial pigments are included in this exercise (Figure 7–7).

In addition to the agar plates, solid medium preparations can be made into slants. Such preparations are made by pouring melted agar medium into test tubes and allowing the material to solidify at an angle. Agar slants are valuable in studying growth characteristics and in maintaining pure cultures.

Broth media generally are contained in test tubes closed by nonabsorbent plugs or special closure caps. Broth can be used to grow large numbers of organisms and to perform various biochemical tests. Microorganisms growing in a broth frequently demonstrate a particular type of growth in or on the surface of the medium. Examples of some types of growth associated with agar and broth cultures are depicted in Figures 7–2, 7–3, 7–4, and 7–5. Additional growth features will be presented in many of the following exercises.

Certain problems may arise during culture examination, and must be avoided if reliable results are to be obtained. The major problems are contamination or the presence of unwanted organisms, and species variations.

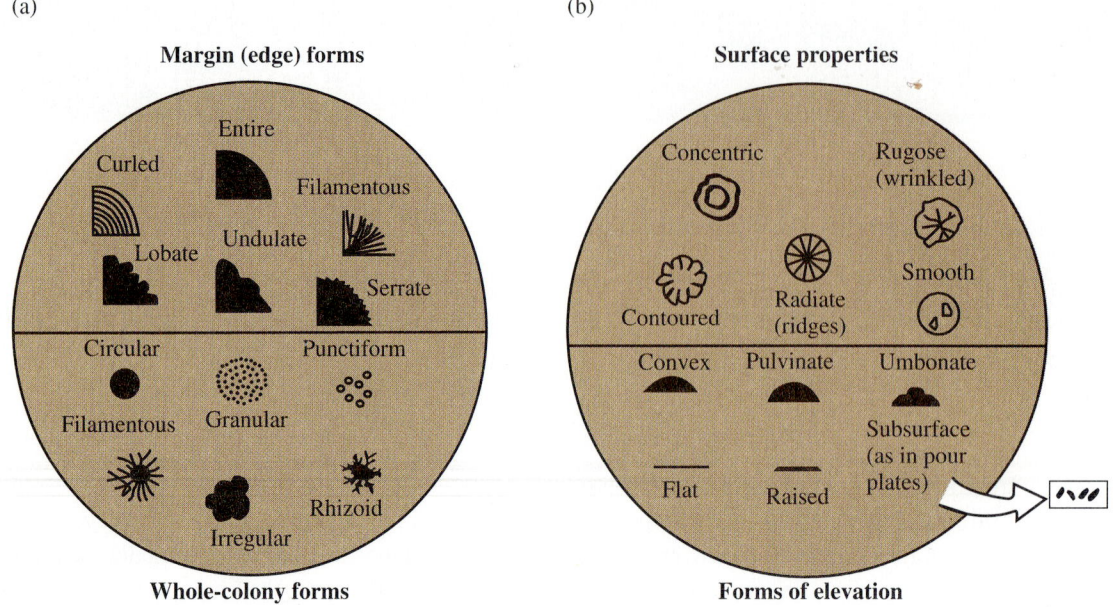

Figure 7–1

Diagrams of bacterial colony characteristics. (a) Whole-colony forms, and margin (edge) forms. (b) Surface properties and elevations.

Species variations can be the result of (1) an environmentally induced change of a genetically controlled characteristic; (2) spontaneous and undirected permanent genetic changes known as *mutations;* and (3) the acquisition of a new assortment of genes as a consequence of *gene transfer.* The first of these factors is generally temporary in its effect. That is, it occurs only as long as the environmental conditions are present. The second and third factors may be permanent.

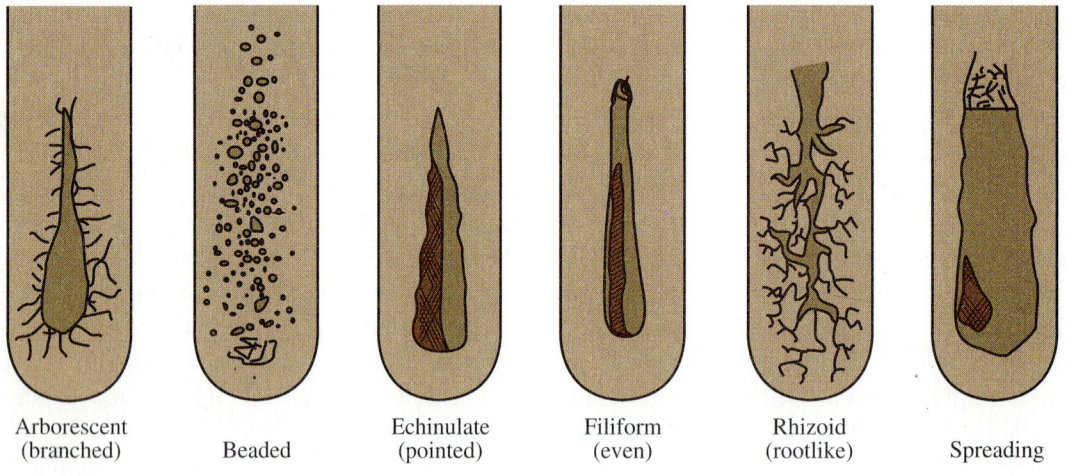

Arborescent (branched) Beaded Echinulate (pointed) Filiform (even) Rhizoid (rootlike) Spreading

Figure 7–2

Selected features of an agar slant stroke culture.

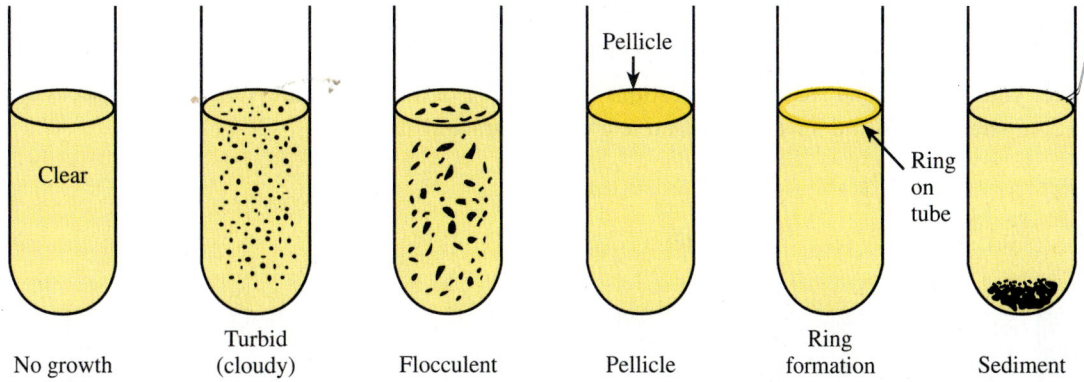

Figure 7–3
Selected features of broth cultures. (Note: Several features may be found in one tube.)

A. Examining Bacterial Culture Characteristics

Materials

All the microorganisms listed below should be prepared by the instructor in accordance with specified incubation periods and temperatures.

❏ 1. Divided streak plates of each microbial combination (1 per 4 students). The cultures should be growth on trypticase soy agar for 24 hours, except as indicated.

    ❏ a. *Micrococcus luteus* and *Serratia marcescens* (25°C)
    ❏ b. *Bacillus subtilis* and *Escherichia coli* (37°C)
    ❏ c. *Enterococcus (Streptococcus) faecalis* (en-ter-ō-KOK-us, fē-KAL-us) and *Pseudomonas fluorescens* (sū-dō-MŌ-nus, flor-ES-senz) (25°C)
    ❏ d. *Mycobacterium phlei* (mī-kō-bak-TE-rē-um, flāy) and *Z. faecalis* (96-hour culture, 37°C)

❏ 2. Twenty-four-hour pour plate preparations of each of the following cultures in trypticase soy agar, grown at 37°C (1 per 4 students):

    ❏ a. *Bacillus subtilis*   ❏ d. *Serratia marcescens*
    ❏ b. *Escherichia coli*   ❏ e. *Enterococcus (Streptococcus) faecalis*
    ❏ c. *Pseudomonas fluorescens*

❏ 3. Twenty-four-hour agar slant stroke cultures on trypticase soy agar incubated at 25°C of each of the following organisms (1 per 4 students):

    ❏ a. *Bacillus subtilis*   ❏ d. *Mycobacterium phlei* (96-hour culture)
    ❏ b. *Escherichia coli*   ❏ e. *Streptococcus faecalis*
    ❏ c. *Serratia marcescens*   ❏ f. *Pseudomonas fluorescens*

❏ 4. Forty-eight-hour trypticase soy broth cultures incubated at 25°C (except where indicated differently) of the following (1 per 4 students):

    ❏ a. *Bacillus subtilis*   ❏ d. *Pseudomonas fluorescens*
    ❏ b. *Escherichia coli*   ❏ e. *Enterococcus faecalis*
    ❏ c. *Mycobacterium phlei* (96-hour culture)   ❏ f. *Serratia marcescens*

Procedure: Examining Bacterial Culture Characteristics

This procedure is to be performed by students in groups of 4.

    ❏ 1. Examine the various cultures provided using the diagrams accompanying this exercise and the selected characteristics of bacterial growth in broth and on agar plates and slants shown in Figures 7–4, 7–5, and 7–6.

❑ 2. Be careful not to disturb (by shaking) the broth preparations, as surface growths may be dislodged from the sides of the cultures tubes and fall to the bottom.

❑ 3. Enter your findings and answer the questions in Table 7–1 of the Results and Observations section. Be certain to compare cultures and look for distinctive growth characteristics.

(a) (b) (c) (d) (e) (f) (g) (h)

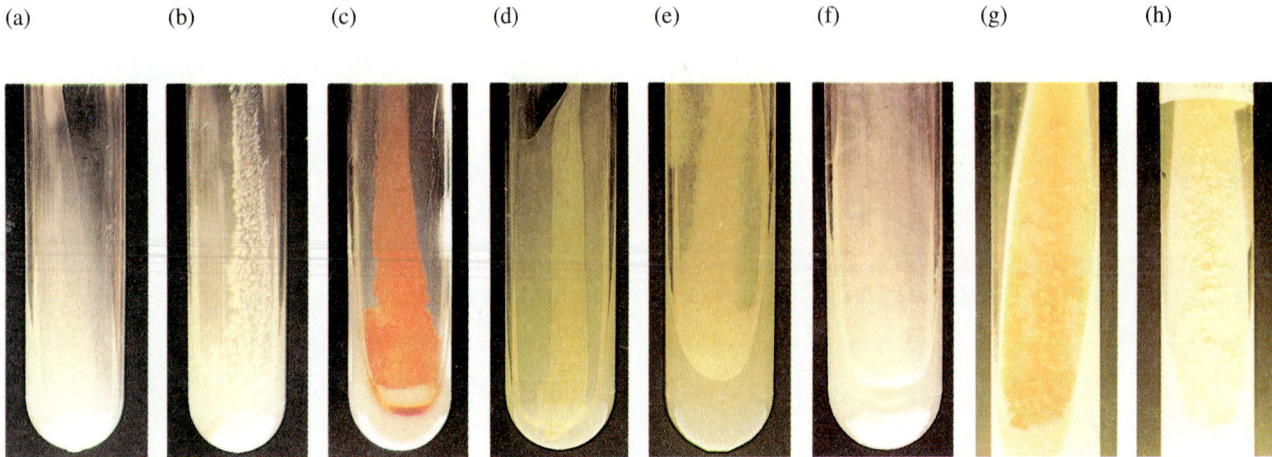

Figure 7–4
Agar stroke patterns. Tube a spreading growth; Tube b, the beaded form of growth; Tube c, flat, red, and filiform; Tube d filiform, nonpigmented growth; Tube e, rhizoid growth; Tube f, aborescent (branched); Tube g, shows the orange growth of *Mycobacterium phlei;* Tube h shows the nonpigmented growth of *M. smegmatis.* Both species are growing on a form of Lowenstein-Jensen medium.

(a) (b) (c) (d) (e) (f)

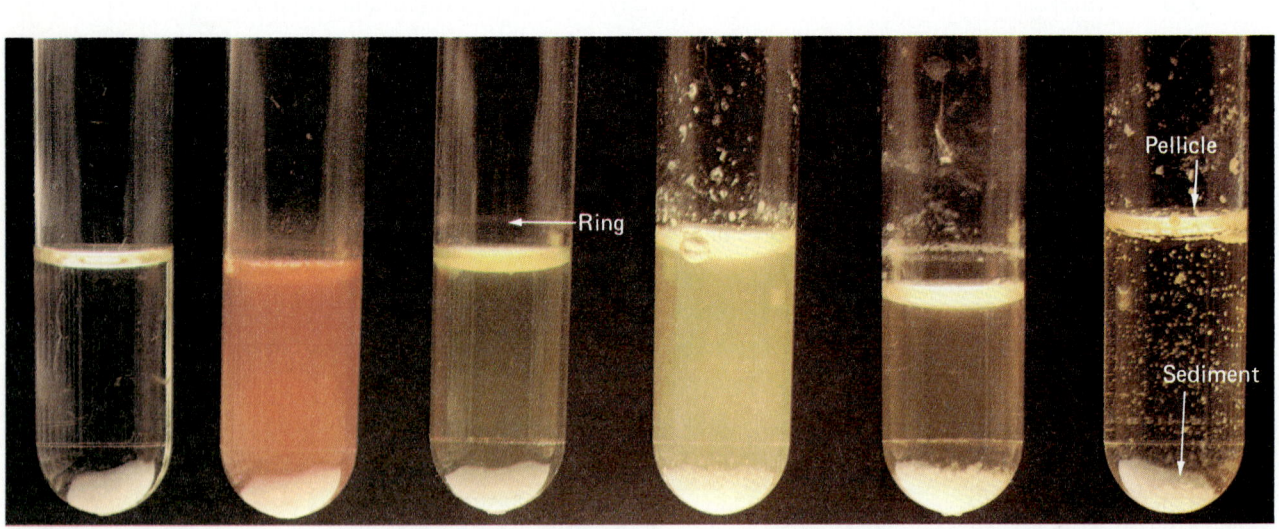

Figure 7–5
Selected broth culture characteristics of bacteria. Tube a, clear, uninoculated broth; Tube b, turbid (cloudy growth) and red pigmentation; Tube c, turbid and green pigmentation in the upper portion of the broth; Tube d, turbid; Tube e, slightly turbid, and ring formation (1 mL of the broth was removed to demonstrate the presence of the ring); Tube f, somewhat clear broth with flocculent growth.

(a)   (b)   (c)   (d)

(e)   (f)   (g)   (h)

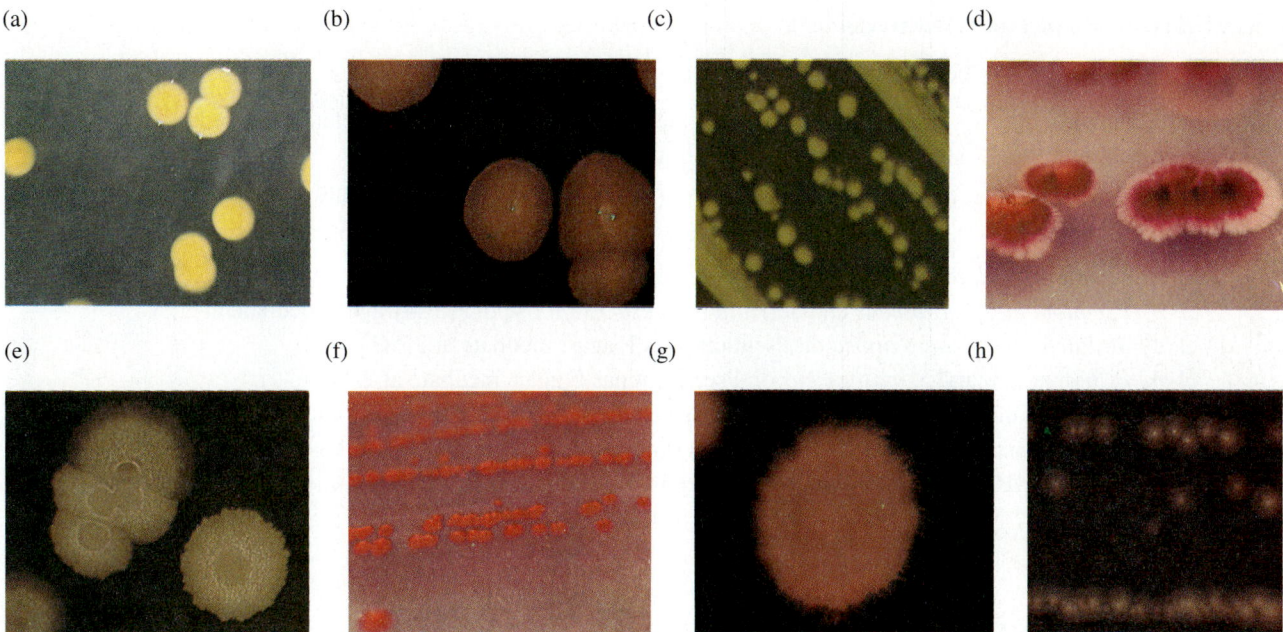

Figure 7–6

Examples of bacterial colony characteristics. (a) Circular, entire, smooth, opaque, yellow-pigmented colonies. (b) Irregular, undulate, umbonate, opaque, cream-colored colonies. (c) Punctiform (less than 1 mm in diameter), smooth, entire, opaque colonies among larger circular ones. (d) Unusual red-and-white-pigmented colonies that also exhibit irregular, lobate, radiate, and opaque properties. (e) Irregular, concentric, undulate, opaque, somewhat raised colonies. (f) Red-pigmented, circular, smooth, entire, opaque, convex colonies. (g) Filamentous, opaque, umbonate colonies. (h) Circular, semitransparent, umbonate colonies.

B. Demonstration of Pigment Production and the Influence of Temperature

Materials

All the microorganisms noted in this section should be prepared by students using specified incubation periods, temperatures, and media. The following items should be provided per 2 students:

❑ 1. Trypticase soy broth 24-hour cultures of the following bacterial species:
   ❑ a. *Micrococcus luteus*   ❑ c. *Pseudomonas fluorescens*
   ❑ b. *Pseudomonas aeruginosa*   ❑ d. *Serratia marcescens*

❑ 2. Two plates of the following media:
   ❑ a. Pseudomonas F agar   ❑ b. Pseudomonas P agar

❑ 3. Three trypticase soy agar plates

❑ 4. Ultraviolet lamps (below 260 nanometers) for general class use to view fluorescent pigments in a darkened room

❑ 5. Marking pens or wax marking pencils

Additional Technique Required for This Portion of the Exercise:

Streak Plate Technique, Procedure Diagram 10, Exercise 5

Procedure 1: Inoculations

The following procedure is to be performed by students in pairs.

❏ 1. With a marking pen or wax marking pencil, divide the underside of each agar plate provided in half and label each half according to the combinations that follow.

❏ 2. Using the streak plate technique, inoculate each of the following bacterial species on the halves of the media indicated (Figure 7–7). Incubate all streaked plates for 48 hours and at the temperatures specified.

   ❏ a. *Micrococcus luteus* and *Pseudomonas aeruginosa* on trypticase soy agar; incubate at 25°C.
   ❏ b. *Pseudomonas fluorescens* and *Serratia marcescens* on trypticase soy agar; incubate at 25°C.
   ❏ c. *M. luteus* and *P. aeruginosa* on Pseudomonas P agar; incubate at 25°C.
   ❏ d. *P. fluorescens* and *S. marcescens* on Pseudomonas F agar; incubate at 25°C.
   ❏ e. *M. luteus* and *S. Marcescens* on trypticase soy agar; incubate at 37°C.
   ❏ f. *P. aeruginosa* and *P. fluorescens* on Pseudomonas P agar; incubate at 37°C.
   ❏ g. *P. aeruginosa* and *P. fluorescens* on Pseudomonas F agar; incubate at 37°C.

❏ 3. Continue on to Procedure 2.

Procedure 2: Demonstration of Pigment Production and the Influence of Temperature and Medium

This procedure is to be performed by students in groups of 4.

❏ 1. Examine the specific cultures provided for the presence or absence of pigmentation. Pay particular attention to the appearance of cultures grown at different temperatures and different media. (Refer to Procedure 1 for the types of media and the incubation temperatures.)

❏ 2. Examine the areas around individual colonies to determine if pigments have diffused into the media. Refer to Figure 7–7.

❏ 3. In addition, examine individual colonies for sectored pigmented areas. Refer to Figure 32–3.

❏ 4. Examine all cultures for the production of a fluorescent pigment. In the area specified by the instructor (e.g., darkened room, drawer), examine each section of the divided plates, with the covers removed, under ultraviolet light. Place the plates so that this light does not shine directly at your face. *Do not look directly at*

(a)   (b)

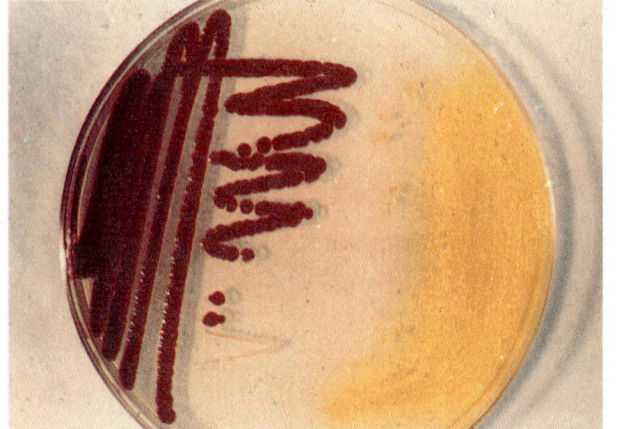 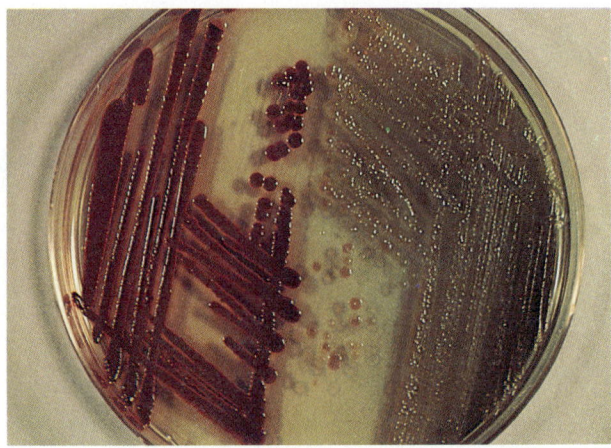

Figure 7–7

Media differentiation of *Pseudomonas* sp. (a) On *Pseudomonas agar F medium.* Left side, *Serratia marcescens;* right side, *Peudomonas fluorescens.* Note the production and diffusion of fluorescein (yellow fluorescent pigment) by the culture on the right-hand side of the plate. (b) On *Pseudomonas agar P medium.* This medium enhances the production of pyocyanin (a bluish-colored pigment) and inhibits the formation of fluorescein. Note the dark blue color of the medium surrounding the growth of *Pseudomonas* species. The red-pigmented *Serratia marcescens* is shown for purposes of comparison.

the ultraviolet light. The detection of a green or yellowish-green fluorescent pigment in the medium surrounding bacterial growth is a positive result.

❑ 5. Complete Table 7–2 and answer the questions in the Results and Observations section. If you recognize features other than those indicated previously, note them in the appropriate column.

Results and Observations

1. Complete Table 7–1 by describing the culture characteristics of each organism according to Figures 7–1 through 7–6.

2. Were other distinctive properties present? Describe them. _____

_____

_____

3. Complete Table 7–2 by indicating whether or not pigmentation was observed on the media and at the temperatures listed. Specify the color of the pigment formed.

4. Which of the bacterial species producing a pigment at 25°C did not do so at 37°C? _____

_____

_____

5. Were any sectored colonies produced? If so, which species produced them? Refer to Figure 32–3._____

_____

Table 7–1

Culture Characteristics

| Microorganisms | Pigmentation | | Colonial Form | | Petri Dish Observations[a] | | | Broth Properties | Agar Slant Stroke Properties |
	25°C	37°C	SP	PP	Margin Characteristics (SP)	Elevation Properties (SP)	Surface Properties SP)		
B. subtilis									
E. coli									
M. phlei				NP					
P. fluorescens									
M. luteus									
S. marcescens									
E. faecalis									
b									
b									

[a]Explanation of symbols: PP = pour plate; SP = streak plate; NP = not provided.
[b]Space for additional organisms.

Table 7-2

Pigment Production and Temperature

Microorganism	Medium — Incubation Temperature	Trypticase Soy Agar 25°C	Trypticase Soy Agar 37°C	Pseudomonas[a] F Agar 25°C	Pseudomonas[a] F Agar 37°C	Pseudomonas[a] P Agar 25°C	Pseudomonas[a] P Agar 37°C
M. luteus							
P. aeruginosa							
P. fluorescens							
S. marcescens							

[a]Specify the color of the pigment formed.

6. Which culture(s) produced diffusible pigments? _____
   _____

7. Were there any differences in pigmentation when the same bacterial species were grown on different media?
   If so, describe the results. _____
   _____
   _____

Laboratory Review 7   A Demonstration of Bacterial Culture Characteristics

1. a. What is a nutrient (culture) medium? _____
   _____
   _____

   b. What factors determine the formulation of a culture medium? _____
   _____
   _____

   c. Why is agar added to media? _____
   _____

2. Which of the culture characteristics studied in this exercise are most useful in identifying a bacterial species?
   _____
   _____
   _____

3. List 2 cultural characteristics presented in this exercise that are subject to mutation.

    a. _____   b. _____

4. Which microbial types cannot be routinely cultured in or on artificial media (e.g., nutrient agar, blood agar, and broths)? _____

    _____

5. With which culture characteristics are the following associated: flocculent, pellicle, ring, sediment, and turbid?

    _____

    _____

6. List the 3 general or basic forms of bacteriological culture media.

    a. _____   b. _____   c. _____

7. List 6 specific margin characteristics of bacterial colonies.

    a. _____   d. _____

    b. _____   e. _____

    c. _____   f. _____

8. Define or explain:

    a. aseptic technique _____

    _____

    b. synthetic (chemically defined) media _____

    _____

    c. anaerobe _____

    _____

    d. facultative anaerobe _____

    _____

    e. pure culture _____

    _____

Key Terms

bacterial colony: a visible mass of bacteria on a solid culture medium

opaque (Ō-pāk): in this exercise, the term refers to colonies that do not allow the transmission of invisible light

pellicle (PEL-i-kel): a thin film formed on broth culture surfaces

ring formation: the deposit of the bacteria on tubes near broth culture surfaces

translucent (trans-LOO-sent): in this exercise, the term refers to colonies that allow partial transmission of visible light

transparent (trans-PĀR-ent): in this exercise, the term refers to colonies that allow full transmission of visible light

Photograph Quiz 2

Give the margin characteristics for each of the labeled colonies shown in Figure 7–8. _____

_____

Photograph Quiz 3

Identify the broth features of the tubes shown in Figure 7–9. _____

_____

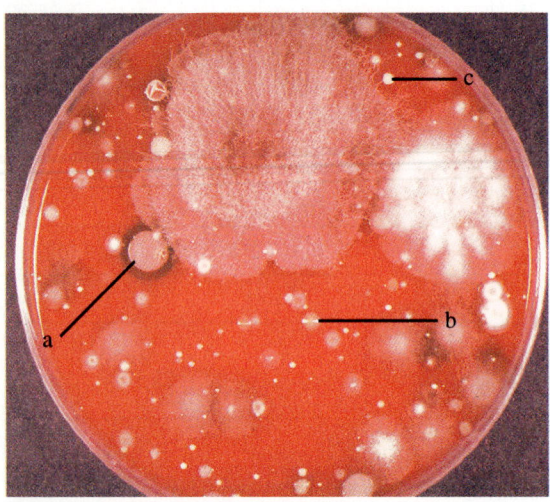

Figure 7–8
Bacterial colonies on an agar medium.

Figure 7–9
Bacterial broth cultures.

After completing this exercise, you should be able to:

1. Perform shake culture and paraffin plug anaerobic techniques.
2. Distinguish between strict (obligate) and facultative anaerobes and between microaerophiles and strict aerobes by means of the shake and paraffin plug cultures.
3. Recognize the value of thioglycollate medium for the isolation of aerobes and anaerobes.
4. Use commercially available devices for the cultivation of anaerobic organisms.

Microorganisms that grow only in an environment containing oxygen (air) are called **aerobes,** while those that grow only in the absence of oxygen are called **anaerobes.** Some microbes can grow in either the presence or absence of oxygen. Such organisms are generally referred to as **facultative anaerobes.** The oxygen requirement of a microorganism is directly related to its metabolism.

Oxygen is toxic for **obligate** or **strict anaerobes,** which can survive and grow only in environments where oxygen is completely excluded. Dissolved oxygen in a medium forms toxic hydrogen peroxide and free radicals, which cannot be detoxified by obligate anaerobes. Because of their extreme susceptibility to the toxic effects of oxygen transfer and/or inoculation, procedures involving anaerobes must be performed in anaerobic chambers (Figure 8–1). Chambers of this type are generally found in specialized laboratories that deal with anaerobes on a regular basis.

In practice, bacteria that are unable to grow on or near the surfaces of semisolid or solid media in air at atmospheric pressure are *anaerobes*. The early bacteriologists believed that successful cultivation of obligate anaerobes could be achieved only through methods that excluded all free oxygen. Recently, it has become apparent that providing adequate anaerobic conditions requires only the removal of free oxygen from the immediate environment of the microorganisms, or simply the maintenance of a low oxidation-reduction (redox) potential, or E_h, in the media. The E_h is a measure of the tendency of a preparation to be oxidized or reduced. In routinely used laboratory media, oxygen is primarily responsible

Figure 8–1
A modern chamber used for transferring and incubating strict anaerobes in an oxygen-free environment.
(Courtesy of Lab-Line Instruments, Inc., Model 6500 PACE.)

for the increasing E_h. Various redox dyes can be used to estimate the E_h of a medium or a culture. Useful dyes are reversibly oxidized or reduced, and are colored in the oxidized state and colorless in the reduced state.

Basically, procedures for cultivating and/or identifying anaerobic bacteria are much like those for aerobes. The difference lies in the incubation atmosphere. Reducing agents, which are nontoxic, are added to most anaerobic media to depress and maintain the redox potential at low levels. Examples of such agents are sodium thioglycollate and cysteine hydrochloride.

Procedures for preparing media and for culturing most anaerobes are not difficult. One procedure employs potassium hydroxide and pyrogallol to absorb the oxygen in the container in which the microorganisms are being incubated. Another method utilizes anaerobic jars. These containers are relatively inexpensive and allow one to examine the surface colonies of anaerobes closely. Among the most widely used are the Brewer anaerobic jar with a heat-activated catalyst and the GasPak Anaerobic System (Figure 8–2a). With the former apparatus, hydrogen or a mixture of gases is introduced into the anaerobic jar after it is sealed. Electrical heat activation of a platinum catalyst present in the lid creates anaerobic conditions.

(b)

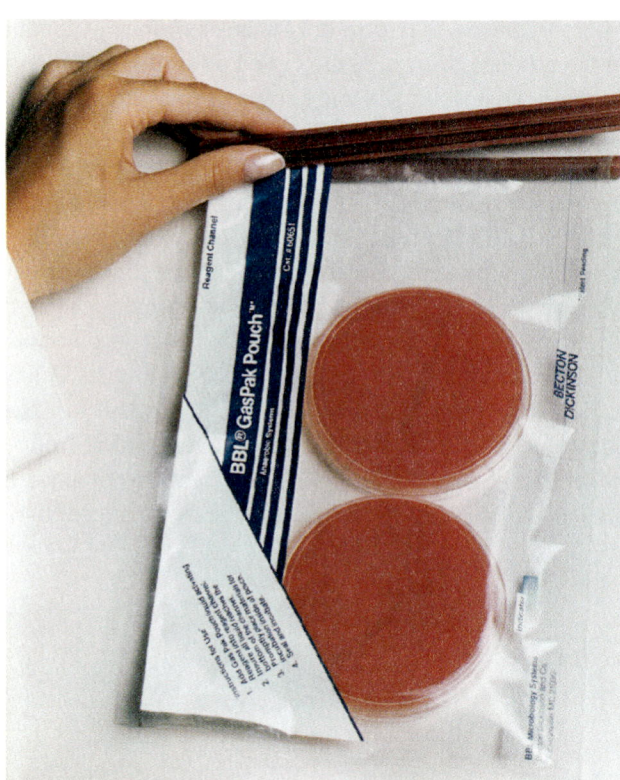

(a)

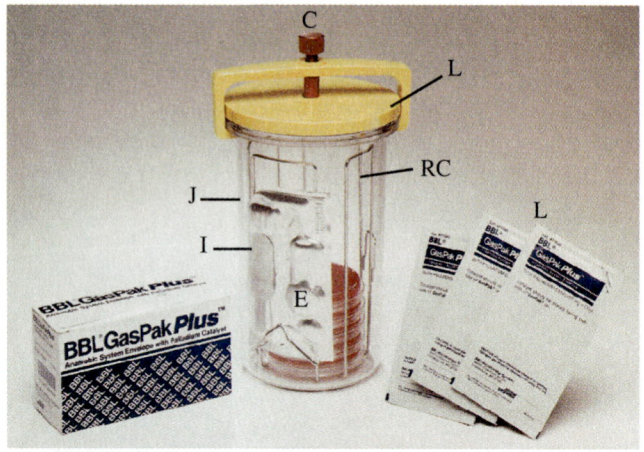

Figure 8–2

A commercial anaerobic cultivation device. (a) The components of the GasPak Anaerobic System: GasPak anaerobic jar (J), lid (L), clamp screw (C), charged catalyst reaction chamber (RC), disposable hydrogen + GasPak carbon dioxide generator envelope (E), and GasPak disposable anaerobic indicator (I). (b) The GasPak Pouch Anaerobic System. This system uses a reagent that contains all the reactants necessary to create an anaerobic environment. The reagent also contains a built-in indicator to show that anaerobiosis has been achieved.
(Photo courtesy of Becton Dickinson Microbiology Systems. GasPak and GasPak Plus are trademarks of Becton, Dickinson and Company.)

The GasPak jar differs from all other anaerobic jars in that it has no external connections and it utilizes a room-temperature catalyst system, thus rendering electrical connections or other means of heating the catalyst unnecessary. This anaerobic jar is specially designed to be used with a disposable hydrogen- and carbon dioxide-generating system. When water is added to a GasPak envelope, hydrogen gas is released. This gas, in turn, reacts with oxygen in the presence of a catalyst to produce anaerobic conditions. Carbon dioxide is also produced in quantities sufficient to support the growth of anaerobes requiring it. A disposable anaerobic indicator containing methylene blue is used to determine whether an anaerobic condition exists within the system.

Another system used for the cultivation of some anaerobic, facultative organisms (which can grow with or without oxygen) and microaerophilic organisms (which prefer low oxygen tensions) is the GasPak Pouch (Figure 8-2*b*). This system is compact, self-contained, and disposable.

Procedures that do not involve complicated pieces of equipment or elaborate techniques are also commonly used. The so-called shake cultures are examples. For shake cultures, nutrient agar is melted, cooled to approximately 45°C, and inoculated with a culture. The mixture is then caused to solidify rapidly. Upon incubation, the different oxygen requirements of organisms may be evidenced as follows: (1) strict anaerobes will be found only in the deeper portions of the culture; (2) facultative anaerobes will be encountered throughout the preparation; (3) microaerophiles will grow near the surface; and (4) aerobes will be located at the surface. Figure 8–3 shows the characteristics of bacteria with different oxygen requirements growing in shake cultures.

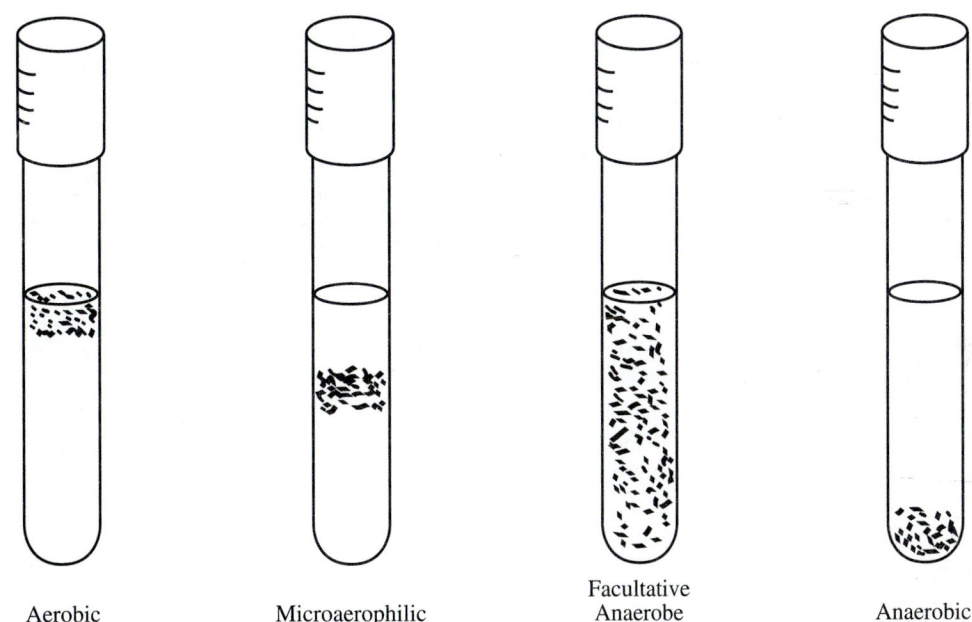

Aerobic   Microaerophilic   Facultative Anaerobe   Anaerobic

Figure 8–3
Growth patterns (distributions) of bacteria in shake cultures.

Other methods include the paraffin plug technique and the use of media containing a reducing compound, such as sodium thioglycollate, to reduce the oxygen content. In the paraffin plug technique, the tube of medium is heat for several minutes, cooled quickly, and then inoculated. A layer of melted paraffin approximately one-quarter of an inch thick is poured onto the top of the medium. The culture is then ready for incubation. Figure 8–4 shows paraffin plug cultures.

Thioglycollate media are used in the cultivation of anaerobic, microaerophilic, and aerobic bacteria and for the detection of bacteria in normally sterile materials. The ingredients of these media support the growth of a wide variety

Figure 8–4
The paraffin-plug technique. Left tube, uninoculated with a paraffin plug; center tube, turbidity (cloudiness) indicating growth; right tube, no growth under anaerobic conditions.

of microorganisms having a broad range of growth requirements. The presence of sodium thioglycollate lowers the oxidation-reduction potential of the media, while resazurin serves as the oxidation-resolution (OR) indicator and indicates the status of oxidation or aerobiosis.

Representative growth characteristics of bacteria in fluid thioglycollate broth are shown in Figure 8–5.

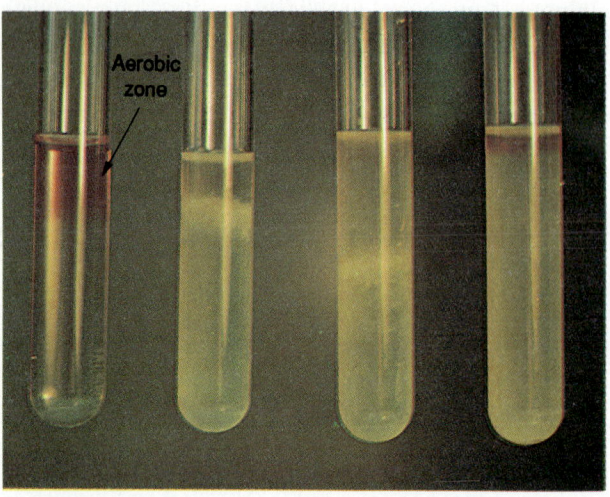

Figure 8–5

Representative growth characteristics of bacteria in fluid thioglycollate broth. Tube a is uninoculated. Note the presence of the upper pink zone (for aerobes) and the lower yellow zone (for anaerobes). Tube b shows the growth of a typical anaerobic organism. Tubes c and d show facultative (adjustable) organisms.

In this exercise, you are to perform the shake culture and paraffin plug techniques and inoculate tubes of thioglycollate broth. Included in the exercise is a demonstration of the use of an anaerobic jar system and of the GasPak Pouch.

Materials

❑ 1. Forty-eight-hour thioglycollate broth cultures of the following (1 per 4 students):
   ❑ a. *Clostridium sporogenes* (klō-STRID-ē-um, spor-a-GEN-ēz)
   ❑ b. *Escherichia coli*
   ❑ c. *Serratia marcescens*

❑ 2. Four divided, freshly streaked trypticase soy agar plates of each of the following microbial combinations:
   ❑ a. *Clostridium sporogenes* and *Escherichia coli*
   ❑ b. *Clostridium sporogenes* and *Serratia marcescens*

   These plates will be prepared during class time and used in the demonstration of the anaerobic jar system and the GasPak Pouch.

❑ 3. Media (3 per student):
   ❑ a. Thioglycollate broth
   ❑ b. Nutrient agar deeps
   ❑ c. Beef heart infusion broth

❑ 4. Melted paraffin (in 1 container to be used by the entire class)

❑ 5. Enough Pasteur pipettes for class use for the dispensing of paraffin

❑ 6. One GasPak Anaerobic System, which includes 1 each of the following: polycarbonate jar, lid, clamp, clamp screw, "O" ring gasket, charged catalyst reaction chamber, catalyst charge, GasPak disposable hydrogen + carbon dioxide generator envelopes, and GasPak disposable anaerobic indicator

❑ 7. Ten mL of sterile distilled water and one 10-mL sterile pipette for use with the GasPak system

❑ 8. Two GasPak Pouch Systems, including liquid activating reagent and sealing bars

❑ 9. Boiling water bath or 500-mL beakers for class use

❑ 10. Safranin stain in dropper bottles for class use

A. Shake Cultures

Procedure

This procedure is to be performed by students in pairs.

❏ 1. Incubate all preparations at room temperature or as indicated by the instructor.

❏ 2. After performing the six steps of the technique shown in Procedure Diagram 12 with each culture provided, examine the preparations and record your findings in the Results and Observations section. (Refer to Figure 8–3.)

Procedure Diagram 12
Shake Cultures

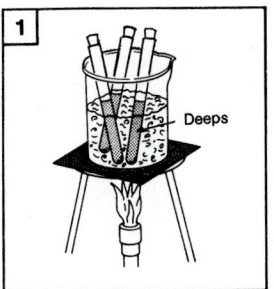

1. Melt 3 nutrient agar deeps.

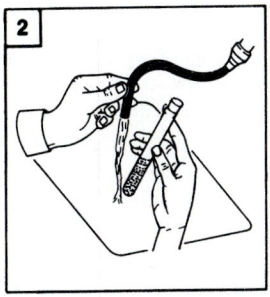

2. Cool the deeps to about 45°C by carefully rotating them under cold running water.

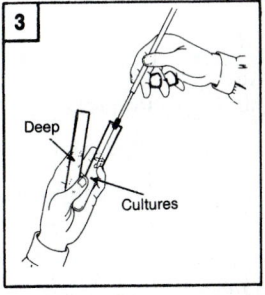

3. Inoculate each tube with one of the cultures provided. (Refer to Procedure Diagram.)

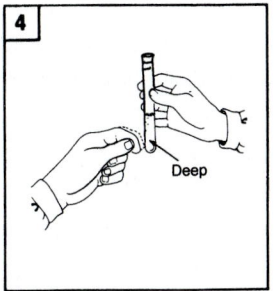

4. Shake each tube by gently striking the preparation with your fingers

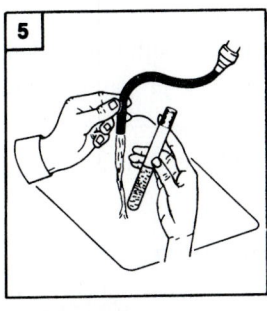

5. Cool the tubes rapidly under cold running water.

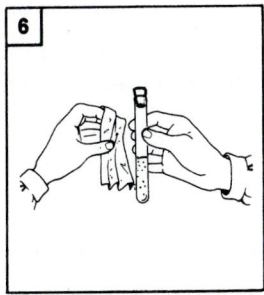

6. Wipe the tubes, mark them to indicate the microorganism used, and incubate as indicated earlier.

B. Paraffin Plug Technique

Additional Technique Required for This Portion of the Exercise:

Broth Transfer, Procedure Diagram 5, Exercise 4

Procedure

This procedure is to be performed by students in pairs.

- ❏ 1. Heat the tubes of sterile beef heart infusion broth for 5 minutes in a boiling water bath.
- ❏ 2. Cool the media and inoculate separate tubes with each of the cultures provided.
- ❏ 3. Remove the plug from the tube aseptically, tilt the tube, and place melted paraffin on top of the medium with the aid of a Pasteur pipette. The layer should be approximately 7mm or 0.25 inch in thickness. Replace the plug in the tube.
- ❏ 4. Incubate at 37°C for 48 hours.
- ❏ 5. After incubation, compare and describe your findings in the Results and Observations section. Consult Figure 8–4.

C. Inoculation of Thioglycollate Medium

Additional Techniques Required for This Portion of the Exercise:

1. Bacterial Smear Preparation, Procedure Diagram 2, Exercise 2
2. Simple Staining, Procedure Diagram 3, Exercise 2
3. Broth Transfer, Procedure Diagram 5, Exercise 4

Procedure

This procedure is to be performed by students in pairs.

- ❏ 1. Using the safranin stain reagent, prepare a smear and simple strain of each culture provided. Examine these preparations along with those to be obtained in step 4.
- ❏ 2. Inoculate individual tubes of thioglycollate medium with each of the cultures used in this exercise.
- ❏ 3. Incubate at 37°C for 48 hours.
- ❏ 4. Observe the regions of growth in each tube. Consult Figure 8–5. Using safranin, prepare a simple stain of material taken from each region. Examine these preparations under oil immersion, along with those made in step 2.
- ❏ 5. Record your findings in the appropriate portion of the Results and Observations section.

D. Anaerobic Jar Demonstration

Additional Techniques Required for This Portion of the Exercise:

1. Streak Plate Technique, Procedure Diagram 10, Exercise 5
2. Bacterial Smear Preparation, Procedure Diagram 2, Exercise 2
3. Simple Staining, Procedure Diagram 3, Exercise 2

Procedure

This procedure is to be performed by students in pairs.

- ❏ 1. The instructor will demonstrate the use of this system.
- ❏ 2. *Note:* Only one set of streaked plates will be incubated at 37°C in the anaerobic jar. The other set will be incubated aerobically at the same temperature for 48 hours or longer.
- ❏ 3. Examine the streak plates after incubation for growth and record your findings in the Results and Observations section.
- ❏ 4. Prepare a simple stain, with safranin, of 1 isolated bacterial colony from each region in which growth occurred. Examine these preparations under oil immersion and record your findings.
- ❏ 5. The 6 steps for the use of the system are shown in Procedure Diagram 13.

Procedure Diagram 13
The Anaerobic Jar System

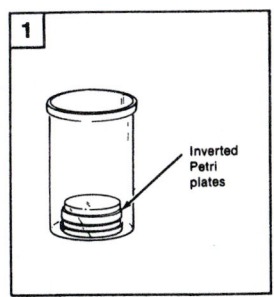

1. Place 1 set of streaked plates into the jar in an inverted position.

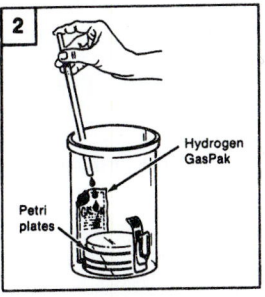

2. Prepare the generator by tearing off a corner of the envelope. Using a pipette, add 10ml of sterile distilled water into the GasPak. Put the preparation into the jar.

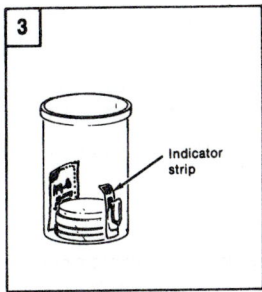

3. Open the indicator envelope to expose the indicator strip and place it in the jar. Be certain that the strip is visible from the outside.

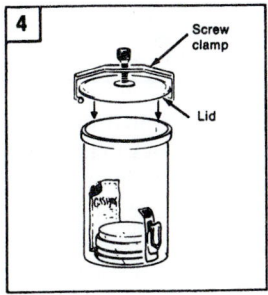

4. Place the lid and screw clamp in position on the jar.

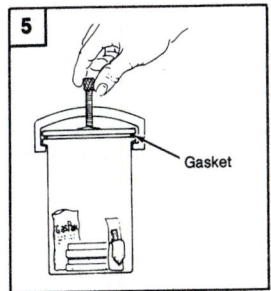

5. Tighten the screw clamp by hand only and incubate the system as directed earlier.

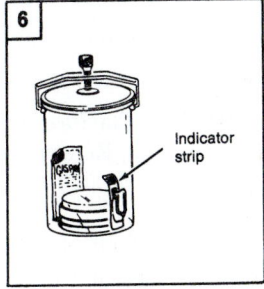

6. Check the indicator strip periodically for a change from a blue color to a colorless state. If this change does not occur within 1 or 2 hours, the generator should be replaced.

E. GasPak Pouch Anaerobic System Demonstration

Procedure

This procedure is to be performed by students in pairs.

❑ 1. The instructor will demonstrate the use of this system.

❑ 2. Examine the streak plates after incubation and compare them to those obtained with the anaerobic jar system.

❑ 3. Using safranin, prepare a simple stain of 1 isolated bacterial colony from each region in which growth occurred. Examine these preparations under oil immersion.

❑ 4. Answer the questions and record your findings in the Results and Observations section.

❑ 5. The 3 steps for the use of the system are shown in Procedure Diagram 14.

Procedure Diagram 14
The GasPak Pouch Anaerobic System

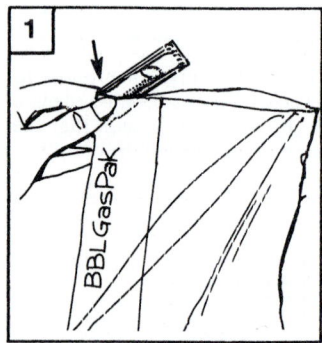 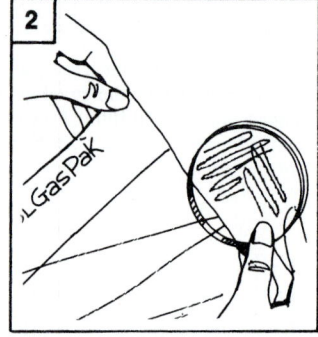 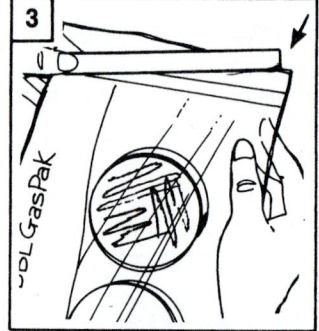

1. Pour the BBL GasPak Pouch Liquid Activating Reagent into the reagent channel (arrow).

2. Place 2 freshly streaked plates into the GasPak pouch.

3. Hold the pouch in an upright position and lock in the anaerobic environment by clamping the open end with the GasPak sealing bar. Incubate the system at 37° for 48 hours.

Results and Observations (A through D)

1. Shake cultures

    a. Appearance of growth

    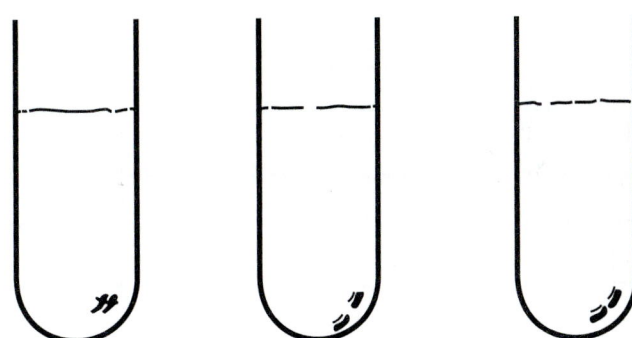

    b. Organism   _____   _____   _____

                    _____   _____   _____

2. Describe the results obtained with the paraffin plug technique.

    a. In which tubes did growth occur? _____

    _____

    b. In which tubes did growth not occur? _____

    _____

3. Thioglycollate medium microscopic appearance

    Organisms

    a. _____

    b. _____

    c. _____

    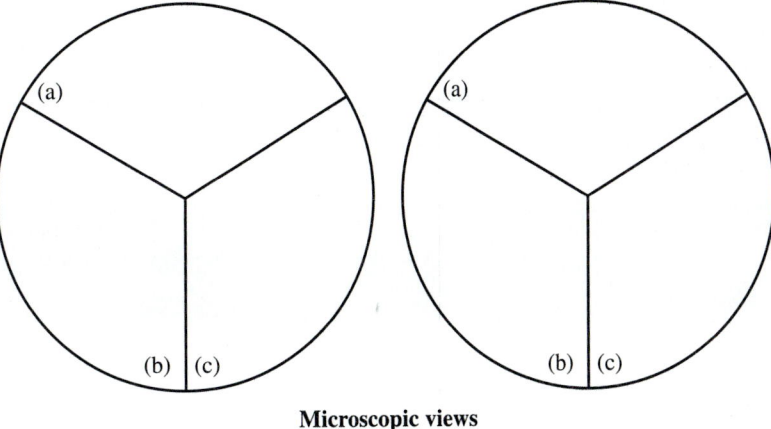

    Microscopic views

    Inoculum   After incubation

4.   Anaerobic jar system. Complete Table 8–1 by indicating the presence (+) or absence (−) of growth under the conditions specified and the morphology of the cultured bacterial species.

Table 8–1

Anaerobic Jar Culture Properties

	Cultures					
	C. sporogenes		E. coli		S. marcescens	
Characteristics	Aerobic	Anaerobic	Aerobic	Anaerobic	Aerobic	Anaerobic
Growth						
Morphology						

5.   GasPak Pouch Anaerobic System. Complete Table 8–2 by indicating the presence (+) or absence (−) of growth under the conditions specified and the morphology of the cultured bacterial species.

Table 8–2

GasPak Pouch Anaerobic System Culture Properties

	Cultures					
	C. sporogenes		E. coli		S. marcescens	
Characteristics	Aerobic	Anaerobic	Aerobic	Anaerobic	Aerobic	Anaerobic
Growth						
Morphology						

Laboratory Review 8   Selected Techniques for the Cultivation of Anaerobes

1.   In Part B, what is the reason for heating the beef heart infusion broth before inoculation?_____

_____

2.   Are all anaerobic bacteria spore formers? Explain. _____

_____

3. List 3 non-spore-forming anaerobic bacteria and the activities associated with each. (Refer to your text.)

    Genus, species   *Activity*

    a. _____   _____

    b. _____   _____

    c. _____   _____

4. a. What is the purpose of the indicator systems used in the anaerobic jar and the GasPak Pouch? _____

    _____

    b. What gas is produced by the generator in the GasPak system? _____

    _____

5. What is a Brewer place? _____

    _____

6. Thioglycollate broth can be used to culture and to differentiate between what 2 general microbial groups? _____

    _____

7. Differentiate between the following types of organisms:

    a. aerobe and anaerobe _____

    _____

    b. facultative and microaerophilic _____

    _____

Key Terms

aerobe (ER-ōb): a microorganism that requires free oxygen for growth and other cellular activities

anaerobe (an- ER-ōb): a microorganism that grows in the absence of air or free oxygen

bacterial spore former: a microorganism capable of forming resistant resting structures

E_h (redox potential): measure of the tendency of a preparation to release electrons (oxidized) or to accept electrons (reduced)

facultative (fak-cul-TAY-tiv) anaerobe: refers to organisms capable of growth under either aerobic or anaerobic conditions

microaerophilic (mī-krō- ER-ō-fil-ik): refers to aerobes that require environments with small amounts or less than atmospheric oxygen levels for growth

redox dye: a dye capable of showing oxidation-reduction reactions in which one compound becomes oxidized (releases electrons), and another compound becomes reduced (takes up the released electrons)

thioglycollate (thī-ō-GLĪ-kō-lāt) medium: a nutrient preparation used for the cultivation of aerobic, anaerobic, and microaerophilic microorganisms

A Survey of the Microbial World

Do there exist many worlds, or is there but a single world? This is one of the most noble and exalted questions in the study of nature.

—St. Albertus Magnus

The basic unit of living organisms is the cell. Nearly all forms of life are single-celled (unicellular) or many-celled (multicellular). In the cells of such forms, specialized structures called *organelles* conduct specific activities.

The living cell is a highly organized, self-directing, and complex system. Each cell is composed of molecules and molecular aggregates. The energy needed for a cell's various activities is obtained from the environment either in the form of chemical bonds or as light energy. The latter form of energy is exclusively utilized by green plants, algae, and certain species of bacteria.

Despite the similarities shared by cellular organisms, vast differences exist as well. For example, plants differ from animals in having the following characteristics: (1) the presence in the cells of a rigid outer layer called the *cell wall;* (2) the possession of chlorophyll and the performance of photosynthesis; (3) the absence of active motility; and (4) the ability to store food in the form of starch.

Traditionally, the various forms of life have been classified in one of two biologic kingdoms—Animalia and Plantae. However, more recent classification schemes include five kingdoms, thus accounting for certain groups that cannot be neatly placed in either the animal or plant kingdoms. In the five-kingdom approach, the prokaryotic bacteria, which include the blue-greens or the cyanobacteria (Figure III–1*a*), chlamydia, and rickettsia (Figure III–1*b*), are placed into a third kingdom, **Prokaryotae**. Many single-celled eukaryotic algae (Figure III–2*a*) and protozoa (Figure III–2*b*) are placed into the kingdom **Protista,** and one kingdom is exclusively composed of fungi, the kingdom **Fungi.**

Bacteria appear to be structurally simpler than the cells of multicellular plants and animals. Their structural components include a plasma membrane, a bacterial *nucleoid* and *ribosomes*. Photosynthetic bacteria have, in addition to these parts, specific structures containing photosynthetic pigments and associated enzymes. The organelles of motion, called *flagella,* and the vesiculated structures, called *mesosomes,* are found in some, although not all, bacteria. Bacteria are grouped by shape as spherical, rodlike, spiral, or square. The different morphological arrangements of bacteria and their various components will be considered in detail in later exercises.

(b)

(a)

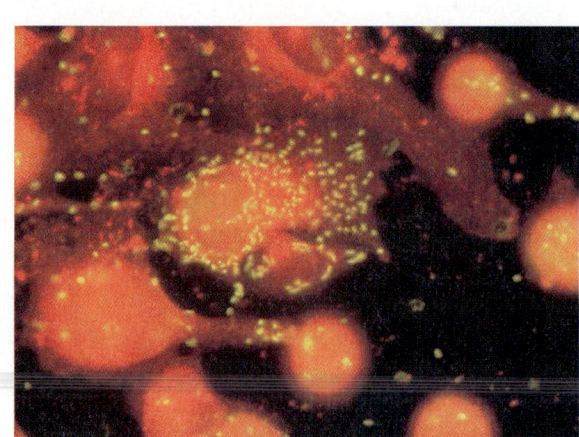

Figure III–1
Examples of prokaryotic microorganisms (a) The cyanobacterium, *Oscillatoria*. (b) A microscopic view of rickettsia. The causative agent of Rocky Mountain spotted fever appears as small greenish yellow rods in this preparation. The special approach known as fluorescent staining was used to distinguish the bacteria from the larger (orange) animal cells shown.

(a)

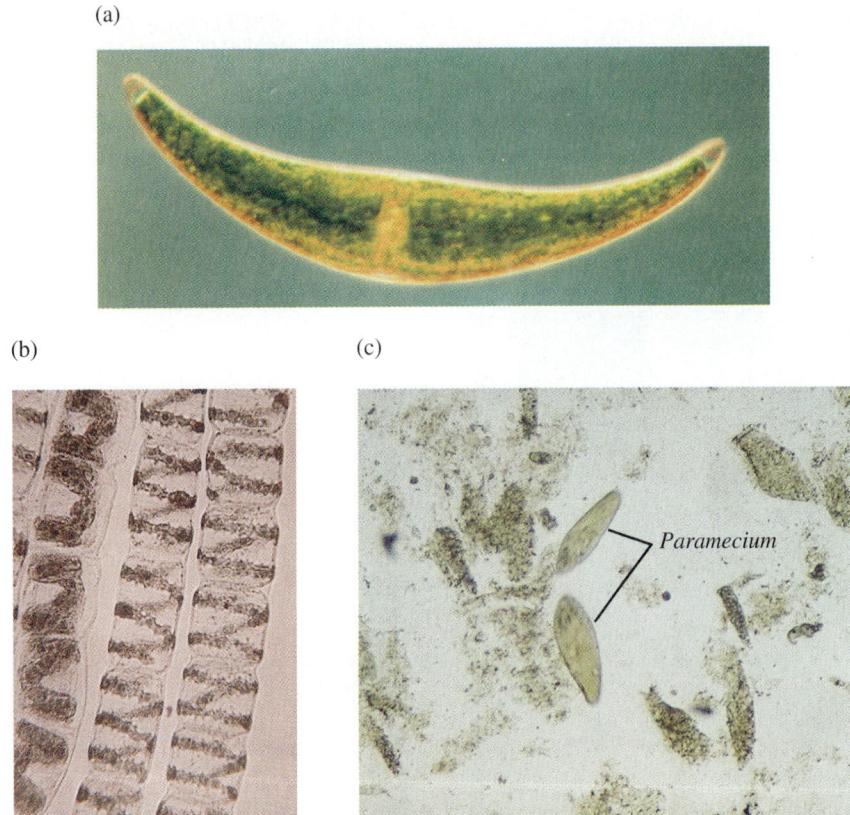

(b)   (c)

Figure III–2

The protists. (a) The colorful alga (desmid) *Closterium ehrenbergii*
(Courtesy Professor Dr. K. Ueda, Women's University, Nara, Japan.)

(b) The green alga, *Spirogyra*. (c) The well known protozoan *Paramecium*.

True fungi are organisms that lack roots, stems, or leaves. The kingdom of Fungi includes both molds and yeasts. These microorganisms are larger and more complex structurally than bacteria. Microscopically, molds exhibit tubular branching cells, called *hyphae,* and reproductive structures known as *spores* (Figure III–3). Yeasts appear as oval, spherical, or elongated cells.

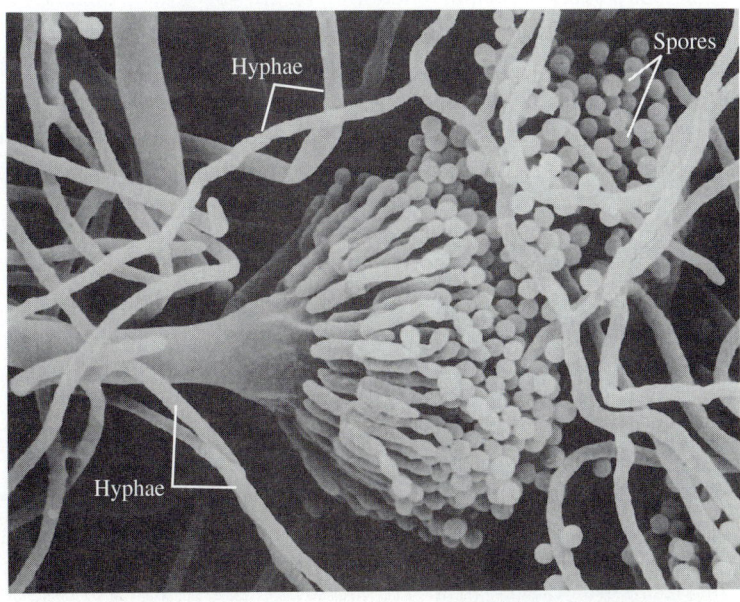

Figure III–3

A scanning micrograph showing the tubelike *hyphae,* and the reproductive structures, *spores,* of fungi.

Fungi do not possess the photosynthetic pigment chlorophyll, and as a consequence they exhibit a heterotrophic form of nutrition. These microorganisms are dependent upon enzyme systems for obtaining energy from organic substances. Most fungi are either obligate saprobes (saprophytes) or parasites. True fungi are divided into several classes on the basis of differences in morphology and modes of reproduction. In this manual, consideration will be given to representative fungi including the gilled, fleshy forms known as the *mushrooms*. These fungi come in a variety of forms, sizes, and colors (Figure III–4). Both macroscopic and microscopic features of fungi used in identification and classification will be presented in this section.

(a)

(b)

Figure III–4

The kingdom of the fungi. These forms of life come in various forms, sizes, and colors. (a) An example of a fleshy mushroom. (b) The appearance of a bush-like mushroom.

Fungi, as a group, are capable of exhibiting a wide range of activities, some beneficial and others harmful to other organisms. Fungi are used in the production of antibiotics, foods, beverages, and chemicals. Like bacteria, fungi degrade many types of organic compounds that thus perform essential functions. The growth of fungi on foods will also be considered.

The kingdom of Protista consists of a great number of eukaryotic organisms. At the present time, as many as 200,000 species are estimated to be in this kingdom. The boundaries of the Protista have been expanding in recent years, largely because there is no universal agreement among biologists as to what properties clearly define a "protist." The kingdom broadly includes protozoa, algae, slime molds (Figure III–5), and water molds. Although these forms of life may superficially resemble animals, plants, or fungi in certain properties, they should not be considered simple forms of animals, plants, or fungi.

Figure III–5

The scrambled egg slime mold, *Fuligo septica.* Both a well-formed yellow plasmodium and the filamentous newly forming plasmodia are shown.

All protozoa consist of a single cell or are independent parts of a cellular group referred to as *colony.* These eukaryotic microorganisms are found in fresh water, oceans, soils, and the bodies of other forms of life. Protozoa are amazingly complex in structure, physiology, and behavior. They exhibit a heterotrophic form of nutrition. Because they cannot manufacture food molecules from inorganic substances, protozoa must obtain previously synthesized molecules, either directly or indirectly, from autotrophs such as algae and plants. Protozoa have various mechanisms for obtaining and digesting nutrients. In addition, many protozoa have internal relationships with other microbial types including bacteria and algae, and harbor them as *endosymbionts* (Figure III–6). Such endosymbionts are organisms found within a eukaryotic cell, contain DNA, and are enclosed by at least one membrane. These relationships, which may be temporary or permanent, are a major point of focus in the cell biology area coming to be known as *endocytobiology.* Several properties of protozoa and other microorganisms are presented in the exercises of this section.

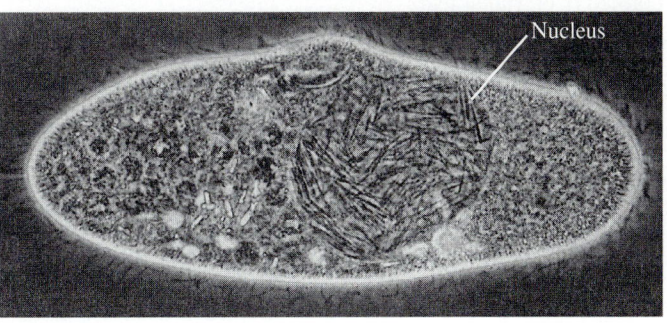

Figure III–6

The protozoon. *Paramecium caudatum* infected with bacteria in its large nucleus (**macronucleus**). These endonuclear symbiotic bacteria (arrows) are resistant to digestion by the protozoon's enzymes. Original magnification, 12,000X.
(From Lee, J.J., A.T. Soldo, W. Reisser, M.J. Lee, K.W. Jeon, and H.D. Gortz, *J. Protozool.* 32:391–403 [1985])

Many microscopic, single-celled eukaryotic algae belong to the aquatic-wandering, or plankton, community. These organisms live suspended in both fresh water and salt water, and are generally restricted to the sunlit zones of the aquatic environment. Algae are also found on and within soil, on moist stone and wood, and in association with various forms of life, such as fungi (Figure III–7). All algae are autotrophs, utilizing the absorbed energy of sunlight and the composite pigment chlorophyll to synthesize sugars from carbon dioxide. This process is known as *photosynthesis.*

Algae are classified in several divisions. Each group has a unique complement of photosynthetic pigments, storage products, cellular characteristics (Figure III–2a), and habitats.

The exercises in this section are designed to demonstrate the morphological details of selected examples of microorganisms. Some exercises are also intended to illustrate the wide environmental distribution of microorganisms as well as the methods used for their isolation and cultivation.

Figure III–7

Several species of lichens on a tree branch, the fibrous form called the "old man's beard," and orange and white crustose lichens.

After completing this exercise, you should be able to:

1. Recognize the basic shapes and arrangements of bacteria.
2. Develop a perspective on size relationships among bacteria and other microorganisms.
3. Locate and examine stained bacterial smears.
4. Calibrate on ocular micrometer.
5. Use an ocular micrometer to measure the sizes of microbial cells.
6. Observe what Anton van Leeuwenhoek saw in peppercorn-soaked water preparations in the late 1660s.

Microbiology is the study of biological entities too small to be seen with the naked eye. Certain algae, bacteria, fungi, protozoa, and viruses constitute the world of microbes with which the science deals. The majority of microorganisms are unicellular and exhibit apparent structural and organizational simplicity. The possession of this property, however, does not necessarily indicate a simplicity of physiological processes. This exercise will deal with characteristic morphological features useful in the identification of selected microorganisms.

The general morphological shapes of bacteria—rod, coccus, and spiral—were described by Anton van Leeuwenhoek in the late 1600s. A new morphological type of prokaryote, the square, was described in 1980 by A. E. Walsby. These cells appear as flat, rectangular boxes with perfectly flat edges. Apart from these differences in cellular shape, definite patterns in cellular numbers and arrangements are known to exist among different bacterial species (Figure 9–1).

In the case of spherical or coccus forms, **five** patterns can be found. These include pairs of bacteria (diplococci), chains of **four** or more organisms (streptococci), **four** bacteria in a square arrangement (tetrads), irregular groups of microorganisms resembling grape clusters (staphylococci), and **eight** cocci grouped into a cuboidal packet (sarcinae formation).

Rod-shaped bacteria, also referred to as *bacilli* (singular, *bacillus*), can be found occasionally in pairs (diplobacilli) and in chains (streptobacilli). However, these morphological patterns of rod-shaped forms are not as constant as in the case of the cocci and therefore should not necessarily be considered as characteristic for a particular species.

Two groups of coiled or spiral-shaped bacteria are known. One group, the *spirochetes* (Figure 9–2), consists of flexible, waving forms with several coils. The second group, the *spirilla,* are rigid bacteria possessing one or several curves. Spiral forms that are short and do not form complete coils are called *vibrios*.

Figure 9–3 summarizes the characteristic cell shapes and morphological arrangements found among bacteria.

Variations in the general shapes and sizes of bacteria are frequently seen and can be explained in terms of environmental factors. *Pleomorphism* is the term used to denote these modifications when they occur under favorable conditions. Under unfavorable conditions, these variations are called *involution forms*.

The determination of the dimensions of *cells* requires a functional microscope, a good light source, and a calibrated ocular micrometer. Measurements are made with the aid of an *ocular micrometer.* This special disk is placed in the *microscope ocular lens,* or "microscope eyepiece." The disk contains a number of equally spaced graduations, or divisions, engraved on one side. The graduations look like a miniature ruler or scale. The divisions in the disk represent different units of measurement, depending on the objective magnification used; hence, it is necessary to calibrate (compare) the ocular micrometer with a special slide called a *stage micrometer.* A stage micrometer also contains an engraved scale. For each objective lens, the ocular micrometer is calibrated by superimposing the ocular scale on the stage scale. Then one determines the number of ocular graduations that coincide with *one* graduation of the stage micrometer scale. The purpose of the calibration procedure is to determine the exact distance between the graduations of the ocular micrometer for each objective lens.

This exercise presents a procedure for the calibration of a microscope. Once calibrated, the microscope will be used for precise measurements of the different smears prepared in the exercise.

(a)

(b)

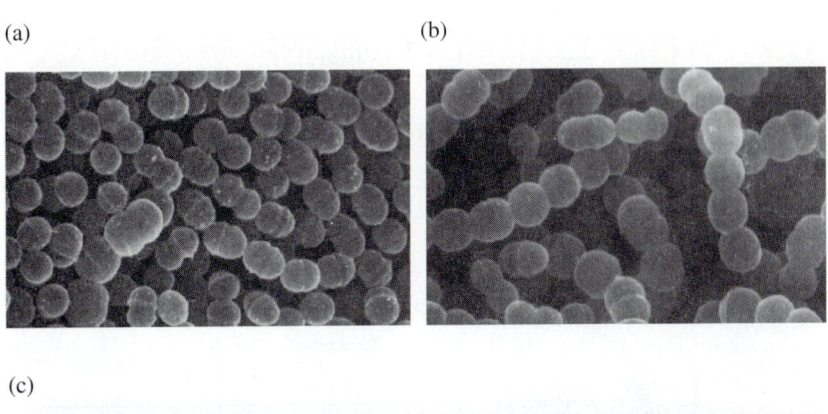

(c)

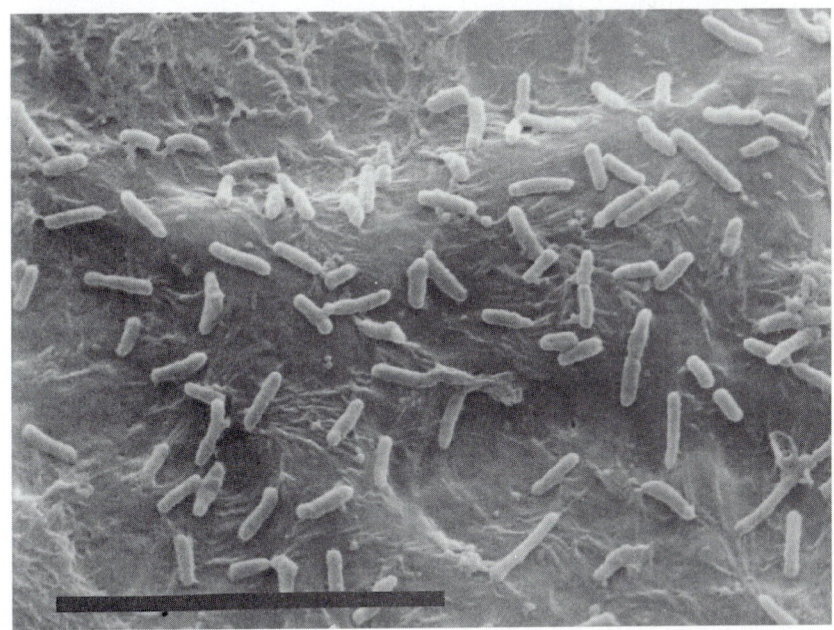

Figure 9–1

Morphological arrangements among bacteria (a) A scanning micrograph showing the appearance of diplococci. (b) Cocci or spherical forms in a streptococcus or chain arrangement.
(From Pincus, S.H. and S.F. Hayes, *J. Bacteriol.* 174:3739 [1992])

(c) The appearance of *Escherichia coli,* a small rod.
(From T. Yamamoto, *J. Inf. Diseases* 166:1295-1310 [1992])

The bar markers represent 1 micrometer (μm).

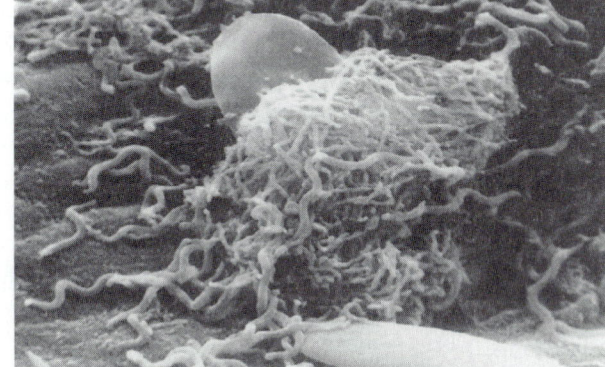

Figure 9–2

A scanning micrograph showing spirochetes in mass. The large oval structures are chicken red blood cells.
(From Swayne, D.E., and Associates. *Avian Dis* 36:776–781 [1992])

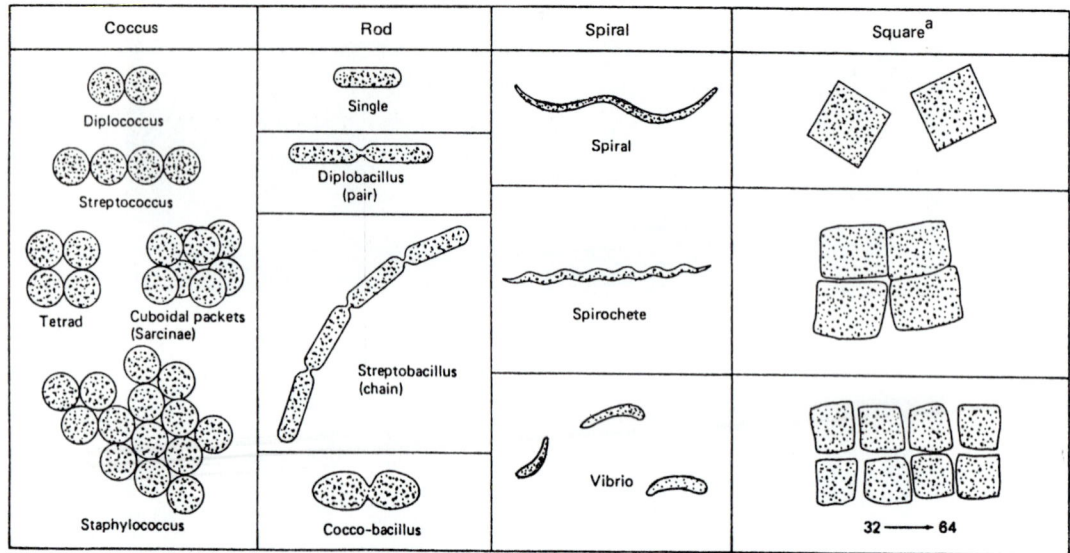

Figure 9–3
Characteristic bacterial cell shapes and arrangements.

This exercise also considers one of Anton van Leeuwenhoek's early experiments concerning his attempts to discover whether the spiciness of pepper could be observed microscopically. While the views of the water-soaked peppercorns contained no surprises, van Leeuwenhoek was amazed by the great number of tiny organisms that swarmed around in a single drop of the peppercorn material. Like many scientists noted for important discoveries, Anton van Leeuwenhoek stumbled on the existence of bacteria and protozoa while looking for something entirely different.

This exercise will be limited to the demonstration of representative morphological features of bacteria.

A. Bacterial Morphology

Materials

❑ 1. Twenty-four-hour trypticase soy broth cultures of the following (1 per 4 students):

    ❑ a. *Moraxella (Branhamella) catarrhalis* (mōr-ax-EL-a kat-TAR-al-is)

    ❑ b. *Enterococcus (Streptococcus) faecalis*

    ❑ c. *Aerococcus viridans* (er-Ō-KOK-us veer-E-danz)

    ❑ d. *Micrococcus luteus*

    ❑ e. *Staphylococcus aureus*

    ❑ f. *Escherichia coli*

    ❑ g. *Rhodospirillum rubrum* (rō-dō-spī-RIL-um ROO-brum)

❑ 2. Containers with crystal violet (1 per 4 students)

❑ 3. Commercially prepared slides of the following organisms for demonstration:

    ❑ a. *Treponema pallidum*

    ❑ b. *Vibrio cholerae* (VIB-rē-ō, KAHL-er-ī)

    ❑ c. *Spirillum volutans* (spī-RIL-um vol-lū-tans)

❑ 4. Glass slides

❑ 5. Immersion oil, xylene, and lens paper

Procedure I: Broth Cultures

This procedure is to be performed by students individually.

❑ 1. Prepare smears of each broth culture according to the pattern illustrated below. Simply spread the inoculum across the width of the glass slide. Place only 4 bacteria on 1 slide.

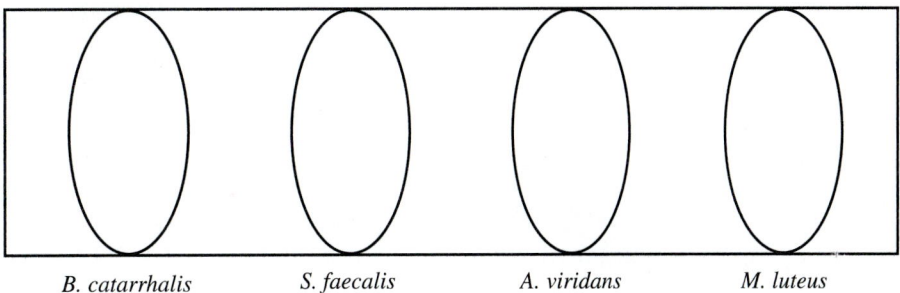

| *B. catarrhalis* | *S. faecalis* | *A. viridans* | *M. luteus* |

❑ 2. Stain each slide with crystal violet. Stain the slides for 1 minute.
❑ 3. Carefully rinse the slides, blot dry, and examine under oil immersion.
❑ 4. Examine the demonstration slides.
❑ 5. Answer the questions and make the sketches called for in the Results and Observations section. Refer to Figure 9–3.

Procedure 2: Stained Preparations

This procedure is to be performed by students individually.

❑ 1. Examine each stained slide under oil immersion. With the aids of Figures 9–3, 9–4a, and 9–4b, identify the morphological features of the stained organisms.
❑ 2. Answer the questions in the Results and Observations section and the Laboratory Review.

B. Microscope Calibration and Measurements

Materials

The following items should be provided 1 per 4 students:

❑ 1. Ocular micrometer or eyepiece containing an ocular micrometer
❑ 2. Stage micrometer
❑ 3. Prepared blood or tissue smears with (refer to figures in Exercise 2 and Figure 9–4):
    ❑ a. *Borrelia recurrentis*
    ❑ b. *Treponema pallidum*
❑ 4. Normal blood smear
❑ 5. Student or commercially prepared stained smears of:
    ❑ a. *Bacillus subtilis*
    ❑ b. *Escherichia coli*
    ❑ c. *Micrococcus luteus*

(a)

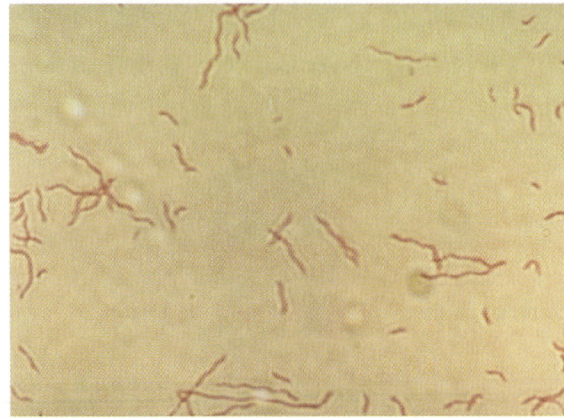

(b)

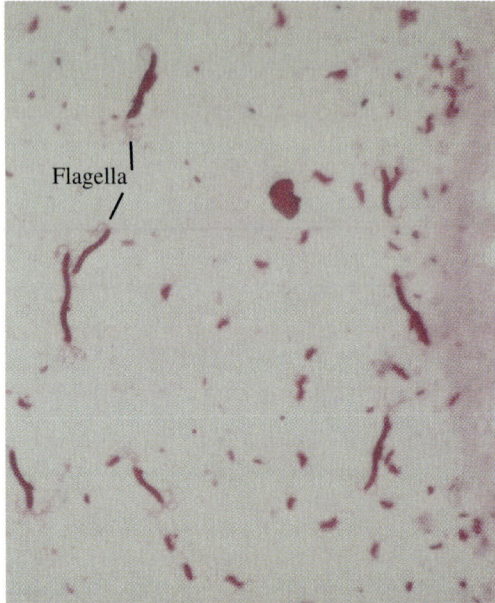

Flagella

(c)

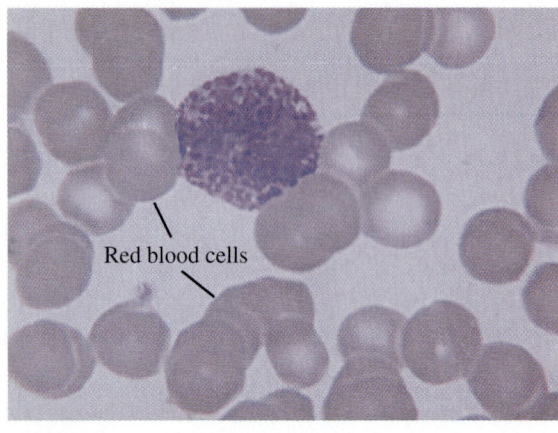

Red blood cells

Figure 9–4

Bacterial morphology and human blood. (a) *Vibrio cholerae,* the causative agent of Asiatic cholera. Note the curved appearance of individual cells. (b) *Spirillum volutans.* In addition to showing the spiral morphology of this, organism, the arrangement of two or more flagella at one or both ends of a cell is evident. Note that special staining procedures are needed to demonstrate these organelles. (c) A blood smear showing red blood cells and a basophil.

Procedure: Microscope Calibration and Measurements

This procedure is to be performed by students individually.

❑ 1. Before beginning the calibration of the microscope, find out if the ocular micrometer is to be inserted below the bottom lens of the ocular or placed between 2 lenses within the ocular.

❑ 2. To calibrate the microscope ocular, refer to the 9 steps in Procedure Diagram 15.

❑ 3. Once the lens of the microscope has been calibrated, remove the stage micrometer and measure the diameters of typical cells on both the bacterial smears and blood smears prepared earlier, or on other materials provided. Use 3 cells of each type and then calculate the average diameter.

❑ 4. *Note:* Each space of most stage micrometers equals 10 μm or 0.01 mm. This multiplication factor must be used to obtain the true value of each ocular division in step 9.

❑ 5. Enter your findings in Table 9–2 in the Results and Observations section.

Procedure Diagram 15
Microscope Calibrations

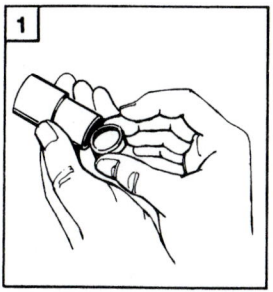

1. Carefully remove the ocular lens from the microscope and either insert the ocular micrometer on the retaining ring or shelf or attach it by screwing it into the ocular. The surface with the scale graduations should be closest to the ocular lens.

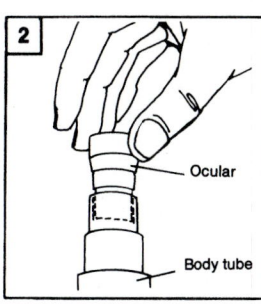

2. Reassemble the ocular lens carefully, if necessary, and place it back in the microscope.

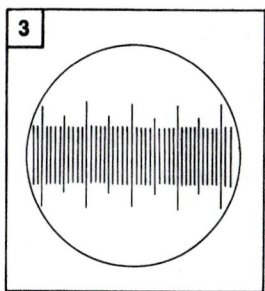

3. Look through the ocular to see the micrometer.

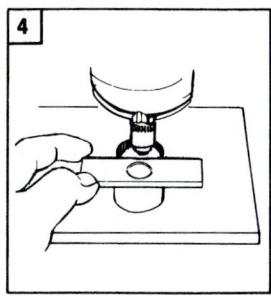

4. Next, mount the stage micrometer on the microscope stage and center it exactly over the light source.

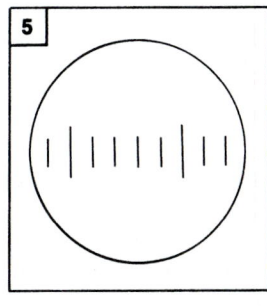

5. Using the coarse-adjustment knob and the low-power objective, bring the scale gradations (lines) of the stage micrometer into focus.

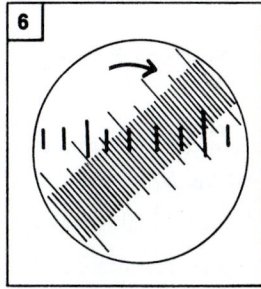

6. Rotate the ocular lens until the lines of the ocular micrometer are superimposed on those of the stage micrometer. If the low-power objective is to be calibrated, proceed to the next step. If the high-dry objective is to be calibrated, swing it into position before proceeding to the next step. If the oil-immersion lens is to be calibrated, place a drop of immersion oil on the stage micrometer and swing the oil-immersion lens into position before continuing to the next step.

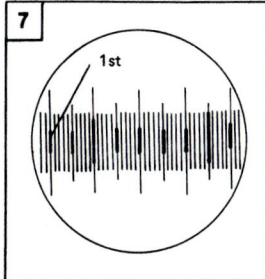

7. Move the stage micrometer laterally until the lines at 1 end of the scale coincide.

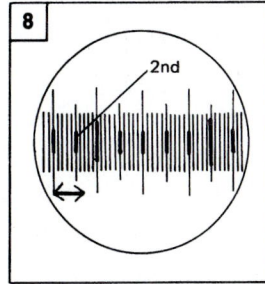

8. Find another line on the ocular micrometer that coincides with one on the stage micrometer. Count the number of ocular divisions that are equivalent to 1 stage division. Enter your findings in the Results and Observations section.

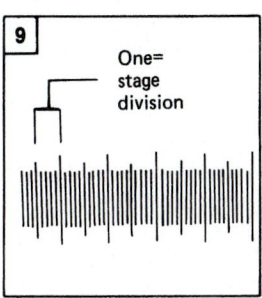

9. Calculate the value of 1 ocular division (OD) by the formula:

$$1 \text{ OD} = \frac{\text{Number SD*}}{\text{Number OD}} \times 0.01 \text{ mm}$$

$$\left(\begin{array}{l} \text{*SD} = \text{Stage Divisions} \\ \text{OD} = \text{Ocular Divisions} \end{array} \right)$$

C. Leeuwenhoek's Peppercorn Water

Materials

The following items should be provided for general class use:

- ❏ 1. Containers with peppercorns soaked in water for two weeks
- ❏ 2. Eye droppers
- ❏ 3. Glass slides and cover slips

Procedure

Additional Technique Required in This Exercise:

Temporary Wet-Mount Technique, Procedure Diagram 1, Exercise 1

This procedure is to be performed by students individually.

- ❏ 1. Prepare a temporary wet mount of the peppercorn water. (Refer to Procedure Diagram 1.)
- ❏ 2. Observe the preparations first under low power and then under high power. **Do not use the oil-immersion objective.** *Note:* Reducing the brightness of the light source will help greatly in locating specimens.
- ❏ 3. Look for typical bacterial cells. Note the morphology and morphological arrangements.
- ❏ 4. Make a representative sketch of your findings in the Results and Observations section.
- ❏ 5. Answer the questions in the Results and Observations section and Laboratory Review.
- ❏ 6. At the conclusion of the procedure, clean your slide and cover slip as directed by your instructor.

Results and Observations

Bacterial Morphology

1. Complete Table 9–1 by inserting a representative sketch of each bacterial species listed.

2. Make a representative sketch of (a) *S. volutans,* (b) *T. pallidum,* and (c) *V. cholerae.*

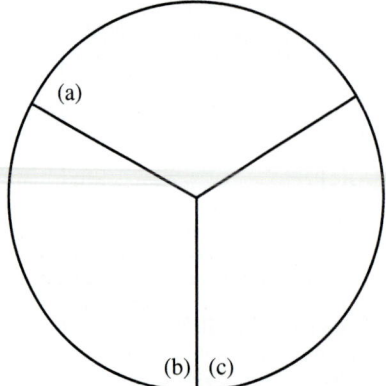

Table 9–1

Bacterial Morphology

Microorganism	Major Morphological Appearance
M. catarrhalis	
E. faecalis	
A. viridans	
M. luteus	
S. aureus	
E. coli	
R. rubrum	

Microscope Calibration and Measurements

1. Enter your findings and calculations for calibration.

   a. The number of ocular divisions equivalent to 1 stage division = _____

   b. One ocular division = _____ μm. Calculations:

2. Complete Table 9–2 by inserting the diameters of the cells measured.

Table 9–2

Cellular Measurements

Cell Type	Diameter (in μm)
B. subtilis	
B. recurrentis	
E. coli	
M. luteus	
T. pallidum	
Red blood cell	
Lymphocyte	
Monocyte	
Neutrophil	
Eosinophil	

aRefer to Figure 1–3 in this manual.

Leeuwenhoek's Peppercorn Water

1. Sketch a representative view of your specimen.

2. What was the predominant morphological type of bacteria in your preparation? _____

_____

3. Were there microorganisms other than bacteria in your preparation? If so, what kinds did you see? _____

_____

Laboratory Review 9   **Morphological Features of Bacteria, Microscopic Measurements, and Leeuwenhoek's Peppercorn Water**

1. List three differential staining procedures used to observe bacteria.

   a. _____   b. _____   c. _____

2. Define or explain:

   a. morphology and morphological arrangement _____

   _____

   b. dye _____

   c. involution form _____

   d. pleomorphism _____

3. Give the specific term for each of the following morphological arrangements of bacterial cells.

   a. cocci appearing in pairs _____   e. cocci in packets of eight _____

   b. four or more cocci in a chain _____   f. a comma-shaped rod _____

   c. cocci in grapelike clusters _____   g. a tightly coiled spiral _____

   d. cocci in packets of four _____

4. List the optical devices used for the calibration of a microscope.

a. _____ b. _____

5. Calculate the number of micrometers in 1 space of an ocular micrometer if 5 divisions of the stage micrometer line up with 20 divisions of the ocular micrometer. Show your equation and give your answer in both millimeters and micrometers.

_____

_____

_____

Key Terms

aseptic (ā-SEP-tik): being free of all life

aseptic technique: precautionary measures used in microbiological methods and in dental and medical practice to prevent contamination of cultures, media, and/or individuals by environmental sources of microorganisms

calibration (kal-i-BRĀ-shun): a procedure used to determine the accuracy of an instrument

dye: a chemical substance used to stain cells

involution (in-vō-LOO-shun): variations in cell shapes formed during unfavorable conditions

micrometer (mī-KROM-e-ter): one-millionth of a meter, or 10^{-6}m (abbreviated μm)

pleomorphism (plē-ō-MŌR-fizm): the ability to change shapes under favorable conditions

smear: a thin film of cells

Photograph Quiz 4

What morphological types of bacteria are apparent in the photograph? _____

_____

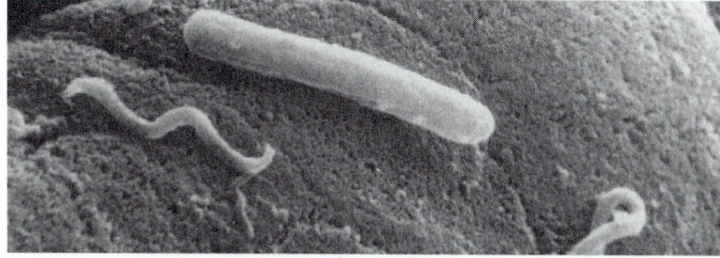

Figure 9–5
Bacterial morphology

EXERCISE 10   Algae and Protozoa: The Protists

After completing this exercise, you should be able to:

1. Recognize characteristic properties of algae and protozoa.
2. Distinguish selected types of movement among the protozoa.
3. Distinguish between prokaryotic and eukaryotic cellular organization in microorganisms.
4. Identify selected organelles of protists.
5. Observe food vacuole formation in certain protozoa.

The protist kingdom includes the heterotrophic protozoa, slime molds, water molds, and many of the autotrophic algae. The major feature of these organisms is their eukaryotic structure, which is shared with the numerous forms of life in three other kingdoms—*animals, plants,* and *fungi.* The eukaryotic cellular organization makes the separation between the protists and the members of the kingdom Prokaryotae (Exercise 9) quite distinct. Eukaryotic cells have true nuclei and other membrane-bounded organelles such as *mitochondria, Golgi bodies,* and *plastids.*

While protists are simple eukaryotic organisms their cellular organization is more complex than that of individual animal, plant, or fungal cells (Exercise 11). Animals, plants, and fungi have different cells, tissues, organs, and organ systems to perform the various functions of a living organism. Single-celled protists accomplish all of these functions within one cell. This exercise will survey the protists by presenting several representatives of the kingdom. In addition, the formation of food vacuoles using the protozoan *Paramecium* will be demonstrated (Figure 10–1).

Protozoa are unicellular microorganisms with nuclei and other specialized cytoplasmic parts or organelles. Many of these eukaryotic microbes exhibit a high degree of specialization with respect to their form and function. At least 45,000 species of protozoa have been described to date. Many of them are parasitic and are of major medical importance because they cause some extremely disabling diseases. This is especially true in persons with acquired immune deficiency syndrome (AIDS). Several of the exercises in Section XIII present representative protozoa known to cause human infections.

As a group, protozoa can readily adapt to and survive in a variety of environments. To a large extent, their success is the result of their remarkable development of organelles, which perform the same functions as organs in higher forms of life. Contractile vacuoles, which regulate water and salt content in many protozoa, are examples of such structures.

Several types of nutrition are found among the protozoa. These include *holozoic nutrition,* in which entire organisms or parts thereof are ingested (Figure 10–2), and *saprozoic nutrition,* in which nutrients are assimilated by diffusion through the cell membrane.

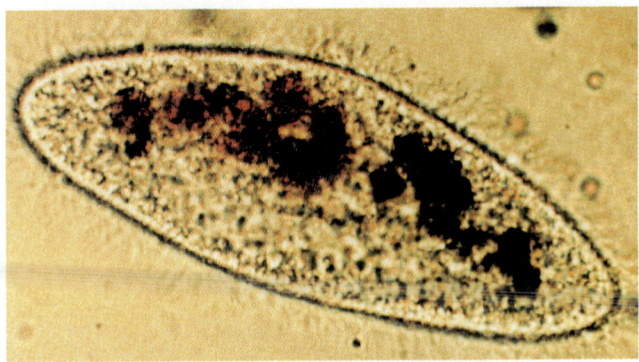

Figure 10–1

A micrograph showing a *Paramecium* that formed food vacuoles in the presence of India ink (black deposits). The India ink after ingestion accumulated in the posterior end of the cell.
[From Kaneshiro, E.S. and associates. *J. Protozool,* **39:**713–718, (1992).]

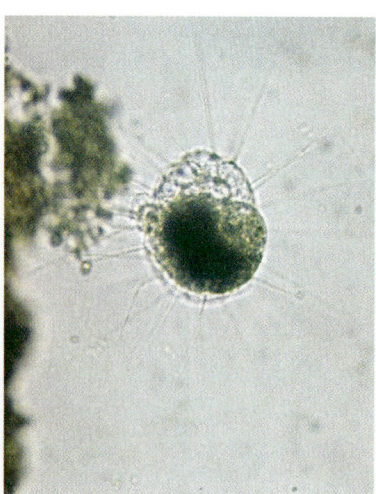

Figure 10–2

An actinopod with an ingested cell. These protozoans have long filamentous projections that protrude through pores in their outer covering or skeletons. These projections or **axopods** are used to trap other protozoa.

Movements by protozoa generally involve four basic types of organelles: flagella (long, whiplike structures), cilia (miniature flagella), pseudopodia (temporary extensions of the cell's cytoplasm), and undulating ridges (waving motions of the plasma membrane). Figure 10–3 shows several examples of protozoa using these organelles.

Reproduction in protozoa may be either asexual or sexual, although many species alternate between the two in their life cycles. In the usual method of asexual reproduction, protozoa divide into two, break into several parts, or give rise to

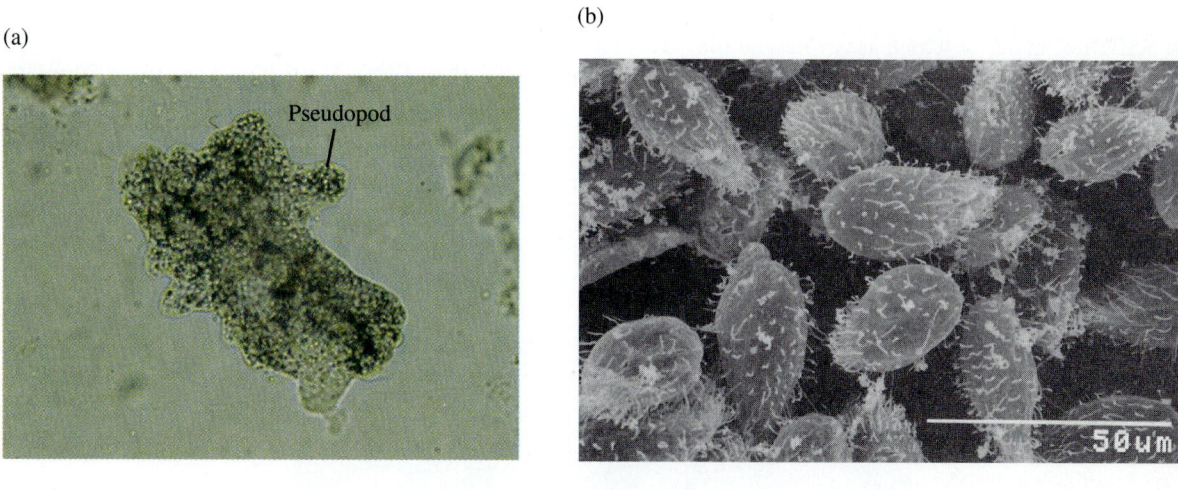

(a)

(b)

Pseudopod

50 um

(c)

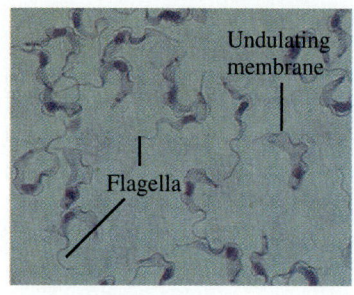

Undulating
membrane

Flagella

Figure 10–3

Examples of protozoa and their organelles of locomotion (a) The ameba showing several **pseudopodia.** (b) A scanning electron micrograph showing the protozoan *Tetrahymena* and its cilia.
(From Kiy, T., and A. Tiedtke, *FEMS Microbiology Letters* 106:117–122, [1993].)

(c) A smear showing *Trypanosoma gambiense,* the causative agent of African sleeping sickness. The undulating membrane and flagellum used by these organisms are shown.

buds that develop into another cell. Two basic forms of sexual reproduction have already been noted. One of these is *conjugation,* in which a temporary union of two protozoa occurs for the exchange of nuclear material. The second form is *sexual fusion,* or *syngamy,* in which a union of two sex cells takes place.

   Algae are of interest in a wide variety of fields, including aquatic biology, limnology, oceanography, and phycology. The study of these organisms, known as *phycology* or *algology,* involves the examination of their cellular arrangements, cell walls, and pigmentation, and the nature of their food reserves, distribution, and habitats (Figure 10–4).

Figure 10–4
Green and brown algae in a coastline environment.

   Algae are found in a variety of environments, including both fresh and salt water, soil, and moist surfaces. They are also found in association with certain species of fungi, animals, and protozoa. Algae are involved in a variety of basic biological activities, many of which are of human importance.

   Algae range from microscopic simple forms, such as the unicellular forms (Figure 10–5), to very large and complex ones, such as the well-known kelps, or brown algae, seen along coastlines. This exercise will deal only with the unicellular forms, including diatoms (Figure 10–6a), desmids (Figure 10–6b), and green algae. The cellular organization in algae does not differ fundamentally from that found in plants. Naturally, variations exist among the specific groups of these organisms. The conspicuous organelles of algae include the cell wall (Figure 10–5) and chloroplasts. All algae possess chlorophyll *a* and yellow and orange carotenoid pigments. However, a number of algae contain a variety of other pig-

Figure 10–5
A scanning micrograph showing the characteristics features of diatom cell walls. The large diatom is a member of the genus
Campylodiscus
(Courtesy of Dr. C.M. Pringle, U.C. Berkeley.)

(a)

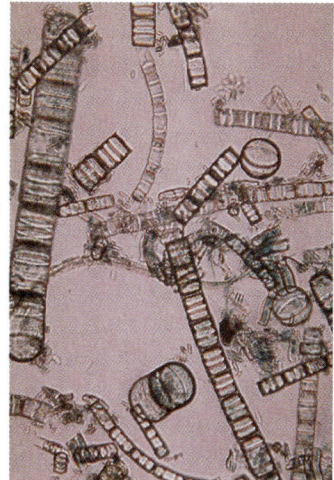

(b)

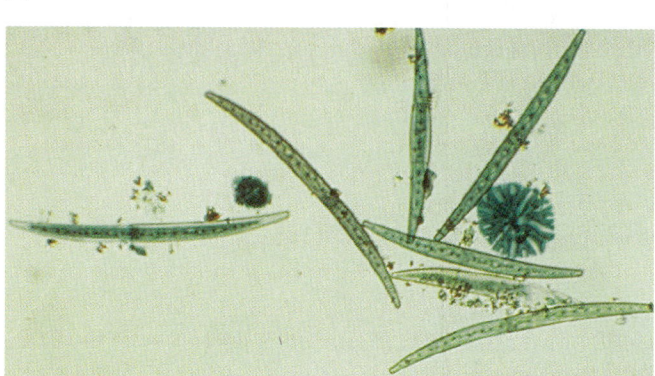

Figure 10–6

Photomicrographs of representative algae commonly found in natural waters. (a) Diatoms. (b) Desmids. Note the characteristic symmetry exhibited by these microorganisms.

ments. The nuclei of individual algal cells may be concealed by pigment-containing *plastids*. Classification schemes for algae make use of their pigment composition and the various products produced by their photosynthetic and related activities.

Reproduction in unicellular algae occurs both asexually and sexually. These organisms are known to produce several specialized reproductive and related types of cells and structures (Figure 10–7).

This exercise will be limited to the demonstration of representative morphological features of protozoa and selected algae. Consideration of other microbial forms, such as bacteria, fungi, viruses and rickettsiae, and of additional features of organisms covered in this exercise are presented in other portions of the manual. Selected species of cyanobacteria also are included in this exercise for purposes of comparison.

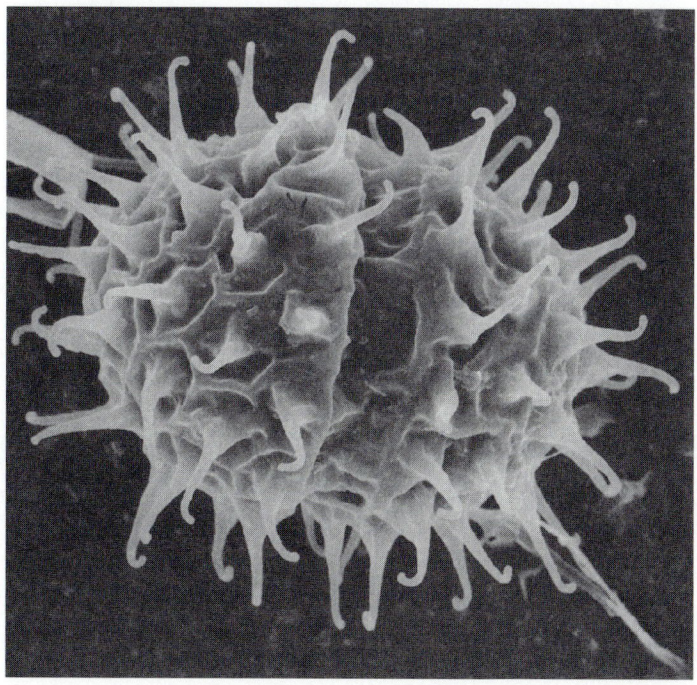

Figure 10–7

A scanning micrograph of a dinoflagellate cyst or **hypnozygote** from Antarctic sea ice. This thick-walled cyst represents a stage in the sexual part of dinoflagellate life cycles.
(Courtesy of Dr. Kurt Buck, Monterey Bay Aquarium Research Institute, Pacific Grove, CA [J. Phycol 28:15–18 1992].)

A. Protozoan Morphology

Materials

❏ 1. Live cultures of the following protozoa for class use:
  ❏ a. *Amoeba* (a-MĒ-ba) spp.
  ❏ b. *Paramecium* (par-a-MĒ-sē-um) spp.
  ❏ c. *Vorticella* (vor-TĒ-sel-a) spp.
  ❏ b. *Euplotes* (Ū-plōy-tēz) spp.
  ❏ b. *Phacus* (FAK̄-us) spp.
  ❏ b. *Stentor* (STEN-tōr)

❏ 2. Pond water samples

❏ 3. Stained prepared slides of the following:
  ❏ a. A blood smear with *Trypanosoma* variety *brucei gambiense* (one of the causative agents of African sleeping sickness)
  ❏ b. A blood smear with *Plasmodium vivax* (one of the causative agents of human malaria)
  ❏ c. Conjugating paramecia

❏ 4. Glass slides and cover slips

❏ 5. Pasteur pipettes or eye droppers (1 per culture)

Procedure I: Live Cultures

This procedure is to be performed by students individually.

❏ 1. Prepare temporary wet mounts of each protozoan culture and pond water sample provided. Refer to Procedure Diagram 1 in Exercise 1, if you are in doubt about the technique.

❏ 2. Examine each preparation under the low- and high-power objectives. With the aid of the figures in the Results and Observations section, Figures 10–1, 10–2, 10–3, and your textbook, attempt to identify the protozoa and their respective organelles.

❏ 3. Answer the questions in the Result and Observations section and the Laboratory Review.

Procedure 2: Stained Preparations

This procedure is to be performed by students individually.

❏ 1. Examine each stained slide under low- and high-power objectives and oil immersion. With the aid of Figures 10–3 and 10–8 and your textbook, identify the paramecia, their organelles, and *Plasmodium vivax* and *Trypanosoma gambiense*.

❏ 2. Answer the questions in the Results and Observations section and the Laboratory Review.

Results and Observations

Live Cultures

1. Which of the protozoa was the largest? _____

_____

2. a. Did the pond-water specimens contain protozoa that you recognized? If so, which ones? _____

_____

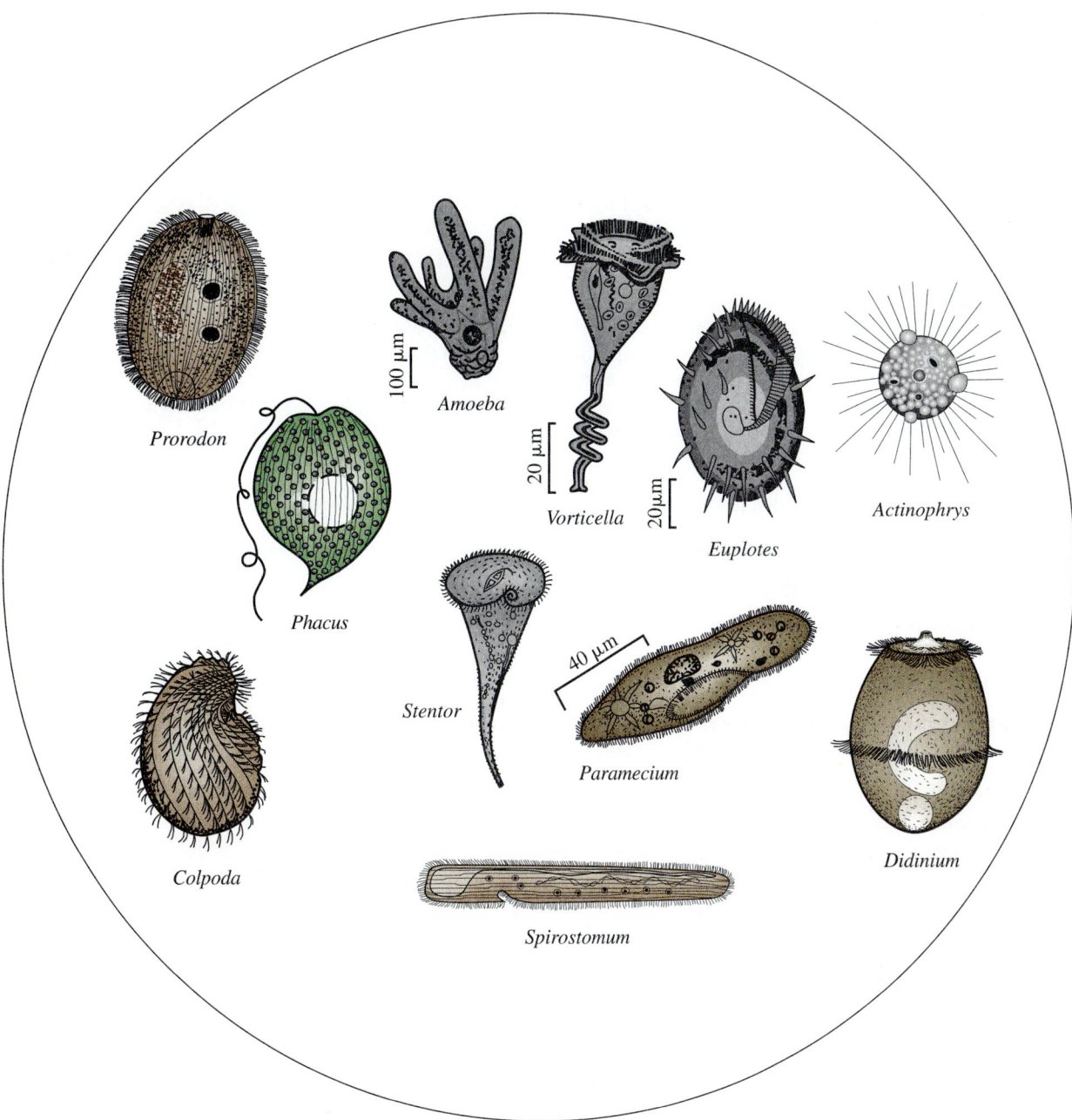

Figure 10–8
Common protozoa. Note the relative sizes.

      b. What other forms of life did you see in the pond-water specimens? _____

_____

3. Which protozoan organelles were apparent in the various preparations you examined? Do all protozoa move?

_____

_____

Stained Preparations

  1. Which protozoan organelles were apparent in the stained preparations? _____

_____

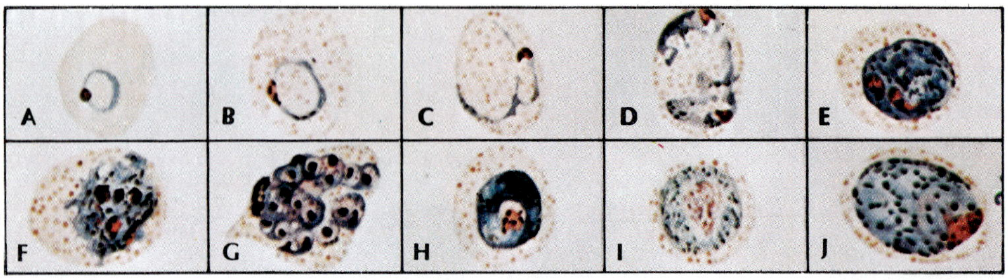

Figure 10–9

Plasmodium vivax. (a) A typical "signet ring" stage. (b) An enlarged erythrocyte with a ring form of trophozoite. The cell also contains Schüffner's stippling. It should be noted that such stippling may not always be present in infected red blood cells. (c) Another trophozoite form. (d) The presence of two amoeboid-shaped trophozoites. (e) and (f) Erythrocytes showing progressive stages in schizont division. (g) A mature schizont. Note the number of infective small units called **merozoites.** (h) A developing sex cell or gametocyte (i) and (j) Mature micro- and macrogametocytes.

2. In the blood smear preparations provided, which protozoan species generally appears outside of blood cells?

   Which one is found intracellularly? _____

   _____

3. What stages of Plasmodium were obvious in the blood smear? _____

   _____

B. Algal and Cyanobacterial Morphology

Materials

❑ 1. Live cultures of the following for class use:
- ❑ a. Mixed diatoms
- ❑ b. *Euglena* (ū-GLĒ-na)
- ❑ c. *Spirogyra* (spī-rō-JĪ-ra)
- ❑ d. *Chlorella* (klor-EL-la)
- ❑ e. *Anabaena* (an-na-BĒ-na), a cyanobacterium
- ❑ f. *Oscillatoria* (os-sil-a-TŌR-ē-a), a cyanobacterium

❑ 2. Pond-water samples

❑ 3. Pasteur pipettes or eye droppers (1 per culture)

❑ 4. Glass slides and cover slips

Procedure

This procedure is to be performed by students individually.

❑ 1. Prepare temporary wet mounts (TWMs) of each culture and pond-water sample provided. Refer to Procedure Diagram 1 in Exercise 1 if you are in doubt as to the technique.

❑ 2. Examine each preparation under the low- and high-power objectives. With the aid of the figures in the Results and Observations section and your textbook, identify the algae and their respective organelles. Note that the cyanobacteria are included here for purposes of a comparison between prokaryotes and eukaryotes.

❑ 3. Answer the questions in the Results and Observations section.

Results and Observations

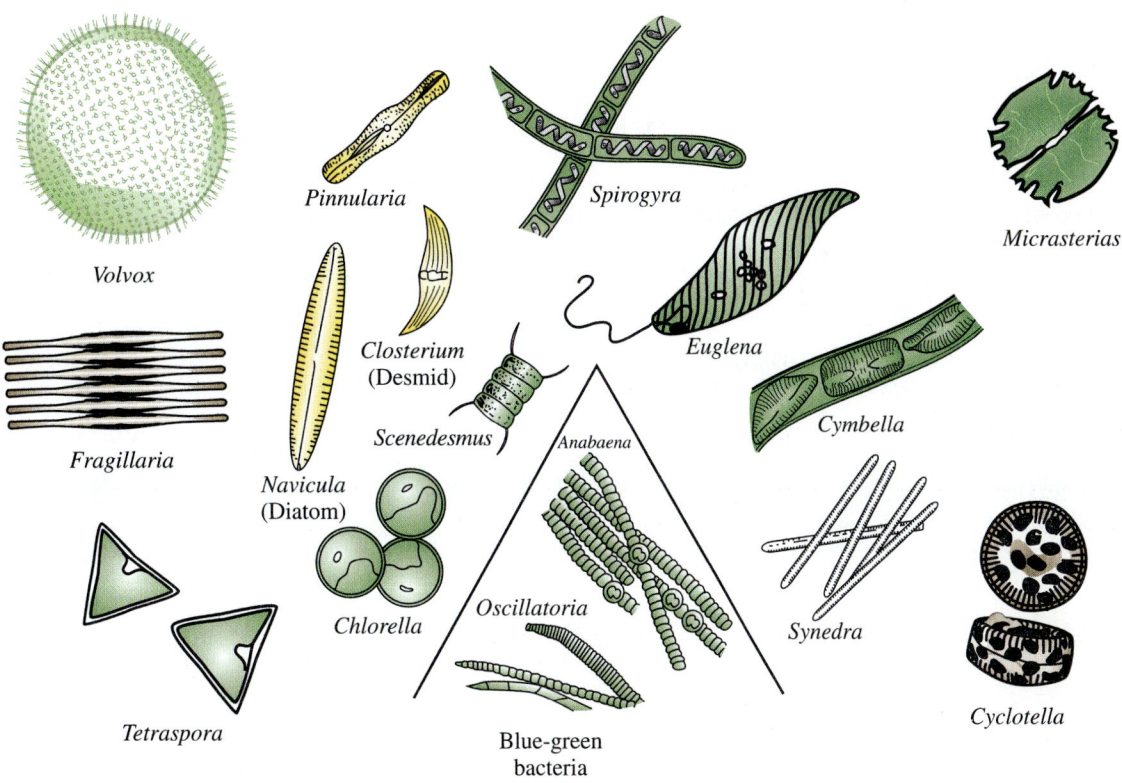

Figure 10–10
Common algae. Note the relative sizes and shapes.

1. What specific structures were evident among the algae studied? _____

   _____

2. Were the cyanobacteria larger or smaller than *Spirogyra?* _____

   _____

3. Which of the algae exhibited vital movement? _____

   _____

Laboratory Review 10   Algae and Protozoa: The Protists

1. What is the outermost cellular structure of the following organisms?

   a. bacteria _____   c. algae _____

   b. protozoa _____   d. fungi _____

2. List 4 characteristics used for the classification of algae.

    a. _____   c. _____

    b. _____   d. _____

3. Are cyanobacteria more closely related to bacteria or to algae? _____

    _____

4. List 4 properties of algae that distinguish them from protozoa.

    a. _____   c. _____

    b. _____   d. _____

5. List 4 properties of bacteria that distinguish them from algae.

    a. _____   c. _____

    b. _____   d. _____

6. Complete the following table by listing at least 2 beneficial and 2 harmful activities for each microbial type indicated.

Table 10–1

Beneficial and Harmful Activities

Microbial Type	Beneficial Activities	Harmful Activities
Algae	1. 2.	1. 2.
Protozoa	1. 2.	1. 2.

7. Give the function or functions of the following:

    a. contractile vacuole _____

    b. chloroplast _____

    c. cilium _____

    d. vacuole _____

Key Terms

cell wall: the relatively rigid structure found outside of the cell membranes of algae, fungi, and most prokaryotes

cyanobacteria (sī-an-ō-bak-TĒ-rē-a): photosynthesizing prokaryotes that contain chlorophyll and phycocyanin

morphology (mor-FOL-ō-jē): the study of structure and form

plastid (PLAS-tid): a cytoplasmic organelle found in plants and algae; usually contains pigments or metabolic products

Photograph Quiz 5

Identify the organism in Figure 10–11. Indicate if the organism is an algae, bacterium, or protozoan. _____

Figure 10–11
Microscopic view of a pond water specimen.

After completing this exercise, you should be able to:

1. Recognize the macroscopic (mycelial phase) and microscopic features of common molds and yeasts (i.e., fungi), with particular reference to asexual reproductive structures.

2. Prepare a slide fungus culture.

3. Recognize the major structures of fleshy mushrooms.

4. Prepare a spore print.

5. Identify fungal structures in foods and other materials.

The study of fungi (molds, yeasts, mushrooms, and related forms), referred to as *mycology,* involves both examination of specific morphological components and determination of their respective functions. In general, fungi are characterized by vegetative structures that elongate into branching filaments called *hyphae* (singular, *hypha*). These structures originate from specialized reproductive cells called spores (propagules) and may or may not be divided into sections by cross walls called *septa* (singular, *septum*). These structures are shown in Figure 11–1. Hyphae containing cross walls are called *septate,* while those lacking them are referred to as *nonseptate*. In the latter situation, the cellular material flows uninterrupted to a limited extent within the confines of the filamentous structure. The term *coenocytic* is used to describe this property. All fungi have cell walls and produce propagules of some type. These organisms are nonmotile throughout their life cycles.

A large group of general interest, and perhaps the most familiar, consists of the fleshy fungi, including the gilled mushrooms (Figure 11–2). We will consider representatives of the basidiomycontina (Basidiomycetes). The common mushroom found in markets, *Agaricus* sp., is a member of this group. A major characteristic of the basidiomycetes is the *basidium* (Figure 11–3a). This club-shaped structure produces basidiospores, which are formed externally at the tip of delicate projections known as *sterigmata* (singular, *sterigma*). Usually, four *basidiospores* are borne on each basidium.

The hyphae of basidiomycetes are septate and often have specialized structures known as *clamp connections* (Figure 11–3b). These connections are temporary bridges over a septum formed during cell division. They function as a bypass mechanism to ensure the distribution of two genetically different types of nuclei in hyphal cells.

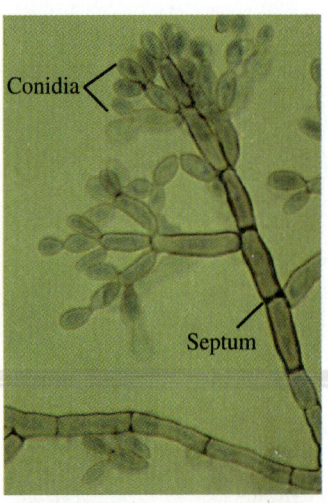

Figure 11–1

A microscopic view of a fungus showing septate hyphae and a type of spore called **conidia** (*singular,* **conidium**).
(Courtesy of Dr. Tadahiko Matsumoto.)

Figure 11–2

A number of the common mushroom *Mycena* growing on a lawn.

(a)   (b)

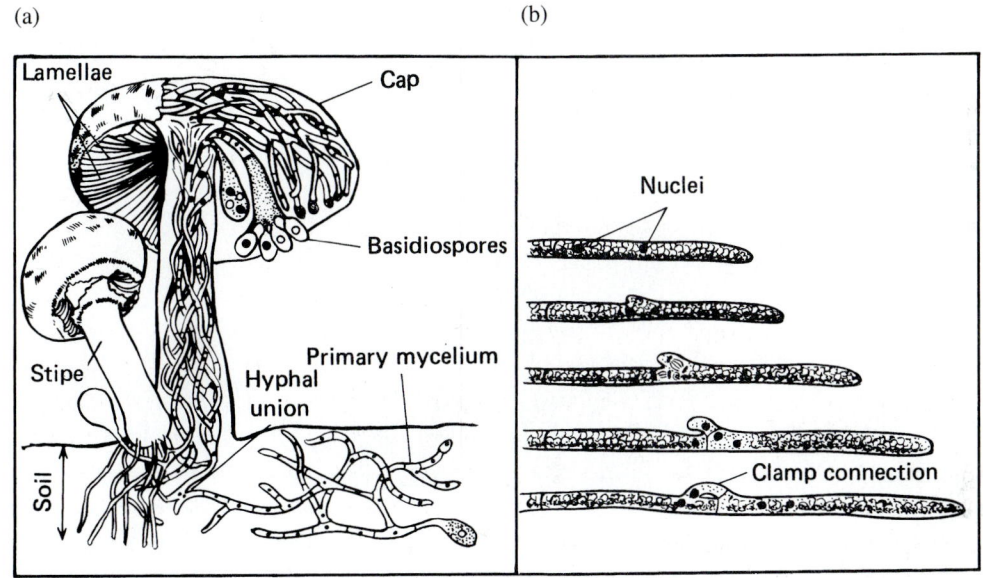

Figure 11–3

Properties of the Basidiomycotina. (a) A diagrammatic view of the internal and external properties of a fruiting body. (b) The formation of a clamp connection.

The *mushroom,* or *toadstool* (which is the term used to designate poisonous species), is a visible, plantlike structure representing the reproductive phase of the fungus that formed it. It is called a *fruiting body,* or *basidiocarp,* and develops from a mycelium that is hidden from view in soil or another substrate. A mushroom is a specialized body that ensures the maximum production of spores. There are many variations of the basic mushroom structure. However, in general, a mushroom consists of a *cap* or *pileus,* radiating layers of *lamellae* found on the underside of the cap and representing the spore-producing region, and a *stipe* or *stalk* on which the cap sits. Many mushrooms may have one or two layers of tissue, known as *veils,* on young fruiting bodies. Patches of veil material may be found on the cap and on portions of the stipe. A persistent ring of tissue on the stipe is called an *annulus.* Figure 11–4 shows the characteristic parts of a mushroom.

The identification of a particular mushroom involves noting the presence or absence of these parts and other properties including the size, shape, cap and stipe, color, surface appearance and texture, odor, color changes of structures following injury, habitat, and spore color. Determination of spore color is one of the most essential steps in mushroom identification. A spore print (Figure 11–5) usually is the technique used to determine spore characteristics, as well as certain features of gills. This procedure will be used in this exercise.

Figure 11–4

The structures of a mature mushroom. The specific parts shown include the *cap* (**pileus**) attached to the *stalk* (**stipe**), a *ring* (**annulus**), and the lower surface of a cap showing the radiating strips of tissue called *gills* (**lamellae**).

Figure 11–5

The results of the spore print technique. With the spore-bearing surface placed down on a piece of white paper for 1 or 2 hours, the gill pattern and spore color will become evident. Both of these properties are important in the identification of fleshy fungi.

In the laboratory, molds and yeasts are studied by cultivation methods similar to those used for bacteria. Because most fungi require oxygen for growth, either solid media or small amounts of broths are used for their cultivation. Once inoculated into or onto media, most fungi tend to grow more slowly than bacteria. Such slow growth can present problems, such as a competition with bacteria for food. To prevent such situations, selective media favoring fungi can be used. Sabouraud's dextrose agar is one example of this type of medium. It contains about a 4% concentration of sugar and its pH ranges from 5 to 6. This medium is widely used for the isolation and cultivation of molds and yeasts.

Another approach to the study of molds is the slide fungus culture. It is a valuable tool and provides a way to study a fungus as it grows, without disturbing any of its vegetative or reproductive structures. Permanent mounts for later use and study can be prepared from such cultures.

Selected properties of several representative fungi will be considered, including members of the Ascomycotina *(Aspergillus flavus, A. niger, Penicillium notatum, Rhodotorula sp.,* and *Saccharomyces cerevisiae),* Basidiomycontina *(Agaricus* sp. or other genera), Deuteromycontina *(Alternaria* spp., *Fusarium* sp., and *Scopulariopsis* sp.), and Zygomycotina *(Rhizopus nigrans).* (See Figures 11–6 through 11–12.)

The continual branching and intertwining of fungal filaments result in the formation of a visible structure called a *mycelium* (plural, *mycelia*) on the surfaces of substances (substrates) used as food by the fungus. The portion of the mycelium that develops on surfaces is called the *reproductive* or *aerial mycelium,* while the portion that penetrates the surface is called the *vegetative mycelium.*

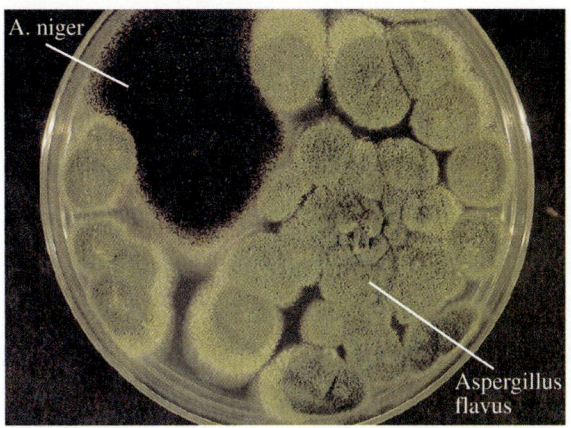

Figure 11–6

Two species of the genus *Aspergillus, A. flavus* (olive colored mycelia) and *A. niger* (black mycelium). These molds are growing on Sabouraud's dextrose agar.

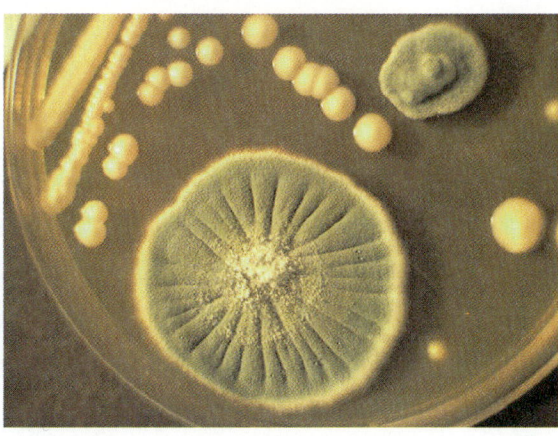

Figure 11–7

The appearance of the mold *Penicillium notatum* (green mycelium), and the cream colored yeast *Saccharomyces cerevisiae*. Both fungi are growing on Sabouraud's dextrose agar.

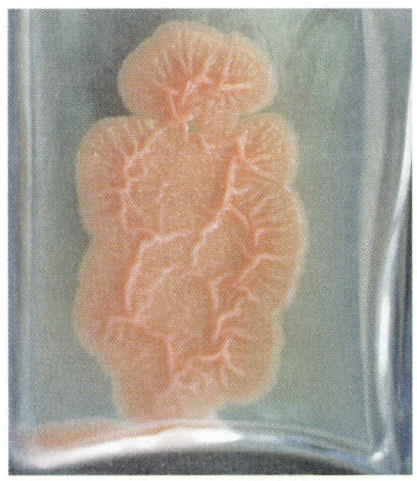

Figure 11–8

Rhodoturula sp. This genus includes nonfermenting yeasts usually forming mucoid, yellow to red colonies.

Figure 11–9

The mycelium of the fungus *Alternaria* sp. Growth is on Sabouraud's dextrose agar.

Figure 11–10

Fusarium sp. This mold forms a fluffy, cottony mycelium, which may be accompanied by a diffusible, purple or yellow pigment.

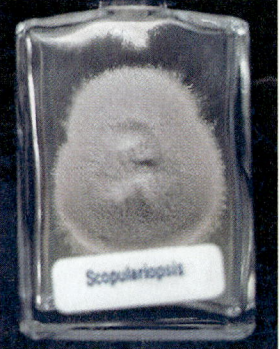

Figure 11–11

Scopulariopsis sp. This mold forms white to light beige or dull grayish mycelia.

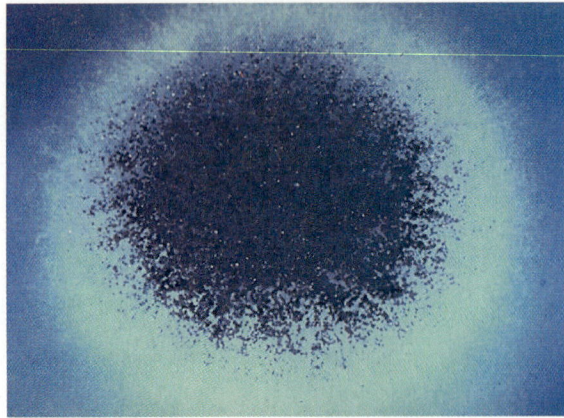

Figure 11–12
The bread mold, *Rhizopus nigricans,* also grown on Sabouraud's dextrose agar. Note the different pattern of growth. The black dotlike structures are sporangia.

A fungus cannot manufacture essential organic nutrients from the inorganic substances in its environment. All fungi are heterotrophic. Certain fungal species form specialized structures to obtain food. *Rhizoids* (Figure 11–13a), for example, which are associated with organisms that live on dead organic matter (saprobes), anchor fungi to a substrate and provide the means to absorb nourishment. Parasitic plant fungi sometimes have *haustoria,* which penetrate and absorb nutrients directly from living host cells.

Many fungi have both sexual and asexual (vegetative) reproductive phases in their life cycles. In the sexual mode of reproduction, two different hyphal structures called *gametangia* fuse or intertwine and result in the formation of a variety of sexual spores. This process is not very common and usually involves special requirements. The usual method of asexual reproduction for fungi is by the formation of reproductive units called **spores** or **conidia.** Certain of these reproductive units are not produced from specialized spore-bearing hyphae. Spores may result from the breaking up of a hypha into separate cells, known as *arthroconidia,* or form thick-walled cells known as *chlamydoconidia,* or develop as buds known as *blastoconidia.* In other situations, spores are produced or formed on specialized asexual propagules. Spore development and the morphological features of these structures are important criteria used for classifying fungi. Spores come in many forms and develop by genetically controlled pathways. In this exercise, we will be concerned with the asexual or vegetative form of multiplication.

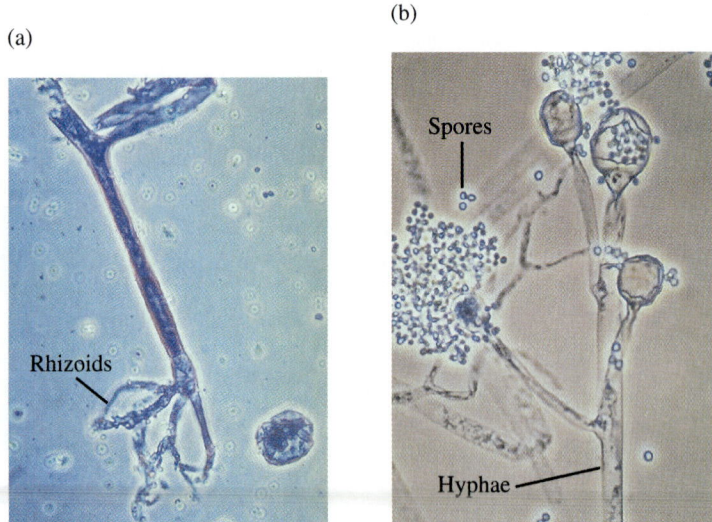

(a)

(b)

Figure 11–13
Selected microscopic parts of a mold (a) The rootlike **rhizoids.** (b) The sac-like **sporangia** and **sporangiospores.** Nonseptate hyphae also can be seen in these micrographs.
(From St.-Germain G., A. Robert, M. Ishak, C. Tremblay and S. Claveau. *Clin. Inf. Dis.* 16:640–5, [1993].)

Some spores are produced in swollen sac-like structures called *sporangia* (singular, *sporangium*) and are consequently called *sporangiospores* (Figure 11–13*b*). Others arise on club-shaped structures known as *basidia* (singular, *basidium*) and are called *basidiospores*. Still others grow out from another type of structure, the *conidiophore,* and are named *conidia* (singular, *conidium*). Specialized arrangements of hypha and/or conidia cells exist among the fungi to package and/or spread conidia more effectively. An example of this arrangement is the *pycnidium,* a flask-shaped reproductive structure containing large numbers of *pycnidiospores.* This exercise presents a number of examples of asexual propagules.

Yeasts, another form of fungi, are single-celled, lack chlorophyll, and exhibit characteristic shapes including spherical and ellipsoidal forms (Figure 11–14*a*). They have a typical eukaryotic organization and are larger than bacteria. Yeasts reproduce asexually by a fission (splitting) process or by budding. In budding, daughter cells, referred to as *buds* (Figure 11–14*b*), pinch off from a parent cell. At times, these daughter cells may themselves begin to bud during their formation, and thus form chains of cells that do not separate. Such chains may be quite long and are called *pseudomycelia.* Sexual reproduction can also occur with yeasts.

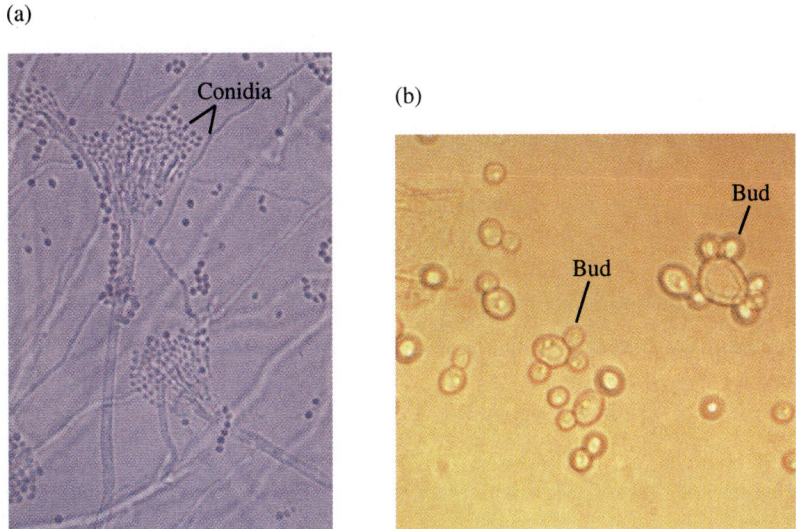

Figure 11–14
Two microscopic views of fungi, showing certain differences between molds and yeast. (a) The mold *Penicillium* with its **conidiophores** and **conidia.** (b) The yeast *Saccharomyces cerevisiae* showing parent cells and buds.

A. Molds and Yeasts

Materials

The following materials should be provided per 4 students:

❏ 1. One-week-old cultures of the following fungi grown on Sabouraud dextrose agar plates. (One culture should be sealed and used only for the study of mycelial morphology.)

  ❏ a. *Alternaria spp. (awl-ter-NĀ-rē-a)*

  ❏ b. *Aspergillus* flavus (as-per-JIL-us, FLĀ-vus)

  ❏ c. *A. niger (A. NI-jer)*

  ❏ d. *Fusarium sp. (fū-ZĀ-rē-um)*

  ❏ e. *Penicillium notatum* (pen-ī-ŠIL-ē-um, nō-TĀY-tum)

  ❏ f. *Rhizopus nigricans* (rī-ZŌ-pus, NĪ-grē-kans)

  ❏ g. *Rhodotorula* (RŌ-dō-tōr-ū-la)

  ❏ h. *Saccharomyces cerevisiae* (sak-a-rō-MĪ-sēz, sār-ā-VIS-ē-ī)

  ❏ i. *Scopulariopsis* (SKŌP-ū-lar-i-op-sis)

❏ 2. One ripe, sweet apple such as the Delicious or Jonathan variety

❏ 3. One tube of sterile distilled water (5 mL)

❏ 4. One hand lens

❏ 5. Four dissecting needles

❏ 6. One sterile scalpel or single-edge razor blade

❏ 7. One 500-mL beaker

The following items should be provided for general class use:

❏ 1. Lactophenol cotton blue mounting solution

❏ 2. Aluminum wrap or similar covering material

❏ 3. Glass slides and cover slips

❏ 4. Microscopes

Procedure 1: Macroscopic and Microscopic Examination

This procedure is to be performed by students individually.

❏ 1. Examine the plate cultures of the fungi provided. Observe both the top and back surfaces of each species. Refer to Figures 11–6 through 11–12.

❏ 2. Remove a portion of the fungal culture using the dissecting needles and prepare a wet mount using lactophenol cotton blue solution as the suspending material. Avoid removing large amounts of agar. Before applying the cover slip to the specimen, tease the fungal growth apart with the aid of dissecting needles.

❏ 3. Diagrams of the respective fungi can be found in the Results and Observations section. Use them to identify the various fungal components—type of hyphae, spore characteristics, and so on. (Refer to your text and Figures 11–15 through 11–23.)

Results and Observations

Macroscopic and Microscopic Examination

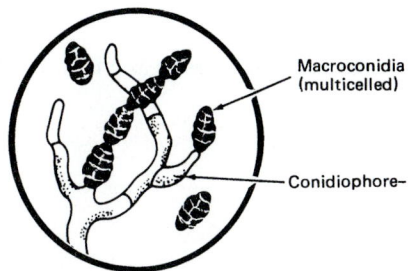

Figure 11–15
Alternaria spp.

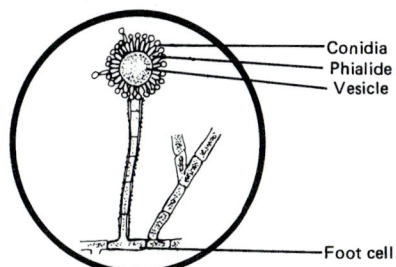

Figure 11–16
Aspergillus flavus.

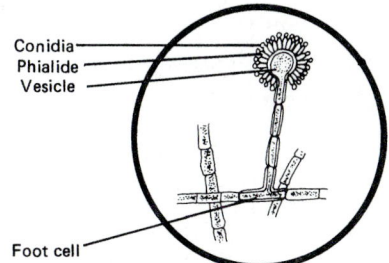

Figure 11–17
Aspergillus niger.

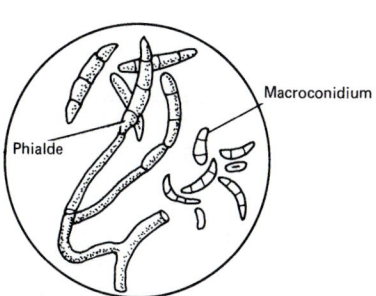

Figure 11–18
Fusarium sp.

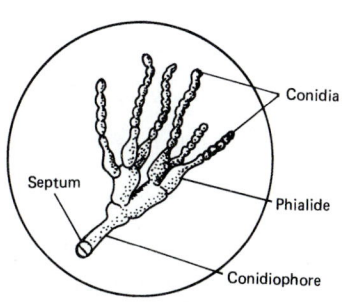

Figure 11–19
Penicillium notatum

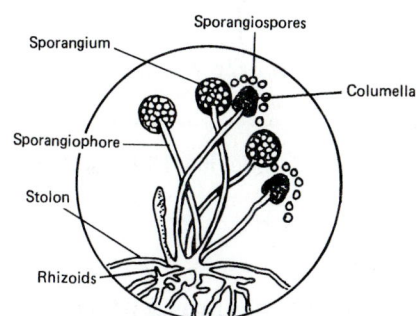

Figure 11–20
Rhizopus nigricans.

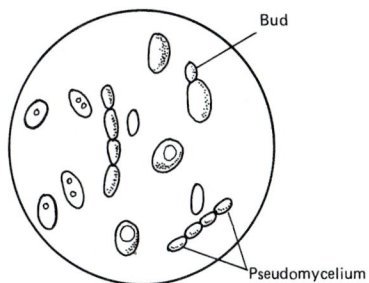

Figure 11–21
Rhodoturula sp.

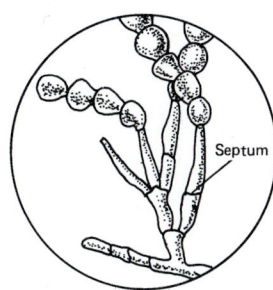

Figure 11–22
Scopulariopsis sp.

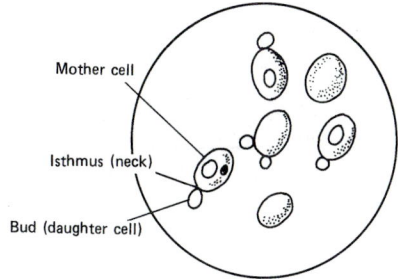

Figure 11–23
Saccharomyces cerevisiae.

Procedure 2: Demonstration of Spoilage

This procedure is to be performed by students in groups of four.

❏ 1. Make several 1.5-cm-deep punctures with a dissecting needle in one of the ripe apples provided.

❏ 2. Flame a dissecting needle, allow it to cool, and then use it to inoculate the punctured areas with spores of mycelia from an available *Penicillium* culture.

❏ 3. Pour 5 mL of distilled water into the sterile 500-mL beaker provided.

❏ 4. Place the inoculated apple into the beaker and cover the container with aluminum wrap.

❏ 5. Incubate the apple in an area specified by your instructor for 10 days or until definite signs of rot appear in and around the inoculated sites.

❏ 6. Remove a small piece of rotted apple, cut the apple into thin slices, and prepare a temporary wet mount. Use lactophenol cotton blue as the mounting solution, and tease the specimen apart with the dissecting needle before applying the cover slip.

❏ 7. Examine the preparation under both low and high power. Look for the presence of hyphae, conidiophores, and spores among the apple cells. (Refer to Figure 11–19.)

❏ 8. Sketch representative fields of your specimen, and answer the questions in the Results and Observations section.

Results and Observations

Demonstration of Spoilage

1. When did the first signs of fungal growth and/or spoilage appear? _____

_____

2. Did all inoculated sites show evidence of spoilage? If not, how would you explain this finding? _____

_____

_____

3. a. Were the hyphae in the specimens examined growing between cells, or were they penetrating apple cells? _____

_____

   b. Were the hyphae septate or coenocytic? _____

_____

4. Sketch and label representative views in the areas provided.

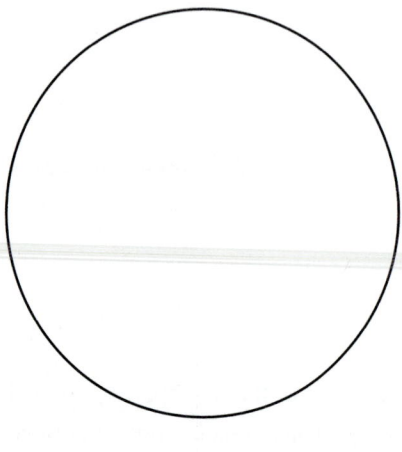

Low power

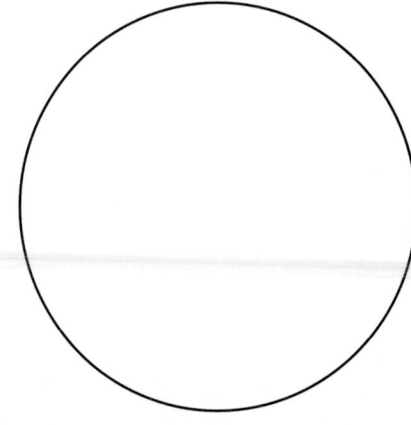

High power

B. Slide Fungus Culture

Materials

The following items should be provided per student:

❑ 1. Two standard sterile plastic Petri dishes (15 × 100 mm)

❑ 2. Two sterile filter papers, cut if necessary, to fit the Petri dish bottom

❑ 3. Two sterile depression slides with 2 sterile cover slips (22 × 40 mm)

❑ 4. Four sterile wooden applicator sticks about 75 mm in length

❑ 5. Two dissecting (teasing) needles

❑ 6. Sterile forceps

❑ 7. One-week-old cultures provided for Section A

The following items should be provided for general class use:

❑ 1. Sterile distilled water in tubes (1.0 mL/per tube)

❑ 2. Sterile Sabouraud's dextrose agar in tubes (1.0 mL/per tube)

❑ 3. Sterile, cotton-plugged Pasteur pipettes with rubber bulbs

❑ 4. Vaspar (1 volume each of vaseline and paraffin)

❑ 5. Boiling water bath

❑ 6. Container for glass disposal

Procedure 1: Preparation of Slide Culture Chamber

This procedure is to be performed by students individually.

❑ 1. Obtain a sterile Petri dish with a sterile filter paper in the bottom portion.

❑ 2. Lift the Petri dish top enough so that with a pair of sterile forceps, 2 sterile wooden applicator sticks can be placed in the dish bottom, as shown in Figure 11–24.

❑ 3. Next, flame the forceps and after allowing them to cool, place a sterile depression slide onto the applicator sticks, as shown in Figure 11–24.

❑ 4. Repeat steps 1 through 3.

❑ 5. Continue on to Procedure 2.

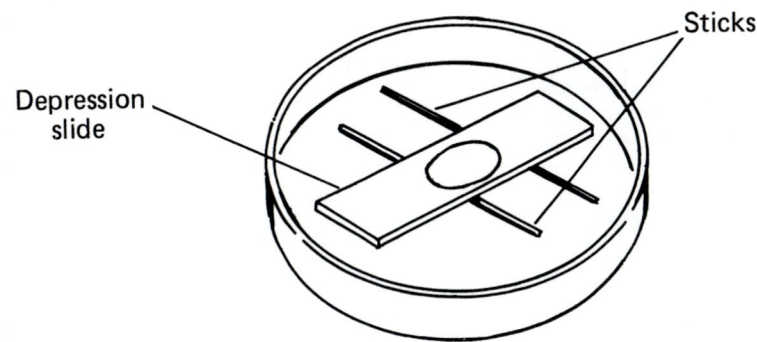

Figure 11–24
Slide culture chamber preparation.

Procedure 2: Inoculation

This procedure is to be performed by students individually.

❑ 1. Obtain 1 tube of melted Sabouraud's dextrose (SD) agar and select 2 mold cultures from those provided earlier in this exercise.

❑ 2. With a Pasteur pipette, place 2 drops of the SD agar into the depression of the slide and allow it to harden.

❑ 3. Perform the 9 steps of Procedure Diagram 16 with each of the cultures.

❑ 4. Incubate the slide cultures at room temperature and examine them under low power daily after the first day of incubation. Compare the microscopic appearance of your culture to the appropriate figures provided earlier. Record your findings and answer the questions in the Results and Observations section.

❑ 5. To examine a slide culture, remove it from the Petri dish, wipe the bottom of the depression slide, and place it on the microscope stage.

❑ 6. Dispose of all cultures after 2 weeks, as directed.

**Procedure Diagram 16
Slide Fungus Culture Technique**

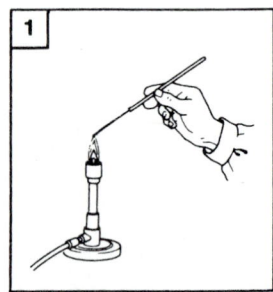

1. Flame the tip of a teasing needle and allow it to cool.

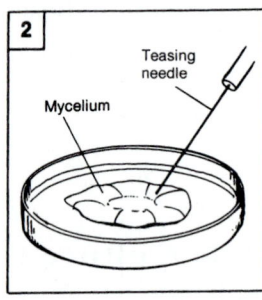

2. Remove a small amount of mycelium with the cooled needle.

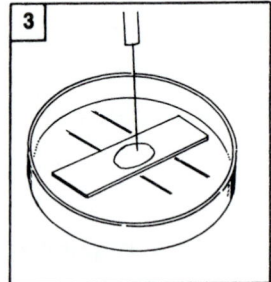

3. Inoculate the slide culture medium. Flame and cool the teasing needle.

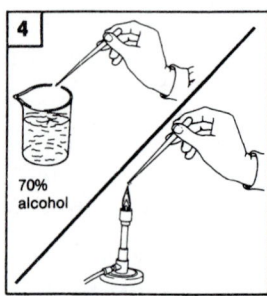

4. Heat-sterilize the forceps by dipping the tips into the beaker containing 70% alcohol and flaming. *Hold the forceps with the tips in a downward position and allow them to cool.*

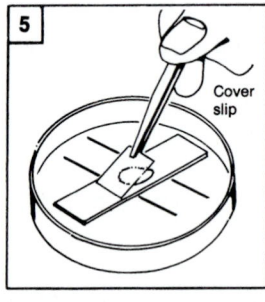

5. Place a coverslip over the agar with the cooled forceps.

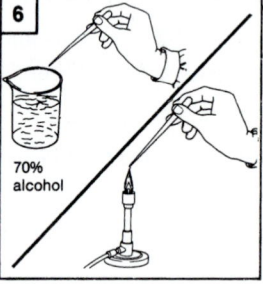

6. Flame the forceps as in step #4 and place them in a suitable place.

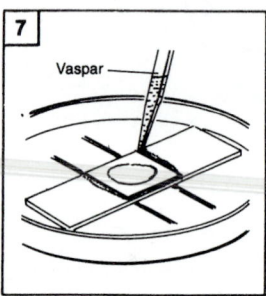

7. Seal 3 sides of the coverslip with melted vaspar.

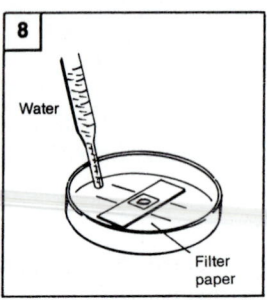

8. Pour 1.0 ml of distilled sterile water onto the filter paper.

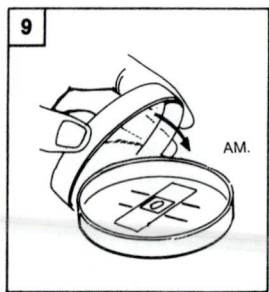

9. Close the Petri dish and incubate as directed.

Results and Observations

Slide Fungus Culture

1. On which day of incubation did recognizable growth occur? _____

2. When did the culture reach the stage of development so that it could be identified according to the figures in this exercise? _____

3. Were all structures of the fungus clearly visible after 1 week? _____ After 2 weeks? _____

C. The Mushrooms (Gill Fungi)

Materials

The following items should be provided for general class use:

❏ 1. Fresh, edible mushrooms (*Agaricus* spp.)
❏ 2. Fresh or preserved poisonous mushrooms (*Amanita* spp. or other specimens), if available
❏ 3. Fresh, thin slices of the cap (pileus) and stipe of edible and other mushrooms to show clamp connections
❏ 4. Lactophenol cotton blue mounting solution
❏ 5. Scalpels or single-edge razor blades
❏ 6. Dissecting needles
❏ 7. White paper board or paper squares cut to 12.6 × 12.6 cm (5 × 5 inches)
❏ 8   Glass slides and cover slips
❏ 9. Microscopes

Procedure 1: Mushroom Macroscopic and Microscopic Features

This procedure is to be performed by students individually.

❏ 1. Obtain and examine the fruiting bodies of the mushroom species provided. Locate the basic structures—cap, gills, stipe, veil, and annulus. (Refer to Figure 11–4.)
❏ 2. Look for the following features of the fruiting bodies and enter your findings in Table 11–1 of the Result and Observations section:
  ❏ a. Cap color and color distribution: even, uneven, etc.
  ❏ b. Cap texture: velvety, dry, sticky, moist, slimy, etc.
  ❏ c. Gills: color
  ❏ d. Gill attachment: connected to stalk
  ❏ e. Stipe color
  ❏ f. Stipe location: center, off-center, other arrangements
  ❏ g. Veil: present or absent
  ❏ h. Annulus: present or absent
❏ 3. With the aid of a scalpel or single-edge razor blade, cut very thin slices of the mushroom cap and stipe specimens provided and prepare temporary wet mounts. Use lactophenol cotton blue as the mounting fluid.
❏ 4. Examine each temporary wet mount under low and high power. Look for the presence of clamp connections on the surface layers of the cap and stipe cortex. (Refer to Figure 11–3*b*.)

❏ 5. In addition, examine the gill section of the cap for the presence of spores. Sketch a representative view in Table 11–1 of the Results and Observations section.

❏ 6. Enter your findings in Table 11–1 and answer the questions in the Results and Observations section.

❏ 7. Dispose of all used materials as indicated by your instructor.

Procedure 2: Spore Print

The following procedure is to be performed by students in pairs.

❏ 1. Obtain a mature mushroom (*Agaricus* sp.) and with the aid of a scalpel or single-edge razor blade, *carefully* remove the stem.

❏ 2. Place the pileus with its gills (spore-bearing surface) down on a white poster board or paper square.

❏ 3. Put the spore print square in an area where it will not be disturbed for 2 hours. A clear spore print usually takes 1 to 2 hours to appear. (Refer to Figure 11–5.)

❏ 4. After 2 hours, remove the pileus and examine the spore prints, noting the following:

   ❏ a. Color of different regions of the spore print (e.g., the interior, margin)
   ❏ b. Spacing between spore lines (e.g., close, distinct, absent)
   ❏ c. Margin properties (e.g., smooth, ragged)

❏ 5. Next, with a clean scalpel or single-edge razor blade, remove a small amount of the deposited spores and make a temporary wet mount. Use lactophenol blue cotton as the mounting fluid.

❏ 6. Examine your preparation under low and high power. Pay particular attention to spore color, shape, and other distinctive features. Sketch a representative field in the space provided in Table 11–2.

❏ 7. Enter all of your findings and answer the questions in the Results and Observations section.

❏ 8. Repeat steps 1 through 7 with other mushroom species that are available.

Results and Observations

Mushroom Macroscopic and Microscopic Features

1. Enter your findings in Table 11–1.

2. Do the spores found with the basidiomycetes differ from those observed with the molds? If yes, list at least 2 distinguishing properties. _____

_____

_____

Table 11–1

Mushroom Properties

Features	Mushroom Species		
	Agaricus Species	a	a
Cap: a. Color			
b. Color distribution			
c. Texture			
Gills: a. Color			
b. Type of attachment			
Stipe: a. Color			
b. Location (attachment)			
Veil: (present/absent)			
Annulus: (present/absent)			
Clamp connection (present/absent): a. Cap			
b. Stipe cortex			
Spore color			
Spore arrangement (sketch)	Low High	Low High	Low High

aFor other mushroom species.

3. Why are clamp connections generally located in the cap? _____

_____

_____

Spore Print

1. Enter your findings in Table 11–2.

Table 11–2

Spore Printing Observations

Features of Spore Printing and Spores	Mushroom Species					
	Agaricus Species		a		a	
Color of print						
Spacing between spore lines						
Margin properties						
Spore color						
Spore shape (sketch)	Low	High	Low	High	Low	High

aFor other mushroom species.

2. Of what value is a spore print to the identification of mushroom species? _____
_____
_____

Laboratory Review 11   The Fungi: Molds, Yeasts, and Mushrooms

1. Distinguish between a hypha and a mycelium. _____
_____
_____

2. What type of cell organization do fungi exhibit? _____
_____

3. List 4 types of asexual spores.

   a. _____   c. _____

   b. _____   d. _____

4. List 2 forms of asexual reproduction associated with yeasts.

   a. _____   b. _____

5. What is a pseudomycelium? _____
_____

6. List 2 microscopic structures generally found with the basidiomycetes.

   a. _____   b. _____

7. Distinguish between a mushroom and a toadstool. _____

_____

_____

8. List 4 properties of mushrooms used in their identification.

   a. _____   c. _____

   b. _____   d. _____

9. What is a spore print? _____

_____

_____

10. a. What is an aflatoxin? _____

_____

_____

   b. What species of fungi produces aflatoxins? _____

_____

11. a. Which gill fungus genis is known for its extremely poisonous nature? _____

_____

   b. To what genus does the commercially cultivated mushroom belong? _____

_____

12. Define or explain:

   a. septate hyphae _____

   b. rhizoid _____

   c. haustoria _____

   d. propagule _____

Key Terms

aerial mycelium (ĀR-i-al, mī-SĒ-li-um): the portion of a mycelium that develops on surfaces

annulus (AN-ū-lus): a ring-shaped or collar structure found on the stalk of some mushrooms

basidiospore (ba-SID-ē-ō-spōr): a specialized asexual reproductive unit found with members of the basidiomyastina

coenocytic (SĒ-nō-sit-ik): a condition in which cross walls are absent, thus allowing cellular material to flow uninterrupted in certain hyphae

conidium (kō-NID-i-um): an asexual fungal spore that may be one- or multi-celled and of many sizes and shapes

haustoria (HAWS-tōr-ē-a): a specialized type of hypha that is used by some parasitic plant fungi to penetrate and absorb nutrients from living host cells

hypha (HĪ-fa): the structural unit of a fungal mycelium; one filament or thread

lamella (la-MEL-a): one of several radiating membranes or layers found on the underside of a mushroom cap's spore-producing region

mycelium (mī-SĒ-lē-um): a mass of threadlike filaments (hyphae) branching and intertwining to form a network of the visible structure of a fungus

nonseptate (non-SEP-tāt): having no dividing walls (septa) in hyphae (filaments)

pileus (PĪ-lē-us): the cap of a mushroom

Sabouraud's agar: a selective medium used for the cultivation of molds and yeasts; the medium contains a higher than usual concentration of carbohydrate, and has a pH ranging between 5 and 6

septate (SEP-tāt): having dividing walls (septa) in hyphae

spore (spōr): a reproductive unit (propagule) that may be formed asexually and/or sexually

spore (spōr) print: a technique used to determine the properties of spores and lamellae

stipe (stȳp): the stalk or structure that supports the cap of a mushroom

vegetative mycelium (VEJ-e-tā-tiv, mī-SĒ-li-um): the portion of a mycelium that obtains nutrients and may penetrate the surface on which a mold is growing

Photograph Quiz 6

    a. Give the genus of the microorganisms shown. _____

_____

    b. Label the parts of the organism indicated in the photograph (Figure 11–25).

1. _____ 2. _____ 3. _____

Photograph Quiz 7

    a. Identify the labeled parts of the fruiting body shown (Figure 11–26).

      1. _____ 3. _____

      2. _____ 4. _____

    b. What specific structures are associated with part 2? _____

    c. Where is the mycelium in this illustration? _____

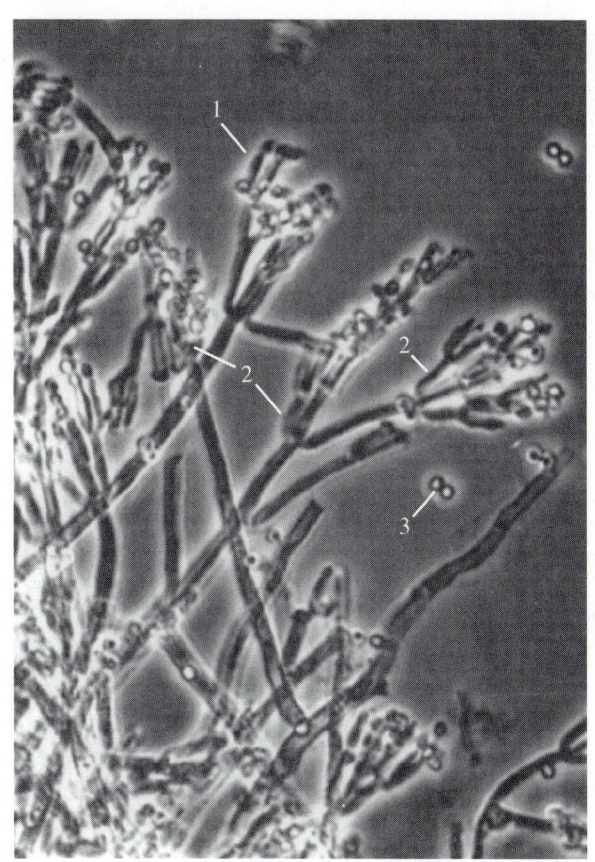

Figure 11–25
Microscopic view of a mycelium
(From Ciegler, A., D.I. Fennel, G.A. Sansing, R.W. Detroy, and G.A. Bennett. *Appl. Microbiol.* 26:217–278, [1973].)

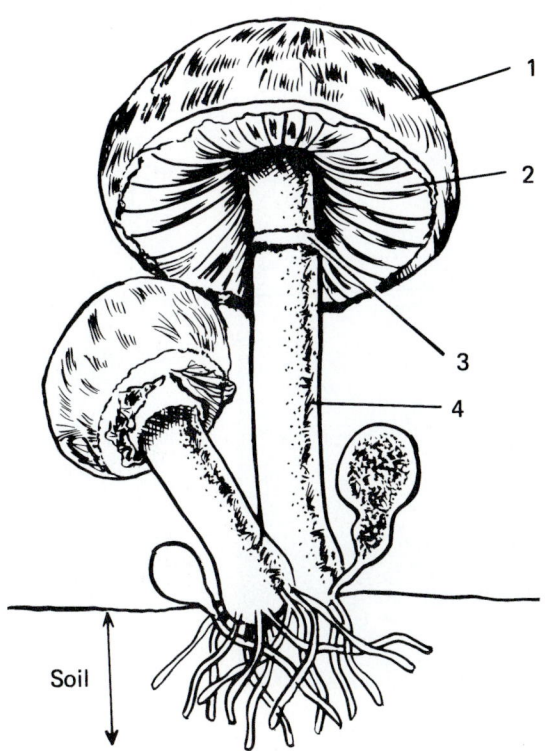

Soil

Figure 11–26
Mushroom parts.

The Distribution of Microorganisms in the Environment

After completing this exercise, you should be able to:

1. Collect environmental samples.
2. Realize the wide distribution of microorganisms in the environment.
3. Recognize common microorganisms isolated from or present in environmental samples.
4. Understand the experimental basis of the historical arguments against the doctrine of the spontaneous generation of life.

Microorganisms thrive in a wide variety of natural environments. They are ubiquitous on land and water, having even been found at altitudes above 30,000 m. In no environment where higher forms of life are present have microbes been shown to be absent. Moreover, in several environments devoid of higher organisms such as animals and plants, microorganisms exist and even flourish. Because individual microorganisms are usually microscopic, the existence of these forms of life in a given area may be unsuspected. Yet, strangely enough, without microorganisms and their associated activities, higher organisms would experience great difficulties. The environments in soil and water that provide the life support conditions for microbes vary, and the types of organisms making up microbial populations vary as well. Such populations can include algae (Figure 12–1), bacteria, fungi, and protozoa.

Many factors play a decisive role in establishing the population of a particular region. For example, the numbers and kinds of microorganisms present in soil are determined by the chemical and physical properties of the soil, the presence of plants, and the degree of contamination by the remains of animals and plants. The conditions of a soil environment control fluctuations in both numbers and kinds of microorganisms. The interactions of all living organisms in a given area (the community) with the nonliving environment of that area form an ecological system, or ecosystem. Many different

(a)   (b)

 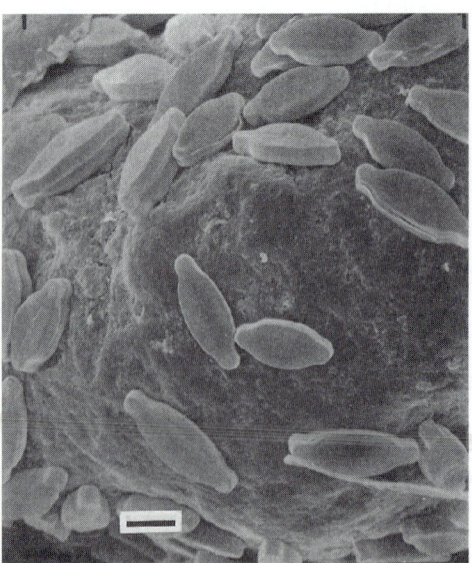

Figure 12–1

The distribution of diatoms within aquatic environments. (a) A wide assortment of diatoms on a grain of sand. The scale marker = 25 μm. (b) Microcolonies of the diatom *Achnanthes exigua* on the smooth portion of a sand grain. Scale bar = 5 μm. (From Pringle, C.M., *J. Phycol.* 21:185–194 [1985].)

ecosystems exist, each with its particular physical, or nonliving, environment and its range of living organisms. *Ecology* is the specialty of the biological sciences that deals with the relationships between organisms and their respective environments.

This exercise consists of several parts: (1) the microscopic examination of specimens from different environments: (2) the isolation of microorganisms from an environment chosen by the student; and (3) consideration of the question, "Where do microorganisms come from?" That life appeared spontaneously from inanimate or decomposing organic matter was a commonly held belief since 346 B.C. Despite various convincing experiences through the centuries, the issue of spontaneous generation was not settled until the work of Louis Pasteur and others appeared in the mid-nineteenth century. The origin of microorganisms had to be understood before microbiology could be established as an experimental science.

Among the various experiments performed by Pasteur was that involving swan- or gooseneck flasks to show how microbes gained access to or could be prevented from entering fermentable and related types of nutrients. No plugs of any type were used to prevent the passage of microorganisms into these systems. The flasks and their contents were first sterilized by boiling. Despite the fact that these systems were open to the external environment, growth did not usually occur. The occasional failure was found to be caused by bacterial spores in the broth that could resist boiling for short periods of time. The length and the bend of a flask's gooseneck prevented microorganisms present in the air from entering the flask proper. However, when the top of such a system was broken off, or when the flask was tilted so that the sterile liquid nutrient ran into the exposed part of the neck and then returned, microbes appeared in the fluids after a brief incubation period. Pasteur also used these and other flasks to show the distribution of microorganisms in the environment.

Materials: Pond-Water Examination and Environmental Isolations

❏ 1. Beaker containing pond water

❏ 2. Nutrient agar plates (1 per student)

❏ 3. One blood agar plate for class demonstration

❏ 4. Sterile cotton-tipped applicator sticks (2 per student)

❏ 5. Glass slides and cover slips

❏ 6. Aqueous methylene blue stain

❏ 7. Pasteur pipettes for pond water and additional specimens

❏ 8. Charts or reference texts showing representative microorganisms (e.g., algae and protozoa)

A. Examination of Pond Water

Procedure

This procedure is to be performed by students individually.

❏ 1. Prepare a wet mount of each specimen provided.

❏ 2. Examine each preparation with the low- and high-power objectives. (Refer to Figures 12–2a through d and Exercise 10.)

❏ 3. With the aid of these figures, attempt to identify the microorganisms present.

❏ 4. Answer the questions in the Results and Observations section.

B. Isolations from the Environment

Procedure

This procedure is to be performed by students individually.

❏ 1. The instructor will remove the lid and expose a blood agar plate to the general laboratory environment for 30 minutes.

❏ 2. In choosing an environment from which to gather specimens for your microbial isolations, select a location within or close to the laboratory—for example, a water faucet, laboratory table, door handle, staining rack—and obtain a specimen, using the sterile (cotton-tipped) applicator sticks provided.

(a) (b) (c) (d)

Figure 12–2
Photomicrographs of representative protists commonly found in natural waters. (a) The algae, diatoms, and (b) desmids. (c) The protozoa, *Stentor* and (d) *Euplotes.*

❏ 3. Inoculate a nutrient agar plate with your specimen by running the cotton-tipped applicator stick over the surface of the medium several times.

❏ 4. Mark the bottom of the plate, indicating the location from which the specimen was taken.

❏ 5. Make several temporary wet mounts and/or simple stains of the organisms growing on the inoculated plate.

❏ 6. Answer the questions in the Results and Observations section. (Refer to Figure 12–3.)

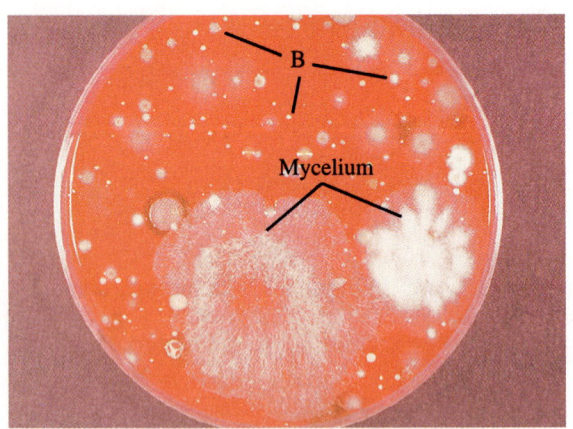

Figure 12–3
The result of exposing a blood agar plate to a laboratory environment. Note the variety of smooth bacterial (b) colonies and the dramatic features of the cottony fungal mycelia.

Results and Observations: A and B

1. Were the major types of organisms encountered in the pond water and environmental specimen different? Explain.

_____

_____

2. What factors would account for a lack of growth on your nutrient agar plates?

3. Is blood agar better than nutrient for the isolation of microorganisms?

C. Pasteur's Swan-Neck Flask Demonstration

Materials

❏ 1. Three cotton-stoppered, 250-mL Erienmeyer flasks containing 75 mL of sterile phenol red glucose broth

❏ 2. Three sterile 1-hole rubber stoppers with glass tubing approximately in the configuration shown in Figure 12–4a (the tube opening that will be exposed to the external environment will be covered by aluminum foil). The entire unit will be kept wrapped until ready for use.

❏ 3. One mL of a 72-hour nutrient broth culture of *Bacillus subtilis*

❏ 4. One sterile Pasteur pipette with rubber bulb

❏ 5. One beaker containing disinfectant

❏ 6. One wax marking pencil or similar marker

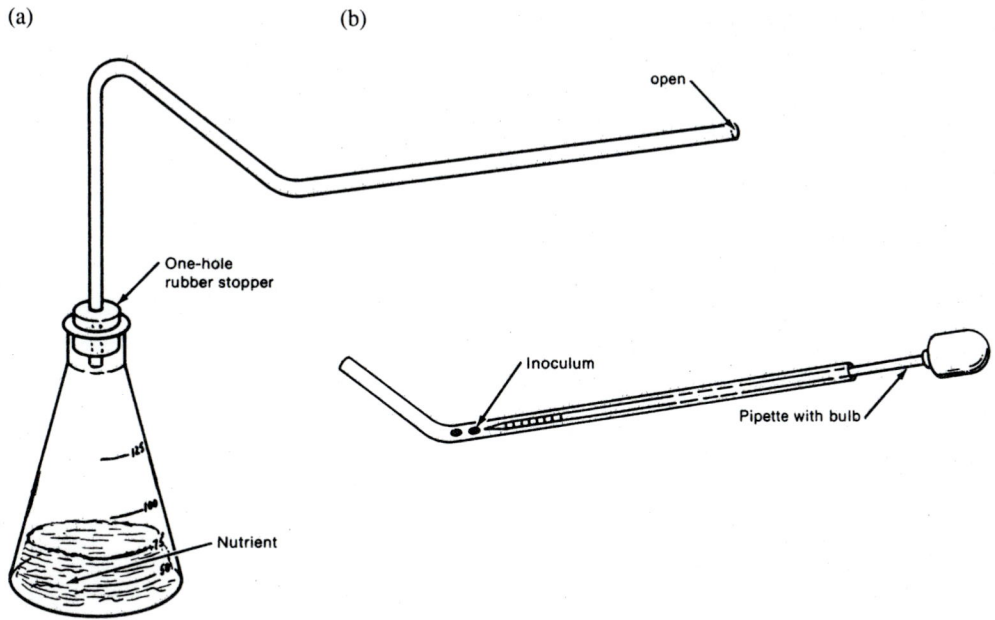

Figure 12–4

Diagrammatic representation of an environment sampling flask setup. (a) This device approximates the swan-neck system used by Louis Pasteur. (b) The introduction of an inoculum into the system by a pipette.

Procedure

This portion of the exercise will be a laboratory demonstration.

❏ 1. The instructor will assemble each flask system separately as follows:

    ❏ a. Carefully uncover the wrapped rubber stopper and glass tubing unit.

    ❏ b. Remove the cotton stopper from one of the Erlenmeyer flasks, flame the mouth of the flask, and quickly insert the rubber stopper unit.

    ❏ c. Assemble the other two systems.

❏ 2. Examine the assembled systems. Label each flask system A, B, and C, respectively. Remove foil from the tube ends of flasks B and C.

❏ 3. Flask A will be used as a control; nothing will be done to it. Flask B will be an environmental sampler.

❏ 4. With the aid of a sterile Pasteur pipette, aseptically remove about half of the broth culture provided and introduce it into the open end of the tubing unit in flask C. Insert the pipette as far as it will go into the tubing before releasing the culture. This procedure is illustrated in Figure 12–4b.

❏ 5. Incubate the three systems at room temperature for 1 month or until the first sign of growth occurs in an area that is available for periodic observation.

❏ 6. Examine each flask system weekly for the presence of growth and/or color change in the medium. If organisms are present and are able to break down the glucose present in the phenol red broth, a color change will occur. (Refer to Figure 12–5.)

Figure 12–5
Environmental sampling flasks. The flask on the left shows the color change that can be expected if growth occurs and the microorganisms are glucose fermenters. If microbial growth occurs and the organisms involved are not capable of using glucose, the medium will be cloudy and red.

❏ 7. If, after 1 month, growth and/or a color change does not occur in the medium, tilt flasks B and C so that the medium will pass into the last bend in the tubing. Then return the flasks to their right positions.

❏ 8. Examine the flasks for 1 week, and record your findings in the Results and Observations section.

❏ 9. After growth appears in flasks B and C, remove the stopper tubing unit from flask A and keep the system exposed for 1 hour. Replace the unit, continue to incubate at room temperature, and examine the flasks for growth after 1 week. Record your findings in the Results and Observations section.

Results and Observations

Complete the following table. Use "−" to show the absence of growth (G) and/or a color change (C) and "+" to show the presence of growth (G) and/or a color change (C).

Table 12-1

Pasteur's Swan-Neck Flask Observations

Week of Observation

Flask	1		2		3		4		5		6		7		8	
	G	C	G	C	G	C	G	C	G	C	G	C	G	C	G	C
A																
B																
C																

Laboratory Review 12   The Distribution of Microorganisms in the Environment

1. Would viruses and/or anaerobic microorganisms be found in any of the samples or systems used in this exercise? Explain your answer.

   _____

   _____

2. Distinguish between the following:

   a. diatoms and cyanobacteria _____

   _____

   b. ecology and ecosystem _____

   _____

   c. symbiosis and parasitism _____

   _____

Key Terms

community (ko-MEW-ni-te): all the interacting populations within an ecosystem

ecology (ē-KOL-ō-jē): the study of the interrelationships between organisms and their respective environments

ecosystem (Ē-kō-sis-tem): the total community of living organisms together with their chemical and physical environments

population (pop-ū-LAY-shun): a group of interbreeding individuals of the same species within an ecosystem

spontaneous generation: an ancient point of view which held that lower forms of life develop from nonliving materials

Photograph Quiz 8

Identify the general microbial types shown in the photograph of this environmental sample.

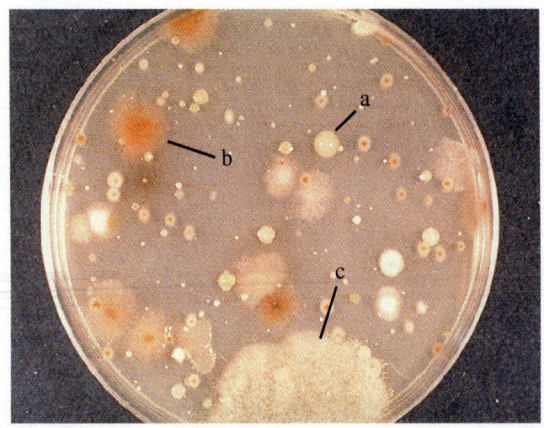

Figure 12–6
An environmental plate.

a. _____

b. _____

c. _____

After completing this exercise, you should be able to:

1. Construct a Winogradsky column and explain the role of each component.

2. List and discuss biogeochemical cycles.

3. Recognize the morphological types of microorganisms and typical genera found in the Winogradsky artificial ecological system and their respective roles in color development.

Microorganisms possess a remarkable ability to degrade an enormous number of organic compounds. The catalytic powers of these organisms play major roles in the chemical transformations occurring on the surface of the earth. While all types of organisms participate in the turnover of the chemical elements that make up living forms, the contributions of microbes are not only important but in certain situations essential. These natural or biogeochemical cycles include the cyclic conversions (use and reuse) of carbon, nitrogen, oxygen, phosphorus, and sulfur. The integration of the reactions in such cycles results in a balanced production and consumption of biologically important elements. For example, the use and reuse of carbon and oxygen are brought about primarily by the reactions involved with two processes: *oxygenic photosynthesis* and *respiration.*

In this exercise, the Winogradsky column, a simple means of encouraging the growth of a wide variety of soil microorganisms, will be used to simulate microbial processes occurring in the water and mud of a pond. This system was developed by Sergei Nikoliavich Winogradsky in the 1880s to study the complex interactions between environmental conditions and microbial activities and the role of soil enrichment in the isolation of pure bacterial cultures. In his initial studies, Winogradsky used the system for the isolation of photosynthetic bacteria. He was able to isolate and study these microorganisms because of their visible brown, green, and purple pigmentations.

The Winogradsky column is an artificial ecological system, or "minipond," that provides an oxygen gradient from top to bottom. When it is sealed and exposed to a light source, suitable environments are created for both aerobic and anaerobic photosynthetic organisms. Actually, a succession of microbes will develop according to the concentrations of oxygen and the nutrients and light available at different regions in the column. The types of soil and nutrients used to prepare the column also influence the varieties of organisms that will appear. A typical column contains a mud layer with sources of cellulose, sulfur, and carbon dioxide. The Winogradsky column and the types of organisms that could appear at different levels are shown in Figure 13–1.

The artificial minipond for this exercise has been designed to test various sources of substances necessary for the development of a wide variety of microorganisms, in particular the colorful photosynthetic bacteria.

Procedure

This exercise will be performed in groups the size of which will be determined by your instructor and based upon the variety of materials available. Each group will be assigned different materials to test their ability to assist in the development of a functional ecological system.

❑ 1. Check the assigned soil or mud sample for clumps that will prevent packing of the column. Although the samples should have been sifted prior to class, some clumping may occur on storage. Remove any clumps present.

❑ 2. The amount of mud or moistened soil needed will depend upon the actual size of the container chosen by your instructor, who will indicate how much is appropriate. The only caution at this point is that the water you use must be appropriate to the mud or soil type (i.e., fresh water with soils and fresh water muds and marine water with marine muds).

❑ 3. Mix the sulfur source and calcium carbonate with the mud or moistened soil so that each will be 1 to 2% by weight. Then mix this mass with an equal volume of the cellulose source.

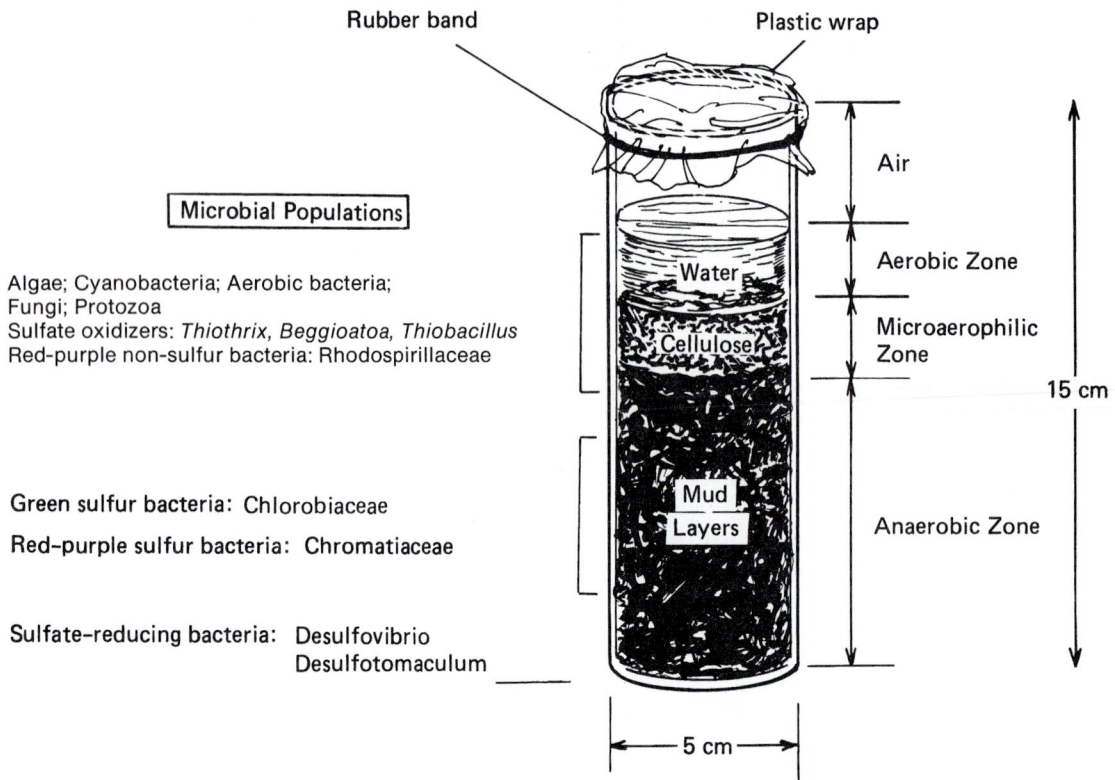

Microbial Populations

Algae; Cyanobacteria; Aerobic bacteria;
Fungi; Protozoa
Sulfate oxidizers: *Thiothrix, Beggioatoa, Thiobacillus*
Red-purple non-sulfur bacteria: Rhodospirillaceae

Green sulfur bacteria: Chlorobiaceae

Red-purple sulfur bacteria: Chromatiaceae

Sulfate-reducing bacteria: Desulfovibrio
Desulfotomaculum

Figure 13–1

The general appearance of a *Winogradsky column.* Environmental conditions and the respective locations of possible microbial populations are also shown.

Materials

The following items are to be provided per group of students:

❑ 1. A glass jar approximately 15 cm in height and 5 cm in diameter

❑ 2. Clear plastic film and a rubber band

❑ 3. A 20-cm wooden dowel approximately 1 cm in diameter with a cork at one end

❑ 4. A selection of fresh water and marine muds, if available

❑ 5. A selection of soils

❑ 6. A selection of water samples, including fresh, marine, and aquarium sources

❑ 7. *Cellulose source:* shredded papers of different types, shredded grass, leaves, lettuce, or other types of vegetation, and sawdust or wood shavings

❑ 8. *Sulfur source:* calcium sulfate, magnesium sulfate, the yolk of a hard-boiled egg, cooked meat, or cheese

❑ 9. *Immediate carbon dioxide source:* calcium carbonate

❑ 10. A 250-mL beaker

❑ 11. Plastic ruler

❑ 12. Balance, spatula, and weighing papers or boats

❑ 13. Graduated cylinders

❑ 14. Pasteur pipettes and a rubber bulb

❑ 4. Pack this material into the container in layers of 2 to 3 cm. Use the cork end of the dowel to force out trapped air, adding small amounts of the same water appropriate to the soil or mud to displace air bubbles. Your last layer should stop approximately 5 cm from the top.

❑ 5. Add 2 to 3 cm of your water sample to cover the column and cover the jar with plastic film held in place by a rubber band.

❑ 6. Place your column near low-intensity incandescent lighting so that it will not overheat.

❑ 7. If possible turn the columns one-half turn daily. This will maximize light exposure and encourage growth.

❑ 8. Examine your column weekly for at least 1 month for pigmentation changes as shown in Figure 13–1, and record these changes in the Results and Observations section. Notice any other changes (e.g., gas production).

❑ 9. After an incubation period of 4 weeks or longer at the discretion of your instructor, sample different layers in the column with a Pasteur pipette and examine by the temporary wet mount technique. Record observations of different morphological types for each region of the column in the Results and Observations section.

Results and Observations

1. Record color changes in the areas provided on a weekly basis.

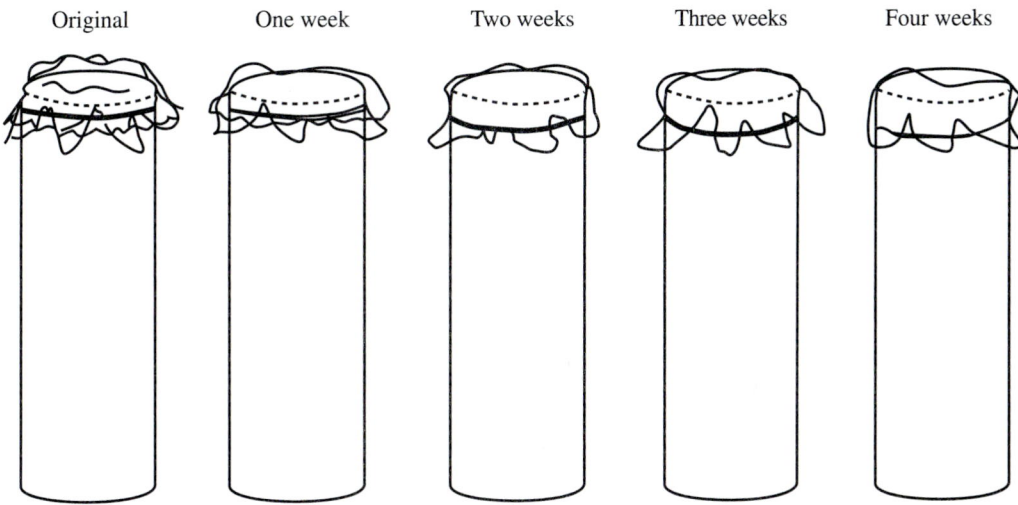

| Original | One week | Two weeks | Three weeks | Four weeks |

2. Complete the following table by recording the observed morphological types of organisms in the specific layer of the column.

3. Was gas production in the system noticeable? _____ When? _____

Table 13–1

Microscopic Observations

Layer	Morphological Types of Organisms Observed
Water	
Cellulose	
Mud	

Laboratory Review 13   An Artificial Ecological System: The Winogradsky Column

1. What is the role of the sulfate-reducing bacteria in this ecological system?

   _____

   _____

2. Describe and sketch the sulfur cycle and correlate observed organisms, where possible, and using colors with portions of that cycle. (Refer to your text.) _____

   _____

   _____

   _____

   _____

   _____

3. Is there evidence of other natural cycles in the column? If so, which ones? _____

   _____

   _____

4. What is meant by *anoxygenic photosynthesis?* _____

   _____

5. Gas production should have occurred during the exercise. What gaseous substances would you normally expect to be formed? _____

   _____

6. What might occur if an identical column were prepared and covered entirely with foil to prevent exposure to light? _____

   _____

   _____

7. Define the following terms:

   a. ecosystem _____

   b. habitat _____

   c. niche _____

   d. synergism _____

Key Terms

biogeochemical: chemical activities performed by living organisms *(bio)*, and associated with the earth *(geo)*

combustion: an act of burning

oxygenic: producing oxygen

respiration: cellular energy-yielding reactions in which the last step involves reduction of an inorganic chemical

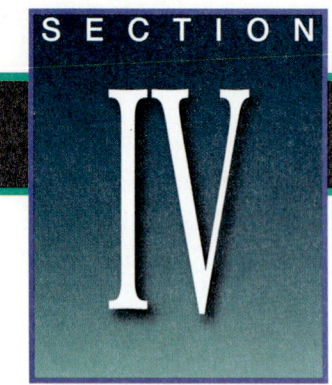

The Differentiation of Bacterial Groups by Staining Reactions

If the number of victims which a disease claims is the measure of its significance, then all diseases, particularly the most dreaded infectious diseases such as bubonic plague, Asiatic cholera, etc. must rank far behind tuberculosis.

—Robert Koch, 1882

A preliminary grouping of bacteria is usually based upon gross morphology and the manner in which the bacteria react to staining procedures. The purpose of this section is to demonstrate two major procedures by which most bacteria can be categorized.

Properly performed, the Gram stain differentiates nearly all bacteria into two major subgroups of prokaryotes that differ with respect to the nature of the cell wall (Figure IV–1). One subgroup, the *gram-positive bacteria,* includes non-pathogenic organisms as well as various causative agents of such diseases as anthrax, boils, diphtheria, rheumatic fever, and certain forms of pneumonia. The second subgroup, the *gram-negative* bacteria, includes all photosynthetic prokaryotes; pathogenic organisms that cause cholera, dysentery, and plague; and various nonpathogenic species. In addition to these two major categories of bacteria, two other types of reactions are known, namely, the *gram-variables* and the *gram-nonreactives.* Gram-variables may include bacteria belonging to the genera *Branhamella* and *Neisseria,* which under ordinary conditions may be either gram-positive or gram-negative. This type of variation in staining is caused by a tendency of certain microorganisms to resist decolorization. It may also be caused by inadequate technique. It should be noted that the walls of these microorganisms contain lipopolysaccharide, phospholipid, protein, and peptidoglycan organized according to the usual gram-negative form of architecture.

(a)

(b)

 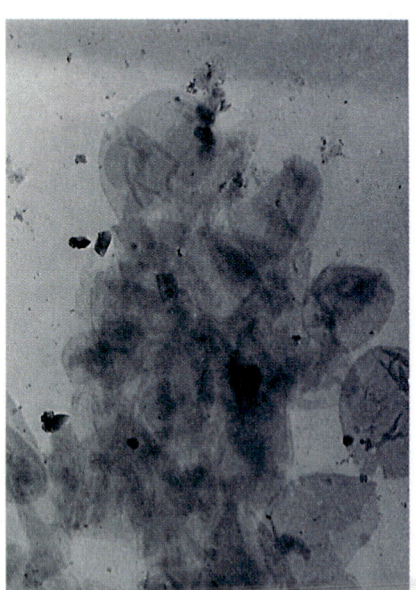

Figure IV–1

Electron micrographs of prokaryotic cell walls. Both electron micrographs were prepared by the zeolite-disruption procedure developed by G.A. Wistreich, M.D. Lechtman, J.W. Bartholemew, and R.F. Bils. (a) The cell walls of the gram-positive *Bacillus megaterium.* (b) A cell-wall preparation of the gram-negative *Enterobacter aerogenes.*

The gram-nonreactive group of microorganisms includes those that do not stain or that stain very poorly. Various spirochetes fall into this category.

The Gram stain is a relatively simple procedure to perform. It is of diagnostic value only when applied to prokaryocytes having cell walls. Differentiation of the two major subgroups is possible by other more reliable means, such as examination by electron microscopic or chemical analysis. These methods, however, are neither as simple nor as inexpensive to apply. Modern technology has also been applied to the Gram stain procedure. An automated device known as the **Aerospray Stainer** can stain more than 180 slides per hour. This stainer (Figure IV–2) sprays the Gram staining reagents in proper order onto the surfaces of prepared bacterial smears.

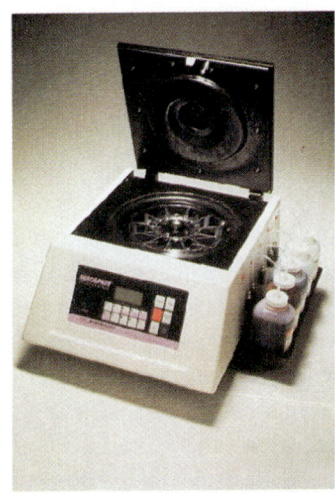

Figure IV–2

The Aerospray Gram Stainer. A relatively new concept for the Gram Stain. This instrument is an automated device capable of staining more than 180 smears per hour.
(Illustrations courtesy of Wescor, Inc.)

Another method used for bacterial differentiation is the acid-fast staining procedure. Here again, bacteria in general are divided into two groups, namely, *acid-fast* and *non-acid-fast.* Two of the better-known acid-fast organisms of medical significance are *Mycobacterium tuberculosis* and *M. leprae,* the etiological agents of tuberculosis and leprosy, respectively. Most other pathogens are generally non-acid-fast and are differentiated primarily by the Gram stain reaction.

In addition to presenting both differential staining procedures, Exercises 14 and 15 include unknown specimens. Such specimens are used to test the ability of students to apply and interpret the results of the respective procedures.

After completing this exercise, you should be able to:

1. Carry out the Gram stain procedure correctly.
2. Differentiate between gram-positive and gram-negative reactions.
3. Interpret Gram stain reactions with unknown specimens.
4. Recognize the importance of the Gram stain in disease detection and diagnosis.
5. List at least two bacterial species to be gram-positive and two species to be gram-negative.

The Gram staining reaction was discovered in 1883 by Christian Gram, who chanced on it while he was trying to stain biopsy specimens so that microorganisms could be differentiated from surrounding tissue. His procedure had an important shortcoming: some bacteria could be easily distinguished, but others could not be seen at all. In 1886, it was realized that his technique could be used to divide bacteria into two major groups.

The Gram differentiation is based upon the color reactions exhibited by bacteria when they are treated with crystal violet dye (the **primary stain**) followed by an iodine-potassium iodide solution (a **mordant**). Certain organisms lose the violet color rapidly when ethyl alcohol is applied, while others lose their color more slowly. After the decolorization step, a **counterstain**—usually safranin—is used. Bacteria resistant to decolorization will retain a blue or purple color and will not take the counterstain. Such organisms are referred to as *gram-positive*. Those microorganisms unable to retain the crystal violet stain will take the counterstain and consequently exhibit a pink or red color. The term *gram-negative* is used to describe these organisms. It is important to note that this differentiation is based on the rate at which the dye leaves the cell. For this reason, the procedure must be performed with great care.

Since the original work of Gram, many investigators have attempted to determine the mechanism of the Gram stain reaction. Various studies have shown that the response of an organism in the procedure is a consequence of its cell wall structure (Figure 14–1) rather than the presence of any particular biochemical. Table 14–1 lists the respective reagents and the color results of the Gram stain procedure.

Figure 14–1 shows a comparison of the cell walls of gram-positives and gram-negatives. Treatment of cells first with crystal violet and then with Gram's iodine (the mordant) results in the formation of large, insoluble dark purple, almost black, dye-iodine complexes inside cells. Such dye-iodine complexes are retained in gram-positives by the thick **pepti-**

Table 14–1

The Gram Stain Procedure and Results

Reagent	Cell Color Results	
	Gram-Positive Cells	Gram-Negative Cells
Crystal violet (primary stain)	(purple)	(purple)
Gram's iodine (mordant)	(purple)	(purple)
Acetone-alcohol (decolorizer)	(purple)	(no color)
Safranin (counterstain)	(purple)	(red)

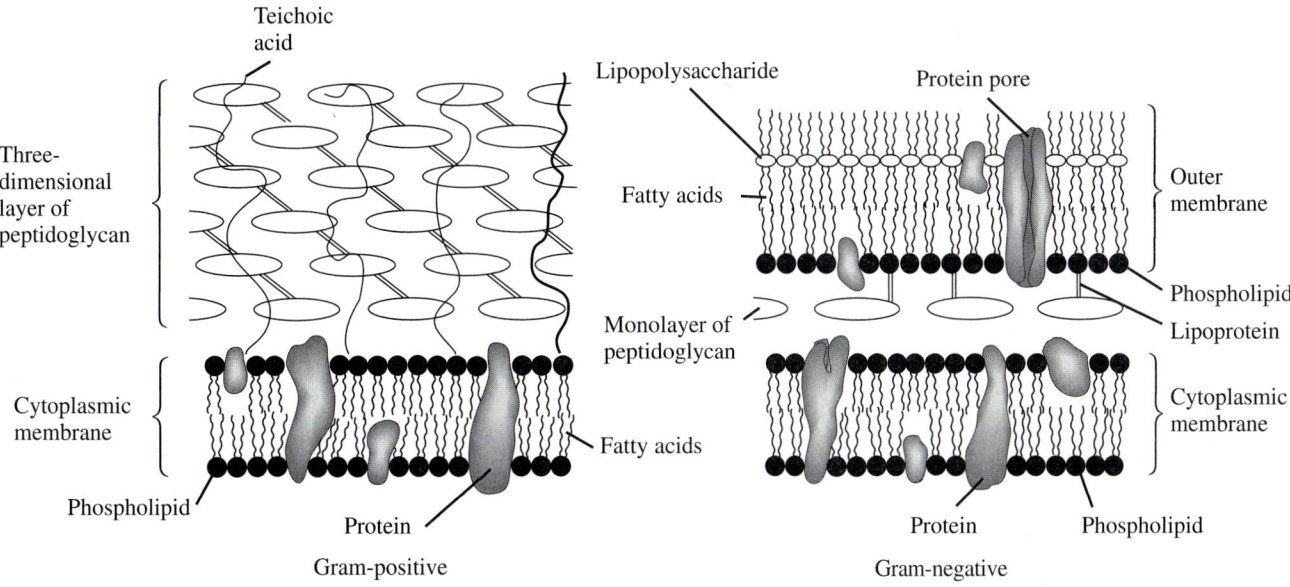

Figure 14—1
A comparison of gram-positive and gram-negative cell walls.

doglycan meshwork in their cell walls. In gram-negative cells the crystal violet-iodine complexes are readily released through the very thin peptidolayer remaining after their membranes have been dissolved during the decolorization step.

Several factors can affect the Gram reaction, including cell age, autolysin (self-dissolving enzyme) levels, and growth conditions. As cultures become old, bacterial walls become more permeable to the crystal violet-iodine complex and therefore tend to exhibit gram-negativity more readily. This shift is believed to be the direct result of the lytic (dissolving) action of autolysins. Situations of this type would have significant effects on gram-positives because cell walls would become leaky and allow the cells to be easily decolorized.

The general procedure is quite simple—perhaps deceptively so. It involves the application of the primary dye, next the iodine solution, and then alcohol. It is customary to wash off excess reagent after each step. But excessive washing can remove the dye or dye-iodine complexes within the cells and consequently greatly affect the overall staining reaction. The final step in the procedure, the application of the counterstain safranin, must be performed very carefully. As the decolorization step is probably the most critical, precautions should be taken to guard against misleading results. The most commonly employed control is the use of a mixed smear of a gram-positive coccus and a slender gram-negative rod on a portion of the same slide used for the unknown culture.

Materials

❑ 1. Twenty-four-hour trypticase soy broth cultures of the following microorganisms (per pair of students):
　❑ a. *Escherichia coli*
　❑ b. *Bacillus subtilis*
　❑ c. *Staphylococcus aureus*
❑ 2. Unknown cultures (2 per student)

❑ 3. Microscope slides
❑ 4. Gram stain reagent sets (which include acetone alcohol, 95% ethanol, or other reagent as the decolorizer)
❑ 5. Immersion oil
❑ 6. Lens paper

A. Known Cultures

Procedure

This procedure is to be performed by students individually.

❑ 1. Prepare separate thin, air-dried, heat-fixed smears of the bacterial cultures provided. (If necessary, refer to Procedure Diagram 2 in Exercise 2.)

❏ 2. With each of the prepared smears, perform the 9 steps as described in Procedure Diagram 17. (*Note:* The rate of decolorization depends on several factors, including alcohol concentration, smear thickness, and whether the smear is wet or dry.)

❏ 3. Observe your results under oil immersion. (Refer to Figure 14–2.)

❏ 4. In the Results and Observations section, sketch the morphology of and indicate the Gram reaction with the color for each culture used.

Procedure Diagram 17
The Gram Stain

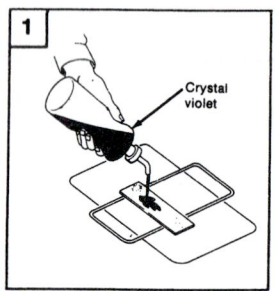

1. Cover the smear with crystal violet reagent for 1 min. Add more stain if drying occurs.

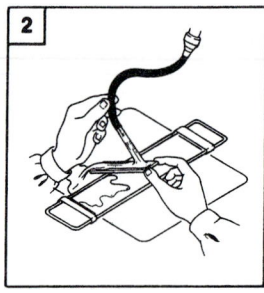

2. Rinse the slide (front and back) in slowly running water for 5 sec.

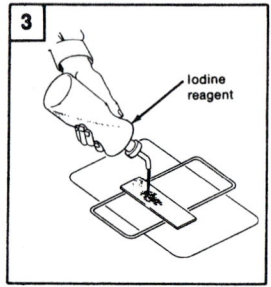

3. Rinse the smear with iodine reagent and pour-off the excess. Cover the smear again with the iodine and allow it to remain for 1 min.

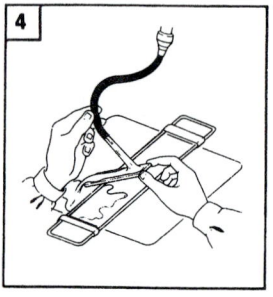

4. Rinse the smear with running water as in step 2.

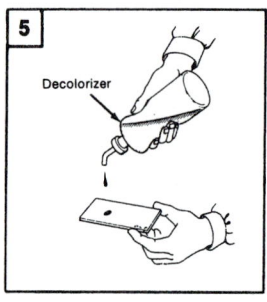

5. Apply the alcohol decolorizer dropwise and slowly. Continue to add this reagent until dye no longer runs off from the smear.

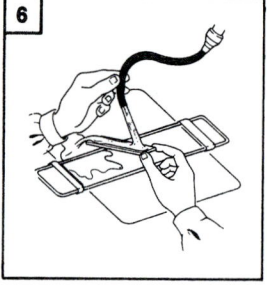

6. Rinse with running water again as in step 2, to prevent additional decolorization.

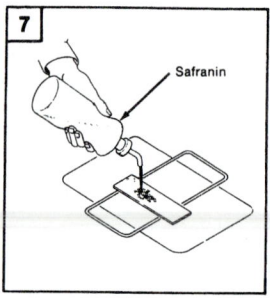

7. Cover the smear with safranin for 1 min.

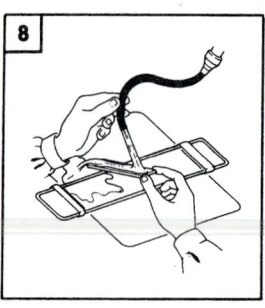

8. Rinse again with water as in step 2.

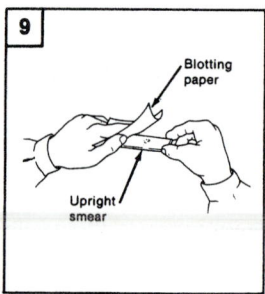

9. Blot dry but do not rub. Observe as directed.

Results and Observations

Record the results of your Gram stain of known cultures.

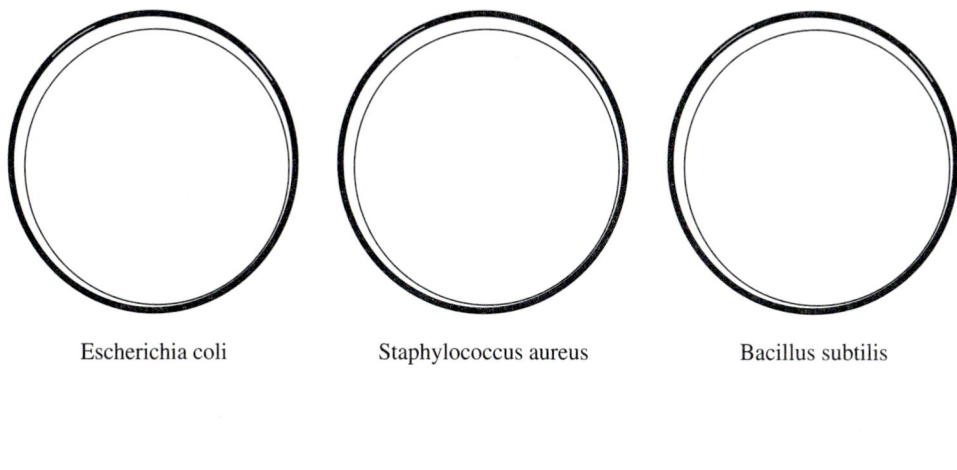

Escherichia coli   Staphylococcus aureus   Bacillus subtilis

(a)   (b)

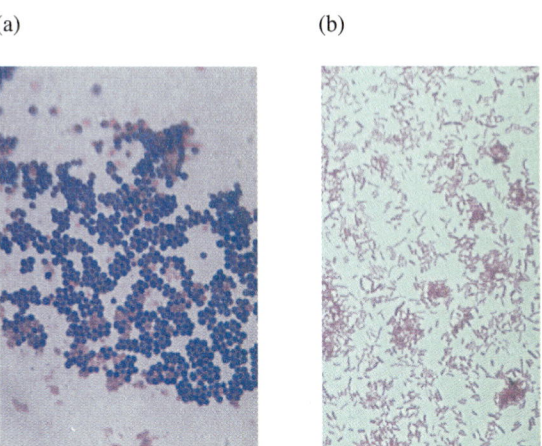

Figure 14–2
Differential staining—the Gram stain procedure. a. A typical gram-positive reaction exhibited by *Staphylococcus aureus*. These organisms characteristically retain the primary dye, usually crystal violet, and appear dark purple in color. This photomicrograph demonstrates the staphylococcal arrangement of cells. b. The appearance of a gram-negative bacterial species. In this case, the primary dye was not retained following the decolorization step. Consequently, the organisms take the counterstain safranin.

B. Unknown Cultures

Procedure

❏ 1. Each student will receive an unknown culture containing either a gram-positive or a gram-negative microorganism.
❏ 2. Perform the Gram stain procedure as described earlier. It is suggested that a mixed smear of *E. coli* and *S. aureus* be included on a portion of the slide as a control. (Refer to Figure 14–2.)
❏ 3. Record your results on the Unknown Form, page 150. In addition, include a brief description of the morphological features observed.

Laboratory Review 14   The Gram Stain

1. What are the chemical differences between the cell walls of gram-positive and gram-negative bacteria that might explain differences in the rate of decolorization?

_____

_____

2. a. Does the age of a culture affect the Gram stain reaction? _____

_____

    b. Why or why not? _____

_____

3. List 2 factors, in addition to the age of the culture, that can affect the Gram stain reaction.

    a. _____   b. _____

4. List 5 gram-positive bacteria and the diseases they cause.

Table 14–2

Gram-Positive Pathogens

	Genus/Species	Disease
a.		
b.		
c.		
d.		
e.		

5. List 5 gram-negative bacteria and the diseases they cause.

Table 14–3

Gram-Negative Pathogens

	Genus/Species	Disease
a.		
b.		
c.		
d.		
e.		

6. List 4 chemical and/or structural differences that distinguish gram-positive from gram-negative bacteria.

   a. _____

   b. _____

   c. _____

   d. _____

Key Terms

autolysin (aw-TOL-i-sin): an enzyme produced or originating from within a cell, and involved with the digestion of cellular structures

differential (dif-er-EN-shal) staining: a procedure in which more than one stain is used to distinguish the chemical composition of cells or their parts (e.g., Gram stain, acid-fast stain, and spore stain)

Gram stain: a differential staining procedure used to divide bacteria into 2 groups based on their staining reactions, gram-positive and gram-negative

mordant (MŌR-dant): a substance that fixes (holds firmly) a stain or dye

peptidoglycan (pep-ti-dō-GLĪ-kan): the structural molecule of bacterial cell walls consisting of the molecules N-acetylglucosamine, N-acetymuramic acid, tetra-peptide side chain, and peptide side chain

permeability (per-mē-a-BIL-i-tē): the physical property of membranes that permits the passage of molecules and ions in solution across the membrane

Photograph Quiz 9

The smear prepared from the gram-positive control shown gave gram-negative results. Give 3 reasons for this incorrect reaction.

a. _____

b. _____

c. _____

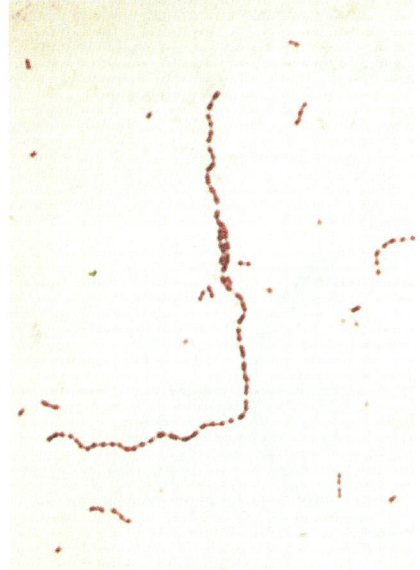

Figure 14–3
A gram stain reaction.

Photograph Quiz 10

What is the Gram reaction and morphological arrangement shown in the following figure? _____

_____

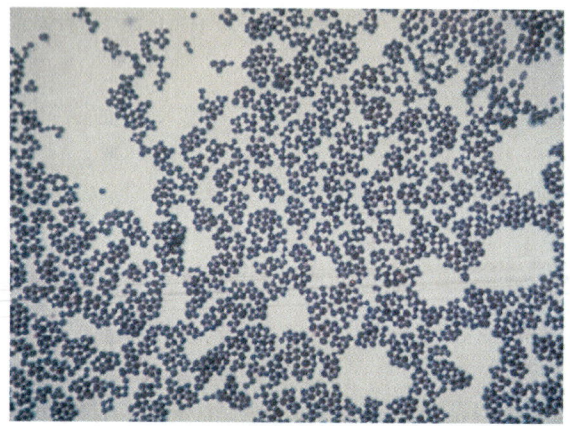

Figure 14–4
A gram stain unknown.

_____

Unknown Form for Gram Stain Reaction

Student's Name _____   Score _____

Date _____   Laboratory Section _____

Unknown Number _____   Result _____

Sketch of oil immersion view (in color)

_____

Unknown Form for Gram Stain Reaction

Student's Name _____   Score _____

Date _____   Laboratory Section _____

Unknown Number _____   Result _____

Sketch of oil immersion view (in color)

_____

After completing this exercise, you should be able to:

1. Perform the acid-fast procedure.

2. Distinguish between acid-fast and non-acid-fast reactions.

3. Interpret acid-fast reactions with unknown specimens.

4. Recognize the importance of the acid-fast procedure in disease detection and diagnosis.

5. List at least two bacterial species known to be acid-fast.

The acid-fast staining procedure was developed by Paul Ehrlich in 1882. He found that tubercle bacilli (*Mycobacterium tuberculosis*) retained a dye reagent composed of crystal violet and aniline in water even after a wash treatment with acidified ethanol solution. After the development of this initial technique, changes in methodology resulted in the formation of the Ziehl-Neelsen procedure, which will be used in this exercise. The acid-fast procedure is another example of a differential stain used in microbiology to distinguish one group of bacteria from another.

The cell walls of certain bacteria and the outer structures of protozoa contain long chain fatty acids called *mycolic acids*. These chemicals give to these microorganisms the property of resistance to destaining of basic dyes by acid alcohol, the hydrochloric acid-containing decolorizer used in the acid-fast procedure. These organisms are referred to as *acid-fast* (Figure 15–1*a*) and include the species of bacterial genera such as *Mycobacterium* and *Nocardia,* and the protozoan genus of *Cryptosporidum*. Nearly all other bacterial and protozoan genera are non-acid-fast. Acid-fast organisms stain red, while non-acid-fast organisms stain blue (Figure 15–1*b*).

Within the last 10 years a dramatic resurgence of tuberculosis (**TB**) has appeared and created a major worldwide health threat. Public health officials are particularly concerned about the increasing outbreaks of multidrug-resistant *Mycobacterium tuberculosis*. Infections caused by such organisms are difficult to treat and unfortunately are associated with a high mortality rate.

(a)   (b)

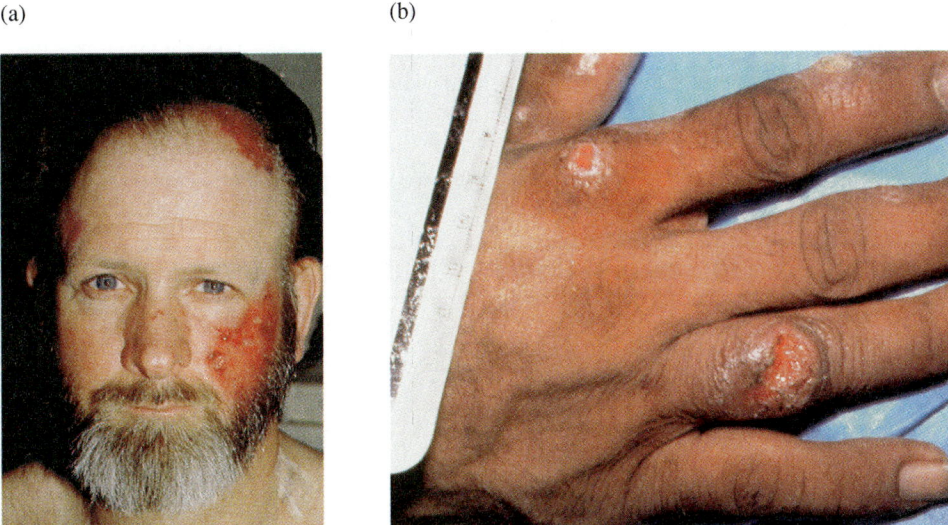

Figure 15–1

(a) The effects of *Mycobacterium avium-intracellulare.*
(From Elewski, Boni *et al. Inter. J. Dermatology* 30:491, [1991].)

(b) An infection caused by *M. haemophilum*
(From Jorgansen, J.H., *et al. Clin. Inf. Dis.* 14:1195–1200, [1992].)

In the United States, TB is primarily caused by *Mycobacterium tuberculosis. M. tuberculosis* and other very closely related TB-causing strains are grouped together as the *M. tuberculosis complex.* Tuberculosis generally is contracted by inhaling particles containing TB-causing mycobacteria. These particles are dispersed through the air by coughing.

In addition to tuberculosis, other infections have come to the attention of the medical community, especially in relation to individuals who have poorly functioning immune systems. Such *immunosuppressed* persons are not only more likely to develop tuberculosis but are of a much greater risk to develop infections caused by mycobacteria that normally do not cause human infections (Figure 15–2). In patients with AIDS, *M. avium* and *M. intracellulare* cause deadly opportunistic infections and are known as *M. avium complex* (MAC) or as *M. avium-intracellulare* (MAI). These are among the most common opportunistic infections in people with AIDS. Without treatment, AIDS patients with MAC infections usually die in less than six months.

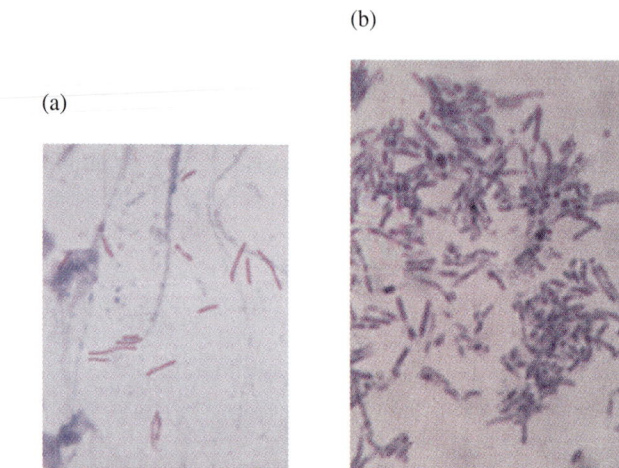

(b)

(a)

Figure 15–2
Differential staining—the acid-fast stain procedure. (a) Typical acid-fast reaction. Microorganisms exhibiting this type of result show a red coloration. (b) Bacterial cells exhibiting a non-acid-fast (blue) reaction.

While *M. tuberculosis* is capable of infecting any tissue of the body, it, as well as *Nocardia asteroides,* is most commonly found in sputum. *M. leprae,* the causative agent of leprosy, may be found in skin scrapings, nasal mucosa, or tissue biopsy material of skin taken from infected persons. Other mycobacteria can be found in these and related tissues. The acid-fast procedure and other procedures are useful in the detection and/or diagnosis of mycobacterial infections.

As mentioned earlier, the Ziehl-Neelsen acid-fast procedure will be used. The reagents and their effects are indicated in Table 15–1A. In this staining procedure, the primary dye *carbol fuchsin* (basic fuchsin) is formulated with

Table 15–1A

Acid-Fast Staining Procedures and Results Ziehl-Neelson Procedure

Reagent	Cell Color Results	
	Acid-Fast Cells	Non-Acid-Fast Cells
Carbol Fuchsin (primary stain)	(red)	(red)
Acid-alcohol (decolorizer)	(red)	(colorless)
Methylene Blue (counterstain stain)	(red)	(blue)

phenol to allow permeation through the waxlike cell walls of the mycobacteria. The slide is usually heated in order to facilitate permeation. Ethyl alcohol is employed as the decolorizer. However, this reagent is prepared with hydrochloric acid to aid in the decolorization of non-acid-fast cells. It is called *acid alcohol*. The final step in the procedure is the application of the counterstain methylene blue. Acid-fast microorganisms stain red, while non-acid-fast organisms appear blue. In addition, acid-fast organisms range in morphology from coccobacilli to long, slender, slightly curved rods.

As is the case of the Gram stain reaction, the integrity of the cell wall (or, more specifically, structurally intact cells) is necessary to demonstrate the acid-fast nature of microorganisms. Because of this property, the mechanism of the acid-fast reaction may also be associated with permeability factors.

Several modifications of the acid-fast procedure exist. One of these is the Kinyoun method, which uses carbol fuchsin (without the application of heat) and brilliant green as a counterstain. The Kinyoun method is also referred to as the cold method since it doesn't use heat. This exercise presents the Kinyoun procedure as an alternate method. Table 15–1B lists the reagents for this procedure together with effects of each on the cells being stained.

Table 15–1B

Kinyoun (Cool) Procedure

Reagent	Cell Color Results	
	Acid-Fast Cells	Non-Acid Fast Cells
Carbol Fuchsin (primary stain)	(red	(red)
Acid Alcohol (decolorizer)	(red)	(colorless)
Brilliant Green (counterstain stain)	(red)	(green)

For tissues, the Fite-Faraco procedure may be used. It is identical to the Ziehl-Neelsen method, except for the application of hematoxylin as a counterstain (Figure 15–3).

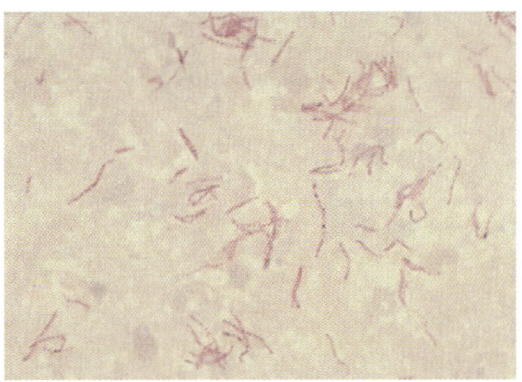

Figure 15–3

A section of a lymph node containing *M. avium-intracellulare* and stained by the Fite stain. The bacterial rods as shown under oil immersion are short and beaded in appearance.
(From Jannottie, F.S. and Sidaway, M.K. *Arch Pathol Lab Med* 113:1120, [1989].)

Materials

❑ 1. Cultures of each of the following microorganisms (1 per 2 students):

    ❑ a. *Mycobacterium smegmatis,* 48-hour trypticase soy agar slant or broth cultures

    ❑ b. *Staphylococcus aureus,* 24-hour trypticase soy agar slant or broth cultures.

❑ 2. Unknown culture (1 per student)

The following materials should be provided per 4 students:

❑ 1. Ziehl-Neelsen acid-fast reagent set

    ❑ a. Carbol fuchsin

    ❑ b. Methylene blue

    ❑ c. Acid alcohol

❑ 2. Kinyoun acid-fast reagent set

    ❑ a. Carbol fuchsin

    ❑ b. Brilliant green

    ❑ c. Acid alcohol

❑ 3. One staining rack designed for heating slides, or 4 large clothespins

❑ 4. One bottle of immersion oil

❑ 5. A sufficient number of glass slides

❑ 6. An adequate number of paper towel squares (approximately 1.5 × 1.5 cm)

A. Known Cultures

Procedure 1: The Ziehl-Neelsen Acid-Fast Technique

This procedure is to be performed by students individually.

❑ 1. Prepare separate air-dried, thin, heat-fixed smears of the bacterial cultures provided. (If necessary, refer to Procedure Diagram 2 in Exercise 2.)

❑ 2. With each of the prepared smears, perform the 9 steps as described in Procedure Diagram 18. (*Note:* The rate of decolorization [step 5] depends on several factors, including alcohol concentration, smear thickness, and whether the smear is wet or dry.)

❑ 3. Observe your results under oil immersion. (Refer to Figure 15–2.)

❑ 4. In the Results and Observations section, sketch the morphology of and indicate the acid-fast reaction with the color for each culture used.

Procedure Diagram 18
The Acid-Fast Stain

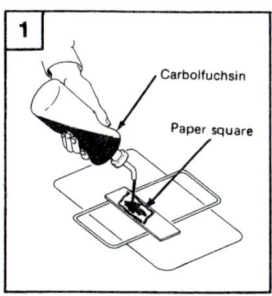

1. Place the slide on the staining rack. Cover the smear with a small paper towel square and then with carbolfuchsin. Allow the stain to stand for 30 to 60 sec. before heating.

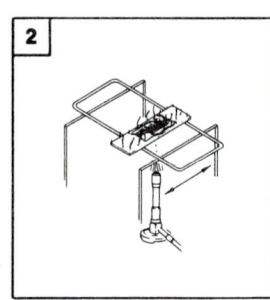

2. Heat the preparation gently by passing the Bunsen burner flame under the slide on the staining rack, or passing the slide over the flame if a clothespin is used. Continue heating until you observe a slight steaming when the flame is removed.

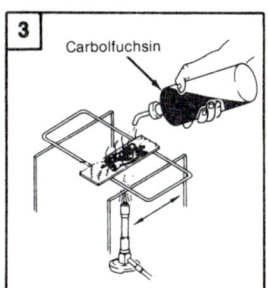

3. Maintain steaming for 5 min. Add more dye as needed to prevent the smear from drying out.

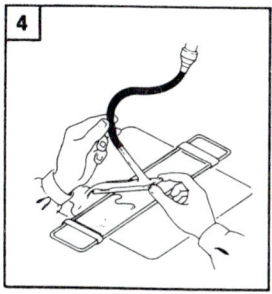

4. Allow the slide to cool on the staining rack for 15 sec. Remove the square piece of paper toweling and then rinse with running water.

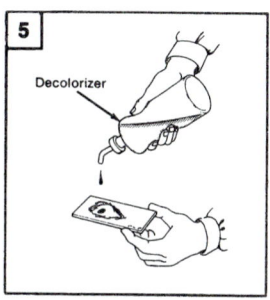

5. Apply the acid-alcohol decolorizer dropwise and slowly. Continue to add this reagent until the dye no longer runs off from the smear.

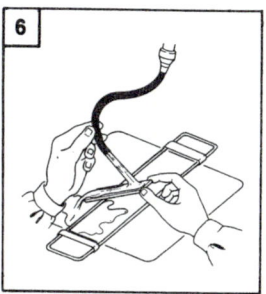

6. Rinse with running water again, as in step 4.

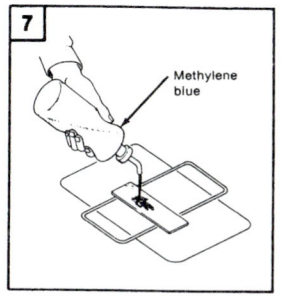

7. Cover the smear with methylene blue for 1 min.

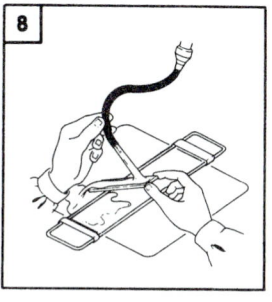

8. Rinse again with water.

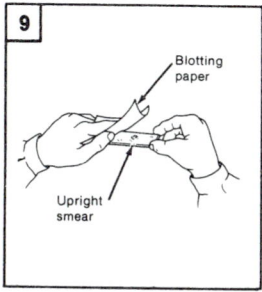

9. Blot dry but do not rub the smear. Observe as directed earlier.

Procedure 2: The Kinyoun Technique

This procedure is to be performed by students individually.

❏ 1. Prepare separate air-dried, thin, heat-fixed smears of the bacterial cultures provided. (If necessary, refer to Procedure Diagram 2 in Exercise 2.)

❏ 2. With each of the prepared smears, perform the 9 steps as described in Procedure Diagram 19. (*Note:* The rate of decolorization [step 5] depends on several factors, including alcohol concentration, smear thickness, and whether the smear is wet or dry.)

❏ 3. Observe your results under oil immersion.

❏ 4. In the Results and Observations section, sketch the morphology of and indicate the acid-fast reaction with the color for each culture used.

Procedure Diagram 19
The Kinyoun (Cold) Stain

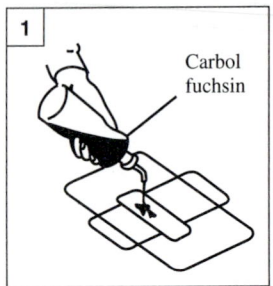
1. Cover the smear with carbol fuchsin for 5 min. Add more stain if drying occurs.

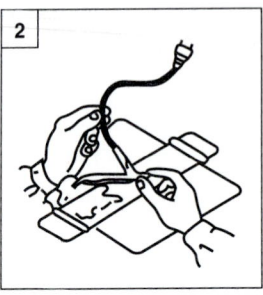
2. Rinse the slide (front and back) in slowly running water for 5 sec.

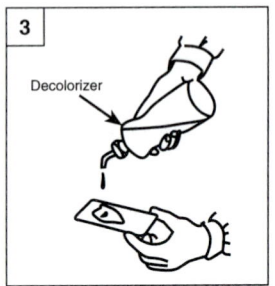
3. Apply acid-alcohol for 3 min.

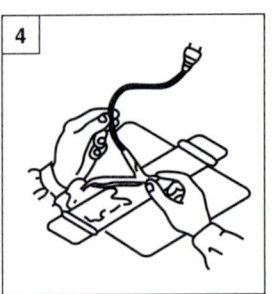
4. Rinse with running water again as in step 2.

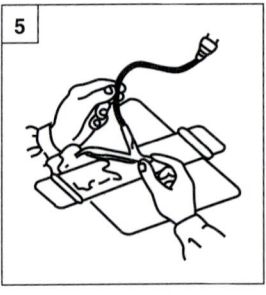
5. Apply the acid-alcohol decolorizer again for 1 min. or until no more red color runs from the **smear.**

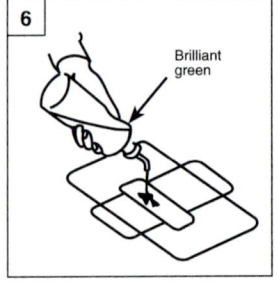
6. Cover the smear with brilliant green and allow to stain for 4 min.

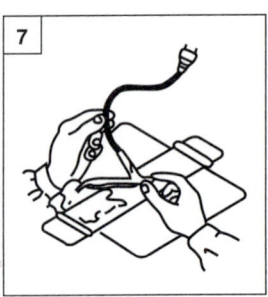
7. Rinse again with water.

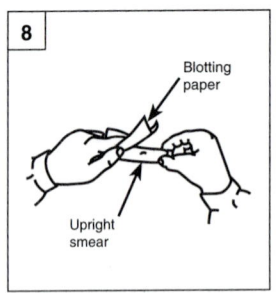
8. Blot dry but do not rub the smear. Observe as directed earlier.

Results and Observations

1. Sketch the results obtained with the Ziehl-Neelsen technique here.

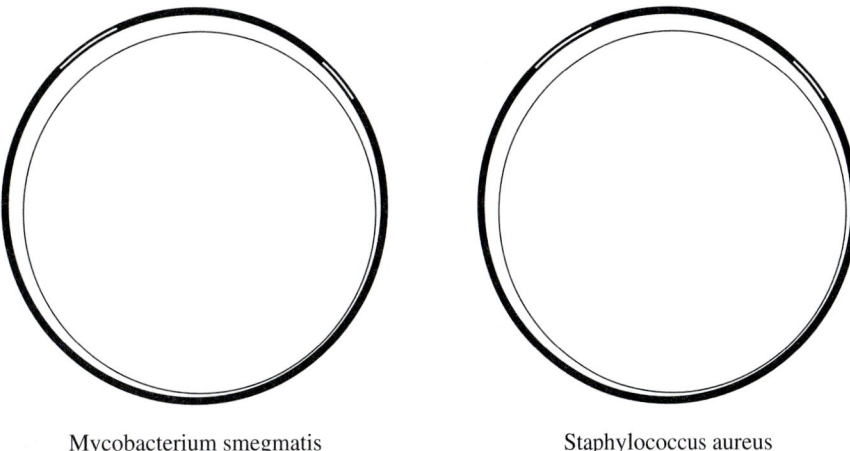

Mycobacterium smegmatis   Staphylococcus aureus

2. Sketch the results obtained with the Kinyoun technique here.

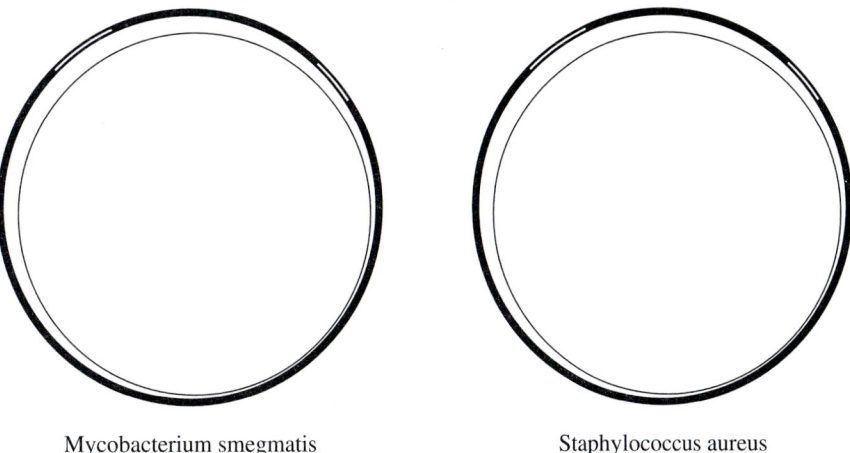

Mycobacterium smegmatis   Staphylococcus aureus

B. Unknown Cultures

Procedure

  ❏ 1. Each student will receive an unknown culture.

  ❏ 2. Perform the acid-fast staining procedure as described earlier. (Your instructor will indicate which of the techniques to use with your unknown.)

  ❏ 3. It is suggested that a mixed smear of *M. smegmatis* and *S. aureus* be included on the same slide as the unknown. (This smear will serve as a check on your technique.)

  ❏ 4. Enter your results obtained with your unknown on the Unknown Form (page 159), and turn it in to the instructor for grading.

Laboratory Review 15   The Acid-Fast Stain

1. Were there any non-acid-fast forms in the *M. smegmatis* culture? Explain. _____

_____

_____

2. What are the major chemical differences among bacteria that might explain the acid-fast state? _____

_____

_____

3. How many genera contain acid-fast organisms? List them. _____

_____

4. List 3 acid-fast pathogens and the diseases they cause.

   a. _____

   b. _____

   c. _____

5. Distinguish between the following combinations:

   a. Gram stain and acid-fast procedures: _____

   _____

   _____

   b. Non-acid-fast and acid-fast cells: _____

   _____

   _____

   c. Acetone alcohol and acid alcohol: _____

   _____

   _____

6. What type of acid-fast staining procedure would you use for tissues suspected of containing *Mycobacterium*

   tuberculosis? _____

   _____

Key Terms

acid-fast reaction: a staining reaction in which organisms resist decolorization with acid alcohol and retain the primary dye; acid-fast bacteria contain large amounts of wax in their cell walls

mycolic (mī-KOL-ik) acids: long-chain fatty acids found in the cell wall of bacteria, such as *Mycobacterium* species and outer protozoan structures

non-acid-fast reaction: a staining reaction in which organisms are susceptible to decolorization with acid alcohol and retain the counterstain

Ziehl-Neelsen (zēl-NĒL-sen) method: a differential staining procedure usually used to distinguish species of *Mycobacteria* and *Nocardia;* these microorganisms are acid-fast

Photograph Quiz 11

What is the acid-fast reaction and morphological arrangement shown in the following figure? _____

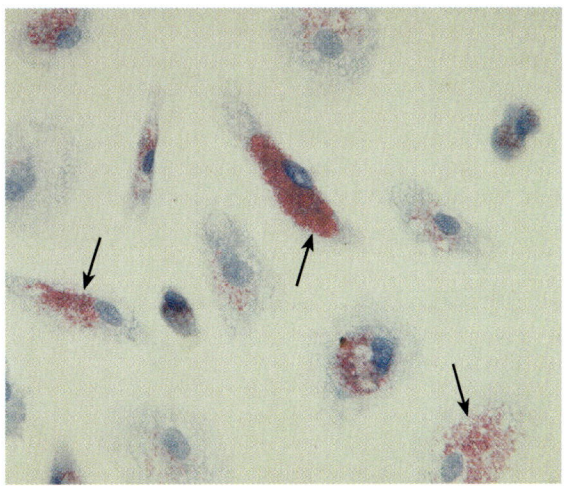

Figure 15–4
An acid-fast reaction from a skin specimen.
(Courtesy of Hoffner, S.E., G. Kallenius, and S.B. Svenson, *Res Microbiol* 143:391–398, [1992].)

Unknown Form for Acid-Fast Stain Reaction

Student's Name _____   Score _____

Date _____   Laboratory Section _____

Unknown Number _____   Result _____

Sketch of oil immersion view (in color)

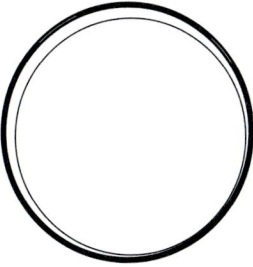

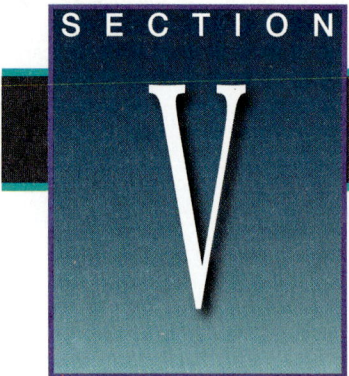

And the little things that our naked eye can not penetrate into, have in them a greatness not to be seen without astonishment . . . how exquisite, how stupendous must the structures of them be!

—Cotton Mather, Thanksgiving sermon, 1689

After Leeuwenhoek first observed bacteria in the seventeenth century, about 200 years elapsed before the development of the procedures and instruments required to give them careful study. Early observations of bacteria were limited in value because of unstained or insufficiently stained specimens and the poor resolving power of microscopes. With modern improvements in specimen preparation and microscopes, various morphological features and structures have been discovered (Figure V–1).

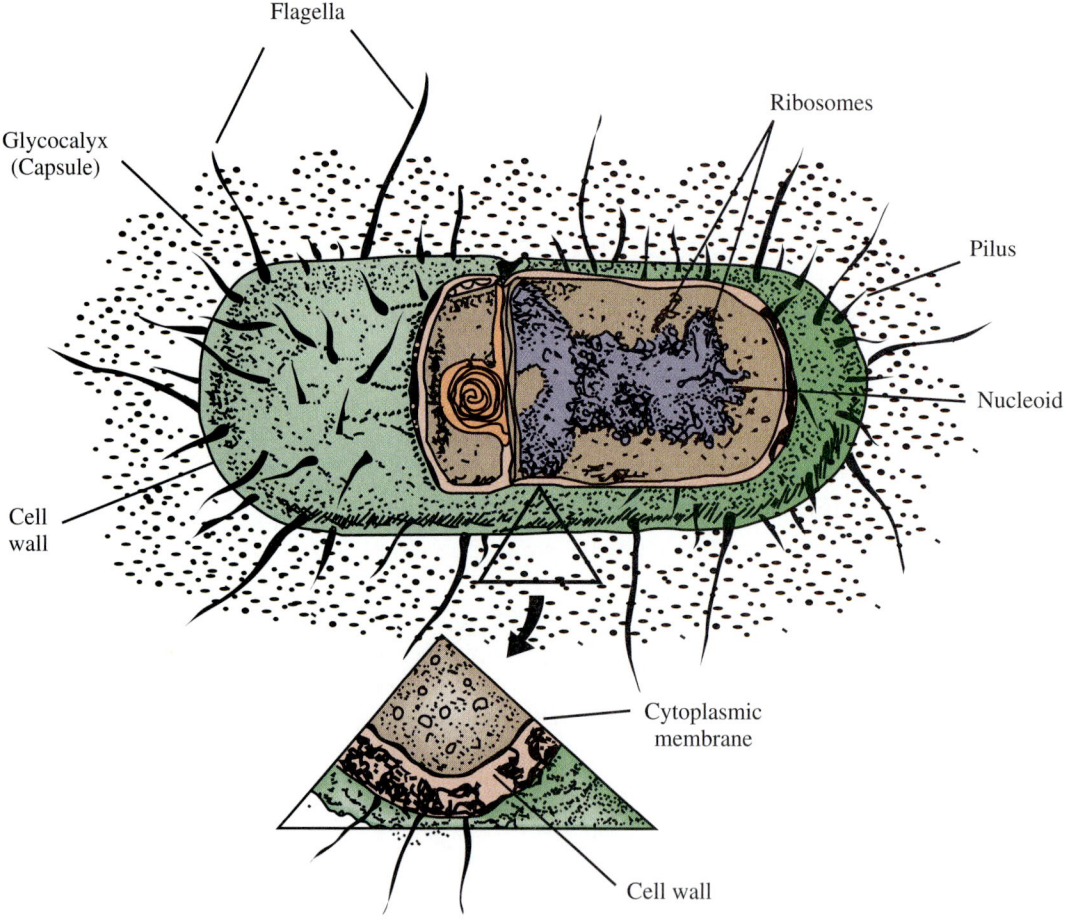

Figure V–1

The structure of a prokaryote. Several of the parts typically found in bacterial cells are shown. Not all of these structures are found in a single cell type.

The hundreds of bacterial species generally are alike in that the individual cell is composed of a rigid cell wall surrounding a cytoplasmic membrane or sac (Figure V–2), which contains several fluid or semifluid components. Within the cytoplasm is an area containing some of the characteristics of a nucleus. This area contains deoxyribonucleic acid (DNA) and is referred to as the *bacterial nucleoid*. The cytoplasm also contains a variety of granules composed of such substances as ribonucleic acid (RNA) and protein, poly-beta-hydroxybutyrate, sulfur, and carbohydrates. Specific granules

(a)

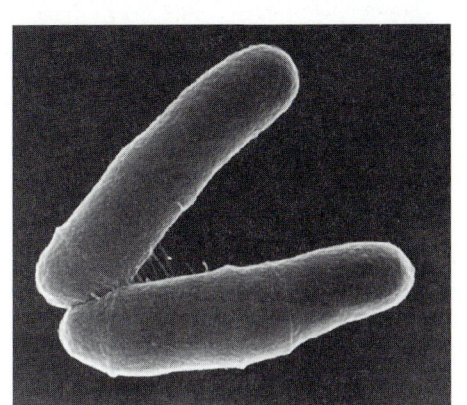

(b)

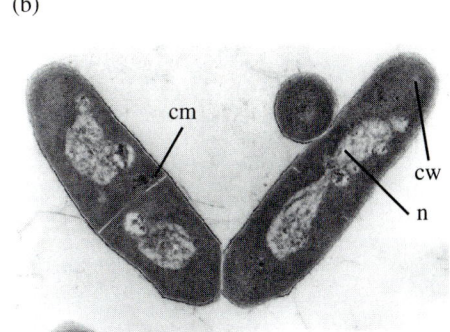

Figure V–2

The anatomy of a prokaryotic cell. (a) An outer view of two cells bending to separate from each other shown by scanning electron microscopy. (b) The internal view of similar cells showing various parts, including the cell wall (cw), *nucleoid* (n), and cell membrane (cm). New cell wall formation (arrows) also can be seen.
(From Takade A., K. Takeya, H. Taniguchi, and Y. Mizguchi, *J. Gen. Microbiol.* 129:2315, [1983].)

are characteristic of certain bacterial species. Ribosomes, a major submicroscopic component of bacterial cells associated with protein synthesis, are found in large numbers (Figure V–1*b*).

Some bacterial cell walls are surrounded by a layer composed of polysaccharide or a combination of polysaccharide and polypeptide, generally referred to as a *glycocalyx*. If the covering has a structural organization, it is called a *capsule*. If such organization is lacking, the layer is referred to as *slime*. Capsular material also serves as a basis for antigenic differentiation.

Many bacteria possess either one or two kinds of surface appendages. These structures are peripheral to the rigid wall and are known as *flagella* (singular, *flagellum*) and *pili* (singular, *pilus*). Flagella, the organelles of locomotion, originate within the cytoplasm but are primarily external structures. Bacterial flagella and movements influenced or controlled by them have been used as models for studying protein synthesis, organelle development, and molecular aspects of sensory reception (detecting environmental stimuli) and information processing. Flagella consist of protein subunits and have specific antigenic properties that aid in the identification of bacterial species. *Pili* are generally considerably thinner (submicroscopic) and more numerous than flagella on the bacterial species that possess them. Like flagella, pili consist of protein subunits and originate within the cytoplasm. From a functional standpoint, pili can be divided into two categories. One group, referred to as the *sex pili*, is associated with the transfer of nucleic acid (DNA) from one cell to another. The second group serves to attach bacteria to various surfaces and objects.

Endospores and *exospores* are distinctive structures seen with certain bacterial species. The capacity to form these structures is limited to members of certain genera. Endospore formation takes place within the vegetative cell, while exospore formation involves a budding process at one end of the vegetative cell. Highly ordered sequences of reactions and morphological changes occur during the spore formation process, which is referred to as *sporogenesis* or *sporulation*. Most bacterial spores are unusually resistant to environments and agents—including chemicals, drying, radiation, and temperature extremes—that would normally kill vegetative cells.

The exercises in this section include brief discussions and demonstrations of the more characteristic bacterial structures.

Table V–1

Prokaryotic Structures

Structure	Chemical Composition	Function(s)
capsule	polysaccharide or polypeptide	provides protection against phagocytosis and penetration of antibiotics
cell wall (generalized)	peptidoglycan, -acetylglucosamine, N-acetyl-muramic acid, polypeptides	protects against osmotic lysis; provide shape or form to cell
chromosome	DNA (single molecule)	information center for cellular activities, carrier of cell's genetic information
chromatophore	protein, lipid, and photosynthetic pigments	photosynthesis
cytoplasm	nucleic acids, proteins, lipids, carbohydrates, components of marcomolecules, and inorganic ions	serves as site of primary activities of cell
plasma (cytoplasmic) membrane	lipoprotein	regulates passage of materials, these include waste, and building materials
cyst	protein, polysaccharide	formed as a part of certain cells' life cycles
endospore	protein, polysaccharide, calcium, dipicolinic acid	formed as a part of certain life cycles
flagella	protein	provides for motility (independent movement)
metachromatic granules (volutin)	polymetaphosphate	reserve material
pili	protein	a. one type enables bacteria to attach to surfaces b. second type involved with the transfer of DNA
plasmid	DNA (circular)	contains information for nonessential functions, such as antibiotic resistance
poly-β-hydroxybutyrate	poly-β-hydroxybutyrate	reserve carbon and energy source
ribosome	RNA and protein	protein synthesis

After completing this exercise, you should be able to:

1. Carry out a standard procedure for the demonstration of bacterial spores.
2. Detect the presence of bacterial spores in a culture.
3. Distinguish between vegetative cells and bacterial spores.
4. Identify endospores in unknown cultures.

Various types of specialized, resistant, resting cells are produced by different bacterial species as part of their respective reproductive cycles (Figure 16–1). These include: the *cyst,* a dormant cell sometimes enclosed in a tubelike structure called a sheath; the *endospore,* a heat-resistant structure formed within a cell; the *exospore,* another heat-resistant cell formed by a budding process and found outside the cell; and the *myxospore,* a resting cell formed within the fruiting bodies of the myxobacteria. The primary focus of this exercise is the endospore (Figure 16–2).

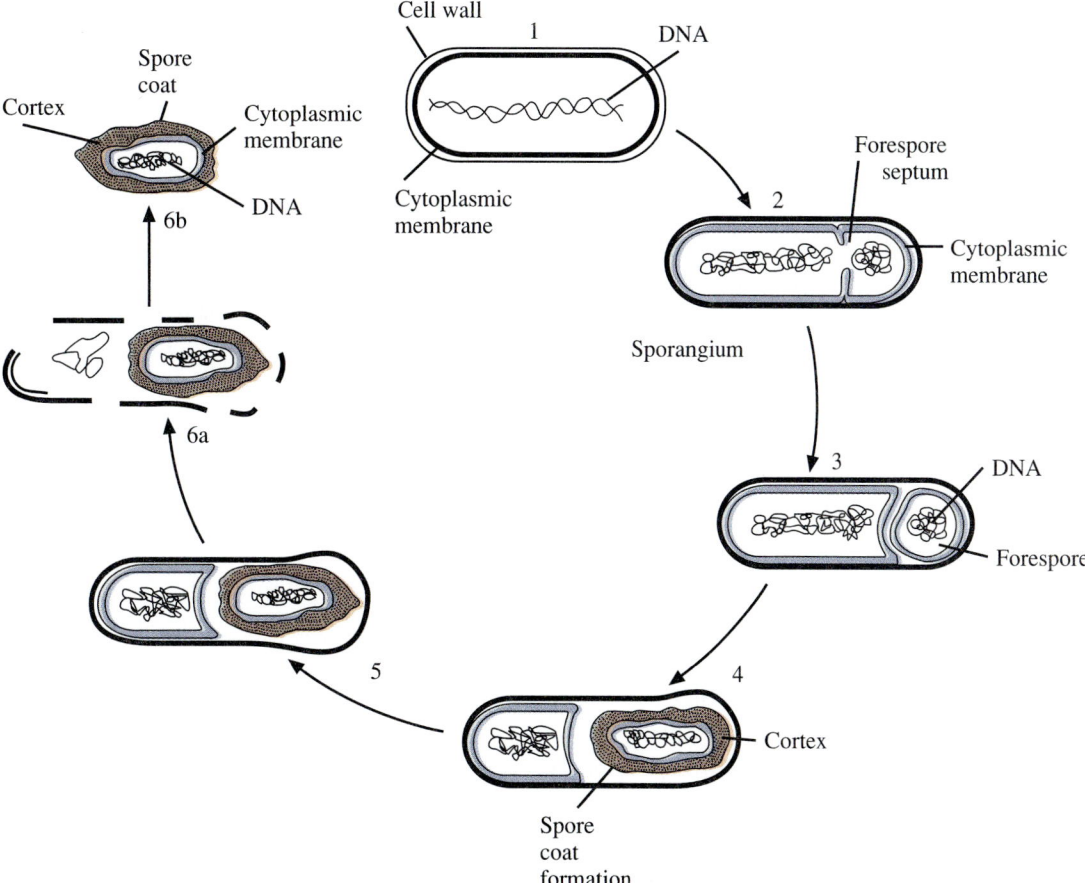

Figure 16–1

The steps of a sporeforming or sporulation process. 1. The vegetative cell; 2. DNA splits into individual chromosomes, the cytoplasmic membrane begins to invaginate, and a forespore septum begins to form; 3. DNA becomes completely surrounded by membrane and a forespore is formed; 4. Cortex and spore coat formation; 5. Spore cortex and coat formation complete; 6a. Sporagium lyses thereby releasing the endospore; 6b. The general parts of a mature spore.

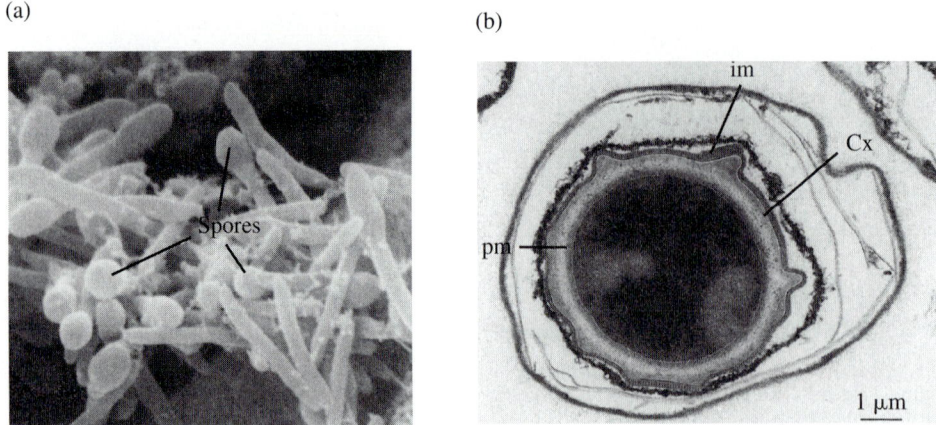

(a)   (b)

Figure 16–2

Endospores. (a) A scanning micrograph showing spore formers and non-spore formers. The terminal spores are quite evident.
(From Ducluzeau, R., M. Ladire, C. Calut, P. Raiband and G.D. Abrams, *Infect. Immun.* 17:415 [1971].)
(b) An electron micrograph showing the relationship of the endospore and the parent cell. The structures shown include the *inner spore membrane* (im); *cortex* (Cx); and *parent cell membrane* (pm).
(From Robinson, R.W., and C.R. Spotts, *Can. J. Microbiol.* 29:807–814, [1983].)

Bacterial endospores are so well known for their resistance to high temperature, radiation, desiccation, and chemical disinfection that one is always tempted to consider sporulation as a major factor in the perpetuation of bacteria. If spore-forming bacteria were the only microbial forms in existence, this statement would be correct. Furthermore, it was thought for many years that sporulation occurred when the cell was "confronted" with a "sporulate-or-perish" situation—threatened by starvation or by a toxic environment. Studies have shown the nutritional and environmental requirements and enzymatic activities for sporulation to be not too different from those initially needed for normal vegetative cell division. Later events in the endospore-forming cycle require the formation of biochemical compounds that are not formed during the vegetative cell cycle. These compounds are utilized in the formation of spore parts such as the cortex and coat, and to establish the highly environmentally resistant nature of the spore.

Spores aid in the survival of those organisms that can produce them. In the 1930s, Cook succinctly stated that "bacteria form spores, because they form spores." Essentially, the process may be considered as a primitive mode of differentiation and as a normal part of a life cycle. The latter is characteristic of certain higher plants. The development of a completed bacterial endospore involves the formation of inner and outer spore coats and a spore cortex (Figure 16–2b) within a parent cell. Once it is formed, the endospore can be released from the interior of the parent cell into the environment by cell lysis. In its cell-free state, the highly resistant spore can remain metabolically inactive for a prolonged period.

Bacterial endospore formation occurs in several genera, including *Bacillus, Clostridium, Desulfotomaculum, Sporosarcina,* and *Sporolactobacillus.* The species within these genera are, for the most part, rod-shaped. The genus of *Sporosarcina* contains coccus-shaped cells.

Exospore formation occurs with the purple, nonsulfur, photosynthetic bacterium *Rhodomicrobium vannielii* and members of the genus *Methylosinus.* While these structures are resistant to heat and drying, they are distinguished from endospores by their chemical composition, structural features, and process of formation.

This exercise will demonstrate the Schaeffer-Fulton technique for the differentiation of endospores and vegetative cells. In this procedure the aqueous primary dye, *malachite green,* is applied to a specimen and steamed to force the penetration of the dye through the relatively impermeable spore coats. Malachite green does not bind strongly to bacterial cell surfaces or interiors. Moreover, it easily penetrates bacteria cells and is easily removed by washing stained cells with water. The spore presents another situation. Once inside the spore (Figure 16–3a), malachite green can not be easily removed. After cells are steamed with the dye, then cooled, and followed with a water rinse, the spores remain green while the other cell parts and non-spore-forming cells become colorless. In order to demonstrate the presence of spores, a counterstain such as *safranin* is used to provide a contrasting color. The colorless cell parts of the sporangia and nonspore-forming cells appear red (Figure 16–3b). Adequately prepared spores will resist replacement by the counterstain. Often, one can observe green spores inside as well as outside red cells. The intracellular location of the spores of certain organisms may be determined and used for taxonomic purposes.

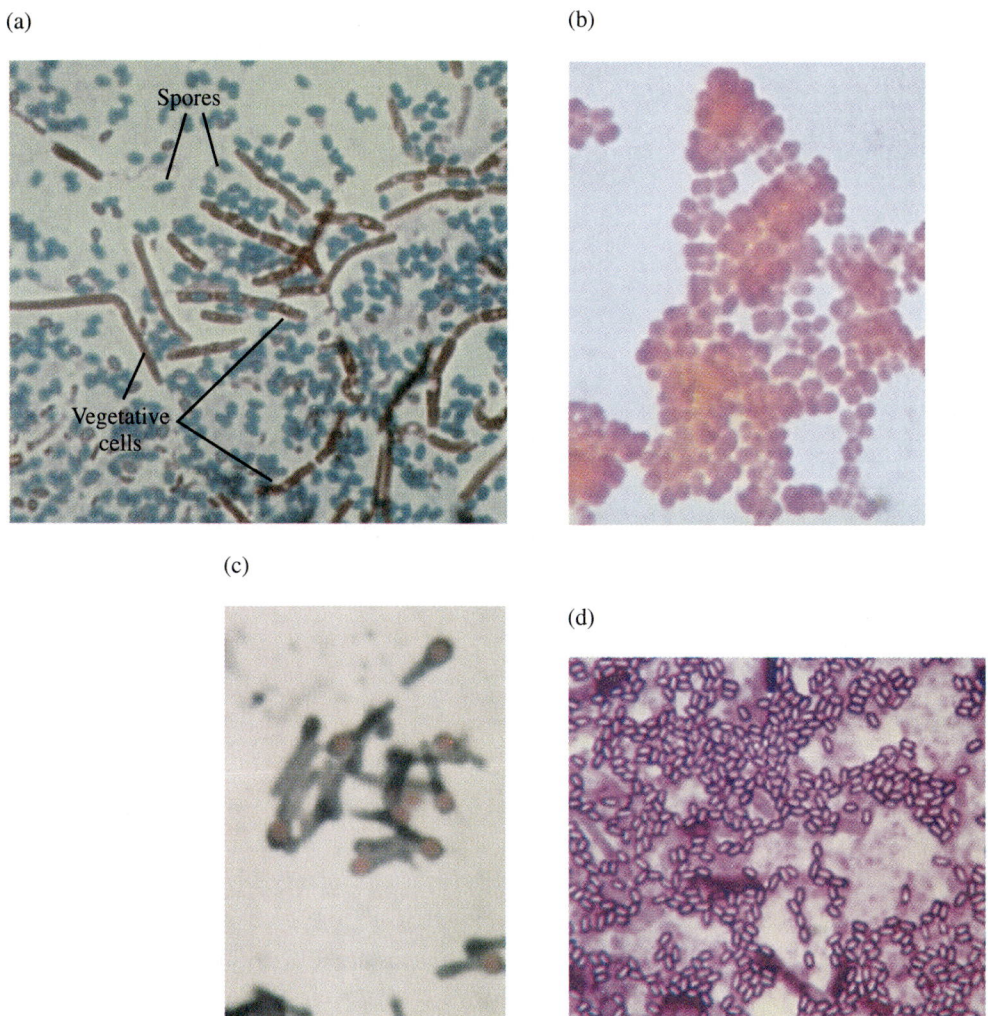

(a) (b)

(c)

(d)

Figure 16–3

Bacterial spore stains. a. The spores and vegetative cells of *Bacillus megaterium.* The preparation shown was stained by the Schaeffer-Fulton technique. Spores appear green, while vegetative portions of cells are red. Note the presence of spores (endospores) inside their respective vegetative cells. b. The appearance of a nonsporeforming culture. c. The spores of the anaerobic *Clostridium tetrani.* These heat-resistant structures are red and located terminally (end of cells). Vegetative portions of cells are stained blue. The primary and secondary stains used were carbol fuchsin and methylene blue, respectively. d. A simple stained spore culture.

Other dye combinations such as the primary stain carbol fuchsin and the counterstain methylene blue show the presence of spores. In this case, spores are red and sporangia stain blue (Figure 16–3c). Simple staining with crystal violet or safranin can be used as well to detect endospores. In this case, spores appear as intracellular or extracellular clear structures surrounded by stained borders (Figure 16–3d).

The species of *Bacillus* and *Clostridium* are gram-positive rods that are commonly found in the environment. Both genera contain members that play a pathogenic role in human infections and diseases of livestock and wildlife. *B. anthracis* (the cause of anthrax), is a well-known pathogen in the genus *Bacillus*. Several clostridial species are noted for their exotoxin production and associated diseases. Examples of this group include: *Clostridium botulinum*, the cause of botulism; *C. difficile*, the cause of certain cases of toxic enterocolitis; *C. perfringens*, one cause of gas gangrene, and some cases of food poisoning; and *C. tetani*, the cause of tetanus or lockjaw. Other clostridia are nonpathogenic and are found in soil and in the lower intestinal tracts of humans and lower animals.

Materials

The following materials should be provided per 4 students:

❑ 1. Forty-eight-hour nutrient agar slant cultures of the following:
  ❑ a. *Bacillus subtilis*
  ❑ b. *Staphylococcus aureus*

❑ 2. Four unknown cultures

❑ 3. One Schaeffer-Fulton staining set
  ❑ a. Malachite green
  ❑ b. Safranin

❑ 4. One staining rack or 4 large clothespins for heating slides

❑ 5. A sufficient number of paper towel squares (approximately 1.5 × 1.5 cm)

❑ 6. Forceps (for the removal of the paper squares)

A. Known Cultures

Procedure

This procedure is to be performed by students individually.

❑ 1. Prepare separate thin, air-dried, heat-fixed smears of the bacterial cultures provided. (If necessary, refer to Procedure Diagram 2 in Exercise 2.)

❑ 2. With each of the prepared smears, perform the 7 steps as described in Procedure Diagram 20.

❑ 3. Observe your results under oil immersion. (Refer to Figure 16–3.)

❑ 4. In the Results and Observations section, sketch the appearance, with the colors, of each culture.

Procedure Diagram 20
The Schaeffer-Fulton Spore Stain

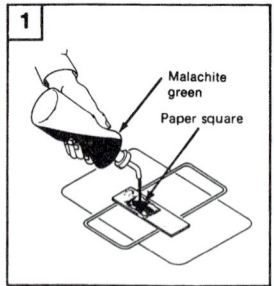

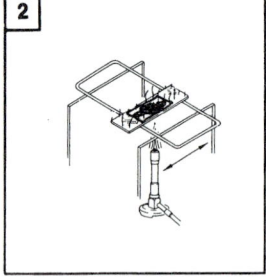

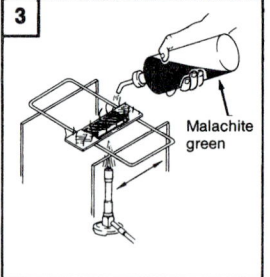

1. Place the slide on the staining rack. Cover the smear with a small paper towel square and then with malachite green. Allow the stain to stand for 30 to 60 sec. before heating.

2. Heat the preparation gently by passing the Bunsen burner under the slide. Continue heating until you see a slight steaming when the flame is removed.

3. Maintain steaming for 5 min. Add more dye as needed to prevent the smear from drying out. Be careful not to overheat the slide. Overheated slides may crack.

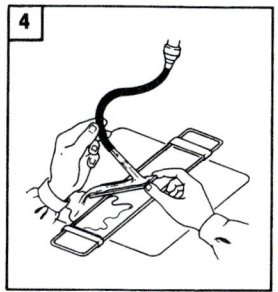

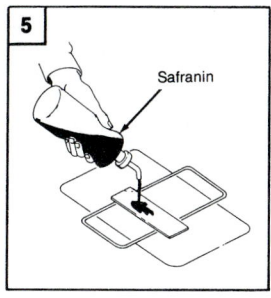

Safranin

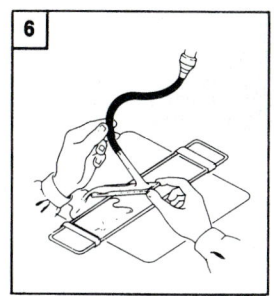

4. Allow the slide to cool, remove the paper towel square, and rinse the slide (front and back) in slowly running water.

5. Apply the counterstain, safranin, for 1 min.

6 Rinse the smear with running water as in step 4.

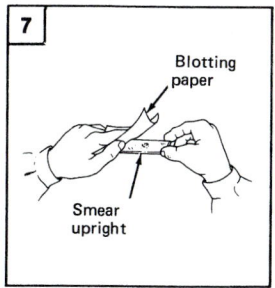

Blotting paper

Smear upright

7. Blot dry and observe as directed earlier.

B. Unknown Cultures

Procedure

This procedure is to be performed by students individually.

❑ 1. Prepare a heat-fixed smear of the unknown culture provided.

❑ 2. Process this smear in the same manner as for known cultures in Part A.

❑ 3. Enter your results on the Unknown Form (page 170) and give it to your instructor for grading.

Results and Observations

Sketch your results here.

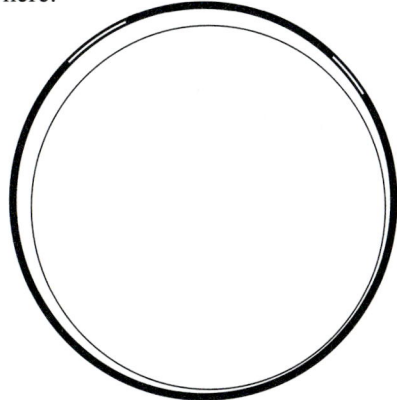

Bacillus subtilis

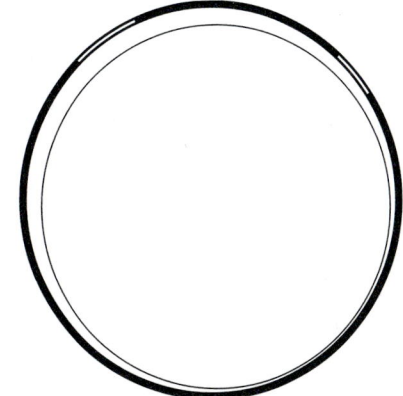

Staphylococcus aureus

Laboratory Review 16   Spores

1. a. What is the function of a bacterial spore? _____
    _____
    _____

    b. Briefly outline the general features of a bacterial spore cycle. Sketches may be used.

2. Are all anaerobic bacteria spore formers? _____
    _____

3. List 1 aerobic and 3 anaerobic spore-forming pathogenic bacteria and the disease states they cause.

    a. aerobic: _____

    b. anaerobic: _____
    _____
    _____

4. The first microorganism proven to be the etiologic (causative) agent of a disease was a spore former.

    a. What was the organism? _____
    _____

    b. How did Robert Koch demonstrate his proof? _____
    _____
    _____
    _____

5. a. Do nonpathogenic clostridia exist? _____
    _____

    b. Where can they be found? _____
    _____

6. List and briefly describe 3 other types of specialized, resistant, resting cells produced by bacteria.

a. _____

b. _____

c. _____

Key Terms

cyst (sist): a resting or dormant cell formed by some microorganisms during specific stages of their life cycles

endospore (en-DŌ-spōr): a thick-walled, generally highly heat-resistant, resting or dormant form developed within the cell by certain microorganisms during specific stages of their life cycles

exospore (eks-Ō-spōr): generally heat-resistant, resting or dormant forms developed externally from the cell through a budding process during specific stages of their life cycles

myxospore (miks-Ō-spōr): resting cells formed in the fruiting bodies of the myxobacteria

sporangium (spō-RAN-jē-um): a cell giving rise to a spore

sporulation (spōr-ū-LĀ-shun): the process of spore formation

vegetative (vej-e-TĀ-tiv) cell: an actively growing and/or metabolizing cell that does not serve as a specialized reproductive or resting form

---

Photograph Quiz 12

A bacterial culture was simple-stained. The results are shown. Did the culture contain spores? How can the finding be confirmed? _____

_____

_____

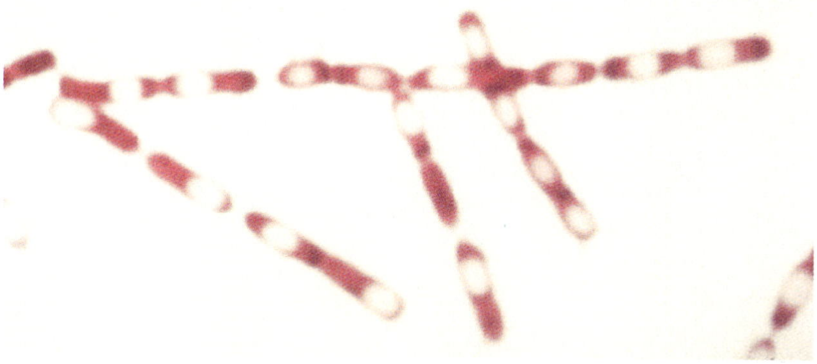

Figure 16–4
A simple stained specimen.

Photograph Quiz 13

In the transmission electron micrograph shown in Figure 16–5, identify the labelled structures.

a. _____   b. _____

c. _____   d. _____

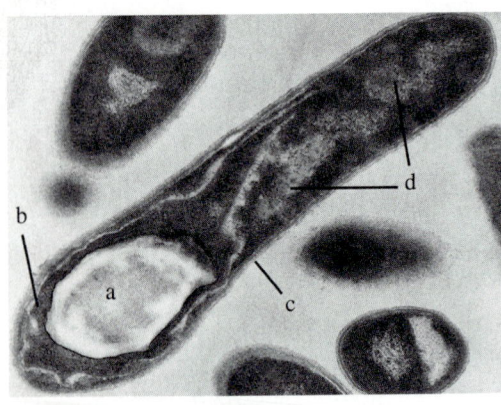

Figure 16–5
A transmission electron micrograph of a bacterial cell.

Unknown Form for Spores

Student's Name _____

Date _____

Unknown Number _____

Score _____

Laboratory Section _____

Results

(spores present or absent) _____

After completing this exercise, you should be able to:

1. Carry out a negative staining technique.

2. Differentiate between bacterial capsules and artifacts.

3. Identify capsules, if present, in unknown cultures.

The *glycocalyx* is currently defined as any polysaccharide-containing structure outside of the bacterial cell wall. It may be composed of fibrous polysaccharides or globular glycoproteins. (A few bacterial species produce such extracellular structures entirely composed of protein.) The term **glycocalyx** is generally used for a structured mass of extracellular polysaccharide that attaches tightly to the cell wall. Less organized and loosely attached extracellular polysaccharide is referred to as a *slime*. One type of glycocalyx is the **capsule** (Figure 17–1). Capsules vary in thickness, may be rigid or flexible, and may or may not be closely associated with the bacterial cell surface.

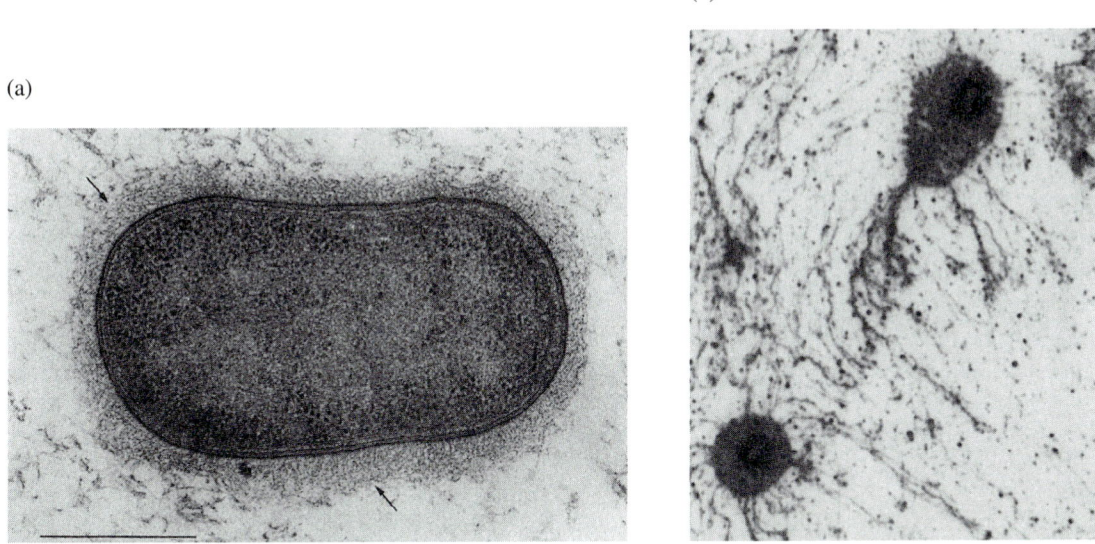

(a)

(b)

Figure 17–1

The bacterial capsule. (a) An electron micrograph showing the bacterial surface covered with a thick capsule. Note that capsule contains a number of fine fibers. The bar marker = 0.5 μm.
(From Meno, Y., and Associates, Kyushu Univ., *Inf. Imm. 58:*1421–1428, [1990].)

(b) A special technique using immunoglobulins (antibodies) to demonstrate the presence of capsules. Capsules can be seen as clear areas around individual cells. The magnification used = 10,000X.
(From Wacharotayankuh, R., Y. Arakawa, M. Ohta, K. Tanaka, J. Akashi, M. Mori, and N. Kato *Inf. Imm.* 61:3164–3174, [1993].)

The glycocalyx has several important functions. With this structure, bacteria can adhere to other bacteria and to the surface of inert materials (soil, sand, etc.) and of animal and plant cells. In this way, bacteria form microcolonies that are the major type of bacterial growth in nature and in several diseases. The glycocalyx also provides bacteria with some protection from antibacterial agents such as antibiotics, bacteriocins, bacterial viruses, immunoglobulins, and phagocytes. This exercise is concerned with one type of glycocalyx: the bacterial capsule.

Bacterial capsules are indistinct in some organisms and well developed in others. The latter include *Streptococcus pneumoniae, Clostridium perfringens,* and *Klebsiella pneumoniae.* Capsules appear to increase the virulence of organisms by protecting them from the defense mechanisms of their hosts. In addition, capsules impart specific immunologic properties to some microorganisms. Pneumococci, for example, are differentiated on the basis of the antigenic character-istics of their capsules, which are polysaccharide in nature. The presence of pneumococcus capsules is demonstrated by means of a serologic test known as the *Quellung reaction,* first described by Neufeld in 1902. The reaction is character-ized by the occurrence of capsular swelling when pneumococci of a specific type are mixed with homologous antisera. In this exercise, the presence of capsules will be demonstrated by a simpler, nonspecific method called the *negative stain* (Figure 17–2).

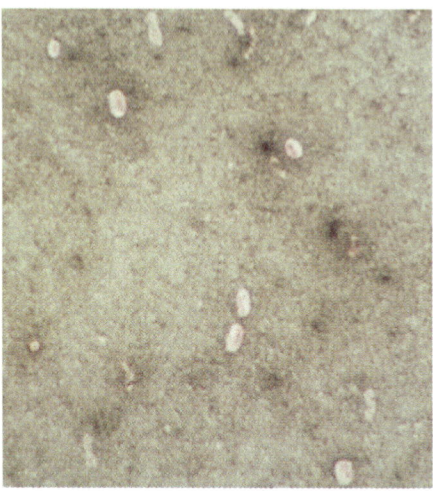

Figure 17–2

The capsules of *Enterobacter aerogenes.* In the procedure used to demonstrate this bacterial structure, a combination of India ink and safranin was used. The capsules (clear areas) are seen against the dark background. Individual organisms are red.

The negative stain technique incorporates material such as **India ink,** which is composed of particles too large to enter a cell. A small amount of culture is mixed with India ink, and the resulting smear is stained with a dye, **safranin,** which penetrates the bacterial cell. Upon examination of the preparation, the capsule will appear as a clear zone surrounding the cell wall. One cannot state with certainty that all the clear zones observed are capsules, be-cause shrinkage of cells or withdrawal and cracking of the India ink may cause irregular results. In general, however, when a treated smear contains many uniformly shaped clear zones, it is probable that they are actual capsules and not artifacts.

Materials

❏ 1. Twenty-four-hour glucose broth cultures of:
    ❏ a. *Enterobacter aerogenes* (en-ter-ō-BAK-ter ER-ō-jen-ēz) (1 per 4 students)
    ❏ b. *Staphylococcus aureus* (1 per 4 students)
    ❏ c. Unknown culture (1 per student)

❏ 2. India ink or nigrosin

❏ 3. Safranin

❏ 4. Glass slides

❏ 5. Inoculating loop

A. Known Culture

Procedure

This procedure is to be performed by students individually.

❑ 1. With each culture provided, perform the 9 steps described in Procedure Diagram 21.

❑ 2. Observe your results under oil immersion. (See Figure 17–2 to help you in the identification of capsulated organisms.)

❑ 3. In the Results and Observations section, sketch the appearance with the color for each culture used.

Procedure Diagram 21
The Capsule Stain

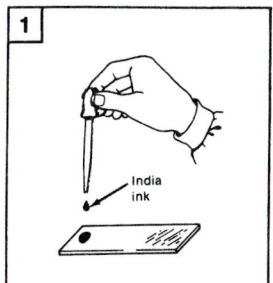

1. Place a small drop of India ink or nigrosin near one edge of a clean slide.

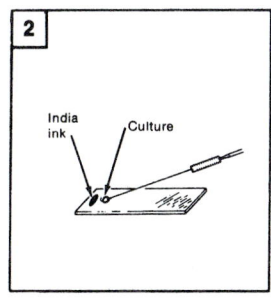

2. Aseptically remove a loopful of culture and mix it well with the India ink.

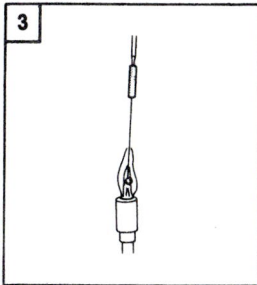

3. Flame the loop and put it in an appropriate place.

4. Take a fresh, clean slide, touch one edge to the mixture, and allow the mixture to run across the width of the slide.

5. Hold the second slide (spreader) at about a 30-degree angle and push the mixture toward the opposite end of the first slide.

6. Allow the newly formed smear to air-dry.

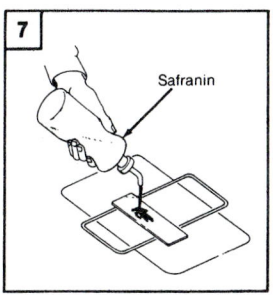

7. Cover the smear with safranin for 30 sec.

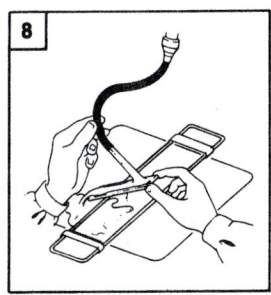

8. Pour-off the safranin and gently rinse the smear with running water. (*Note:* excess rinsing will remove the mixture.)

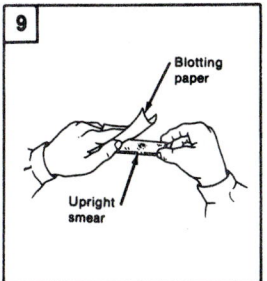

9. Blot dry and examine as indicated earlier.

B. Unknown Cultures

Procedure

This procedure is to be performed by students individually.

- ❑ 1. Perform this procedure with the unknown culture provided.
- ❑ 2. Enter your results on the Unknown Form (page 176) and give it to your instructor for grading.

Results and Observations

1. Sketch your results here.

   a. Known cultures

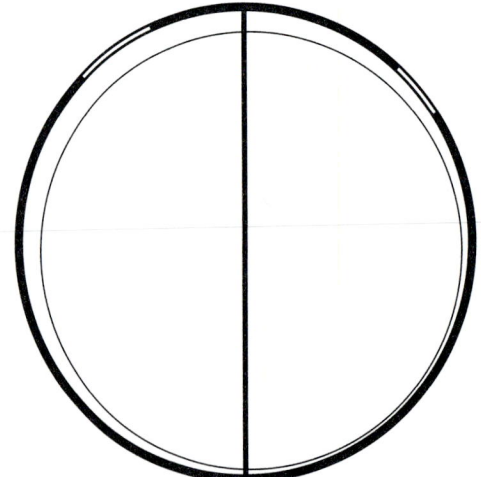

Enterobacter aerogenes   Staphylococcus aureus

   b. Unknown culture

2. Why is heat not used in this capsule staining procedure? _____

_____

Laboratory Review 17   Capsules

1. How important are capsules in establishing a disease process? _____
_____
_____

2. Are capsules found with both gram-positive and gram-negative bacteria? _____
_____

3. List 3 specific functions of capsules.

   a. _____

   b. _____

   c. _____

4. Can the presence of a capsule always be correlated with virulence? If not, give an example.
_____
_____

5. a. What is a bacterial glycocalyx? _____
_____

   b. What is a bacterial slime? _____
_____

Key Terms

bacteriocin (bak-TĒ-rē-ā-sin): a bacterial antimicrobial product that kills sensitive cells of related strains

capsule (CAP-sul): generally, a compact polysaccharide layer found exterior to the cell walls of certain bacteria; represents one form of glycocalyx

glycocalyx (GLĪ-kō-KĀL-iks): a general term for polysaccharide and/or protein cell components found outside of the cell wall

negative stain: a staining procedure in which the background surrounding a specimen is stained, but the specimen is not

virulence (VIR-ū-lens): the relative ability of an organism to cause disease

Photograph Quiz 14

Identify the labeled structures shown in the micrograph.

a. _____ b. _____

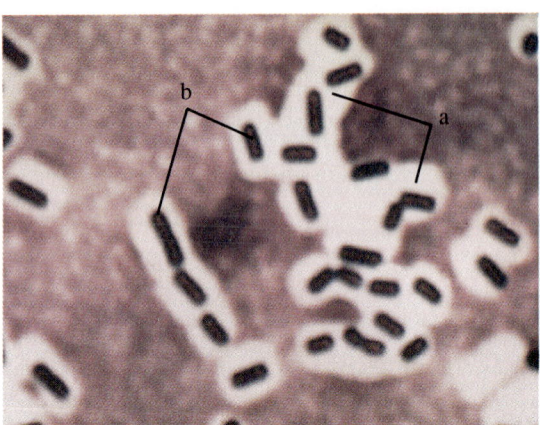

Figure 17–3

A capsule stain preparation.
(Courtesy of Bradshaw-Rouse, J.J., M.H. Watley, D.L. Coplin, A. Woods, L. Sequeira, and A. Kelman.)

Unknown Form for Capsules

Student's Name _____   Score _____

Date _____   Laboratory Section _____

Unknown Number _____   Result _____

After completing this exercise, you should be able to:

1. Demonstrate the presence of flagella on bacteria.
2. Distinguish among the bacterial arrangements of flagella.
3. Use motility media to show bacterial motility.
4. Interpret motility media reactions.
5. Interpret flagella staining reactions with unknown cultures.

Two general types of surface filamentous appendages can be found with several bacterial species. These are the submicroscopic pili (also known as *fimbriae*) and flagella. Pili are thinner, shorter, and straighter than flagella. Figure 18–1*a* shows these differences, while Figure 18–1*b* demonstrates two types of pili, type F_1 and F_S. The F_S type is also known as the sex pili and can be detected by the attachment of specific RNA bacterial viruses or bacteriophages. At the present time, several different pili types are recognized. Chemically, they are protein, and function in the transfer of genetic material or aid in the attachment of bacteria to surfaces.

Flagella are thin, protein, hairlike organelles of locomotion originating in the cytoplasmic region just beneath the cell wall. The wide use of electron microscopy has shown the bacterial flagellum to consist of three morphologically and chemically distinct parts: the *filament, hook,* and *basal structure* (Figure 18–2). Not all organisms are motile or have these structures. Motility is not essential to survival. Nevertheless, important functions are served by the movement of flagella. These functions include escaping from predators or unfavorable environments, and entering into symbiotic relationships.

(a)

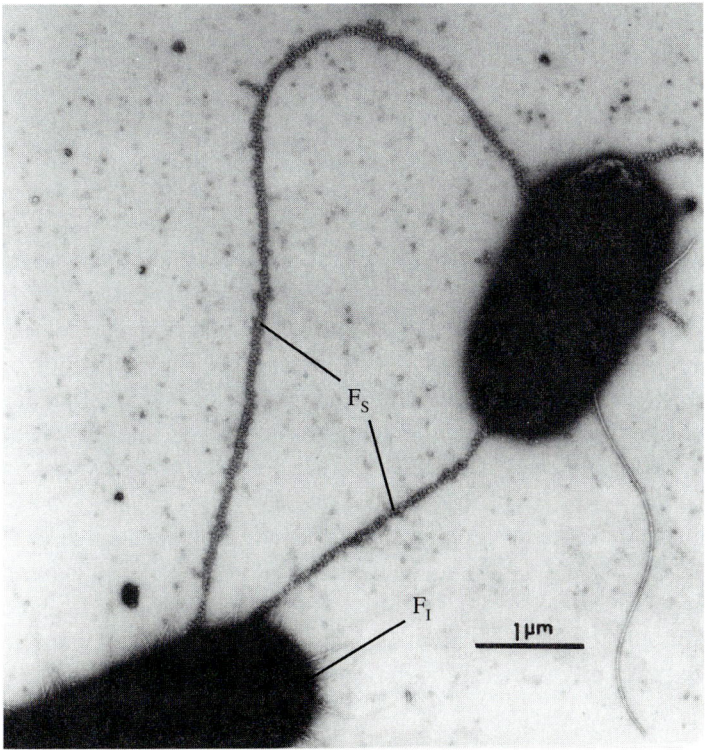

Figure 18–1

Pili and flagella. (a) A negatively stained preparation of *Escherichia coli* showing type 1 (F_1) and sex (F_S) pili. The latter have spherical RNA bacteriophages attached to their surfaces.
(Courtesy of Curtiss, R., III, L.G. Caro, D.P. Allison, and D.R. Stallions, *J. Bacteriol.* 100:1091–1104 [1969].) (Continued)

(b)

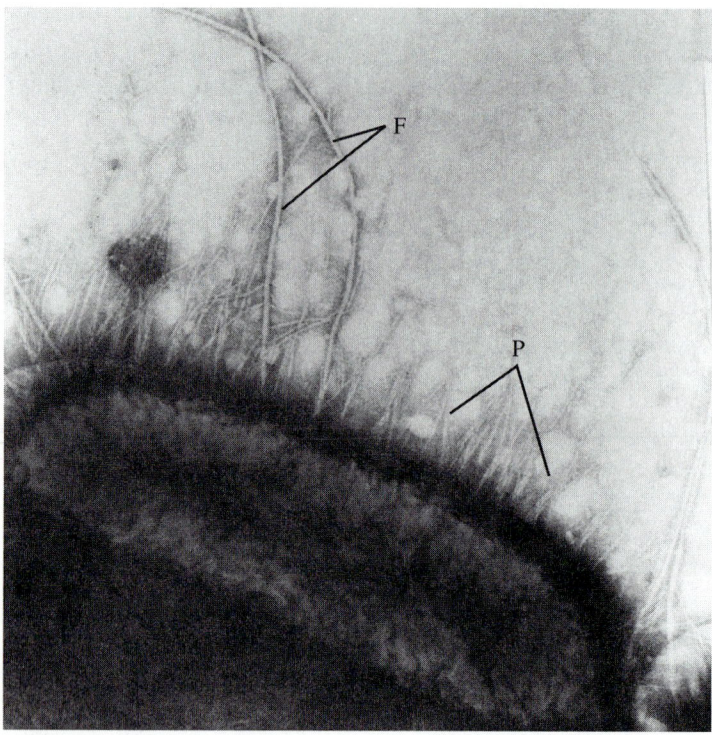

Figure 18–1 (continued)

Pili and flagella. (b) An electron micrograph of a negatively stained preparation of *Proteus rettgeri* showing the differences in appearance of pili (P) and flagella (F).

Movement toward a favorable environment, such as a source of nutrients, is referred to as *positive chemotaxis,* while movement away from an unfavorable environment, such as harmful chemicals, is referred to as *negative chemotaxis.* The presence of flagella, as well as the pattern of attachment and the number possessed by microorganisms, are useful characteristics for classification. Particular patterns of flagellation are shown in Figure 18–3 and include:

1. *atrichous:* no flagella present
2. *monotrichous:* a single flagellum at one end of a cell
3. *lophotrichous:* a cluster of flagella at one end of a cell
4. *amphitrichous:* either single flagella or clusters at both ends of a cell
5. *peritrichous:* flagella inserted over the entire surface of a cell

While it is difficult to observe bacterial flagella in their natural state, they can be seen using light microscopy when dye and mordant are deposited on them, thereby increasing their diameter. The lengths of bacterial flagella are often found to be several times the length of the cells possessing them. However, their diameters range from approximately 0.01 to 0.05 μm, thereby making them too small to be seen with the ordinary laboratory microscope. The application of the staining procedure developed by Einar Leifson in 1954 can make flagella visible in laboratory microscopes. The procedure uses **tannic acid,** which serves as a mordant for the red dye, **basic fuchsin.** The mordant-bound dye makes the flagella appear thicker so that they can be seen with the light microscope. This technique can be used to demonstrate a single flagellum or bundles of flagella called *fasicles* (Figure 18–4).

Bacterial motility also can be shown with the aid of several different types of **motility media.** The composition of these preparations is such that they offer no more resistance to movement during incubation than would a broth culture. Motility Medium S, which contains 2,3,5-triphenyltetrazolium chloride (TTC), is used in this exercise. Motility can be recognized by the presence of diffuse red growth away from the line of inoculation or stabline (Figure 18–5). Nonmotile organisms grow only along the stabline. Media of this type generally work best with facultative anaerobes.

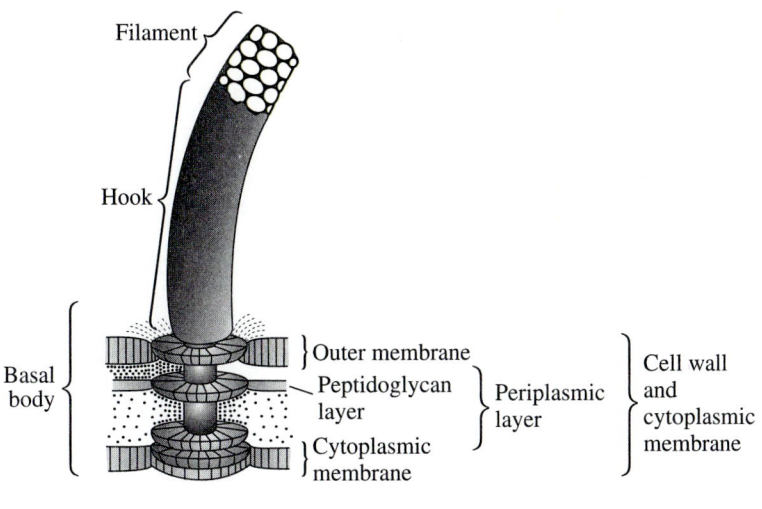

Filament

Hook

Basal body

Outer membrane
Peptidoglycan layer
Cytoplasmic membrane

Periplasmic layer

Cell wall and cytoplasmic membrane

10 nm

Figure 18–2

A diagrammatic view of flagellar structure based on electron microscopic examination. This arrangement is found with gram-negative bacteria.

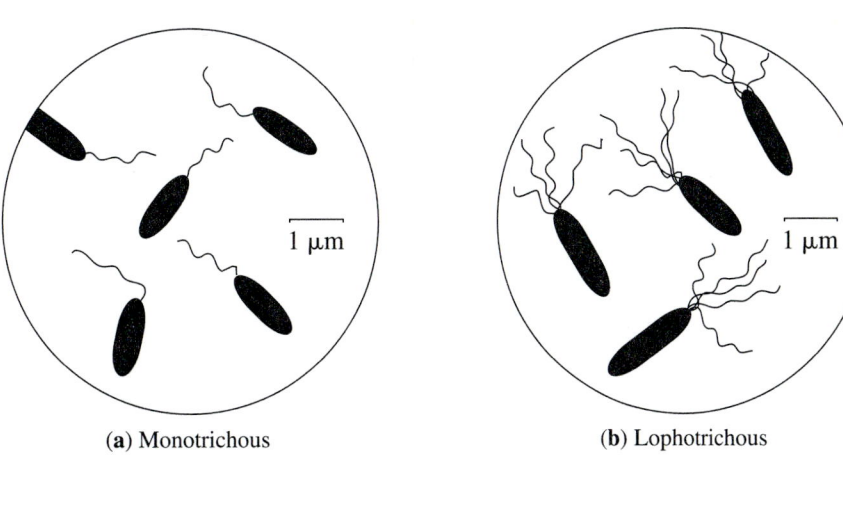

1 μm

(a) Monotrichous

1 μm

(b) Lophotrichous

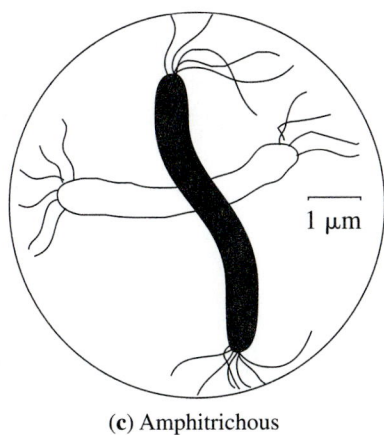

1 μm

(c) Amphitrichous

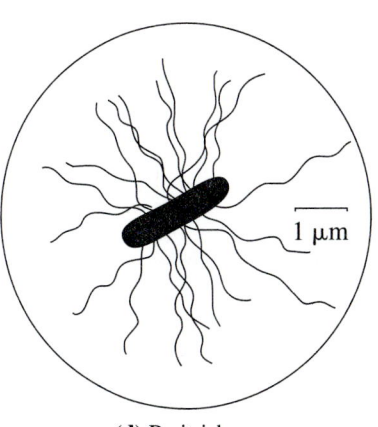

1 μm

(d) Peritrichous

Figure 18–3

Bacterial flagellar arrangements. (a) *Monotrichous,* a single flagellum. (b) *Lophotrichous,* a cluster of flagella. (c) *Amphitrichous* flagella, either a single flagellum or clusters of flagella, at both ends. (d) *Peritrichous,* flagella inserted over entire cell surface.

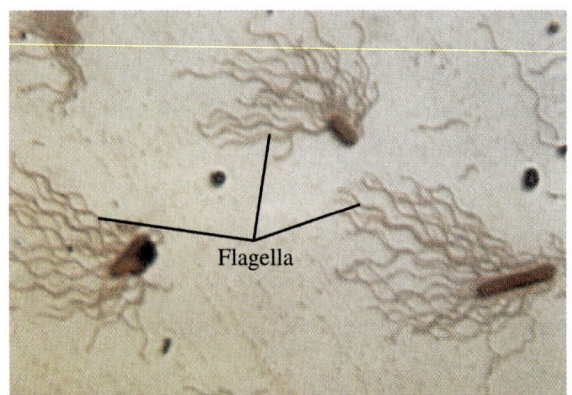

Figure 18–4

Salmonella typhi, stained especially to demonstrate flagella. These organelles are not invisible in ordinary stained preparations.

(a)

Figure 18–5

Motility medium. This preparation, to which a dye called *2,3,5-triphenyltetrazolium chloride* is added, can be used to detect motile organisms. (a) Examples of patterns that can be obtained with bacterial cultures include: (1) nonmotile culture, growth only appears along the line of inoculation; and (2) motile culture, growth spreads from the line of inoculation making most of the medium turbid or cloudy. (b) The use of the motility medium. Motility is indicated by a reddish-brown pattern spreading away from the line of inoculation. Nonmotile organisms grow along the stab line. One uninoculated and two inoculated tubes are shown.

Line of inoculation

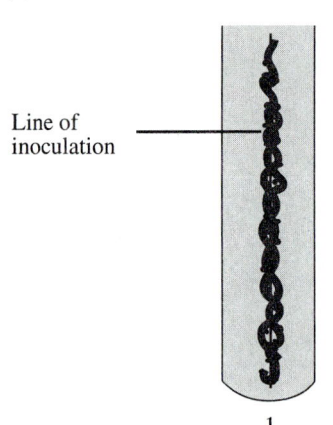

1   2

(b)

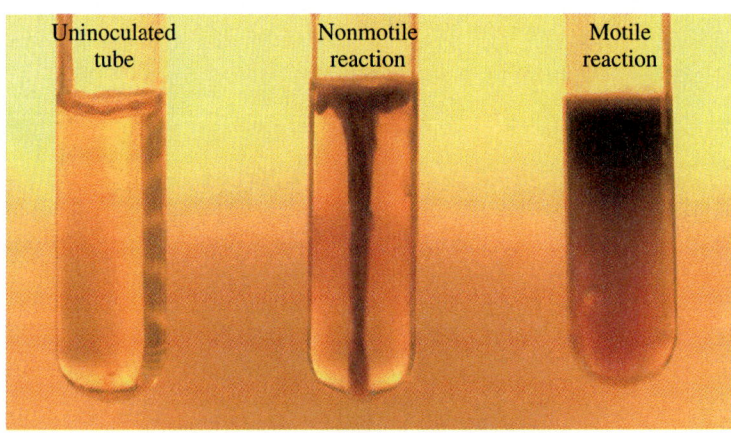

A. Observing Flagella

Materials

The following items should be provided for general class use:

❑ 1. Prepared slides for demonstration of flagella
  ❑ a. *Salmonella typhi* (flagella stain)

Procedure

This procedure is to be performed by students individually.

❏ 1. Examine and study the demonstration slides.

❏ 2. Sketch a representative field in the Results and Observations section. Refer to Figure 18–4 for a representative view of bacterial flagella.

B. Flagella Staining of Known Cultures

Materials

The following items should be provided per 4 students:

❏ 1. Twenty-four-hour trypticase soy agar slant cultures of the following:
   ❏ a. *Bacillus subtilis*
   ❏ b. *Proteus vulgaris*
   ❏ c. *Pseudomonas aeruginosa*

❏ 2. Three tubes each containing 5 mL of sterile distilled water

❏ 3. One staining bottle of Leifson stain

❏ 4. Acid-cleaned slides

❏ 5. One staining rack or 4 large clothespins

Procedure

❏ 1. With the aid of a loop, aseptically remove a small amount of growth from one of the culture slants and introduce it into a tube of sterile distilled water. The suspension should be only slightly cloudy.

❏ 2. Next, place 1 drop of the suspension on the center of 3 different slides. Allow the preparation to run down the length of each slide. Use your loop to move the suspension, if necessary.

❏ 3. Slant the slides and allow the preparations to air-dry. *Do not fix.*

❏ 4. Place 1 preparation on a level staining rack or hold it with a clothespin and cover the smear with the staining solution.

❏ 5. Allow the stain to remain for 5 minutes.

❏ 6. Gently rinse the slide with tap water.

❏ 7. Air-dry the slide in an upright position. *Do not blot dry.*

❏ 8. When the slide is dry, examine the preparation under the oil-immersion objective for the presence of flagella.

❏ 9. Repeat the staining procedure with the remaining 2 slides. Stain 1 slide for 10 minutes and the other for 15 minutes.

❏ 10. Repeat steps 1 through 9 with the other bacterial cultures provided.

❏ 11. Enter your findings and answer the questions in the Results and Observations section.

❏ 12. Sketch a representative view of all 10-minute stained slides in the Results and Observations section.

C. Flagella Staining of Unknown Cultures

Materials

The following items should be provided per student:

- ❏ 1. One sterile distilled water suspension of *Bacillus subtilis*
- ❏ 2. One unknown bacterial culture suspended in sterile distilled water
- ❏ 3. One bottle of Leifson stain
- ❏ 4. Acid-cleaned slides
- ❏ 5. One staining rack or one large clothespin

Procedure

- ❏ 1. Each student will receive an unknown culture.
- ❏ 2. Perform the flagella staining procedure as described earlier in this exercise.
- ❏ 3. It is suggested that a separate smear of a known flagellated organism be included on the same slide as the unknown. (This approach will serve as a check on your technique and a guide to identifying flagellated cells.)
- ❏ 4. Enter the results obtained with your unknown culture on the Unknown Form on page 185, and give it to your instructor for grading.

D. Motility Agar

Materials

- ❏ 1. Forty-eight-hour nutrient broth cultures (in 1-mL quantities) of the following bacterial species (1 tube per 4 students):
  - ❏ a. *Eacillus subtilis*
  - ❏ b. *Enterobacter aerogenes*
  - ❏ c. *Escherichia coli*
  - ❏ d. *Micrococcus luteus*
- ❏ 2. Bacto Motility Medium S with 2,3,5-triphenyltetrazolium chloride (1 tube per student)

Procedure

This procedure is to be performed by students in groups of 4.

- ❏ 1. Inoculate the motility medium as shown in Procedure Diagram 22.
- ❏ 2. Repeat the procedure with each culture provided.
- ❏ 3. Incubate the preparations at 37°C for 24 hours or as directed.
- ❏ 4. After incubation, examine the tubes and note the growth patterns. (Refer to Figure 18–5.)
- ❏ 5. Answer the questions in the Results and Observations section.

**Procedure Diagram 22
Motility Agar Inoculation**

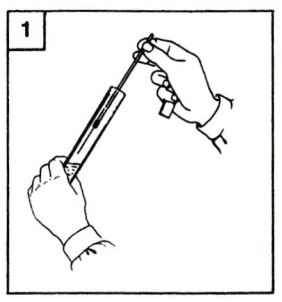

1. Using an inoculating needle, aseptically remove an inoculum from one of the cultures provided. (Refer to Procedure Diagram 2.)

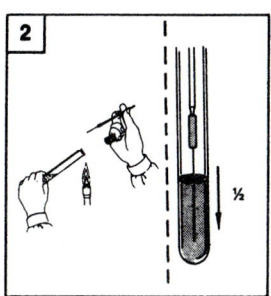

2. Flame the lip of the motility agar tube and stab the needle into the center of the medium to about one-half of the way down. Motile organisms will move into the uninoculated region.

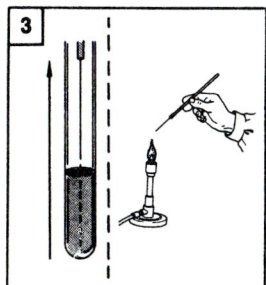

3. Remove the needle through the same stab line and flame the tube and inoculating needle in the usual manner.

Results and Observations

1. Observing flagella. Sketch your observations with 10-minute stained preparations in the spaces provided.

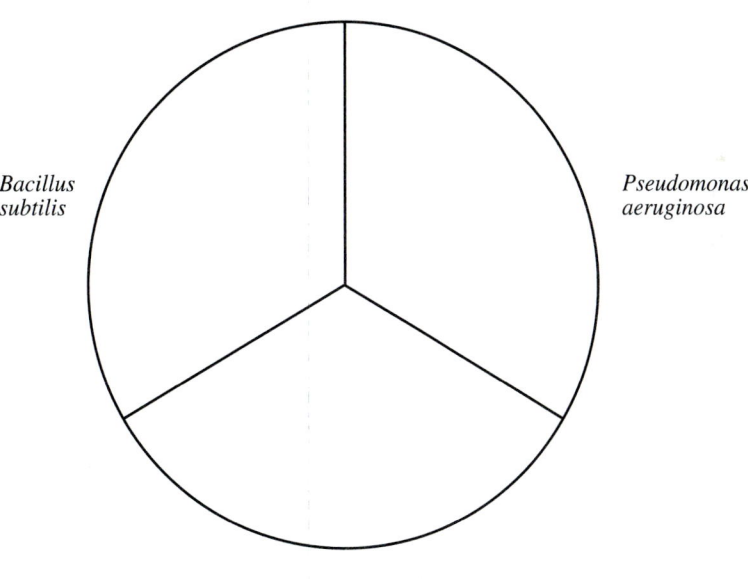

2. Flagella staining of known cultures.

    a. Enter your findings in Table 18–1. Indicate the presence of flagella by a "+" and the absence by a "−".

Table 18–1

Flagella Staining

Organism	Staining Time (in Minutes)		
	5	10	15
B. subtilis			
P. vulgaris			
P. aeruginosa			

b. Were flagella visible in all stained preparations? If not, did longer staining times increase flagella detection?

_____

_____

_____

3. Motility agar reactions.

a. Why was the motility medium inoculated only one-half of the way down the tube? _____

_____

b. Which of the cultures were motile? _____

_____

c. What other methods are available to demonstrate motility? _____

_____

Laboratory Review 18   Pili, Flagella, and Motility

1. Distinguish between bacterial pili and flagella. _____

_____

2. List and briefly explain four different flagella arrangements found among bacteria. (Sketches may be used.)

a. _____

b. _____

c. _____

d. _____

3. What is chemotaxis? _____

_____

4. Differentiate between vital and Brownian movements. _____

_____

_____

5. Can the possession of flagella by bacteria be utilized for classification purposes? Explain. _____

_____

_____

6. Briefly explain how the motility medium demonstrates motility. _____

_____

_____

Key Terms

Brownian movement: a vibrating movement resulting from a bombardment of surrounding molecules

chemoreceptors (kē-mō-rē-SEP-tors): in prokaryotes, this term refers to specific proteins within plasma (cell) membranes that respond to environmental factors (chemicals) and direct chemotactic activity

Leifson staining technique: a procedure for the staining of flagella, which uses tannic acid as a mordant and safranin or related dye as the stain

mordant (MOR-dant): a substance that fixes (holds firmly) a stain or dye

motility: the property of a cell to move a definite distance by its own power

negative chemotaxis (kē-mō-TAK-sis): movement away from a favorable environment

positive chemotaxis: movement toward a favorable environment

Unknown Form for Flagella Staining

Student's Name _____   Score _____

Date _____   Laboratory Section _____

Unknown Number _____   Result _____

Photograph Quiz 15

Determine the flagellar arrangements shown in the following micrographs.

a. _____   b. _____

(a)

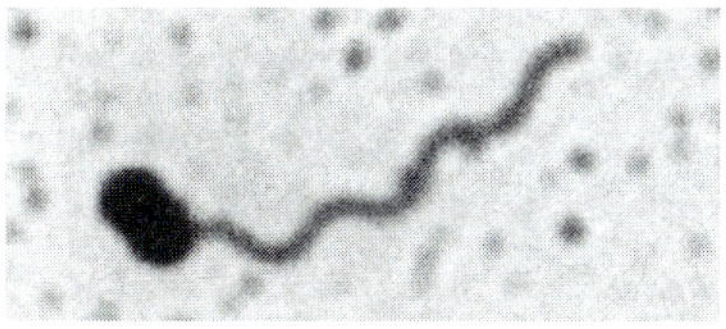

(b)

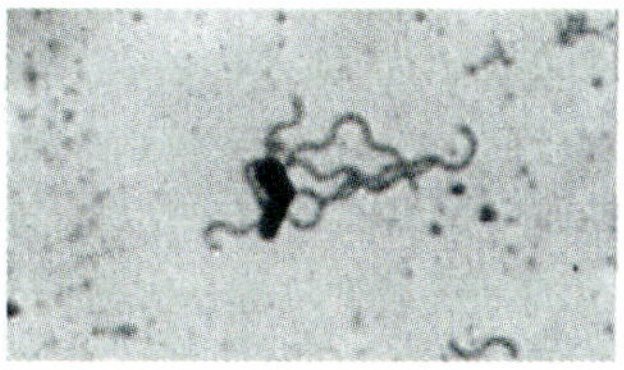

Figure 18–6
Flagellar arrangements.

After completing this exercise, you should be able to:

1. Identify both microscopic and submicroscopic features of bacteria cells.

2. Identify the chemical composition and functions of major bacterial cell components.

Many structures of bacteria can be observed under the proper conditions, particularly when staining procedures are performed by an experienced bacteriologist. Some of these cellular parts are common to all cells, while others are restricted to certain species. The electron microscope and specialized techniques have provided additional means for studying bacterial structures.

The cell wall (Figure 19–1a) is the structure responsible for the physical integrity of the cell. The wall's durability can be seen in its ability to resist considerable osmotic pressures before rupturing. Staining procedures have been developed in which a precipitate of dyes is deposited around the cell wall to make it visible under a microscope. It was thought for a long time that this procedure actually stained the structure. But the nature of the technique and the nature of the walls are now seen to contradict this view.

Bacterial cells accumulate several kinds of water-insoluble reserve materials, including metachromatic granules and lipid inclusions. The choice of methods used for staining these substances is based on their chemical nature. The actual chemical composition of metachromatic granules (or volutin) has been in dispute for some time. Certain investigators hold that volutin is polymerized inorganic phosphate. Others contend that the granules are composed of nucleic acid, lipid, and protein. In any case, when volutin-containing cells are stained with aged methylene blue, the granules take on reddish-blue and violet tones, thus producing the phenomenon called *metachromasia*. The age of the dye preparation is important. As the dye ages, dimers and trimers of methylene blue molecules are formed. These are responsible for the multiple-staining effect. Metachromatic granules are used to identify *Corynebacterium diphtheriae,* the causative agent of diphtheria (Figure 19–2).

Another substance found with certain bacteria is the lipid material called *poly-β-hydroxybutyrate* (**PHB**). PHB serves as a reserve carbon and energy source, and can be easily stained with fat-soluble dyes such as the Sudan series.

(a)

(b)

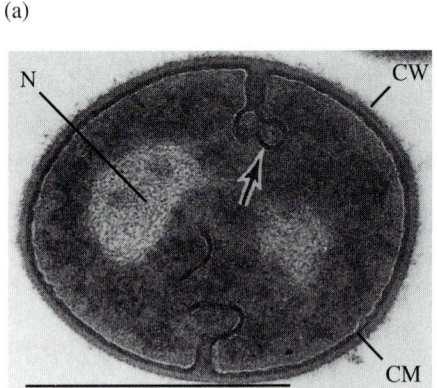

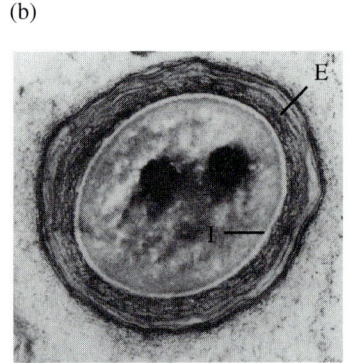

Figure 19–1

Micrographs of various bacterial structures. (a) An ultrathin section of *Staphylococcus aureus* with septal vesicular **mesosomes** (large arrow). These structures are believed to be artefacts produced during specimen preparation. Other structures shown include the **cell wall** (CW), **cell membrane** (CM), and **nuclear area, or nucleoid** (N). Bar marker represents 1.0 μm.
(Courtesy of Popkin, T.J., T.S. Theodore, and R.M. Cole, J. Bacteriol 107:907 [1971].)

(b) A thin section of an *Azotobacter vinelandii* **cyst.** The **intine** (I) and **extine** (E) layers are shown.
(Courtesy of Pope, L.M., and O. Wyss, J. Bacteriol. 102:234–239 [1970].)

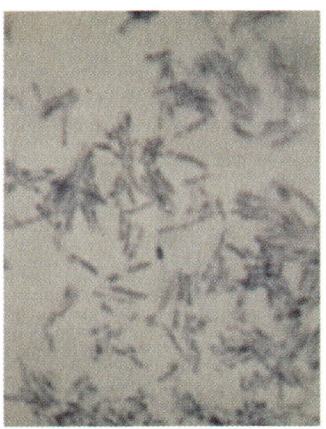

Figure 19–2
The **metachromatic granules** of *Corynebacterium diphtheriae*. Note the rather granular appearance of these cells.

Members of the genus *Azotobacter* (well-known nonsymbiotic, nitrogen-fixing organisms) during their cyclic process of cellular differentiation form an unusual type of resting structure called a *cyst* (Figure 19–1*b*). Cysts, in contrast to bacterial endospores (Exercise 16) are not heat-resistant, nor are they totally dormant. The Vela-Wyss staining procedure is an effective means of differentiating cysts from vegetative cells. Generally, a cyst can be recognized by its green center and red outer region.

Several of the bacterial structures mentioned will be demonstrated in this exercise.

Observing Bacterial Structures

Materials

❑ 1. Prepared slides for demonstration of:
    ❑ a. *Corynebacterium* (kō-rī-nē-bak-TĒ-nē-um) *diphtheriae,* alkaline methylene blue stain for metachromatic granules
    ❑ b. *Spirillum volutans* (spī-RIL-um, vol-Ū-tanz) stained with Sudan black B for lipid granules
    ❑ c. *Azotobacter* (a-zō-tō-BAK-ter) stained by the Vela-Wyss procedure

Procedure

This procedure is to be performed by students individually.

❑ 1. Examine and study the demonstration slides.

❑ 2. Sketch a representative field in the Results and Observations section. (See Figure 19–2 for a representative photograph of metachromatic granules.)

❑ 3. Answer the questions in the Laboratory Review section.

Results and Observations

1. Observing bacterial structures. Sketch your observations in the spaces provided.

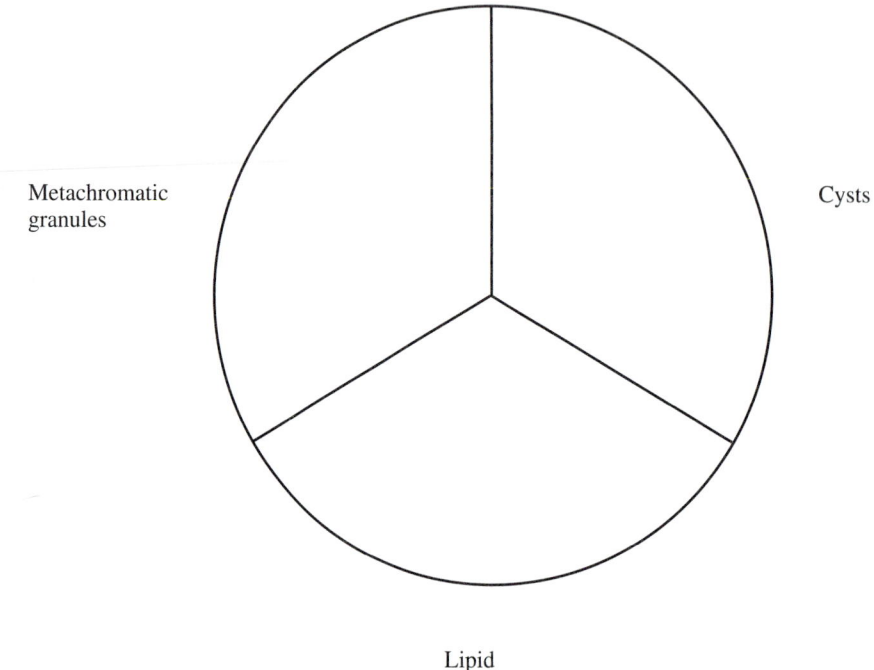

Metachromatic granules

Cysts

Lipid inclusions

Laboratory Review 19   A Demonstration of Selected Bacterial Structures

1. List 2 major differences between bacterial endospores and cysts.

   a. _____   b. _____

2. List 4 major differences between typical prokaryotic and eukaryotic cells. (Refer to your textbook.)

   a. _____   c. _____

   b. _____   d. _____

3. Complete the following table. (Refer to your textbook and the exercises in this section.)

Table 19–1

Bacterial Structures and Functions

Bacterial Structures	Bacterial Function(s)	Chemical Composition
Capsule		
Cell wall		
Cytoplasmic membrane		
Endospore		
Flagellum		
Pilus		
Ribosome		

Key Terms

bacterial cyst (sist): a resistant form of soil bacteria that differs from endo- and exospores in lacking dipicolinic acid and having less heat resistance

cell wall: the rigid, outer, layered structure that encloses the cell of most algae, bacteria, fungi, and plants

fascicle (FAS-i-kl): a bundle

flagellum (fla-JEL-um): long, slender protein structure originating from within the cell's cytoplasm and used for movement

metachromatic (met-a-krō-MAT-ik) granules: a reservoir of inorganic phosphate within a bacterium that is stainable by basic dyes

pilus (PĪ-lus): a submicroscopic tubular surface projection, located on bacterial cells, that functions for purposes of attachment or the transfer of DNA

poly-beta-hydroxybutyrate (pol-ē-BĀ-ta-hī-droxs-i-byou-te-rāt) granules: an energy-storage compound of prokaryotes that consists of several molecules (polymer) of beta-hydroxybutyric acid

ultrathin section: refers to an extremely thin slice of cell

Photograph Quiz 16

Identify the lettered structures shown in the following figure.

a. _____   b. _____   c. _____

d. _____   e. _____

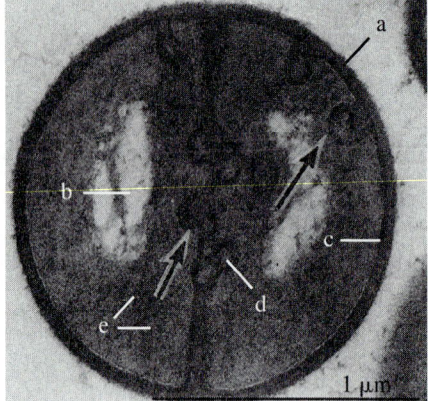

Figure 19–3

A transmission electron micrograph of *Staphylococcus aureus.*
(Courtesy of Popkin, T.J., T.S. Theodore, and R.M. Cole *J. Bacteriol.* 107:907 [1971].)

SECTION VI

Biochemical Activities of Microorganisms

. . . there is a difference between bacterial physiology and bacterial biochemistry. Biochemistry is concerned with the chemical composition and the chemical reactions of living organisms; physiology is concerned, in addition, with the functions of these reactions in the life of the organism.

—E. L. Oginsky and W. W. Umbreit

In any natural ecological environment, a constant and dynamic situation exists in which each member competes with the others for a favorable niche. The destructive effects of microorganisms in the process of causing disease are, in most cases, the results of their particular needs for survival. The disease is a by-product of their metabolic activities. Microbes must compete for nutrients, so they must attempt to adjust the environment (temperature, pH, and oxidation-reduction potentials) to their optimal requirements. In certain cases, the pathogenesis of the disease can involve (1) digestion of tissues by extracellular enzymes, (2) physical blocking of capillaries caused by unlimited growth, or (3) competitive inhibition of normal respiratory chain reactions caused by release of a respiratory pigment that interferes with the analogous pigment of the host. Obviously, the study of microbial metabolism is essential to a complete understanding of the host-parasite relationship.

Metabolism is the sum total of the biochemical reactions required to maintain adequate nutritional levels and functional cellular activities. These reactions include the degradative, or breakdown, process called *catabolism* and the synthetic process known as *anabolism.* Both processes occur simultaneously and complement each other to provide for the essential needs of the cell. All metabolic reactions occur as a series of steps leading from one compound to another. Such reaction series are referred to as *metabolic pathways* (Figure VI–1).

Microorganisms break down nutrients to obtain energy in two ways: **respiratory catabolism** and **fermentative catabolism.** Respiratory catabolism, which is the oxidative, energy-yielding part of metabolism, is much more efficient than fermentative catabolism in converting nutrients into the usable form of energy known as **adenosine triphosphate** (ATP). Respiratory catabolism involves a number of enzymes known to rapidly oxidize (break down) compounds such as sugars by means of the metabolic pathways: **Embden-Meyerhoff-Parnas (EMP) pathway** and the **tricarboxylic acid (TCA) cycle.**

In respiratory catabolism, organic compounds are rapidly and completely oxidized to carbon dioxide. In addition, there is a subsequent removal of many electrons that are no longer of any use to the cell. This is accomplished by passing the energy-spent electrons through an electron transport chain to a terminal electron acceptor. If the terminal electron acceptor is oxygen, the process is referred to as **aerobic respiration.** However, if the electron acceptor is any other inorganic compound, such as thiosulfate, the process is called **anaerobic respiration.** Figure VI–2 shows a comparison of metabolic oxidations linked to these forms of respiration. Exercise 24 demonstrates the activities of two enzymes,

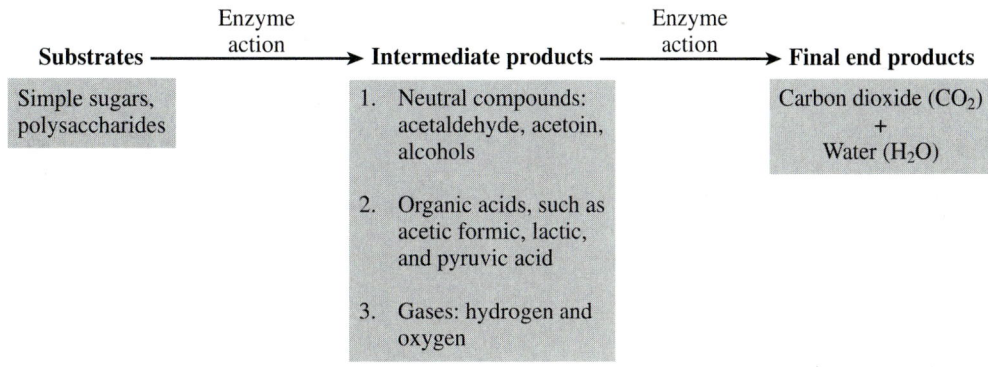

Figure VI–1

Diagram of the basic reactions associated with carbohydrate metabolism. These reactions can be demonstrated by the inoculation of various carbohydrate-containing media.

cytochrome c and **catalase,** that are involved with the microbial utilization of oxygen in the process of aerobic respiration.

Microorganisms, like other forms of life, may alter their environment to a certain extent and use chemical compounds in solution as sources of energy and as components for growth and reproduction. All cellular activities are mediated by enzymes. In order to observe microbial enzymatic activity, various kinds of specifically prepared media are inoculated with pure cultures of microorganisms. Such microbes multiply during a suitable incubation period

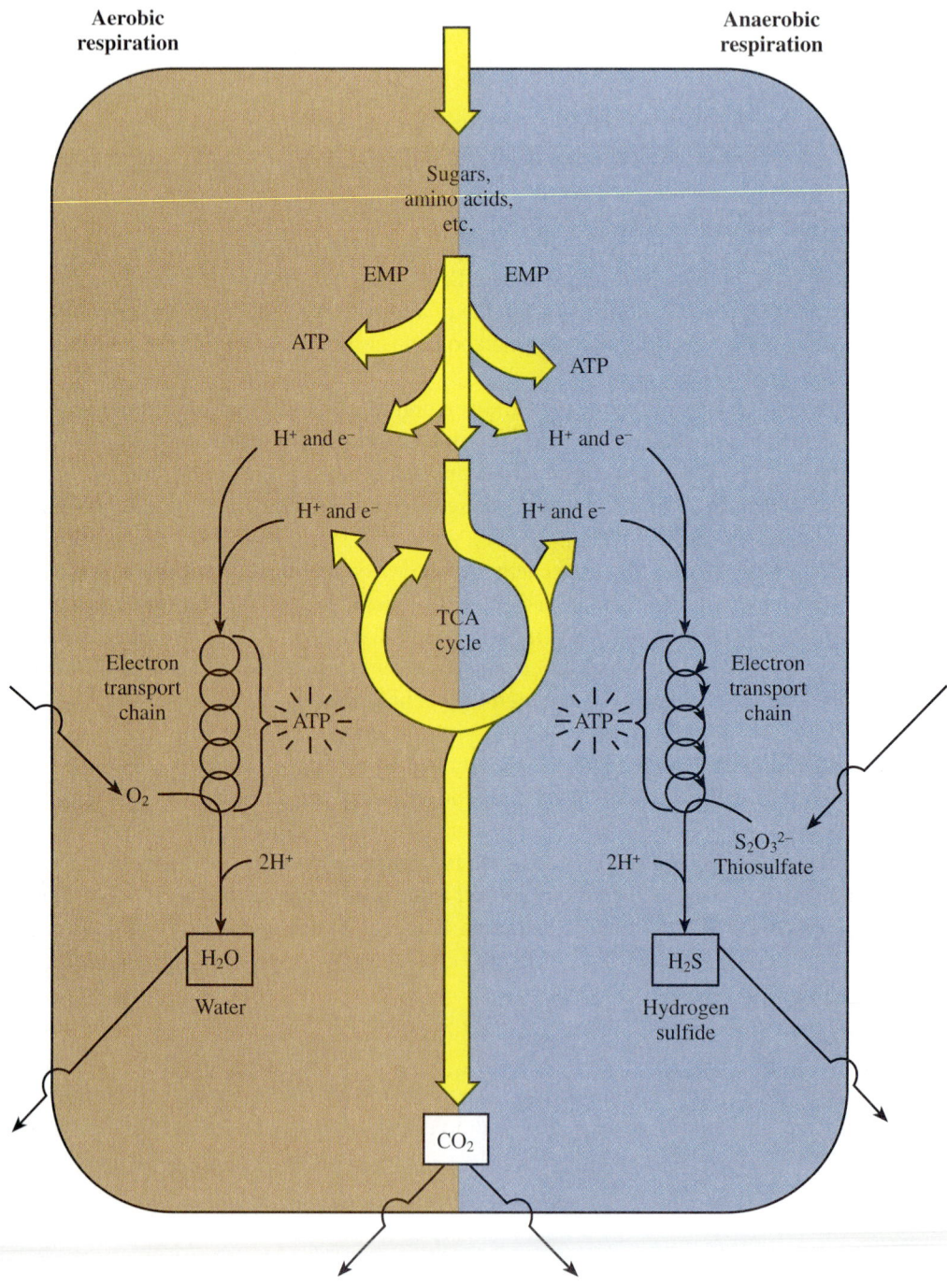

Figure VI–2

Comparison of metabolic oxidation associated with aerobic and to anaerobic respiration. Aerobic respiration reduces oxygen to water by means of the energy-spent electrons (plus positive charged particles) at the end of the electron transport system. Anaerobic respiration reduces thiosulfate (instead of oxygen) to hydrogen sulfide (instead of water) by use of the energy-spent electrons (plus protons) from the electron transport system. Explanation of symbols: ATP, adenosine triphosphate; EMP, Embden-Meyerhoff-Parnas pathway; TCA, tricarboxylic acid cycle.

and exhibit their respective enzymatic activities. Many distinctive enzyme activities can be demonstrated by testing for the by-products resulting from the action of enzymes on specific substances within the specially prepared media. Microbial species can differ greatly and may be identified by the actions of their enzymes. By using a series of different media (biochemical tests), an investigator can establish a pattern of activity for specific species. Such reaction (enzymatic) patterns can be employed in the identification and differentiation of microorganisms. This section will present an overall view of microbial reactions associated with carbohydrates, proteins, other nitrogenous materials, and lipids. Several of the following exercises describe selected aspects of microbial metabolism. Exercise 28 shows how various combinations of biochemical activities establish useable patterns for the identification of unknown bacterial species.

Carbohydrate metabolism includes the processes by which microorganisms utilize various sugars and polysaccharides for energy and for the synthesis of cellular components (Figure VI–1). While different sets of reactions exist for the degradation of glucose and other sugars, the pathways of general concern are mainly the EMP pathway and the TCA cycle. In the former, glucose is converted into pyruvic acid and constitutes the major portion of the fermentative activities of microorganisms. If a particular organism also contains the enzymes necessary for the TCA cycle, the pyruvate that is metabolized to acetate enters the cycle and results in the production of carbon dioxide, water, and (ultimately) energy. If the organism is not able to metabolize via the TCA cycle, the pyruvate may be converted into lactic acid, ethyl alcohol, and/or other fermentation products.

In conjunction with carbohydrate metabolism, mention should be made of the metabolism of lipids, because the enzymatic cleavage of these organic compounds yields acetate, which is the same compound as that produced during glycolysis. Thus, by-products or metabolites of lipids can be used in the TCA cycle.

Carbohydrate metabolism yields energy and carbon skeletons for various cellular components. Metabolism of nitrogenous materials, on the other hand, produces amino acids and subsequently proteins, which are used to build and repair structures and to promote enzymatic activities (Figure VI–3). Heterotrophic organisms (which require organic compounds) either degrade proteins or utilize existing amino acids for their protein-synthesizing processes.

Many bacteria are able to use inorganic nitrogenous compounds such as ammonia, nitrates, and gaseous nitrogen for their nutritional requirements. Similarly, certain bacteria are able to fix carbon dioxide, much as green plants do. A particularly interesting aspect of nitrogenous metabolism involves the set of reactions by which an amino group is transferred from one amino acid to the precursor of a different amino acid, subsequently resulting in the formation of a new amino acid. Two intermediates of the TCA cycle, as well as pyruvic acid, are amino acid precursors.

Thus, the three seemingly separate and distinct metabolic processes involved with carbohydrates, proteins, and lipids are really closely interconnected. All three can yield compounds that are active in the TCA cycle, providing energy for synthesis and furnishing building blocks for cellular components.

A significant number of procedures have been devised to isolate specific bacterial species that are based on the various combinations of these and related metabolic activities. Many different types of media are used in laboratories as well as in other areas utilizing microbiological procedures to isolate and differentiate bacterial species on the basis of the presence or absence of a given metabolic (enzymatic) activity. Exercise 20 describes representative media used for bacterial isolation and differentiation.

Numerous ingenious improvements have been made in procedures used for the identification of bacterial species, among which are miniaturized and rapid-test systems. With the former, as many as 21 different biochemical tests can be performed by a simple inoculation operation. In the case of rapid procedures, the results can be read within a matter of a few hours of incubation. Exercise 27 demonstrates representative systems and associated procedures.

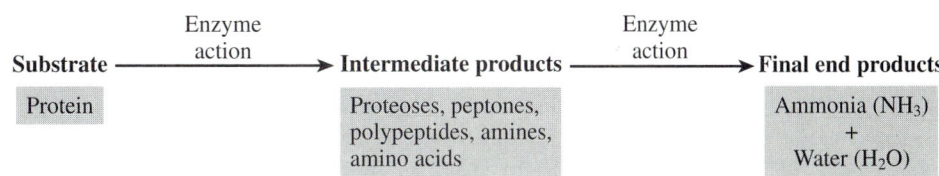

Figure VI–3

Diagram of the basic reactions associated with nitrogen metabolism. Various media in the exercises presented in this section can be used to demonstrate these metabolic reactions.

A Demonstration of Selective and Differential Media

After completing this exercise, you should be able to:

1. Understand the basic differences among differential, enriched, selective, and selective and differential media.

2. Recognize the role of such media in the isolation and identification of microorganisms.

Generally speaking, a *selective medium* is defined as one that permits the growth of certain organisms while preventing or retarding the growth of others. Selection, in general, can be carried out through (1) control of ingredients of the medium, (2) alteration of atmospheric components, or (3) adjustment of incubation temperature. On the other hand, a *differential medium,* which does not prevent the growth of organisms, will cause certain colonies to develop differently (differentiated) from other organisms present. Several media in use today incorporate both selective and differential substances (Table 20–1). One example of such preparations is Bacto-Brilliant Green gar (Figure 20–1). This medium is a highly selective preparation used for the isolation of *Salmonella* species other than *S. typhi,* the causative agent of typhoid fever from stools or other specimens suspected of containing these organisms. The growth of other bacteria is almost completely inhibited by the presence of brilliant green dye. This medium also contains lactose and sucrose as substrates for enzymatic action. Typical *Salmonella* form slightly pink-white opaque colonies surrounded by brilliant red areas. The few lactose- or sucrose-fermenting organisms that can grow on the medium are easily differentiated from *Salmonella* by the formation of yellow-green colonies surrounded by intense yellow-green zones. This exercise will demonstrate the effectiveness and general properties of commonly used differential, selective, and differential media (Tables 20–1 and 20–4).

In the next section of this manual (Exercise 36), the dye, crystal violet, will be shown to be selectively bacteriostatic for gram-positive bacteria. It is not too surprising to find this dye incorporated into a selective medium for the examination of enteric pathogens, which are predominantly gram-negative. Furthermore, since these pathogens are usually incapable of fermenting lactose, a medium that would allow differentiation of these organisms from the usually nonpathogenic, lactose-fermenting bacteria is also possible. Such a medium would be both selective and differential. A medium commonly used for the examination of stool specimens is MacConkey agar. This medium incorporates 1 mg per liter of crystal violet to prevent the growth of gram-positive bacteria and 30 mg per liter of neutral red as a pH indicator. The gram-negative bacteria that ferment lactose (the only sugar in the medium) produce acid and their colonies appear red (Figure 20–2), while the nonlactose-fermenting bacteria produce colorless colonies.

Another selective and differential medium also noted for better bacterial colonial differentiation is Hektoen (HEK-tō-en) enteric agar. This preparation is effective in uncovering rapid lactose-fermenting organisms in fecal specimens. In

Table 20–1

Selective and Differential Media Properties

Medium	Selective Agent(s)	Organisms Encouraged to Grow
Brilliant green agar	Brilliant green	Gram-negative rods[a]
Eosin-methylene blue agar	Eosin Y, methylene blue	Gram-negative rods
Hektoen enteric agar	Bile salts	Gram-negative rods
MacConkey agar	Bile salts, crystal violet	Gram-negative rods
Mannitol-salt agar	Sodium chloride	*Staphylococcus aureus*

[a]This medium is not used for the isolation of *Salmonella typhi.*

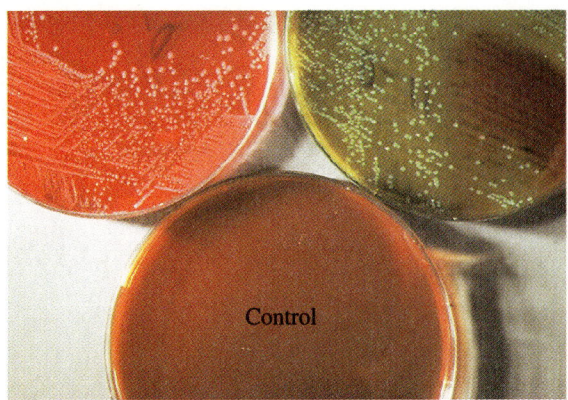

Figure 20–1

Bacto-Brilliant Green agar, a useful medium for the isolation and identification of *Salmonella* species. Typical nonlactose fermenting *Salmonella* colonies appear as slightly pink-white opaque colonies surrounded by a brilliant red zone. Colonies of lactose fermenting organisms form yellow-green colonies surrounded by intense yellow-green zones on the medium. An uninoculated plate also is shown as a control.

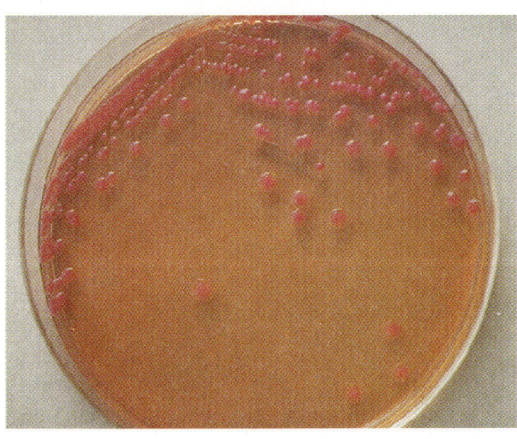

Figure 20–2

The appearance of MacConkey agar reactions. Lactose fermenting microorganisms form red colonies.

addition, hydrogen sulfide (H_2S) producers can easily be detected (Figure 20–3*b*). Among its various ingredients, Hektoen enteric (HE) agar contains 3 carbohydrates—lactose, salicin, and sucrose—and two indicator dyes—Bromthymol blue (65 mg per liter) and acid fuchsin (100 mg per liter). Rapid lactose fermenters appear as bright yellow to salmon-pink glistening drops against a faint pink to red background of the medium (Figure 20–2). Nonlactose fermenters generally are blue-green to green in color. Sodium thiosulfate and ferric ammonium citrate are used in the medium to show the presence of H_2S producers. These organisms exhibit black centers in their colonies (Figure 20–4*c*). It is important to note that HE agar is clear and green in color prior to use.

This exercise will also demonstrate the properties of one selective and differential medium designed for certain gram-positive bacteria. Mannitol-salt agar was formulated for the isolation and differentiation of pathogenic *Staphylococcus aureus*. The selective agent here is sodium chloride in a concentration of 7.5%. This ingredient prevents the growth of most, but not all, other bacteria. In addition, the medium contains 10 g of mannitol per liter for the detection of mannitol fermentation (Figure 20–4). Phenol red in the concentration of 25 mg per liter is incorporated into the medium

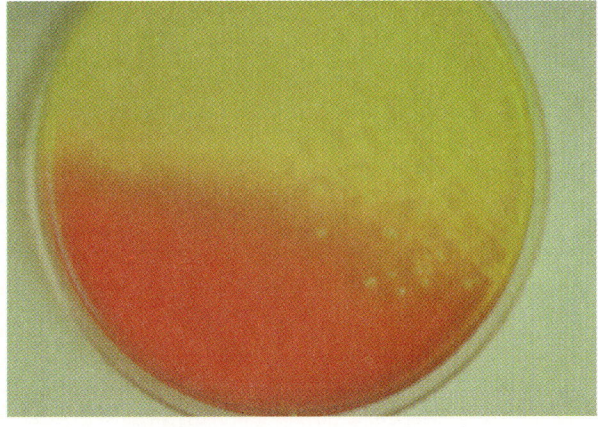

Figure 20–3

A positive (yellow) reaction on mannitol salt agar.

(a)   (b)   (c)

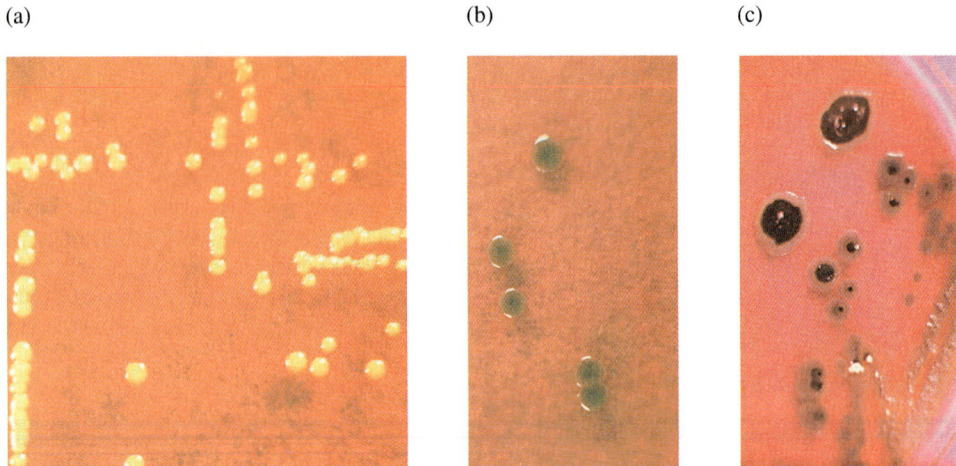

Figure 20–4

Hektoen enteric agar reactions. (a) Lactose fermenters. (b) Nonlactose fermenters. (c) Hydrogen sulfide producers.

as a pH indicator. There are far too many media to present here. Table 20–1 presents the general properties of a number of selective and differential media. Another medium category not considered in this exercise is the one-purpose medium. These preparations are highly selective and usually are designed to isolate a specific organism.

Table 20-4 at the end of this exercise summarizes the biochemical reactions that occur with the media used in this exercise.

For a comparison of the growth response of the organisms selected for this exercise, blood agar was chosen as a good example of an all-purpose medium. Blood agar, which is both an enriched and a differential medium, can be used to differentiate among some organisms in relation to their actions on the medium (Figure 20–5). Several different types of hemolytic reactions can occur on this medium. The incomplete breakdown of hemoglobin results in the formation of a green zone around colonies. Clear zones around colonies indicate a complete breakdown of hemoglobin and is referred to as beta-hemolysis. The absence of any visible reaction around bacterial colonies is referred to as gamma-hemolysis.

The various reactions of bacteria used in this exercise will be demonstrated with the use of **replica plating.** This technique was developed by Joshua and Esther Lederberg in the early 1950s. In this technique, a master plate, containing the isolated colonies of a specific bacterial species, is pressed onto the surface of a supported velveteen surface. The velveteen is a fabric that has thin threads that stand upright like tiny bristles. Each thread acts like an inoculation needle. Pressing onto the velveteen transfers a portion of each bacterial colony. Next, sterile media are oriented in the same manner as the master plate, and pressed onto the same velveteen. In this way the cells imprinted on the velveteen from the master plate are transferred to the various media used to be tested.

(a)   (b)

                                                              (c)

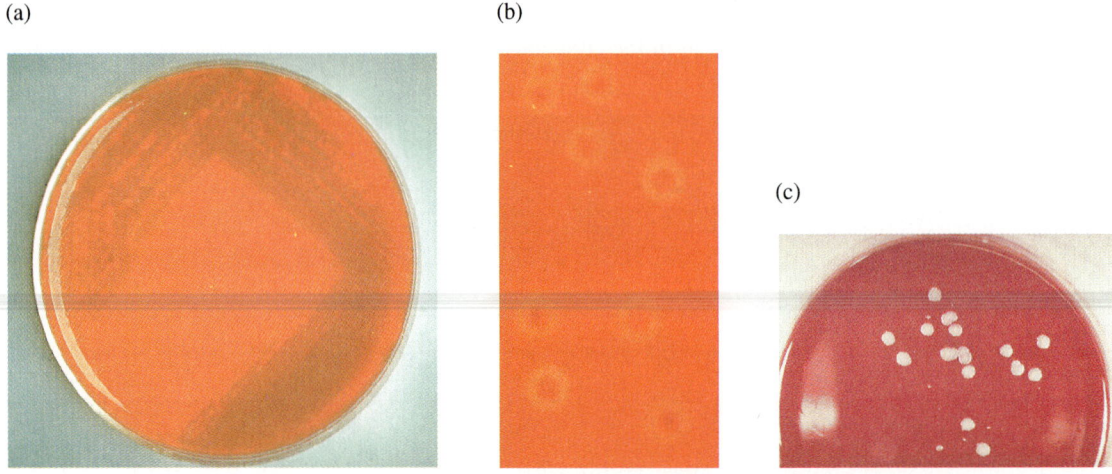

Figure 20–5

Blood agar plates showing different hemolytic reactions. (a) Alpha hemolysis. (b) Beta hemolysis. (c) Gamma hemolysis.

Materials

The following items are to be provided per 4 students:

❑ 1. Divided streak plate of each microbial combination listed (1 per 4 students). The cultures should be grown on trypticase soy agar for 24 hours prior to class use:
  ❑ a. *Escherichia coli* and *Staphylococcus aureus*
  ❑ b. *E. coli* and *Enterobacter aerogenes*
  ❑ c. *S. aureus* and *Enterococcus (Streptococcus) faecalis* (strep-tō-kok-us, FĒ-kal-is)
  ❑ d. *Proteus vulgaris* and *Micrococcus luteus*

❑ 2. Four plates of the following:
  ❑ a. Blood agar   ❑ d. Hektoen enteric agar
  ❑ b. MacConkey agar   ❑ e. Trypticase soy agar
  ❑ c. Mannitol-salt agar

❑ 3. One replicating apparatus marked with a reference line on one side extending from top to bottom. (This is a circular support with a diameter of approximately 81 mm. The replicating structure should be smaller in diameter than the interior of the bottom of a Petri dish.)

❑ 4. Four sterile velveteen pads, approximately 5 x 5 inches

❑ 5. Four rubber bands to anchor the velveteen pads to the replicating apparatus

❑ 6. One marking pen or pencil

Procedure

❑ 1. The instructor will demonstrate the anchoring of the velveteen pad to the replicating block as well as the general replicating rechnique.

❑ 2. Each student will be assigned 1 of the microbial combinations provided for the exercise. Sketch the positions of the bacterial colonies or surface of this reference plate in the designated circle in the Results and Observations section.

❑ 3. Follow the 9 steps shown in Procedure Diagram 23 with your assigned organisms.

❑ 4. Incubate the inoculated plates at 37°C for 48 hours or as directed by your instructor.

❑ 5. After incubation, compare the growth pattern on each medium with the original pattern for your microbial combination and record your findings in the Results and Observations section. (Refer to Figures 20-1–20-6.)

❑ 6. After consulting with your laboratory partners as to their results, tabulate the differential growth characteristics observed on each medium whenever appropriate.

(a)   (c)   (e)

(b)   (d)   (f)

Figure 20–6

Characteristics of selected differential and selective media shown by means of the replica plating technique. *Plate* a, master plate; b, Blood agar; c, Eosin methylene blue agar; d, MacConkey agar; e, Mannitol-salt agar; f, Trypticase Soy Agar. The master plate contains a gram-positive bacterium on the right side and a gram-negative bacterium on the left side of the medium. Note that both organisms grow on the master plate, blood agar, and the trypticase soy agar. Refer to the specific reactions of these media described in this exercise.

Procedure Diagram 23
Replica-Plating Technique

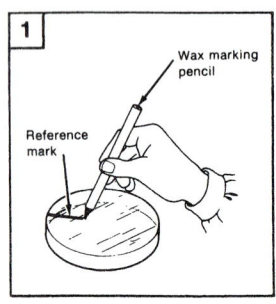

1. Mark the trypticase soy agar plate bottom of the assigned microbial combination with a short reference mark.

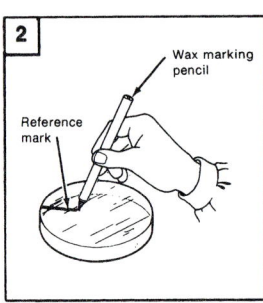

2. Repeat this step with each of the various types of media to be used.

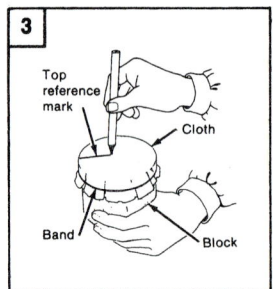

3. Anchor the velveteen pad to the replicating block with the rubber band. Make a small reference mark at the top of the pad.

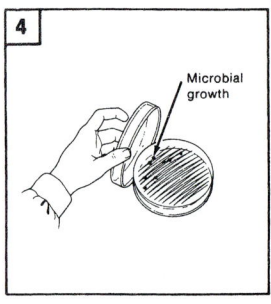

4. Remove the Petri plate cover from the assigned microbial combination preparation.

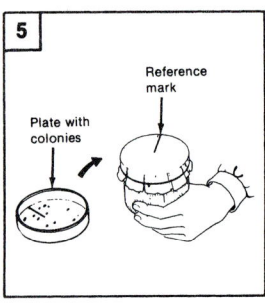

5. Place the inoculated surface on the velveteen pad so that the 2 reference marks are aligned.

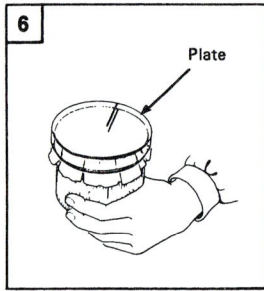

6. Press the streaked surface onto the velveteen so that it comes into close contact with the colonies. *Do not rotate the block or the streaked plate.*

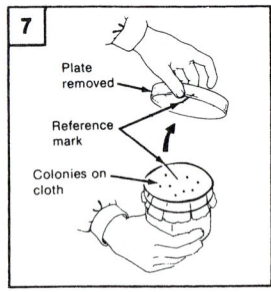

7. Remove the plate from the block, replace the cover of this reference plate, and incubate.

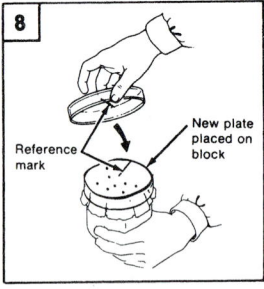

8. Remove the plate covers of the uninoculated media (one at a time) in the order as listed in the **Materials section.** Inoculate each one by pressing the surface of the agar medium onto the velveteen. Align the reference markers before bringing them into contact with the surface of the reference plate.

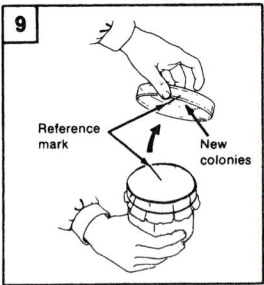

9. Remove the plate from the block, replace the cover of the newly inoculated plate, and incubate as directed.

Results and Observations

1. Growth patterns.

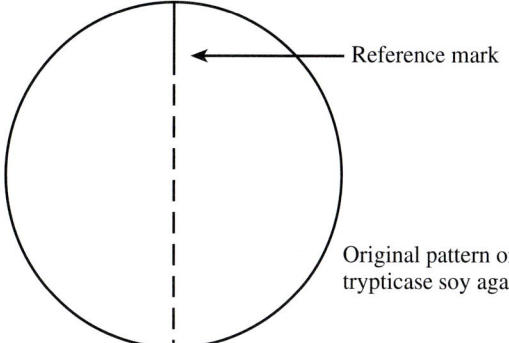

Reference mark

Original pattern on
trypticase soy agar

Microorganism: _____

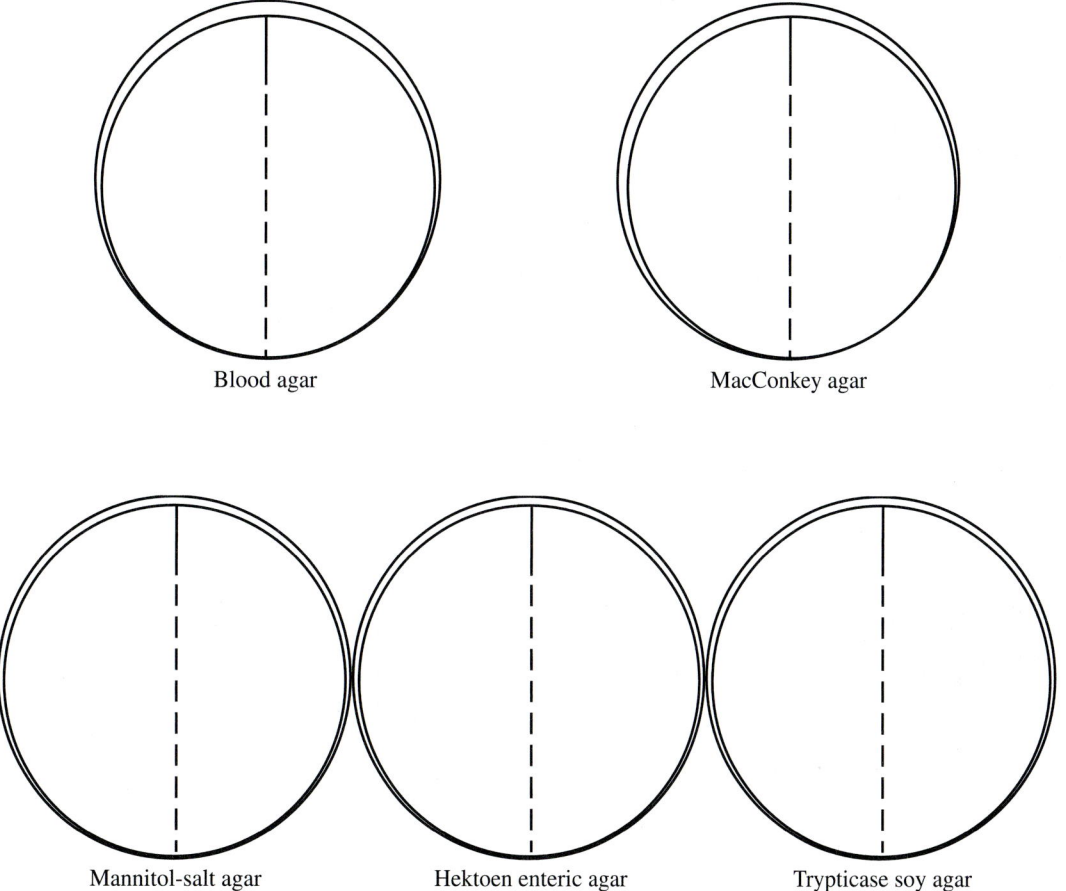

Blood agar

MacConkey agar

Mannitol-salt agar

Hektoen enteric agar

Trypticase soy agar

Microorganism: _____

2. Differential characteristics. Complete the following table by entering the color of colonies, the color of media, hemolytic reactions, H_2S production, and specific carbohydrate fermentations for each microorganism listed.

Table 20–2

Selective and Differential Media Reactions

Organism	Blood Agar	MacConkey Agar	Mannitol-Salt Agar	Hektoen Enteric Agar
E. aerogenes				
E. coli				
P. vulgaris				
M. luteus				
S. aureus				
E. faecalis				

Laboratory Review 20   A Demonstration of Selective and Differential Media

1. Complete the following table by indicating for each plating medium listed the selective agent or agents, the appearance of a positive reaction, and the general types of bacteria that would *not* grow on the preparation.

Table 20–3

Actions of Selective and Differential Media

Medium	Selective Agent(s)	Positive Result	Bacteria Inhibited
Blood agar			
Brilliant green agar			
Eosin-methylene blue agar			
Hektoen enteric agar			
MacConkey agar			
Mannitol-salt agar			
Salmonella-Shigella agar			

2. Distinguish between selective and differential plating media. _____

_____

_____

3. a. What happens to blood agar when an alpha-hemolytic reaction occurs? _____

_____

   b. What happens when a beta-hemolytic reaction occurs? _____

_____

Key Terms

differential (dif-er-EN-shal) *medium:* a medium that contains one or more ingredients designed to stimulate a characteristic biochemical response, which will differentiate individual or groups of microorganisms from one another

enriched medium: a medium containing one or more substances, such as blood, to encourage microbial growth

one-purpose medium: a highly selective medium designed to isolate a specific organism

selective and differential medium: a preparation that incorporates the features of both selective and differential media

selective medium: a medium containing one or more substances that permit the growth of certain organisms *while interfering with the growth* of others

---

Table 20–4

Summary of Biochemical Reactions

Medium	Substrate(s)	Reactions and Descriptions
Blood agar	Hemoglobin	1. alpha hemolysis (green zones around colonies) 2. beta hemolysis (clear zones around colonies) 3. gamma hemolysis (no zone around colonies)
Brilliant green agar	Lactose, sucrose	1. lactose fermenter (yellow-green colonies surrounded by yellow-green zones) 2. lactose nonfermenter (pink to white colonies surrounded by brilliant red zones)
Eosin methylene blue agar	Lactose, sucrose	1. lactose fermenter (dark purple or colonies or colonies with dark centers and transparent colorless borders.) 2. lactose or sucrose nonfermenters (colorlesscolonies)
Hektoen enteric agar	Lactose, sucrose, salicin, and amino acids containing sulfur	1. lactose fermenter (salmon-pink colonies) 2. lactose nonfermenters (green, most colonies) 3. salicin fermenters (pink zones around colonies) 4. salicin nonfermenters (no change) 5. H$_2$S producers (colonies with black centers)
MacConkey agar	Lactose	1. lactose fermenter (pink-red colonies surrounded by pink zones due to precipitated bile 2. lactose nonfermenter (colorless and translucent colonies)
Mannitol salt agar	Mannitol	1. mannitol fermenter (colonies surrounded by yellow zones) 2. mannitol nonfermenter (small colonies with no color change)

Photograph Quiz 17

Identify the reactions taking place on this Hektoen enteric agar plate. _____

_____

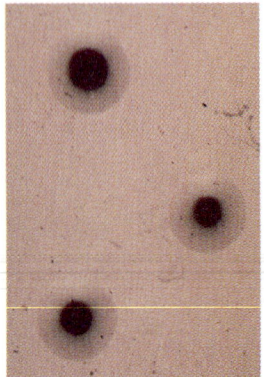

Figure 20–7
Hektoen Enteric agar reaction.

Extracellular Degradation of Polysaccharides, Proteins, Lipids, and DNA

After completing this exercise, you should be able to:

1. Perform and interpret tests for the hydrolysis of each of the following substances: starch, casein, lipid, and deoxyribonucleic acid.
2. Explain how these test results can be used in microbial identification.
3. Explain the role of extracellular enzymes in providing nutrients for cellular metabolism.

Metabolism is a general term denoting the sum of all chemical activities required in living cells. These activities include digestion of complex organic compounds with the production of simpler organic and inorganic chemicals. These by-products of digestion can then be used as building blocks to make the organism's own required complex organic materials. Some of these by-products must also be used to provide the energy for cellular function, driving the many catabolic and anabolic processes, as well as for motility in some organisms.

With animals, metabolism is based upon digestion of complex foodstuffs in two phases: **extracellular** and **intracellular.** Extracellular digestion involves the secretion of amylase in saliva and pancreatic juices, proteases in gastric and pancreatic juices, and lipases in pancreatic juices. The various breakdown products, which include maltose from starch, amino acids from proteins, and glycerol and fatty acids from simple lipids, then become available to be absorbed, circulated to animal cells for transport across their membranes, and subsequently additional catabolic activities. Many microorganisms utilize similar extracellular enzymes to obtain their nutrients from the environment.

Several microorganisms form and then release enzymes into their respective environments such as a medium. Some of these microbial extracellular enzymes can degrade or break down large molecules in the environments surrounding the cells. Most microbial extracellular enzymes are referred to as being **hydrolytic.** These enzymes are so named because they break large molecules into smaller ones and are generally classified according to the large types of molecules they can degrade. Such enzymes include *esterases,* which decompose fats and lipids, *glycosidases,* which break apart polysaccharides, and *proteinases,* which degrade proteins.

Most of the carbohydrates available to microorganisms are in the form of polymers known as *polysaccharides.* Two common examples of these compounds are cellulose and starch, both of which are polymers of glucose. The basic differences between the chemical and physical characteristics of these natural substances depend upon the structural arrangement of their glucose units. If this were not the case, then cellulase, the enzyme that degrades cellulose into simple sugar units, would also degrade starch. The enzyme responsible for hydrolysis of starch is called amylase. Cellulase and amylase are examples of extracellular enzymes (exoenzymes); they are secreted through the cell wall in order to degrade complex substances into units that can readily enter the cell. An example of this type of metabolic pattern is the action of amylase on starch, (Figure 21–1), yielding maltose (a disaccharide composed of two glucose molecules).

The degradative action of an organism on an intact protein is analogous to such action on carbohydrates. In the case of starch, the enzyme amylase degrades the polysaccharide into units that can be readily absorbed by the cell. With a protein, such as casein (milk protein), the enzyme caseinase accomplishes much the same result in yielding polypeptides (Figure 21–2).

Many microorganisms are able to degrade fats and oils (in the process called *lipolysis*) and thus obtain acetate for carbohydrate metabolism and amino acid synthesis. The presence of a lipase in the enzymatic functions of a microorganism can be considered a potential indication of its invasiveness, because animal cell membranes are largely composed of lipid. This degradative ability is yet another characteristic of microorganisms that can be useful in classification. In this exercise, the lipolytic activity of microorganisms will be determined by using a plate containing agar emulsified with an oil and a dye. A clearing of the blue medium will appear in the area of lipolysis (Figure 21–3).

The production of deoxyribonuclease (DNAse), another extracellular enzyme, is useful for the isolation and differentiation of several bacterial species. DNAse agar is used to demonstrate such activity. Flooding inoculated plates after incubation with 0.1 N HCl will show clear areas around the colonies that degraded the DNA in the medium (Figure 21–4).

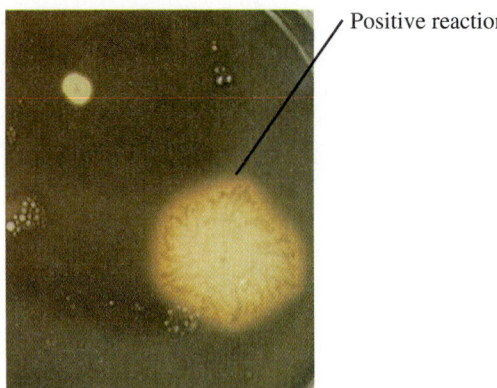

Figure 21–1

Starch hydrolysis (amylase production). A typical starch hydrolytic reaction as produced by the bacterium *Bacillus subtilis*. The complete absence of starch (hydrolysis) upon the addition of an iodine reagent is indicated by the yellow background surrounding the bacterial colonies. A negative reaction, the presence of starch, is indicated by a purple background.

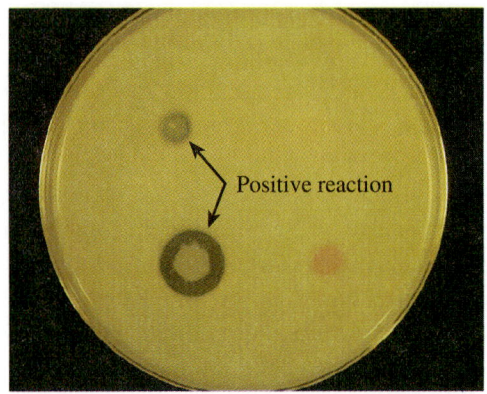

Figure 21–2

Caseinase activity. Casein digestion can be seen as clear zones around bacterial growths. One clearly negative result also is shown.

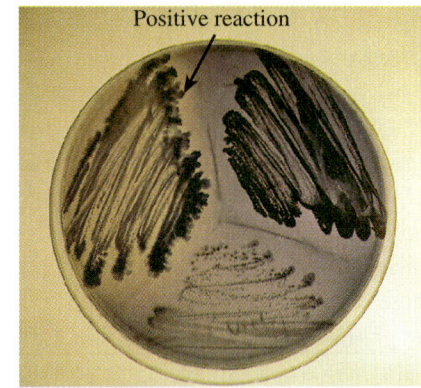

Figure 21–3

The demonstration of lipid hydrolysis using Bacto-spirit blue agar containing Bacto-lipase reagent. The lipolytic activity of a bacterial species is recognized by the deep blue color or clearing that develops in the medium surrounding the test organism. A Comparable color change is not observed with nonlipolytic microorganisms.

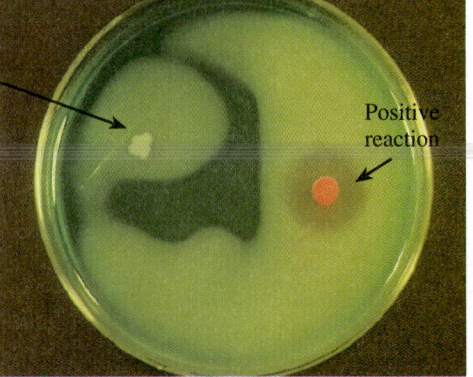

Figure 21–4

DNAse reactions. The presence of the enzyme is indicated by a clear zone around the bacterial growth. The cloudy areas are caused by the addition of hydrochloric acid which precipitates the DNA.

Since microbial species vary in the kinds of extracellular enzymes they possess, demonstrating the presence or absence of a particular enzyme is of value in the identification of unknown organisms. Such applications of metabolic or biochemical tests are explored in Exercise 22 and later expanded in Exercises 28 and 65, where methods of identifying unknown bacterial species are presented.

In this exercise, four tests will be used to demonstrate the presence of extracellular enzymes associated with specific organic macromolecules of importance in metabolism. These molecules are casein, starch, deoxyribonucleic acid, and lipids. Selected aspects of intracellular phases of metabolism are discussed in Exercise 22.

Materials

The following items are to be provided per 2 students:

❑ 1. One 24-hour nutrient broth culture of each of the following:
  ❑ a. *Bacillus subtilis*
  ❑ b. *Serratia marcescens*
  ❑ c. *Staphylococcus aureus*

❑ 2. One plate of the following media:
  ❑ a. Starch agar
  ❑ b. Casein agar
  ❑ c. DNAse agar
  ❑ d. Spirit blue agar

❑ 3. One wax marking pencil or similar device

The following reagents should be in dispenser bottles or accompanied by appropriate droppers and provided after incubation:

❑ 1. 0.1 *N* hydrochloric acid

❑ 2. Lugol's or Gram's iodine solution

Procedure

This procedure is to be performed by students in pairs.

❑ 1. Using a wax marking pencil, mark 3 sectors on the bottom of each plate. Label 1 sector for each of the bacterial cultures provided.

❑ 2. Place a loopful of each culture into the center of its respective sector.

❑ 3. Invert and incubate all plates at 35–37°C for 48 hours.

❑ 4. After incubation, determine the test results as follows, and record your findings in the Results and Observations section:
  ❑ a. **Amylase activity:** Carefully flood the surface of the starch agar plate with iodine solution. Examine the areas surrounding the bacterial growth for starch hydrolysis. A yellow zone indicates the breakdown of starch. (Refer to Figure 21–1.)
  ❑ b. **Caseinase activity:** Examine the casein agar plate and look for an obvious clearing of the area around colonies growing on casein agar. (Refer to Figure 21–2.)
  ❑ c. **DNAse activity:** Carefully flood the surface of the DNAse agar plate with 0.1 *N* HCl. Look for the clear areas surrounding the bacterial growths. (Refer to Figure 21–4.)
  ❑ d. **Lipase activity:** Refrigerate the plate for 30 minutes and then examine it for the presence of a clear, somewhat darker blue color around the bacterial growths. This is the characteristic appearance of lipase activity. (Refer to Figure 21–3.)

Table 21–2 summarizes the reactions that occur with the media used in this exercise.

Results and Observations

1. Diagram the results for each of the extracellular enzymes tested. Label both positive and negative reactions.

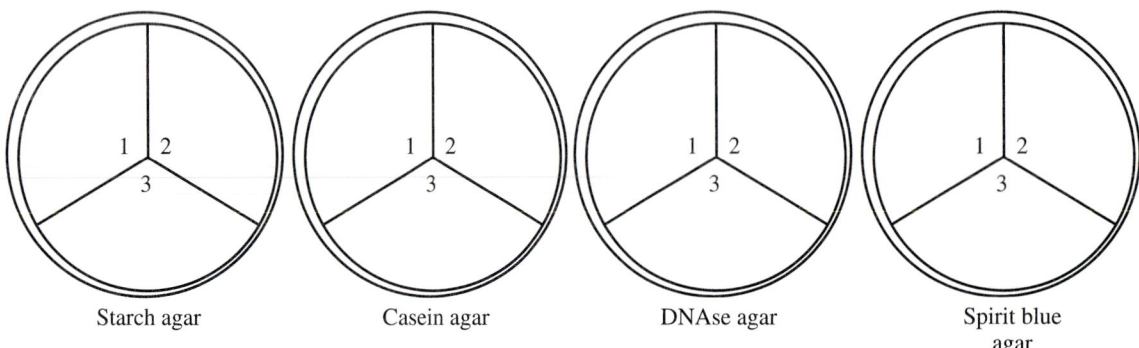

| Starch agar | Casein agar | DNAse agar | Spirit blue agar |

2. Complete the following table, using a "+" for positive results and a "−" for negative results.

Table 21–1

Differential Plate Reactions

Organism	Caseinase	Amylase	Lipase	DNAse
Bacillus subtilis				
Serratia marcescens				
Staphylococcus aureus				

Laboratory Review 21   **Extracellular Degradation of Polysaccharides, Proteins, Lipids, and DNA**

1. What is the role of extracellular enzymes in bacterial metabolism? In an ecosystem? _____

_____

_____

_____

2. How might the results obtained in this exercise be used for the identification and/or classification of bacteria?

_____

_____

_____

_____

3. Design a dichotomous identification key using the results obtained in this exercise using the fewest tests necessary to distinguish among the 3 bacterial species. _____

_____

_____

_____

Table 21–2

Summary of Biochemical Reactions

Test	Substrate(s)	Reagent(s) and Times	Positive Reactions	Negative Reactions
Casein degradation	casein (milk protein)	no reagent/ 24–48 hours	clear zones around bacterial colonies	no clear zones around colonies
DNAse production	DNA	0.I N hydrochloric acid/24–48 hrs	clear zones around colonies after the addition of HCl	cloudy zones around colonies after addition of HCl
Lipid hydrolysis	lipid	spirit blue dye/24–48 hrs	clear and/or dark areas surrounding growth	no clear or dark areas around colonies
Starch hydrolysis	starch	Lugol's or Gram's iodine/ 24–48 hrs	yellow zones around colonies after the addition of the iodine	purple or dark zones around colonies after the addition of iodine

Key Terms

anabolic: refers to those metabolic reactions concerned with synthesis

catabolic: refers to those metabolic reactions concerned with degradation or decomposition of substrates that generate energy

esterase (ES-ter-ās): an enzyme that degrades fats and lipids

extracellular enzymes: catabolic enzymes secreted by the cell into the environment for the purpose of decomposing complex organic nutrients

glycosidase (glī-kō-SĪ-dās): an enzyme capable of degrading polysaccharides into smaller carbohydrate molecules

hydrolysis (hī-DROL-iss-is): a chemical process of decomposition involving the splitting of a bond by the addition of a water molecule

polymer (POL-i-mer): a natural or synthetic product formed by a combination of two or more molecules of the same substance

proteinase (PRŌ-tē-in-ās): an enzyme capable of breaking apart proteins into amino acids and related compounds

Carbohydrate Metabolism: An Introduction to Intracellular Metabolism

After completing this exercise, you should be able to:

1. Perform and interpret tests for selected aspects of carbohydrate metabolism.
2. Determine whether a carbohydrate is oxidized or fermented.
3. Explain the role, in microbial metabolism, of the intracellular enzymes involved in carbohydrate metabolism.

For a microorganism to grow, it must have a source of energy. Whatever the form of energy, it must get into the cell. As shown in Exercise 21 some microorganisms can break down large molecules such as proteins, polysaccharides, nucleic acids, and lipids by producing extracellular enzymes. However, just because such large molecules are broken down does not necessarily insure their passage into cells.

One of the major functions of a microbial cell's plasma membrane is to select which small molecules are to be passed into the cell. Thus, this cellular structure is selectively permeable since it permits only certain kinds of molecules into the cell. For a small molecule to get into a microbial cell it must first bind to a specific binding protein located at the outer plasma membrane surface. After this binding step, the molecule then must be transported or *translocated* across the membrane and into the cell. Thus, after small molecules are bound and translocated, they can be metabolized by the cell and used as energy sources, or used directly as building parts for cell synthesis. This exercise will emphasize the fact that cells must produce many other types of enzymes besides extracellular ones in order to use the variety of large molecules in their environments.

Carbohydrates are the prime sources of energy and carbon skeletons for the synthesis of cellular substances. Simple sugars such as glucose and galactose have been described as the initiators of metabolic reactions. In the case of microorganisms, most of the carbohydrates are available to them in the form of polysaccharides. Intracellular enzymes are used by cells for further metabolic degradation of carbohydrates produced by extracellular enzymes or to synthesize more complex cellular molecules. Two examples of such enzymes are maltase and lactase. The former acts upon maltose to yield two molecules of glucose, while the latter decomposes the carbohydrate lactose and yields one molecule of glucose and one of galactose. All the end products of these metabolic reactions are used in glycolysis.

In this exercise, the reactions of several microorganisms with a disaccharide (lactose) and a monosaccharide (glucose) will be studied. Detecting acid formation from these sugars can be done easily by including a pH indicator in the test medium (Table 22–1). A pH indicator commonly used is phenol red, which is yellow at pH 6.9 (acid) and red at pH 8.5 (alkaline). Gas production in a carbohydrate medium may be detected through the use of an inverted (Durham) tube that serves to trap any gas formed (Figure 22–1).

Table 22–1

Summary of Biochemical Reactions

Test	Substrate(s)	Reagent	Positive Reactions	Negative Reactions
Carbohydrate fermentation (Durham fermentation system)	glucose, lactose	phenol red indicator	acid (yellow color) gas (gas bubble in inverted Durham tube)	alkaline (red color) no gas (no bubble in Durham tube)

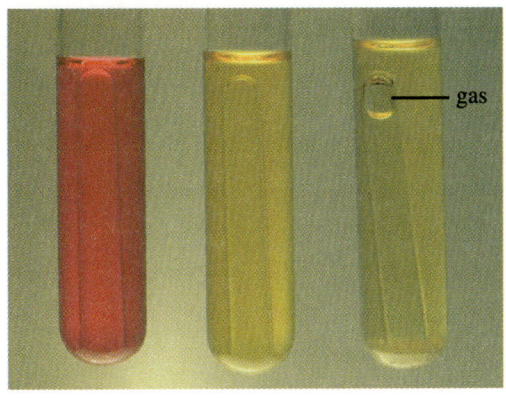

Figure 22–1

Carbohydrate fermentation using Durham fermentation tubes. The indicator used is phenol red. Tube **a,** uninoculated. Tube **b,** acid production. Tube **c,** acid and gas production. Note the collection of gas in the inverted vial.

A portion of the exercise will be used to determine if a carbohydrate is oxidatively or fermentatively utilized. Oxidation-fermentation agar containing a specific carbohydrate and Bromthymol blue as the pH indicator will be used. This indicator is blue at pH 6.8 and yellow at pH 5.2. Mineral oil, vaspar, or another similar material is used before incubation to layer media to show carbohydrate fermentation. In media of this type, acid formation by oxidative organisms occurs at the agar surface, or not at all, if the preparation is layered with mineral oil or similar material. With fermentative organisms, the entire medium, whether layered with oil or not, will show acid production. Determining whether carbohydrate oxidation or fermentation occurs is important in the identification of certain bacterial species. This medium also permits determination of motility by the culture. If the medium becomes cloudy, then the organism is motile (Figure 22–2). Table 22–2 summarizes selected reactions that can be obtained with oxidation-fermentation (O/F) agar.

Additional aspects of intracellular metabolism are presented in other exercises of this section.

(a)   (b)   (c)   (d)   (e)

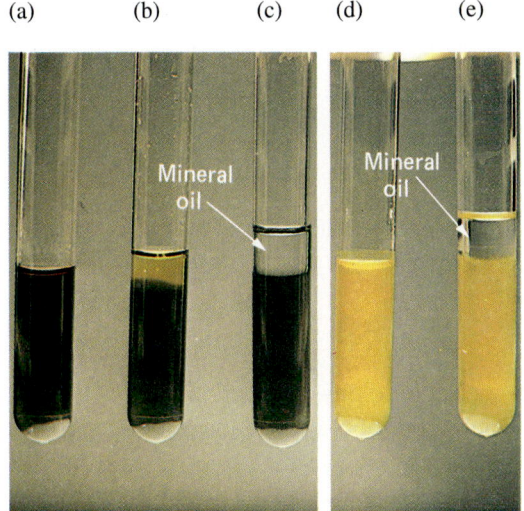

Figure 22–2

Carbohydrate oxidation or fermentation using tubes of glucose O/F agar. The indicator used is bromothymol blue. Tube a, is uninoculated; Tube b, shows glucose oxidation; Tube c, shows no glucose oxidation or fermentation; Tubes d and e show glucose fermentation. Note that the fermentation reaction is detected specifically in the mineral oil-covered medium. Refer to Table 23–1 for an interpretation of reactions.

Table 22–2

Oxidation/Fermentation Reactions

Type of Metabolism	Aerobic Conditions (No Mineral Oil Layer)	Anaerobic Conditions (Mineral Oil Layer)
Oxidative	acid (yellow)	alkaline (green)
Fermentative	acid (yellow)	acid (yellow)
Non-saccharolytic	alkaline (green)	alkaline (green)

Carbohydrate Metabolism

Materials

The following materials should be provided per 4 students:

❏ 1. Twenty-four-hour nutrient broth cultures of the following:
  ❏ a. *Bacillus subtilis*
  ❏ b. *Escherichia coli*
  ❏ c. *Micrococcus luteus*
  ❏ d. *Proteus vulgaris*
  ❏ e. *Pseudomonas aeruginosa*

❏ 2. Five tubes of Durham fermentation media with Durham tubes
  ❏ a. Phenol red glucose broth
  ❏ b. Phenol red lactose broth

❏ 3. Ten tubes of glucose oxidation/fermentation agar medium

The following items should be provided for general class use:

❏ 1. Sterile mineral oil

❏ 2. Sterile Pasteur pipettes and droppers

Additional Techniques Required for This Exercise:

Broth Transfer, Procedure Diagram 5, Exercise 4

Procedure 1: Durham Fermentation

This procedure is to be performed by students in groups of 4.

❏ 1. Inoculate 1 set of fermentation tubes with each of the organisms provided.

❏ 2. Incubate all tubes at 37°C for 24 hours.

❏ 3. Observe the reactions that develop in the 2 fermentation media.
  ❏ a. The acid reaction (A) will be yellow; the alkaline reaction (ALK) will be a deeper red when compared to an uninoculated tube; a no-change reaction (NC) will be the same color as the control tube.
  ❏ b. The presence of gas (G) will be shown by a bubble at the top of the inverted tube.
  ❏ c. Record your findings in Table 22–3 of the Results and Observations section by indicating whether acid only or acid and gas were produced. (Refer to Table 22–1 and Figure 22–1.)

Procedure 2: Carbohydrate Oxidation-Fermentation

❏ 1. Inoculate 2 tubes of glucose oxidation-fermentation agar with each of the bacterial cultures provided. Stab the medium down to the bottom of the tube.

❏ 2. Cover 1 tube of each pair of inoculated media with a thin layer of sterile mineral oil.

❏ 3. Incubate all tubes at 37°C for 48 hours.

❏ 4. After incubation, examine each tube for the production of acid (yellow color). Record your findings in Table 22–3 of the Results and Observations section. (Refer to Figure 22–2 and Table 22–2.)

❏ 5. Examine each culture for the presence of turbidity or cloudiness. This medium will turn cloudy with a motile culture because of the movement of organisms through the soft agar. The medium remains clear with nonmotile cultures.

Results and Observations

Complete the following table by indicating the presence "+" or absence "−" of a specific enzyme for each organism listed.

Table 22–3

Carbohydrate Metabolism Reactions

Organism	Durham Fermentation[a]		O/F[a]		
	G	L	O	F	M
Bacillus subtilis					
Escherichia coli					
Micrococcus luteus					
Proteus vulgaris					
Pseudomonas aeruginosa					

[a]Legend: G = glucose; L = lactose; O = oxidative; F = fermentative; M = motility;

Laboratory Review **22**	Carbohydrate Metabolism: An Introduction to Intracellular Metabolism

1. What are the roles of intracellular enzymes in bacterial metabolism? _____

_____

_____

2. How might the results obtained in this exercise be used for the identification and/or classification of bacteria? ___

_____

_____

Key Terms

fermentation: the anaerobic reduction of organic substances

intracellular enzymes: catabolic and anabolic enzymes that perform their functions within the cell

oxidation: the removal of electrons from a chemical; in this exercise, the term also refers to aerobic catabolism of carbohydrates

reduction: the gaining of electrons by a chemical

Photograph Quiz 18

Which of the phenol red glucose broth tubes shown is exhibiting an acid and gas reaction? _____

_____

Figure 22–3
Durham fermentation reactions.

(a)   (b)

After completing this exercise, you should be able to:

1. Perform and interpret tests for selected aspects of nitrogen metabolism.
2. Explain how selected test results can be used in microbial identification.
3. Explain the role of intracellular and extracellular enzymes in nitrogen metabolism.

Many different nitrogen-containing compounds are involved in the metabolism of a living cell. Proteins are involved in enzymatic and structural activities, purines and pyrimidines of nucleic acids are concerned with heredity mechanisms and certain syntheses, and inorganic compounds act as electron donors or acceptors. This exercise will consider several reactions of microorganisms—protein degradation (gelatin), urease activity, and nitrate reduction—with representative nitrogen-containing compounds. Gelatin hydrolysis and urease action involve extracellular enzymes, while nitrate reduction utilizes intracellular enzymatic activity. Table 23–2 summarizes these reactions.

Gelatin, a protein formed from collagen (an animal protein), was used by Robert Koch in the early 1880s as a solidifying agent for various media preparations. Unfortunately, gelatin was soon found to have certain disadvantages. While gelatin remained solid at temperatures generally below 28°C, it turned to liquid at higher temperatures. In addition, because of its protein nature, it could be hydrolyzed by extracellular enzymes (proteinases) produced by several microbial species. As microorganisms hydrolyze this protein, it changes from a solid state to one that is liquid, thus destroying its value as a solidifying agent. Several tests for proteinases are based on gelatin liquefaction. Robert Koch eventually found a much more suitable solidifying agent for his media, namely agar. This exercise will use gelatin agar deeps to demonstrate gelatin hydrolysis (Figure 23–1).

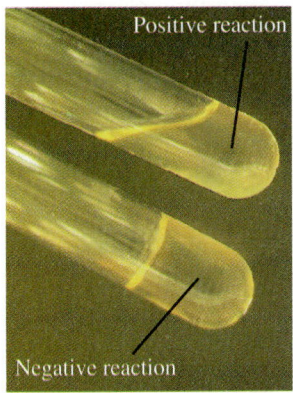

Figure 23–1

Nitrogen metabolism. Gelatin hydrolysis, an example of an extracellular enzyme reaction. A positive result is indicated by a liquefaction of the gelatin agar (upper tube). A negative reaction is indicated by the persistent solid form of the medium (lower tube).

Many bacteria are capable of synthesizing amino acids from by-products of carbohydrate and lipid metabolism when provided with ammonia as a nitrogen source. Some of these organisms are able to split the compound urea, a major organic waste product of animal metabolism, into ammonia and carbon dioxide (Figure 23–2). The product, carbon dioxide, is incorporated into carbohydrate and nitrogen metabolism through a variety of important reactions. The medium becomes highly alkaline due to ammonia production. In this exercise, a medium containing urea as the substrate and the pH indicator phenol red are used to demonstrate urease activity (Figure 23–3).

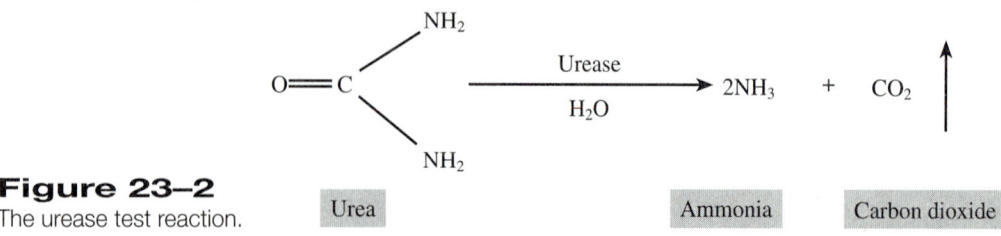

Figure 23–2
The urease test reaction.

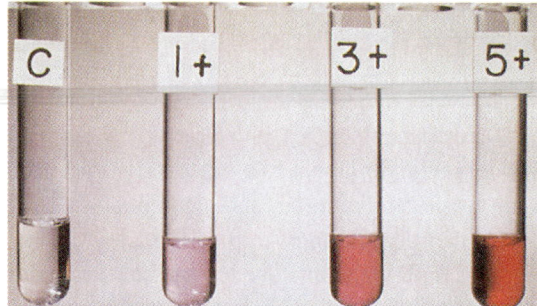

Figure 23–3
Urease activity. A positive reaction is indicated by the red color of the medium. A negative reaction is represented by an orange color. *Proteus spp.* generally produce positive results. The indicator in this medium is phenol red.

A major aspect of metabolism is the generation and transport of electrons, yielding the energy required for metabolism. One facet of this electron transport system is the ability to reduce nitrate to nitrite (Figure 23–4). The enzyme responsible for the reaction is nitrate reductase. In this exercise, several organisms are inoculated into tubes of nitrate broth. After incubation, a test for the presence of *nitrite* is performed by the addition of *sulfanilic acid* first and then *dimethyl-alpha-naphthylamine*. The appearance of a red color is a positive test (Figure 23–5). If a red color does not occur, an additional test is necessary to determine whether nitrate was not reduced to nitrite (the absence of nitrate reductase) or whether nitrate was converted to gaseous nitrogen by means of *denitrification* (the presence of nitrite reductase). To determine what actually occurred, a small amount of powdered zinc is added to the culture. The zinc will reduce nitrate to nitrite, producing a positive (red) reaction. This reaction will show that the organism did not possess the necessary enzyme. A negative result at this point will confirm the presence of *nitrite reduction*. The presence of gaseous nitrogen can be demonstrated with the use of an inverted, medium-filled vial in the test during incubation. Gas formed during the test is trapped in the vial and appears as a bubble.

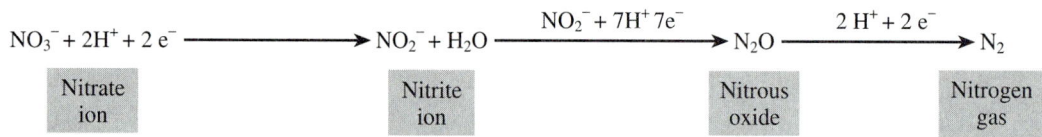

Figure 23–4
Nitrate reduction.

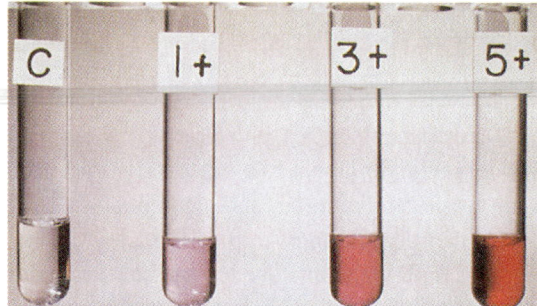

Figure 23–5
Nitrate reduction. A control tube (C) is included with the nitrate reduction reactions to emphasize the range of color changes that can occur.

A Nitrogen Metabolism

Materials

The following items should be provided per 4 students:

❏ 1. Twenty-four-hour nutrient broth cultures of the following:
- ❏ a. *Bacillus subtilis*
- ❏ b. *Escherichia coli*
- ❏ c. *Micrococcus luteus*
- ❏ d. *Proteus vulgaris*
- ❏ e. *Pseudomonas aeruginosa*

❏ 2. Five tubes of the following media:
- ❏ a. Nutrient gelatin agar tubes
- ❏ b. Nitrate broth
- ❏ c. Urea broth with phenol red indicator

The following items will be provided after incubation:

❏ 1. Sulfanilic acid
❏ 2. Dimethyl-alpha-naphthylamine solution
❏ 3. Powdered zinc

Procedure

This procedure is to be performed by students in groups of 4.

❏ 1. Inoculate 1 set of media containing gelatin agar, tryptone broth, nitrate broth, and urea broth with each bacterial culture. Use the inoculating needle to stab the gelatin medium down to the bottom of the tube. Refer to Procedure Diagram 5, Exercise (4) as a guide to broth inoculations.

❏ 2. Incubate at 37°C for 24 hours.

❏ 3. **After incubation,** determine the results as indicated with the respective tests, and record them in Table 23–1 as follows:

- ❏ a. Gelatin hydrolysis
  - ❏ i. Place the tube of nutrient gelatin into a refrigerator for 15 minutes.
  - ❏ ii. Gelatinase activity is indicated when the medium remains liquid after refrigeration.
  - ❏ iii. Record your findings in Table 23–1 of the Results and Observations section. (Refer to Figure 23–1.)
- ❏ b. Urease activity
  - ❏ i. A red color in the urea broth is a positive test. This color is caused by a pH indicator in the medium, demonstrating the presence of ammonia.
  - ❏ ii. Record your findings in the appropriate spaces of Table 23–1 in the Results and Observations section. (Refer to Figure 23–3.)
- ❏ c. Nitrate reductase activity
  - ❏ i. Add 2 drops of sulfanilic acid to each tube of nitrate broth and shake gently.
  - ❏ ii. Next, add 2 drops of dimethyl-alpha-naphthylamine to each tube and shake gently.
  - ❏ iii. The formation of a red or reddish-brown color is a positive test for nitrite.
  - ❏ iv. If the test is negative, add a small amount of powdered zinc. After a few minutes, test for nitrite. A red color at this point demonstrates a true negative nitrate reductase test. (Refer to Figure 23–5.)
  - ❏ v. Record your findings in Table 23–1 of the Results and Observations section.

Results and Observations

Oxidase Test

1. Complete the following table.

Table 24–1

Comparison of Oxidase Test Reactions

Microorganism	Filter Paper Reactions		Agar Plate Reactions		Oxy-Swab Reactions	
	Positive	Negative	Positive	Negative	Positive	Negative
M. catarrhalis						
E. coli						
P. aeruginosa						

2. Were the results obtained with the filter paper test identical to those obtained with the agar plate colonies and Oxy-Swabs? If not, explain your findings. _____

_____

_____

Catalase Test

1. Which of the microorganisms used produced a positive catalase reaction? _____

_____

2. Can this test be performed with broth cultures? _____

_____

Laboratory Review 24 Oxygen Utilization: Oxidase and Catalase Activities

1. What is the substrate acted upon by catalase? _____

2. What type(s) of microorganisms produce catalase? _____

_____

3. Give the reagents used in the following tests for:

a. oxidase: _____

b. catalase: _____

energy to the phosphorylation of adenosine diphosphate (ADP) to form adenosine triphosphate (ATP). This process is called **oxidative phosphorylation.** This exercise deals with cytochrome c and *catalase,* two enzymes involved in the utilization of oxygen by aerobically respiring bacteria.

Cytochromes are heme-containing enzymes that are tightly bound in the plasma membranes of prokaryotes. The oxidase test is used to detect the presence of cytochrome c. The test uses the reagent tetramethyl-para-phenylenediamine, which can give its electrons (electron donor) to the oxidized form of cytochrome c. The oxidized form of this reagent forms a dark violet or purple color (Figure 24–2*a*), while the reduced form is colorless. Commercially available reagent-containing swabs such as Oxy-Swab make testing for cytochrome c relatively simple (Figure 24–2*b*).

(a) (b)

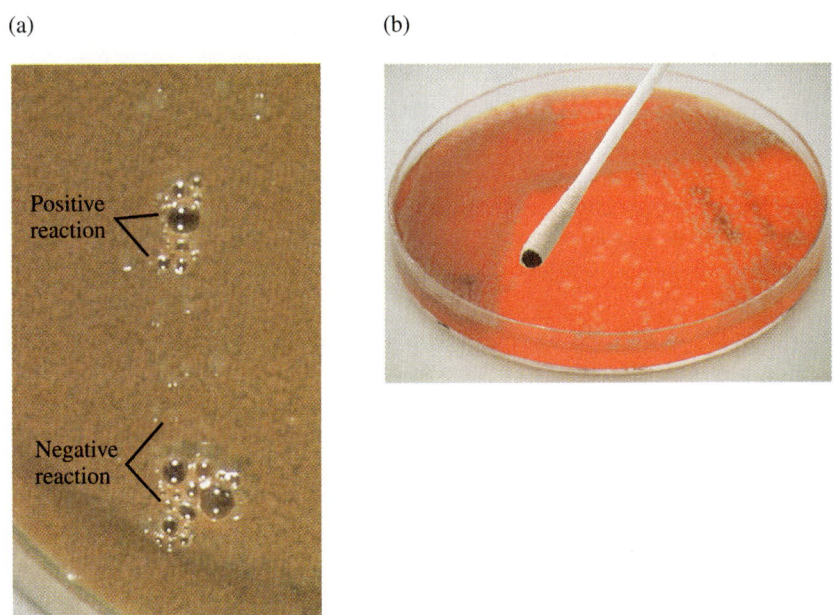

Figure 24–2
The oxidase test. (a) Colonies demonstrating a positive reaction are pink and turn dark red later. Negative reactions are indicated by colorless colonies. (b) A positive oxy-swab test. The reagent is contained in the swab material.
(Courtesy Remel, Lenexa, KS.)

The oxidase test is of value in detecting the presence of gram-negative cocci belonging to the genera of *Moraxella (Branhamella)* and *Neisseria.* In addition, the test can distinguish between oxidase-negative gram-negative enteric bacterial rods, and gram-negative rods belonging to the genera of *Pseudomonas* and *Aeromonas.*

Toxic forms of oxygen such as the *superoxide radical* (O_2) are found in all aqueous environments containing dissolved oxygen. Since these chemicals are toxic to living cells, microorganisms in aerobic environments must be able to detoxify them. Such microbes produce an enzyme, *superoxide dismutase* (SD), which adds protons to the superoxide radical to form hydrogen peroxide:

$$2\,O_2 \xrightarrow{\;2\,e^-\;} 2\,O_2^{\cdot} \xrightarrow[\text{(SD)}]{\;4\,H^+\;} 2H_2O_2$$

| **Superoxide** | **Hydrogen peroxide** |

Since hydrogen peroxide also is toxic, microorganisms also must produce the enzyme *catalase,* which is capable of decomposing the chemical into water and molecular oxygen in order to survive.

$$2H_2O_2 \xrightarrow{\;O_2\;} 2H_2O$$
Catalase

Catalase is produced by all actively growing aerobic microbes. Since strict anaerobes lack the ability to use oxygen in their respiration, they also lack catalase.

(a)

(b)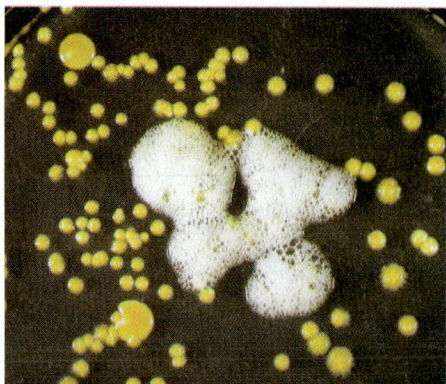

Figure 24–3

The catalase reaction. Catalase is an enzyme that catalyzes the breakdown of hydrogen peroxide (H_2O_2), thereby releasing the oxygen gas. The formation of a white froth when a few drops of 3% H_2O_2 are added to a microbial colony or to a broth culture is a positive reaction. (a) A positive reaction in a broth culture. (b) The agar plate reaction.

The test for catalase is simple to perform. It only requires the adding of a few drops of 3% hydrogen peroxide directly to a young broth culture (Figure 24–3a), to colonies on an agar surface (Figure 24–3b), or to a clump of cells on a glass slide. The vigorous evolution of oxygen bubbles is a positive result. Cultures growing on blood agar should not be used since the agar itself may produce a positive reaction.

The catalase test is very useful in differentiating among bacteria with similar morphological features, but differing in their metabolic activities. See Table 24–3 for a summary of reactions.

A. The Oxidase Reaction: A Comparison of Tests

Materials

The following materials should be provided per 4 students:

❑ 1. Twenty-four-hour trypticase soy agar plate cultures of:
   ❑ a. *Moraxella (Branhamella) catarrhalis*
   ❑ b. *Escherichia coli*
   ❑ c. *Pseudomonas aeruginosa*
❑ 2. One filter paper sheet cut to fit the inside of a Petri plate bottom
❑ 3. One Petri plate
❑ 4. One container of a freshly prepared 1% aqueous solution of tetramethyl-para-phenylenediamine (oxidase reagent) with a dropper
❑ 5. One marking pen or similar device
❑ 6. Three Remel Oxy-Swabs®
❑ 7. Container for disposal of contaminated items

Procedure 1: The Filter Paper Test

This procedure is to be performed by students in groups of 4.

❑ 1. With a marking pen, divide the filter paper sheet provided into three equal sectors. Label one sector for each organism provided.
❑ 2. Insert the filter paper inside one-half of a disposable Petri plate, and place a few drops of the oxidase reagent into the center of one marked filter paper sector.
❑ 3. Aseptically remove a colony from the agar plate of the culture for which the sector was marked. Rub the sample on the reagent moistened filter paper.

❏ 4. Examine the treated area, and record the presence of any color appearing within 10 seconds in the Results and Observations section.

❏ 5. Repeat steps 3 and 4 with each of the other two cultures provided.

❏ 6. Do not discard the cultures.

Procedure 2: The Plate Test

This procedure should be performed by students in groups of 4.

❏ 1. Place 1 drop of the oxidase reagent on two isolated colonies at each plate culture provided.

❏ 2. Examine the plates for color changes. (Refer to Figure 24–2.) If the test is positive, colonies should first turn red and then a deep violet color.

❏ 3. Record your findings in the Results and Observations section and answer the questions in the Laboratory Review.

Procedure 3: The Oxy-Swab Test

This procedure may be performed as a class demonstration or by students in groups of 4.

❏ 1. Remove one Oxy-Swab from its package.

❏ 2. Touch the swab to the surface of one colony growing on one of the agar plate cultures provided.

❏ 3. Look for the formation of a dark violet or black color on the swab. (Refer to Figure 24–2*b*.)

❏ 4. Record your findings in the Results and Observations section and answer the questions in the Laboratory Review.

❏ 5. Repeat steps 1 through 4 with the other cultures provided.

B. The Catalase Reaction

Materials

The following materials should be provided per 2 students:

❏ 1. Forty-eight-hour trypticase soy broth culture of the following:
   ❏ a. *Staphylococcus aureus*
   ❏ b. *Streptococcus faecalis*

❏ 2. One trypticase soy agar plate

❏ 3. One container of a 3% hydrogen peroxide solution and a pipette for dispensing (to be provided after incubation).

Procedure

This procedure is to be performed by students in pairs.

❏ 1. With a wax marking pencil, divide the bottom of the trypticase soy agar plate. Label one sector for each organism provided.

❏ 2. Streak one half of the plate with *S. aureus* and the other half with *S. faecalis*.

❏ 3. Incubate the inoculated plate at 37°C for 24 hours.

❏ 4. Add 2 or 3 drops of the 3% hydrogen peroxide solution to an isolated colony of each organism used.

❏ 5. A positive test is indicated by the production of gas bubbles. (Refer to Figure 24–3.)

❏ 6. Answer the questions in the Results and Observations section and the Laboratory Review.

Results and Observations

Oxidase Test

1. Complete the following table.

Table 24–1

Comparison of Oxidase Test Reactions

Microorganism	Filter Paper Reactions		Agar Plate Reactions		Oxy-Swab Reactions	
	Positive	Negative	Positive	Negative	Positive	Negative
M. catarrhalis						
E. coli						
P. aeruginosa						

2. Were the results obtained with the filter paper test identical to those obtained with the agar plate colonies and Oxy-Swabs? If not, explain your findings. _____

_____

_____

Catalase Test

1. Which of the microorganisms used produced a positive catalase reaction? _____

_____

2. Can this test be performed with broth cultures? _____

_____

Laboratory Review 24   Oxygen Utilization: Oxidase and Catalase Activities

1. What is the substrate acted upon by catalase? _____

2. What type(s) of microorganisms produce catalase? _____

_____

3. Give the reagents used in the following tests for:

a. oxidase: _____

b. catalase: _____

4. Define or explain the following terms:

   a. cytochrome *c:* _____

   b. superoxide dismutase: _____

   c. respiratory catabolism: _____

   d. tricarboxylic acid cycle: _____

Key Terms

ADP: adenosine diphosphate

ATP: adenosine triphosphate

catalase (KAT-a-lās): the enzyme that catalyzes the decomposition of hydrogen peroxide to water and oxygen

cytochrome c: a heme-containing enzyme that functions in the transfer of electrons to oxygen, which combines with hydrogen to form water

Embden-Meyerhoff-Parnas pathway: the metabolic pathway used by certain bacteria to obtain the energy required for growth and other activities

oxidative phosphorylation (fos-for-i-LĀ-shun): refers to the reactions associated with the transfer of electrons from one cytochrome to another along with the transfer of some energy the combining of phosphate to ADP to form ATP

respiratory catabolism: The oxidative, energy-yielding portion of respiratory metabolism

tricarboxylic acid cycle: a series of enzymatic cellular reactions involving the oxidative metabolism of pyruvic acid and the release of energy

---

Table 24–2

Summary of Biochemical Reactions

Test	Substrate or Detects Presence of	Regent and Times	Positive Reactions	Negative Reactions
Oxidase (cytochrome c)	no substrate; detects electron transport enzyme cytochrome c	tetramethyl-*R* phenylenediamine depending on procedure may be necessary 24-48 hours	formation of a deep violet or purple color within 10 seconds after additon of reagent	no major color change after addition of reagent
Catalase	no substrate; detects the presence of the enzyme catalase	hydrogen peroxide cultures must be less than 24 hours old	formation of visible bubbles after addition of reagent	no visible bubbles formed after addition of reagent

Differential Test Patterns: The IMViC Test

After completing this exercise, you should be able to:

1. Perform and interpret the IMViC set of tests.
2. List the biochemical basis of each IMViC reaction.
3. List the specific products detected in positive IMViC reactions.
4. Distinguish *Escherichia coli* from *Enterobacter aerogenes* on the basis of IMViC test results.
5. List the specific IMViC test reagents.

Bacteria, like all other forms of life, may change their respective environments to some degree and use a variety of chemical substances as sources of energy and as building material for cellular growth, repair, and reproduction. All such cellular activities are under the direction of enzymes. The tests in this exercise will clearly show that the chemical end products formed by certain enzymatic actions or the disappearance of specific substances from a test medium can be detected. This exercise will also show that by grouping a series of different tests, a pattern of activity can be determined for a specific bacterial species, which can then be used in the identification and/or differentiation of the microorganism from closely related species. The pattern of reaction results reflects the enzymatic makeup or biochemical fingerprint of the microorganism being studied. For example, *Escherichia coli*, a normal inhabitant of the intestinal tracts of humans and lower animals, resembles *Enterobacter aerogenes,* a bacterial species that is widely distributed in nature, especially on plants and plant products. Both organisms are gram-negative and are similar in morphological and cultural characteristics. Differentiating these bacteria from one another can be achieved by means of the **IMViC** set of tests. Each letter in the series (except *i*) refers to a separate procedure. *I* stands for the *indole test,* the *M* represents the *methyl red test,* the *Vi* refers to the *Voges-Proskauer reaction,* and the *C* is for the *citrate test.* The small *i* after the *V* is used to make pronunciation of the series easier.

The indole test is used to determine the ability of an organism to cleave the amino acid tryptophan into *indole, ammonia,* and *pyruvic acid* (Figure 25–1). During the metabolic activities of microorganisms, amino acids enter cells and are subsequently degraded by specific intracellular enzymes, the *decarboxylases* and *deaminases.* The former remove the carboxyl group (—COOH) from an amino acid, while the latter remove the amine group (—NH_2) from an amino acid. Tests for these enzymatic reactions are routinely used in the identification of certain bacterial species. The indole test will be used in this exercise to detect *deamination.* The procedure here involves the inoculation of a medium containing the amino acid *tryptophan.* If the organism being studied has the enzyme *tryptophanase,* removal of the amino group and indole formation from tryptophan will occur. Indole is easily detected by the addition of Kovacs' reagent and the appearance of a cherry red-colored layer (Figure 25–2).

The methyl-red (MR) and Voges-Proskauer (VP) tests are used to differentiate two major types of facultatively anaerobic enteric bacterial species. These organisms first metabolize glucose aerobically, thus exhausting all of the avail-

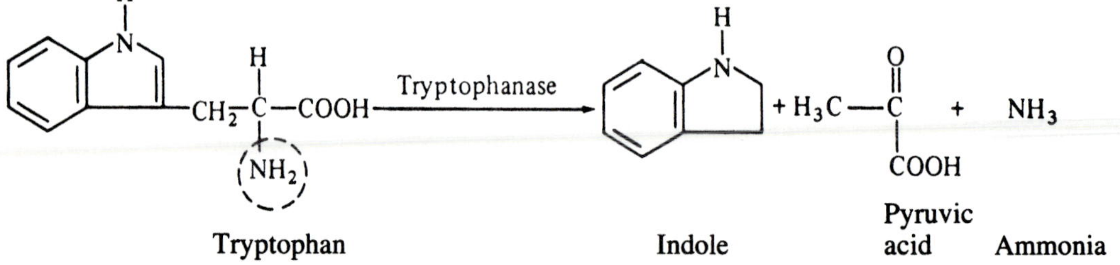

Figure 25–1

The indole test reaction. After incubation, Kovac's solution or other appropriate reagents are added to the test medium. The formation of a red-colored layer is a positive test.

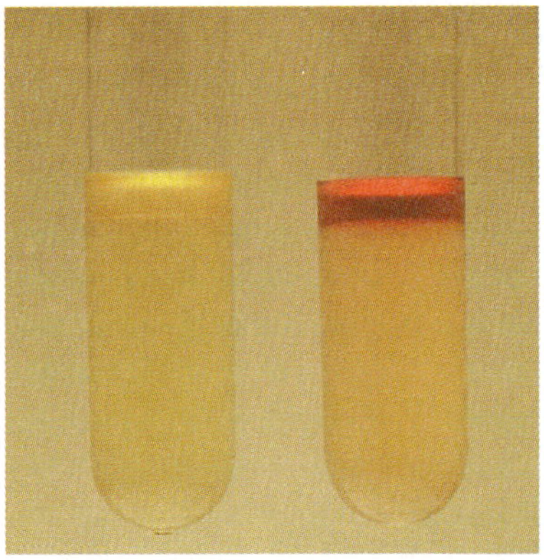

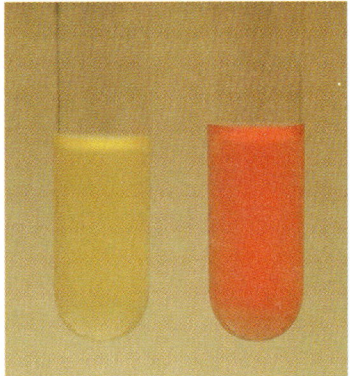

Figure 25–2
The Indole test. *Left tube:* negative. *Right tube:* positive.

able oxygen by means of respiratory metabolism. This is then followed by one of two types of glucose fermentation, *mixed-acid* or *butylene glycol*. The type of fermentation and the end-products formed are of value in species identification. Enteric bacterial species that use mixed-acid fermentation of glucose excrete large quantities of *acetic, formic, lactic,* and *succinic acids,* and *ethanol*. These excreted metabolic products lower pH significantly to approximately 4.2, which is detectable by the methyl red indicator (Figure 25–3).

Figure 25–3
Methyl Red Test. *Left tube:* negative. *Right tube:* positive.

The Voges-Proskauer (VP) test is used to detect *acetoin,* also known as *acetylmethyl carbinol*. This is an intermediate compound produced by organisms carrying out the butylene glycol type of glucose fermentation. A positive reaction is indicated by the formation of a pink or cherry-red color upon the addition of alpha-naphthol and potassium hydroxide (KOH) creatine solutions (Figure 25–4). The VP test is of major importance differentiating *E. coli* from *E. aerogenes* and *Klebsiella* species.

Figure 25–4
The Voges-Proskauer Test. *Left tube:* negative. *Right tube:* positive.

The final test in the series, the *citrate test,* determines the presence of enzymes that enable citrate to enter the cell and then be used as a sole source of carbon for its metabolism and growth. Two different media can be used for citrate test, *Simmons citrate* and *Koser's citrate. Simmons citrate agar* contains the pH indicator bromothymol blue, which is green under acidic conditions, and dark blue when the medium becomes alkaline (Figure 25–5*a*). Organisms utilizing citrate produce an alkaline reaction. The positive reaction in Koser's citrate medium (a clear colorless liquid) is indicated by the preparation becoming cloudy (Figure 25–5*b*) after incubation. A light inoculum must be used with Koser's citrate to avoid false positive results. Both types of media will be used in this exercise. Table 25–1 summarizes all of these reactions.

(a)   (b)

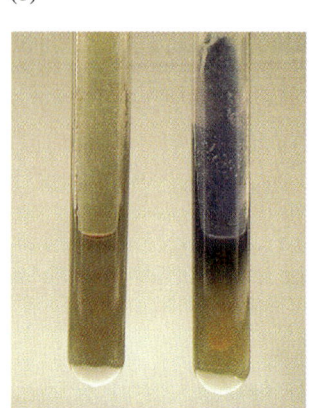

Figure 25–5
The Citrate Test (a) Kosers' citrate. *Left tube:* negative. *Right tube:* positive. (b) Simmons' citrate. *Left tube:* negative. *Right tube:* positive.

A. The IMViC Set of Reactions

Materials

The following items should be provided per 4 students:

❑ 1. Trypticase soy broth 24-hour cultures of the following bacterial species:
   ❑ a. *Enterobacter aerogenes*   ❑ c. *Proteus vulgaris*
   ❑ b. *Escherichia coli*   ❑ d. *Staphylococcus aureus*

❑ 2. Four tubes (except where noted differently) of the following media:
   ❑ a. Tryptone broth   ❑ c. Koser's citrate medium
   ❑ b. Methyl red—Voges-Proskauer broth (8 tubes)   ❑ d. Simmon's citrate medium

The following reagents listed should be in dispenser bottles or accompanied by eye droppers and provided after incubation:

❑ 1. Kovac's reagent   ❑ 3. Alpha-naphthol solution

❑ 2. Methyl red indicator   ❑ 4. Potassium hydroxide-creatine solution

Procedure

The IMViC tests are to be carried out by students in pairs.

❑ 1. Inoculate each organism provided into 1 tube of tryptone broth, 2 methyl red—Voges-Proskauer broth tubes, and 1 tube each of the Koser's citrate and Simmon's citrate media.

❑ 2. Incubate all tubes at 37°C for 24 hours, and then perform the following procedures:
   ❑ a. **Indole test**
      ❑ i. Add 5 drops of Kovac's reagent to each tube.
      ❑ ii. The development of a red color is a positive test. (Refer to Figure 25–2.)

❏ b. Methyl red test
  ❏ i. Add 4 or 5 drops of the indicator to 1 set of methyl red broth cultures.
  ❏ ii. The development of a red color is a positive test. A yellow color is a negative reaction. (Refer to Figure 25–3.)
  ❏ iii. Record questionable color reactions as +.
❏ c. Voges-Proskauer test
  ❏ i. Add 5 drops of the alpha-naphthol solution.
  ❏ ii. Add 5 drops of the potassium hydroxide-creatine solution, then shake well for 1 minute.
  ❏ iii. A reddish color (which may develop slowly) is a positive reaction. (Refer to Figure 25–4.)
❏ d. Citrate test
  ❏ i. The presence of growth in Koser's citrate medium is a positive reaction. With Simmon's citrate medium, the development of a blue color is a positive reaction. (Refer to Figure 25–5.)
  ❏ ii. Record your data and answer the questions in the Results and Observations section.

Table 25–1

Summary of Biochemical Reactions

Test	Substrate(s)	Reagent(s) and Times	Positive Reactions	Negative Reactions
Indole	tryptophan	Kovacs' reagent/ 24–48 hours	indole production (dark red) color in layer on top of broth	negative indole production (any color other than dark red)
Methyl red	glucose	methyl red indicator/ 48 hours	acid production (red color)	no acid production (any color other than red)
Voges-Proskauer	glucose	alpha-naphthol, and KOH plus creatine/48 hours	acetoin produced (red color)	no acetoin produced (any color other than red)
Citrate immons citrate agar	sodium citrate	Bromothymol blue/ 24–48 hours	growth (agar turns blue)	no growth (medium remains green)
Koser's citrate medium	sodium citrate	none/24–48 hours	growth (medium becomes cloudy)	no growth medium is clear

Results and Observations

1. IMViC reactions. Complete the following table. Indicate a positive test by "+" and a negative test by "−."

2. Were there any differences between the results observed in the Koser's and Simmon's citrate media? _____

_____

3. Can a false positive result be obtained with Koser's citrate medium? Explain. _____

_____

4. In which test does deamination occur? _____

Table 25–2

IMViC Reactions

Microorganism	Test			C^a	
	I	M	Vi	K	S
E. coli					
E. aerogenes					
P. vulgaris					
S. aureus					

aK = Koser's citrate; S = Simmons' citrate.

Laboratory Review 25   **Differential Test Patterns: The IMViC Test**

1. What is an enteric organism? _____

2. Indicate the reagents used in the following tests.

   a. indole: _____

   b. Voges-Proskauer: _____

   c. Simmon's citrate: _____

   d. Koser's citrate: _____

3. What are the IMViC test patterns for the following bacteria?

   a. *Escherichia coli:* _____

   b. *Enterobacter aerogenes:* _____

4. In which IMViC test is acetoin formed? _____

Key Terms

deaminase: an enzyme that catalyzes the removal of an amino group (NH_2) from organic compounds

deamination: loss of an amino group from amino-containing compounds

decarboxylase (dē-kar-BOK-si-lās): an enzyme that catalyzes the release of carbon dioxide (CO_2) from compounds such as amino acids

enteric (en-TER-ik): pertains to the intestines and related structures

intracellular enzymes: catabolic and anabolic enzymes that perform their functions within the cell

Photograph Quiz 19

Interpret the IMViC set of reactions shown.

I: _____

MR: _____

VP: _____

C: _____

C: _____

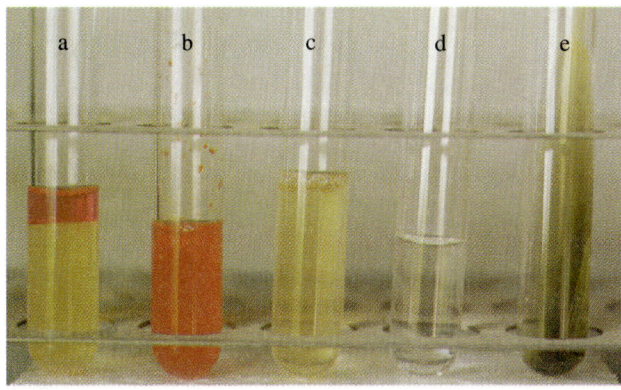

Figure 25–6

IMViC reaction results. The tests are from left to right: (a) Indole, (b) Methyl Red, (c) Voges-Proskauer, (d) Citrate (Kosers' medium), (e) Citrate (Simmons' medium).

Microbial Reactions in Multiple-Test Media: Litmus Milk, Triple Sugar Ion Agar, and Sulfide Indole Motility Medium

After completing this exercise, you should be able to:

1. Inoculate various types of multiple-test media.
2. Identify and interpret the characteristic reactions produced by microorganisms in multiple-test media.
3. Recognize the value of multiple-test media in the identification of microorganisms.

The identification of bacterial species is based on schemes, or keys, that take into consideration a variety of characteristics, including differential staining reactions, colonial properties, and biochemical tests that determine the ability of a particular organism to utilize or attack certain substances (substrates) and to produce chemical products that can be analyzed. Traditionally, the use of a variety of well-chosen biochemical tests has offered the best means of specific, or near specific, identification of unknown bacterial cultures. Assistance in this type of activity is provided by numerous reference works and identification schemes containing the results of key reactions obtained with a known bacterial species.

In many cases, the biochemical tests used in the identification of unknown cultures are performed in separate Petri plates or test tubes. These rather standardized procedures are not only costly but time-consuming. It is no small wonder, then, that investigators have long been interested in reducing the expense and drudgery associated with microbiological methods. The efforts of many individuals have resulted in the development of at least three general categories of improved biochemical testing procedures and materials: (1) several test substrates combined in one or two tubes that are inoculated and incubated in the conventional manner; (2) separate biochemical tests contained in miniaturized, multi-compartment devices that are inoculated by other than conventional methods but incubated according to standard practices; (3) biochemical substrates, impregnated in paper strips, that are inoculated by other than conventional methods and produce reactions in a significantly less time than conventional tests.

The first category includes several widely accepted media such as litmus milk (Figures 26–1 and 26–2), triple sugar iron agar (Figure 26–3), and sulfide indole motility medium (Figure 26–4). This exercise will consider these media, and the following exercise will deal with the tests of the second category.

Very few media, after inoculation and incubation, can yield as much information on microbial reactions as litmus milk. The medium includes materials that most bacteria require for growth, and specifically the substrates *lactose* and the protein *casein*. The milk is composed of materials that most bacteria require for growth. Addition of litmus provides an indicator of the organism's acid or alkali production and oxidation and reduction activities. (Characteristic reactions observed with litmus milk are as follows [Figures 26–1 and 26–2].)

1. Acid and acid curd formation
2. Alkaline conditions
3. Rennin curd production
4. Peptonization
5. Litmus reduction
6. Gas formation

Production of acid by an organism is a function of its ability to utilize lactose, one of the most abundant sugars in milk. Acid production is demonstrated when a litmus medium changes from blue to pink. If the organism is able to produce a considerable amount of acid, an insoluble complex of calcium and casein may be formed, resulting in a curd. An organism that cannot attack lactose might well utilize the milk proteins as a source of nitrogen and carbon, and an alkaline reaction might result. This reaction is indicated by an intensification of the blue color in the litmus medium. If the amount of acid or alkali produced is low, no apparent change in the medium may be observed. However, a rennin curd

(a)   (b)   (c)   (d)   (e)   (f)

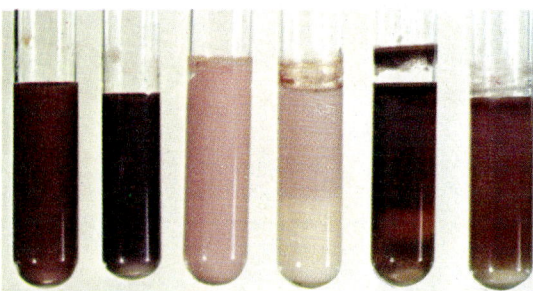

Figure 26–1

Selected litmus milk reactions. Tube a, **uninoculated.** Tube b, an **alkaline** reactions (blue to purple), indicating utilization by the inoculated organism of the milk proteins as a source of carbon and nitrogen. Tube c, an **acid** result (pink to red), demonstrating the production of a considerable quantity of acid. Tube d the beginning of **litmus reduction** (white portion of the medium). The reaction is caused by the oxidation-reduction activities of the inoculated bacteria. This reaction begins at the bottom of the tube and spreads upward. Tubes e and f, **casein curd formation** and **peptonization.** With peptonization, a reduction in curd size occurs and a brownish or light clear supernatant, called **whey,** is formed.

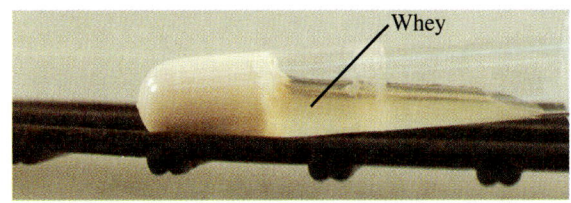

Whey

Figure 26–2

Coagulation and **peptonization.** Note the solidification of the milk protein and the clear liquid or whey.

may develop even with low acid or alkali production if the organism contains enzymes that can produce the insoluble calcium-casein complex. This type of curd usually retracts and yields a grayish liquid known as *whey.* Acid or rennin curds are quite palatable dairy products known as *cottage cheese.* Either type of curd can be digested by bacterial proteinases, a process known as *peptonization* (or *solubilization*). Peptonization is usually characterized by a reduction in curd size and the formation of a brownish liquid. This reaction can be observed easily when it occurs alongside the curd. Reduction of the litmus by oxidation-reduction activities of the bacteria can be shown by loss of the litmus color. This reaction is quite apparent when a pink acid curd begins to turn white. The change develops at the bottom of the tube and moves upward. Another readily apparent characteristic is gas production, which can be observed as holes or tears in curds or as separated curd strands.

Triple sugar iron agar is a characteristic preparation used to differentiate gram-negative enteric organisms by their ability to ferment dextrose (glucose), lactose, or saccharose (sucrose) and to reduce sulfites to sulfides. The medium is dispensed as agar slants. It is inoculated through the insertion of an inoculating needle into the bottom or butt region of the tube: as the needle is withdrawn from this area, the slant surface is streaked. Several types of reactions can be recognized. (See Figure 26–3 and Table 26–1.)

Table 26–1

Triple Sugar Iron Agar Reactions

Appearance of	Positive Only for Dextrose Fermentation*	Positive Lactose and/or Saccarose Fermentation*	Negative Fermentation*	H₂S Production	Gas Production from Sugars
Slant	Red (ALK)	Yellow (A)	Red (ALK)	Blackening of the medium	Holes or disrupted medium
Butt	Yellow (A)	Yellow (A)	Red (ALK)		

*ALK, alkaline reaction; A, acid.

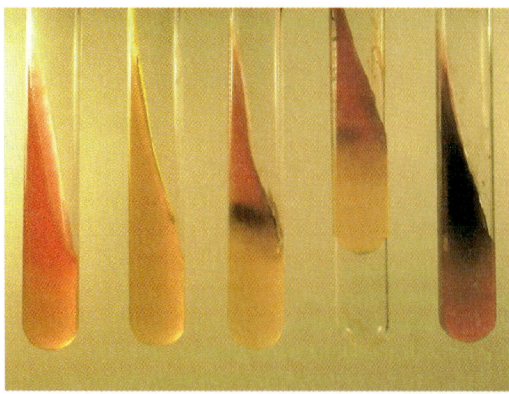

Figure 26–3

Selected triple sugar iron agar medium reactions. This medium is an example of a preparation used in the identification of gram-negative enteric pathogenic bacteria. Alkaline reactions (**Alk**) are indicated by a red coloration, acid (**A**) production by a yellowing of the medium, gas formation by the presence of air pockets in the preparation, and the presence of **H₂S** by a blackening of the media. A summary of the reactions shown is as follows.

	Tube a	Tube b	Tube c	Tube d	Tube e
Slant	Alk	Acid	Alk	Alk	Alk
Butt (bottom)	A	A	A	A	Alk
Gas	Absent	Absent	Absent	Absent	Absent
H₂S	Absent	Absent	Present	Present	Absent[a]

[a]The dark color here is caused by the purple pigment of the organism. Note acid in glucose (bottom) must first be produced before H_2S can occur.

When a culture ferments only dextrose (whose concentration is one-tenth that of lactose or sucrose), the entire medium will turn yellow from the acid produced. Since the concentration of glucose is low, it is quickly utilized. In the event that the organisms in the culture cannot ferment either of the other two carbohydrates in the medium, growth in the butt stops. However, the organisms on the slant continue to utilize, as a source of energy, either the peptone in the medium, the intermediate products of dextrose fermentation, or both. Use of these intermediate products results in their reduction, and the use of peptone results in the eventual secretion of ammonia into the medium. The ammonia will neutralize the intermediate products, causing the slant to turn red.

The entire medium will turn yellow and remain so if the culture can utilize lactose, sucrose, or both, in addition to dextrose. Apparently, the organisms in such cases will not exhaust the available fermentable carbohydrates. The peptones, or the intermediate end products, are not utilized.

Hydrogen sulfide (H_2S) production by a culture results in the blackening of the medium. This color is caused by the production of H_2S from an ingredient of the medium, sodium thiosulfate, which, when combined with another component of the medium, ferrous ammonium sulfate, results in the formation of the black insoluble compound ferrous sulfide.

Sulfide indole motility medium can be used to determine hydrogen sulfide production, indole production, and motility. The presence of H_2S is indicated by a blackening along the line of inoculation in the medium. Quite often, the dark color diffuses when a motile organism is involved. The presence of indole is detected by the use of Kovac's reagent. Motility is recognized by the appearance of a diffuse growth or turbidity from the inoculation site. (See Figure 26–4.)

(a)   (b)   (c)   (d)

Figure 26–4

Bacto-SIM medium reactions. Tube a, uninoculated, Tube b, positive for motility (general turbidity in the medium). Tube c, positive for hydrogen sulfide production (blackening of the medium), positive for indole (red layer), and positive for motility. Tube d, positive for indole, and motility.

All procedures in this exercise are to be performed by students working in groups of 4. Each student is to select one of the organisms provided and use it for all inoculations. After incubation, all the results obtained with the different organisms and multiple-test systems are to be compared. (See Table 26–7.)

A. Litmus Milk

Materials

❑ 1. Twenty-four-hour nutrient broth cultures of the following (1 per 4 students):
    ❑ a. *Escherichia coli*   ❑ c. *Bacillus subtilis*
    ❑ b. *Proteus vulgaris*
❑ 2. Twenty-four-hour thioglycollate broth culture of *Clostridium sporogenes* (klō-STRID-ē-um, spor-a-GEN-ēz) (1 per 4 students)
❑ 3. Tubes of litmus milk (5 per student, 1 tube to be used as a sterile control)
❑ 4. Sterile pipettes (cotton plugged) and rubber bulbs (1 per student)
❑ 5. Melted paraffin and a disposable pipette for dispensing

Additional Technique Required for This Portion of the Exercise:

Broth Transfer, Procedure Diagram 5, Exercise 4

Procedure

This procedure is to be performed by students individually.

❑ 1. Inoculate each of the 4 tubes of litmus milk medium with 1 of the bacterial cultures provided. Inoculate *C. sporogenes* as indicated in step 2. (Refer to Procedure Diagram 5.)
❑ 2. Use a sterile pipette to inoculate the remaining litmus milk medium with *C. sporogenes*. In order to do this procedure, obtain an inoculum from the bottom of the stock culture and transfer it to the bottom of the litmus medium.
❑ 3. Place approximately 2 mL of melted paraffin over the medium inoculated with *C. sporogenes*.
❑ 4. Incubate all 5 tubes at 37°C for 24 hours.
❑ 5. Observe each culture using the sterile litmus milk as a color control and record your results below. (See Figures 26–1 and 26–2.)

Results and Observations

1. Complete the following table by entering your findings.

Table 26–2

Litmus Milk Test Reactions

Organism	Acid	Alkaline	No Change	Curd	Peptonization	Reduction	Gas
E. coli							
P. vulgaris							
C. sporogenes							
B. subtilis							

B. Triple Sugar Iron Agar (TSIA)

Materials

The following items should be provided per 4 students:

❑ 1. Divide streak plates of each bacterial combination listed. The cultures should be grown on trypticase soy agar plates for 24 hours at 37°C.
    ❑ a. *Enterobacter aerogenes* and *Escherichia coli*
    ❑ b. *Proteus vulgaris* and *Serratia marcescens*
❑ 2. Four triple sugar iron agar slants

Procedure

❑ 1. With each culture specified by the instructor, perform the 4 steps shown in Procedure Diagram 24.

❑ 2. Label each tube with the organism used, and incubate after inoculation at 37°C for 18 to 24 hours.

❑ 3. After incubation, examine the medium for pH changes in the butt and on the slant, and for gas formation within the agar. Yellowing of the medium indicates acid production, and formation of H_2S results in blackening of the agar. In the Results and Observations section, record your findings using "A" for acid, "NC" for no change, and "G" for gas in the case of carbohydrate fermentations. Use "+" or "−" to express the presence or absence, respectively, of H_2S. Refer to Figure 26–3 for typical TSIA reactions.

**Procedure Diagram 24
Triple Sugar Iron Agar (TSIA)
Inoculation Technique**

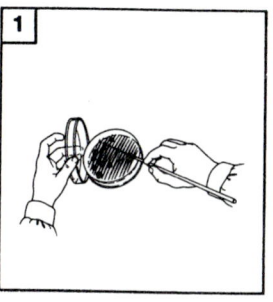

1. Using a flamed inoculating needle, pick a colony from one of the culture plates provided. Close the lid.

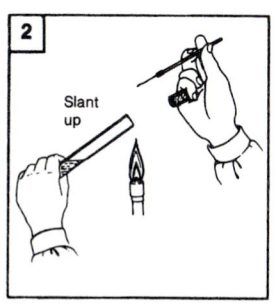

2. Remove the plug or cap of the TSIA tube and flame the lip of the tube.

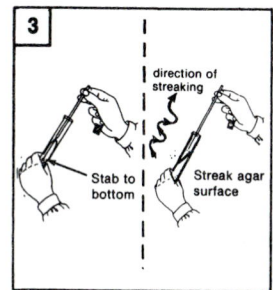

3. Inoculate the TSIA slant by inserting the needle through the medium down to the bottom or butt of the tube. Before removing the needle from the tube, streak the surface of the agar slant. Note that this type of inoculation is performed in one operation.

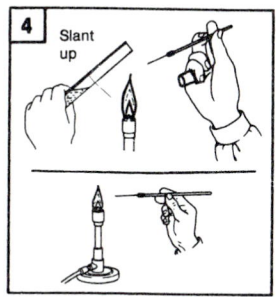

4. Flame the lip of the tube, replace the plug or cap of the tube, and flame the inoculating needle.

Results and Observations

1. Enter your findings in the following table.

Table 26–3

Triple Sugar Iron Agar Test Reactions

Organism	Butt (Tube Bottom)	Slant	Gas Production	H_2S Production
E. aerogenes				
E. coli				
P. vulgaris				
S. marcescens				

2. Was the agar slant moved significantly in any of the tubes? If so, with which one(s)? _____

C. Sulfide Indole Motility (SIM) Agar

Materials

❑ 1. Forty-eight-hour trypticase soy broth in 2-mL quantities of the following (1 each per 4 students):
   ❑ a. *Enterobacter aerogenes*   ❑ c. *Proteus vulgaris*
   ❑ b. *Escherichia coli*   ❑ d. *Staphylococcus aureus*

❑ 2. Sulfide indole motility agar medium (1 per student)

The following items will be provided after incubation for the determination of test results:

❑ 1. Kovac's reagent   ❑ 2. One-mL pipette with rubber bulb

Procedure

This procedure is to be performed by students in groups of four.

❑ 1. Inoculate this semisolid medium as shown in Procedure Diagram 25.

❑ 2. Repeat the procedure with each culture provided.

❑ 3. Incubate the preparations at 37°C for 24 hours or as directed.

❑ 4. After incubation, examine the tubes and record your findings in the Results and Observations section.

❑ 5. Carry out the indole test using sulfide indole motility agar as follows. Add several drops of Kovac's reagent. The formation of a pink color is a positive test for indole. (Refer to Figure 26–4.)

❑ 6. With this procedure, motile organisms will move into the uninoculated region. The medium will become cloudy with such movement.

Procedure Diagram 25
Sulfide Indole Motility (SIM)
Semisolid Medium Inoculation

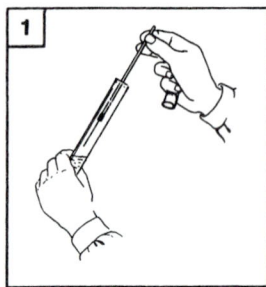 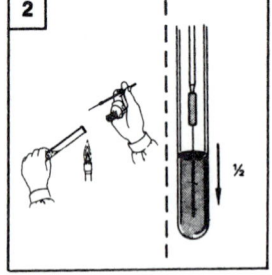 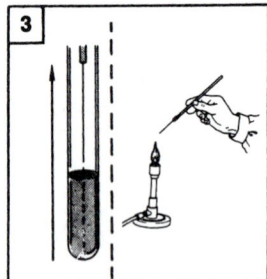

1. Using an inoculating needle aseptically remove an inoculum from one of the cultures provided. (Refer to Procedure Diagram 10.)

2. Flame the lip of the SIM agar tube and stab the needle into the center of the medium to about one-half of the way down.

3. Remove the needle through the same stab line and flame the tube and inoculating needle in the usual manner.

Results and Observations

Sulfide Indole Motility Agar Reactions

1. Enter your findings in the following table.

Table 26–4

SIM Reactions

| Organism | Sulfide Indole Motility Medium | | |
	H_2S	Indole	Motility
E. aerogenes			
E. coli			
P. vulgaris			
S. aureus			

Laboratory Review 26 — Microbial Reactions in Multiple-Test Media: Litmus Milk, Triple Sugar Iron Agar, and Sulfide Indole Motility Medium

1. Complete the following table by indicating the specific purpose(s) of each multiple-test medium listed.

Table 26–5

Multiple-Test Media Comparison

Medium	Purpose
Litmus milk	
Sulfide indole motility	
Triple sugar iron agar	

2. Interpret the reactions described for each multiple-test medium listed.

Table 26–6

Interpretations of Multiple-Test Media Reactions

Medium	Reaction	Interpretation
Litmus milk	a. Pink color throughout	
	b. Curd formation	
	c. Solid mass formed	
	d. White color throughout	
Sulfide indole motility	a. Blackening of the medium	
	b. Red color formed with addition of Kovac's reagent	
	c. Growth only on line of inoculation	
Triple sugar iron agar	a. Yellow butt	
	b. Yellowing of entire medium	
	c. Large holes in agar	

3. List 2 advantages of a multiple-test system such as triple sugar iron agar.

    a. _____

    b. _____

4. Can a mixed culture be used to inoculate a differential tube medium? Explain your answer.

    _____

    _____

Table 26-7

Summary of Biochemical Reactions

Test	Substrate(s) and Times	Reagent(s)	Positive Reactions	Negative Reactions
Litmus milk	Lactose and/or casein/24–48 hours	Litmus indicator	**acid** (pink) **alkaline** (light to dark blue) **litmus reduction** (white) **curd** (milk changes to solid state with fluid [whey]) **proteolysis** (decrease in milk turbidity and eventual formation of clear brownish or purple fluid)	No change (light blue)
Triple sugar iron agar	Dextrose, lactose, sucrose/24–48 hours	phenol red	**acid from dextrose** (yellow bottom) **acid from lactose and sucrose** (yellow slant)	**alkaline,** no change (red bottom) **alkaline** (red slant)
	sodium thiosulfate/ 24–48 hours	Ferrous ammonium sulfate	**H$_2$S-positive** (blackening of medium)	**H$_2$S-negative** (no blackening of medium)
Sulfide, indole, motility medium	Sodium thiosulfate/ 24–48 hours	Peptonized iron	**H$_2$S production** (blackening of medium)	**H$_2$S-negative** (no blackening of medium)
	Tryptone/24–48 hours	Kovac's reagent	**Indole production** (red layer)	**Indole-negative** (no red layer)
	None/18–24 hours	None	**Motility** (turbidity throughout medium)	**No motility** (growth only on line of inoculation)

Key Terms

peptone (PEP-ton): nitrogen-containing compounds resulting from the action of protein-digesting (proteolytic) enzymes

peptonization (pep-to-ni-ZĀ-shun): the process of changing proteins into peptones by proteolytic enzymes

rennin (REN-in) curd: an enzyme- (rennin) produced insoluble calcium-casein complex

whey (wāy): the liquid formed during curd formation

Photograph Quiz 20

Describe the litmus milk reaction(s) shown in the following figure. _____

_____

_____

(a)   (b)

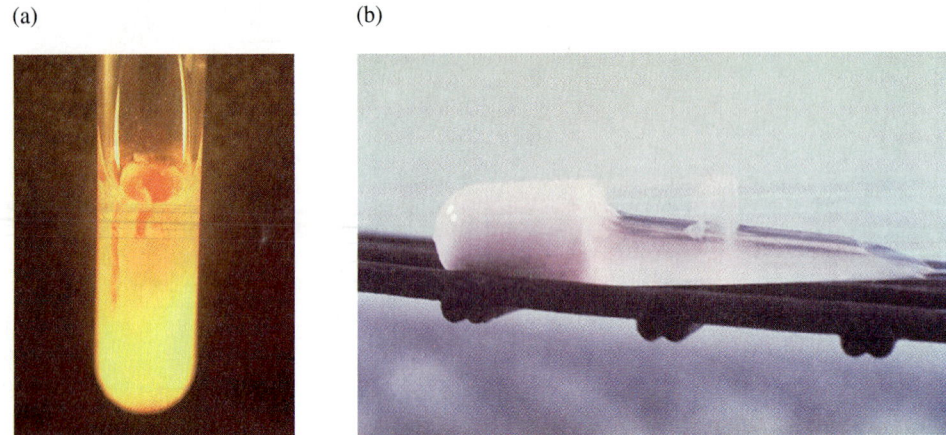

Figure 26–5
Litmus milk reactions.

After completing this exercise, you should be able to:

1. Inoculate miniaturized and rapid microbiological systems.

2. Interpret reactions possible with API Enteric and Enterotube II systems.

3. Explain the role of miniaturized and rapid systems in the laboratory identification of unknown bacterial cultures.

Improved biochemical procedures include several commercially available miniaturized multiple-test systems and devices designed to facilitate the identification of unknown bacterial cultures. Complete instructions together with accurate identification keys based on established biochemical reactions are provided by the respective manufacturers. Two systems, API 20 Enteric (Analytab Products, Inc.) and the Enterotube II (Becton Dickinson Microbiology Systems), consist of manufacturer-selected combinations of substrates that not only are ready for use but require minimal storage space and are stable at either room or refrigerator temperatures for significant time periods. The carefully selected tests in each case also form the bases for computer-developed identification systems. These two systems emphasize the importance placed on the need for the rapid analysis of test results for the identification of clinically isolated bacteria.

The API system employs a plastic strip holding 20 miniaturized compartments, or **"cupules,"** each containing a dehydrated substrate for a different test (Figure 27–1). At least 20 standard biochemical tests can be performed, including o-nitrophenyl-β-D-galactosidase (ONPG), arginine dihydrolase, lysine and ornithine decarboxylase, citrate utilization, hydrogen sulfide (H_2S) production, urease, tryptophan deaminase, indole production, acetoin production, gelatinase, and fermentation of glucose, mannitol, inositol, sorbitol, rhamnose, sucrose, melibiose, amygdalin, and arabinose. A brief listing of possible reactions can be found with an accompanying explanation in Table 27–1. The dehydrated substrates are inoculated with a bacterial suspension and subsequently incubated according to a procedure described by the manufacturer.

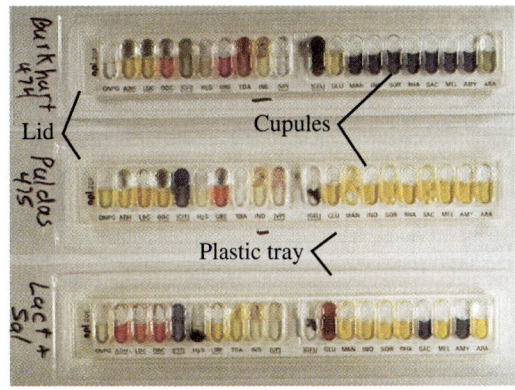

Figure 27–1

Examples of the API 20 Enteric (E) System (Analytab Products, Inc.). This rapid and ready-to-use system uses a combination of substrates which allows the performance of 22 standard biochemical tests. The various substrates are contained in individual compartments known as *cupules*. Note the following components of the system: the plastic strip with the cupules, and the plastic incubation tray and lid. The system can be labeled on the end of the incubation tray.

The Enterotube II E incorporates conventional media into a single, ready-to-use, multicompartment tube with an enclosed inoculating needle (Figures 27–2a and 27–2b). The unique inoculating arrangement permits simultaneously inoculation of all compartments and the performance of 15 biochemical tests. The Enterotube II, which is used for the identification of lactose fermenters, is used by first unscrewing the end caps of the system (Figure 27–2a). This exposes a long inoculating rod, which is used to pick an isolated colony from a selective and differential agar or related plate

Table 27–1

API Reactions

| Test | Positive Result Appearance Test Symbols (TS) | | | Condition(s) Indicated by Positive Result |
| | Enterotube | API | | |
		TS	Rx[c]	
Beta-galactosidase	n/a[a]	ONPG	Yellow	Hydrolysis of orthonitrophenol-beta-D-galactose
Arginine dihydrolase	n/a	ADH	Red	Transformation of arginine into ornithine, ammonia, and CO_2
Lysine decarboxylase	Purple	LDC	Red or orange	Cadaverine formation by removal of CO_2 (decarboxylalation) from lysine
Ornithine decarboxylase	Purple	ODC	Red	Putrescine formation by removal of CO_2 from ornithine
Citrate	Blue	CIT	Blue-green	Utilization of sodium citrate as sole source of carbon
H₂S formation	Black	H₂S	Black deposit	Reduction of sulfur-containing compounds such as peptones and sodium thiosulfate
Urease	Pink to red	IND	Red ring	Ammonia is released from urea.
Tryptophane deaminase	n/a	TDA	Brown	Formation of indolepyruvic acid from tryptophan
Indole	Pink to red	IND	Red ring	Indole formation from tryptophan
Voges-Proskauer	n/a	VP	Red	Acetoin production
Gelatin	n/a	GEL	Diffusion of black material	Gelatin liquefaction
Glucose (dextrose)	Yellow[b]	GLU	Yellow	Glucose fermentation
Mannitol	n/a	MAN	Yellow	Mannitol fermentation
Inositol	n/a	INO	Yellow	Inositol fermentation
Sorbitol	n/a	SOR	Yellow	Sorbitol fermentation
Rhamnose	n/a	RHA	Yellow	Rhamnose fermentation
Sucrose	n/a	SAC	Yellow	Sucrose fermentation
Melibiose	n/a	MEL	Yellow	Melibiose fermentation
Amygdalin	n/a	AMY	Yellow	Amygdalin fermentation
Arabinose	n/a	ARA	Yellow	Arabinose fermentation
Lactose	Yellow		n/a	Lactose fermentation
Dulcitol	Yellow		n/a	Dulcitol fermentation

Continued

Table 27–1 continued

API Reactions

| Test | Positive Result Appearance Test Symbols (TS) | | | Condition(s) Indicated by Positive Result |
	Enterotube	TS	Rxᶜ	
Esculin hydrolysis	Gray to black		n/a	Hydrolysis of esculin to esculetin
Malonate utilization	Green to blue		n/a	Formation of alkaline end-products from sodium malonate
Phenylalanine deaminase	Brown		n/a	Pyruvic acid formation by the removal of an amino group (deamination) from phenyla-lanine
Nitrate reduction	n/a		Red	Nitrate formation
Cytochrome oxidase	n/a		Dark purple	Cytochrome C oxidation
Catalase	n/a		Bubbles	Release of oxygen gas on the addition of hydrogen peroxide

ᵃn/a = no application ᶜRx = reaction
ᵇGas production can occur. This reaction is indicated by the separation of the wax overlay from the dextrose agar surface.

(a)

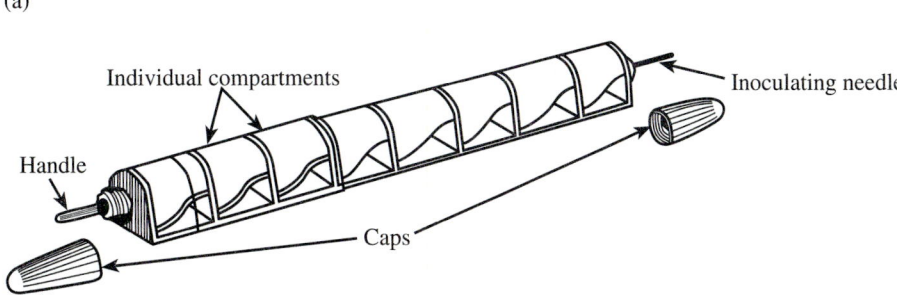

(b) 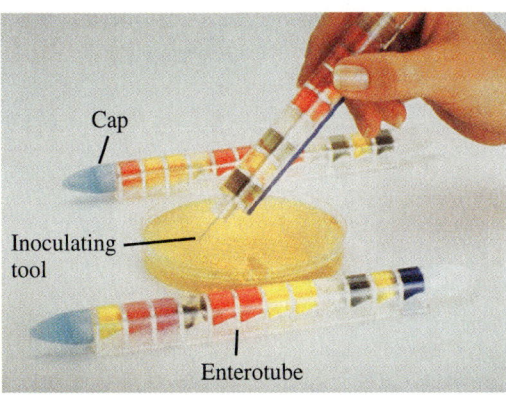 (c)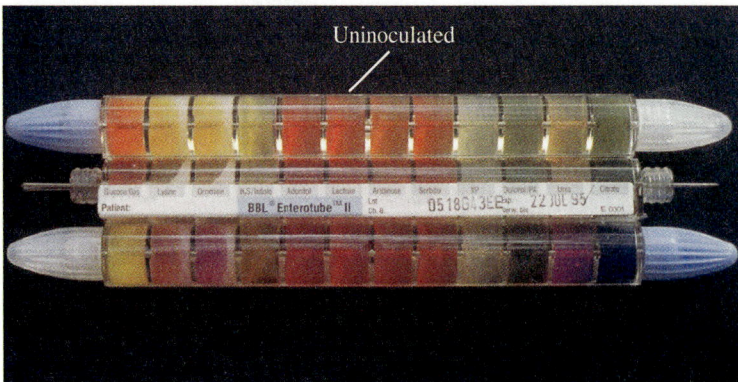

Figure 27–2

The Enterotube System (a) A diagram of the components of the multiple-test Enterotube system. (b) The inoculation procedure. The use of the inoculating instrument to obtain an inoculum. Two additional Enterotubes also are shown. (c) The results after inoculation and incubation. The top and center systems are uninoculated. Note the color changes in the bottom tube. (Courtesy Becton Dickinson Microbiology Systems.)

medium (Figure 27–2*b*). The rod is then drawn through a series of substrate-containing agar compartments in the plastic tube comprising the system. The end caps are then replaced and the system is incubated. After incubation, appropriate reagents are introduced into the chambers requiring them. The color changes resulting from the biochemical activity of unknown bacterial cultures used as inocula are interpreted and used in their identification.

The various systems and techniques described provide convenient means for performing the traditional biochemical tests used in the identification of unknown bacterial cultures. Each has certain advantages and disadvantages that should be considered before the decision is made to adopt one particular system. This exercise will cover a representative selection of systems to show their potential value in culture identification. The specific tests of the API 20E and the Enterotube II systems, together with the descriptions of expected positive results, are listed in Table 27–1.

All procedures in this exercise are to be performed by students working in groups of 4. Each student is to select one of the organisms provided as the source of inoculation material for all miniaturized and rapid-test systems provided. After incubation, all the results obtained with the different organisms are to be compared. Your instructor will indicate the procedure to follow for the disposal of all systems used.

Materials: For All Systems

The following materials should be provided per 4 students:

❑ 1. One divided streak plate of each of the bacterial combinations listed (1 per 4 students). The cultures should be grown on tryptic soy agar or trypticase soy agar for 24 hours at 37°C.
    ❑ a. *Enterobacter aerogenes* and *Escherichia coli*
    ❑ b. *Proteus mirabilis* and *Serratia marcescens*

❑ 2. Miniaturized and rapid-test systems:
    ❑ a. Four API 20E systems complete with test strips, incubation trays, and lids
    ❑ b. Four Enterotube II systems

❑ 3. Eight tubes, each containing 5 mL of sterile 0.85% saline, pH 7.0

❑ 4. Four-mL sterile mineral oil

❑ 5. Eight sterile 5-mL pipettes for the inoculation of API systems

❑ 6. Eight sterile Pasteur pipettes with bulbs

❑ 7. One wax marking pencil or other marking device

❑ 8. Container with disinfectant

The following items should be provided for class use:

❑ 1. Reagents to be provided after incubation for class use with the API 20E and/or Enterotube II systems:
    ❑ a. A 20% aqueous ferric chloride solution
    ❑ b. Hydrogen peroxide
    ❑ c. Kovac's reagent
    ❑ d. Nitrate reduction solutions
        ❑ i. Sulfanilic acid
        ❑ ii. Dimethyl-α-naphthylamine
    ❑ e. Oxidase solution

❑ f. Voges-Proskauer solutions
    ❑ i. Alpha-naphthol
    ❑ ii. Potassium hydroxide

❑ 2. Pasteur pipettes with rubber bulbs for use with test reagents

❑ 3. Distilled water in a plastic squeeze bottle

❑ 4. Filter paper strips

❑ 5. Coding manuals for API and Enterotube systems

❑ 6. Color pencils

A. The API Enteric System

Procedure 1:

This procedure is to be performed by students in groups of 4.

❑ 1. Obtain and examine the API system using step 1 in Procedure Diagram 26 and Figure 27–1 as guides. Note the following components: (a) the strip itself with its 20 microtubes, (b) an individual microtube, which consists of a bottom tube and an upper cupule, and (c) the plastic incubation lid and tray.

❑ 2. Label each API system with your assigned or selected organism. This is an extremely important step, because the results obtained will be used to confirm the identity of organisms according to the API system standard codes (Table 27–3).

❑ 3. Perform the oxidase test by removing a small amount of growth of the test organism from the agar plate used and rubbing it onto a filter paper strip to which a drop of oxidase reagent has been added. A dark blue or purple reaction is a positive result.

❑ 4. Prepare for the inoculation of the strip by performing the following steps:
   ❑ a. Place about 5 mL of distilled water into the plastic incubation tray. (A plastic squeeze bottle can be used.)
   ❑ b. Remove the API strip from its sealed pouch and place it into its incubation tray.

❑ 5. Inoculate the API strip as follows:
   ❑ a. With the aid of a sterile pipette and bulb or other device, fill the microtubes by placing the pipette against the side of the cupule. (Your instructor will demonstrate the procedure.)
   ❑ b. Do not completely fill the microtubes for ADH, LDC, ODC, H$_2$S and URE. After all microtubules have been inoculated, fill these microtubes with mineral oil.

❑ 6. Place the lid on the incubation tray and incubate the system for 18–24 hours or as directed by your instructor.

❑ 7. After incubation, examine the API strip and perform the steps shown in Procedure Diagram 26. Add the reagents to the microtubes in the following order and as follows:
   ❑ a. Add the reagents to the TDA and VP tubes. (TDA reactions develop quickly whereas VP reactions may take 10 minutes to appear.)
   ❑ b. Next add Kovac's reagent to the IND tube.
   ❑ c. Perform the nitrate reduction test after the IND test. Add the nitrate reagents to the GLU tube.

❑ 8. Record all reactions not requiring the addition of reagents. **Refer to Figure 27–3 and Table 27–1 for the interpretation of all reactions.** Table 27–2 lists the respective values for each test. The tests of the API test system are arranged in specific groups to yield a 7-digit code number from which an organism's identity can be made. This arrangement is shown in the Results and Observations section. Table 27–3 lists several examples of the 7-digit code.

❑ 9. Record your findings in the Results and Observations section in Table 27–6.

❑ 10. Using positive and negative reactions as the basis for an organism's identification, convert the test results to digital form for each organism using the diagrams provided. (*Note:* A negative reaction counts as a zero. The digit code provides a profile or identification number. An example of this step is described in Procedure 2.)

❑ 11. Record the reactions of your laboratory partners in Table 27–6 also.

Procedure Diagram 26
API Inoculation

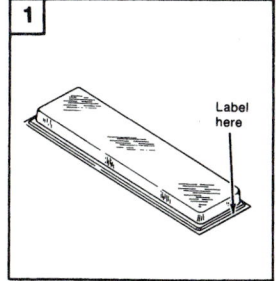

1. With a wax marking pencil, label the elongated flap of the API incubation tray with the culture to be used.

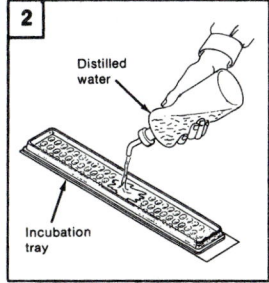

2. Put about 5 ml of water from a plastic squeeze bottle into the incubation tray.

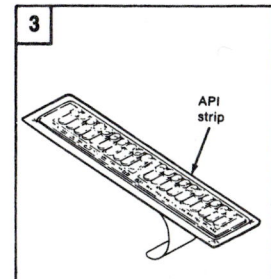

3. Remove the API strip from its sealed envelope and place it in the incubation tray.

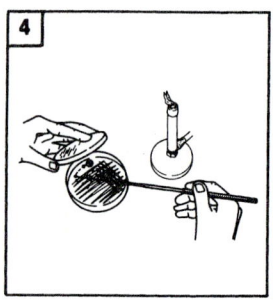

4. With a flamed inoculating loop, touch the center of a well-isolated colony of your assigned culture.

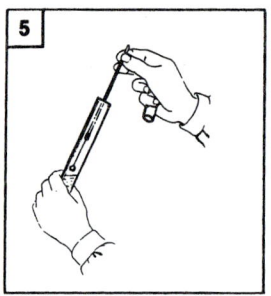

5. Aseptically insert the inoculum into a tube of saline.

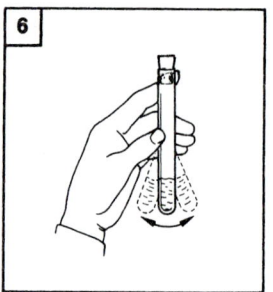

6. Mix the contents by shaking the tube gently.

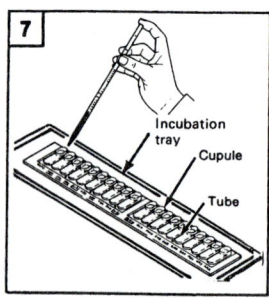

7. Tilt the API incubation tray. Using a Pasteur pipette, fill the tube sections of each compartment by placing the pipette tip against the side of the cupule.

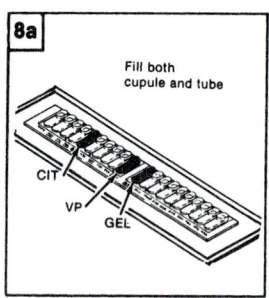

8a. Fill both the tube and cupule sections of the CIT, VP, and GEL compartments.

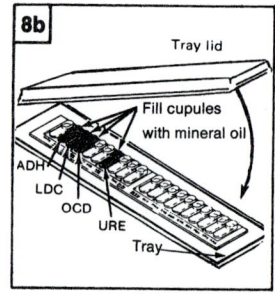

8b. After inoculation, fill the cupules of ADH, LDC, ODC and URE tubes with sterile mineral oil. Apply the tray lid and incubate as directed.

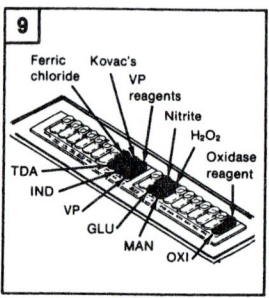

9. After incubation, add reagents in order as follows: (a) Kovac's to IND; (b) VP reagents to VP; (c) ferric chloride to TDA; (d) nitrate reagents to GLU; (e) oxidase reagent to OXI; (f) hydrogen peroxide to MAN. Record your findings as directed earlier.

Procedure 2: Determining an Organism's Identity with the API System

❏ 1. Place the API system in line with the cupules shown in Figure 27–3 as best you can.

❏ 2. Circle the reaction values shown below each cupule for all positive reactions. (Refer to Tables 27–1 and 27–2.)

❏ 3. Enter your results and reaction values for the respective reactions obtained with your assigned organism in Table 27–7. Add the sum of each group of three tests and enter the number in the space provided.

❏ 4. Enter the 7-digit number in the designated box and refer to Table 27–3 (or to the appropriate code book if one is provided) to identify the test organism. Find the 7-digit identification number closest to the results obtained.

(a)

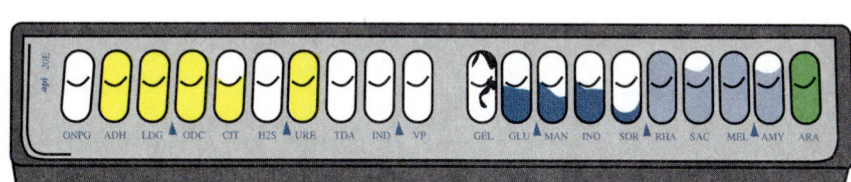

(b)

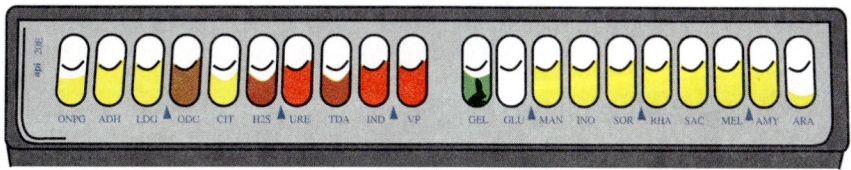

Figure 27–3

The API 20 system. (a) The colors and appearance of an inoculated system. (b) The colors and appearance of reactions after incubation of an inoculated system. Refer to Table 27–1 for additional details of these reactions.

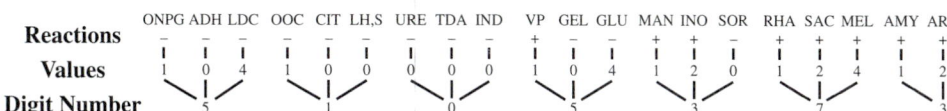

	ONPG	ADH	LDC	OOC	CIT	LH,S	URE	TDA	IND		VP	GEL	GLU	MAN	INO	SOR		RHA	SAC	MEL	AMY	ARA
Reactions	–	–	–	–	–	–	–	–	–		+	–	–	+	+	–		+	+	+	+	+
Values	1	0	4	1	0	0	0	0	0		1	0	4	1	2	0		1	2	4	1	2
Digit Number		5			1			0				5			3			7			3	

7 DIGIT NUMBER:

5105373

IDENTITY OF UNKNOWN BACTERIUM:

Enterobacter aerogenes

Figure 27–4

Determining an organism's identity using the 7 digit number arrangement of the API system.

Table 27-2

API 20E Tests and Respective Values

Test	Value	Test	Value
ONPC	(1)	GLU	(4)
ADH	(2)	MAN	(1)
LCD	(4)	INO	(2)
ODC	(1)	SOR	(4)
CIT	(2)	RHA	(1)
H_2S	(4)	SAC	(2)
URE	(1)	MEL	(4)
TDA	(2)	AMY	(1)
IND	(4)	ARA	(2)
VP	(1)	OXI	(4)
GEL	(2)	NO_3	(−)

Table 27-3

API 20E ID Values

API (Profile) Values[a] (7 Digits)	Bacterial Species
3244104	*Aeromonas hydrophila*
5304552	*Arizona* species
0104562	*Citrobacter freundii*
5105373	*Enterobacter aerogenes*
1105563	*Enterobacter cloacae*
7144542	*Escherichia coli*
4115112	*Hafnia alvei*
5204773	*Klebsiella pneumoniae*
0336000	*Proteus mirabilis*
0076021	*Proteus vulgaris*
0064200	*Providencia stuartii*
0216004	*Pseudomonas aeruginosa*
4704552	*Salmonella enteriditis*
4404540	*Salmonella typhi*
5306561	*Serratia marcescens*
1104012	*Shigella sonnei*
4044104	*Vibrio parahaemolyticus*
0054523	*Yersinia enterocolitica*

[a]NOTE: Several species have more than one API 20E value. Refer to the Code Book for additional values.

B. Enterotube System

Procedure 1: Inoculation

This procedure is to be performed by students in groups of 4.

❏ 1. Examine the Enterotube using Figure 27–2a. Note the following components: (1) caps on each end of the test system; (b) the different compartments; (c) the inoculating rod within the system; and, if present, (d) blue tape; and (e) sliding clear band.

❏ 2. Perform the steps shown in Procedure Diagram 27 with your selected culture or one assigned by the instructor.

❏ 3. Observe the following steps in particular.

    ❏ a. Inoculate the Enterotube by gently pulling the inoculating rod through all of the compartments with a twisting motion. DO NOT PULL THE INOCULATING ROD COMPLETELY OUT OF THE LAST COMPARTMENT.

    ❏ b. Push the inoculating rod back through the compartments carefully until the tip reaches the H_2S/indole compartment. A notch in the rod will be visible.

    ❏ c. Break the inoculating rod at the notch by bending it. (The portion of the tool remaining in the Enterotube provides the anaerobic conditions needed for certain anaerobic reactions to occur.)

    ❏ d. Punch holes with the broken-off part of the inoculating rod through the thin tapelike material on the bottom of the Enterotube covering the last 8 compartments (adonitol, lactose, arabinose, sorbitol, Voges-Proskauer, dulcitol/PA, urease, and citrate). The holes will provide for aerobic growth in these compartments.

    ❏ e. Replace both caps.

❏ 4. Label each Enterotube with the organism used, and incubate at 35°–37°C for 18–24 hours or as indicated by your instructor.

❏ 5. After incubation, add the necessary reagents as described in Procedure Diagram 27. However, interpret and record all reactions with the exception of the indole and Voges-Proskauer tests. These tests must be done last because they may affect the other results obtained with the Enterotube system. **Refer to Table 27–1 for the interpretation of test results.** Table 27–4 lists the respective values for each test. Figure 27–3 shows the color reactions possible with the Enterotube.

❏ 6. The reactions of the Enterotube test system are arranged in specific groups to yield a 5-digit code number from which an organism's identity can be determined. This arrangement is shown in Procedure 2 and in the Results and Observations section. Table 27–5 lists several examples of the 5-digit code.

❏ 7. Record your findings and those of your laboratory partners in Table 27–8 in the Results and Observations section.

❏ 8. Using positive and negative reactions as the basis for an organism's identification, convert the test results obtained with your organism to digital form using the diagram provided. An example of this step is shown in Figure 27–6.

Procedure Diagram 27
Enterotube Inoculation

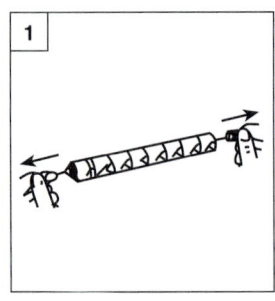

1. Remove the caps from the Enterotube, and keep them for later use.

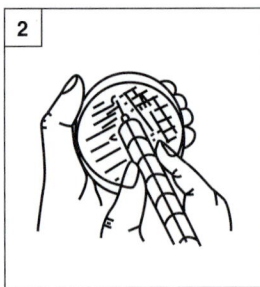

2. Pick a well-isolated colony from your assigned plate with the Enterotube inoculating needle.

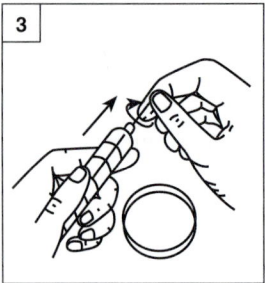

3. Inoculate the tube by first twisting the inoculating wire and then withdrawing the needle through all 8 compartments using a turning motion.

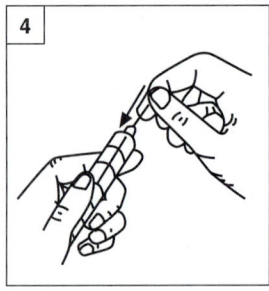

4. Put the needle back into the tube (without sterilizing it) and, using a turning motion, push it through the dextrose and orinthine compartments. You should be able to see the tip of the needle in the H_2S/indole compartment.

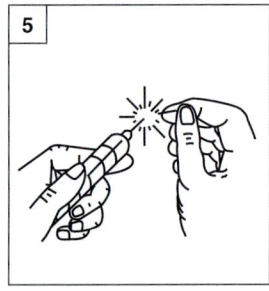

5. Next, break the needle at the notch by bending, discard the handle, and replace the caps on the tube. The portion of the needle remaining in the tube will maintain the anaerobic conditions necessary for the true fermentation of dextrose, gas production of gas, and decarboxylation of lysine and ornithine.

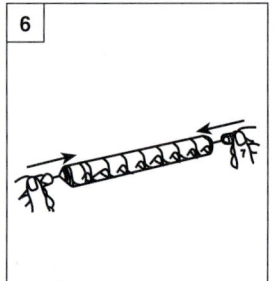

6. Replace the caps on the Enterotube.

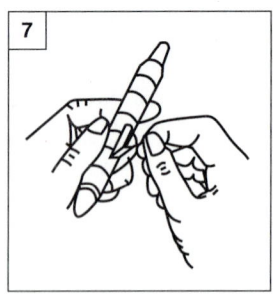

7. Remove the blue tape on the tube to provide aerobic conditions in the lactose, dulcitol, phenylalanine, urea, and citrate compartments.

Procedure 2: Determining an Organism's Identity with the Enterotube

❏ 1. Place the incubated Enterotube in line with Figure 27–5 so that the appropriate compartments are next to it.

❏ 2. Circle the number appearing below the positive test compartment in the figure.

❏ 3. Add the number in the bracketed sections and enter the totals in the spaces provided below each respective arrow.

❏ 4. Locate the 5-digit number either in Table 27–5 or in the appropriate code book if one is available. Find the best or closest answer.

❏ 5. Enter the 5-digit number and the identity of the organism it represents.

(a)

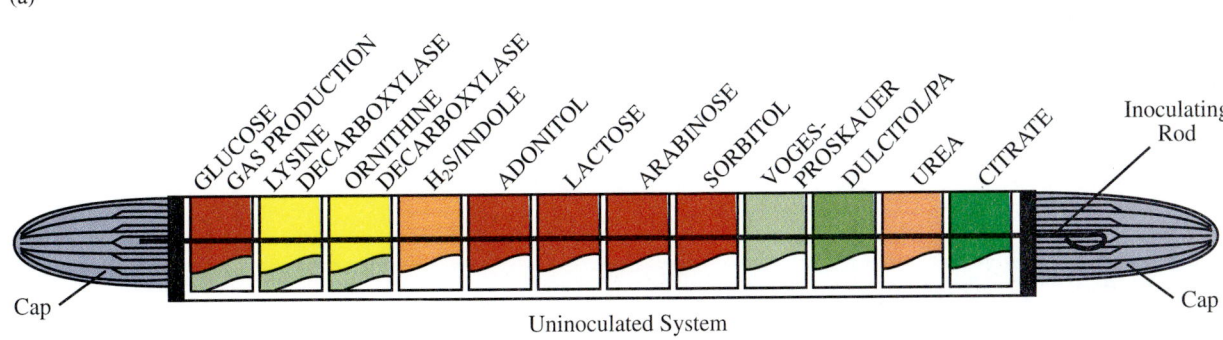

(b)

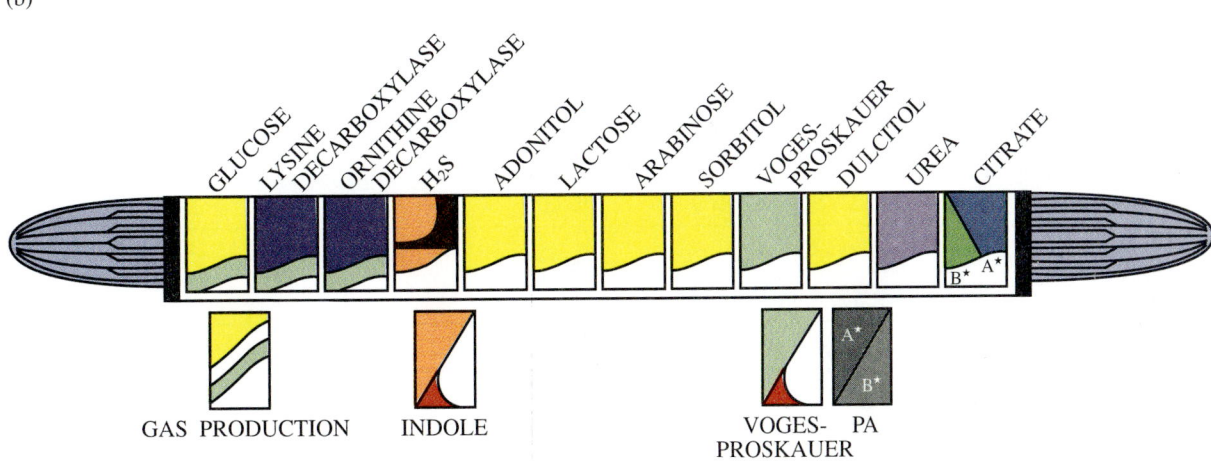

Figure 27–5

The Enterotube II (a) The colors and appearance of an uninoculated system. (b) The colors and appearance of reactions after incubation of an inoculated system. Refer to Table 27–1 for additional descriptions of reactions. (After Roche Diagnostic Systems, Division of Hoffman-Laroche, Inc., Nutley, NJ 07110).

Table 27–4

Enterotube II System Tests and Respective Values

Test	Value	Test	Value
Glucose-acid	(1)	Arabinose	(4)
Gas production	(2)	Sorbitol	(1)
Lysine	(4)	Voges-Proskauer	(2)
Ornithine	(1)	Dulcitol	(4)
H_2S	(2)	Phenylalanine deaminase	(1)
Indole	(4)	Urea	(2)
Adonitol	(1)	Citrate	(4)
Lactose	(2)		

Table 27–5

Enterotube II ID Values

Enterotube II ID Value	Bacterial Species	Enterotube II ID Value	Bacterial Species
#40673	*Klebsiella pneumoniae*	#56001	*Arizona sp.*
#41447	*Providencia stuartii*	#56150	*Salmonella enteriditis*
#42226	*Proteus mirabilis*	#60050	*Shigella flexneri*
#45162	*Yersinia enterocolitica*	#60371	*Enterobacter cloacae*
#50040	*Salmonella typhi*	#62006; 36007, 16007	*Proteus vulgaris*
#50063	*Serratia marcescens*	#62302	*Citrobacter freundii*
#55241; 55650; 55601	*Escherichia coli*	#70020	*Hafnia alvei*
#55760	*Enterobacter aerogenes*		

Results and Observations

API Reactions

1. Enter your findings in the following table, using a "+" for positive reactions and a "−" for negative reactions. (Refer to Table 27–1 for descriptions of reactions.)

Table 27–6

API Reactions

Test[a]	Organisms			
	E. aerogenes	E. coli	P. mirabilis	S. marcescens
ONPG				
ADH				
LDC				
ODC				
CIT				
H₂S				
URE				
TDA				
IND				
VP				
GEL				
GLU				
MAN				
INO				
SOR				
RHA				
SAC				
MEL				
AMY				
ARA				
OXI				
NO₃				

[a]Test symbols; ONPG = beta-glactosidase; ADH = arginine dihydrolase; LDC = lysine decarboxylase; ODC = ornithine decarboxylase; CIT = citrate; H₂S = hydrogen sulfide production; URE = urease; TDA = tryptophan deaminase; IND = indole; VP = acetoin; GEL = gelatin; GLU = glucose; MAN = mannitol; INO = inositol; SOR = sorbitol; RHA = rhamnose; SAC = sucrose; MEL = melibiose; AMY = amygdalin; ARA = arabinose; OXI = oxidase; NO₃ = nitrate reduction.

Continued

Table 27–6 continued

API Reactions

Test[a]	Organisms			
	E. aerogenes	E. coli	P. mirabilis	S. marcescens
ONPG				
ADH				
LDC				
ODC				
CIT				
H_2S				
URE				
TDA				
IND				
VP				
GEL				
GLU				
MAN				
INO				
SOR				
RHA				
SAC				
MEL				
AMY				
ARA				
OXI				
NO_3				

[a]Test symbols; ONPG = beta-glactosidase; ADH = arginine dihydrolase; LDC = lysine decarboxylase; ODC = ornithine decarboxylase; CIT = citrate; H_2S = hydrogen sulfide production; URE = urease; TDA = tryptophan deaminase; IND = indole; VP = acetoin; GEL = gelatin; GLU = glucose; MAN = mannitol; INO = inositol; SOR = sorbitol; RHA = rhamnose; SAC = sucrose; MEL = melibiose; AMY = amygdalin; ARA = arabinose; OXI = oxidase; NO_3 = nitrate reduction.

2. Use colored pencils or pens to color the cupules in Figure 27–6 and confirm the identity of the organism you used. Use the procedure described earlier for the API system, and enter your determinations in Table 28–7. Use Tables 27–1 through 27–3 and Figures 27–3 and 27–4 as guides.

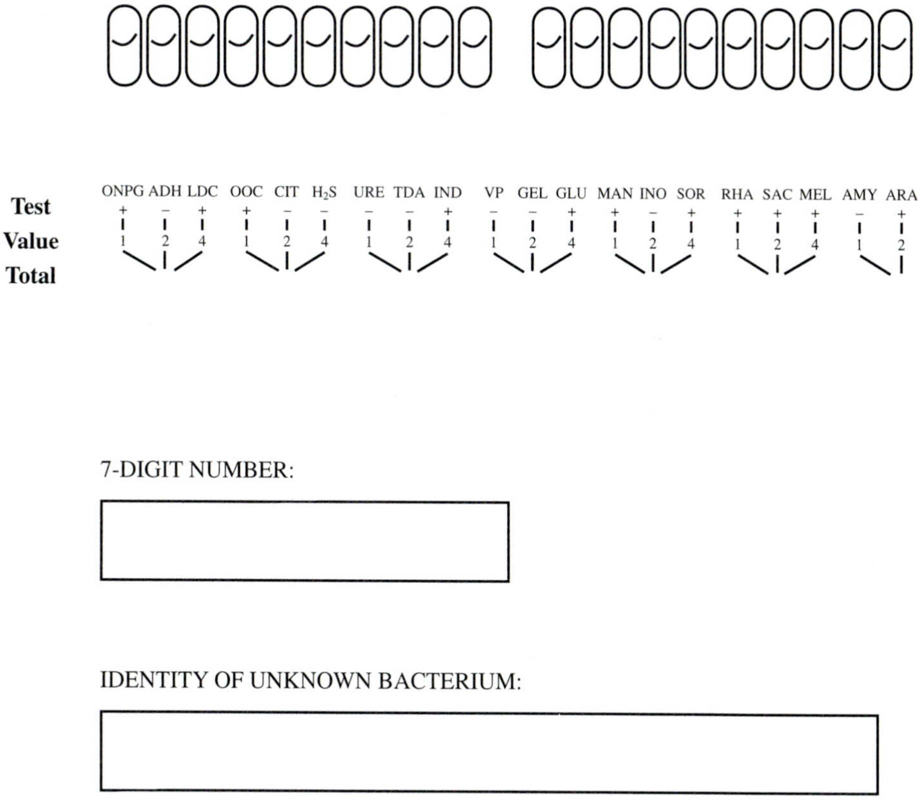

	ONPG	ADH	LDC	OOC	CIT	H$_2$S	URE	TDA	IND	VP	GEL	GLU	MAN	INO	SOR	RHA	SAC	MEL	AMY	ARA
Test	+	–	+	+	–	–	–	–	+	–	–	+	+	–	+	+	+	+	–	+
Value	1	2	4	1	2	4	1	2	4	1	2	4	1	2	4	1	2	4	1	2
Total																				

7-DIGIT NUMBER:

IDENTITY OF UNKNOWN BACTERIUM:

Figure 27–6
API determination.

3. Did your test results correspond to the digit code for the organism you used? _____ If not, why? _____
_____

4. Did the results of your laboratory partners confirm the identity of their organisms? _____

Enterotube Reactions

1. Enter your findings in the following table, using a "+" for positive reactions and a "−" for negative reactions. (Refer to Table 27–1 for descriptions of reactions.)

2. Use colored pencils or pens to color the compartments in Figure 27–7 and confirm the identity of the organism you used. Use the procedure described earlier for the APT system. Use Tables 27–1 and 27–2 and Figure 27–5 as guides.

3. Did your test results correspond to the digit code for the organism you used? _____ If not, why? _____
_____

4. Enter the reactions of your laboratory partners in Table 27–8.

5. Did the results of your laboratory partners confirm the identity of their organisms? _____

Table 27–7

API System Test Reactions

Test	Value	Reaction (±)	Numerical Value	Section Value	Test	Value	Reaction (±)	Numerical Value	Section Value
ONPC	(1)				GLU	(4)			
ADH	(2)				MAN	(1)			
LDC	(4)				INO	(2)			
ODC	(1)				SOR	(4)			
CIT	(2)				RHA	(1)			
H_2S	(4)				SAC	(2)			
URE	(1)				MEL	(4)			
TDA	(2)				AMY	(1)			
IND	(4)				ARA	(2)			
VP	(1)				OXI[a]	(4)			
GEL	(2)				NO_3[a]	(−)			

[a]Note these tests are not in the test strip but must be done separately. Refer to the inoculation procedure.

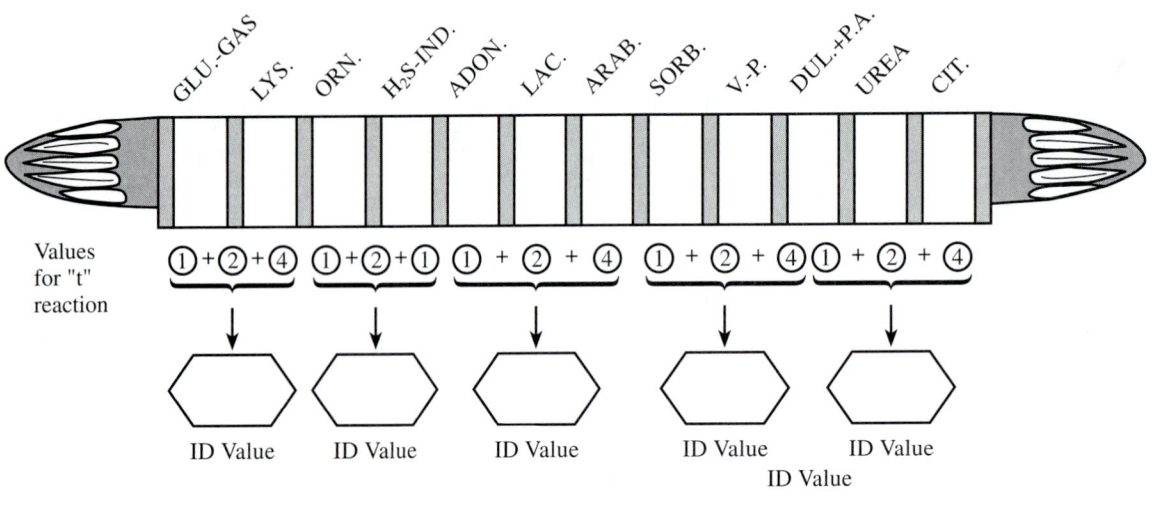

ID Value   ID Value   ID Value   ID Value   ID Value

ID Value

Organism

7 DIGIT NUMBER:

IDENTITY OF UNKNOWN BACTERIUM

Figure 27–7

Enterotube determination.

Table 27–8

Enterotube Reactions

Test[a]	Organisms			
	E. aerogenes	E. coli	P. mirabilis	S. marcescens
D[a]				
Gas				
Lys				
Orn				
H_2S				
I				
Lac				
Phenyl				
Dul				
U				
Cit				

[a]Test symbols: D = dextrose; Gas = gas production from dextrose; Lys = lysine; Orn = ornithine; H_2 = hydrogen sulfide production; I = indole; Lac = lactose; Phenyl = phenylalanine; Dul = dulcitol; U = urease; Cit = citrate.

Laboratory Review 27   Miniaturized and Rapid Microbiological Systems

1. a. List 2 advantages of a miniaturized test system. _____

    _____

    _____

    b. List 2 disadvantages of a miniaturized test system. _____

    _____

    _____

2. Complete the table on the following page by indicating the specific purpose and the appearance of a positive reaction for each test or medium listed. Compare the systems used in this exercise.

3. List 4 carbohydrates routinely used to determine the fermentative ability of bacterial species.

    a. _____   c. _____

    b. _____   d. _____

Table 27–9

Test Reaction Interpretations

Test or Medium	Purpose	Appearance of Positive Reaction	
		API System	Enterotube
Citrate			
Glucose			
H$_2$S			
Indole			
Lysine decarboxylase			
Ornithine decarboxylase			
Urease			

After completing this experimental exercise, you should be able to:

1. Correctly interpret the reaction obtained with selective and/or differential plate and tube media.

2. Use a biochemical flow chart for the identification of unknown bacterial cultures.

3. Isolate and identify organisms within a mixed bacterial culture.

4. Outline and explain a functional approach to the identification of unknown bacterial cultures.

The closely related bacterial species inhabiting the large intestine of mammals are called *enterobacteria*. This group contains well-known and studied species of the genera *Citrobacter, Enterobacter, Escherichia,* and *Proteus.* While some of these microorganisms are members of the normal flora or residents of the gastrointestinal system, certain species may be opportunists waiting for an opening to establish a disease process. The potential disease-causing group includes species of *Citrobacter, Escherichia, Klebsiella, Proteus, Salmonella, Serratia,* and *Shigella.* Identification of many of these organisms can only be achieved through the careful combined use of several media and biochemical tests, because they are morphologically similar.

Some members of the gastrointestinal normal flora, frequently referred to as *coliforms* because of their lactose-fermenting capability, are used as indicator organisms to check for fecal contamination of water, food, or other types of materials. Such coliforms are universally found in feces and include *Enterobacter* species and *Escherichia coli.*

The differentiation and identification of various bacterial species is primarily based on demonstrating their biochemical (enzymatic) activity. Exercises 20 through 26 in this section emphasized the properties of several media, tests, and reagents used in the identification, separation, and differentiation of bacteria. How these materials can be used effectively in combination to identify unknown gram-negative microorganisms is the main feature of this exercise. The value of tests and unknown gram-negative microorganisms is the main feature of this exercise. The value of tests and procedures used in combination frequently is not appreciated when they are performed individually or simply used to demonstrate a particular microbial activity. A flowchart of biochemical reactions is provided to aid in the separation and eventual identification of unknown bacterial species within a mixed culture. Examination of the chart also will show that "unknown" identification requires attention to detail, interpretation of results, and more than one laboratory session.

One additional test is introduced here, namely the *deamination of the amino acid phenylalanine.* Detecting the deamination of phenylalanine requires the use of phenylalanine agar. After incubation, ferric chloride is added directly to a growing culture. The development of a green color indicates the presence of phenylpyruvic acid, a product of phenylalanine deamination (see Figure 28–1).

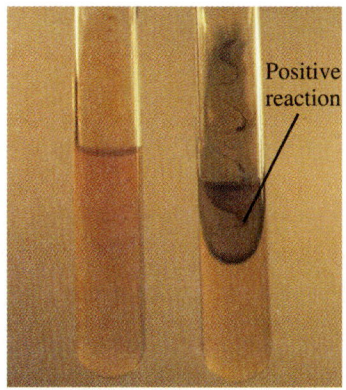

Figure 28–1

Phenylalanine agar (PA). This medium is used to detect the presence of phenylalanine deaminase. The formation of a green or dark brown color in the medium upon the addition of ferric chloride is a positive reaction.

Later exercises introduce additional media and tests that are applicable to isolation and identification procedures.

In this exercise, each student will be provided with a mixed culture containing 2 bacterial unknowns. These organisms will exhibit the same Gram stain reaction, but may have different morphological properties. Separation of the organisms in the unknown preparation is the first major challenge.

Be certain and careful in your approach to the problem. Identification of organisms requires careful application of the techniques previously studied. To avoid problems of limited time and supplies of microorganisms, complex media requirements, and danger to the experimenter, a relatively simple yet thorough exercise has been constructed. However simple it may be, carrying it out should give you some understanding of the problems encountered in the isolation and identification of the enterobacteria.

Hundreds of media and tests have been developed for the identification of unknown organisms. From a practical point of view, however, it is important that the number of tests used for identification be well chosen and kept to a minimum. A test pattern, or biochemical key, based on reactions with known organisms can serve as a rapid and efficient approach to the identification of an unknown organism. A key to this type is provided in this exercise. Table 28–1 provides explanations of the symbols used in the biochemical key.

Table 28–1

Explanation of Symbols

M	medium	±	slightly or weakly positive
R	result	O	no growth
T	test	A	acid
+	positive reaction	Alk	alkaline
−	negative reaction	NC	no change

A. Bacterial Enteric Unknown

Materials

❑ 1. The following items are to be provided per pair of students:
    ❑ a. One each of 24-hour trypticase soy agar slant cultures of *Escherichia coli* and *Staphylococcus aureus* (*Note:* These microorganisms are the controls for the Gram-staining reactions.)
    ❑ b. One unknown bacterial mixed culture
    ❑ c. Four trypticase soy agar slants
❑ 2. The following media and reagents will be provided for this exercise only upon authorization by the instructor:

Media	*Reagents*
❑ a. Eosin-methylene blue agar plates	❑ a. Gram stain reaction
❑ b. Trypticase soy agar plates	❑ b. Ferric chloride
❑ c. Pseudomonas P agar plates	❑ c. Methyl red: methyl red indicator
❑ d. Motility agar	❑ d. Oxidase: dimethyl-p-phenylenediamine
❑ e. Phenylalanine agar slants	❑ e. Oxy-Swab (Remel), if available
❑ f. Triple sugar iron agar slants	
❑ g. Carbohydrate fermentation media	
❑ h. Phenol red lactose broth	
❑ i. Koser's citrate broth	
❑ j. Methyl red broth	
❑ k. Urease broth	

Procedure

This procedure is to be performed by students individually.

Additional Techniques Required for This Exercise:

❑ 1. Streak Plate Technique, Procedure Diagram 10, Exercise 5

❑ 2. Inoculation of an Agar Slant, Procedure Diagram 7, Exercise 4

❑ 3. The Gram Stain, Procedure Diagram 17, Exercise 14

❑ 4. Broth Transfer, Procedure Diagram 5, Exercise 4

❑ 5. Triple Sugar Iron Agar (TSIA) Inoculation Technique, Procedure Diagram 24, Exercise 26

❑ 1. Obtain a mixed-culture unknown, and enter its number in the appropriate space of the Report Form in the Results and Observations section.

❑ 2. Streak an Eosin-Methylene Blue plate with the unknown, and incubate the preparation at 37°C for 24 hours, or as directed.

❑ 3a. After incubation, observe the plate. Select 1 lactose fermenter colony and inoculate a trypticase soy agar slant with a small inoculum from this colony. Label the slant with an identification number and indicate that the culture is a lactose fermenter. Enter this number into one of the columns on the Report Form.
(*Note:* This slant culture, after incubation, will serve as the only source of organisms for all tests and/or subsequent media inoculations.)

❑ 3b. Use a small inoculum from your selected colony and streak a trypticase soy agar plate with it. Label the plate as you did with the slant. After incubation this plate should be used to determine colonial characteristics and to perform the oxidase test.

❑ 4. Repeat step 3, but this time select one nonlactose fermenter colony.

❑ 5. After incubation, perform the Gram stain with each slant culture. Record the Gram stain and morphology on the Report Form. This step is performed as a check for purity.
(*Note:* In making a smear from an agar slant or plate, always mix the culture with a loopful of water.)

❑ 6. After incubation of your plate culture determine the colonial characteristics of culture and perform the oxidase test. Record your findings on your unknown report form.

❑ 7. Based on the lactose fermentation reaction, select the portion of the biochemical identification key (pages 262 or 263) to follow for each of the cultures obtained.
(*Note:* Incubation for at least 18–24 hours is needed for cell media inoculations.)

❑ 8. Record each test and/or medium used, together with the results obtained for each step taken toward the identification of each unknown culture, on the Report Form.

❑ 9. Successful identification of any unknown culture depends on common sense and the correct application of basic procedures and the identification key. If you have difficulty, perform the following unknown check:
   ❑ a. Check the purity of your stock culture.
   ❑ b. Go over the identification key to see if you performed the correct tests.
   ❑ c. Check if the correct reagents were used.
   ❑ d. Make certain that tests have been made carefully and interpreted correctly. Use the appropriate figures in Exercises 20 through 26.

❑ 10. After all tests and/or media reactions have been interpreted and recorded, the results should point to the identification of each unknown culture. Enter the names of the unknown in the appropriate spaces and give the report to your instructor for evaluation.

Identification Key

In the identification key that follows, an approach is given for the identification of important enteric bacteria. Certain symbols used in the identification key are explained in Table 28–1.

Biochemical Identification Key
for Selected Gram-Negative Bacteria

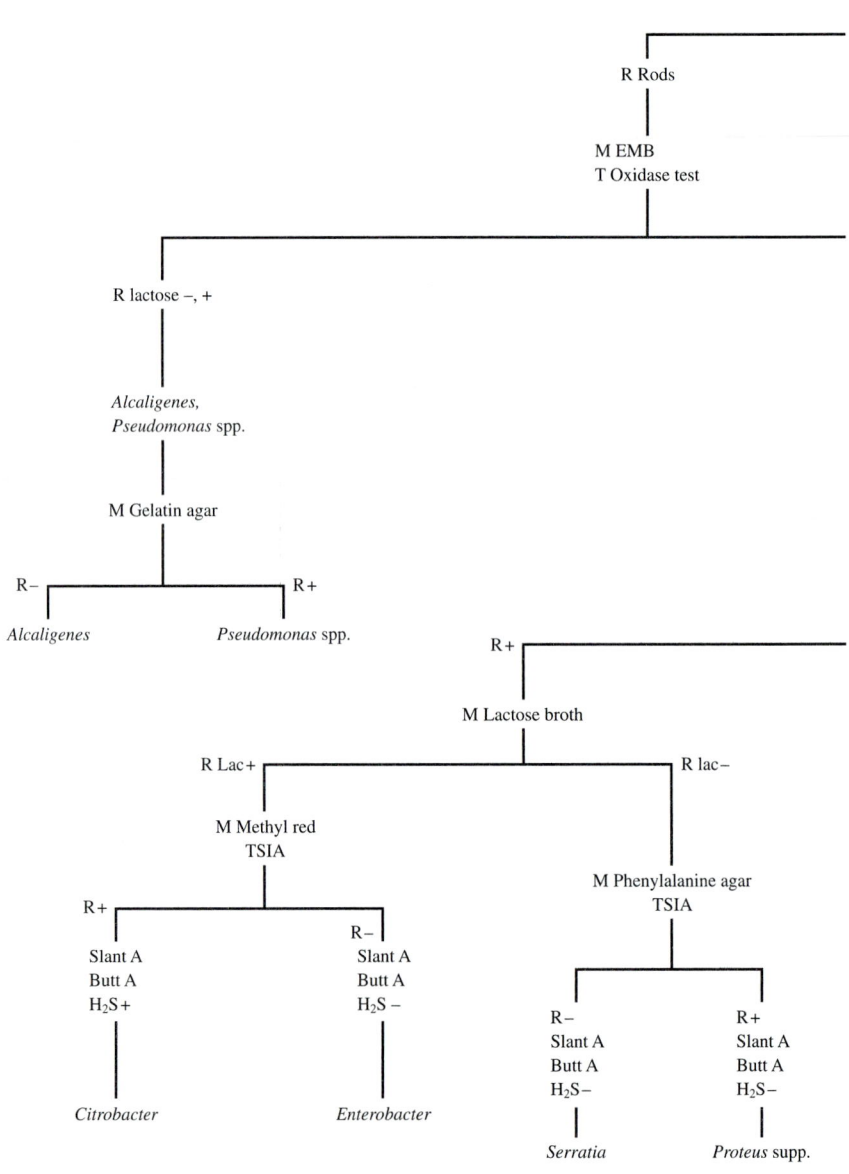

Biochemical Identification Key
for Selected Gram-Negative Bacteria

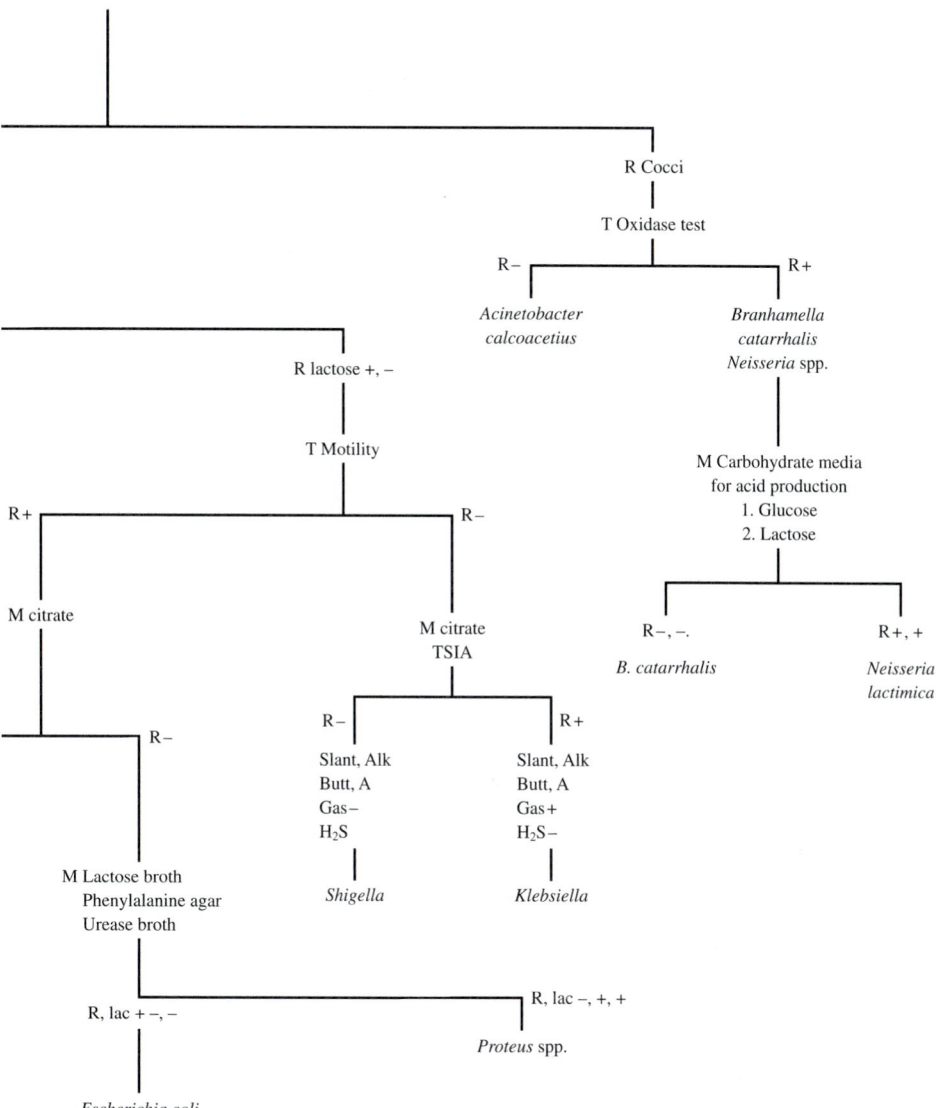

Results and Observations

Enter your findings on the form below.

Name _____ Date _____

Report Form

Unknown Culture 1 Code No. _____	Unknown Culture 2 Code No. _____
Description of colony:	Description of colony:
1. Pigment _____	1. Pigment _____
2. Margin _____	2. Margin _____
3. Colonial growth _____	3. Colonial growth _____
4. Elevation _____	4. Elevation _____
5. Odor _____	5. Odor _____
6. Other _____	6. Other _____
Gram reaction and morphology	Gram reaction and morphology
_____	_____

Tests (approved by _____)   Tests (approved by _____)

	Result:		Result:
1.	1.	1.	1.
2.	2.	2.	2.
3.	3.	3.	3.
4.	4.	4.	4.
5.	5.	5.	5.
6.	6.	6.	6.
7.	7.	7.	7.
8.	8.	8.	8.
9.	9.	9.	9.
10.	10.	10.	10.

This bacterium is probably	This bacterium is probably
_____ _____	_____ _____
genus species	genus species
Student's name _____	Student's name _____
_____	_____
Score _____	Score _____
Remarks (for instructor only)	Remarks (for instructor only)
_____	_____

Laboratory Review 28   An Introduction to Bacterial Identification: An Unknown Challenge

1. What are coliforms? _____

_____

2. List 7 disease-causing bacteria belonging to the enteric group.

a. _____   e. _____

b. _____   f. _____

c. _____   g. _____

d. _____

3. Identify the reaction from the following culture media:

a. A green coloration of phenylalanine agar upon the addition of ferric chloride. _____

_____

b. The formation of bubbles upon the addition of hydrogen peroxide. _____

_____

c. The formation of dark purple colonies on EMB agar. _____

_____

d. The formation of a red color in methyl red broth upon the addition of methyl red indicator. _____

_____

e. A cloudy citrate medium after incubation. _____

_____

f. Black coloration of triple sugar iron agar after incubation. _____

_____

The Control of Microorganisms by Chemical and Physical Factors

Control of water and food supplies, better ventilation of dwellings, and various procedures of sanitation have greatly reduced the spread of many killing infectious agents. . . .

René Dubois

All life forms require a particular set of environmental factors for their metabolic and reproductive activities. A variety of chemical and physical factors—including nutrients, water, temperature, pH, and osmotic pressure—are involved in these reactions. The interaction of such factors causes conditions to develop that can (1) be optimum for growth, (2) prevent growth, or (3) cause the death of the organism.

Bacteriostatic and bactericidal effects (2 and 3 above) may involve osmotic pressure. When the materials in a medium or in a suspending liquid exert little or no osmotic pressure, the solution is considered to be isotonic. But if the dissolved substances are increased, a hypertonic situation occurs and water is drawn out of the organism. This causes shrinkage of the cytoplasm, or *plasmolysis,* usually a bacteriostatic state. On the other hand, when the amount of dissolved substances in the medium decreases and a hypotonic solution is formed, water enters the cell and the cell bursts. This process, called *plasmoptysis,* is an example of bactericidal activity, and can occur when a bacterial cell wall is damaged by enzymes or antibiotics.

Salt water is hypertonic for most microorganisms, which undergo plasmolysis when exposed to it. Because it prevents microbial growth, brine has been used for centuries to preserve a wide variety of foods such as fish and beef. Unfortunately, some microorganisms are *halophiles* (salt lovers), which thrive in a 15–30% salt concentration. Spoilage of foods by these organisms is not prevented through use of brine. Some fungi are able to grow in honey or other concentrated sugar solutions and are known as *saccharophiles.* It is not known whether the high sugar content is an actual requirement. Other microorganisms grow in low-water environments. Included are those organisms that degrade petroleum because of their ability to exist in extremely low levels of water. If any of the organisms just discussed experience a substantial increase in the water content of their environments, plasmoptysis will usually occur. A significant increase in dissolved solids would bring about plasmolysis with most organisms, however. The effects of osmotic pressure is considered in this section.

The moisture content of the environment is particularly relevant to the growth—or death—of microorganisms. Drying (desiccation) can be considered generally as a lethal treatment, assuming that the various organic materials that could protect the organisms are not present. Drying is a good method of preserving food materials, because it inhibits the growth of bacteria that cause spoilage in dried foods. However, organisms are also preserved in the dried food that, when given a sufficient amount of water, will grow and flourish once more. An excellent means of spreading respiratory infections is by creating aerosols of pathogens through coughing or sneezing. The organisms are present in droplets of mucus and saliva. Even when these droplets are dry, they are a potential source of infectious microorganisms because spores, certain bacteria, and viruses are resistant to the effects of desiccation. The influence of drying will be considered as it pertains to the handling of clinical specimens in Section XII.

The action of temperature can be demonstrated quite easily. Most organisms of medical significance thrive in a temperature range of approximately 20° to 45°C, with an optimum temperature of 36° to 37°C. Cold temperatures of 5° to 10°C are used for temporary preservation of foods and dairy products in the home. The fact that these temperatures do not prevent the growth of microorganisms is demonstrated by the spoiling of refrigerated milk within a week and the growth of fungi on refrigerated cheeses, fruit, and the like. Freezing will prevent microbial growth, but even freezing will not kill all contaminating organisms. Several years ago, an investigation of a typhoid fever epidemic revealed that ice cream made with cream contaminated by *Salmonella typhi* and then frozen for several months still contained viable microorganisms. It is evident that the use of high temperatures is an effective way to kill organisms. Among nonsporeformers, the heat resistance of microorganisms is related to their optimum growth temperatures. Psychrophilic microorganisms (**psychrophiles**) are the most heat-sensitive of the three temperature groups, followed by the **mesophiles** and **thermophiles.** One exercise in this section examines the thermal resistance of selected microorganisms and the effectiveness of commonly used heating methods on microbial survival.

Most microorganisms have an optimum pH range of 6 to 8, but many deviate from these values. Selective media for fungi, for example, employ a high sugar content and a pH of 5.6 in order to prevent the growth of most bacteria. At the other extreme, the pathogen *Vibrio cholerae* requires a pH of more than 8 for optimal growth. In addition, some bacteria are known to be able to oxidize sulfur and nitrogen compounds to form fairly strong solutions of sulfuric and nitric acids. In general, one can say that placing an organism at a pH outside of its optimal range will prevent its growth and that extreme changes will probably kill it.

Other physical factors normal to the environment that will kill or eliminate microorganisms include ultraviolet radiation and filtration. A substantial amount of ultraviolet radiation permeates the atmosphere during daylight hours and can destroy airborne organisms as well as organisms on exposed surfaces and those in the upper few inches of water. Its lethal effect apparently is caused by radiant energy absorption by nucleoproteins. Ultraviolet radiation does not penetrate very well into solids or liquids. Environmental filtration occurs when water ladened with microorganisms (such as sewage effluents) is allowed to percolate through sand or soil. The organisms may adhere to particles of sand or soil, or will be prevented from passing through layers of materials with an effective pore size smaller than the organism. Water supplies can be purified if allowed to pass through sufficient filtration material.

Chemical considerations for this section must be limited to those compounds that are usually toxic for microorganisms and are considered to be *disinfectants* or *antiseptic* materials. These terms are differentiated according to whether the material is used to decontaminate inanimate materials (disinfection) or living tissue (antisepsis). In some cases, the same substances are used for both purposes. In this section, the bacteriostatic activity of dye materials (crystal violet) and metabolic by-products (antibiotics) will be examined.

Dyes also are examples of biological compounds effective in controlling bacterial growth. This effect can also serve to differentiate gram-positive from gram-negative bacteria when suitable concentrations of the dye are used. Furthermore, dyes can act as more general agents at higher concentrations. In contrast to dyes, the antibiotics are biologically produced compounds, made during the metabolic stages of microorganisms, and are often capable of interfering with the activities of other organisms.

In this section, exercises representing both physical and chemical methods of control will be presented through the consideration of the activities and applications of common disinfectants, antiseptics, dyes, and antibiotics. One exercise will consider certain microbe-control methods and an evaluation of a procedure used to disinfect selected instruments.

After completing this exercise, you should be able to:

1. Compare the effectiveness of dry-heating processes and moist-heating methods.
2. List the advantages and disadvantages of methods using high temperatures for microbial killing.
3. Give examples of the application of high-temperature methods for control of microorganisms.
4. Define and explain the importance of *thermal death time* (TDT), and *thermal death point* (TDP).

High temperatures are used in a variety of ways for the effective killing and/or control of microorganisms (Table 29–1). Heat may be applied in either a dry form (hot air) or a moist form (steam or water). The effectiveness of the most extreme form of high temperature, *incineration,* is clearly evident in several exercises in this manual.

Determining the susceptibility of microorganisms to high temperatures is extremely important in controlling microbial populations. This susceptibility can be expressed in terms of the ***thermal death time*** **(TDT).** The TDT is the shortest period of time required to kill **all** *microorganisms,* when exposed to a specific temperature under standard conditions. Such standard conditions include the nature of the medium, pH, and the initial concentration of the microorganisms. Another important measurement is the ***thermal death point*** **(TDP).** The TDP is the temperature at which a suspension of organisms is kindled after a 10-minute exposure.

Dry heat, or hot air, used at sufficiently high temperatures and for adequate exposure times will kill microorganisms. However, this type of heating process is not as effective as certain moist-heat methods and also has a number of disadvantages. These include the requirement of long exposure times, and not all materials lend themselves to being sterilized by dry-heating temperatures. (See Table 29–1.)

Table 29–1

High-Temperature Methods Used for Microbial Control

Method	Temperature	Application	Limitations
Incineration	100°C + (combustible temperatures)	Sterilization of inoculation loops and needles; disposal of carcasses of infected animals; disposal of contaminated objects that cannot be reused or salvaged	Size of incinerator must be adequate to burn largest load promptly and completely and in accordance with local ordinances; potential for air pollution exists
Dry heat	170–180°C for 1–2 hours	Sterilizing materials impermeable to or damaged by moisture (e.g., oils, glass, sharp instruments, metals)	Destructive to materials that cannot withstand extreme high temperatures
Moist heat Autoclave	121.6°C at 15 lb/in^2 pressure, 15–30 minutes	Sterilizing instruments, linens, utensils and treatment trays, media, and other liquids	Ineffective against organisms in materials not penetrable by steam; cannot be used for heat-sensitive articles to be reused
Boiling water	100°C	Killing vegetative cells on instruments, containers	Endospores are not killed; cannot be relied upon to sterilize
Pasteurization	62.8°C for 30 minutes, or 71°C for 15 seconds	Killing vegetative cells of disease-causing microorganisms and of many other microorganisms in milk, fruit juices, and other beverages	Note this is not a method for sterilization

Moist heating methods are more effective for microbial killing than those using dry heat. This is largely because moist heat causes *denaturation* and *coagulation of proteins* and *nucleic acids,* which are vital to the functioning of all life. Dry heat on the other hand brings about an oxidation of such cell components as well as other organic cellular constituents. Examples of **moist-heating methods** include *boiling, subboiling,* and the use of the *autoclave* (steam under pressure). Boiling water to 100°C (the boiling point of water) will kill vegetative cells of various microorganisms but will not guarantee complete killing of all life (sterility). Therefore, boiling water is not used as a preferred method of sterilization.

Because the temperatures needed for sterilization are known to have destructive effects on various foods and beverages, other approaches are used to reduce microbial numbers and/or contamination. One of the better known methods is *pasteurization.* This controlled application of moist heat, developed by Louis Pasteur in the 1860s to prevent the spoilage of French wines, kills most microbial vegetative cells, but does not sterilize. Applying pasteurization to milk and milk products kills the vegetative cells of pathogens and those of most other microorganisms, and thus prolongs the keeping quality of the product. Modern dairies generally use *flash pasteurization* methods, which involve passing milk continuously through a heat exchanger where its temperature is quickly raised to 71°C and held there for 15 seconds, and then is quickly cooled.

The *autoclave,* an equipment item that is quite commonly seen and used in microbiology laboratories, uses steam under pressure to kill microorganisms. Autoclaving, when used correctly, kills all forms of life including cells, viruses, and microbial spores. The autoclave commonly employs a temperature of 121°C. The sterilization time needed is determined by the amount of materials to be sterilized; while autoclaving is very effective for sterilization, it does have limitations (see Table 29–1).

Sterilization methods require some means of monitoring their effectiveness. These include thermometers, chemicals that darken on exposure to specific heat treatments, and biological monitors (Figure 29–1). Biological monitors consist of viable bacterial spores that are resistant to steam on dry heat. After undergoing the sterilization procedure, the monitor is incubated at an appropriate temperature. Growth after 24 to 48 hours indicates a failure of the methods.

This exercise will be used to demonstrate several high-temperature methods and to compare their effectiveness.

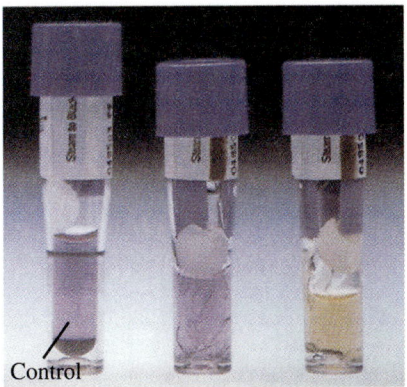

Control

Figure 29–1

An example of a biological monitor, the AMSCO sterility "Proof Plus" monitoring system. No color change from the control after incubation indicates proper sterilization, whereas a yellow color indicates sterilization failure.

A. High-Temperature Methods

Materials

The following items should be provided per 4 students:

❑ 1. Six mL of a twenty-four-hour nutrient broth culture of *Escherichia coli*

❑ 2. Six mL of a six-day-old nutrient broth culture of *Bacillus subtilis*

❑ 3. Two nutrient agar plates

❑ 4. Ten tubes of glucose phenol red broth in 5-mL amounts

❑ 5. Ten sterile 1.0-mL pipettes with rubber bulbs and/or pipetters

❑ 6. Two sterile 5.0-mL pipettes

❑ 7. One container with disinfectant for pipette disposal

❑ 8. One standard thermometer

The following items should be provided for general class use:

❑ 1. Water bath set at 63°C

❑ 2. Dry-heat oven set at 160°C

❑ 3. Test tube racks

❑ 4. Two hundred-and-fifty-mL Pyrex beakers

❑ 5. Sterile glass test tubes

❑ 6. Metal tripods and wire mesh squares for boiling water baths

❑ 7. Marking pens or similar devices

Procedure 1: Incineration

This procedure is to be performed by students in pairs.

❑ 1. Obtain two nutrient agar plates, and mark one for each of the bacterial cultures provided.

❑ 2. With the marker, section each plate into two parts. Label one section of each plate "control" and the other as "test."

❑ 3. Streak *Escherichia coli* on the control half of the plate labeled for the organism.

❑ 4. Flame the inoculating loop and obtain another loopful of the *E. coli* culture.

❑ 5. Flame the loop again and touch it at one end of the surface "test" section to cool it.

❑ 6. Next, streak the "test" or second half of the plate.

❑ 7. Repeat steps 3 through 6 with the *Bacillus subtilis* culture.

❑ 8. Incubate both plates at 37°C for 24 hours.

❑ 9. Examine each plate for growth and record your findings in Table 29–2 in the Results and Observations section.

Procedure 2: Pasteurization, Boiling, and Autoclaving

This procedure is to be performed by students in pairs.

Additional Techniques Required for This Portion of the Exercise:

❑ 1. Broth Transfer, Procedure Diagram 5, Exercise 4

❑ 2. The Use of a Pipette Pump, Procedure Diagram 11, Exercise 6

❏ 1. Using a metal tripod, wire mesh square, and beaker with sufficient water, set up a boiling water. Maintain it at the boiling temperature.

❏ 2. Label 5 tubes for each bacterial culture provided with the initials of the organism being used and with the letter representing one of the following heating methods:
   ❏ a. *P* = pasteurization
   ❏ b. *D* = dry heat
   ❏ c. *B* = boiling
   ❏ d. *A* = autoclave
   ❏ e. *C* = control

❏ 3. With the aid of a pipettor and a 5-mL sterile pipette, aseptically transfer 1 mL of the *E. coli* culture to each of the marked tubes.

❏ 4. Discard the pipette into the container with disinfectant or as directed by the instructor.

❏ 5. Repeat steps 2 through 4 with the *B. subtilis* culture.

❏ 6. Place the labeled tubes into racks or containers into the following heating systems, and for the time periods indicated.

Tube	Heating Systems	Time Period
P	63°C water bath	30 minutes
D	Hot air oven set at 160°C	5 minutes
B	Boiling water bath	10 minutes
A	Autoclave (121°C)	15 minutes
C	37°C incubator or room temperature	30 minutes

❏ 7. After each heating process has been completed, allow the cultures to cool. This can be done easily by holding the bottom of the culture tubes under running tap water. (DO NOT ALLOW THE WATER TO CONTAMINATE THE CULTURES.)

❏ 8. Obtain and label 1 tube of phenol red glucose broth for each culture. (This step will require *10* tubes.)

❏ 9. With a pipetting aid (pipettor) and separate sterile 1-mL pipettes, septically transfer 0.1 mL of each heated and control culture to the appropriate labeled phenol red glucose broth. *(Note the color of broth.)*

❏ 10. Discard the pipette into the container of disinfectant or as indicated by your instructor.

❏ 11. Incubate all tubes of 37°C for 24–48 hours or as specified by your instructor.

❏ 12. After incubation, examine all tubes as to a color change in the phenol red glucose broth and the presence of growth as indicated by turbidity. (*Note:* Phenol red glucose broth will change color if growth occurs and the enzymes of the organism are functional. A yellow color indicates the breakdown of glucose. Refer to Exercise 22 and Figure 22–1 for additional details.)

❏ 13. Enter your findings and answer the questions in the Results and Observations section.

B. Demonstration of the Autoclave and the Use of Biological Monitors

Materials

The following items should be provided for a class demonstration:

❏ 1. Two commercial biological monitor systems

❏ 2. One water bath or incubator set at 56°C

❏ 3. One standard autoclave

Procedure 1: Demonstration of the Autoclave

Your instructor will demonstrate the use of the autoclave. Specific features to be emphasized in the demonstration should include:

❏ 1. Air-tight doors

❏ 2. Autoclave chamber

❏ 3. Temperature and pressure indicators

❏ 4. Exhaust system

❏ 5. Timer and cycle indicator

Procedure 2: The Use of Biological Monitors

This procedure will be performed by your instructor or students assigned by the instructor.

❏ 1. Obtain two biological indicator systems and label one "control" and the second "test."

❏ 2. Place the "control" system into the 56°C water bath or incubator for 48 hours.

❏ 3. Place the "test" system into the autoclave and set the instruments for a full-sterilizing cycle.

❏ 4. After the cycle has been completed, incubate the "test" system for 48 hours in the 56°C water bath or incubator.

❏ 5. After incubation, examine each system for the presence or absence of growth.

❏ 6. Answer the questions in the Results and Observations and Laboratory Review sections.

Results and Observations

Incineration

1. Enter your findings in Table 29–2. Use a "+" for growth and "−" for no growth.

Table 29–2

Results and Incineration

Microorganism	Control Section	Test Section
Bacillus subtilis		
Escherichia coli		

Pasteurization, Boiling, and Autoclaving

1. Enter your findings in Table 29–3. Indicate an acid *(yellow color)* reaction by "A," and an alkaline *(red color)* reaction by "Alk" for the results obtained with the phenol red glucose red broth. Use a "+" for the presence of growth and turbidity, and "−" for no growth and no turbidity.

Table 29–3

Results of Heating Methods

Microorganism	Pasteurization Phenol Red Glucose Broth[a]			Dry Heat P R G B [a]			Boiling P R G B [a]			Autoclave P R G B [a]		
	C	T	G	C	T	G	C	T	G	C	T	G
B. subtilis												
E. coli												

[a] PRGB = Phenol Red Glucose Broth, C = color, T = turbidity, and G = growth

2. Were all heating methods effective in killing the bacterial cultures used? _____

    a. If not, which ones were not effective? Offer an explanation for the failure. _____

_____

The Use of Biological Monitors

1. Was growth present in both systems? _____

    a. If not, which one did not exhibit growth? _____

    b. On the basis of your findings, what conclusion can be made about the autoclave cycle? _____

_____

2. What is the temperature and time period needed to insure complete sterilization with the autoclave?

_____

Laboratory Review 29   The Effectiveness of Heating Methods on Microbial Survival

1. Complete the following table by listing four heating methods used for microbial control, the temperatures used, and examples of their applications.

Table 29–4

Heating Methods Comparison

Method	Temperature(s) Used	Applications
1.		
2.		
3.		
4.		

2. What pressure, temperature, and sterilization times are used in routine autoclaving? (Refer to your text.) _____

    _____

3. Why is pasteurization used with milk and milk products rather than autoclaving or other forms of heat-sterilization? _____

    _____

4. Define or explain the following:

    a. thermal death time _____

    _____

    b. thermal death point _____

    _____

    c. biological monitor _____

    _____

Key Terms

coagulation (kō-ag-ū-LĀ-shun): the process of curdling or clotting

denatured: destruction of the usual or natural form of a substance causing it to lose some of its physical and chemical properties

pasteurization (pas-tūr-i-ZĀ-shun): a heating method designed to destroy pathogens and food-spoilage organisms without changing the chemical composition of materials undergoing the process

sterilization (ster-il-i-ZĀ-shun): process of completely removing or destroying all life on or in a substance

vegetative cell: a growing and metabolizing cell

The Use of Ultraviolet Radiation for Sterilization and Its Mutagenic Effects

After completing this exercise, you should be able to:

1. Explain radiation resistance differences among selected bacterial cultures.
2. Interpret laboratory data concerning the ability of ultraviolet radiation to penetrate clear plastic.
3. Recognize selected mutations by physical appearance.
4. Compare the frequency of occurrence of natural (spontaneous) mutations with that of induced mutations for pigmentation of bacterial culture.
5. Relate mutation frequency to the duration of exposure to ultraviolet light.
6. Test the penetrating capacity of ultraviolet light.

Because of its apparent ease of handling, ultraviolet radiation has been considered for sterilizing a wide variety of materials. But it has been found to have a very limited application, because it has a very poor penetration capacity. It is effectively used to sterilize the air in hospital operating rooms; ultraviolet wavelengths for this purpose are in the range of 250 to 265 nanometers. To secure maximum penetration, several banks of ultraviolet lamps are arranged so that a thin layer of air can pass around and between them. Only in this manner can ultraviolet radiation be effective for air sterilization.

Ultraviolet light is absorbed by various compounds in bacterial cells, most significantly by the purine and pyrimidine bases in deoxyribonucleic acid (DNA). The binding of two adjacent thymine bases, resulting in the production of thymine dimers, is a major outcome of exposure to ultraviolet radiation. The formation of such dimers, unless repaired, may cause significant damage or death to a cell because it cannot correctly transcribe or replicate the dimer-containing DNA without producing errors.

Actively multiplying microorganisms are the most easily killed by ultraviolet radiation. Pigmented cells and endospores are the most resistant and require longer ultraviolet exposure radiation in order for the lethal waves to penetrate the organisms (Figure 30–1).

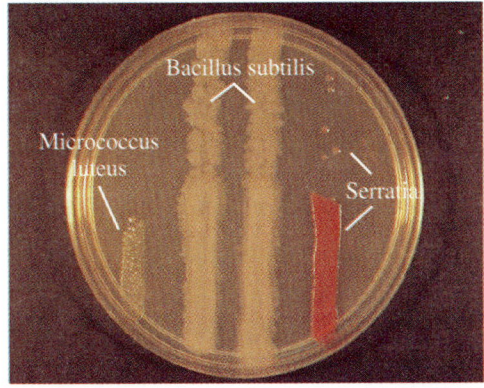

Figure 30–1

The lethal effect of ultraviolet (UV) light exposure. This plate was partially covered by the Petri plate top during exposure. The organisms shown are *Micrococcus luteus* (yellow growth), *Bacillus subtilis* (2 streaks), and *Serratia marcescens.*

The degree of injury is also related to the efficiency of repair mechanisms of exposed organisms. Several bacterial species, for example, have photoreactivation enzymes that can repair the damage caused by ultraviolet light. These enzymes, activated by exposure to visible light, repair the DNA by the process known as *photoreactivation.* The process involves the enzymes binding to the thymine dimers and with the aid of light energy, split the dimer-forming bonds, thereby restoring the thymine nucleotides to their original state. *DNA-polymerase* and *DNA-ligase* are enzymes that repair ultraviolet radiation-caused damage in the dark. This process known as *dark repair* requires the enzymatic removal of the distorted thymine dimers. This excision step creates a wide gap that DNA polymerase fills in by synthesizing new DNA is complementary to the undamaged DNA strand. The DNA ligase completes the process by joining the old (undamaged) DNA and the newly formed DNA. Photoreactivation is not totally effective for all cells in a population exposed to and damaged by ultraviolet light (UVL).

A major disadvantage of ultraviolet light is its limited penetrating power. Unless microorganisms are directly exposed to the source of radiation, they are likely to escape its destructive effects (Figure 30-2). This exercise includes a demonstration of the limited penetrating capacity of UVL.

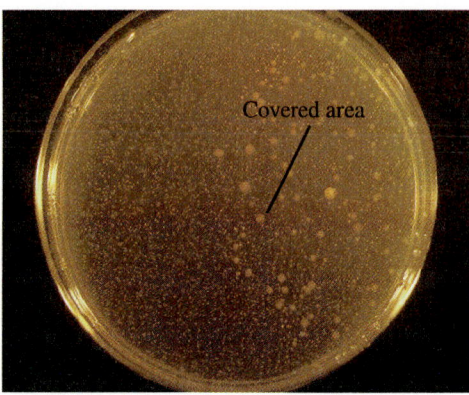

Figure 30–2
The limited penetrating capacity of ultraviolet light (UVL). One-half of this plate was covered with a Petri plate top during UVL exposure. Surface growth occurred only on the covered area.

Specific genes found along DNA strands are responsible for genetic characteristics of bacteria such as antibiotic resistance, capsule formation, colony size, motility, pigment production, and the synthesis of various biochemical compounds, including amino acids and vitamins. Any alterations in the nucleotide base sequences of DNA molecules can result in permanent genetic changes or *mutations,* which in turn cause changes or losses of associated properties of the organism. Some mutations are believed to occur as the result of damages of nucleic acids caused by ultraviolet light, X-rays, or other types of irradiation. Examples of such mutations include changes in colony size, loss of the capsule, loss of pigment (Figure 30–3), formation of colonies with different pigment, pie-shaped sectors, and the most serious effect, a significant reduction in viable cells (lethal effect). The rate of mutation can be increased by exposing a bacterial culture to such forms of radiation. Any chemical or physical agent that increases the mutation rate is called a *mutagen.* Mutations brought about by the laboratory application of a mutagen are said to be induced rather than spontaneous.

This exercise will demonstrate various effects that can develop in a bacterial culture exposed to ultraviolet light.

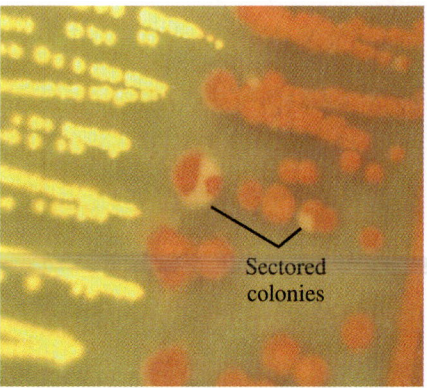

Figure 30–3
The appearance of a sectored colony of *Serratia marcescens.* Such colonies result when a mutation causing the absence of pigmentation occurs during the formation of a colony of pigmented organisms.

A. Lethal Effects of Ultraviolet Light

Materials

❑ 1. Forty-eight-hour nutrient broth cultures of the following (1 per 4 students):
    ❑ a. *Bacillus subtilis*
    ❑ b. *Micrococcus luteus*
    ❑ c. *Serratia marcescens*

❑ 2. Nutrient agar deeps (2 per student)

❑ 3. Sterile Petri plates (2 per student)

❑ 4. Ultraviolet lamps for class use (2 per class)

❑ 5. Aluminum wrap

Procedure

This procedure is to be performed by students in pairs.

❑ 1. Pour and prepare 2 nutrient agar plates for streaking.

❑ 2. Make streak lines of each organism on the agar surface of each plate as shown in Figure 30–4. Note that 2 streaks of *B. subtilis* are needed.

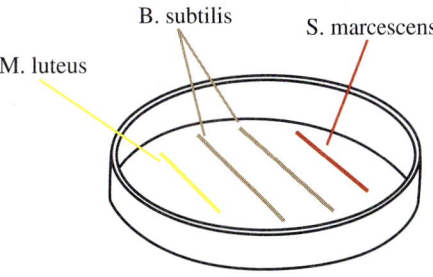

M. luteus   B. subtilis   S. marcescens

Figure 30–4
A diagram of a streak plate ready for lethal ultraviolet irradiation.

❑ 3. Label the bottom of each plate to indicate the location of the respective organisms.

❑ 4. Allow the UV lamp to warm up for 5 minutes and then expose the prepared plates under the ultraviolet lamp for 10 minutes. One plate should be totally uncovered, and the second plate should be partially covered with its Petri dish top so that one-half of each of the parallel streaks is exposed. The instructor will set the distance, which may vary from a few inches to 10 to 12 inches, depending upon the intensity of the radiation. *Do not look directly at the lamp, as severe eye damage can result.*

❑ 5. Close the plates, wrap completely in aluminum foil, and incubate them at room temperature for 24–48 hours.

❑ 6. After incubation, sketch your observations and answer the questions in the Results and Observations section. (Refer to Figure 30–1.)

B. Penetrating Capacity of Ultraviolet Light

Materials

❑ 1. The following materials should be provided per 4 students:
    ❑ a. Two mL of a 48-hour trypticase soy broth culture of *Serratia marcescens*
    ❑ b. Four nutrient agar deeps
    ❑ c. Four sterile Petri plates
    ❑ d. Two sheets of aluminum foil or wrap, about 8–11 inches each

❑ 2. Two or more ultraviolet lamps (germicidal or with a wavelength of 200 to 300 nanometers)

Procedure

Additional Technique Required for This Portion of the Exercise:

❑ Pour Plate Technique, Procedure Diagram 9, Exercise 5

This exercise is to be performed by students in groups of 4.

❑ 1. Turn on the ultraviolet lamp. Allow it to warm up for at least 15 minutes before use. Set up the lamp so that there is about 10 mm, or 4 inches, of space between the lamp and the surface on which the agar plate will rest. The instructor will set the distance, which may vary from a few inches to 10 to 12 inches, depending upon the intensity of the radiation. *Do not look directly at the lamp, as severe eye damage can result.*

❑ 2. Prepare 4 pour plates with the 48-hour *S. marcescens* culture provided.

❑ 3. After the media has hardened, label both the covers and bottoms of each plate with exposure UVL exposure times as follows: 2 plates for **5 minutes** and 2 plates for **10 minutes.** In addition, draw a line dividing the bottom of each plate into 2 sections. One-half of each plate will be irradiated and the other will not.

❑ 4. Expose the 2 plates designated for 5 minutes to ultraviolet light with the Petri plate tops partially covering the respective agar surfaces (Figure 30–5).

❑ 5. Close the plates, wrap completely in aluminum foil, and incubate one plate at room temperature and the other at 37°C for 24–48 hours.

❑ 6. Repeat steps 4 and 5 with the pour plates designated for a 10-minute exposure to ultraviolet light.

❑ 7. After incubation, examine the plates and look for the presence or absence of bacterial colonies on and in the respective agar plates. In addition, look for differences in pigmentation. (Refer to Figure 30–2.)

❑ 8. Enter your findings and answer the questions in the Results and Observations section.

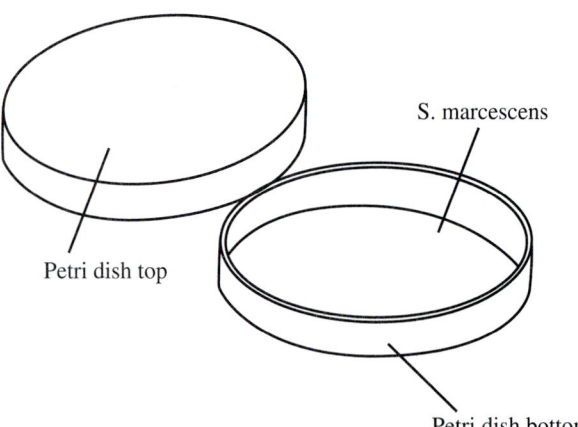

S. marcescens

Petri dish top

Petri dish bottom

Figure 30–5
A diagram of a streak plate ready for ultraviolet irradiation.

C. Mutagenic Effects of Ultraviolet Light

Materials

❑ 1. The following materials should be provided per 4 students:
   ❑ a. Two mL of a 24-hour trypticase soy broth culture of *Serratia marcescens*
   ❑ b. Four trypticase soy agar plates
   ❑ c. Four 3 × 5-inch cards
   ❑ d. One sheet of aluminum foil or wrap, about 8–11 inches

❑ 2. Two or more ultraviolet lamps (germicidal or with a wavelength of 200 to 300 nanometers)

❑ 3. Tape and tape dispenser

Procedure

This exercise is to be performed by students in groups of 4.

❑ 1. Turn on the ultraviolet lamp. Allow it to warm up for at least 15 minutes before use. Set up the lamp so that there is about 10 mm, or 4 inches, of space between the lamp and the surface on which the agar plate will rest.

❑ 2. Streak 4 agar plates with the 24-hour *S. marcescens* culture provided. Streak back and forth across the plate so that most of the surface will be utilized.

❑ 3. Label both the cover and the bottom of each streaked plate with one of the following exposure times: 5, 15, 30, and 60 seconds. In addition, draw a line dividing the bottom of each plate into 2 sections. One-half of each plate will be irradiated and the other will not.

❑ 4. Remove the cover of the first plate to be used. As shown in Figure 30–6, place one of the cards over the section of the plate that is not to be irradiated. Secure the card to the side of the Petri plate with a small piece of tape. Expose the preparation for the designated time.

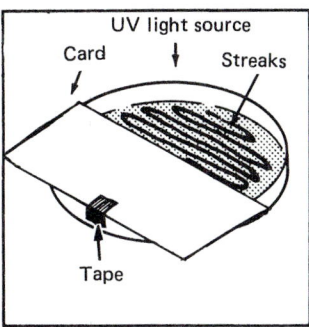

Figure 30–6
A diagram of a pour plate ready to determine the penetrating capacity of ultraviolet light.

❑ 5. After exposure, replace the cover. Put the plate in an inverted position in your desk drawer, and cover it with a sheet of aluminum wrap. Dispose of the card as directed by your instructor.

❑ 6. Repeat steps 4 and 5 with each remaining plate.

❑ 7. Incubate all plates in your desk drawer under the sheet of aluminum wrap.

❑ 8. After incubation, examine all plates for the presence of white and sectored colonies (refer to Figure 30–3). Count the number of these colony types on both sides of the exposed plates. Enter your findings in Table 30–1 of the Results and Observations section.

❑ 9. Select the plate containing the most white colonies on the irradiated side and count the *total* number of colonies on both the irradiated and nonirradiated sides. Use this information to calculate the effects of ultraviolet radiation:
    ❑ a. The percent frequency of total mutations after irradiation
    ❑ b. The percent frequency of spontaneous mutation
    ❑ c. The percent of induced mutation
Formulas for these calculations are as follows:

$$\text{Percent frequency of total mutations after irradiation} = \frac{\text{Number of white colonies on irradiated side}}{\text{Total number of colonies on irradiated side}} \times 100$$

$$\text{Percent frequency of spontaneous mutation} = \frac{\text{Number of white colonies on nonirradiated side}}{\text{Total number of colonies on nonirradiated side}} \times 100$$

$$\text{Percent of induced mutation} = \frac{\text{Percent frequency of total mutations}}{\text{Percent frequency of spontaneous mutations}} \times 100$$

Results and Observations

Lethal Effects of Ultraviolet Light

1. Sketch the growth patterns that were observed.

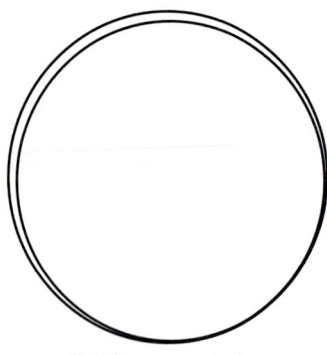   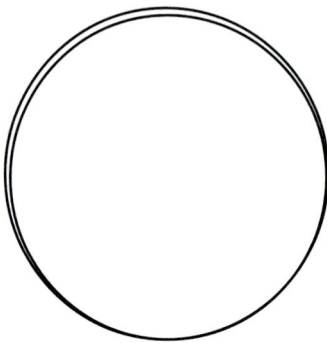

Totally exposed plate   Partially exposed plate

2. Which of the bacterial species used exhibited resistance to the lethal effects of UV irradiation? _____

_____

3. What factors account for UV resistance? _____

_____

Penetrating Capacity of Ultraviolet Light

1. Sketch the surface growth patterns observed on the exposed parts of the agar plates.

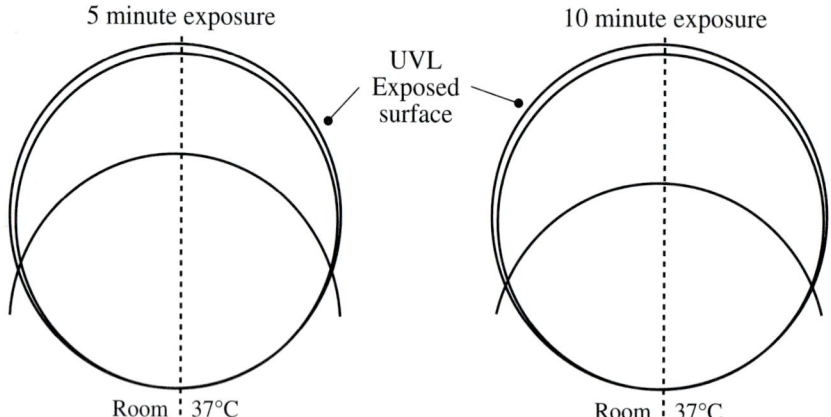

2. Briefly describe any differences with the following:

   a. Plates exposed for the same time period, but incubated at different temperatures. _____

_____

   b. Plates exposed for different times, but incubated at room temperature. _____

_____

   c. Plates exposed for different times, but incubated at 37°C. _____

_____

Mutagenic Effects of Ultraviolet Light

1.   Enter your findings in the following table.

Table 30–1

Mutagenic Effects of Ultraviolet Light

Exposed Plates According to Exposure Time	Nonirradiated Side			Irradiated Side		
	Total Number of White Colonies	Total Number of Sectored Colonies	Total Number of Colonies	Total Number of White Colonies	Total Number of Sectored Colonies	Total Number of Colonies
5 seconds						
15 seconds						
30 seconds						
60 seconds						

2.   a. Did you find more sectored colonies on the irradiated or the nonirradiated sides of the exposed plate? _____

   _____

   b. Why do sectored colonies form? _____

   _____

   _____

Laboratory Review 30   The Use of Ultraviolet Radiation for Sterilization and Its Mutagenic Effects

1.   Would you expect spore formers and pigmented bacteria to be more resistant to the effects of ultraviolet radiation? Explain your answer briefly. _____

   _____

   _____

2.   What are the specific effects of ultraviolet radiation on DNA? _____

   _____

   _____

3.   a. What is photoreactivation? _____

   _____

   b. How effective is it? _____

   _____

4. What functions do DNA-polymerase and DNA-ligase serve? _____

    _____

5. What is a sectored colony? _____

    _____

6. Give 2 clinical applications of ultraviolet light. (Refer to your text.)

    a. _____

    b. _____

7. What is a major disadvantage of ultraviolet light? _____

    _____

Key Terms

DNA-ligase (LĪ-gās): an enzyme that joins short pieces of deoxyribonucleic acid

DNA-polymerase (pol-IM-er-ās): an enzyme that joins the nucleotides of deoxyribonucleic acid

endonuclease (en-dō-NEW-klē-ās): an enzyme that attacks the internal regions of nucleic acids (covalent bonds between nucleotides)

mutagen: any chemical or physical agent that increases the mutation rate

mutation: a permanent genetic change

purine (PŪ-rēn): a nitrogen-containing ring base found in nucleic acids; examples include adenine and guanine

pyrimidine (pī-RIM-id-in): a nitrogen-containing ring base found in nucleic acids; examples include cytosine, thymine, and duracil

thymine dimer: a compound resulting from the joining of two adjacent thymine molecules; causes distortions in the DNA molecule

After completing this exercise, you should be able to:

1. Recognize the influence of the hydrogen ion concentration on microbial growth.
2. Recognize that an optimum concentration of hydrogen ions exists for each organism in which it grows best.
3. Identify the general pH range in which microbial growth can occur.

In addition to incubation temperature, osmotic conditions, and media composition, the *pH* of an organism's environment exerts a significant influence on that organism's growth and metabolism. The development of a bacterial population may be severely limited by pH changes that result from the growth and related activities of organisms themselves. The pH value of a given solution is the logarithm of the reciprocal of the hydrogen ion concentration. A convenient method used to express hydrogen ion concentrations utilizes the well-known pH scale. The acidic range of the scale extends from 0 to 6.9, while the alkaline, or basic, range extends from 7.1 to 14.

It has been well established that most microorganisms grow best at pH values around 7.0, or neutrality. Few organisms grow below 4.0. It is important to note that for each organism, an optimum, or most favorable, concentration of hydrogen ions exists. Bacteria tend to be more sensitive than fungi to pH changes in their environments. This susceptibility is evident in the keeping quality of certain foods. It is quite common for fruits to undergo mold or yeast spoilage, or both. Such spoilage is caused by the capacity or tolerance of these organisms for growing at pH values below 3.5, which is considerably below the minimum pH value for most food-spoilage and all food-poisoning bacteria.

The detection of acid production and/or acidic conditions in culture media is most often accomplished with the use of **acid-base (pH) indicators** (Table 31–1). These indicators are themselves either weak acids or bases. They do not change color suddenly at a specific pH value, but do so gradually over a range of pH. Phenol red, which has a specific pH range of 6.9 to 8.5 (see Table 31–1), is used in this exercise.

This exercise will demonstrate the influence of pH on microbial growth and metabolism with media containing different hydrogen ion concentrations.

Table 31–1

Some Examples of pH Indication Used in Microbiological Media

pH Indicator	pH Range	Color Change (Acid→Alkaline)
Bromocresol purple	5.4⟷7.0	yellow→purple
Bromophenol blue	3.1⟷4.7	yellow→blue
Bromothymol	6.1⟷7.7	yellow→blue
Methyl red	4.2⟷6.3	red→yellow
Phenol red	6.0⟷8.5	yellow→red

Materials

The following materials should be provided per 4 students:

❏ 1. Forty-eight-hour nutrient broth cultures of the following:
    ❏ a. *Bacillus subtilis*
    ❏ b. *Escherichia coli*
    ❏ c. *Serratia marcescens*

❏ 2. Forty-eight-hour sucrose broth culture of the yeast *Saccharomyces cerevisiae*

❏ 3. Four each of glucose phenol red broths in 5-mL amounts, with pH values of 3, 5, 7, and 9

❏ 4. Four nutrient agar plates, to be used after incubation of broths

❏ 5. One wax marking pencil

Procedure

This exercise is to be performed by students in groups of 4.

❏ 1. Note the variations in color among the glucose phenol red broths with different pH values: the pH indicator phenol red is orange at pH 7, yellow at pH below 7, and red at pH above 7. Enter the colors in the table in the Results and Observations section.

❏ 2. Inoculate the separate sets of phenol red glucose broths with each of the microbial cultures provided.

❏ 3. Label each tube with its pH value and the organisms used.

❏ 4. Incubate all tubes at 37°C until the next laboratory period.

❏ 5. Examine each tube for the amount of growth. Record your findings in the Results and Observations section using 0 for no growth, + for slight turbidity, 2+ for definite turbidity, 3+ for quite turbid, and 4+ for extreme turbidity. Note also any color changes in the media.

❏ 6. With a wax marking pencil, divide 1 nutrient agar plate into 4 sectors for each set of inoculations, and label accordingly.

❏ 7. Streak 1 sector with a loopful of each of the phenol red broths.

❏ 8. Incubate the plate preparations at 37°C until the next laboratory period.

❏ 9. Examine each plate for growth and record your findings in Table 31–2 in the Results and Observations section.

Results and Observations

1. Complete the following table.

Table 31–2

pH Effects on Microbial Growth

Organism	Medium Color Before Incubation	Growth at pH				Growth on Nutrient Agar from pH			
		3	5	7	9	3	5	7	9
B. subtilis									
E. coli									
S. cerevisiae									
S. marcescens									

2. Did all organisms provided grow at each hydrogen ion concentration? If not, which ones did not? _____

_____

3. Did any of the tubes of media exhibit changes in color after incubation? Explain any changes. _____

_____

_____

Laboratory Review 31 — The Effects of pH on Microorganisms

1. a. What is pH? _____

_____

   b. What microbial activities does pH affect? _____

_____

2. a. Can individual bacterial species tolerate a pH range from 4 to 8? _____

_____

   b. What types of microorganisms grow best in an acid medium? _____

_____

3. Why are pH indicators frequently incorporated into culture media? _____

_____

4. Complete the following table by listing 3 pH indicators commonly used in media. Indicate their useful pH ranges and the colors formed at these ranges.

Table 31–3

pH Indicators Used in Media

Indicator	pH Range	Colors Formed

5. a. What substances are incorporated into culture media to control the large change in pH that may occur during growth? _____

_____

   b. Give 2 examples of these substances. _____

_____

6.   Why do high-acid foods require less treatment for preservation in canning? _____

_____

_____

Key Term

pH: the degree of acidity or alkalinity of a solution. It is expressed in units based on the concentration of hydrogen ions. These units in the pH scale range from 0 (the most acid) to 14 (the most alkaline).

After completing this experimental exercise, you should be able to:

1. Explain the relationship between solute concentration and osmotic pressure.

2. Test the effect of changes in environmental osmotic pressure on microorganisms.

3. Support or reject the hypothesis that changes in environmental osmotic pressure will affect microbial growth and cultural characteristics.

4. Explain the practical role of increased osmotic pressure in the preservation of food.

Solutions consist of two parts, the **solute,** the substance that is dissolved in a solution, and the **solvent,** the fluid (solution) in which the solute is dissolved. The phenomenon of *osmosis* occurs with the passage of solvent through a semipermeable membrane from a region of low solute concentration to one of higher concentration. **Osmotic pressure** is the physical phenomenon to which the solute concentration of a solution is related, and it develops when two solutions of different concentrations are separated by a semipermeable membrane. This pressure varies with respect to the concentration of the solution. Animals cells such as red blood cells or aquatic plant cells (Figure 32–1*a*) have an osmotic pressure approximately equal to that of the circulating fluid in their respective environments. Solutions exerting this osmotic pressure are referred to as being **isotonic.** Solutions with greater osmotic pressure cause cell membranes to shrink (Figure 32–1*b*), and are referred to as being **hypertonic.** Still other solutions with lesser osmotic pressure cause cells to swell. Such solutions are called **hypotonic.** Cells having only a cell membrane as their outermost structure undergo *crenation* in hypertonic solutions, and *lysis* in hypotonic solutions. Plant or plant-like cells also have cell membranes, but have cell walls as their outermost structure. Such cells exhibit **plasmolysis** in hypertonic solutions (Figure 32–1*b*) and **plasmoptysis** in hypotonic solutions.

(a)   (b)

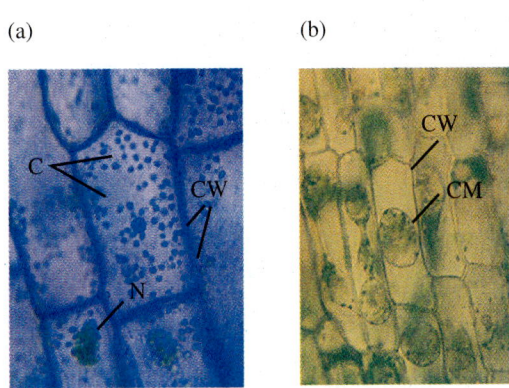

Figure 32–1
The effects of osmotic pressure on plant cells. (a) A photomicrograph of Elodea (*Anacharis* sp.) a common aquarium plant. Note the presence of the large number of chloroplasts (C), the nucleus (N), the cell wall (CW), and the general shape of the cells. (b) The appearance of Elodea after exposure to a highly concentrated salt (hypertonic) solution. The cell (plasma) membranes (CM) have separated from their positions next to the cell walls. The cells clearly show plasmolysis.

The growth and survival of a microorganism can be affected drastically by the amount of water that is allowed to leave or enter its cytoplasm. In the aqueous environments inhabited by microorganisms, the presence of dissolved solids, such as salts, creates solutions that may be isotonic, hypotonic, or hypertonic. In isotonic solutions, the majority of organisms grow naturally. In hypotonic environments, water enters the cytoplasm, thus causing the process of plasmoptysis and leading to the eventual bursting of the cell if the osmotic imbalance is too extreme. Plasmolysis occurs in hypertonic environments where water is drawn out of the cell, causing it to shrink, resulting in metabolic inhibition or cell death (Figure 32–2). Exceptions are known to these situations. These include organisms that grow in the concentrated salt solutions of the Great Salt Lake and the Dead Sea, as well as those that grow and/or survive in a heavy sugar solution such as honey, and in distilled water.

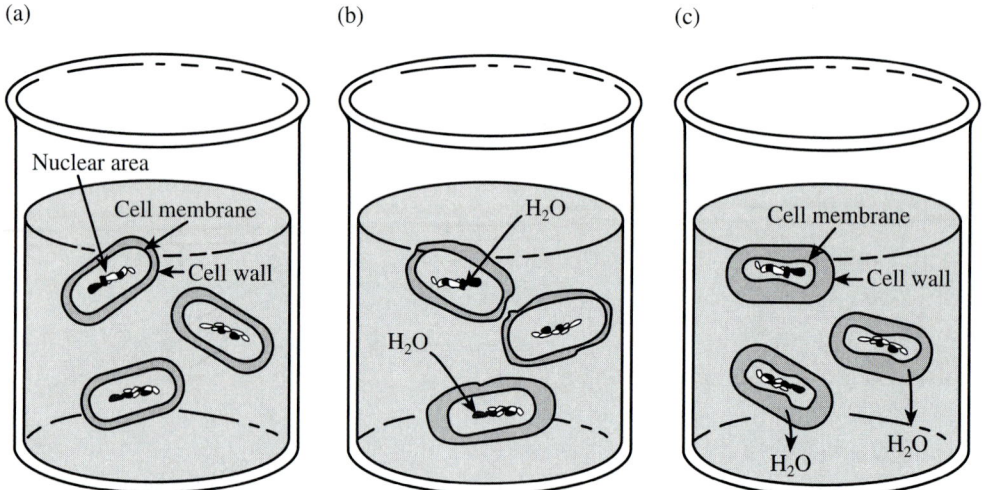

(a)   (b)   (c)

Figure 32–2

Osmotic pressure effects on cell wall-containing microorganisms. (a) The **isotonic** environment. No changes in cell size occur. (b) The **hypotonic** environment. Water molecules enter cells causing them to swell. The presence of a cell wall enables the cell to withstand the increase in osmotic pressure. (c) The **hypertonic** environment. Despite the presence of a cell wall, water loss and shrinkage of cellular content occurs. This process is called plasmolysis.

The osmotic pressure phenomenon is effectively used in food preservation methods. For example, if any plant or animal tissue used as food is dried, solutes become concentrated causing the plasmolysis of bacteria cells that are always present. These tissues are preserved because microbial chemical degrading (decay) is prevented. During the drying process, bacterial cells lose water, thus slowing their metabolism to such a low level that growth and cell division stop. This type of situation is an example of a *bacteriostatic effect.* The same bacteriostatic effect is achieved by the addition of sugar to fruits to make jams and jellies. It is important to note that the osmotic effect on microbial cells does not destroy them, but simply prevents normal microbial activity.

The use of large amounts of salt (sodium chloride) for the preservation of meats is well known. High concentrations of salt removes water from tissues *(dehydration),* thus causing the osmotic effect plasmolysis with any microorganism present. In contrast to the effects caused by large concentrations of sugar in food preservation methods, salt may be not only bacteriostatic, but toxic to certain microorganisms.

While most bacteria are sensitive to high osmotic pressure environments, yeasts and molds appear to be more tolerant of such situations. This is why it is not unusual to find them growing on or in foods and/or materials such as jams, honey, dried fruit and meats, and even clothing. Organisms that can grow and reproduce under conditions of high osmotic pressure are referred to as being **osmophilic** (osmotic pressure-loving).

In this exercise, the effects of varying concentrations of dextrose and sodium chloride on microbial growth and cultural properties, including pigment production and colonial characteristics, will be studied. A modification of the gradient plate techniques described in 1952 by Bryson and Szbalski also will be used to test the salt tolerance of several microbial species. A gradient plate is prepared by pouring nutrient agar into a Petri plate and placing the plate at an angle while the agar solidifies. After solidification of the agar, the plate is placed on a level surface and nutrient agar containing a test substance is poured onto the solid agar wedge. When the agar preparation solidifies, a concentration gradient forms with little or no test substance at one end to full concentration at the other end because of its diffusion into the underlying agar.

Materials

The following items should be provided per 4 students:

❑ 1. One 24-hour trypticase soy agar broth culture of each of the following:
  ❑ a. *Bacillus subtilis*
  ❑ b. *Escherichia coli*
  ❑ c. *Micrococcus luteus*
  ❑ d. *Serratia marcescens*
  ❑ e. *Staphylococcus aureus*

❑ 2. One 48-hour glucose broth culture of *Saccharomyces cerevisiae*

❑ 3. One 5-day Sabouraud's dextrose agar slant culture of each of the following:
  ❑ a. *Penicillium notatum*
  ❑ b. *Rhizopus nigricans*

❑ 4. Two each of the following nutrient agar plate preparations:
  ❑ a. Nutrient agar, 0% NaCl
  ❑ b. Nutrient agar plus 5% NaCl
  ❑ c. Nutrient agar plus 10% NaCl
  ❑ d. Nutrient agar plus 15% NaCl

❑ 5. One marking pen or pencil

Additional Technique Required for This Portion of the Exercise:

❑ The Gram Stain, Procedure Diagram 17, Exercise 14

Procedure 1: The Effect of Osmotic Pressure on Bacterial Cultural Properties

This procedure should be performed by students in pairs.

❑ 1. Using a marking pen or pencil, mark 5 sectors on the bottom of 1 of each nutrient agar plate preparation. Label 1 sector for each of the 5 bacterial cultures provided.

❑ 2. Streak each sector on each plate with the appropriate organism in the pattern shown in Figure 32–3.

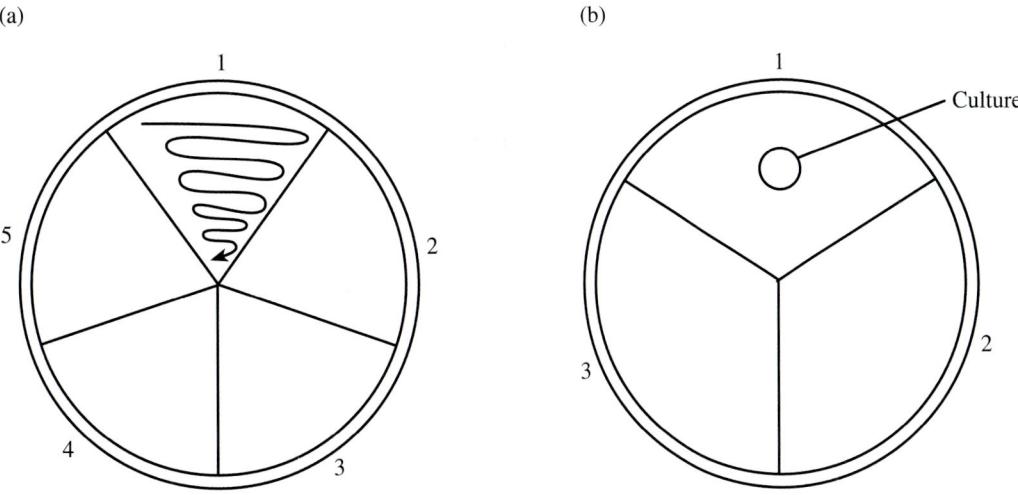

Figure 32–3
Streaking patterns. (a) For bacteria. (b) For fungi.

❑ 3. Invert and incubate the set of plates at room temperature for 48 hours.

❑ 4. After incubation, compare the growth of each species on the different plates. Record your findings in Table 32–1 of the Results and Observations section.

❏ 5. Prepare Gram stains of representative organisms from each plate. Examine and record your findings as to Gram reaction and morphology.

❏ 6. Examine the growth of each species on each plate as to the colonial characteristics of elevation, margins, whole colony appearance, and pigment production, if present (refer to Exercise 7). Answer the questions in the Results and Observations section.

Procedure 2: The Effect of Osmotic Pressure on Fungal Properties

This procedure should be performed by students in pairs.

❏ 1. Using a marking pen or pencil, mark 3 sectors on the bottom of 1 of each nutrient agar plate preparation. Label 1 sector for each of the 3 fungal cultures provided. (Refer to Figure 32–3b.)

❏ 2. Aseptically place a loopful of each fungal culture provided into its appropriately labeled sector on each plate.

❏ 3. Invert and incubate the set of plates at room temperature for 4 days.

❏ 4. After incubation, compare the growth of each culture on the different plates. Examine each culture as to change in pigmentation and mycelial size. Record your findings and answer the questions in the Results and Observations section. (Refer to Figure 32–4.)

(a)   (b)   (c)

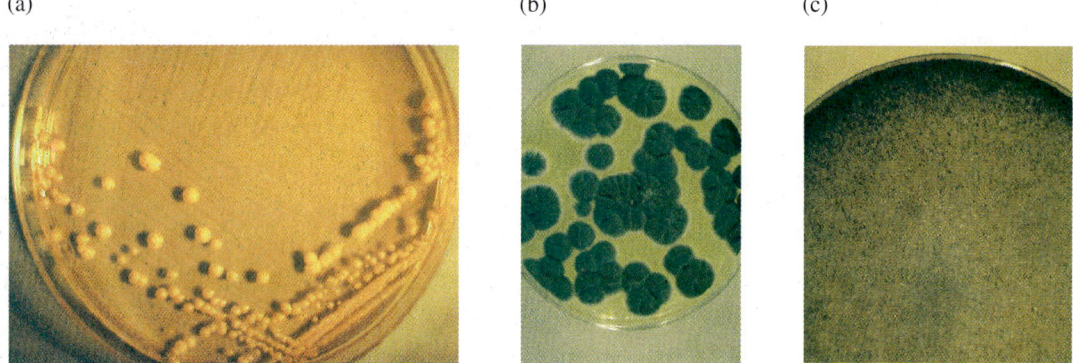

Figure 32–4
The appearance of normal fungal growth. (a) The yeast *Saccharomyces cerevisiae.* The molds, (b) *Penicillium notatum;* and (c) *Rhizopus nigricans.*

Results and Observations

Effect of Osmotic Pressure on Bacterial Cultural Properties

1. Enter your findings in the following table. Indicate the quantity of growth by "4+" for the largest amount, "1+" for the least amount, and "0" for no growth.

2. Did any changes in Gram reactions and/or morphology occur with the cultures used? If so, which ones? _____

_____

3. Did all organisms grow on all plates? If not, which ones did not? _____

_____

4. What explanation can be offered for a lack of growth on any of the plates? _____

_____

Table 32–1

Osmotic Pressure Effects on Microbial Cultural Properties

Organism	0% NaCl		5% NaCl		10% NaCl		15% NaCl	
	Growth	Gram Rx[a] and Morphology	Growth	Gram Rx[a] and Morphology	Growth	Gram Rx[a] and Morphology	Growth	Gram Rx[a] and Morphology
B. subtilis								
E. coli								
M. luteus								
S. marcescens								
S. aureus								

[a]Rx = reactions

5. Were colonial characteristics affected by any of the NaCl concentrations? If so, indicate which concentrations, and which characteristics. _____

_____

_____

Effect of Osmotic Pressure on Fungal Properties

1. Did all fungi grow on the NaCl plates? If not, indicate which ones did not. _____

_____

_____

2. Were mycelial properties affected by the higher concentrations of NaCl? (Refer to Figure 32–4.)

_____

_____

B. Gradient Plate

Materials

The following items should be provided per pair of students:

❑ 1. One 24-hour trypticase soy agar broth culture of each of the following:
  ❑ a. *Bacillus subtilis*   ❑ c. *Serratia marcescens*
  ❑ b. *Escherichia coli*

❑ 2. One 48-hour glucose broth culture of *Saccharomyces cerevisiae*

❑ 3. Two sterile Petri plates

❑ 4. Two nutrient agar melted deeps (15 mL per tube)

❑ 5. One nutrient agar melted deep containing 25% NaCl (15 mL per tube)

❑ 6. One nutrient agar melted deep containing 50% sucrose (15 mL per tube)

❑ 7. One marking pen or pencil

Procedure

This procedure is to be performed by students in pairs.

❏ 1. Prepare 2 gradient plates according to the steps shown in Procedure Diagram 28. One plate should contain NaCl and the second sucrose. Label each to identify the test substance used.

❏ 2. On the bottom of each plate, with a marking pen or pencil, draw an orientation line across the end which corresponds to the high side of the nutrient agar wedge (step 5). This is the area of lowest concentration.

❏ 3. Next, make 4 separate lines perpendicular to the line (step 5). Label one line for each culture provided.

❏ 4. Streak each culture over the area of its labeled line, starting at the end next to the orientation line (step 5).

❏ 5. Invert and incubate both plates at 25°C for 48 hours.

❏ 6. After incubation, examine the plates to see the extent of growth along the inoculation lines. Sketch the general growth pattern and answer the questions in the Results and Observations section.

Procedure Diagram 28
Gradient Plate Preparation

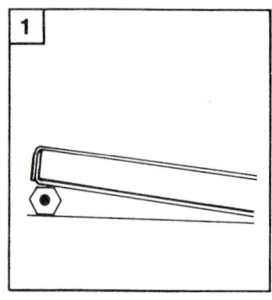

1. Place a sterile Petri dish on a pencil or pen.

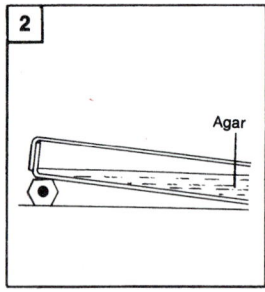

2. Aseptically pour 15 ml of nutrient agar into the plate, and allow it to harden.

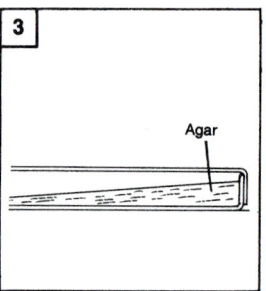

3. Place plate with hardened wedge on a level surface.

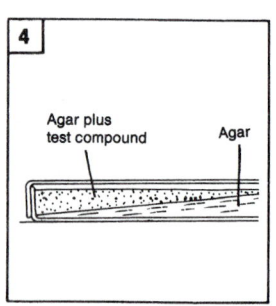

4. Pour 15 ml of nutrient agar containing test compound into the plate, and allow it to harden.

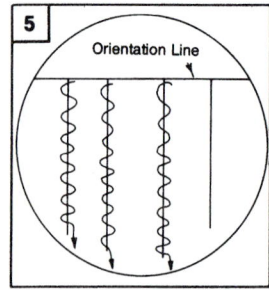

5. Streak a loopful of each culture provided over its labeled sector.

Results and Observations

1. Sketch the growth patterns on the gradient plates in the space provided.

2. Did all organisms grow along the length of their individual streaks on both preparations? If not, which ones did

    not? _____

    _____

3. Was pigment production affected? If so, on which preparation? _____

    _____

Laboratory Review 32 — The Effects of Osmotic Factors on Microbial Growth

1. Why is an NaCl-resistant strain of bacteria likely to be useful for physiological studies? _____

    _____

    _____

2. On a gradient plate, if a colony develops about three-fourths of the way along the streak, by what simple

    procedure can you determine whether or not the organism is resistant to even higher concentrations of sucrose or

    salt? _____

    _____

    _____

    _____

3. Is the appearance of an NaCl-resistant colony on a gradient plate caused by the action of salt (the test substance) or does it represent merely the selecting of a mutant already present in the population before the culture was exposed to the selective agent? _____

_____

_____

4. What types of microorganisms would be capable of survival in foods containing high sucrose concentrations?

_____

_____

Key Terms

bacteriostatic (bak-te-rē-ō-STAT-ik): inhibiting bacterial growth

crenation (krē-NĀ-shun): osmotically changing cells, such as red blood cells, to shrunken, knobbed forms

hypertonic: refers to a solution having a greater amount of solute (dissolved solids) as compared with another, usually on different sides of a semi-permeable membrane

hypotonic: refers to a solution having a lesser amount of solute as compared with another, usually on different sides of a semi-permeable membrane

isotonic: refers to solutions on both sides of a semi-permeable membrane having similar levels of solute

lysis (LĪ-sis): destruction

osmophilic (oz-mō-FIL-ik): favoring higher than normal osmotic pressure

osmotic pressure: the pressure that develops when 2 solutions of different concentrations of dissolved solids are separated by a semi-permeable membrane

plasmolysis (plaz-MOL-ē-sis): the shrinking of living cells' cytoplasm due to excessive water loss by osmosis; crenation in cells without walls

plasmoptysis (plaz-MOP-te-sis): the swelling and eventual bursting of cells due to excessive water intake through osmosis

solute (SOL-ūt): the substance dissolved in a solution (solvent)

The Inhibitory Action of Heavy Metals and Other Chemicals

After completing this exercise, you should be able to:

1. Recognize the inhibitory actions of heavy metals.
2. Compare the inhibitory effects of heavy metal compounds.
3. Use a filter paper disk method for testing purposes.

In 1893, Naegeli observed that silver in very high dilutions killed green algae. Other heavy metals, such as copper and mercury, either alone or in certain compounds, have also long been known to produce harmful effects on microorganisms. The ability of extremely small quantities of certain metals to exert toxic effects on microorganisms is referred to as *oligodynamic action.* The term *oligodynamic* is a compound formed from the Greek words *oligo,* meaning small, and *dynamis,* meaning power.

The phenomenon of oligodynamic action can be demonstrated by placing a freshly cleaned copper or silver coin in a Petri plate and covering it with an inoculated agar preparation. After suitable incubation, you will see a clear, or oligodynamic, zone surrounding the metal coin (Figure 33–1). This effect is believed to be caused by the affinity of certain cellular proteins for metallic ions, which results in the accumulation of large amounts of these ions. Denaturation of these cellular proteins results. The amount of metal needed for this effect amounts to only a few parts per million.

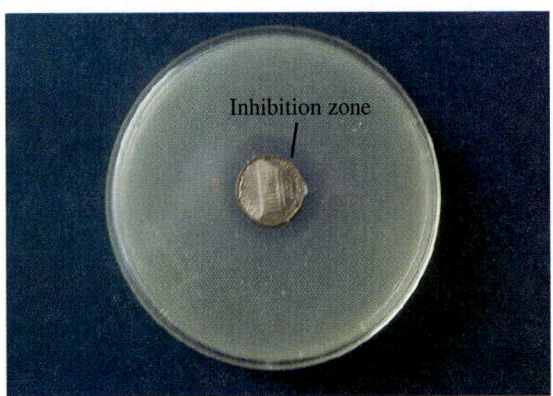

Figure 33–1
The effects of a copper-containing coin on *Serratia marcescens.*

Several compounds of heavy metals have antiseptic or germicidal activity and are used to control microorganisms in a variety of circumstances. Such applications include the application of such compounds in the treatment of various alcoholic beverages, milk, and water; the preparation of antiseptic agents; and the impregnation of various fabrics.

Among the most commonly used antimicrobial compounds of heavy metals are those containing copper, mercury, and silver. Copper compounds are widely used as fungicides in agriculture, and as algicides. Several organic mercury compounds, including Mercurochrome, Merthiolate, and Metaphen, are applied as antiseptics. Silver nitrate has long been used in the prevention of blindness in newborns in cases of gonorrhea and syphilis.

This exercise will demonstrate oligodynamic action and the individual reactions of selected microorganisms to the kinds and concentrations of heavy metal ions.

A. Filter Paper Disk Technique

Materials

The following materials should be provided per 4 students:

❏ 1. Forty-eight-hour nutrient broth cultures of the following:
  ❏ a. *Bacillus subtilis*
  ❏ b. *Escherichia coli*
  ❏ c. *Micrococcus luteus*
  ❏ d. *Serratia marcescens*

❏ 2. Four nutrient agar deeps

❏ 3. Four sterile Petri plates

❏ 4. Sterile filter paper disks in a sterile Petri plate

❏ 5. One pair of sterile forceps

❏ 6. One 50-mL beaker of 70% alcohol

❏ 7. Approximately 10 mL of each of the following 1% solutions in 50-mL beakers:
  ❏ a. Mercuric chloride
  ❏ b. Silver nitrate
  ❏ c. Stannous chloride

❏ 8. One ruler with a millimeter scale

Additional Technique Required for This Portion of the Exercise:

❏ Pour Plate Technique, Procedure Diagram 9, Exercise 5

Procedure

This portion of the exercise is to be performed in groups of 4.

❏ 1. Prepare a pour plate with each culture provided. (Refer to Procedure Diagram 9.) Label each plate with the organism used.

❏ 2. With each pour plate, perform the 6 steps shown in Procedure Diagram 29.

❏ 3. Incubate all preparations at room temperature for 48 hours, or as directed.

❏ 4. After incubation, measure the zone of inhibition with the ruler provided and record your findings in the Results and Observations section. (Refer to Figure 33–2.)

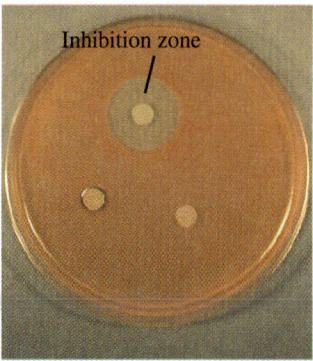

Figure 33–2

Oligodynamic action. The inhibitory effects of mercuric chloride on *Serratia marcescens* are indicated by the absence of growth (clear area) around a disk containing this heavy metal ion.

Procedure Diagram 29
Filter Paper Disk Technique

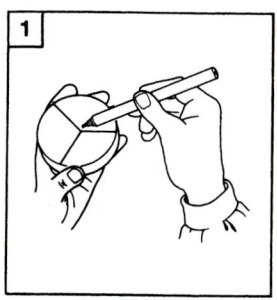

1. Using a wax marking pencil, divide the bottom of a plate into 3 sections and label each section for one of the chemicals provided.

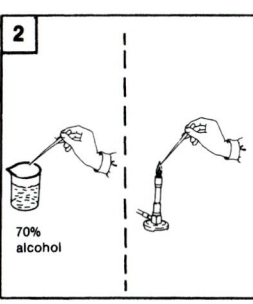

2. Heat-sterilize the forceps by dipping the tips into the beaker containing 70% alcohol and flaming. *Hold the forceps with the tips in a downward position.*

3. Allow the forceps to air-cool, and then aseptically remove a sterile filter paper disk from the container.

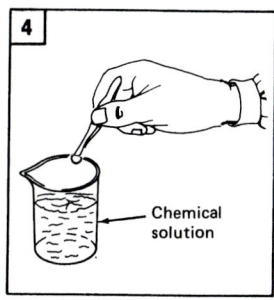

4. Dip the disk *halfway* into one of the chemical solutions.

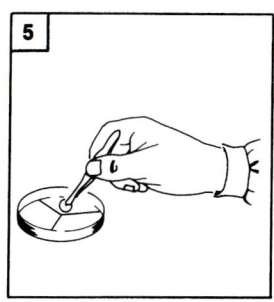

5. Place the disk in the center of its labeled section.

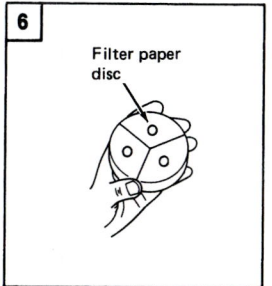

6. Repeat the procedure for each chemical solution provided.

B. Demonstration of Oligodynamic Action by Metal Coins

Materials

The following materials should be provided per 4 students:

❑ 1. Forty-eight-hour nutrient broth cultures of the following (the cultures used in Part A can also be used here):
  ❑ a. *Bacillus subtilis*
  ❑ b. *Escherichia coli*
  ❑ c. *Micrococcus luteus*
  ❑ d. *Serratia marcescens*
❑ 2. Four nutrient agar deeps

❑ 3. Four sterile Petri plates
❑ 4. Four copper or silver coins
❑ 5. One pair of sterile forceps
❑ 6. One container of cleanser powder
❑ 7. One 50-mL beaker containing about 10 mL of 10% nitric acid
❑ 8. One 50-mL beaker containing 20 mL of distilled water

Additional Technique Required for This Portion of the Exercise:

❑ Pour Plate Technique, Procedure Diagram 9, Exercise 5

Procedure

Each student is to carry out this portion of the exercise using only 1 of the cultures provided.

❏ 1. Clean a test coin first by scrubbing with the cleanser powder and then rinsing thoroughly in running water. It is important that all parts of the coin be free of cleanser.

❏ 2. Using the forceps provided, dip the clean coin *carefully* first into the beaker of nitric acid for a few seconds and then into the beaker of water; rinse the coin under running water.

❏ 3. Carefully shake the coin to remove any excess water and place it in the center of a sterile Petri plate.

❏ 4. Prepare a pour plate with your organism. (Refer to Procedure Diagram 9.) Pour the preparation over the freshly cleaned coin.

❏ 5. Allow the medium to harden.

❏ 6. Incubate at room temperature for 48 hours, or as directed.

❏ 7. Examine your plate, as well as those of others. (Refer to Figure 33–1.)

❏ 8. Measure the zones of inhibition and record your findings in the Results and Observations section.

Results and Observations

1. Complete the following table.

Table 33–1

Oligodynamic Actions

Organism	Diameters of Zones Produced (in mm)			
	Mercuric Chloride	Silver Nitrate	Stannous Chloride	Copper or Silver Coin
B. subtilis				
E. coli				
M. luteus				
S. marcescens				

2. Did gram-positive organisms differ from gram-negatives in their sensitivities to heavy metals? _____

_____

_____

3. Was pigment production altered in any way by the metal ions used? _____

_____

Laboratory Review **33** The Inhibitory Action of Heavy Metals and Other Chemicals

1. What is oligodynamic action? _____

    _____

    _____

2. List 2 compounds of heavy metals that are used as antiseptic agents.

    a. _____   b. _____

3. What is the mechanism of action responsible for the inhibitory effects of heavy metals and associated

    compounds? _____

    _____

    _____

Key Terms

antiseptic (an-ti-SEP-tik): an agent that inhibits the growth of or kills microorganisms; it is used only on skin or mucous membranes, and never internally

denaturation (dē-nā-chur-Ā-shun): the destruction of the usual form of a substance; usually results in the loss of some chemical and physical properties

fungicide (FUN-ji-sīd): an agent that kills fungi

germicide (JER-mi-sīd): an agent that kills microorganisms

oligodynamic (ol-i-gō-dī-NAM-ik) action: microbial growth inhibition by heavy metal ions

Photographic Quiz 21

A foreign coin was tested for oligodynamic activity against *Micrococcus luteus*. The results are shown in Figure 33–3. Is there any indication of oligodynamic activity? _____

_____

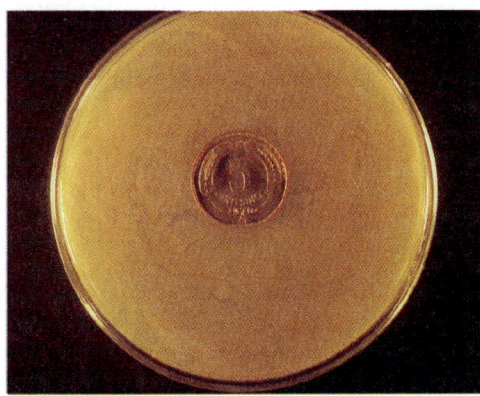

Figure 33–3
Results with an unusual coin.

Disinfection of Selected Instruments and Equipment

After completing this exercise, you should be able to:

1. Perform a disinfectant-use test with clinical instruments, selected disinfectants, and bacteria.
2. Interpret laboratory data obtained with a disinfectant-use test.
3. Compare the relative effectiveness of a variety of antimicrobial chemicals.

Rapid disinfection and sterilization at room temperature are often desirable, particularly for various items that would be damaged by the application of heat, such as clothing with fabrics made of natural and some synthetic fibers, many plastic materials, and cutting instruments that would be dulled if exposed to heat sufficient for sterilization. Hospitals, clinics, dental and medical offices, and related health care facilities frequently find it necessary to sterilize such heat-sensitive items. *Cold sterilization* is commonly used for these materials. Antimicrobial chemical agents, known as *germicides,* have wide application in situations where it is impractical to use heat for sterilization.

Antimicrobial chemicals range from **disinfectants** and **antiseptics** to **sterilants** and **preservatives.** *Disinfectants* are chemicals that kill microorganisms and are applied to inanimate objects. *Antiseptics* are chemicals that kill or inhibit the growth of microorganisms and because of their general nontoxic properties can be applied to living tissue. Several chemicals—phenolic compounds, formaldehyde, glutaraldehyde, alcohols, halogens, and detergents—are available for room-temperature disinfection. Each type has its own advantages and disadvantages. Although the phenols and glutaraldehyde are best for sterilization, they are generally corrosive and toxic; detergents have limited effects on certain types of microorganisms.

Antimicrobial chemicals may be liquid, gaseous, or solid. Liquid chemicals generally consist of a water, alcohol, or a mixture of water and alcohol base into which various substances (solutes) are dissolved. Solutions having pure water as the solvent are referred to as being *aqueous,* while those solutions with pure alcohol or water-and-alcoholic mixtures as the solvents are called *tinctures.*

Cationic detergents such as **quaternary ammonium compounds** (usually referred to as *quats*) are commonly used for purposes of clinic and office disinfection. Quats include benzalkonium chloride, Roccal, and Zephiran. Some of these chemicals when diluted appropriately are mixed with cleaning agents to disinfect and clean at the same time. While quaternary ammonium compounds when used in medium concentrations are effective against a number of algae, gram-positive bacteria, fungi, and viruses, they are ineffective against *Mycobacterium tuberculosis, Pseudomonas species,* bacterial spores, and hepatitis A virus.

An antimicrobial chemical agent's concentration (strength) can be expressed in several ways including percentage and dilution ratios. In dilutions, a small volume of a liquid detergent (solute) is diluted in a larger volume of water to obtain a specific ratio. For example, when 1 part (volume) of Roccal, a common disinfectant, is added to 200 parts of water by volume, a ratio of 1:200 is achieved. Other agents such as the halogen chlorine may be used in very high dilutions and are expressed in parts per million or **ppm.**

A common practice has been to regard 70% ethanol or isopropanol as universal sterilizing solutions. While these chemicals are good disinfectants, they have only slight activity against bacterial spores and certain viruses, and can be inactivated by body fluids.

Several factors control germicide effectiveness. These include: the degree of contamination **(microbial load),** the time of exposure, the concentration of the germicide, the kinds of microorganisms being exposed, and the type of material being treated.

This exercise is designed to demonstrate the properties of selected disinfectants and how certain factors can influence their effectiveness.

Materials

The following materials should be provided for each pair of students:

❏ 1. Twenty-four-hour nutrient broth cultures of the following organisms, in 5-mL aliquots:
   ❏ a. *Serratia marcescens*   ❏ b. *Staphylococcus epidermidis*

❏ 2. Forty-eight-hour nutrient broth cultures of the following organisms, in 5-mL aliquots:
   ❏ a. *Bacillus subtilis*   ❏ b. *Mycobacterium smegmatis*

❏ 3. One set of test instruments, including:
   ❏ a. Thermometer (oral or rectal)   ❏ c. Mouthpiece from physiological equipment
   ❏ b. Small surgical scissors   ❏ d. Scalpel or hypodermic needle

❏ 4. One set of disinfectants and wash solutions, including:
   ❏ a. One suitable container with Roccal solution, or other commercial antimicrobial liquid agent   ❏ c. One suitable container with sterile soap solution (Ivory soap is anionic and will neutralize the cationic detergent)
   ❏ b. One suitable container with 70% isopropyl alcohol   ❏ d. Two jars with sterile, distilled water
                                    ❏ e. Three jars for additional disinfectants

❏ 5. The following materials should be provided for each student:
   ❏ a. Two nutrient agar plates   ❏ f. Five 1-mL sterile cotton-plugged pipettes
   ❏ b. One nutrient agar deep (15-mL)   ❏ g. Gram stain reagents and glass slides
   ❏ c. One large sterile tube containing 6 swabs   ❏ h. Container with disinfectant for the disposal of contaminated swabs and pipettes
   ❏ d. One empty sterile test tube
   ❏ e. One empty sterile Petri dish

Procedure 1

Additional Technique Required for This Portion of the Exercise:

❏ Streak Plate Technique, Procedure Diagram 10, Exercise 5

All procedures in this exercise are to be performed by students in pairs. The instructor will demonstrate the proper use of the pipette before the start of this exercise. Note the correct technique for holding a pipette and pump (Figure 34–1). Refer to Procedure Diagram 11.

❏ 1. Pipette 1 mL of each culture into the sterile test tube, and mix well by rolling the tube between both palms.

❏ 2. Select two instruments and wipe a portion of each with a swab wetted with the mixed bacteria. Place these items on a paper towel to air-dry for a few minutes.

❏ 3. Put swabs into the disposal container provided.

❏ 4. Each student will then select one of the disinfectant solutions in a container or jar and place the instruments into the solution for 20 minutes.

❏ 5. Melt and cool the nutrient agar deeps during this waiting period.

❏ 6. After the instruments have been in the disinfectant for 20 minutes, remove and treat as follows:
   ❏ a. Roccal solution
      ❏ i. Choose an instrument and swirl it gently twice in neutral soap solution.
      ❏ ii. Rinse in distilled water as above. Label the distilled water jar with an **R** for Roccal, or other letter if another antimicrobial agent is provided.
   ❏ b. Isopropyl alcohol
      ❏ i. Rinse instruments in distilled water, as above.
      ❏ ii. Label this distilled-water jar with an **I.**

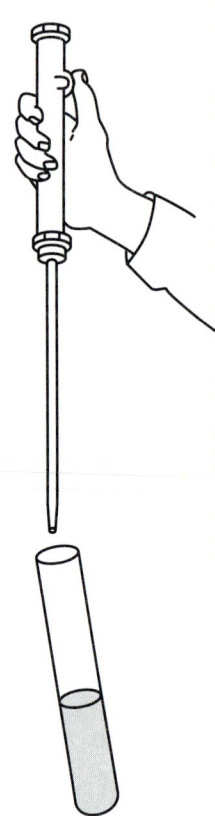

Figure 34–1
Hold the Pi-Pump as shown and insert the pipet into the fluid. Rapidly draw the required volume of fluid into the pipette by rotating the Pi-Pump knob with your thumb. (Refer to Procedure Diagram 11.)

Procedure 2

❑ 1. When the instruments have been rinsed, wipe each with a sterile swab, being certain to take specimens from all areas that might harbor microorganisms.

❑ 2. Inoculate a properly labeled nutrient agar plate with each swab as follows:

   ❑ a. Swab a small area approximately one-third of the way from 1 edge of the plate.

   ❑ b. With an inoculating needle or loop, streak this area and spread the specimens back to the edge prior to proceeding with the streaking method indicated in Procedure Diagram 10.

❑ 3. Incubate the streaked plates at 37° C for 48 hours.

Procedure 3

Additional Technique Required for This Portion of the Exercise:

❑ Pour Plate Technique, Procedure Diagram 9, Exercise 5

❑ 1. Label the empty Petri plate according to the disinfectant studied.

❑ 2. Pipette 1 mL of the distilled water that was used for instrument rinsing into the melted and cooled nutrient agar deep. (Refer to Procedure Diagram 9.)

❑ 3. Mix well by rolling the tube between your palms and pour the contents of the tube into the empty Petri plate.

❑ 4. Incubate the plates at 37° C for 48 hours.

Procedure 4

Additional Techniques Required for This Portion of the Exercise:

❏ 1. Bacterial Smear Preparation, Procedure Diagram 2, Exercise 2

❏ 2. The Gram Stain, Procedure Diagram 17, Exercise 14

❏ 1. Examine the various Petri plates and determine whether or not the instruments were sterilized and whether or not the distilled water showed evidence of contamination. Record your findings in the Results and Observations section.

❏ 2. Perform Gram stains on at least two of the colonial types observed. Record the Gram stain reaction and the morphology of the organisms in Table 34–1 of the Results and Observations section.

Results and Observations

1. Enter your findings in Table 34–1.

Table 34–1

The Effects of Disinfectants

Disinfectant	Instruments	Effective Sterilization		Distilled-Water Contamination		Microorganisms Studied	
		Yes	No	Yes	No	Gram Reactions	Morphology
Roccal	1.						
	2.						
Isopropyl alcohol	1.						
	2.						
Others	1.						
	2.						
	3.						
	4.						
	5.						
Additional Observations							

2. a. Which bacteria survived disinfection? Why? _____

_____

_____

_____

   b. How could one modify procedures or solutions to destroy these organisms? _____

_____

_____

_____

   c. What is the significance of the survival of these organisms with regard to pathogenic bacteria? _____

_____

_____

3. Based upon your results, could the distilled water have been contaminated? If it was not, why not? What is the significance of distilled-water contamination? _____

_____

_____

_____

Laboratory Review 34 — Disinfection of Selected Instruments and Equipment

1. Distinguish between disinfectants and antiseptics. _____

_____

_____

2. What is cold sterilization? _____

_____

3. Distinguish between aqueous mixtures of disinfectants and tinctures. _____

_____

_____

4. List two ways an antimicrobial chemical agent's concentration can be expressed.

   a. _____   b. _____

5. What does ppm mean? _____

6. List four factors that can control a germicide's effectiveness.

   a. _____

   b. _____

   c. _____

   d. _____

Key Terms

cationic: positively charged

detergent: synthetic water-soluble or liquid organic preparation able to emulsify oils and acts as a wetting agent

disinfection: a process, usually chemical, that kills the pathogenic microorganisms usually associated with a particular item or fluid

sterilization: any process that kills or removes all microorganisms

After completing this exercise, you should be able to:

1. Explain how filters sterilize.
2. Explain how a membrane filter can be used to determine the bacterial count in any fluid material.
3. Filter an unknown water sample.
4. Detect contamination of a water sample using a membrane filter system.

Filtration has been used for the removal of microorganisms for over 100 years. In 1884, one of Louis Pasteur's associates, Chamberland, designed and fabricated a filter candle of unglazed porcelain. This filter usually is in the form of a thin-walled tube, closed at one end, and placed in a sterile flask so that the open end is within the flask. The entire candle is enclosed in a funnel or tube, and the solution to be sterilized is poured into the area around the candle. The resulting sterile fluid is drawn into the flask by suction.

Today, other types of filters use diatomaceous earth, fused porcelain, sintered powdered glass, and disks of compressed asbestos. In these filters, pore size is determined by the amount of heat or pressure used in fabrication. In recent years, a membrane filter has been developed, and it is much more versatile than those previously described. Membranes are made from *cellulose nitrate* or *cellulose acetate.* When simple adjustments are made in the preparation procedure, membrane filters of reasonably specific pore size can be obtained. While most of the filter materials mentioned earlier absorb or adsorb components of the solution, membrane filters usually do not. Because of their versatility, these filters are used routinely to sterilize beer and other reasonably clear beverages, as well as to trap organisms in fluids and airstreams (Figure 35–1). When a membrane filter is placed on the surface of a nutrient medium after having been used in the filtering of a fluid containing microorganisms (e.g., bacteria and yeast), colonies of the microorganisms will develop after a suitable incubation period (Figure 35–2). A colony count indicates the number of viable microorganisms that were present in the fluid. Variations of the membrane filtering procedure used in this exercise have wide applications in the areas of industry, public health, and pharmaceutical preparation. It should be noted, however, that while microorganisms such as bacteria and yeasts can be removed from fluids, viruses cannot be eliminated. Figure 35–3 shows the relative sizes of small particles and microorganisms.

As the demands of industry and the general population for water supplies have increased through the years, so too have the problems associated with obtaining and treating water for domestic use. Surface waters, such as lakes, reservoirs, rivers, and streams, contain numerous microorganisms that pose hazards both to aquatic life and to the human population. Sources of such organisms include agricultural wastes, industrial wastes, and raw sewage.

Water pollution caused by sewage is an ever-present hazard in densely populated areas. The sources of various infectious disease agents are found in such situations. Because actual disease-causing microorganisms can be quite difficult to detect, public health personnel and others routinely check for the presence of certain bacteria that can serve as indicators of raw sewage contamination. These microorganisms, known as *coliforms,* are normally present in the intestinal tracts of humans and other warm-blooded animals. Coliforms are introduced into raw sewage by excreted fecal matter. Thus, these organisms are always found in sewage. They are also always found in the presence of pathogens, such as those in the genus *Salmonella.* The value of using coliforms as indicators of pollution lies in the fact that these organisms are very hardy and some are more resistant to chlorination than bacterial pathogens. Thus, if coliforms are not found in a water test sample, one can be reasonably certain that other organisms associated with sewage pollution are not present either. It is important to note here, however, that the mere presence of coliforms alone does not mean that water pollution exists. The concentration of such organisms is also extremely important. When the number of coliforms exceeds the standards established for specific areas and types of water, water pollution is understood to exist. In such situations, additional tests are performed to isolate and culture pathogens.

In this exercise, a procedure similar to the coliform test performed regularly in pollution testing laboratories will be demonstrated. Selective media that promote the growth of coliforms and discourage the growth of most other bacteria

(a)

(b)

(c)

Funnel

Filter

Figure 35–1

Examples of membrane-type filters. (a) Acrodisc PF syringe filters. This membrane filter has a combination of membrane pore sizes, which increases its effectiveness and prevents clogging.
(Courtesy Gelman Sciences).

(b) A variety of filter holders with receivers (bottom portions). (c) Components of disposable, preassembled, presterilized filter units and funnels with membranes.
(*b* and *c* Courtesy of Nalge Company.)

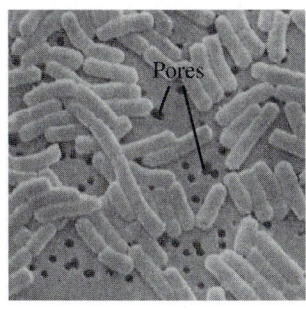

Pores

Figure 35–2

A scanning micrograph of bacteria on the surface of a membrane filter. Note the relative sizes of the bacterial cells and the pores.
(From Todd, R.L., and T.J. Kerr, *Appl. Microbiol.* 23:1160–1162, [1972].)

will be used. One example is MF-Endo medium, which contains the disaccharide lactose, various other nutrients, and the dye, basic fuchsin (Figure 35–4). Coliforms enzymatically break down lactose, thereby forming a group of simpler chemicals known as *aldehydes.* These compounds, in turn, react with basic fuchsin, causing a characteristic shiny green coating to form on coliform colonies.

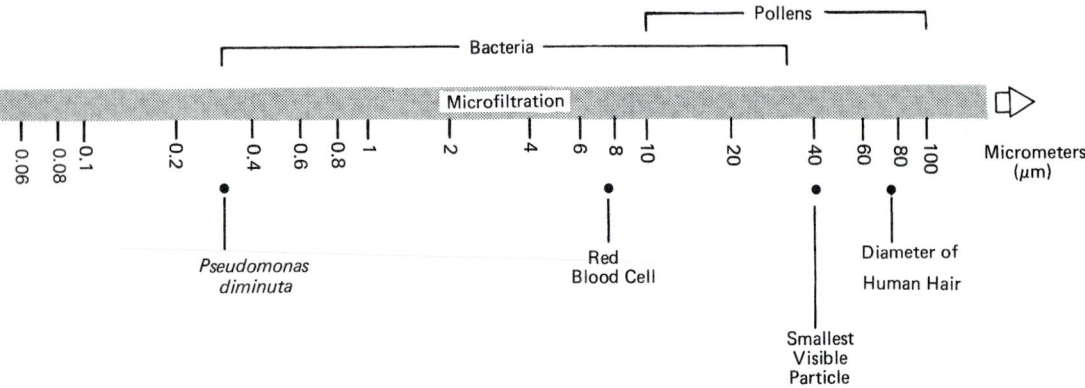

Figure 35–3
The relative sizes of small particles and microorganisms.

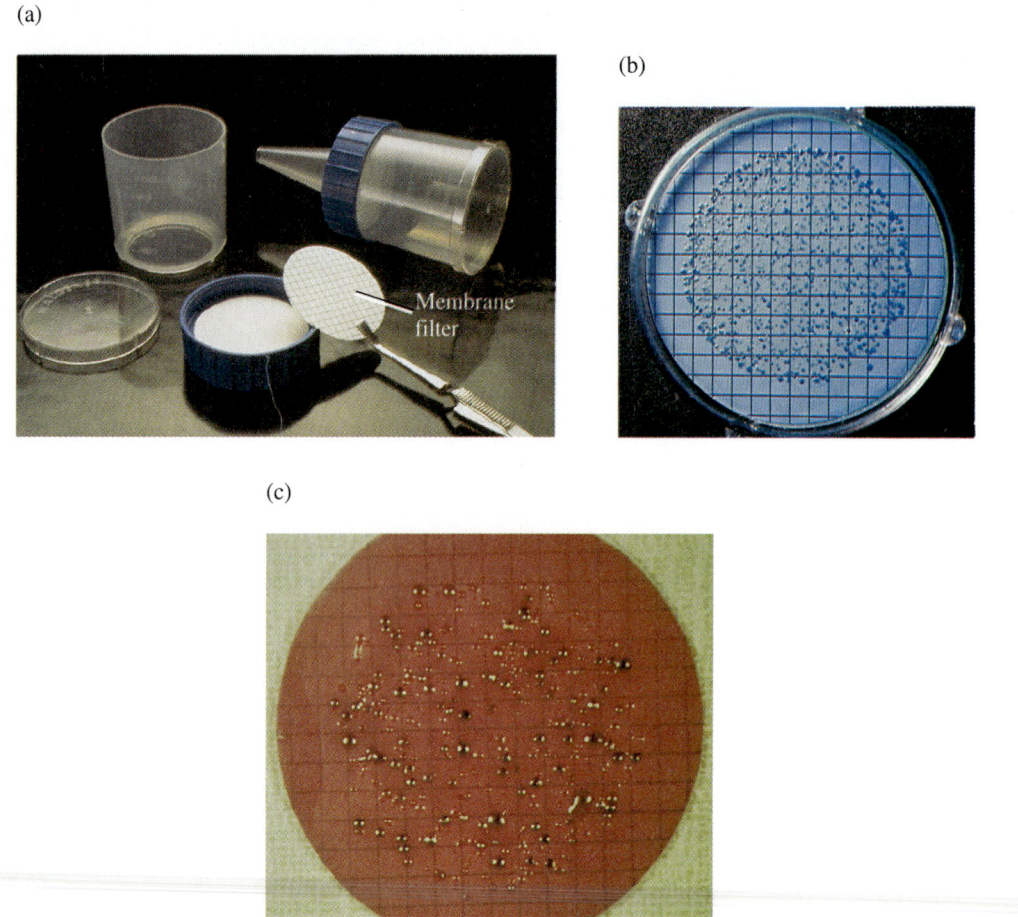

Figure 35–4
Membrane filters. (a) A membrane filter being put into place with filter forceps.
(Courtesy, Nalgene company.)

Various bacteria recovered from surface-water samples. (b) Fecal streptococci grown on a cellulose membrane filter saturated with KF medium. (c) Coliform bacterial colonies (lactose fermenters) grown on a membrane filter saturated with MF-Endo medium.
(Courtesy of Nucleopore Corporation.)

A. Membrane Filter Demonstration

Materials

- ❏ 1. One sterile Millipore filter assembly, or similar filter system
- ❏ 2. Twenty-four-hour nutrient broth culture of *Escherichia coli*
- ❏ 3. One hundred mL of sterile, distilled water
- ❏ 4. One tube of nutrient broth
- ❏ 5. One eosin-methylene blue agar plate
- ❏ 6. Two sterile 1-mL pipettes
- ❏ 7. One pipette pump or rubber bulb
- ❏ 8. One pair of sterile, flat-bladed (nonserrated) forceps
- ❏ 9. A vacuum system such as a mechanical pump or a similar device, and sufficient high-pressure tubing
- ❏ 10. Container with disinfectant for pipette disposal

Procedure

- ❏ 1. The instructor will assemble the Millipore filter assembly or other filter system and describe the individual components and their respective functions. (Refer to Figure 35–5.)
- ❏ 2. Using a 1-mL pipette and pipette pump, the instructor will inoculate the sterile, distilled water with 1 mL of the bacterial culture and set the filter into operation. (Refer to Procedure Diagram 11.) This water sample will be drawn through the filter by suction.
- ❏ 3. With a fresh, sterile pipette and pipette pump, the instructor will introduce 1 mL of the filtrate into the tube of nutrient broth.
- ❏ 4. The pipette should be discarded into the container provided.
- ❏ 5. The filter system will be disassembled and the membrane filter carefully removed and placed on the surface of the eosin-methylene blue agar plate.
- ❏ 6. Both preparations will be incubated at 37°C for 24 hours.
- ❏ 7. After incubation, examine the resulting cultures. (See Figure 35–4b.)
- ❏ 8. Record your findings and answer the questions in the Results and Observations section.

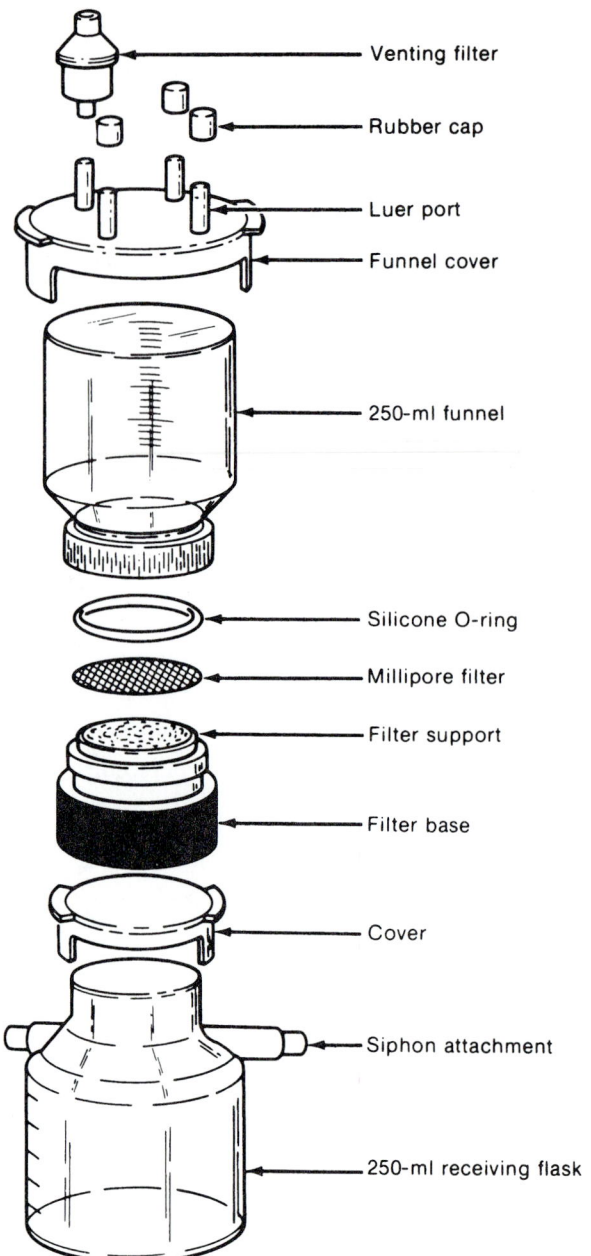

Figure 35–5

The components of a filter membrane system.

B. Detection of Bacterial Water Contamination

Materials

❏ 1. One unassembled Millipore, Nalgene, or similar filter system for class observation

❏ 2. The following materials should be provided per 2 students:

    ❏ a. One sterile Millipore, Nalgene, or similar filter system, ready for assembly

    ❏ b. One sterile membrane filter

    ❏ c. Two pairs of sterile, flat-bladed (nonserrated) forceps

    ❏ d. One 50-mL beaker containing 70% alcohol

    ❏ e. One sterile 47-mm Petri dish

    ❏ f. One sterile absorbent pad

    ❏ g. One ampule of MFC or MF-Endo liquid medium

    ❏ h. Ten mL sterile water

    ❏ i. One hundred mL of an unknown water sample

    ❏ j. One hand vacuum pump or other suitable vacuum source, such as a water aspirator

Procedure

After the instructor's demonstration, unknown water samples will be provided for membrane filtration. This portion of the exercise will be performed by students in groups of 2 or 4.

❏ 1. Examine an assembled and an unassembled filter system.

❏ 2. Review the instructor's demonstration of assembling and using the membrane filter.

❏ 3. Carry out the 9 steps in Procedure Diagram 30.

❏ 4. Incubate your preparation at 37°C until the next laboratory period, or as directed by your instructor.

❏ 5. Examine your membrane filter for the presence of coliforms. (Refer to Figures 35–4*b* and 35–4*c*.)

❏ 6. Count the number of coliforms. These organisms will generally produce colonies with a green sheen. Calculate the number of coliforms in your water sample according to the following formula:

$$\frac{\text{Number of coliform colonies} \times 100}{\text{Milliliters of water sample filtered}} = \text{Number of coliforms/100 mL}$$

❏ 7. Enter your results in the Results and Observations section.

Procedure Diagram 30
Membrane Filter Assembly and Use

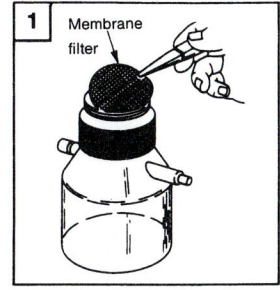

1. Using sterilized forceps center a membrane filter on the filter apparatus support.

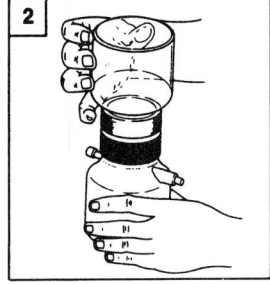

2. Place the filter funnel over the filter support and screw it down so that the O-ring seals the filter in place.

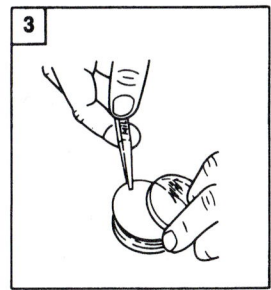

3. Using alcohol-flamed forceps (refer to Procedure Diagram 29) place a sterile absorbent pad into a 47-mm Petri plate.

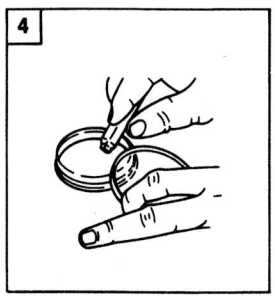

4. Break open an ampule of MF-Endo medium or any other medium provided and pour the entire contents on the absorbent pad. Close the dish and set it aside until step 9.

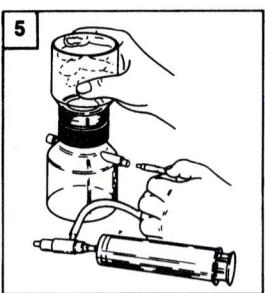

5. Attach the hand vacuum pump or other vacuum source, such as a water aspirator.

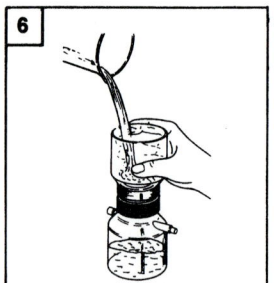

6. Add 10 ml of sterile water followed by 100 ml of the unknown water sample to the filter funnel.

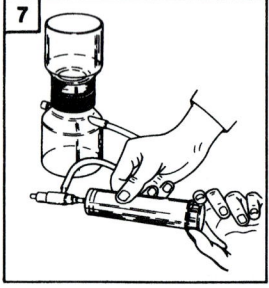

7. Apply suction with the vacuum device. This procedure will cause the water to flow through the filter, leaving any bacteria trapped on the filter membrane surface.

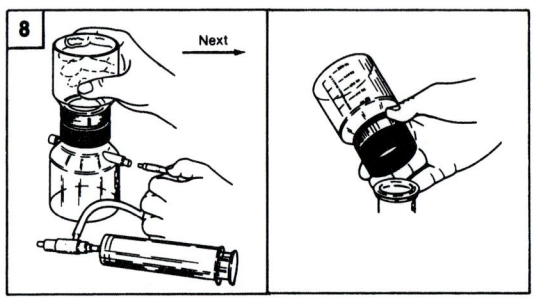

8. After filtration, carefully release the vacuum in the system by removing the vacuum pump tubing from the side arm of the filter receive flask. Unscrew the funnel.

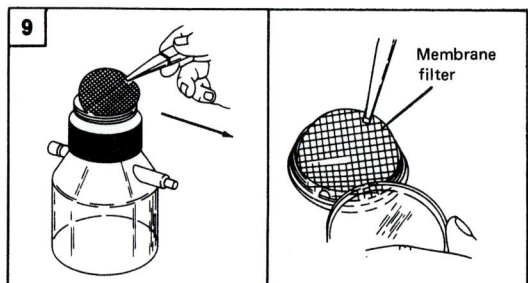

9. With alcohol-sterilized forceps, lift the membrane filter from the base and place it on the previously prepared (step 4) saturated pad in the Petri dish. Set the filter membrane down with a slight rolling movement so that it is evenly centered. Close the cover and incubate.

Results and Observations

1. Describe the appearance of the bacterial colonies obtained from the membrane filter demonstration. _____

_____

_____

2. a. Did your water sample contain any coliforms? _____

    b. If so, calculate how many coliforms were in the sample. Show your calculations. _____

    _____

3. What was the pore size of the membrane filter used, in micrometers? _____ _μ_m

---

Laboratory Review 35   A Demonstration of Sterilization by Filtration and Water Pollution Detection

1. Explain how filters work in the sterilization of solutions. _____

    _____

    _____

2. Explain why the filtering of a solution does not prevent the development of disease symptoms or a disease state
    when ingestion of, or contact with, the resulting filtrate (filtered fluid) occurs. _____

    _____

    _____

    _____

3. List 4 types of commercially important materials that can be sterilized by filtration.

    a. _____   c. _____

    b. _____   d. _____

4. List 2 advantages of membrane filtration.

    a. _____

    b. _____

5. List 2 disadvantages of membrane filtration.

    a. _____

    b. _____

Key Terms

coliforms (KŌ-lē-forms): a term used to describe gram-negative fermentative bacteria found in the large
intestine; includes *Enterobacter aerogenes* and *Escherichia coli*

membrane filter: thin, porous sheets composed of cellulose esters or related materials

membrane filtration: a physical method to remove cellular organisms from fluids with the aid of thin,
membranous materials; large proteins and viruses are not necessarily removed

Salmonella **(sal-mō-NEL-a):** a genus of gram-negative rods known to be capable of causing diseases in both
humans and lower animals; diseases include typhoid and paratyphoid fevers and food poisoning

The Bacteriostatic Activity of Dyes

After completing this exercise, you should be able to:

1. Recognize the inhibitory action of crystal violet and other dyes.
2. Understand the selective and inhibitory nature of certain media.

In microbiology, a wide variety of preparations used to grow microorganisms such as bacteria and fungi are *general* or *all-purpose media.* Such media contain a variety of nutrients and can thus support the growth of a number of microorganisms having different nutritional requirements. All-purpose media, however, are not appropriate for all situations. This is especially true when an organism present in low numbers in a specimen or in a mixed culture is to be isolated. This challenge can often be met with the use of a *selective medium.* Generally speaking, a selective medium contains at least one ingredient that can inhibit the growth of more numerous and/or unwanted organisms in a specimen without preventing the growth of less numerous and/or the wanted organisms. Substances commonly used as selective agents include antibiotics, chemicals such as acids, and certain types of dyes such as crystal violet and malachite green.

In studying the activities of microorganisms, other types of media in addition to selective media also are used. These include *differential* and *selective and differential* preparations. Differential media are used to distinguish one type of microorganism from all others in a mixed culture. Such preparations contain one or more ingredients (substrates) that change during the growth of only certain organisms causing their colonies and/or the medium to take on a distinctive appearance. Selective and differential media combine the features of both differential and selective media. Such media are especially important in the isolation, detection, and identification of disease-causing microorganisms. Several of the exercises in Sections VI and XIII present a variety of these media. This exercise will demonstrate the selective properties of dyes.

Crystal violet has long been employed therapeutically for candidiasis, or thrush, a fungal disease of the mouth. It was once fairly common to see young children undergoing treatment for the infection with violet tongues. At low concentrations, crystal violet has also been found to be selectively bacteriostatic for gram-positive, but not gram-negative, bacteria. This effect can be explained in terms of differences in the cell walls of these two groups of bacteria. Experiments have shown that crystal violet interferes with the synthesis of *Staphylococcus aureus* cell walls. If much higher concentrations of the dye are used, both types of bacteria are prevented from growing. This result can be explained on the basis of cell wall chemistry. It would appear that only high concentrations of the dye are able to diffuse sufficiently through the outer layers of the gram-negative cells in order to reach the area of cell wall synthesis.

Malachite green also has a history of uses as a local antiseptic for application to wounds and burns. It has a limited effectiveness against gram-positive and gram-negative bacteria. The bacteriostatic activity of this dye also will be examined in this exercise.

A. Comparing the Bacteriostatic Activity of Dyes

Materials

The following items should be provided per 4 students:

❏ 1. Twenty-four-hour nutrient broth cultures of the following microorganisms:
  ❏ a. *Enterobacter aerogenes*
  ❏ b. *Staphylococcus aureus*
❏ 2. Two nutrient agar deeps
❏ 3. Two sterile Petri plates
❏ 4. Sterile filter paper disks (5-mm diameter) in a sterile Petri plate
❏ 5. One pair of sterile forceps

❏ 6. One 50-mL beaker of 70% alcohol
❏ 7. Approximately 5–10 mL of each of the following 1% solutions in 50-mL beakers:
  ❏ a. Crystal violet
  ❏ b. Malachite green
❏ 8. One ruler with a millimeter (mm) scale
❏ 9. One wax pencil or other marking device

The following item should be provided for class use:

❏ 1. Water bath set at 60°C for melted agar deeps.

Procedure

Additional Technique Required for This Portion of the Exercise:

❏ Pour Plate Technique, Procedure Diagram 9, Exercise 5

This procedure should be performed by students in groups of 4.

❏ 1. Prepare a pour plate with each culture provided. Label each plate with the organism used.
❏ 2. Using a wax marking pencil, mark 2 sectors on the bottom of each plate. Label 1 sector for each dye provided.
❏ 3. Holding a filter paper disk with the forceps, touch it to the surface of the crystal violet solution until the dye penetrates throughout the disk. (Refer to Procedure Diagram 29.)
❏ 4. Remove the Petri plate top of the *S. aureus* pour plate and place the disk in the center of one of the marked sectors. Press the disk gently with the forceps, so that a good contact is made with the agar surface. Close the plate.
❏ 5. Flame the forceps as shown in Procedure Diagram 29 and repeat steps 3 and 4 with the malachite green solution placing the disk in the center of its marked sector.
❏ 6. Repeat steps 3, 4, and 5 with the *E. aerogenes* pour plate.
❏ 7. Incubate both preparations at room temperature for 48 hours, or as directed.
❏ 8. After incubation, measure the zone of inhibition with the ruler provided and record your findings in the Results and Observations section. (Refer to Figure 36–1.)

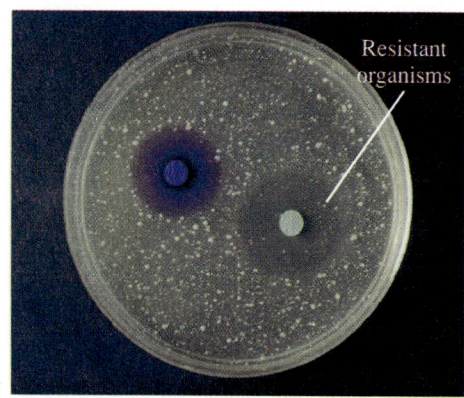

Figure 36–1
The relative bacteriostatic effects of crystal violet (purple) and malachite green.
Note the presence of the resistant organisms.

B. The Use of Dyes in a Selective Medium

Materials

❑ 1. Twenty-four-hour nutrient broth cultures of the following (1 per 4 students):
    ❑ a. *Enterobacter aerogenes*
    ❑ b. *Staphylococcus aureus*

❑ 2. MacConkey agar plates (1 per student; this medium contains 0.001 g of crystal violet per liter)

❑ 3. Nutrient agar plates (1 per student)

❑ 4. Sterile Petri plates (2 per student)

❑ 5. One water bath (for class use)

❑ 6. Wax pencil or marking pen

Procedure

This procedure is to be performed by students in pairs.

❑ 1. Using a wax marking pencil, mark 2 sectors on the bottom of each plate. Label 1 sector for each organism provided.

❑ 2. Streak *E. aerogenes* and *S. aureus* onto each of their designated sectors of MacConkey and nutrient agars.

❑ 3. Incubate at 37°C for 24 hours.

❑ 4. Observe and record your findings in the Results and Observations section. (See Figure 36–2.)

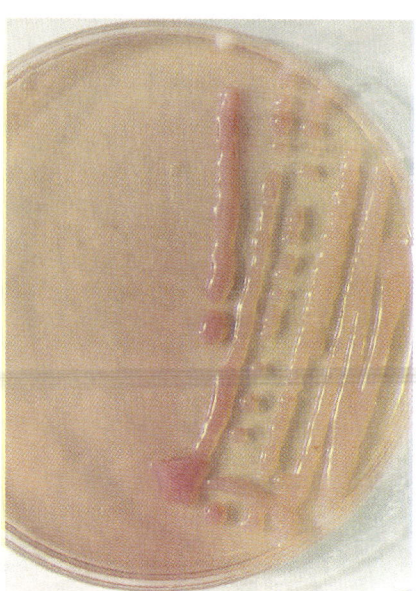

Figure 36–2
The bacteriostatic activity of dyes. Two bacterial species grown on MacConkey agar.
Left side streaked with *Staphylococcus aureus* (a gram-positive microorganism). Right
side streaked with a gram-negative microorganism.

Results and Observations

Comparing the Bacteriostatic Activity of Dyes

1. Complete the following table.

Table 36–1

Dye Bacteriostatic Activity

Organism	Diameters of Zones Produced (in mm)	
	Crystal Violet	Malachite Green
E. aerogenes		
S. aureus		

2. Did the gram-positive culture differ from the gram-negative one in sensitivity to either of the dyes? _____

_____

The Use of Dyes in a Selective Medium

1. Which medium can be considered a selective medium? _____

_____

2. Sketch your results here.

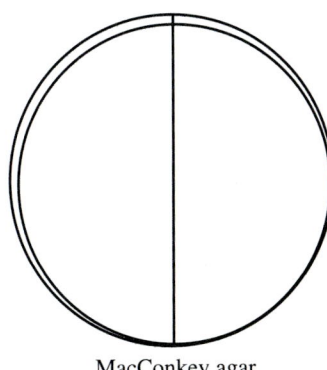

MacConkey agar

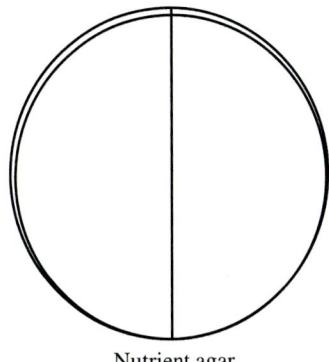

Nutrient agar

Laboratory Review 36

The Bacteriostatic Activity of Dyes

1. Distinguish between differential and selective media. _____

_____

_____

2. List 3 substances incorporated into culture media as selective substances for gram-negative bacteria.

a. _____   b. _____   c. _____

3. List 3 substances used to make media differential.

a. _____   b. _____   c. _____

4. What type of medium is MacConkey agar? _____

_____

5. List 2 physical factors or methods that may be used for enrichment purposes (i.e., to favor the growth, survival, or spatial separation of one specific microorganism from another).

a. _____   b. _____

6. Define *bacteriostasis*. _____

_____

Key Terms

bacteriostatic: growth inhibition of bacteria

differential (dif-er-EN-shal) medium: a medium that contains one or more ingredients designed to stimulate a characteristic biochemical response, which will differentiate individual or groups of microorganisms from one another

enriched medium: a medium containing one or more substances such as blood, to encourage microbial growth

selective medium: a medium containing one or more substances that permit the growth of certain organisms *while preventing the growth* of others

selective and differential medium: a preparation that incorporates the features of both selective and differential media

After completing this exercise, you should be able to:

1. Perform an antibiotic sensitivity test on a bacterial culture.
2. Distinguish the relative resistance and sensitivity of bacterial cultures to selected antibiotics.
3. Interpret the results of the E-test.
4. Demonstrate the presence of antibiotic-resistant bacteria with commercially available devices.

Antibiotics (*anti,* against; *bios,* life) form an interesting—and therapeutically useful—group of compounds. These chemicals are characterized as metabolic by-products of one microorganism that are able to kill other microorganisms (*bactericidal),* or inhibit growth (*bacteriostatic).* The production of antibiotics by microorganisms is common. One of the first reported discoveries of this microbial relationship was made in 1927 in England. During the course of an investigation, bacteriologist Alexander Fleming accidentally left open a Petri plate containing a culture of *Staphylococcus aureus.* A few days later, he noticed that a fungus had contaminated the plate, and he observed that the *S. aureus* grew everywhere but in the area immediately surrounding the fungus. Further, he discovered that filtrates from broth cultures of the fungus were effective in preventing bacterial growth. Such filtrates were also effective in preventing the growth of many types of bacteria. The fungus subsequently was identified as *Penicillium notatum,* and the crude antimicrobial extract obtained from it was named *penicillin.*

Since the 1940s, a vast array of antibiotics has been isolated, synthesized, and subsequently used in the treatment of disease states caused by a variety of pathogenic microorganisms. It is important to note that the sensitivity of microorganisms to an antibiotic varies. Therefore, the selection of the appropriate *chemotherapeutic* agent is extremely important. Decisions of this kind are facilitated by the investigator's knowledge of the antimicrobial activity of specific antibiotics.

The antimicrobial activity of an antibiotic may be tested by several methods designed to determine the smallest amount of the agent needed to inhibit the growth of a microorganism. The resulting value is known as the *minimal inhibitory concentration (MIC).* This exercise will demonstrate the features of the commonly used agar diffusion method (Figure 37–1). Variations on the procedure are known and include the disk diffusion test of Bauer, Kirby, Sherris, and Turck, and the agar overlay method of Barry, Garcia, and Thrupp.

In the basic agar diffusion procedure, a Petri plate containing a suitable medium is heavily inoculated with the microorganism whose antibiotic sensitivity is to be determined. Commercially available filter paper disks, each containing defined concentrations of specific antibiotics, are released from an automatic disk dispenser (Figure 37–2) or removed from individual containers and placed on the agar surface. The preparation is incubated for a definite period of time, during which the antibiotics diffuse from the disks into the agar. At a particular distance from each disk, the MIC for the an-

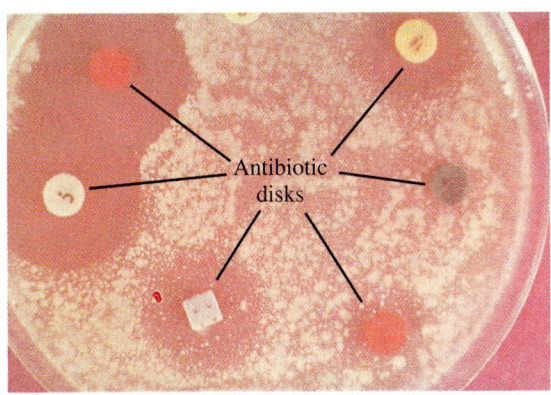

Figure 37–1

Antibiotic disc sensitivity testing (antibiogram). The greater the inhibitory effect of an antibiotic, the larger the clear zone surrounding the specific disc containing it. Note the relative effects of the antibiotics shown.

(a)

(b)

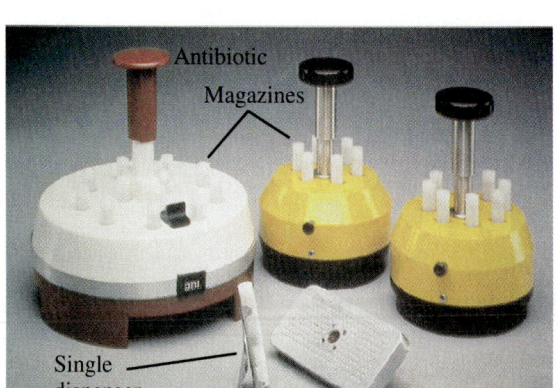

Figure 37–2

Antibiotic dispensers. (a) A single antibiotic dispenser. Depressing the plastic side arm will release an antibiotic disc. (b) A Difco Dispens-O-Disc system. The dispenser shown holds 8–10 antibiotic (magazines). The entire device is first placed over the surface of an inoculated plate. Then the plunger at the top is depressed once and individual antibiotic discs are released. Various versions of antibiotic disc dispensers are in use.
(Photos courtesy of Becton Dickinson Microbiology Systems.)

tibiotic is reached. The MICs are recognized by the presence of growth inhibition (clear) zones surrounding the various antibiotic disks used. The diameters of such zones can be measured with a ruler. The results constitute an *antibiogram*. It should be noted that the relative diameters of the zones do not necessarily indicate the relative activities of the chemotherapeutic agents used in the test. The size of the growth inhibition zones can be affected by several factors, including (1) the culture medium used; (2) incubation conditions; (3) the rate of diffusion of the antibiotic; (4) the concentrations of the antibiotics used; and (5) the antibiotic sensitivity of the organism being tested.

A novel approach to test antimicrobial activity is provided by the *E-test*. This *in vitro* susceptibility method is designed to determine minimum inhibitory concentrations (MICs) of antimicrobial agents. The E-test involves the use of a thin, plastic strip, one side of which contains a continuous concentration gradient (range) of a stabilized and dried antibiotic (Figure 37–3). The drug diffuses from the strip producing both qualitative and quantitative results showing the antibiotic susceptibility of the organism being test.

The effective concentrations of an antibiotic can be readily determined from the growth inhibitory zone produced after incubation. The E-test has been used for several purposes including studies of antibiotic-resistance mechanisms and quality control of media used for susceptibility testing.

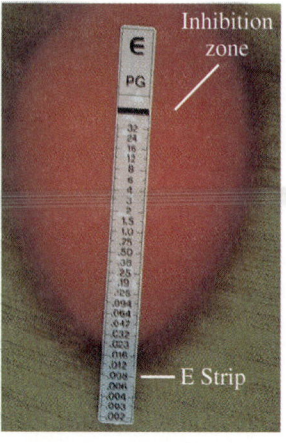

Figure 37–3

The appearance of an E-test. Note the shape of the growth of inhibition.
(Courtesy Remel.)

The relative effectiveness of different antibiotics provides the basis for a sensitivity spectrum of the organism. This information, together with various pharmacological considerations, is used in the selection of an antibiotic for treatment. It should be emphasized that chemotherapeutic agents are not chosen simply on the basis of the drug producing the widest growth inhibition zone.

Within the past few years, antibiotic-resistant bacteria have been isolated with increasing frequency (Figure 37–4). Examples of such organisms include ampicillin- and penicillin-resistant strains of *Bacteroides* species, *Haemophilus influenzae, Neisseria gonorrhoeae,* and *Staphylococcus aureus.* This resistance has been shown to be due to the production of the enzyme beta (β)-lactamase. Detection of this enzyme is crucial for effective antibiotic therapy. One way to demonstrate the presence of beta-lactamase is to use a commercially available disk containing benzipenicillin as a substrate for the enzyme. The disk, which is purple because of the presence of brom-cresol purple, a pH indicator, will change to yellow when a beta-lactamase-producing microorganism is applied to its surface. Rapid-detection tests will be demonstrated in this exercise.

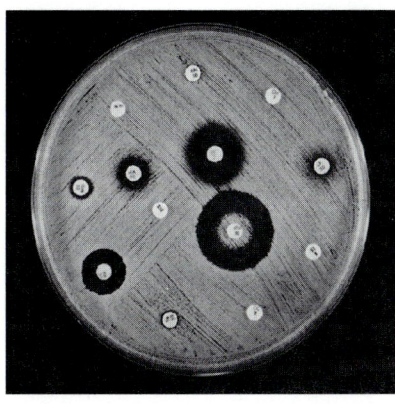

Figure 37–4

Antibiotic disc susceptibilities of a *Staphylococcus haemolyticus* clinical isolate that contained vancomycin-resistant subpopulations. Note the double zone of growth around a standard disc.
(Courtesy of Schwalbe, R.S., Univ of Maryland School of Medicine, Baltimore, MD., *J Infect Dis* 161:45–51, [1990].)

A. Antibiotic Susceptibility (Sensitivity) Testing

Materials

Items 1 through 8 should be provided in quantities of 1 per 4 students:

❏ 1. Twenty-four-hour nutrient broth cultures of the following:
　❏ a. *Staphylococcus aureus*
　❏ b. *Proteus vulgaris*

❏ 2. Individual antibiotic disks in separate containers (Figure 37–1a), or a commercial type of multiple antibiotic disk dispenser (Figure 37–1b)

❏ 3. One ruler with a millimeter scale

❏ 4. One pair of forceps

❏ 5. One beaker containing 70% ethanol

❏ 6. Four sterile 1-mL pipettes

❏ 7. Two nutrient agar deeps and 2 sterile Petri plates

❏ 8. Wax marking pencil

❏ 9. Water bath for class use

Procedure

Additional Technique Required for This Portion of the Exercise:

❏ Pour Plate Technique, Procedure Diagram 9, Exercise 5

This procedure is to be performed by students in groups of 4.

❏ 1. Prepare 1 pour plate each with *S. aureus* and 1 with *P. vulgaris,* as shown in Procedure Diagram 9.

❏ 2. Perform the steps shown in Procedure Diagrams 31 or 32, and as directed by your instructor.

❏ 3. Repeat the procedure with each prepared pour plate.

❏ 4. Invert and incubate all plates at 37°C for 24 hours.

Procedure Diagram 31
Individual Antibiotic Disks

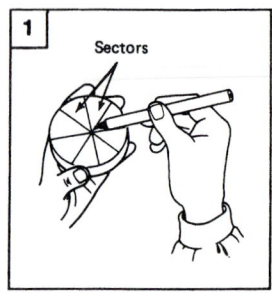

1. Using a wax marking pencil, divide the prepared plate into 8 sectors, and indicate the organism to be used.

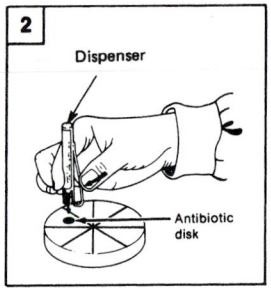

2. Deposit individual antibiotic disks on the surface of the agar plate by pushing in the lever on the side of the dispenser.

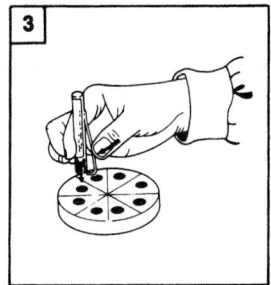

3. Repeat step 2 until 1 of each antibiotic disk provided has been placed on the agar surface.

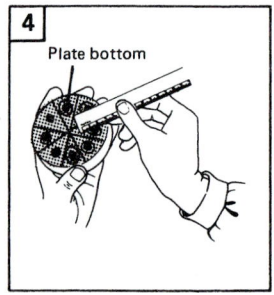

4. After incubation, measure the zones of inhibition with a millimeter ruler.

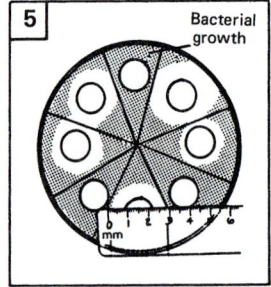

5. Refer to table 29-1 for inhibition zone interpretations.

Procedure Diagram 32
Commercial-Type Multiple Antibiotic Disk Dispenser

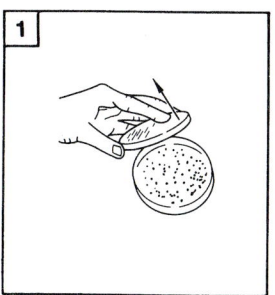

1. Remove the top of one of the seeded plates.

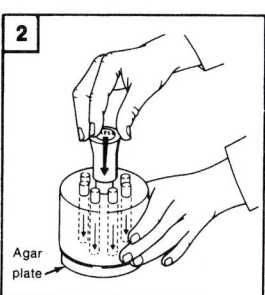

2. Place the dispenser over the agar surface and press firmly on the plunger device, thus depositing several antibiotic disks simultaneously on the agar.

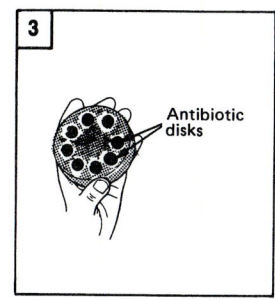

3. Repeat the procedure with each seeded plate.

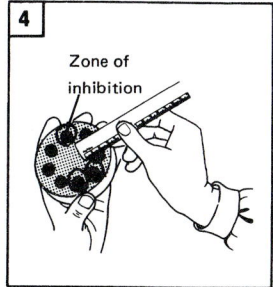

4. After incubation, measure the zones of inhibition with a millimeter ruler.

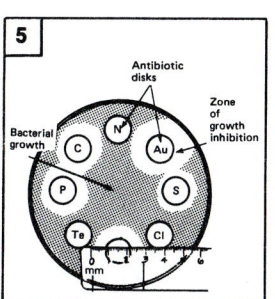

5. Refer to Table 29-1 for inhibition zone interpretations.

❑ 5. Examine each plate and look for the presence of growth inhibition zones. Using the millimeter scale, measure the diameters of such zones. In the Results and Observations section, note whether the test microorganism is susceptible, intermediate, or resistant to specific antibiotics. Figure 37–5 shows the respective sizes of these zones. (See Figure 37–1). Table 37–1 provides a guide to the interpretation of inhibition zones.

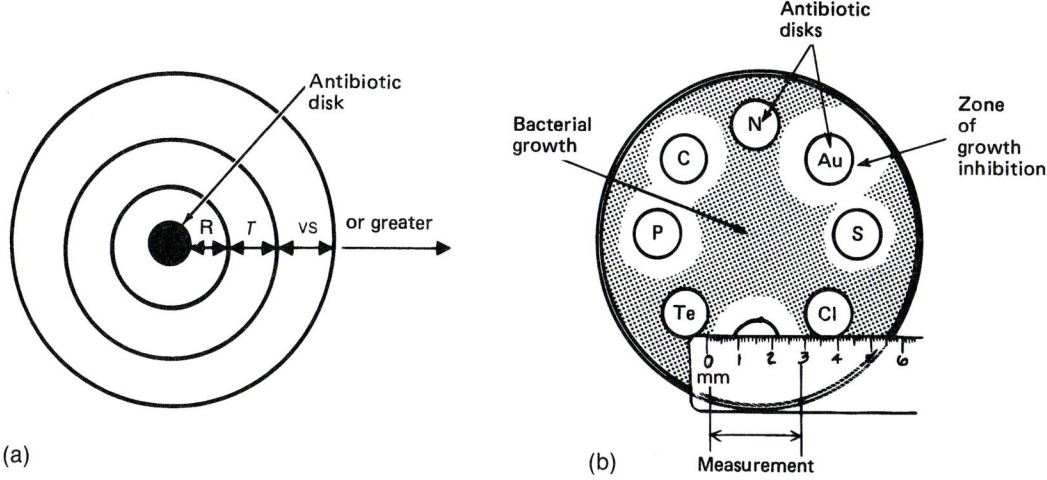

(a)   (b)

Figure 37–5
(a) Zones of an antibiogram, R = resistant; I = intermediate; and S = susceptible. (Note: This figure is only a general guide. Zone sizes will vary according to the antibiotic and the microorganism being tested.) (b) Measuring the diameter of a growth inhibition zone. The zone extends from the rim of the clear ring on one side of the disc to the same area on the opposite side.

Table 37–1

Guide to Interpretation of Antibiogram Inhibition Zones

Antibiotic	Disk Content[b]	Zones of Inhibition Diameter (in mm)[a]		
		Resistant	Intermediate	Susceptible
Ampicillin with gram-negatives and enterococci	10 mcg	13 or less	14–16	17 or greater
Ampicillin with staphylococci and Penicillin G-susceptible microorganisms	10 mcg	28 or less	—	29 or greater
Bacitracin	10 units	8 or less	9–12	13 or greater
Carbenicillin with *Proteus* species and *Escherichia coli*	50 mcg	19 or less	20–22	23 or greater
Cephalothin when reporting susceptibility to cephalothin, cephaloridine, and cephalexin	30 mcg	14 or less	15–17	15–21
Chloramphenicol	30 mcg	13–17	—	18 or greater
Clindamycin	2 mcg	14 or less	15–16	17 or greater
Clindamycin when reporting susceptibility to lincomycin	2 mcg	14	15–20	21 or greater
Colistin	10 mcg	8 or less	9–10	11 or greater
Erythromycin	15 mcg	13 or less	14–22	23 or greater
Gentamicin	10 mcg	12 or less	13–14	15 or greater
Kanamycin	30 mcg	13 or less	14–17	18 or greater
Methicillin	5 mcg	9 or less	10–13	14 or greater
Neomycin	30 mcg	12 or less	13–16	17 or greater
Novobiocin	30 mcg	17 or less	18–21	22 or greater
Penicillin G, with staphylococci	10 units	28	—	29 or greater
Penicillin G, with other microorganisms	10 units	14–26	27–46	47 or greater
Polymyxin B	300 units	8 or less	9–11	12 or greater
Streptomycin	10 mcg	11 or less	12–14	15 or greater
Tetracycline	30 mcg	14 or less	15–18	19 or greater
Vancomycin when reporting susceptibility for gram-positives other than enterococci	30 mcg	9	10–11	12 or greater

[a]Source: Difco Laboratories, Detroit, Michigan
[b]mcg = microgram

B. E-Test Demonstration

Materials

The following items should be provided for an instructor demonstration:

❏ 1. Five mL of 48-hour cultures of the following bacteria:
    ❏ a. Ampicillin-resistant *Staphylococcus aureus*
    ❏ b. *Escherichia coli*
    ❏ c. *Haemophilus influenzae*
❏ 2. Six blood agar plates
❏ 3. Six sterile cotton swabs

❏ 4. Six of each of the following E-test strips:
    ❏ a. Doxycycline
    ❏ b. Penicillin V
❏ 5. One marker pen
❏ 6. Container with disinfectant

Procedure

The following procedure will be demonstrated by the instructor or by students assigned.

❏ 1. Label two agar plates for each of the bacterial cultures provided.

❏ 2. Obtain one sterile cotton swab, and dip it into one of the cultures provided.

❏ 3. Remove any excess fluid by gently pressing the swab against the side of the tube.

❏ 4. Next swab the agar plate completely by covering the surface with the culture material on the swab.

❏ 5. Allow the agar surface to dry.

❏ 6. Place an E-test strip in the center on the swabbed agar surface.

❏ 7. Repeat steps 2 through 6 with the same culture and the second E-test strip provided.

❏ 8. Repeat steps 2 through 7 with the remaining bacterial cultures.

❏ 9. Incubate all plates at 37° C for 24–48 hours.

❏ 10. Read the minimal inhibitory concentration (MIC) values on each plate. The MIC value is read where the zone of growth inhibition intersects the E-test strip (see Figures 37–3 and 37–6).

❏ 11. Enter the MIC values in the Results and Observations section and answer the questions in the Results and Observations section and Laboratory Review.

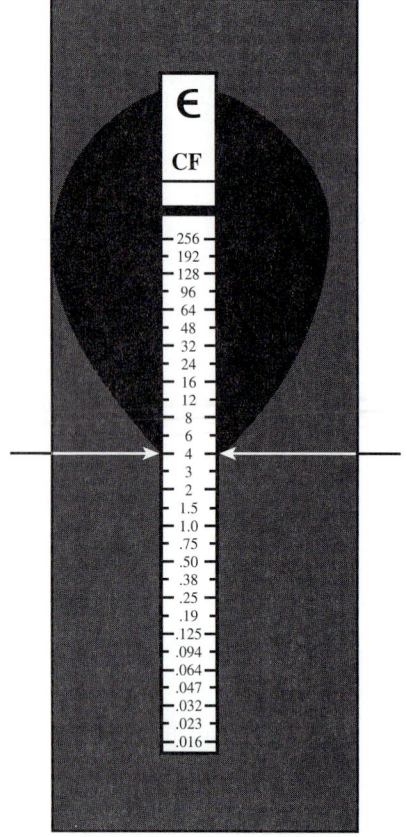

Figure 37–6
Reading the MIC value. Read the value at
the point where the zone of growth
inhibition intersects the E-test strip.

C. Demonstration of Beta-Lactamase Activity

Materials

❏ 1. Forty-eight-hour streak plate preparations of each
of the following bacterial species grown on
chocolate agar:
   ❏ a. *Escherichia coli*
   ❏ b. *Haemophilus influenzae*
   ❏ c. Ampicillin-resistant *Staphylococcus
   epidermidis*

❏ 2. One sterile Petri plate
❏ 3. Two mL sterile, distilled water
❏ 4. Three beta-lactamase reagent disks (Marion
Scientific Corporation)
❏ 5. One wax marking pencil, or felt-tip marker

Procedure

This procedure will be performed by your instructor.

❏ 1. Using a wax marking pencil or felt-tip marker, divide the empty Petri plate into 4 sectors and indicate the
organisms for which each sector is to be used. One sector will serve as an inoculated control.

❏ 2. With the 1-mL pipette, place 1 small drop of sterile water into each sector.

❏ 3. Place 1 beta-lactamase reagent disk into each drop of water. Wait a few seconds for the disks to rehydrate.

❏ 4. Aseptically remove a portion of an isolated colony from 1 of the cultures provided and streak it across the
surface of the rehydrated disk in the appropriately marked sector.

❏ 5. Repeat step 4 with each remaining culture.

❏ 6. The appearance of a yellow zone on the disk is a positive test.

❏ 7. Examine each disk within a 10-minute period and answer the questions in the Results and Observations section.

D. Methicillin-Resistant *Staphyloccus aureus* (MRSA) Detection

Materials

The following items are to be provided for an instructor demonstration:

❏ 1. Two BBL® Crystal™ MRSA ID systems (one of these systems will be used for examination only)
❏ 2. One 24-hour trypticase soy broth culture of each of the following:
    ❏ a. *Staphylococcus aureus* (methicillin-resistant strain)
    ❏ b. *Staphylococcus aureus* (methicillin-sensitive strain)
    ❏ c. Unknown
❏ 3. Container for disposal of used test system
❏ 4. One hand-held long-wave ultraviolet violet (**UV**) light source
❏ 5. One marking pen or similar device

Procedure

The following inoculation is to be performed by the instructor or student specified by the instructor.

❏ 1. Examine the demonstration test system. Note the wells in the device and its cover (Figure 37–7).
❏ 2. Starting from the left-hand side, label the wells, **MRSA, SA,** and **X** (unknown), respectively.
❏ 3. Inoculate the different wells with the cultures indicated.
❏ 4. Cover the system by snapping on the lid.
❏ 5. Incubate the closed system for 4 hours at 35°C.
❏ 6. After incubation, turn on the hand-held long-wave UV light source. Allow it to warm up for 4–5 minutes.
❏ 7. Unsnap the lip and expose the system to the ultraviolet light.
❏ 8. Note any fluorescence. The presence of a bright orange color indicates a positive result.
❏ 9. Record your findings in Table 37–4 in the Results and Observations section.

(a)

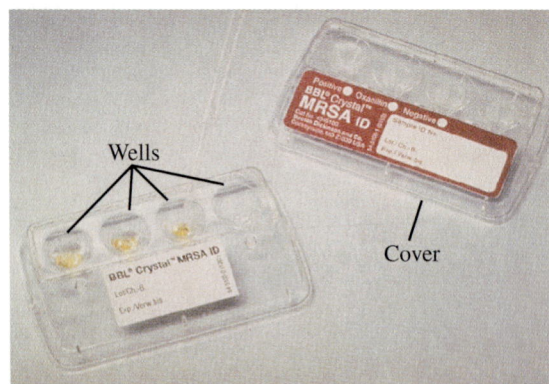

(b)

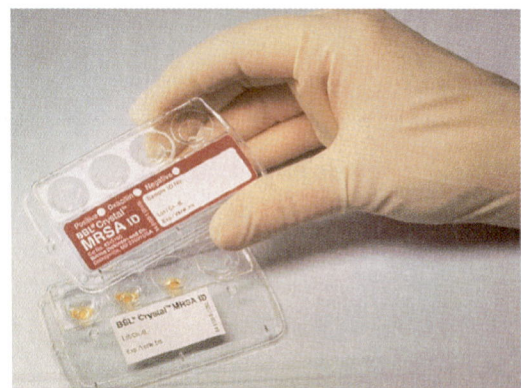

(c)

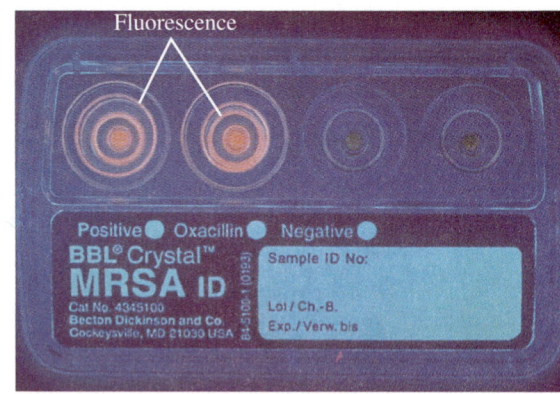

Figure 37–7
The BBL® Crystal™ methicillin resistant *Staphylococcus aureus* (MRSA)
system. (a) The components of the system. (b) Inoculation. (c) The results
with an MRSA strain.
(Photos courtesy of Becton Dickinson Microbiology Systems.)

Results and Observations

Antibiotic Susceptibility (Sensitivity) Testing

1. Antibiotic susceptibility (sensitivity) testing. Indicate the degree of susceptibility as determined by zone diameters measured in millimeters (mm) in the following table.

2. Were the antibiotic susceptibilities the same for both cultures? _____

3. a. To which antibiotics were both cultures resistant? _____

   b. To which antibiotic(s) were both cultures susceptible? _____

E-Test Demonstration

1. Enter the MIC values in Table 37–3.

2. Were MIC values easily determined? _____

   If not what type(s) of reactions interfered with the interpretation? _____

   _____

Table 37–2

Antibiotic Susceptibility Test Results

Antibiotic	Concentration	Diameter of Inhibition Zone	
		S. aureus	P. vulgaris

Table 37–3

MIC Determinations

Microorganism	MIC Values		
	Doxycycline	Penicillin V	Other Antibiotics
Ampicillin-resistant S. aureus			
E. coli			
H. influenzae			

Demonstration of Beta-Lactamase Activity

1. Which of the cultures showed beta-lactamase activity? _____

   _____

2. What type of reaction (color change) occurred in the control? _____

   _____

Methicillin-Resistance Detection

1. Record your findings in the following table.

Table 37–4

Methicillin Resistance

Strains	Fluorescence	No Fluorescence
Methicillin-Resistant S. aureus		
Non-methicillin-resistant S. aureus		
Unknown		

2. What color was the fluorescence observed? _____

3. Was the unknown a methicillin resistant? _____

Laboratory Review 37   The Antibiogram, Antibiotic Testing, and Antibiotic Resistance Detection

1. List 5 microbial diseases and the antibiotic currently used for treatment. (Refer to your text.)

   a. _____

   b. _____

   c. _____

   d. _____

   e. _____

2. Are antibiotics considered to be effective against virus infections? Explain. (Refer to your text.) _____

   _____

   _____

3. Is penicillin usually considered to be nontoxic to humans? Explain. (Refer to your text.) _____

   _____

   _____

4. What is penicillin hypersensitivity? (Refer to your text.) _____

   _____

   _____

5. a. What is beta-lactamase? _____

    _____

    b. Where are beta-lactamases found in a bacterial cell? _____

    _____

6. List 4 bacterial species known to produce beta-lactamase.

    a. _____   c. _____

    b. _____   d. _____

7. What is a monobactam? (Refer to your text.) _____

    _____

Key Terms

antibiogram (an-ti-BĪ-ō-gram): the results of an antibiotic sensitivity test

antibiotic (an-ti-bī-OT-ik): a chemical compound produced by microorganisms and/or synthesized commercially that can inhibit the growth of or kill other microbes

bactericidal (bak-ter-i-SĪ-dal): an antimicrobial agent that kills bacteria, whereas a bacteriostatic agent inhibits growth

beta-lactamase (BĀ-ta LAK-ta-māc): an enzyme that attacks β (beta)-lactam ring-containing antibiotics, such as the penicillins and related antibiotics

chemotherapeutic (kē-mō-THER-a-pū-tik) agent: a chemical used in the treatment of a disease

minimal inhibitory concentration (MIC): the concentration of a drug that will inhibit the growth of microorganisms

Selected Activities of Viruses

Do there exist many worlds or is there but a single world? This is one of the most noble and exalted questions in the study of nature.

St. Albertus Magnus

At the beginning of this century, when viruses were first cultured in the laboratory, several of their unique properties were uncovered. They were found to pass through filters that normally held back bacteria (hence the original designation filterable viruses) and to grow only in living cells, referred to as *hosts*. These findings served to emphasize the submicroscopic and obligate-intracellular-parasitic nature of viruses. With the aid of various subsequent technological achievements and extensive investigations, additional properties of these infectious agents were determined. Several of the properties of viruses that are pertinent to the exercises in this section are briefly described here.

Extensive work with many viruses has shown that individual viral particles contain a nucleic acid core consisting of a single type of nucleic acid, either deoxyribonucleic acid (DNA) or ribonucleic acid (RNA), and a protein coat that surrounds the nucleic acid. The protein coat is called a *capsid* and is composed of subunits referred to as *capsomeres*. The nucleic acid core and capsid combination is called a *nucleocapsid*. More complex forms of viruses also exist. Only those virus particles that have the components mentioned above are considered to be complete. These are referred to as *virions*.

Viruses exist in a range of highly characteristic morphological types that can provide one functional basis for classification schemes (Figure VIII–1). The majority of viruses are either *polyhedral* (many-sided) or *helical* in shape, or display a *binal* form that combines these two shapes (see Figure 38–2a).

Viruses multiply (under normal conditions) only in particular host cells. Accordingly, the viral host range is used for classification purposes. Thus, there are three main classes: animal viruses, plant viruses, and bacterial viruses, or bacteriophages. The successful infection of host cells by virions depends on several factors, including the properties of the host's surface and the virus particle's coat.

An infectious cycle must occur within a living cellular host for viruses to replicate (propagate). Depending on the properties of the virus-host system, this cycle has several stages. Viruses utilize the host's cellular organelles for the syn-

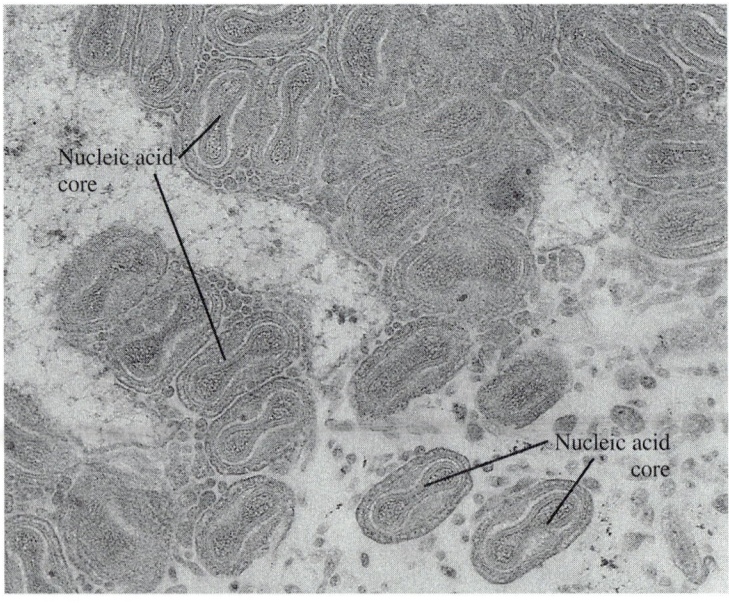

Figure VIII–1
An electron micrograph of a poxvirus. The nucleic acid core of these particles is clearly shown.

thesis of viral proteins and the replication of viral nucleic acid. These products are then accurately assembled into new viral particles and released in a pattern characteristic of the infecting virus. This mode of propagation accounts for the obligate-intracellular-parasitic nature of viruses. As briefly described here, viruses do not increase their numbers by simple divisions (splitting in half), but by an involved assembly process known as *replication.*

The viral infection cycle, or even a portion of it, may have a variety of effects on the host. Depending on the virus-host system, the effects include local lesions—such as cell lysis (destruction) and abnormal cellular development—or general reactions such as antibody production and death. Additional effects are described in the exercises of this section.

In order to study viruses, one must have effective means of detecting and identifying them. The presence of viruses in a host is generally recognized by the appearance of some abnormal effect, either a natural disease or an experimentally induced disease state. These outcomes can be expected, because viruses in their extracellular states neither multiply nor carry out active exchanges with their environments. Most significant viral activities are expressed within hosts and are revealed in the form of disease symptoms.

In the laboratory, viruses can be studied with the aid of laboratory animals and plants, cells and tissues of animals grown in various containers (tissue cultures), and microbial cultures. An infected cell in such cultures serves as a local site from which the virus spreads. Depending on the infecting virus, additionally infected cells either die or grow abnormally. The destructive effects at times are obvious. However, at other times, staining and/or immunological methods are needed to demonstrate viral presence.

The exercises in this section present selected general activities of bacteriophages and animal viruses. Several of the techniques presented here are used for the quantitative determination of viral activity, called *titration.* A viral titration is performed to determine the smallest amount of virus suspension capable of producing a recognizable effect, such as the localized lesions or generalized reactions mentioned earlier.

After completing this exercise, you should be able to:

1. Perform standard procedures to detect, identify, and demonstrate bacterial virus activities.
2. Recognize the value of standard procedures to the study of viral growth cycle.
3. Recognize the epidemiological significance of bacteriophage typing.

In 1915 and 1917, respectively, F. Twort and F. d'Herelle independently discovered that bacteria are susceptible to infection with viruses called *bacteriophages* (ϕ). These eaters of bacteria, also referred to as *phages,* are widely distributed in nature and have been isolated from feces, sewage, human gastrointestinal tracts, nasopharyngeal areas, and sputum specimens (Figure 38–1). The finding of bacteriophages in commercial fetal calf serum used for human vaccines as well as in the vaccines themselves emphasizes the widespread occurrence of these viruses in our environment. This discovery has also created interest and concern regarding the possible effects of bacteriophages on humans.

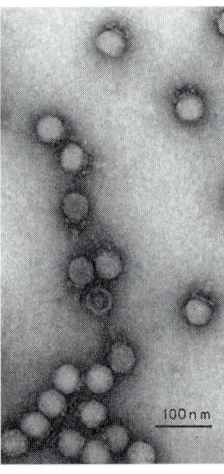

Figure 38–1
An electron micrograph of *Staphylococcus carnosus* bacteriophages isolated from fermented meat.
(From Burtin, A., *et al;* J. Appl. Bacteriol. *73:*401–406, [1992].)

Like viruses affecting humans, lower animals, and plants, bacteriophages are obligate, intracellular parasites that may alter the host cells they infect. In addition, specific bacteriophages may bring about devastating effects in the mammals harboring these viruses and their bacterial hosts. For example, they may (1) cause bacterial toxin production, as in diphtheria and botulism; (2) mediate the antibiotic resistance of invading bacterial pathogens; and (3) change bacterial surface properties.

The bacteriophage-host interaction provides a model system for the study of the infection cycle at a cellular level. Bacteriophages lend themselves to such studies because they can be obtained in large quantities and can be manipulated genetically. Among the viruses studied most extensively are those affecting *Escherichia coli,* strain B. These bacteriophages are known as the *T (type) coliphages.* Seven strains, numbered T_1 to T_7, are recognized and classified into specific groups on the basis of biologic, morphologic, and serologic properties. Portions of this exercise will be concerned with

(a)

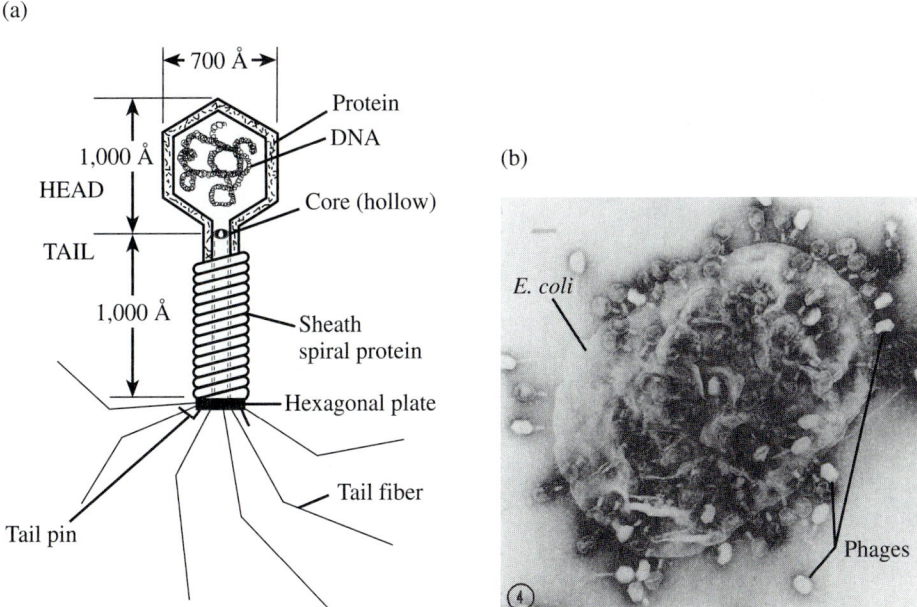

Figure 38–2
Bacteriophages. (a) The components of a T$_2$ bacteriophage. (b) An electron micrograph showing T$_4$r bacteriophages normally absorbing onto the bacterial host *Escherichia coli*. The cell wall of the bacterium has been severely damaged. Note that the viral tail sheaths are contracted and that several phages have injected their nucleic acid components (the clear phage heads) into the host.
(Courtesy of Simon, L.D., J.G. Swan, and J.E. Flatgaard, *Virology* 41:77 [1970].)

the T phages, specifically T$_7$, and their lytic or reproductive cycle (replication). The morphologic features of this strain are shown in Figure 38–2a. Phages exhibit a variety of shapes and sizes. Some are helical (spring-shaped), while others are polyhedral (many-sided). Certain bacteriophages are binal in shape. This arrangement is a combination of a polyhedral (head) portion and a helical (tail) structure (Figure 38–1a). Animal and plant viral particles or *virions* do not exhibit this shape. With binal phages, all of the viral nucleic acid is located within the head. The tail functions as an organ of attachment to susceptible host cells.

The first step in the reproductive cycle is the attachment of a fully infectious bacteriophage (virion) to specific receptor sites on the bacterial cell. This process is referred to as *adsorption* (Figure 38–2b). The phage nucleic acid—containing the genetic information of the virus—separates from the viral coat, is injected into the host, and becomes free in the cell. These events are followed by the numerous sequential steps that constitute the process of bacteriophage replication. This reproductive cycle ends with the lysis of the host and the release of newly synthesized progeny bacteriophages. These viruses are called *virulent.* Infections by bacteriophages such as T$_2$ cause a rearrangement of all macromolecular syntheses, which includes replacing host cell deoxyribonucleic acid (DNA) synthesis with viral synthesis.

It should be emphasized that not all bacteriophage infections end in the lysis of host cells. In certain cases, a phage-host interaction known as *lysogeny* may develop. Here the viral nucleic acid persists in the host for many cell generations, even indefinitely, without causing lysis. The bacterial host functions normally and the viral nucleic acid is transmitted through successive bacterial generations. The bacteriophages that produce the phenonemon of lysogeny are known as *temperate,* while bacterial hosts are referred to as *lysogenic* strains or *lysogens.*

The lytic activity of bacteriophages may be observed by the clearing of a cloudy broth suspension of infected bacterial cells or by the formation of clear zones, known as *plaques,* in a dense bacterial culture growing on the surface of an agar medium (Figure 38–3a and b). Plaque formation is the basis for one of the virology procedures used to assay bacteriophage and animal viral suspensions.

Bacteriophage assays or titrations can be performed in the following manner. First, serial dilutions (e.g., 1:10 [10^{-1}], 1:100 [10^{-2}], 1:1,000 [10^{-3}]) of phage-containing specimens are made. Specific aliquots of such dilutions and a standard suspension of a suitable bacterial host are added to a standard quantity of melted soft agar. The mixture is then poured over the surface of an appropriate medium contained in a Petri plate. After the soft agar mixture hardens, the entire prepa-

(a)   (b)

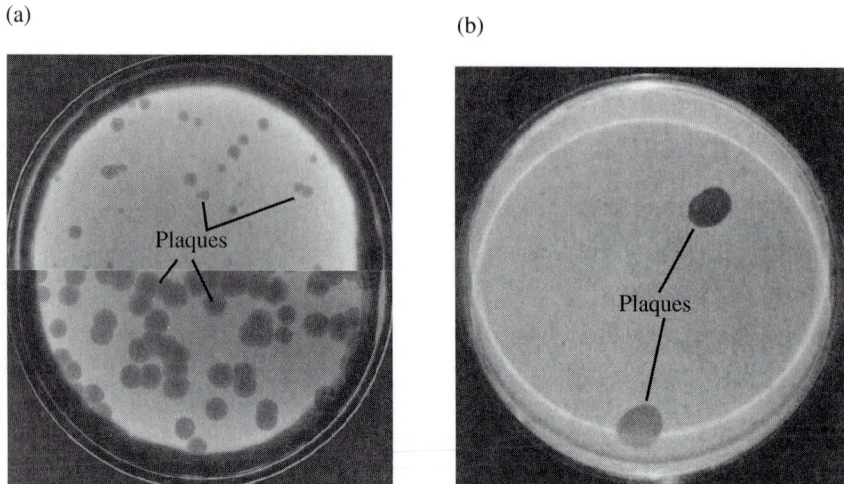

Figure 38–3

Different bacteriophages. (a) The opaque background is composed of a lawn of bacteria. Each clear area, or hole in the lawn, is a *plaque* formed as a result of lysis of the susceptible bacteria by the progeny (offspring) of a single phage. This figure shows the variations in plaque morphology possible with different bacteriophage-host combinations. (b) A typical result in a phage-typing procedure.

ration is inverted and incubated for a specified length of time and temperature. During the incubation period, the reproductive cycle described earlier takes place in the soft agar. In time, infected bacteria lyse and release several hundred progeny phages. These viruses, in turn, attach to and infect neighboring bacterial hosts, which also lyse and release new bacteriophages. Meanwhile, uninfected bacteria grow and reproduce to form a dense, opaque bacterial cell layer called a lawn. While the soft agar permits phage diffusion to nearby cells, it limits their spread to more distant regions of the plate; hence, so-called secondary centers of infection cannot develop. After incubation, the infected zones, or *plaques,* appear as clear areas against the opaque bacterial lawn background (Figure 38–3a).

Bacteriophages are specific in their attraction or affinity for particular bacterial species or strains of such species. This fact serves as the basis for a relatively new technique used to identify certain sources of bacterial strains associated with outbreaks of infections caused by such organisms as *Pseudomonas aeruginosa, Serratia marcescens,* and coagulase-positive *Staphylococcus aureus.* This precise identification procedure is based on the lytic action of specific phages on specific bacterial hosts. The second portion of this exercise will illustrate the phage-typing technique (Figure 38–3b).

A. Bacteriophage Titration

Materials

The following materials should be provided per 4 students:

❏ 1. Six-tenths (0.6) mL of T_2 bacteriophage suspension (1:1,000, or 10^{-3} dilution)
❏ 2. Three mL of an 18- to 24-hour tryptose phosphate broth culture of *Escherichia coli,* strain B
❏ 3. Twenty-five mL sterile tryptose phosphate broth in an Erlenmeyer flask
❏ 4. Ten tryptose agar plates
❏ 5. Ten tubes of soft agar in 2.5-mL quantities
❏ 6. Ten sterile 1-mL pipettes graduated in tenths
❏ 7. One sterile 5-mL pipette graduated in tenths
❏ 8. One sterile 25-mL pipette graduated in tenths
❏ 9. Five sterile 13 × 100-mm tubes
❏ 10. One Pi-Pump or similar device for pipetting small volumes
❏ 11. One Brinkman Pipet Helper or similar device for pipetting larger volumes
❏ 12. A test tube rack
❏ 13. A wax marking pencil or washable-ink felt-tip pen
❏ 14. Container of disinfectant for used pipettes

The following items should be provided for general class use:

❏ 1. A Quebec Colony Counter
❏ 2. Water baths, one set for boiling and the other set at 45° C
❏ 3. An incubator set for the temperature range 35° to 37° C

Procedure

Additional Technique Required for This Portion of the Exercise:

❏ The Use of a Pipette Pump, Procedure Diagram 11, Exercise 6

This assay is to be done in duplicate and performed by students in groups of 4. (*Note:* Pipetting should be carried out as shown in Procedure Diagrams 11 and 33. Pipetting should not be done by mouth. Your instructor will demonstrate the proper use of pipetting aids.)

❏ 1. Perform the steps shown in Procedure Diagram 34. (*Note:* Pipettes should be disposed of in the container provided.) Refer to Procedure Diagrams 11 and 33 for pipetting techniques.
❏ 2. Invert all plates and incubate at 37°C for 24 hours.
❏ 3. After incubation, using the colony counter provided, examine the plates and count the number of plaques—the holes in the lawn of bacteria.
❏ 4. Enter your findings in Table 38–1 in the Results and Observations section. (*Note:* Count only those plates having 30 to 300 plaques. If more than 300 plaques are formed, indicate this finding by the letters TNTC [too numerous to count].)
❏ 5. Average the number of plaques for each dilution and calculate the number of plaque-forming units for 1 mL of the original bacteriophage suspension. The formula for this assay procedure is shown in the Results and Observations section.

Procedure Diagram 33
The Brinkmann Pipet Helper

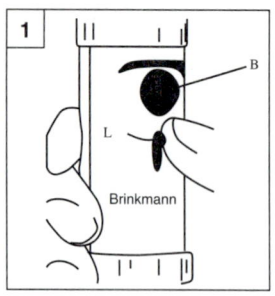

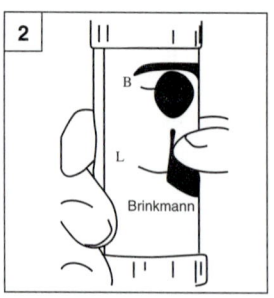

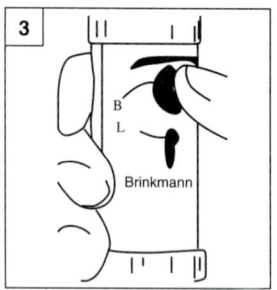

1. Filling. Slowly push the pipetting lever (**L**) upward. The further up the lever is moved, the faster the fluid will rise. Draw the level of the liquid (**meniscus**) slightly above the desired volume mark.

2. Volume adjustment and delivery. Gently press the lever (**L**) down to adjust the meniscus to the required level. To release or deliver the liquid press the lever completely down.

3. Blowout. When using "blow-out" pipets, allow liquid to run out until it comes to a stop. Then press the blow-out button (**B**) to deliver the remaining drops of liquid.

Procedure Diagram 34
Bacteriophage Titration

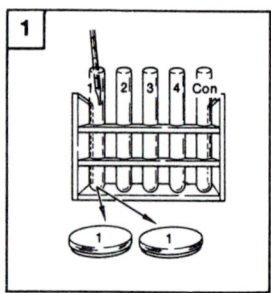

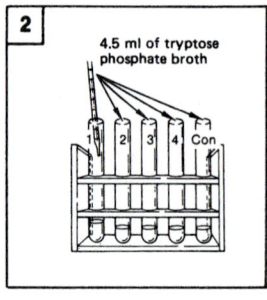

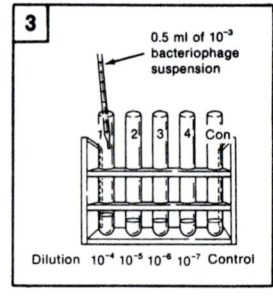

1. Place 5 sterile tubes in a row in the test tube rack and label the tubes consecutively from 10^{-4} through 10^{-7} and control. Also label 2 tryptose agar plates to correspond with each of the labeled tubes.

2. Using the sterile 25-ml pipette, aseptically place 4.5 ml of the tryptose phosphate broth into each tube.

3. Prepare serial 10-fold dilutions of the 10^{-3} bacteriophage preparation provided. To do this aseptically transfer 0.5 ml of the bacteriophage suspension to the 10^{-4} labeled tube (1) with a 1-ml pipette.

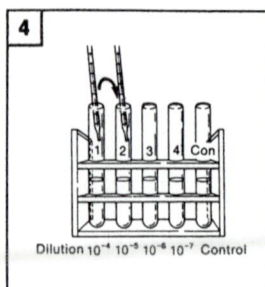

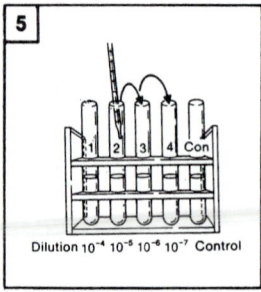

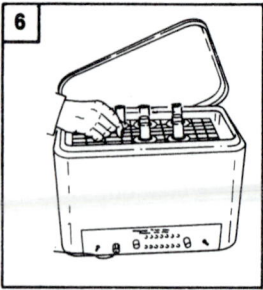

4. Shake the tube well and, with a new 1-ml pipette, transfer 0.5 ml from the contents of the 10^{-4} tube (1) to the tube labeled 10^{-5} (2).

5. Repeat this diluting procedure through tube 10^{-7} (4), using a new pipette for each transfer. However, discard 0.5 ml from the last tube after the ingredients have been mixed.

6. Place 10 tubes of soft agar into the boiling water bath. After the contents have been melted, cool the tubes and keep them in the 45° C water bath.

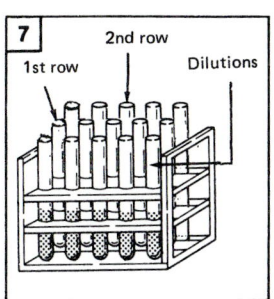

7. Remove the soft-agar, wipe, and place in rows parallel to the bacteriophage-dilution tubes. *Note*: You must perform the following steps quickly.

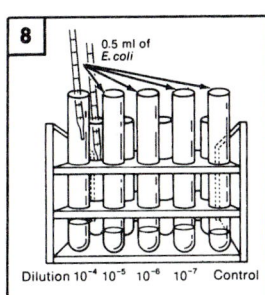

8. With the sterile 5-ml pipette, aseptically add 0.5-ml of the *E. coli*, strain B suspension to each of the tubes of soft agar.

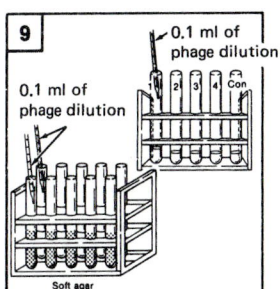

9. To each of the tubes of soft agar, aseptically transfer 0.1 ml of the respective phage dilutions and control. Remember, there are 2 tubes of soft agar for each dilution. Use a separate 1-ml pipette for each dilution and control.

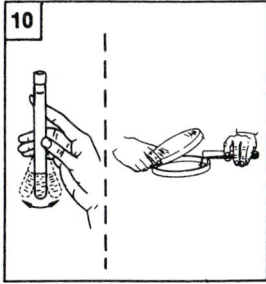

10. Quickly mix the contents of each of the soft-agar tubes and pour onto the surface of the appropriately labeled plate (for example, 10^{-4} agar plate.)

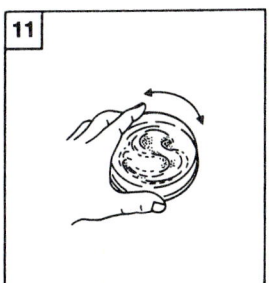

11. Rotate the plate to ensure the even distribution of the soft agar. Allow the soft agar to harden. Incubate as directed.

Results and Observations

1. Complete the following table.

Table 38–1

Bacteriophage Titration Results

	10^{-4}		10^{-5}		10^{-6}		10^{-7}		Control	
Dilution of Virus	1	2	1	2	1	2	1	2	1	2
Number of plaques										
Average number of plaques										

2. Using the following formula, calculate the number of plaque-forming units present in the original sample used.

$$\text{Plaque-forming units} = \frac{\text{Number of plaques formed by original specimen}}{\text{Dilution of original specimen used} \times \text{volume used}}$$

B. Bacteriophage Typing

Materials

The following materials should be provided per 4 students:

❏ 1. One tube containing 5 mL of a 24-hour trypticase soy broth unknown bacterial culture

❏ 2. One trypticase soy agar plate

❏ 3. One sterile cotton swab

❏ 4. One 250-mL beaker containing disinfectant

The following materials should be provided for general class use:

❏ 1. Four individual Pasteur pipettes with rubber bulbs, or tuberculin syringes with 27-gauge needles containing 1 of the following bacteriophages:
   ❏ a. *Escherichia coli* T_2
   ❏ b. *E. coli* T_3
   ❏ c. *Staphylococcus aureus* P_1
   ❏ d. *Streptococcus pyogenes* A_1

❏ 2. A wax marking pencil or washable-ink felt-tip pen (1 per class)

❏ 3. An incubator set at the temperature range of 35° to 37°C

Procedure

This portion of the exercise is to be performed by students in groups of 4.

❏ 1. Perform the steps shown in Procedure Diagram 35.

❏ 2. Incubate your plate without inverting it for 24 hours at 35° to 37°C.

❏ 3. Observe the different squares for plaque formation (phage strains that lysed the unknown bacterial cells). Use Figure 38–3b as a guide

❏ 4. Record your findings in the pattern diagram in the Results and Observations section. On the basis of your findings, you should be able to identify the unknown bacterial culture.

Procedure Diagram 35
Bacteriophage Typing

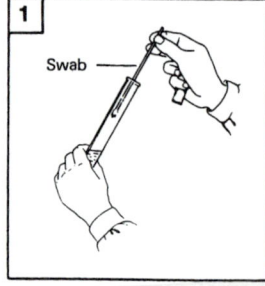 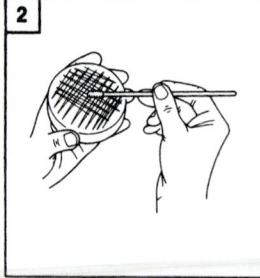 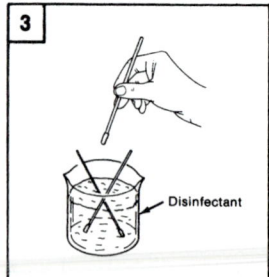

1. Moisten the cotton swab by dipping it into the unknown bacterial broth culture provided.

2. Rub the swab over the surface of the trypticase soy agar plate. Cover the surface thoroughly with the culture.

3. Place the swab into the beaker containing disinfectant.

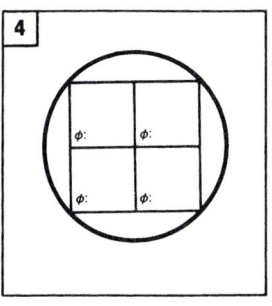

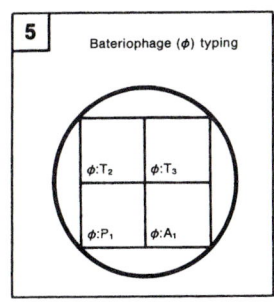

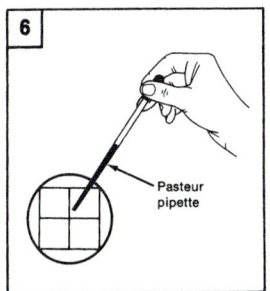

4. Invert the inoculated plate and, using a wax marking pencil or other instrument, draw a square pattern on the bottom similar to that shown.

5. Label the squares with the numbers corresponding to the phage strains provided: T_2, T_3, P_1, and A_1.

6. Turn the plate right side up and place 1 small drop of each of the different bacteriophage preparations provided into their labeled squares. If the virus preparations are in the tuberculin syringes, exert care in pushing the plunger portion, since it may move quickly and discharge more than 1 drop.

Results and Observations

1. Record your findings in the following diagram. Indicate plaque formation by a "+".

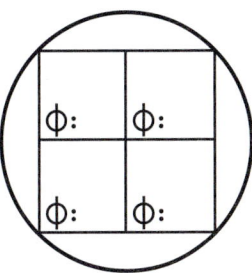

2. What was the unknown bacterial culture used? _____

Laboratory Review 38   Characteristics of Bacteriophages

1. What procedural factors can affect the outcome of a bacteriophage titration? _____

_____

_____

2. What is the resulting dilution of a bacteriophage suspension if 0.5 mL of a 10^{-6} dilution is placed into 4.5 mL of sterile broth? _____

_____

_____

3. Are bacteriophages specific for certain bacterial species? Explain your answer. _____

_____

_____

4. How does a viral plaque compare to a bacterial colony? _____

_____

_____

5. Define or explain the following terms:

   a. prophage _____

   b. lysogenic host _____

   c. temperature phage _____

   d. polyhedral _____

   e. virulent phage _____

Key Terms

adsorption (ad-SORP-shun): the process of attachment

bacteriophage (bak-TE-rē-ō-fāj): a bacterial virus; phage

coliphage (KŌ-lē-fāj): one of several bacterial viruses affecting *Escherichia coli*, strain B, usually designated as a T (type) phage

helical (HĒL-i-kal): resembling a spiral or coil

lysogen (lī-sō-JEN): a bacterium that harbors a temperate bacteriophage

lysogeny (lī-SOJ-eh-nē): the relationship between a temperate bacteriophage and a bacterium, in which the bacterium is not destroyed by the phage

phage (FĀJ): shortened form for bacteriophage

plaque (PLAK): virus-containing zones that appear as clear areas in a lawn of bacteria or tissue cultures

The Uses of the Embryonated Chicken Egg (Chick Embryo) and Cell Systems to Demonstrate Animal Viruses

After completing this exercise, you should be able to:

1. Understand the basic methods used for the inoculation of chick embryos with microorganisms.
2. Recognize the general types of reactions caused by microorganisms that are indicative of pathogenicity.
3. Recognize the value of the chick embryo and cell systems as tools with which to grow, study, and identify certain microbial agents.

The use of animals to solve laboratory problems in the various specialties has steadily increased, especially in the areas of virology and immunology. Laboratory animals are used in (1) investigating the pathogenesis and mode of multiplication of infectious agents; (2) making the primary isolation and identification of infectious disease agents; (3) testing chemotherapeutic agents for their effectiveness against pathogens or for toxicity; (4) producing viruses when large quantities are required; (5) producing and testing vaccines and immune sera; and (6) studying tissue transplant techniques. The animals most often used are mice, rats, guinea pigs, dogs, cats, monkeys, and ferrets. The chick embryo is very frequently employed in virology experiments. In this exercise, we will utilize the chick embryo as a representative laboratory animal, bearing in mind that our overall approach applies to experimentation with other animals as well.

A schematic diagram of a typical chick embryo approximately 12 days old is shown in Figure 39–1. The egg has several cavities into which viruses may be inoculated and cultivated. The site and method of inoculation chosen depend on the purpose of the procedure. For example, the *chorioallantoic membrane* (**CAM**) is customarily selected as an inoculation site for the (1) titration of viruses, (2) investigation of viral morphology, and (3) identification of viruses. Several viruses are capable of producing local blister-like lesions known as *pocks* on the CAM (Figure 39–2). When large

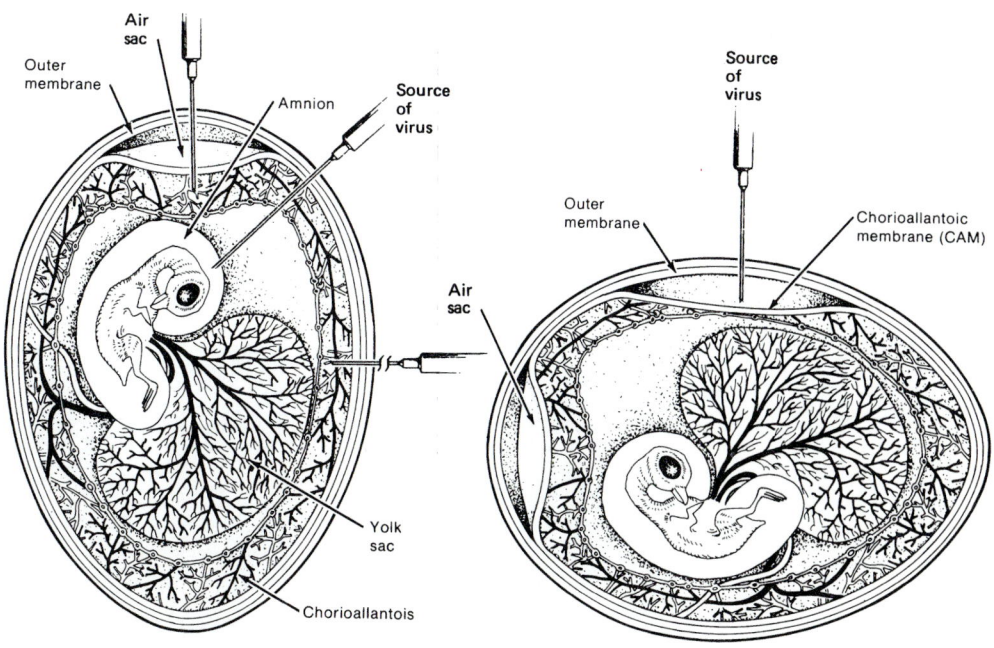

Figure 39–1
Cutaway diagram of a chicken embryo. The different parts of this animal and the sites used for inoculation are shown.

343

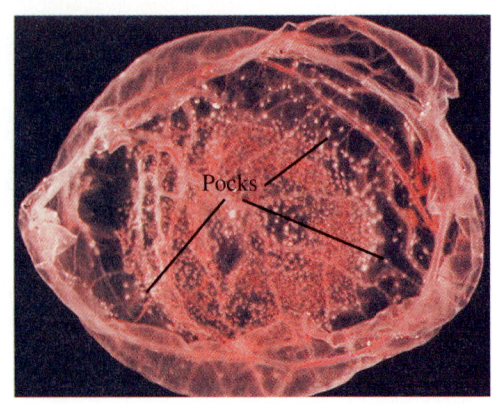

Figure 39–2
Pock formation on the chorioallantoic membrane of a chick embryo. Pocks can be recognized as small, whitish, dense areas.
(Courtesy of U.S. Public Health Service.)

quantities of viruses are required for vaccine production or chemical studies, the allantoic sac is utilized. Inoculations of the yolk and amniotic sacs are used for the isolation of various viral agents.

The detection, identification, and propagation of well-known animal viruses can be a routine procedure. However, characterizing a new virus, present only in extremely small amounts in a tissue sample or biological fluid, is often difficult. Several viruses are very sensitive and rapidly decompose during transport to a laboratory or during procedures used to isolate them.

In addition to the cultivation of viruses in embryonated bird eggs, the use of tissue cell cultures has proved to be an economical and readily available means for the isolation and cultivation of most animal viruses. The cells of most body organs can be grown *in vitro* in a variety of glass or plastic containers. Such cells, when introduced into a container, attach to the available surface and continue to divide until most or all of the surface area is occupied. When the surfaces of all cell membranes make contact with adjacent cells, division stops. This situation, referred to as *contact inhibition,* results in the formation of a single layer or sheet (monolayer) of cells. Such cell cultures are used for viral studies.

Two types of cell cultures are in use. These are *primary* and *continuous.* Primary cultures consist of cells obtained from normal tissue, and which cannot divide indefinitely. Continuous cultures generally are obtained from malignant (cancerous) tissues, which can undergo an unlimited number of divisions. These cells have lost the contact inhibition property.

Inoculation of susceptible animal cell cultures with many, but not all, animal viruses results in a pattern of cell destruction and death, which often is characteristic for a given host-virus combination. These changes can sometimes be detected by the naked eye, or more often microscopically, with or without the use of stains. The most frequent effect of virus infection is lysis (destruction) of cells. Such cytolytic viruses produce cytopathic effects, the most recognizable of which is the *plaque* (Figure 39–3). Plaques are clear, infectious centers in a monolayer resulting from the death of cells. Microscopically, cytopathic effects include chromosomal damage, shrinkage of the nucleus, loss of contact inhibition, and giant cell formation (several cells fuse, resulting in the accumulation of their nuclei into one cell). Several animal virus assays utilize plaque formation to determine the number of infectious *virions* (virus particles) in an original

(a) (b)

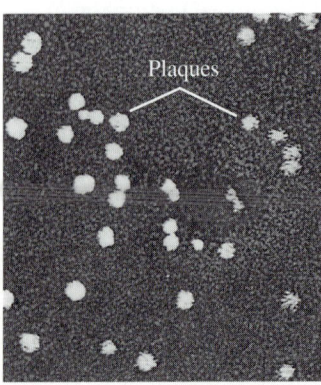 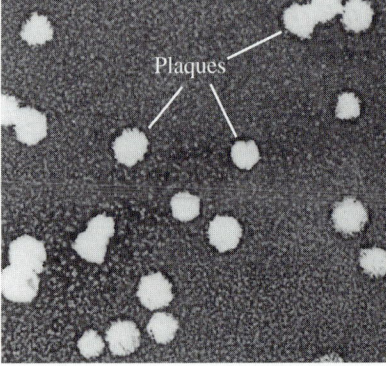

Figure 39–3

The appearance of various-sized animal virus (vaccinia) plaques.
(Courtesy of M. Esteban, Department of Biochemistry, SUNY.)

(a)   (b)

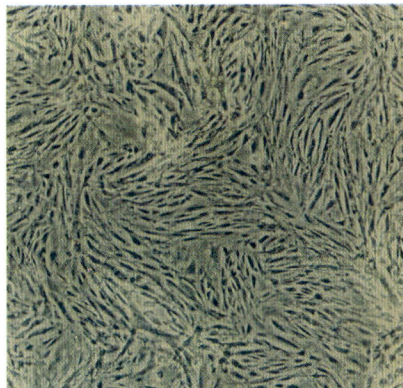 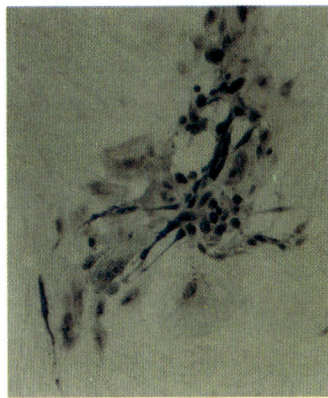

Figure 39—4

Detection of viral infection of tissue cultures. (a) Virus-infected baby rabbit kidney tissue, unstained. (b) Virus-infected baby rabbit kidney tissue, stained. Original magnification, 100 X for both micrographs.

specimen. The value obtained from counting plaques, multiplied by the dilution of the specimen used, represents the number of infectious virions and is expressed as plaque-forming units.

Microscopically, areas of viral growth produce cellular damage that can be demonstrated with the use of a *vital dye,* such as neutral red. Living cells take up the dye, whereas dead cells remain colorless. In this case, infected areas appear as a *colorless plaque.* Viruses that infect cells without causing cytopathic effects are detected by other techniques. Among these is the use of antibodies for the identification of viral antigens (Figure 39–4).

In this exercise, preparations with and without viruses will be used to inoculate cell culture systems. After allowing sufficient time for viral adsorption, these cultures will be treated with a modified neutral red agar overlay to show evidence of viral presence and the signs of their cytopathic effects.

A. Chick Embryo Inoculation

Materials

The following materials should be provided per 4 students:

- ❑ 1. Two 10-day-old fertilized chicken embryos
- ❑ 2. Two tubes of canary-pox, cowpox, or vaccinia virus in approximately 1.0-mL quantities. One tube should contain undiluted vaccine, while the other should contain a 1:2 dilution of the virus-containing material. Both suspensions should contain 500 units of both penicillin and streptomycin
- ❑ 3. Two bleach-soaked towels, rung dry
- ❑ 4. Ten cotton squares (2 × 2 inches)
- ❑ 5. One bottle of 70% alcohol
- ❑ 6. One bottle of tincture of iodine
- ❑ 7. Two rubber bulbs
- ❑ 8. Two sterile test tubes, each containing a 1-mL tuberculin syringe
- ❑ 9. Two sterile test tubes, each with one 27-gauge, half-inch hypodermic needle
- ❑ 10. One metal drill bit (quarter-inch) or a similar type of instrument
- ❑ 11. Two 50-mL beakers of bleach (to be provided when the eggs are examined after incubation)
- ❑ 12. Two paper bags lined with wax paper for disposal of the inoculated eggs after examination
- ❑ 13. One egg-candling device
- ❑ 14. Transparent tape and dispenser
- ❑ 15. Disposable surgical gloves

Procedure

This procedure is to be performed by students in groups of 4. Gloves should be worn.

❑ 1. Using the candling apparatus provided, place the embryonated egg carefully and expose it to the light to determine if the embryo is alive. A live embryo will respond to the heat of the light by jerking. Once movement is observed, continue with step 2 of this procedure. Dead or unfertile eggs should be discarded as directed by the instructor.

❑ 2. Locate the air sac and areas on the upper surface where large blood vessels are present. Mark the latter in order to avoid puncturing these structures when the virus suspension is injected.

❑ 3. Perform the steps in Procedure Diagram 36.

❑ 4. Incubate the inoculated egg (or eggs) at 37° to 38°C for 72 hours. See the photograph of a chorioallantoic membrane (CAM) showing a typical reaction in Figure 39–2.

❑ 5. Place the infected eggs into the wax paper-lined paper bag provided. At the conclusion of the experiment, the entire contents of the bag should be autoclaved.

❑ 6. Answer the questions in the Results and Observations section.

Procedure Diagram 36
Chorioallantoic Membrane Inoculation

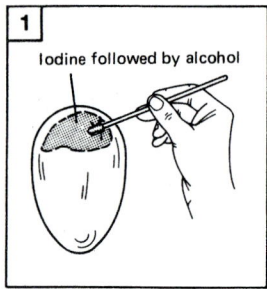

1. Apply the tincture of iodine solution to the area over the air-space region. After the iodine dries, wipe the region with an alcohol soaked square.

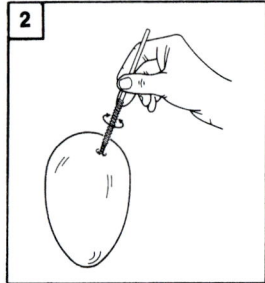

2. Make a hole in the shell over the air sac by gently striking the shell with the pointed end of the drill. Make a slight penetration. Then rotate the tool back and forth between the fingers until the air sac membrane is fully exposed. *Do not puncture the air sac membrane.*

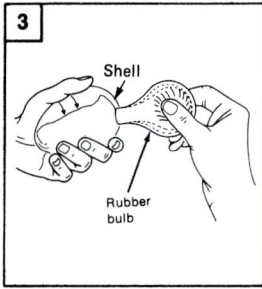

3. Force out the air from the rubber bulb provided by squeezing it. With the bulb fully squeezed in, place the open end tightly over the freshly made shell hole. Release your hold on the bulb. Suction should cause the CAM to recede from its normal position.

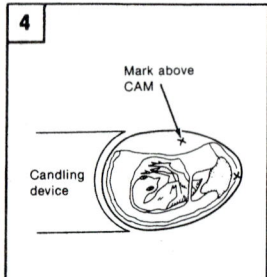

4. Examine the egg again by candling it, and note the region where the CAM has dropped. Before proceeding to the next step, drape the bleach-soaked towel over the area where the egg inoculation is to occur.

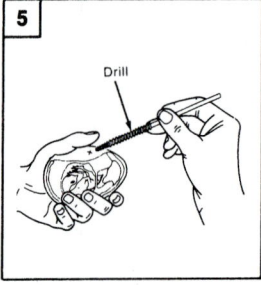

5. Next, surface-sterilize the area noted in step 6 and make a hole over this area, being careful not to injure any of the previously marked blood vessels.

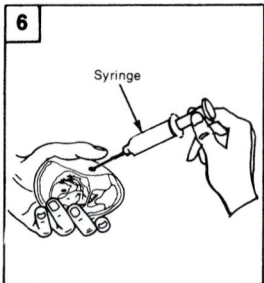

6. One student should inject 0.5 ml of 1 inoculum dilution into 1 egg, while a second student inoculates the second egg with the second dilution.

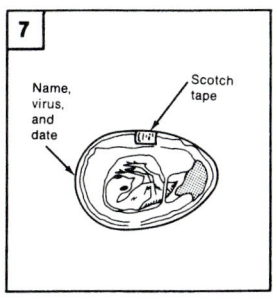

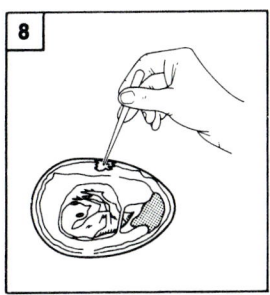

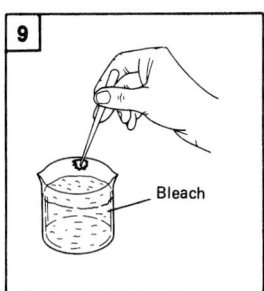

7. Seal each hole with transparent tape. With a lead pencil, print your name, the virus dilution injected, and the date on the egg.

8. *After incubation,* remove the tape covering the area of the CAM and, with a dissecting probe and forceps, remove the overlying shell.

9. Place all contaminated materials into the 50-ml beakers of bleach provided. Examine the CAM for the presence of pocks (local swellings).

B. Cell System Infection, Cytopathic Effects, and Staining

Materials

The following items should be provided per 4 students:

☐ 1. Five tubes containing primary rabbit kidney cells

☐ 2. One test tube rack

☐ 3. Two tubes of equine cytomegalovirus (ATCC VR-536) or equine herpesvirus type 2 (ATCC VR-701) in approximately 0.5-mL quantities. One tube should contain an undiluted virus suspension, while the other should contain a 1:2 dilution of the virus-containing material. Both preparations should contain 500 units of both penicillin and streptomycin

☐ 4. One bottle phosphate buffered saline (approximately 15 mL)

☐ 5. Two bleach-soaked towels (rung dry)

☐ 6. Five sterile 1-mL pipettes with bulbs, or pipette helper

☐ 7. Neutral red agar overlay (approximately 10 mL)

☐ 8. Disposable surgical gloves

☐ 9. Container with disinfectant for pipette disposal

☐ 10. One marking pen or pencil

The following items should be available for general class use during the tissue culture inoculation period:

☐ 1. Water bath set at 48°C.

☐ 2. Sterile 1- and 5-mL pipettes

Procedure 1: Virus Inoculation

All procedures are to be performed by students in groups of 4. The instructor will demonstrate the inoculation procedure.

☐ 1. A culture tube will be on demonstration. Examine it. If the tube contains a cover slip, note that the cover slip holds the tissues.

☐ 2. Label 2 tubes for each virus preparation provided, and one tissue culture system without virus, which will be a control. Include the dilution used and your initials.

☐ 3. Examine the cell monolayer in each of the tubes, microscopically, under low power and with the naked eye. Look for the presence of a cell layer. Refer to Figure 39–4*a*.

❏ 4. With a sterile 1-mL pipette and bulb, introduce 0.2 mL of the undiluted virus preparation into each of the 2 labeled tubes. Be certain to close the caps tightly on each tube.

❏ 5. Place the tubes into the rack provided.

❏ 6. Repeat steps 2 and 3 with the other preparations provided.

Procedure 2: Virus Adsorption and Incubation

❏ 1. Allow the virus to absorb for 30 minutes in a 37°C incubator. Rotate the tubes every 10 minutes to insure even virus distribution.

❏ 2. Place a disinfectant-soaked towel over the work area.

❏ 3. Carefully remove each tube in turn from the rack, and aseptically pour the contents into the container with disinfectant or as directed by the instructor.

❏ 4. Next, aseptically pipette with the aid of a pipette helper and introduce 1 mL of the phosphate buffer saline (PBS). Rotate the tubes to wash the tissue culture cells adequately.

❏ 5. Aseptically pour the PBS wash solution into the container with disinfectant. This step removes any unabsorbed virus.

❏ 6. During the last 5 minutes of the absorption period, remove the neutral red agar overlay preparation from the water bath. Allow it to cool slightly.

❏ 7. With a 1-mL pipette, aseptically add 1 mL or a sufficient volume of the agar overlay preparation to each tube. Be certain that the agar preparation covers the tissue culture. Allow the overlay to harden (about 5–10 minutes).

❏ 8. Incubate all tubes at 37°C for 1 week.

Procedure 3: Observations for CPE

❏ 1. From the second day and through the seventh day of incubation, examine the cultures on a daily basis for cytopathic effects. Score each tube on a scale of 1 to 4, with 4 being the most severe cytopathic effect produced by the virus. The examination should take place in a darkened room with the aid of a desk lamp. (Refer to the introduction of this exercise for a description of cytopathic effects [CPEs].)

❏ 2. Record your findings and describe the appearance of each virus-containing tube and control in the Results and Observations section.

Results and Observations

Chick Embryo Inoculation

1. Describe the general appearance of a pock on the CAM. _____

_____

_____

2. What is the purpose of the antibiotics in the viral inoculum? _____

_____

3. If pock (local blistering) formation did not take place, what could account for this reaction? _____

_____

_____

Cell System Infection, Cytopathic Effects, and Staining

1. Enter your findings in the following table. Indicate the presence of cytopathic effects (CPE) and the demonstration of virus through staining by "+" or "−".

Table 39–1

Cytopathic Effects

Findings	Virus Undiluted	Virus Undiluted	Virus Diluted	Virus Diluted	Control
CPE					
General CPE description					
Virus presence by staining					

2. What was the most common CPE seen? _____

_____

3. Was the presence of virus demonstrated in all cases with the use of stains? If not, which cultures did not? Explain your answer. _____

_____

_____

_____

Laboratory Review 39   The Use of the Embryonated Chicken Egg (Chick Embryo) and Cell Systems to Demonstrate Animal Viruses

1. List four microbial disease agents and the types of animals and/or tissue culture systems that are used for their cultivation. (Refer to your text.)

   a. _____   c. _____

   b. _____   d. _____

2. List two vaccines currently used for immunizations that are prepared from eggs.

   a. _____   b. _____

3. List two vaccines currently used for immunization that are prepared from tissue cultures.

   a. _____   b. _____

4. What complications can be associated with egg vaccine immunizations? _____

_____

_____

5.   What is a synthetic vaccine? (Refer to your text.) _____

_____

_____

6.   List two general methods used for the cultivation of animal viruses.

a. _____   b. _____

7.   What is contact inhibition? _____

_____

8.   What is one major feature that distinguishes a primary cell culture from a continuous one? _____

_____

9.   What is a plaque in a cell culture system? _____

_____

10. List four cytopathic effects associated with viral infections. (Refer to your text.)

a. _____   c. _____

b. _____   d. _____

11. Do cytopathic effects always develop in a virus-infected cell culture system? If not, how can virus infection be

demonstrated? (Refer to your text, if necessary.) _____

_____

_____

_____

Key Terms

contact inhibition: the inhibition of cellular division upon contact with the membranes of adjacent cells

continuous cell culture: cell culture derived from cancerous tissues that has lost the contact inhibition property

plaque (PLAK): in cell culture systems, a plaque is a clear (virus-containing, infectious) area formed in a monolayer of cells; it results from the death of infected cells

pock: a local blister-like formation caused by virus infection of the chorioallantoic membrane

primary cell culture: cell culture derived from normal tissue that has a limited capability to divide

SECTION IX

An Introduction to Microbial Genetics

. . . everything that we discover about microorganisms applies also to elephants, only more so!

F. Jacob

Acquiring genetic traits that favor the survival and reproduction of individuals and populations produce biological variations. These variations result from the new and potentially beneficial combinations of genetic properties. Such genetic variation is a major requirement for the successful evolution of both lower and higher organisms living in different and changing environments. Basic processes responsible for the creation of genetic variation are *mutations* (permanent changes of the deoxyribonucleic acid [DNA] within the genes of a single individual) and *genetic recombination* (the formation of a new chromosome resulting from exchanges of DNA between two different parental cells).

Mutations involve changes in the sequences of the building blocks of DNA, the *nucleotides,* which ultimately can influence an organism's biological characteristics. These changes can occur spontaneously. However, the rates at which such mutations take place are not constant, and can be affected by many environmental conditions. For example, a number of chemical and physical agents, called *mutagens,* can dramatically increase the normal rate of mutation.

Genetic variation in higher organisms primarily occurs by the reshuffling of alternate forms of genes, called *alleles,* that have accumulated within a population over several generations. In higher forms of life, this reshuffling takes place through the process of sexual reproduction. Genetic exchanges occur between two different sets of chromosomes, each set being inherited from two different parental cells. With microorganisms such as bacteria, which have their genes on one very large, circular DNA molecule (the bacterial chromosome), different alleles of a gene are normally not present. Thus, the reshuffling of genes necessary to form a recombinant chromosome requires a mechanism different from that of higher forms of life. Genetic exchange among prokaryotes appears to be an occasional process rather than a requirement in the completion of an organism's life cycle. Three quite different types of gene transfer processes are recognized in bacteria. In order of their discovery, these processes are *transformation, conjugation,* and *transduction.*

In 1928, the English microbiologist Fred Griffith discovered a model system that was to have a profound influence on the future of basic and applied science. He demonstrated the *in vitro* capsular-type transformation between pneumococcus bacteria and the *in vitro* confirmation of pneumococcal-type transformation. Later in 1944, a research team headed by O.T. Avery at the Rockefellar Institute reported that the substance responsible for the transformation, the *transforming principle,* was none other than deoxyribonucleic acid (DNA). The DNA was shown to account for the specific immunologic property of the pneumococcal capsular polysaccharides, which in turn played an important role in the disease-causing capability of a specific type of pneumococcus. These findings, and much of the related research that followed, was a part of the beginning investigations into microbial molecular biology.

The exercises in this section will present the general features of the processes of transformation, conjugation, and transduction. In addition, a special exercise involving the rapid transformation of bacterial colony with commercially available plasmids also will be considered. The Ames test, a relatively inexpensive, very sensitive, and reliable simple procedure for the detection of cancer-causing agents as mutagens using bacteria concludes this section.

Certain experimental exercises presented later in this manual will demonstrate some of the powerful methods used to isolate, manipulate, and analyze DNA.

After completing this exercise, you should be able to:

1. Demonstrate genetic recombination resulting from the conjugation of two different bacterial strains.
2. List two conditions needed for conjugation to occur.
3. List three properties of donor bacteria specified and/or controlled by transfer genes.
4. Outline an experiment to show bacterial conjugation and recombination.

Conjugation is a gene transfer process in which one strain of bacterium, the *donor,* comes into direct physical contact with a different strain of bacterium, the *recipient,* and transfers genetic material. This form of genetic exchange was discovered by J. Lederberg and E. L. Tatum in 1946. The mechanism involved not only proved to be different, but much more efficient than transformation or transduction. Moreover, the donor organism is not destroyed in the process (Figure 40–1).

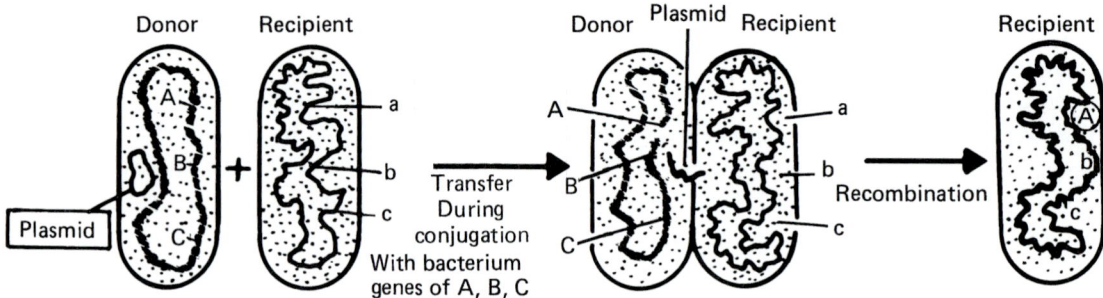

Figure 40–1

Bacterial conjugation. In this process, cell-to-cell contact is necessary so that a plasmid of one bacterium can transfer a DNA segment from a donor bacterium to a recipient cell. The recipient cell's chromosome contains alternate forms of the genes on the transformed DNA segment. The exchange of gene **A** for gene a by recombination is shown.

Bacterial conjugation requires that the donor have a *conjugative plasmid.* Plasmids of this type are extrachromosomal genetic elements that encode for a number of DNA transfer functions. They possess transfer *(tra)* genes that specify and/or control (1) the formation of donor pili that are necessary to allow donor cells to make contact with recipient cells; (2) substances to minimize donor-donor mating; and (3) conjugational transfer of plasmid or chromosomal deoxyribonucleic acid (DNA) for a specific site on a conjugative plasmid. Conjugative plasmids also have genes to control (1) the replication of conjugative plasmids and (2) the number of conjugative plasmids per chromosomal equivalent. Genes associated with antibiotic resistance, bacteriocin production, enterotoxin production, heavy metal ion resistance, and surface antigen production may also be found on conjugative plasmids. Bacterial conjugation is quite common in gram-negative bacteria and occurs less frequently in gram-positives. The mechanism for genetic exchange differs in certain aspects with gram-positive bacteria.

The best-studied conjugative plasmid is the fertility factor, F, of *Escherichia coli* strain K-12. Two general mating types of *E. coli* are genetically determined by the presence or absence of this F factor. Recipient strains do not have the F factor and are termed F^-. Cells carrying the F factor are donors and are termed F^+. Two types of donors are recognized: the *F^+ cell,* in which the F plasmid is detached from the bacterial chromosome and exists in an autonomous or independent cytoplasmic state, and the *Hfr cell,* in which the F plasmid is part of the bacterial chromosome. Hfr donors transfer the genetic markers on their chromosome at higher frequencies and in a specific order from a genetically determined site.

In this exercise, two strains of *E coli* will be used to demonstrate the process of genetic transfer. Recipient cells that have acquired donor genetic information will be isolated (Figure 40–2). The two original strains are called *transconjugants.* The F⁻ strain for the exercise requires the amino acids leucine (L⁻) and threonine (T⁻) for growth and is resistant to streptomycin (Sʳ). The Hfr strain does not require these amino acids but is sensitive to streptomycin (Sˢ). The F⁻ strain will not grow on a minimal medium lacking leucine and threonine, and the Hfr strain will not grow on a minimal medium with streptomycin. By means of the conjugative process, recombinant bacteria should appear that do not require leucine (L⁺) and threonine (T⁺) for growth and are resistant to streptomycin (Sʳ). Figure 40–2 shows the general outcome of this recombination experiment.

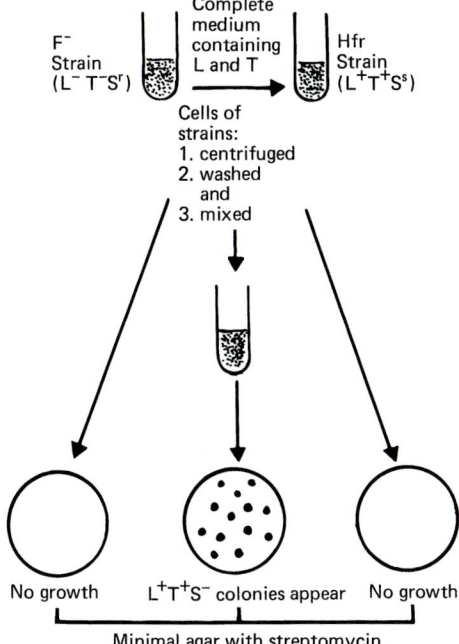

Figure 40–2

The appearance of bacterial strains following conjugation and recombination. The recombinant bacteria exhibit properties of both parent strains, *F⁻* and *Hfr.* (L = leucine; T = threonine; S = streptomycin.)

Materials

The following items should be provided for students in groups of 4:

- ❏ 1. Eighteen-hour tryptose yeast extract glucose (TYEG) broth cultures of the following:
    - ❏ a. *Escherichia coli* (ATCC e-23739). (This Hfr strain is sensitive to the antibiotic streptomycin (Sˢ) and is phototrophic.)
    - ❏ b. *E. coli* (ATCC e-23742). (This F⁻ strain is resistant to streptomycin (Sʳ) and is auxotrophic, and requires threonine (T⁻), leucine (L⁻), and thiamine (Th⁻) for growth.)
- ❏ 2. Five minimal medium agar plates containing thiamine (T), streptomycin (S), and biotin
- ❏ 3. Two trypticase soy agar plates (complete medium)

- ❏ 4. One sterile screwcap test tube
- ❏ 5. Four 9-mL sterile, distilled water blanks
- ❏ 6. Two sterile bent-glass rods (spreaders)
- ❏ 7. Nine sterile 1-mL pipettes and rubber bulbs
- ❏ 8. One 250-mL beaker with alcohol for sterilizing glass rods
- ❏ 9. One marking pencil or pen
- ❏ 10. One container of disinfectant large enough for pipettes

The following items should be provided for general class use:

- ❏ 1. Rotary spreaders
- ❏ 2. One water bath set at 37°C
- ❏ 3. Test tube racks

Procedure 1: The Controls

All procedures should be performed by students in groups of 4. Used pipettes should be placed in the disinfectant container provided. Before beginning this procedure, your instructor will demonstrate the proper methods of pipetting and spreading an inoculum over an agar surface. In addition, refer to Figure 40–3 for a general overview of the procedure.

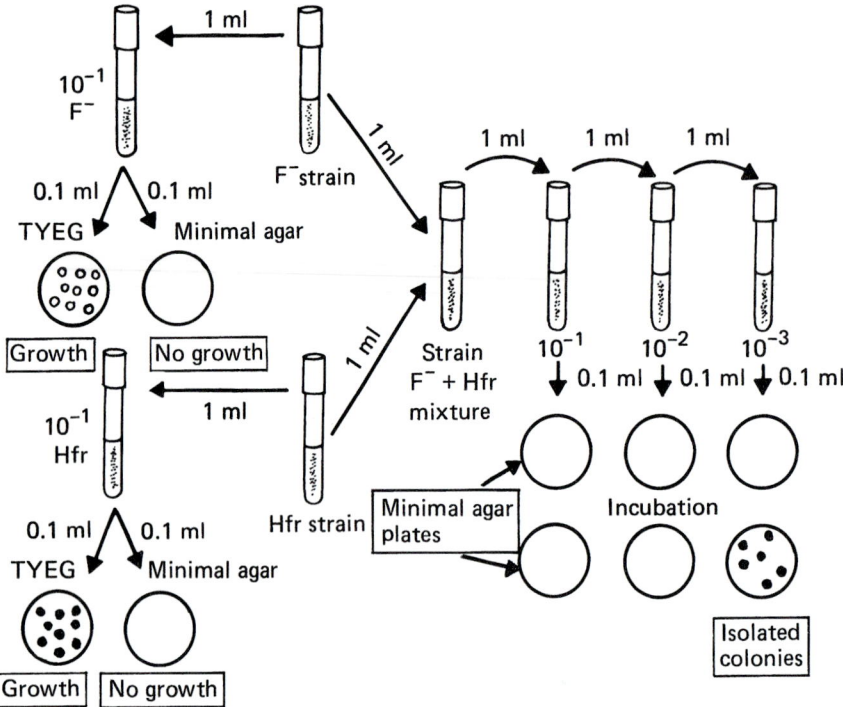

Figure 40–3

An overview of the dilution and plating procedure for the bacterial conjugation exercise. The results obtained should clearly demonstrate that only recombinant bacteria will grow on a minimal medium containing the antibiotic streptomycin.

❏ 1. With a sterile 1.0-mL pipette, aseptically transfer 1.0 mL of the *E. coli* F⁻ culture into a sterile 9-mL water blank. Label this tube 10^{-1}, F⁻.

❏ 2. With another sterile 1.0-mL pipette, aseptically transfer 1.0 mL of the *E. coli* Hfr culture into a separate 9-mL water blank. Label this tube 10^{-1}, Hfr.

❏ 3. Gently shake the 10^{-1}, F⁻ tube and, with a sterile pipette, aseptically place 0.1 mL of the preparation on the centers of one minimal medium agar plate containing streptomycin and one trypticase soy agar plate.

❏ 4. Sterilize a bent-glass rod as shown in Figure 40–4 and as follows, in order to spread the 0.1-mL culture on the minimal medium agar surface.

    ❏ a. Dip the bottom of the rod into the beaker containing alcohol.
    ❏ b. Hold the rod at a slight angle to allow excess alcohol to drain back into the beaker.
    ❏ c. Pass the rod quickly through a flame to ignite the alcohol. This action will sterilize the rod. (*Note:* If reflaming is necessary, be certain that the flame is completely extinguished before inserting the rod back into the alcohol beaker.)
    ❏ d. Allow the rod to cool for about 10 seconds before proceeding.

❏ 5. With the aid of a rotary spreader (see Figures 40–4 and 40–5), use the sterilized glass rod to spread the 0.1-mL culture evenly over the entire agar surface. If a rotary spreader is not available, use the glass rod with a circular motion to spread the culture.

❏ 6. Label the plate accordingly, invert it, and incubate the plate at 37°C until the next laboratory period or for at least 48 hours.

❏ 7. Resterilize the glass rod.

❏ 8. Repeat steps 3 through 7 with the 10^{-1} dilution of the Hfr strain.

(a)   (b)   (c)

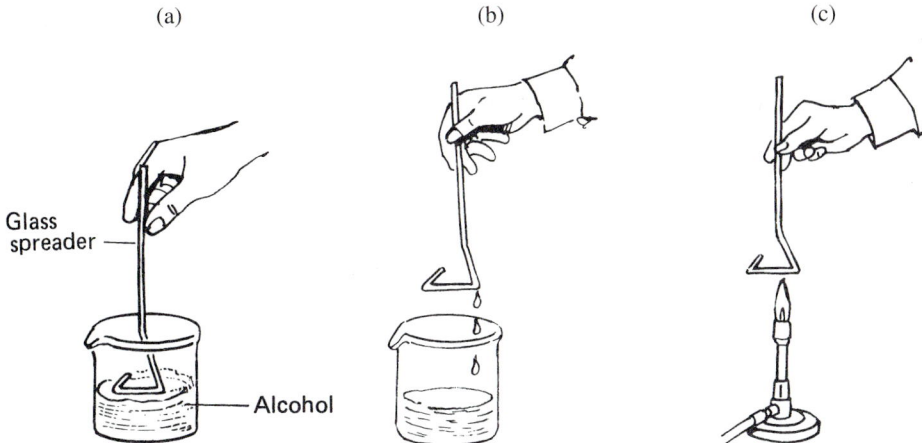

Figure 40–4
Steps in sterilizing a bent glass rod spreader. (a) Inserting the spreader portion of the rod into alcohol. (b) Draining. (c) Flaming and sterilizing.

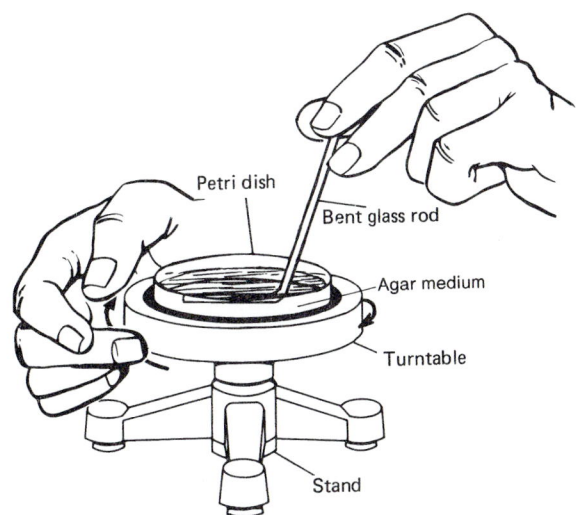

Figure 40–5
The use of a spreader and turntable to inoculate an agar plate. After the inoculum is placed on the agar surface, the spreader is applied to it and the plate gently spun on the turntable. The spreader is held stationary while the plate is being rotated. The inoculum is thus spread easily over the entire surface of the medium.

Procedure 2: Conjugation

❑   1. Label 1 sterile tube F^- + Hfr. With a sterile 1-mL pipette, aseptically transfer 1 mL of the *E. coli* F^- strain into the labeled sterile test tube.

❑   2. With another sterile 1-mL pipette, aseptically transfer 1 mL of the *E. coli* Hfr strain into the same labeled test tube.

❑   3. Gently shake the mixture of strains. Note the time and incubate the tube for 1 hour in a 37°C water bath.

❑   4. While the *E. coli* mixture is incubating, label 3 9-mL water blanks 10^{-1}, 10^{-2}, and 10^{-3}.

❑   5. Label three minimal medium agar plates for each of the dilutions: 10^{-1}, 10^{-2}, and 10^{-3}.

❑   6. After the 1-hour incubation, using a sterile 1-mL pipette, aseptically transfer 1 mL of the *E. coli* mixture to the tube marked 10^{-1} (see Figure 40–3).

❑   7. Gently shake the contents, and with another sterile 1-mL pipette, transfer 1 mL of the 10^{-1} (1:10) dilution to the tube marked 10^{-2} (see Figure 40–3).

❑   8. Again, shake the contents, and with still another sterile 1-mL pipette, transfer 1 mL of the 10^{-2} (1:100) dilution to the tube marked 10^{-3} (1:1,000). Shake the contents of the tube quickly and gently. With the same pipette, aseptically remove 1 mL of the dilution.

❏   9. With individual sterile 1-mL pipettes, remove 0.1 mL of each dilution and place them on the center of separate minimal agar plates.

❏ 10. Sterilize the bent-glass rods and spread the 0.1 mL of each dilution as described in steps 3 through 7 in Procedure 1.

❏ 11. Label each plate to indicate the dilution used. Invert and incubate the plates at 37°C until the next laboratory period or for at least 48 hours.

❏ 12. After incubation, examine each control and conjugation plate for growth. Count the number of colonies.

❏ 13. Enter all findings in Table 40–1 and answer all questions in the Results and Observations section.

Results and Observations

1. Complete Table 40–1 by indicating the colony counts for each medium used.

2. Which of the inoculated plates served as controls? _____

Table 40–1

Colony Counts

Preparation	Medium	
	Minimal Medium	Complete Medium
10^{-1}, F⁻, and Hfr		
10^{-2}, F⁻, and Hfr		
10^{-3}, F⁻, and Hfr		
10^{-1}, F⁻		
10^{-1}, Hfr		

3. Complete Table 40–2 by indicating which of the organisms used in this experiment would be expected to grow on the media described.

Table 40–2

Growth Predictions

Medium	Organism(s) Expected to Grow
Minimal medium without streptomycin	
Minimal medium with leucine, threonine, and streptomycin	
Minimal medium with leucine	
Complete medium with leucine and threonine	

Laboratory Review 40   Bacterial Conjugation

1. List and define three types of gene transfer.

   a. _____

   b. _____

   c. _____

2. What are two major differences between the recombination by conjugation processes of prokaryotic and eukaryotic microorganisms?

   a. _____

   b. _____

3. a. What is a plasmid? _____

   _____

   b. List four plasmid-mediated (controlled) traits.

   i. _____   iii. _____

   ii. _____   iv. _____

4. What types of pili are involved with the cell-to-cell transfer of plasmids? _____

   _____

5. What does the term *plasmid-curing* mean? (Refer to your text.) _____

   _____

Key Terms

conjugation (kon-jū-GĀ-shun): a gene transfer process in which a donor bacterial strain comes into physical contact with a recipient strain and transfers genetic material

F plasmid (PLAZ-mid): an extrachromosomal genetic element that encodes a number of DNA transfer functions, including donor pili formation

F⁻ cell: a recipient cell lacking the F plasmid

F⁺ cell: a donor cell carrying the F plasmid that is not attached to the bacterial chromosome

Hfr (high-frequency recombinant) cell: a bacterial strain that exhibits a high rate (frequency) of gene transfer and recombination during a mating process; the F plasmid is part of the bacterial chromosome

plasmid (PLAZ-mid): an extrachromosomal genetic structure that can replicate independently within a bacterial cell

recombination (rē-kom-bi-NĀ-shun): the exchange and incorporation of genetic information into a single genome resulting in the formation of new combinations of genes

transconjugant (trans-KON-jū-gant): an original parental strain that participates in conjugation

After completing this exercise, you should be able to:

1. Perform a bacterial transformation with plasmid DNA.
2. Understand the role of restriction enzymes (endonucleases).
3. Manipulate a micropipettor.
4. Define or explain the terms *endonuclease, Hind*II, plasmid, and transformation.

In the early 1950s, Salvador Luria, Giuseppe Bertani, and Jean Weigle found certain *Escherichia coli* strains to be resistant to infection by various bacterial viruses *(bacteriophages)*. The observed phenomenon seemed to be a property of the bacterial cell to restrict bacteriophage growth and replication. Some twelve years later, Werner Arber found that the bacterial resistance was due to an enzyme system that selectively recognizes and destroys foreign bacteriophage DNA, and also modifies or protects the bacterial cell's chromosomal DNA to prevent self-destruction. Other later investigations resulted in the isolation of *E. coli* extracts that efficiently cut or cleared bacteriophage DNA. These extracts contained the first known *restriction endonucleases*. These enzymes attack and digest the DNA of invading bacteriophages, but not of the invaded or host bacterial cell.

In 1970, Hamilton Smith and Kent Wilcox isolated a new *restriction endonuclease* from the bacterium *Haemophilus influenzae*. The restriction activity of this enzyme, named *Hind*II (Table 41–1) was found to be separate from the host DNA modification activity. *Hind*II cut or cleaved DNA in a predictable manner, at specific nucleotide sequence sites.

Restriction endonucleases have several applications in molecular biology. Their ability to serve as molecular scalpels to cut DNA in a precise and predictable manner has been useful in producing recombinant (joining) molecules, one type of which is known as *plasmids*. A plasmid, in molecular biological terms, is the simplest DNA molecule that can be used as a vehicle to carry foreign DNA sequences into bacteria or other types of host cells. Plasmids are circular molecules that range in length from 1,000 to 2,000 *nucleotide base-pairs* (**bp**), and are separate from the main chromosome in a bacterial cell.

The first recombinant plasmid was constructed in 1973 by Stanley Cohen and Annie Chang, and involved the combing of selected genes from the two plasmids, *p*SC101 and *p*SC102. (The *p* here represents *plasmid,* while the *SC* stands for Stanley Cohen.) The *p*SC101 was known to contain a gene for resistance to the antibiotic tetracycline, and *p*SC102 a

Table 41–1

Examples of Restriction Enzymes

Microbial Source	Enzyme Abbreviation
Bacillus amyloliquefaciens	*Bam*HI
Escherichia coli	*Eco*RI
Haemophilus aegytius	*Hae*III
Haemophilus influenzae	*Hind*II
Serratia marcescens Sb	*Sma*

gene for resistance to kanamycin. The Cohen and Chang procedure included cutting these two plasmids with a restriction endonuclease *Eco*RI isolated from *Escherichia coli,* mixing the cut elements, and then rejoining the parts with the aid of an enzyme DNA ligase (Figure 41–1). The resulting, experimentally produced plasmid was used for the transformation of *Escherichia coli* cells. *Transformation* refers to inducing the uptake and expression of a foreign DNA sequence (gene). Proof of the successful transformation with the recombinant plasmid was provided by plating the treated *E. coli* cells on a nutrient medium containing both tetracycline and kanamycin, and finding colonies containing cells with resistance to both antibiotics. From the results, it was apparent that a recombinant DNA molecule had been successfully introduced into living bacterial cells.

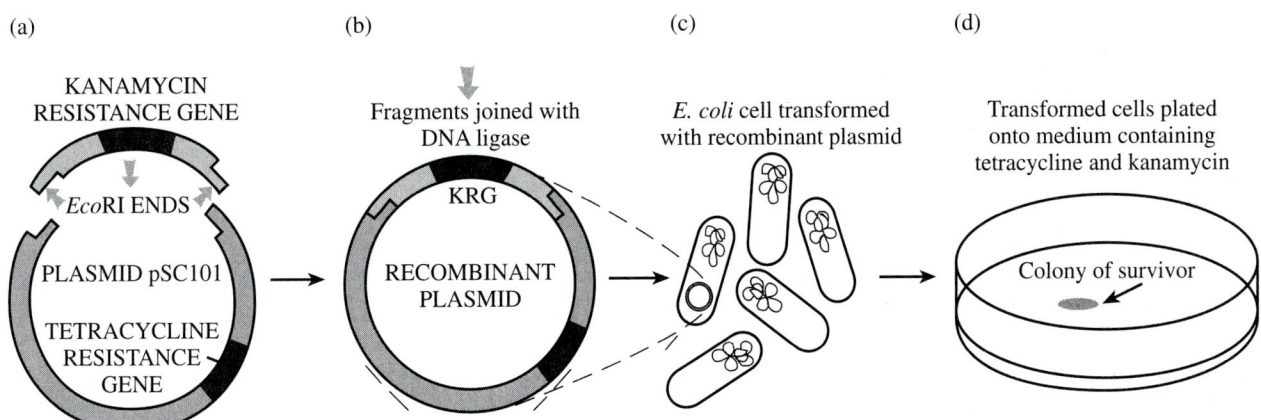

Figure 41–1

Steps involved in the production and expression of a recombinant plasmid. This procedure is based on the Boyer-Cohen-Chang experiment in 1973. (a) The position to be occupied by the kanamycin resistance gene (KRG). (b) The insertion of the KRG and the joining action of DNA ligase. (c) *E. coli* transformation with the recombinant plasmid. (d) Plating of transformed cells. Colonies are formed only by cells containing the recombinant plasmid.

This exercise will demonstrate a rapid method to transform a bacterium such as *Escherichia coli* with a commercially prepared plasmid containing the foreign gene for resistance to the antibiotic ampicillin. The use of a micropipettor and related equipment (Figure 41–2) also will be presented as part of the transformation procedure.

A. Demonstration of Equipment

Materials

The following items should be provided for general class use:

❏ 1. Two 0.5–10-μL *Eppendorf* micropipettors (see Figure 41–2a)

❏ 2. Two 100–1,000-μL *Eppendorf* micropipettors

❏ 3. Micropipettor tips and container

❏ 4. Microtubes or flex-tubes (see Figure 41–2b)

❏ 5. One microfuge (see Figure 41–2c)

(a) (b) (c)

Controls

Figure 41–2

Equipment items used in transformation and related procedures. (a) An assortment of Eppendorf micropipettors. (b) An example of a microtube (flex-tube) used to contain reaction mixtures and for centrifugation (pulsing). (c) A microfuge to spin-down reaction mixture.

(Photos courtesy of Brinkmann Instruments, Inc. Westbury, N.Y.)

Procedure

This procedure should be performed by students in groups of 4.

❏ 1. Examine each of the equipment items placed on demonstration. Pay particular attention to the correct way to hold the instrument while delivering a specific volume. (See Figure 41–3.)

❏ 2. Your instructor will demonstrate the features of the equipment.

❏ 3. Note the following parts of the Eppendorf micropipettors:

    ❏ a. Top-view digital display
    ❏ b. Volume setting dial
    ❏ c. Disposable tips

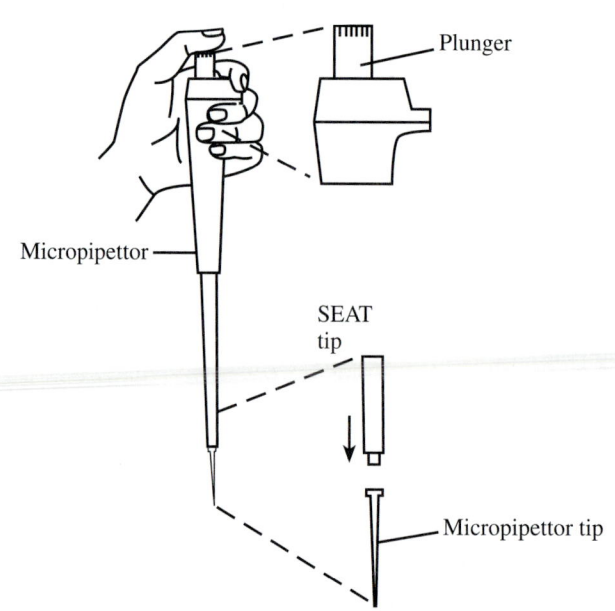

Plunger

Micropipettor

SEAT
tip

Micropipettor tip

Figure 41–3

The position of the hand while using a micropipettor.

B. Manipulation of Micropipettor

Materials

The following items should be provided for pairs of students:

❏ 1. One Eppendorf or related micropipettor with a volume range of 0.5–10 µL or 1–20 µL

❏ 2. One Eppendorf or related micropipettor with a volume range of 100 to 1,000 µL

❏ 3. Ten disposable micropipettor tips

❏ 4. Ten 1.5-mL disposable microtubes

❏ 5. Four test solutions with the volumes as specified:
   ❏ a. Solution 1: 800 µL
   ❏ b. Solution 2: 1,000 µL
   ❏ c. Solution 3: 1,300 µL
   ❏ d. Solution 4 (1% methylene blue): 1,300 µL

❏ 6. One microtube rack

❏ 7. Container for micropipettor and microtube disposal

❏ 8. One permanent marker

Procedure 1: Micropipetting Small Volumes

This procedure should be performed by students individually with each micropipettor. (See Table 41–2.)

❏ 1. Label 3, 1.5-mL microtubes A, B, C, respectively.

❏ 2. Firmly place a fresh micropipettor tip on the end of the micropipettor tip. (See Figure 41–3.)

❏ 3. Set the micropipettor volume dial to 4 µL and withdraw 4 µL from the test solution tube marked *1*. Withdraw the fluid in the following way: (See Procedure Diagram 37, steps 1 to 4.)
   ❏ a. Depress the micropipettor with the thumb plunger to the first stop and hold the pipettor in this position.
   ❏ b. Dip the micropipettor tip into solution 1.
   ❏ c. Gradually release the plunger to allow the fluid to be drawn into the tip.
   ❏ d. Carefully slide the pipette tip along the side of the tube to remove any unwanted droplets of fluid sticking to the tip's surface.

❏ 4. Expel the fluid into the microtube in the following way: (See Procedure Diagram 37, steps 5 to 9.)
   ❏ a. Touch the pipette tip to the inside surface of microtube A.
   ❏ b. Slowly depress the micropipettor plunger to the first stop and hold to expel the fluid.

Table 41–2

The DO NOTS in Using a Micropipettor

1. ***DO NOT*** rotate the volume adjusting knob past the upper or lower ranges specified by the instrument's manufacturer.

2. ***DO NOT*** force the volume adjusting knob. (If there is a problem rotating the volume knob tell your instructor.)

3. ***DO NOT*** use a micropipettor without a tip in place. (The precision piston that measures the volume of fluid can be ruined.)

4. ***DO NOT*** lay the pipettor down with a filled tip. (Fluid can run back into the precision piston.)

5. ***DO NOT*** allow the micropipettor plunger to snap back after fluid is either withdrawn or ejected.

6. ***DO NOT*** immerse the micropipettor barrel into fluid.

7. ***DO NOT*** flame the micropipettor tip.

❑ c. Depress the plunger to the next or second stop and hold to blow out any remaining fluid.

❑ d. Continue to hold the plunger in the depressed position as you remove the micropipettor from the microtube.

❑ 5. Repeat steps 3 and 4 with tubes B and C.

❑ 6. After the solution 1 fluid samples have been introduced into microtubes A, B, and C, eject the micropipettor tip into the container provided in the following way:

❑ a. Depress the micropipettor plunger beyond the second stop or depress a separate *tip-ejection button* if it is present on the pipettor. (See Procedure Diagram 37, step 10.)

❑ 7. Obtain a fresh tip to add solution 2 to tubes A, B, and C, respectively, according to the volumes indicated in the Table 41–3.

❑ 8. Repeat step 7 with solutions 3 and 4.

❑ 9. After all solutions have been added to the respective microtubes, mix the solution by one of the following methods:

❑ a. Sharply tap tube bottom on bench top. Be certain that the individual drops are pooled into one drop at the bottom of the tube.

❑ b. Place the microtubes into a microfuge and apply a short 10-second spin (pulse). Be certain tubes are placed in a balanced configuration in the microfuge rotor. (See Figure 41–4.) (Spinning tubes in an unbalanced position will damage the microfuge motor.)

❑ 10. Obtain a fresh micropipettor tip and set the micropipettor volume dial to 10 μL.

❑ 11. Carefully withdraw the fluid from microtube A. Check to see if the tip is just filled. This finding would indicate that the volumes were correctly added to the microtube.

Table 41–3

Micropipetting Small Volumes

Tube #	Solution 1	Solution 2	Solution 3	Solution 4
1	4 μL	5 μL	1 μL	—
2	4 μL	5 μL	—	1 μL
3	4 μL	4 μL	1 μL	1 μL

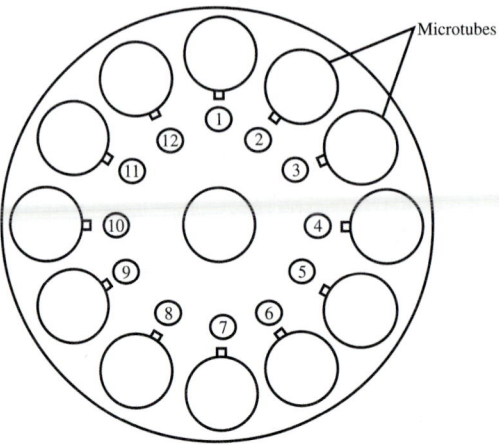

Figure 41–4

Positioning microtubes in a microfuge.

❑ 12. Check to see if any fluid remained in the microtube.

❑ 13. Repeat steps 11 and 12 with microtubes B and C.

❑ 14. Enter all findings and answer the questions in the Results and Observations section and Laboratory Review.

Procedure Diagram 37
The Use of a Digital Pipette

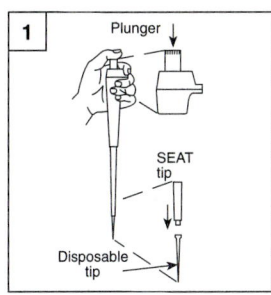

1. The correct way to hold and manipulate a digital pipette. Note the position of the thumb.

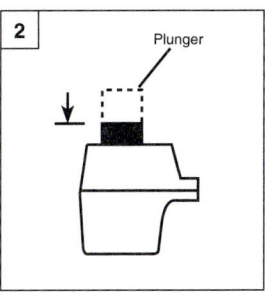

2. Depress the plunger to the first stop.

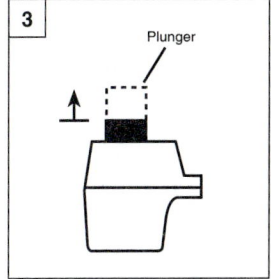

3. Release the plunger slowly.

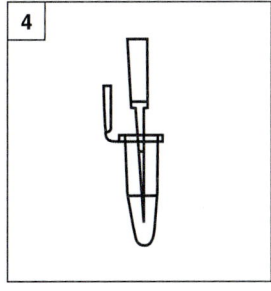

4. Withdraw the sample into the micropipette tip.

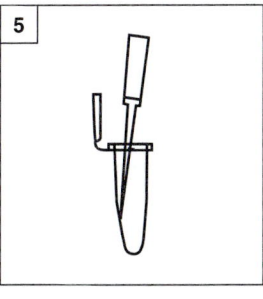

5. Touch the tip to the side of the tube.

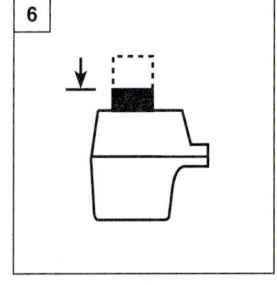

6. Depress the plunger to the first stop.

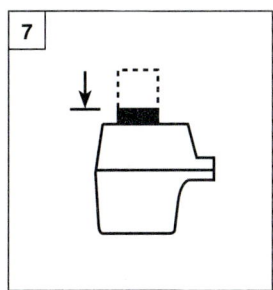

7. Depress the plunger to the second stop.

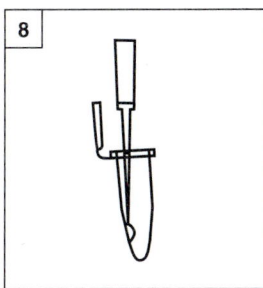

8. Touch the droplet to the side of the tube.

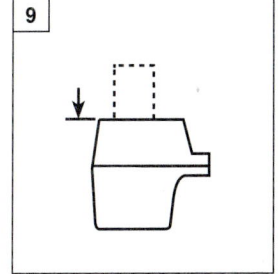

9. Depress the plunger to the third stop.

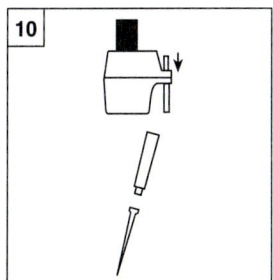

10. Depress the digital pipette tip ejector.

Procedure 2: Micropipetting Large Volumes

This procedure should be performed by students individually.

❏ 1. Label 2, 1.5-mL microtubes A and B, respectively.

❏ 2. Firmly seat the appropriate micropipettor tip on the end of the 100–1,000-μL micropipettor.

❏ 3. Set the micropipettor volume dial to 250 μL and withdraw 250 μL from the test solution marked 1. (Refer to Procedure Diagram 37.)

❏ 4. Expel the fluid into the microtube A. (See Procedure Diagram 37.)

❏ 5. Set the volume dial to 100 μL and micropipette solution 1 into microtube B.

❏ 6. Discard the micropipettor tip.

❏ 7. Obtain a fresh tip to add the volumes of solution 2 to the microtubes A and B as specified in Table 41–4.

❏ 8. Repeat step 7 with the remaining solutions.

Table 41–4

Micropipetting Large Volumes

Tube	Solution 1	Solution 2	Solution 3	Solution 4
A	250 μL	100 μL	200 μL	450 μL
B	100 μL	300 μL	500 μL	100 μL

❏ 9. After all solutions have been added to the respective microtubes, mix the solutions by one of the following methods:

❏ a. Sharply tap tube bottom on bench top. Be certain that the individual drops have pooled into one drop at the bottom of the tube.

❏ b. Place the microtubes into a microfuge and apply a short 10-second pulse. Be certain tubes are placed in a balanced configuration in the microfuge rotor. (See Figure 41–4.) (Spinning tubes in an unbalanced position will damage the microfuge motor.)

❏ 10. Obtain a fresh micropipettor tip and set the micropipettor volume dial to 1,000 μL.

❏ 11. Carefully withdraw the fluid from microtube A. Check to see if the tip is just filled. This finding would indicate that the volumes were correctly added to the microtube.

❏ 12. Check to see if any fluid remained in the microtube.

❏ 13. Repeat steps 11 and 12 with microtube B.

❏ 14. Enter all findings and answer the questions in the Results and Observations section and Laboratory Review.

C. Rapid Colony Transformation

Materials

The following items should be provided for pairs of students:

1. One 0.5–10-μL micropipettor and disposable tips
2. One 100–1,000-μL micropipettor and disposable tips
3. Two 15-mL screwcap culture tubes
4. One 24-hour streak plate of *Escherichia coli*, MM294 (Carolina Biological Supply Company)
5. Six hundred μL 50 mMCaCl$_2$
6. One 250-mL beaker with crushed ice
7. Twenty-five-μL plasmid (p) AMP (0.005 μg/μL)
8. Six hundred-μL Luria-Bertani (LB) broth
9. Two LB plates
10. Two LB/ampicillin (amp) plates
11. Container with disinfectant
12. One permanent marker

The following items should be provided for general class use:

1. Two hundred-and-fifty mL beakers
2. Crushed or cracked ice
3. 95% alcohol in 500-mL beakers for cell spreader sterilization
4. 37°C incubator
5. 42°C incubator
6. Test tube racks
7. Cell spreaders (metal or glass)
8. Quebec Colony Counter or similar instrument

Procedure

The following procedure should be performed by students in pairs.

1. Obtain 2 sterile 15-mL screwcap tubes, and label one "+ pAMP," and the other "− pAMP."
2. With the 100–1,000-μL micropipettor, add 250 μL of a CaCl$_2$ solution to each tube. (Refer to *Procedure Diagram 37*.)
3. Place both tubes into a beaker with ice.
4. Aseptically remove two large bacterial colonies from the streak plate provided and transfer them to the + pAMP tube.
5. Gently shake the tube until no visible clumps of cells are seen. (The suspension should show a uniform cloudiness.)
6. Place the + pAMP tube back into the beaker with ice.
7. Repeat steps 4, 5, and 6 with the − pAMP tube.
8. With a 1–10-μL micropipettor add 10 μL of the pAMP solution to the + pAMP tube (Refer to Procedure Diagram 37.)
9. Discard the micropipettor tip as directed by your instructor and return the + tube to the ice.
10. Keep both tubes on ice for an additional 15 minutes.
11. While the tubes are in ice, obtain 2 LB plates and 2 LB plates with ampicillin (LB/AMP), and label them according to the following scheme:
    a. One LB: "+" (This is a positive control.)
    b. One LB: "−" (This is also a positive control.)
    c. One LB/amp: "+" (This is the experimental plate.)
    d. One LB/amp: "−" (This is a negative control.)
12. Remove both tubes from the ice and immediately place them into a 42°C water bath for 90 seconds.

❏ 13. Remove both tubes and place them into the ice for 1 minute.

❏ 14. Remove the tubes and place them into a rack at room temperature.

❏ 15. With a 100–1,000 μL micropipettor and sterile tip, add 250 μL of LB broth to each tube. Discard the tip as indicated by your instructor.

❏ 16. Gently tap each tube to get a uniform suspension.

❏ 17. With a micropipettor and sterile tip, place 100 μL of the cell suspension from the − pAMP tube onto the centers of the − LB and − LB/amp plates, respectively.

❏ 18. Immediately sterilize the cell spreader and spread the cells on the surface of the − LB plate. (Refer to Figures 40–4 and 40–5.)

❏ 19. Repeat step 18 with the − LB/amp plate.

❏ 20. With a micropipettor and a sterile tip, transfer 100 μL of the cell suspension from the + pAMP tube to the centers of the + LB, and + LB/amp plates, respectively.

❏ 21. Repeat the spreading step described in step 18 with both plates.

❏ 22. Sterilize the spreader before putting it down.

❏ 23. Allow the cell suspensions to become absorbed into agar. This will take a few minutes.

❏ 24. Invert the plates and incubate them at 37°C for 18–24 hours.

❏ 25. After incubation, count the number of individual colonies on each plate. This can be easily done by marking the locations of the colonies or the bottom of each plate. If the transformation was successful, between 50 and 500 colonies should be on the + LB/amp plate. Enter your findings in the Results and Observation section.

❏ 26. Answer the questions in the Results and Observations and Laboratory Review sections.

Results and Observations

Manipulation of Micropipettor

1. Were all micropipettor tips filled accurately, thus indicating a total volume of 1,000 μL? _____

   _____

   a. If not, which ones showed an inaccurate volume(s)? _____

   _____

   b. Explain the finding(s) in 1a. _____

   _____

2. List two possible sources of error that might occur with the use of a micropipettor.

   a. _____

   b. _____

Rapid Colony Transformation

1. Enter your findings in Table 41–5.

2. Which of the plates demonstrated transformation? _____
    _____

3. a. Using a bacterial count of 50–500, was your transformation experiment successful? _____

    b. If not, explain why. _____

Table 41–5

Plate	Number of Colonies
LB/amp"+" (experimental plate)	
LB/amp"−" (negative control)	
LB"+" (positive control)	
LB"−" (negative control)	

Laboratory Review 41   **Rapid Bacterial Colony Transformation with Plasmid DNA**

1. What is a restriction endonuclease? _____
    _____

2. How are endonucleases named? _____
    _____
    _____

3. What is a plasmid? _____
    _____

4. How are plasmids used to transform a bacterium? _____
    _____

5. Explain or define the following:

    a. µL _____

    b. µg _____

    c. mL _____

    d. base pair _____

Key Terms

bacteriophage: a virus that infects bacteria; particular forms such as λ (lambda) are used as vectors for cloning DNA

bp (base pair): a pair of complementary nitrogenous bases in a DNA molecule; also the unit of measurement for DNA sequences

EcoRI: one type of endonuculease obtained from *Escherichia coli*

HindII: an endonuclease obtained from *Haemophilus influenzae*

plasmid (PLAZ-mid): an extrachromosomal genetic structure that can replicate independently within a bacterial cell

transformation: a process in which the genetic constitution (genome) of a cell is changed through the incorporation of DNA from the environment or by artificial means; it was the first mechanism of genetic exchange to be discovered

After completing this exercise, you should be able to:

1. Distinguish bacterial transduction from transformation and conjugation.
2. Perform a generalized transduction procedure.
3. Outline an experiment to show bacterial transduction.

Transduction is a gene transfer process in which a bacteriophage (bacterial virus) established in one bacterial strain, the *donor,* picks up small segments of bacterial genetic information and, upon infection of another bacterial strain, the *recipient,* transfers this genetic information to this strain (Figure 42–1). Recipient cells that acquire a donor trait are called *transductants.* The bacteriophage involved in transduction is a defective virus and is referred to as a *transducing particle.* It contains DNA from the bacterial host's set of genes replacing part or all of the phage's DNA.

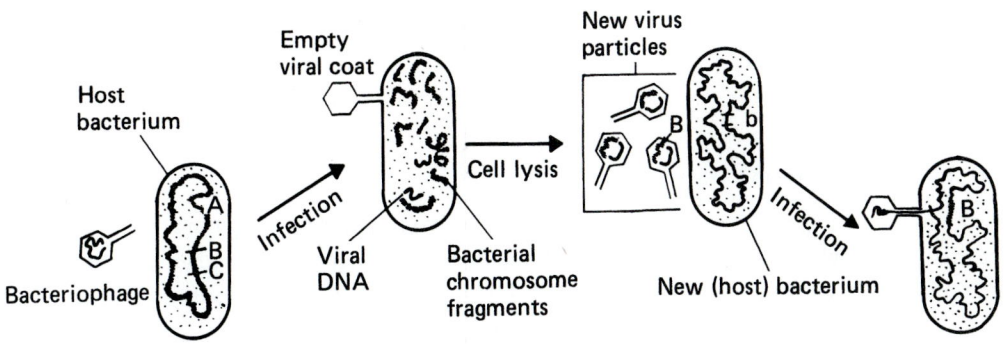

Figure 42–1

Bacterial transduction. In this process, an infecting bacteriophage or bacterial virus introduces a DNA segment from a donor bacterium into a recipient cell. The bacteriophage picks up a DNA fragment carrying an alternate form of a gene from a host bacterium and incorporates it instead of viral DNA *during the formation of new virus particles.* When a virus of this type infects another bacterium (recipient), the bacterial DNA is introduced into the recipient, where it is inserted into its chromosome.

Bacterial viruses are classified as either *virulent* or *temperate.* With virulent viruses, all infected cells undergo a rapid lytic cycle resulting in cellular death and the liberation of new virus particles. A temperate virus can, but ordinarily does not, lyse its host cell. Instead, the viral genetic material is integrated into the host's bacterial chromosome and replicates in synchrony with the bacterial structure. The bacterial host is referred to as being *lysogenic,* and the integrated viral genetic material is called a *prophage.* Temperate bacteriophages capable of establishing this lysogenic state probably account for most transduction-controlled gene flow among bacteria in nature.

Two types of transduction are known: *specialized* and *generalized.* In the specialized type, only donor cell genetic traits next to the integrated prophage are capable of being transduced. With generalized transduction, any donor cell trait, whether it be on a chromosome or an extrachromosomal segment of DNA, as in the case of *plasmids,* is capable of being transduced.

In this exercise, one strain of *Salmonella typhimurium,* ATCC e-23564, will serve as the donor culture for the transducing bacteriophage P-22. This bacterial strain is capable of growth on minimal medium agar. Another strain, *S. typhimurium* ATCC e-23591, will be used as the recipient culture. This strain is auxotrophic and requires the amino acid tryptophan for growth on minimal medium agar. In addition, it is lysogenic, because it carries the P-22 prophage. The prophage provides immunity or protection from lysis by external P-22. The P-22 bacteriophage is capable of transducing the tryptophan genetic marker into the cells of a recipient bacterial strain.

Because *Salmonella typhimurium,* the bacterial species used in this exercise, can cause diarrhea and food poisoning, great care should be used in handling the cultures.

Materials

The following items should be provided per 2 students:

❏ 1. One mL of a 24-hour trypticase soy broth culture of the recipient *Salmonella typhimurium* **try⁻** (ATCC e-23591)

❏ 2. One-half (0.5) mL of the P-22 transducing bacteriophage preparation

❏ 3. Four minimal medium agar plates

❏ 4. Three sterile 1.0-mL cotton-plugged pipettes and rubber bulbs

❏ 5. Two sterile glass spreading rods

❏ 6. One 250-mL beaker of alcohol for glass rod sterilization

❏ 7. One container with disinfectant large enough for pipette disposal

❏ 8. One marking pen or pencil

The following items should be provided for general class use:

❏ 1. Rotary spreaders

❏ 2. One water bath set at 37°C

❏ 3. Test tube racks

Procedure 1: The Controls

All procedures should be performed by students in pairs. Before beginning this procedure, your instructor will demonstrate the proper methods of pipetting and spreading an inoculum over an agar surface. Used pipettes should be placed in the disinfectant container. In addition, refer to Figure 40–4 for the steps involved in sterilizing a glass spreading rod.

❏ 1. With a sterile 1.0-mL pipette and bulb, aseptically transfer 0.1 mL of the recipient *S. typhimurium* culture to the centers of 2 minimal agar plates.

❏ 2. With the aid of a rotary spreader (see Figure 40–5), use a sterilized glass rod to spread the 0.1-mL culture evenly over the entire agar surface. Sterilize the rod as described in Exercise 40 and Figure 40–3.

❏ 3. Sterilize the glass spreading rod again, allow it to cool, and then spread the 0.1 mL of culture over the surface of the second minimal agar plate.

❏ 4. Label these plates as Control Plate 1 and Control Plate 2, respectively. Incubate both plates at 37°C for 48 hours. These plates will show the number of tryptophan reverse mutants present in the auxotrophic recipient culture. No more than 6 colonies should appear after incubation.

❏ 5. After incubation, count the number of colonies on each plate. Record the counts on Table 40–1 and obtain an average value.

Procedure 2: Transduction

❏ 1. With a sterile 1.0-mL pipette and bulb, aseptically add 0.1 mL of the transducing P-22 bacteriophage preparation to the remaining 0.8 mL of the recipient *S. typhimurium* culture.

❏ 2. Gently shake the mixture and incubate it for 15 minutes in the 37°C water bath.

❏ 3. With another sterile 1.0-mL pipette and bulb, transfer 0.1 mL of the *S. typhimurium* P-22 mixture to the centers of two minimal agar plates.

❏ 4. With the aid of a rotary spreader (see Figure 40–5), use a sterilized glass rod to spread the 0.1-mL culture evenly over the entire agar surface. Sterilize the rod as described in Exercise 40 and Figure 40–4.

❏ 5. Sterilize the glass rod once more and use it to spread the mixture over the surface of the second minimal agar plate. Sterilize the glass rod again for safety purposes before putting it down.

❏ 6. Label these plates Experiment 1 and Experiment 2, respectively.

❏ 7. Incubate both plates at 37°C for 48 hours.

❏ 8. After incubation, count the number of colonies formed and enter the findings in Table 42–1.

❏ 9. Determine the average number of colonies and, using the formula in the Results and Observations section, find the total number of *transductants* formed.

❏ 10. Answer the questions in the Results and Observations section.

Results and Observations

1. Enter your findings in the following table.

Table 42–1

Transduction Results

	Minimal Medium Plate					
	Control 1	Control 2	Average 1 and 2	Experiment 1	Experiment 2	Average 1 and 2
Colony Count(s)						

2. How many transductants were produced?

$$\text{Number of transductants} = \frac{\text{Average number of colonies}}{\text{on experimental plates}} - \frac{\text{Average number of colonies}}{\text{on control plates}}$$

Number of transductants = _____ − _____

3. What would have happened if the recipient *S. Typhimurium* strain had been resistant to the transducing P-22 phage? _____

_____

Laboratory Review 42   Bacterial Transduction

1. What are the general types of bacteriophages? _____

   _____

2. Describe two types of transduction known. _____

   _____

   _____

   _____

3. What is phage conversion? (Refer to your text.) _____

   _____

   _____

4. Define *transposon.* (Refer to your text.) _____

   _____

   _____

5. What type of bacteriophage is involved with transduction? _____

   _____

Key Terms

auxotroph (OKS-ō-trōf): a mutant that has a specific growth factor requirement

bacteriophage (bak-TĒ-rē-ō-fāj): a bacterial virus; phage

generalized transduction: the bacteriophage transfer of any bacterial gene that can be incorporated into the viral DNA

lysogeny (lī-SOJ-eh-nē): The relationship between a temperate bacteriophage and a bacterium, in which the bacterium is not destroyed by the phage

plasmid (PLAZ-mid): A circular, double-stranded DNA molecule capable of independent replication within a bacterial cell

prophage (PRŌ-fāj): bacteriophage DNA found within a lysogenic bacterium

specialized transduction: the bacteriophage transfer of only specific bacterial genes

temperate bacteriophage: refers to a bacterial virus that resides in a host lysogenic bacterium and that does not cause the death of its host

transductant (trans-DUK-tant): a recipient bacterium that acquires a donor bacterium's trait

virulent bacteriophage: refers to a bacterial virus that does not cause lysogeny but the lysis of its host bacterium during viral replication

The Ames Test and the Detection of Chemical Carcinogens

After completing this exercise, you should be able to:

1. Perform spot tests for the detection of mutagenic agents.
2. Interpret the results of a spot test for mutagenicity detection.
3. Explain the basis of the Ames test.
4. List two limitations of the spot test technique.

Cancer is recognized as a collection of more than 100 different diseases. The incidence of many cancers is increased by environmental exposure to specific chemicals known as *carcinogens.* Rapid identification of such compounds, and subsequent reduction of human exposure to them, are of major public health importance. Unfortunately, many carcinogens have been recognized from epidemiologic studies in humans, frequently long after the exposure has occurred, or from the results of expensive, time-consuming, and often inconclusive or confusing laboratory animal studies.

Many chemical carcinogens have been shown to cause *mutations.* Moreover, the reverse situation was also found to be true in several situations: several mutagenic chemicals were found to be potentially carcinogenic. Being sufficiently impressed with the correlation between carcinogenesis and mutagenesis, Dr. Bruce N. Ames and his colleagues developed an inexpensive, very sensitive, reliable, and relatively simple test for the detection of possible carcinogens as mutagenic agents using bacterial test systems.

The Ames test, which is another microbiological assay, utilizes three main components: the *tester bacterium,* the *test chemical* to be studied, and *mammalian (usually rat) liver extract.* While in principle any bacterial species in which mutants can be readily detected may be used, the Ames test uses special mutant strains of the gram-negative rod *Salmonella typhimurium.* These organisms, referred to as *his⁻*, are unable to synthesize the amino acid histidine and will not grow unless it is provided in a growth medium. This property makes it relatively simple to detect cells that mutate from a *his⁻* (histidine dependent) cell to a *his⁺* (histidine independent) cell. If a *his⁻* culture is placed onto a minimal medium lacking the amino acid, only those cells that have reverted to the *his⁺* state will form colonies. *S. typhimurium* strains used in the Ames test have other mutations, making them more sensitive to mutagens. One of these is the synthesis of a short lipopolysaccharide that alters bacterial cells walls so that chemicals can penetrate them more readily. The liver extract in the test serves as a source of enzymes that convert nonmutagenic test chemicals into active mutagens that can rapidly enter the tester bacteria and cause mutations within them.

The *standard plate Ames test* is carried out by mixing a tester bacterial culture with the tester compound along with an extract of animal liver. This mixture is then placed onto a medium lacking histidine to determine the frequency of mutation to histidine independence. If the chemical being tested is mutagenic, the reversion rate to histidine independence is greatly increased.

In this exercise, the *spot* or *disk Ames technique* (Figure 43–1) will be used. It is one of the simplest ways to test compounds for mutagenicity and is particularly adaptable for the initial rapid screening of large numbers of compounds in a short period of time. A positive result in a spot test should be considered adequate evidence for mutagenicity only if there is a large increase in colonies (many times the number of spontaneous revertants) around the spot (Figure 43–1*b*).

The spot test is primarily a qualitative test and, although very useful, has certain limitations. It can be used only for the detection of chemicals that are diffusible in the agar. Thus, most water-insoluble chemicals are not easily detected by this procedure. The test is also much less sensitive than the standard plate test since relatively few bacteria on the plate are exposed to the chemical at any particular dose level.

(a)

(b)

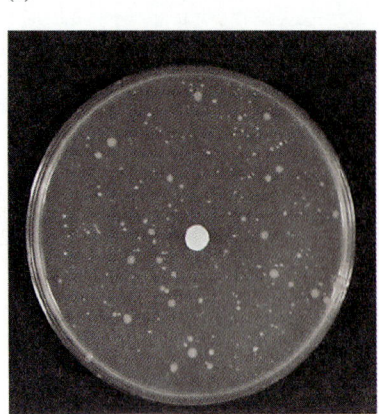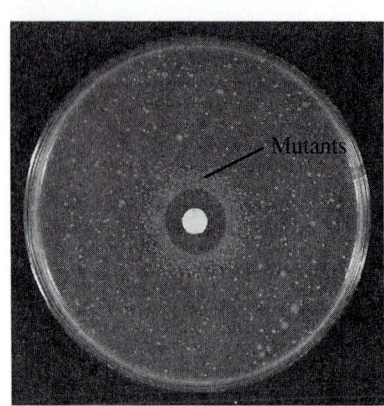

Mutants

Figure 43–1

The results of the Ames test to evaluate the mutagenicity of a chemical. Both plates were inoculated with a histidine requiring bacterial mutant. The medium does not contain histidine, so that the only bacteria that revert back to the wild type are nonhistidine requiring. Spontaneous mutants appear on both plates but the chemical on the filter paper disc in the test plate has caused an increase in mutation rate. This is evident by the large number of bacterial colonies surrounding the disc. (a) A control plate. (b) A mutagenic effect of a mutagen (*tester chemical*).
(From Mamber, S., B. Kolek, K.W. Brookshire, D.P. Bonner, and J. Fung-Tomc. *Antimicro. Agents and Chemother.* 37:213 217, [1993].)

Materials

The following items should be provided per 4 students:

- ❏ 1. Forty-eight-hour trypticase soy broth culture of *Salmonella typhimurium (his⁻)* [ATCC e29630]
- ❏ 2. Five minimal medium agar plates
- ❏ 3. One trypticase soy agar plate
- ❏ 4. One mL of a sterile biotin-histidine solution
- ❏ 5. Three sterile, cotton-plugged, 1-mL pipettes with rubber bulbs, or pipette pump

- ❏ 6. Six sterile disposable Pasteur pipettes with rubber bulbs
- ❏ 7. One sterile microsyringe or a similar delivery device
- ❏ 8. One sterile pair of forceps
- ❏ 9. One marking pen or pencil
- ❏ 10. One container with disinfectant for pipette disposal

The following items should be provided for general class use:

- ❏ 1. Sterile filter paper disks in a sterile Petri plate
- ❏ 2. Water baths set at 45°C for soft agar
- ❏ 3. Desiccator jars for prepared slides
- ❏ 4. Sterile liver extract in screwcap tubes
- ❏ 5. Test chemicals such as the following in screwcap tubes or well-stoppered containers (these chemicals should be kept in a fume hood):
  - ❏ a. Dimethyl-alpha-naphthylamine
  - ❏ b. Para-dimethyl-amine benzaldehyde
  - ❏ c. Tetramethyl-para-phenylene diamine hydrochloride
  - ❏ d. Saturated saccharin solution
  - ❏ e. Physiological saline
  - ❏ f. Sterile distilled water

Procedure

The following procedure is to be performed by students in groups of 4. (*Note:* Certain chemicals used in this exercise should be assumed to be potential carcinogens. *Therefore, take particular care in their handling and disposal.*)

> ## Additional Technique required for this Portion of the Exercise:
>
> ❏ The Use of Pipette Pump, Procedure Diagram 11, Exercise 6

❏ 1. With a marking pen or pencil, prepare a duplicate set of two minimal medium agar plates by dividing the bottom of each plate into 3 sections.

    ❏ a. Label 1 section for each of the test chemicals provided.

    ❏ b. Label 1 set with the letter L for liver extract. These plates will be used to show the effects of this material.

❏ 2. Prepare 4 tubes of soft agar by adding 0.2 mL of the sterile biotin-histidine solution provided with a sterile 1-mL pipette. Dispose of the pipette as directed. Place the tubes in the 45°C water bath until needed.

❏ 3. Take 2 tubes of soft agar. With a microsyringe or a simlar device, add 1 microliter (μL) of the liver extract to each tube.

❏ 4. With a sterile 1-mL pipette and bulb or pipette pump, add 0.5 mL of the *S. typhimurium* culture to each tube. Dispose of the pipette in the disinfectant container. (Refer to Procedure Diagram 11.)

❏ 5. Gently shake these tubes and immediately pour the contents of 1 tube over each surface of a minimal medium agar plate marked with an **"L"**. Rotate the plates using a circular motion so that the soft agar will overlay the minimal agar evenly.

❏ 6. Take the remaining tubes of soft agar and, with a new sterile 1-mL pipette, add 0.5 mL of the *S. typhimurium* culture to each one.

❏ 7. Gently shake these tubes and immediately pour the contents of 1 tube over each surface of a minimal medium agar plate not marked with an **"L"**. Rotate the plates using a circular motion so that the soft agar will overlay the minimal agar evenly.

❏ 8. With sterile forceps, place a sterile filter paper disk on the agar surface of each labeled section of the 4 prepared plates.

❏ 9. Take the 4 plates to the laboratory area containing the test chemicals.

❏ 10. With individual Pasteur pipettes and bulbs, wet each disk with 2 drops of its respective test chemical. Dispose of the pipettes as directed.

❏ 11. Streak 1 minimal medium and 1 trypticase soy agar plate with the *S. typhimurium* culture. Label each plate accordingly.

❏ 12. Place all plates into a desiccator jar or other container and incubate them at 37°C for 48 hours.

❏ 13. After incubation, examine each test chemical plate for the presence of large colonies around individual disks. Record the number in Table 43–1. In addition, note any large zones of growth inhibition around the disks. This type of result indicates a toxic effect.

❏ 14. Count the number of large colonies on the 2 control plates, minimal medium agar, and trypticase soy agar.

❏ 15. Answer the questions in the Results and Observations and Laboratory Review sections.

Results and Observations

1. Complete Table 43–1 by entering the findings of the exercise.

Table 43–1

Colony Counts

| | Large Colony Count | |
Test Chemical	Plate(s) with Liver Extract	Plate(s) without Liver Extract

2. Did any of the test chemicals used inhibit growth? If so, which ones? _____

3. What were the large colony counts found on the following media?

   a. Minimal medium agar _____

   b. Trypticase soy agar _____

4. a. Were the large colony counts with plates containing liver extract greater or less than the counts with plates not containing the extract? _____

   _____

   b. Explain why differences occurred. _____

   _____

   _____

Laboratory Review 43   The Ames Test and the Detection of chemical Carcinogens

1. Distinguish between a carcinogen and a mutagen. _____

    _____

    _____

2. List 4 known examples of carcinogens.

    a. _____   c. _____

    b. _____   d. _____

3. List 4 known examples of mutagens.

    a. _____   c. _____

    b. _____   d. _____

4. a. What is the microorganism used in the Ames test? _____

    _____

    b. What distinctive properties does this bacterium have? _____

    _____

    _____

5. What advantages does the Ames test have over wide-scale animal studies used for the detection of carcinogens?
    List 2.

    a. _____   b. _____

Key Terms

carcinogen: a cancer-producing substance

epidemiological: refers to the study of the causes, incidence, distribution, and control of disease in a population

minimal medium: a medium that allows growth of a parent culture or prototroph and prevents growth of a mutant or auxotrophic culture

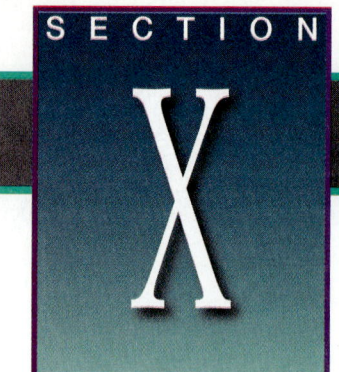

SECTION X

Applied Microbiology

It is characteristic of Science and Progress that they continually open new fields to our vision.

Louis Pasteur

This section is concerned with the use of microorganisms in basic processes that are of industrial importance. As the following exercises will show, the area of applied microbiology involves an intermingling of basic principles and techniques. Consideration will be given to the presence and distribution of microorganisms in foods, and to fermentation.

Nearly all foods contain microorganisms. Fresh and frozen foods whose processing involves numerous steps invariably have high numbers of microorganisms, especially where sanitation is poor and control measures are lacking. Untreated foods may be expected to contain varying numbers of bacteria, molds, or yeasts. The finding of microbes in substantial numbers often raises questions about the safety of a given food product. The numbers and types of microorganisms present in a finished product are affected by the following factors: (1) the general environment from which the food was originally obtained; (2) the microbial content of the food in the unprocessed state; (3) the sanitary conditions during processing; and (4) the adequacy of subsequent packaging, handling, and storage conditions. One exercise in this section deals with the types of microorganisms in foods, their distribution, and methods for their detection.

Over the centuries, fermentation has been a valuable process in the preparation and preservation of foods; the manufacture of alcoholic beverages such as beer, whiskey, and wine; and the production of certain industrially important chemicals. The end products of fermentation processes depend on the type of starting material, or substrate (sugar, starch, and so on), and on the type of microorganism used to perform the fermentation. Exercises in this section cover familiar microbial fermentative processes that are used in the making of sauerkraut, cheese, yogurt, and wine (Figure X–1).

Figure X–1
Each of the foods and beverages shown is a product of microbial fermentation.

The use of microorganisms to perform quantitative analytical tasks, or assays, has become an important area of analytical microbiology. Microorganisms are used when they can provide more specific, more sensitive, or more efficient assays than other known methods. Results obtained by microbial assay systems are also generally more consistent than those obtained in other ways. One such use of microorganisms is the Ames test, a rapid and economical means of detecting cancer- and/or mutation-causing chemicals. The Ames test was presented in Section IX.

The Microbiological Examination and Content of Dairy Products and Selected Foods

After completing this exercise, you should be able to:

1. Perform standard bacteria colony and microscopic counts.
2. Recognize the advantages and limitations of selected standard procedures used in the determination of bacterial loads.
3. Isolate microorganisms from a variety of foods.

One of the more obvious effects of high microbial counts in foods is a shortening of the product's shelf life. Various microbial types, including bacteria and fungi, contribute to such food spoilage. The incidence and types of yeasts and molds are not always reported in food analyses. Nevertheless, these organisms are invariably present unless they are destroyed during processing. This exercise consists of various parts designed to emphasize the presence of microorganisms in various types of commercially available food products and the methods used for their detection.

Because milk contains carbohydrates, fat, minerals, vitamins, and protein, and has a pH of approximately 6.8, it is susceptible to degradation by various species of microorganisms. Milk, as it is drawn from a healthy cow, contains a few bacteria, but unless modern, "closed" methods of milking are used, it may become contaminated by many more microorganisms. Sources of such contamination include the cow itself; the accumulation of dust, dirt, and manure in the milking area; and the various kinds of dairy equipment used—for example, the milking machine, milk cans, and the like—as well as the handlers. While some contamination is unavoidable, it can be substantially reduced by routine measures taken to ensure the cleanliness of the cow, the milking areas, and the various types of milking equipment. In addition, the individuals actually doing the milking and handling of the product should be well informed about the requirements for sanitary milk production. Disease-free dairy personnel and the use of sanitary equipment are important factors in the reduction of contamination from external sources.

Microbial contamination (the **microbial load**) can be reduced or eliminated by means of pasteurization (selective destruction) or heat sterilization. However, it is important to note that the keeping quality and safety of a food product depend on how successfully the reintroduction of microorganisms is prevented and their growth and multiplication are inhibited.

Several different kinds of microorganisms may be present in milk and related products. The numbers and kinds of these organisms fluctuate with the circumstances associated with the production of a particular milk and with the degree and source of contamination. Typically, microorganisms making up the normal flora in milk are gram-positive, nonmotile, microaerophilic, or anaerobic rods or cocci. The genera represented include *Lactobacillus, Microbacterium, Micrococcus,* and *Streptococcus.*

As a result of increased sanitation, pasteurization, and public health control, in recent years milk has been involved in fewer and fewer outbreaks of food-borne illness. However, a variety of diseases are potentially transmissible through milk and related products. Microorganisms belonging to several genera can cause milk-associated illnesses once they gain access to the product, multiply, and are consumed. Brucellosis, diphtheria, dysentery, listeriosis, Q fever, scarlet fever, and other streptococcal diseases, as well as tuberculosis and typhoid fever, are some of the diseases associated with milk and other dairy products.

To protect the consumer, most localities require that a number of standard microbiological tests be carried out periodically on raw and processed milk sold within their boundaries. The results obtained from such microbiological analyses provide useful information regarding the conditions under which the milk was produced and held. Among the tests commonly used are: (1) the standard plate count; (2) the Breed or direct microscopic count; (3) the reductase test; (4) coliform tests; and (5) tests for specific microbial pathogens. These procedures for the examination of milk as well as other products have been carefully evaluated and standardized. The American Public Health Association's *Standard Methods for the Examination of Dairy Products* lists every aspect of the testing procedures used in the examination of milk and

Table 44–1

Selected Dairy Product Standards

Product	Bacterial Count	Colony Count	Coliform Colony Count
Grade A raw milk for pasteurization before mixing with other milk[a]	100,000/mL	—	—
Grade A pasteurized milk products (except cultured products, such as yogurt, buttermilk, etc.)	20,000/mL	—	10/mL
Grade A pasteurized cultured products	—	—	10/mL
Certified raw milk[b]	—	10,000/mL	10/mL
Certified pasteurized milk (before pasteurization)	—	10,000/mL	10/mL
Certified pasteurized milk (after pasteurization)	—	500/mL	1/mL

[a]Recommendations for Grade A milk according to the U. S. Public Health Service.
[b]Recommendation for certified milk according to the American Association of Medical Milk Commissions, Inc.

other dairy items. Table 44–1 lists selected dairy product standards recommended by either the U.S. Public Health Service or the American Association of Medical Milk Commissions, Inc.

In this exercise, three commonly employed procedures will be considered, namely, the standard plate count, the Breed count, and the reductase test.

The **standard plate count** is an agar plate method of estimating bacterial populations. The procedure for milk is carried out in a manner similar to that used to determine the number of bacteria in water samples. Diluted samples of the material to be tested are mixed with standard quantities of melted and partially cooled agar medium. After 48 hours of incubation at 32°C, visible colonies are counted, usually with the aid of a colony counter. To determine the standard plate count milliliter, the total number of colonies is multiplied by the reciprocal of the dilution used. For example, if 30 colonies were found on a plate containing 1 mL of a 1:100 milk dilution, the standard plate count would be 3,000 (30 × 100). Usually, plates containing between 30 and 300 colonies are used for counting, because studies have shown that most accurate results are obtained with these count levels.

The **Breed count** is another procedure for determining the number of bacteria present in a milk sample. An investigator carries out this test by uniformly spreading 0.01 mL of the sample over a 1-cm^2 area. The resulting film is dried and stained with a special preparation. This preparation contains alcohol as a fixative, tetrachlorethane as a defatting agent, methylene blue as the stain, and glacial acetic acid. Bacterial counts are made using the oil-immersion objective. The investigator obtains the final bacterial count, that is, bacteria per milliliter (mL), by determining the reciprocal of the number representing the portion of 1 mL of the sample contained within 1 microscopic field (the microscopic factor) and multiplying it by the average number of bacteria per microscopic field.

The **reductase test** is based upon the oxidation-reduction (O/R) activities of the viable bacteria present in milk. In this procedure, an O/R indicator such as methylene blue is added to the milk. The indicator is blue in the oxidized state and leuco, or white, in the reduced condition. The time required to reduce the methylene blue is taken as an indication of the bacterial load; in other words, the more bacteria present, the faster the reduction. Standards have been established by public health agencies so that a certain degree of correlation can be obtained and the bacterial content determined within one working day. It must be mentioned that different bacterial species have different rates for the reduction of methylene blue. Thus, results of this test can be regarded only as semiquantitative. Figure 44–1 summarizes the procedure and the reduction reactions.

Meats are the most perishable of all important foods. One reason is the availability and abundance in meats of all nutrients required for the growth of bacteria, yeasts, and molds. Ground meats as well as multi-ingredient foods, such as hot dogs, meat pies, and sausages, have higher microbial loads than whole-meat foods such as steak. In general, the number of organisms in the final products reflects the overall microbial quality of the ingredients used. Different meat products will be examined in this exercise.

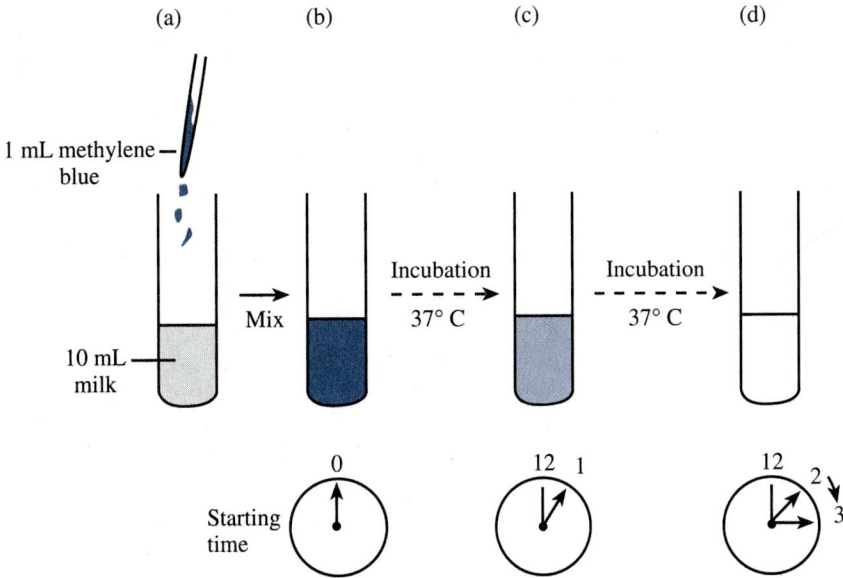

Figure 44–1

The Reductase Test (methylene blue reduction) performed with a milk sample. (a) One mL of dye solution is added to 10 mL of milk and the tube is incubated at 37°C. (b) Initially the milk solution is blue due to the dye solution. (c) After one hour it has become lighter as the dye is partly reduced. (d) In two to three hours (depending on the microbial content) the milk turns white. The end result shows that the milk sample contained a large number of bacteria. The bacteria fermented the lactose and liberated electrons eventually taken up by the dye, causing the sample to become white (reduced).

A. Standard Plate Count: Dairy Products

Materials

❏ 1. One sample of each of the following milk products in individual screwcap tubes per 2 students:
   ❏ a. High-count raw milk (2.5 mL)
   ❏ b. Certified raw milk (2.5 mL)
   ❏ c. Pasteurized milk (2.5 mL)

❏ 2. The following materials should be provided per 2 students:
   ❏ a. Three 99-mL sterile water blanks in either 6-ounce prescription bottles or other suitable stoppered containers
   ❏ b. Three sterile 1.0-mL pipettes
   ❏ c. Three sterile 1.1-mL pipettes
   ❏ d. Six sterile glass or plastic Petri plates
   ❏ e. Six tryptone glucose yeast agar (TGYA) deeps. (Each deep should contain a minimum of 10 mL of the medium.)
   ❏ f. Appropriate pipetting aids

Procedure

This procedure is to be performed by students in pairs. Refer to Figure 44–2 for an overall view of a standard plate count procedure.

❏ 1. Label 1 set of Petri plates 1:100 and 1:1,000 for each milk product provided.

❏ 2. Shake 1 of the milk products well, approximately 25 times, to ensure a well-dispensed specimen; then, using a sterile 1-mL pipette, remove a 1-mL sample.

❏ 3. Introduce this sample into a 99-mL sterile water blank, screw the cap of the bottle on tightly, and shake the contents rapidly (again, approximately 25 times).

❏ 4. With a sterile 1.1-mL pipette, remove 1.1 mL of the new diluted milk sample.

❏ 5. Place 0.1 mL into a sterile Petri plate marked for 1:1,000 dilution, and 1.0 mL into a second sterile Petri plate marked for 1:100 dilution.

❏ 6. Melt the tryptone glucose yeast agar deeps and then cool them to 45°C.

❑ 7. Pour the tryptone glucose yeast agar deep into each of the dishes containing a diluted milk sample. Mix well by gently rocking or rotating the contents of each plate, and allow the agar to harden (approximately 5 to 10 minutes).

❑ 8. Repeat steps 2 through 7 for each of the milk products provided.

❑ 9. Incubate all plates at 35°C for 48 hours.

❑ 10. Count the number of bacterial colonies formed in each plate, multiply this figure by the dilution used, and record your findings in Table 44–2 of the Results and Observations section. Remember to use those plates having only 30 to 300 colonies.

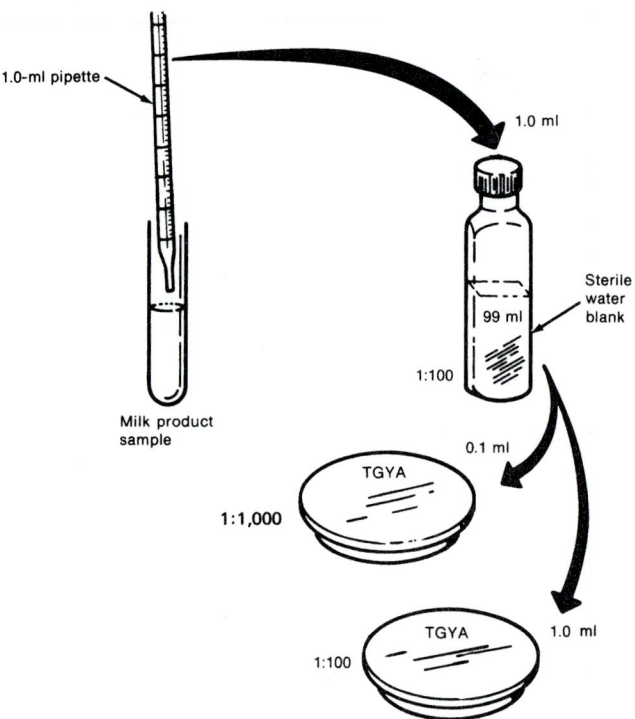

Figure 44–2

The plating of a milk sample. Note that the sample, as well as all diluted materials, should be shaken adequately before being pipetted. (TGYA = tryptone glucose yeast agar.)

B. Breed Count

Materials

❑ 1. One sample (1 mL) of each of the following milk products in individual small containers per 4 students:
    ❑ a. High-count raw milk
    ❑ b. Certified raw milk
    ❑ c. Pasteurized milk
    ❑ d. Buttermilk

❑ 2. One bottle each of the following per 4 students:
    ❑ a. Xylene
    ❑ b. 95% ethanol
    ❑ c. Methylene blue dye solution

❑ 3. Clean, lint-free slides and guide plates containing etched 1-cm squares

❑ 4. Immersion oil

❑ 5. Centimeter rulers (1 per 4 students)

❑ 6. Breed pipettes or similar pipettes to deliver 0.01 mL (1 per student), and appropriate pipetting aids

❑ 7. Boiling water bath with a stainless steel test tube rack

Procedure

This procedure is to be performed by students in groups of 4.

❏ 1. Place a stainless steel test tube rack in the water bath. The water level should be 2 inches lower than the top of the rack. Bring the water to a boil.

❏ 2. With a Breed pipette or a 0.01-mL loop, place 0.01 mL of a milk sample on a carefully cleaned slide and spread it over an area of exactly 1 cm^2. Accomplish this either by placing a guide plate with etched 1-cm squares directly under the slide or marking 1 cm^2 with the aid of a centimeter ruler. (Refer to Figure 44–3).

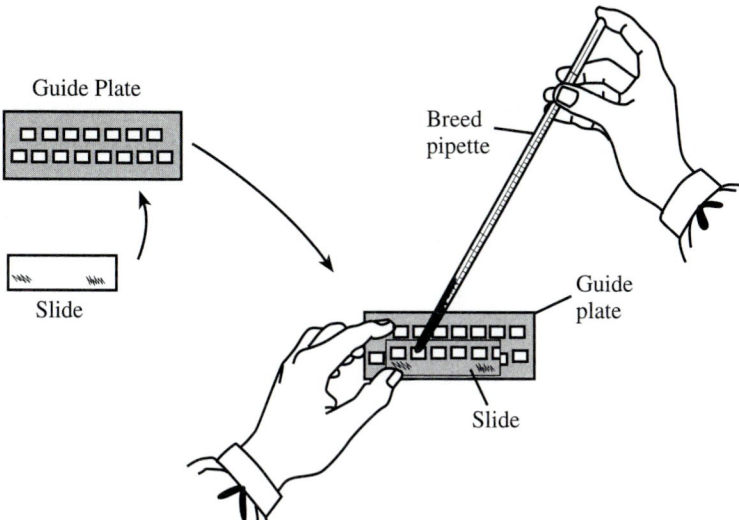

Figure 44–3
The use of a Breed pipette and a guide plate. Note that the delivery of the sample from the Breed pipette is controlled by pressure applied with the index finger.

❏ 3. Air-dry the specimen and fix it by heating it for 5 minutes on the test tube rack in the boiling water bath.

❏ 4. Flood the slide with xylene for 1 minute (to remove the fat in the sample), then wash it with alcohol to remove the xylene.

❏ 5. Dip the slide gently into a beaker of water to remove the alcohol.

❏ 6. Stain with methylene blue for approximately 15 seconds. The smear should be colored a robin's egg blue. If it is darker, it will be difficult to differentiate bacteria from the background of milk solids. If the slide is overstained, flood it with alcohol while it is still wet. The alcohol should remain until the proper color is obtained.

❏ 7. Wash the slide with a gentle stream of water.

❏ 8. Place the slide upright so that the water on it will drain. Allow the preparation to air-dry.

❏ 9. Examine the smear under the oil-immersion objective.

❏ 10. Count 10 separate microscopic fields, and determine the average number of bacteria per field. Multiply this result by the constant of 500,000 and enter your findings in Table 44–2 of the Results and Observations section. The constant used here is generally applicable for most microscopes used in standard microbiology teaching laboratories. The value obtained is an approximation of the number of bacteria per milliliter of the sample examined.

❏ 11. Repeat steps 2 through 10 for each milk sample provided.

C. Reductase Test (Methylene Blue Reduction)

Materials

The following materials should be provided per 4 students:

❑ 1. Ten-mL samples of each of the following milk products:
  ❑ a. High-count raw milk
  ❑ b. Certified raw milk
  ❑ c. Milk freshly pasteurized in class
❑ 2. Three sterile 10-mL pipettes
❑ 3. One sterile 1-mL pipette
❑ 4. Three sterile 15-mL or comparable screwcap test tubes
❑ 5. Four mL of aqueous methylene blue solution (1:10,000 dilution)
❑ 6. One test tube rack
❑ 7. One 37°C incubator or water bath (1 per class)
❑ 8. Appropriate pipetting aids

Procedure

This procedure is to be performed by students in groups of 4.

❑ 1. Pipette 9 mL of each milk sample into a separate sterile test tube. Use a separate pipette for each sample.
❑ 2. Label each tube according to its contents.
❑ 3. With the 1-mL pipette, add 1 mL of the methylene blue solution to each tube containing a milk sample.
❑ 4. Gently shake each tube to disperse the dye solution evenly.
❑ 5. Stopper or close the opening of each tube and place the tubes in the 37°C incubator or water bath.
❑ 6. Observe the tube at 30-minute intervals for 3 hours. The faster the blue color disappears, the greater is the bacterial load. Generally, unspoiled milk should not reduce the methylene blue within the time period for this exercise. (Refer to Figure 44–1.)
❑ 7. Record your findings in Table 44–2 of the Results and Observations section.

Results and Observations

Standard Plate Count, Breed Count, and Reductase Test

1. Complete the following table by entering the findings obtained in Parts A, B, and C.

Table 44–2

Standard Plate and Breed Counts and Reductase Results

	Part A		Part B	Part C
	Colony Counts		Breed Counts	Methylene Blue
Milk Sample Used	1:100	1:1,000		Reduction (in Minutes)
High-count raw				
Certified raw				
Pasteurized				
Buttermilk				

2. Which of the milk dilutions used for standard plate counts produced higher counts per milliliter of milk sample? Does this result agree with the Breed count findings? If not, what factors would account for the differences?

_____

_____

_____

_____

D. Isolation and Activities of Lactic Acid Bacteria

Materials

The following materials should be provided per 4 students:

❑ 1. One sample of the following commercial yogurt products:
  ❑ a. Plain
  ❑ b. Fruit-containing
  ❑ c. Chocolate-flavored
❑ 2. Four tryptone glucose yeast agar plates

❑ 3. Four tomato juice agar plates (a selective medium with a low pH)
❑ 4. Eight tubes of litmus milk
❑ 5. Four tubes containing 3 mL of sterile water
❑ 6. Gram stain sets

Additional Techniques Required for This Portion of the Exercise:

❑ 1. Streak Plate Technique, Procedure Diagram 10, Exercise 5
❑ 2. The Gram Stain, Procedure Diagram 17, Exercise 14
❑ 3. Broth Transfer, Procedure Diagram 5, Exercise 4

Procedure

This procedure is to be performed by students in groups of 4.

❑ 1. Select 1 yogurt sample and aseptically transfer 1 loopful of the material into 1 tube of sterile water.
❑ 2. Label 1 tomato juice agar plate and 1 tryptone glucose yeast agar plate with your name and the sample used.
❑ 3. Mix the contents of the tube to suspend evenly the bacteria and milk solids.
❑ 4. Prepare streak plates on both types of media using the water-diluted yogurt as the inoculum. (Refer to Procedure Diagram 10.)
❑ 5. Repeat steps 1 through 4 with each of the samples provided for the exercise.
❑ 6. Incubate all tomato juice agar plates at 37°C for 48 to 72 hours and the tryptone glucose yeast agar plates at room temperature for the same length of time.
❑ 7. After incubation, examine the plates for colonies that appear to be different from one another. Select 2 from each plate and prepare Gram stains. (Refer to Procedure Diagram 17.)
❑ 8. Examine the stained smears. Note that lactobacilli will be gram-positive long rods, while lactic acid streptococci, although also gram-positive, will appear as elongated cocci in pairs or short chains (Figure 44–4). Record your findings in the Results and Observations section.
❑ 9. Using the 2 different-appearing colonies on the tomato juice agar plates from which the Gram stains were prepared, inoculate separate tubes of litmus milk.

❏ 10. Incubate at 37°C and examine the media at 24, 48, and 72 hours. Refer to Exercise 26 for a description of typical reactions.

❏ 11. Record your findings and answer the questions in the Results and Observations section.

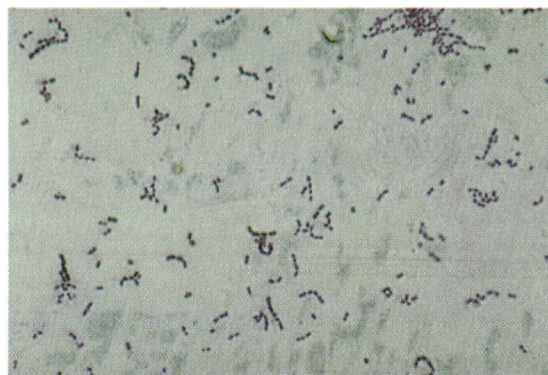

Figure 44–4
The Gram stain reaction of streptococci. Note the number of cells in the various chains.

Results and Observations

1. Gram stain reactions and morphology.

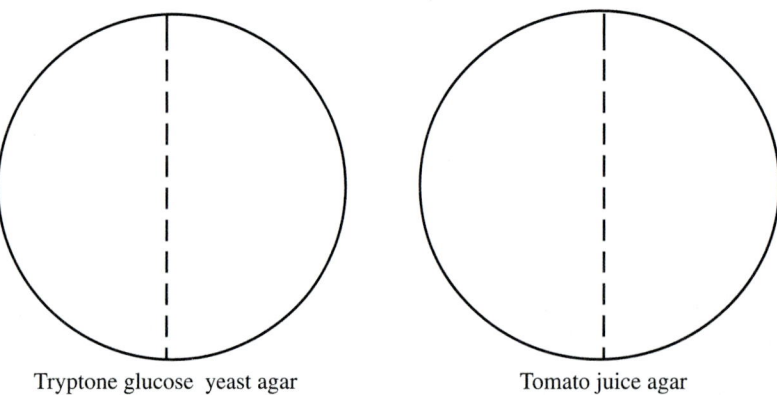

Tryptone glucose yeast agar   Tomato juice agar

2. What litmus-milk reactions occurred with the isolated colonies?

    a. Culture 1 _____

    b. Culture 2 _____

3. Were the litmus-milk reactions the same at 24 and 72 hours? _____

4. Did you find bacteria with all yogurt samples? If not, suggest why bacteria were not isolated._____

_____

_____

E. Microbiological Analysis of Meat Products

Materials

The following materials should be provided for general class use:

❏ 1. Samples of the following meat products in covered containers labeled as follows:
  ❏ a. Fresh ground meat
  ❏ b. Ground beef that has been kept at room temperature in a covered container for 72 hours
  ❏ c. Ground beef that has been kept at refrigerator temperature in a covered container for 72 hours
  ❏ d. A thawed ground beef preparation that has been repeatedly frozen and thawed on 3 consecutive days

❏ e. Uncooked pork sausage links or patties with casings removed
❏ 2. Scale for weighing samples
❏ 3. Wax-coated weighing papers
❏ 4. Sterile scalpels or knives
❏ 5. Sterile forceps for meat transfer
❏ 6. Gram stain sets

The following items should be provided per 4 students:

❏ 1. Five 99-mL sterile water dilution blanks in either 6-ounce prescription bottles or other screwcap containers
❏ 2. Six 9-mL sterile water dilution blanks in screwcap test tubes

❏ 3. Twelve sterile 1-mL pipettes graduated in 0.1 mL, and pipetting aids
❏ 4. Six eosin-methylene blue deeps
❏ 5. Six sterile Petri plates

Additional Techniques Required for This Portion of the Exercise:

❏ 1. The Gram Stain, Procedure Diagram 17, Exercise 14
❏ 2. Pour Plate Technique, Procedure Diagram 9, Exercise 5

Procedure

This procedure is to be performed by students in groups of 4. Refer to Figure 44–5 for an overall view of the procedure.

❏ 1. Select 1 meat sample, place a weighing paper on the scale, and weigh approximately a 1-gram specimen.

❏ 2. Transfer the meat specimen to a sterile 99-mL water dilution bottle. Wipe the surface of a fresh weighing paper with a sterile swab and insert the swab into a sterile 9-mL water dilution blank.

❏ 3. Mix the material thoroughly in the water dilution bottle by shaking the contents vigorously about 25 times.

❏ 4. Allow the coarse particles in the sample to settle.

❏ 5. Aseptically remove a loopful of the mixture and do a Gram stain. Record your findings in the Results and Observations section.

❏ 6. With a sterile 1-mL pipette, transfer 1 mL of the mixture to a sterile 9-mL water dilution blank.

❏ 7. Shake the contents approximately 25 times.

❏ 8. Melt 2 eosin-methylene blue agar deeps and prepare 2 pour plates, 1 with the meat sample and the other with the sample from the fresh weighing paper. Using separate sterile 1-mL pipettes, transfer 0.1 mL of each sample into separate deeps.

❏ 9. Dispose of the pipettes as indicated by your instructor.

❏ 10. Allow the plates to harden, and incubate at 37°C for 24 hours, or as directed.

❏ 11. After incubation, count the number of bacterial colonies formed in each plate, multiply this figure by the dilution used, and record your findings in Table 44–3 of the Results and Observations section. Remember to use only those plates that have between 30 and 300 colonies.

❏ 12. In addition, examine the plates for the presence of lactose-fermenting organisms (see Figure 44–6 for typical reactions).

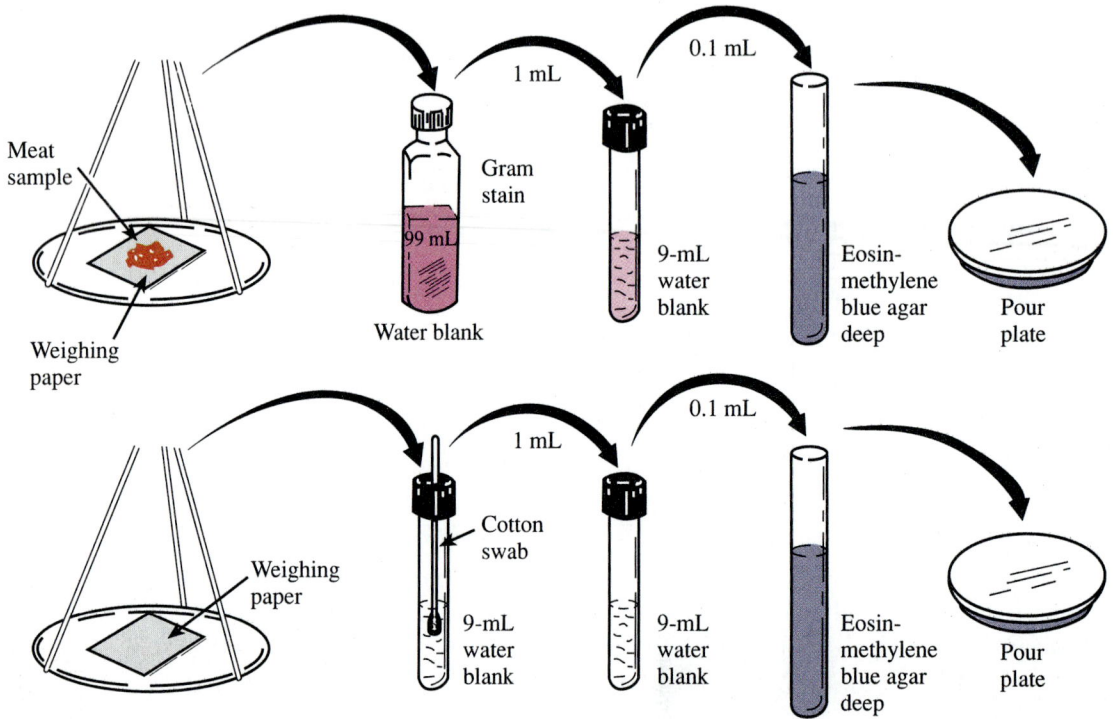

Figure 44–5
An overview of the meat-sampling procedure. Remember to shake all samples well before going on to the next step.

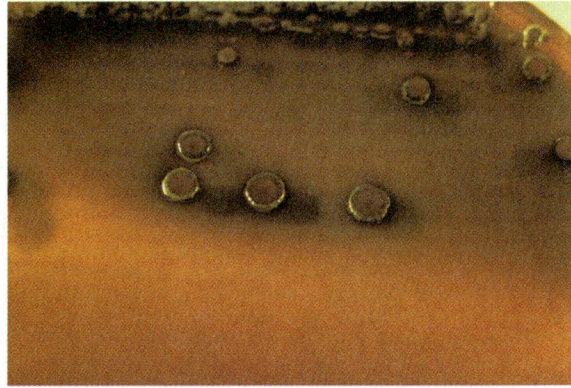

Figure 44–6
The purple colonies of lactose fermenting organisms on an EMB plate.

Results and Observations

1. What was the predominant Gram reaction and the morphology in your sample? _____

_____

2. Complete the following table by entering the findings obtained in the procedure.

Table 44-3

Meat Product Analysis Results

Meat Sample Used	Colony Count	Presence of Lactose Fermenters (+ or −)

3. How many bacteria grew in the plate with the sample from a fresh weighing paper? _____

4. What was the significance of taking a sample from a fresh weighing paper? _____

_____

_____

5. Is there any significance to the presence of lactose fermenters in a meat sample? Explain. _____

_____

_____

Laboratory Review 44　　　The Microbiological Examination and Content of Dairy Products and Selected Foods

Standard Plate Count, Breed Count, and Reductase Test

1. What factors affect the pasteurization process? List 4.

    a. _____　　c. _____

    b. _____　　d. _____

2. List 4 different specific microbial disease agents that can be transmitted by milk or other dairy products.

    a. _____　　c. _____

    b. _____　　d. _____

3. Differentiate between pasteurization and sterilization. _____

_____

_____

4. What methods and devices are currently being used to pasteurize different types of foods? _____

_____

_____

_____

_____

5. Can pasteurized milk sour in time? Why? _____

_____

_____

Isolation and Activities of Lactic Acid Bacteria

1. Can microorganisms be usefully applied in the dairy industry? Explain. _____

_____

_____

2. a. What is a fermented milk product? _____

_____

_____

  b. What are the principal organisms employed in the production of fermented milk products? _____

_____

_____

3. List 6 examples of fermented milk products. (Refer to your text and exercises in this section of the manual.)

  a. _____   d. _____

  b. _____   e. _____

  c. _____   f. _____

4. Briefly outline the process involved in the production of yogurt. (Refer to your text and exercises in this section of the manual.) _____

_____

_____

_____

_____

_____

Microbiological Analysis of Meat Products

1. a. What types of meat products might favor the growth of *Clostridium* species? _____

_____

_____

_____

  b. How can the growth of *Clostridium botulinum* in meat products be prevented? _____

_____

_____

_____

2. List 4 means of preventing food spoilage.

  a. _____   c. _____

  b. _____   d. _____

3.   Which of the methods listed in question 2 is the most effective? _____

_____

_____

_____

4.   What microorganisms would you associate with the following? (Refer to your text.)

a. a jar of spoiled pickles _____

b. tuna salad incriminated in a food poisoning outbreak _____

c. a bulging can of mushrooms _____

d. aflatoxin-contaminated corn _____

e. oysters harvested from fecal-contaminated waters _____

Key Terms

Breed count: a microscopic technique used to determine the number of bacteria present in a milk sample

coliforms (KŌ-li-forms): bacteria normally found in the colon (large intestine): includes *Escherichia coli* and *Enterobacter aerogenes*

microbial load: this term refers to the level of microbial content and/or contamination

pasteurization (pas-tūr-Ī-zā-shun): the process of heating a substance to a temperature high enough to kill all non-spore-forming bacterial pathogens, but not so high as to affect its chemical composition; temperatures used are 63°C (145°F) for 30 minutes, or 72°C (161°F) for 15 seconds

reductase test: a semiquantitative test used to determine the visible bacterial content of a food sample based on oxidation-reduction activities

standard plate count: an agar plate procedure used to estimate the number of living bacteria in a food sample or related material

Cheese Making and Yogurt Production

After completing this exercise, you should be able to:

1. Perform the basic steps and reactions associated with cheese making.
2. Describe the cheese-making process.
3. Recognize the value of microorganisms in the production of certain dairy products.
4. Prepare a yogurt culture.
5. Demonstrate the pH and chemical changes that occur during yogurt production.

The commercial and sometimes the home production of many milk products, such as butter, cultured buttermilk, cheese, cultured sour cream, and yogurt, begin with an appropriate lactic starter culture. For cheese making of all kinds, lactic acid production is essential and the lactic starter is used for this purpose. Lactic starters always include such bacteria as *Streptococcus cremoris* (**strep-tō-KOK-us, krē-MOR-is**), *S. diacetilactis* (**dī-a-sē-ti-LAK-tis),** or *S. lactis* (**LAK-tis**). These organisms are referred to as homolactic acid bacteria because they ferment sugars such as lactose mainly to lactic acid. All homolactic acid bacteria appear to ferment sugars by means of the **Embden-Meyerhoff-Parnas (EMP)** pathway (Figure 45–1). Where flavor and specific aroma compounds are desired, the starter will contain *Leuconostoc citrovorum* (**lū-kon-Ō-stok, sit-rō-VOR-um**), *L. dextranicum* (**DEKS-tran-i-kum**), or *S. diacetilactis.* The discovery, study, and continued applications of the activities of such organisms in the preparation of dairy products have contributed to the establishment of the specialty known as dairy microbiology. This exercise will be concerned with the making of an unripened cheese, cottage cheese, and yogurt.

Cheese, according to the ancient Greeks, was a gift from the Gods. This view may have been a way of saying that nobody could really remember when, where, or how cheese first came into being, or perhaps unaware of the chemical processes involved. The true origins of cheese making are, and probably always will be, a mystery. Like many inventions, cheese was probably discovered by different peoples at the same time. Once it was realized that the milk of certain mammals was both appetizing and nutritious, it was but a short step to the finding that sour milk separates naturally into milk solids known as **curd** and the watery fluid portion known as **whey.**

The simple milk-to-cheese formula has endless possibilities, depending on the source of the milk and partly on the processing by the cheese maker. In addition, microbial activity is central to the production of most cheeses.

Although cheeses differ widely in texture and flavor, they all begin in much the same way. The object is to extract the water from the milk, leaving the milk solids (fat, protein, vitamins, etc.) behind. Cheese-making processes usually start with cow's milk, either whole or skim. The first important step is the curdling of the casein, or milk protein, to form a solid curd. When an appropriate lactic starter is added to the milk, a firm curd and a watery fluid portion, or whey, are formed. The curdling may be exclusively microbiological; however, the enzyme rennin (obtained from calves) is frequently added to hasten the process.

Another important step in the production of many cheeses is ripening. In this phase of the process, the curd is shrunk, pressed, salted, or allowed to ripen under the particular conditions appropriate to the cheese desired. Certain microorganisms play a highly specific role in the ripening process. For example, the blue-green veined appearance and characteristic flavor of Roquefort cheese are a consequence of the metabolic activities of the mold *Penicillum roqueforti* (**pen-i-SIL-ē-um, roq-FOR-te**). Figure 45–2 summarizes the fundamental steps involved with cheese making.

More than 400 varieties of cheese exist. They are grouped according to their texture and whether they are ripened or unripened. Further, ripened cheeses are classified according to whether ripening is caused by bacteria or fungi. Three textural categories of cheeses are recognized: hard, semisoft, and soft. Examples of cheeses classified by textural type are included in Table 45–1. This exercise will demonstrate the basics of cheese making.

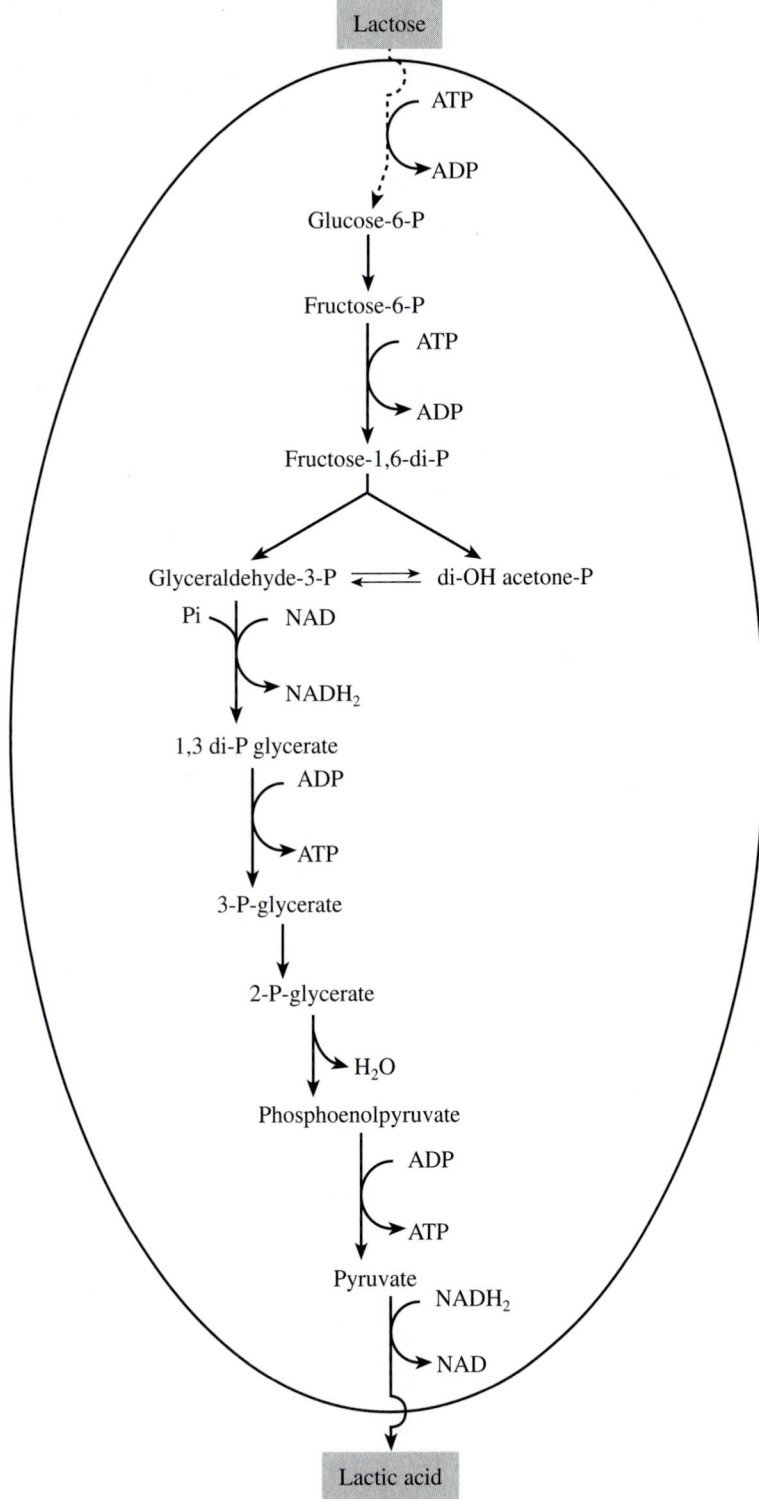

Figure 45–1

The Embden-Meyerhoff-Parnas (EMP) pathway used by the homolactic acid bacteria (such as streptococci and lactobacilli) for fermenting lactose to lactic acid. Most of the lactose fermented by these bacteria is excreted as lactic acid, which increases the hydrogen-ion concentration, thus lowering the pH.

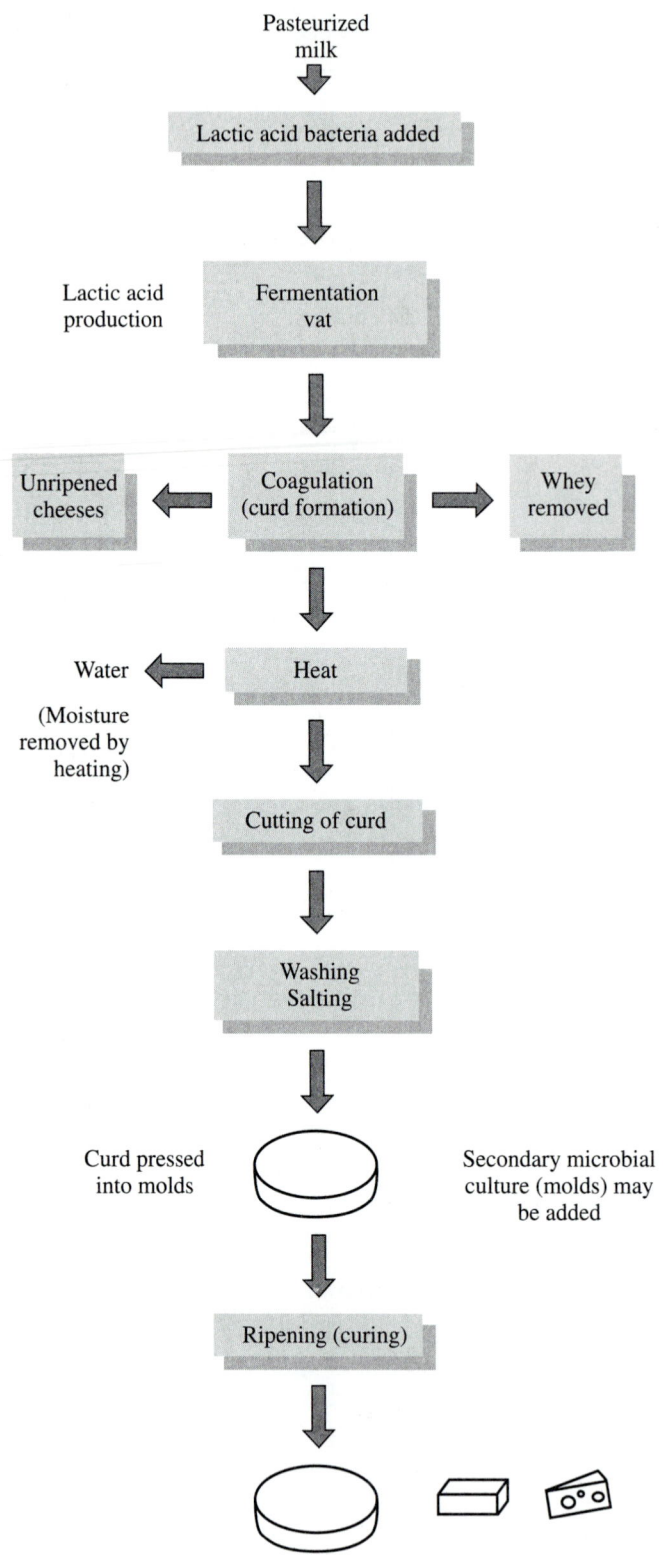

Figure 45–2

The steps in commercial cheese making.

Table 45–1

Examples of Cheese

Cheese Category	Examples
Hard	Cheddar, Edam, Provolone, Romano
Semisoft	Blue Cheese, Gouda, Muenster, Roquefort
Soft	Brie, Camembert, Limburger
Unripened (Unaged)	Cottage cheese, Cream cheese, Neufchatel

Yogurt is one of several well-known fermented dairy products that constitutes an important part of the human diet in many parts of the world (Figure 45–3). A mixture of two lactose-fermenting bacteria, *Lactobacillus bulgaricus* (lak-tō-ba-SIL-us bul-GĀR-i-kus) and *Streptococcus thermophilus* (ther-MŌ-fil-us), is used in the production of this fermented dairy food. Commercially, yogurt generally is made from a standardized mix of whole, partially defatted milk, condensed skim milk, cream, and nonfat dry milk. Milk fat levels in such preparations range from 1.0 to 3.25%. Only products containing a minimum of 3.25% milk fat are labeled *yogurt*. Those with 0.5 to 2.0% or less than 0.5% are labeled *low-fat* and *nonfat yogurt*, respectively. In addition to milk fat, other milk ingredients are found in yogurt. These ingredients include casein, sodium and calcium caseinates, whey, and whey protein concentrates. Additives of several kinds are also permitted in commercially produced yogurt, such as nutritive carbohydrate sweeteners, coloring, stabilizers (to provide smoothness in texture and to increase shelf life), and fruit preparations for flavoring.

Figure 45–3
Yogurt is one of the many types of fermented milk products consumed around the world. Here is one of the commercial advertisements emphasizing the value of the product.

The production of high-quality yogurt is dependent on the starter culture. Equal numbers of *L. bulgaricus* and *S. thermophilus* are desirable for flavor and texture production. The lactobacilli grow first, liberating the amino acids glycine and histidine and stimulating the growth of the streptococci. The production of the characteristic flavor by such cultures is a function of time as well as the sugar content of the yogurt mix. In this exercise, a plain yogurt and a fruit-flavored yogurt will be prepared to show the various changes that occur during yogurt production.

A. Cheese Making

Materials

❏ 1. The following items should be provided per 2 or 4 students:
   ❏ a. One 500-mL (clean) beaker
   ❏ b. One 1-mL sterile pipette
   ❏ c. One mL of Rennilase
   ❏ d. Several sheets of cheesecloth
   ❏ e. One 250-mL or 500-mL graduated cylinder or measuring cup
   ❏ f. One thermometer
   ❏ g. Ordinary table salt
❏ 2. Cultured or ripened pasteurized milk (250 mL per 2 to 4 students). Make the cultured preparation by adding 20 mL of fresh buttermilk per liter of milk and allowing the mixture to stand for at least 10 minutes at 25°C (room temperature)
❏ 3. Hot plates with temperature control (enough for class use)
❏ 4. Markable tape or labels
❏ 5. Plastic storage containers (if available)
❏ 6. A sharp knife for cutting curd

Procedure

This exercise should be performed by students in groups of 2 or 4.

❏ 1. Put 250 mL (1 cup) of the cultured pasteurized milk preparation in the beaker provided and warm it to 32°C.
❏ 2. Using a 1-mL pipette, add 0.5 mL of Rennilase to the warmed milk and stir.
❏ 3. Remove the preparation from the heat and allow it to stand undisturbed for about 15 minutes. Coagulation and curd formation should occur during this period.
❏ 4. Remove the curd, and on a clean surface cut it into small pieces with the knife provided.
❏ 5. Remove as much of the whey (liquid) as possible by placing the curd (cheese) pieces in cheesecloth and gently squeezing them into a ball.
❏ 6. Taste a small amount of your preparation and describe the taste in the Results and Observations section.
❏ 7. Flatten the cheese ball into a hard, solid form and salt it lightly.
❏ 8. Wrap the preparation in the aluminum foil or wax paper and label it.
❏ 9. Place the cheese into a plastic container (if available) refrigerate it for 24 hours, or as directed.
❏ 10. Examine the cheese, pour off any additional liquid, and taste. Enter your findings in Table 45–2.
❏ 11. Continue to examine and taste your cheese periodically for a month. Note differences in taste, texture (softness or hardness), and color. Enter your findings in Table 45–2 and answer the questions in the Results and Observations section.

B. Yogurt Production

Materials

The following items should be provided per 2 students:

- ❏ 1. Pasteurized whole milk or cream (approximately 400mL)
- ❏ 2. One small container of fresh, plain yogurt containing an active culture
- ❏ 3. One small container of fruit preserves
- ❏ 4. One 500-mL Pyrex graduated sterile beaker
- ❏ 5. Three 250-mL sterile beakers or comparable-sized drinking glasses or paper cups
- ❏ 6. One clean tablespoon
- ❏ 7. Three sterile wooden tongue depressors
- ❏ 8. One sterile thermometer
- ❏ 9. Two aluminum foil squares (7×7 cm or 3×3 inches)
- ❏ 10. Three wooden applicator sticks or glass stirring rods
- ❏ 11. Six paper drinking cups

The following items should be provided for general class use:

- ❏ 1. Methylene blue stain in dropper bottles
- ❏ 2. pH paper or a pH meter
- ❏ 3. Hot plates with temperature control or other means of heating such as tripods, wire gauze, or Bunsen burner
- ❏ 4. Ice-water bath
- ❏ 5. Incubator set at 40°C or commercial yogurt makers
- ❏ 6. Glass slides

Additional Techniques Required for This Portion of the Exercise:

- ❏ 1. Bacterial Smear Preparation, Procedure Diagram 2, Exercise 2
- ❏ 2. Simple Staining, Procedure Diagram 3, Exercise 2

Procedure

The following procedure should be performed by students in pairs.

- ❏ 1. Pour a small amount of the milk or cream and yogurt provided into separate glass containers or paper cups. Determine the pH, taste, odor, and texture (e.g., fluid, semisolid, or solid) of both materials. Enter your findings in Table 45–3. In addition, make heat-fixed smears of the milk and yogurt for later staining.
- ❏ 2. Put 400 mL of the milk or cream in the large beaker provided.
- ❏ 3. Carefully place the beaker on the hot plate or other means of heating and heat the milk to approximately 85°C. With the aid of a thermometer, maintain this temperature for 15 minutes.
- ❏ 4. After heating, remove the beaker from the hot plate and allow it to cool. An ice-water bath can be used when the milk is cooled to about 60°C. Cool the heated milk to 43°C.
- ❏ 5. Using the tablespoon provided, add 3 tablespoons of yogurt to the milk. Stir and mix the preparations. Again, remove a small amount of the mixture as described in step 1 and determine the pH, taste, color, and texture. Enter your findings in Table 45–3.

❏ 6. Take 1 of the small sterile beakers or other containers provided, and with a wooden tongue depressor, put enough of the fruit preserves into the container to cover its bottom.

❏ 7. Carefully pour equal amounts of the milk-yogurt mixture into the beaker containing the fruit and another empty sterile container.

❏ 8. Cover both containers with aluminum wrap and place them in an incubator set at 40°C or in a commercial yogurt maker for 8 to 18 hours. Do not disturb or stir the mixture during this time.

❏ 9. After incubation, remove small samples of both preparations for taste, texture, and pH determinations and the preparation of a heat-fixed smear. Enter your findings in Table 45–3.

❏ 10. Refrigerate both new yogurts for 12 to 18 hours.

❏ 11. Taste both preparations and determine the texture of each.

❏ 12. Stain all smears with methylene blue and examine the slides under oil immersion. Look for different shapes and morphological arrangements. Sketch representative fields and answer the questions in the Results and Observations section.

❏ 13. Rinse and clean all glassware, and discard all disposable items as directed by the instructor.

Results and Observations

Cheese Making

1. Enter your findings in the following table.

Table 45–2

Cheese Making

Time of Examination	Characteristic			
	Color	Flavor	Odor	Texture
At preparation				
First 24 hours				
Week 1				
Week 2				
Week 3				
Week 4				

2. Did the taste of the cheese improve with age? _____

_____

3. Did you notice any unusual characteristic in the cheese? _____

_____

Yogurt Production

1. Enter your findings in the following table.

Table 45–3

Time of Examination	Characteristic			
	pH	Flavor	Odor	Texture[a]
At preparation before heating				
At preparation after heating and with yogurt				
Yogurt				
First 18 hours of incubation				
After refrigeration				

[a] *Texture* refers to consistency (e.g., fluid, semisolid, or solid).

2. Representative sketches.

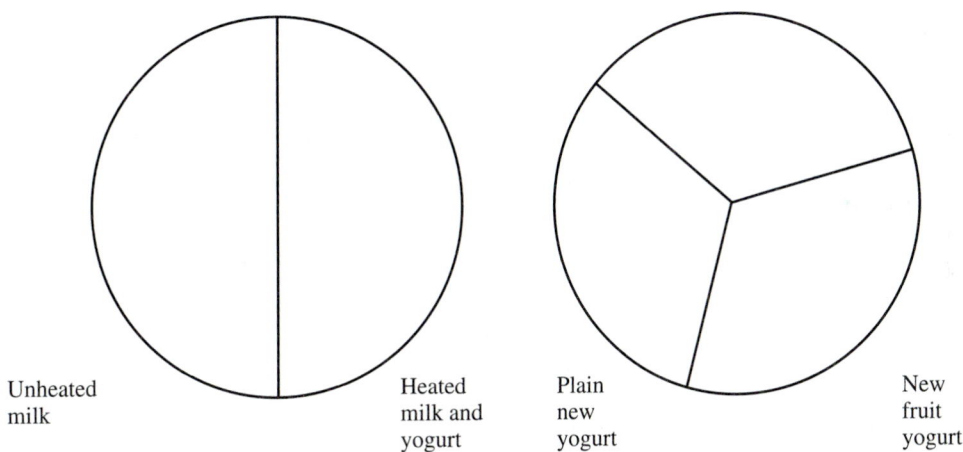

Unheated milk   Heated milk and yogurt   Plain new yogurt   New fruit yogurt

3. Was the microbial content of unheated milk different from that of the plain new yogurt? If so, describe the differences. _____

_____

_____

Laboratory Review 45 — Cheese Making and Yogurt Production

1. From what types of materials are cheeses made? _____
   _____

2. Distinguish between the following:

   a. hard and soft cheese _____
   _____

   b. ripening and salting _____
   _____

   c. curd and whey _____
   _____

3. How are the characteristic holes or "eyes" in Swiss cheese produced? _____
   _____
   _____

4. What microorganisms are specifically involved in yogurt production? _____
   _____

5. Why was the milk heated before the introduction of the yogurt culture? _____
   _____

6. What would happen to a yogurt preparation with the addition of a 2% gelatin solution? _____
   _____

7. What is a homolactic acid bacterium? _____
   _____

8. What is the Embden-Meyerhoff-Parnas pathway? _____
   _____
   _____

Key Terms

casein (KĀ-sēn): milk protein

curd (KURD): the solidified product formed during cheese making: it contains most of the fat and other ingredients suspended in milk

Embden-Meyerhoff-Parnas pathway: the metabolic pathway used by homolactic bacteria to ferment lactose to lactic acid; the pathway provides the energy required for growth. Detection of the enzyme fructose diphosphoaldolase can be used to differentiate between homolactic-acid and heterolactic-acid bacteria (see Figure 49–1)

homolactic acid bacterium: a bacterium that ferments sugars, mostly to lactic acid, which is excreted; this excreted acid increases the hydrogen-ion concentration, thus lowering the pH

rennin (REN-in): an enzyme that coagulated the milk protein, casein; commercially available from fungal cultures, or by extraction from calf stomach linings

salting: a step in cheese making that not only adds flavor, but also continues additional water extraction from the curd

starter culture: a carefully selected pure culture used to inoculate milk or cream to obtain a fermented product that is both uniform in consistency and of high quality

whey (WĀY): the liquid formed during the precipitating of casein from milk during cheese production; it contains water and dissolved materials

EXERCISE 46

The Preparation of a Fermented Food: Sauerkraut

After completing this exercise, you should be able to:

1. Recognize the principal value of microbial activities.
2. Understand the changes caused by microorganisms in the fermentation of foods such as sauerkraut.
3. Outline the stages of microbial succession involved with sauerkraut production.
4. Describe the steps involved in the production of sauerkraut.

Sauerkraut results from the lactic acid fermentation (pickling) of fresh cabbage. During the fermentation process, the cabbage sours mainly as a result of the production of lactic acid by lactic acid bacteria (Figure 46–1). This group of microorganisms includes *Lactobacillus brevis* and *L. plantarum,* which are rods, and *Leuconostoc mesenteroides,* which are cocci. These organisms are normally present in cabbage and are essential to the proper development of the characteristic flavor and smell of sauerkraut.

Basically, the preparation of sauerkraut involves the salting of fresh shredded cabbage. Layers of salted cabbage are packed into appropriate containers, and the preparation is allowed to ferment (Figure 46–2). The salt draws the tissue fluid out of the cabbage, a **brine** (salt-saturated water solution) results that favors the growth of lactic acid bacteria. Sauerkraut production is an example of bacterial or microbial succession because the growth and metabolism of several bacterial species increase and decrease in succession before the final product is formed. Usually the succession process occurs in three stages and requires about three to four weeks incubation.

During the early stage of the sauerkraut production, some naturally occurring, salt-tolerant bacterial species dispose of any dissolved oxygen to make the system anaerobic. The resulting environment enables facultative anaerobes to ferment the plant sugars in the salt-extracted cabbage juice and to produce and excrete various acids (see Figure 46–1), and thereby lower the pH. The gram-negative, facultative anaerobic rods, *Enterobacter cloacae* (**en-ter-Ō-bak-ter, klō-Ā-kē**) and *Erwinia herbicola* (**er-win-Ē-a, er-bik-Ō-la**) are commonly isolated during this stage.

In the **intermediate stage** (about the second or third day of the process, *Leuconostoc* (**lū-kon-Ō-stok**) species are the predominate bacteria in the preparation. These gram-positive cocci are fermentative, salt- and acid-tolerant lactic acid bacteria. *Leuconostoc* species are referred to as **heterolactic fermentors** because they produce almost equal amounts of carbon dioxide, ethanol, and lactic acid by means of the **pentose phosphoketolase (PPK) pathway** (see Figure 46–1). Small quantities of acetic acid and glycerol also are formed.

After four to six days into the fermentation process, the final stage begins. *Lactobacillus* (**lak-tō-ba-SIL-us**) species are the predominate forms found. These gram-positive rods are fermentative, salt-tolerant, lactic acid bacteria. Lactobacilli are **homolactic bacteria,** because they ferment sugar mostly to lactic acid. Their enzymatic activities not only produce a final acid concentration ranging from 1.5 to 2.0%, but also convert a small amount of sugar to aromatic products, which contribute to the flavor of the fermented product.

A combination of several factors contributes to the development of a favorable environment for the cabbage-fermenting microorganisms. These factors include anaerobiosis, decreasing pH, and temperature. The temperature during lactic acid fermentation should be about 21° to 24°C for best results. If the temperature falls below 16°C, the fermentation will be quite slow and incomplete. Good sauerkraut should be light-colored and crisp, and have a pleasant aroma and a clean acid flavor.

Although the finished product has a pH ranging from 3.1 to 3.7, it is still subject to spoilage by bacteria, molds, and yeasts. Sauerkraut is especially subject to spoilage at its surface, where it is exposed to air. These film-forming yeasts and molds destroy acidity, thereby permitting other microorganisms to grow and causing softening, darkening, and unpleasant flavors. Elevated temperatures and changes in pH and salt content are among the environmental factors that bring about abnormal fermentation, inhibit the growth of *Leuconostoc,* and favor the growth of spoilage or undesirable flavor-producing organisms such as *Pediococcus* (**pe-dē-O-KOK-us**).

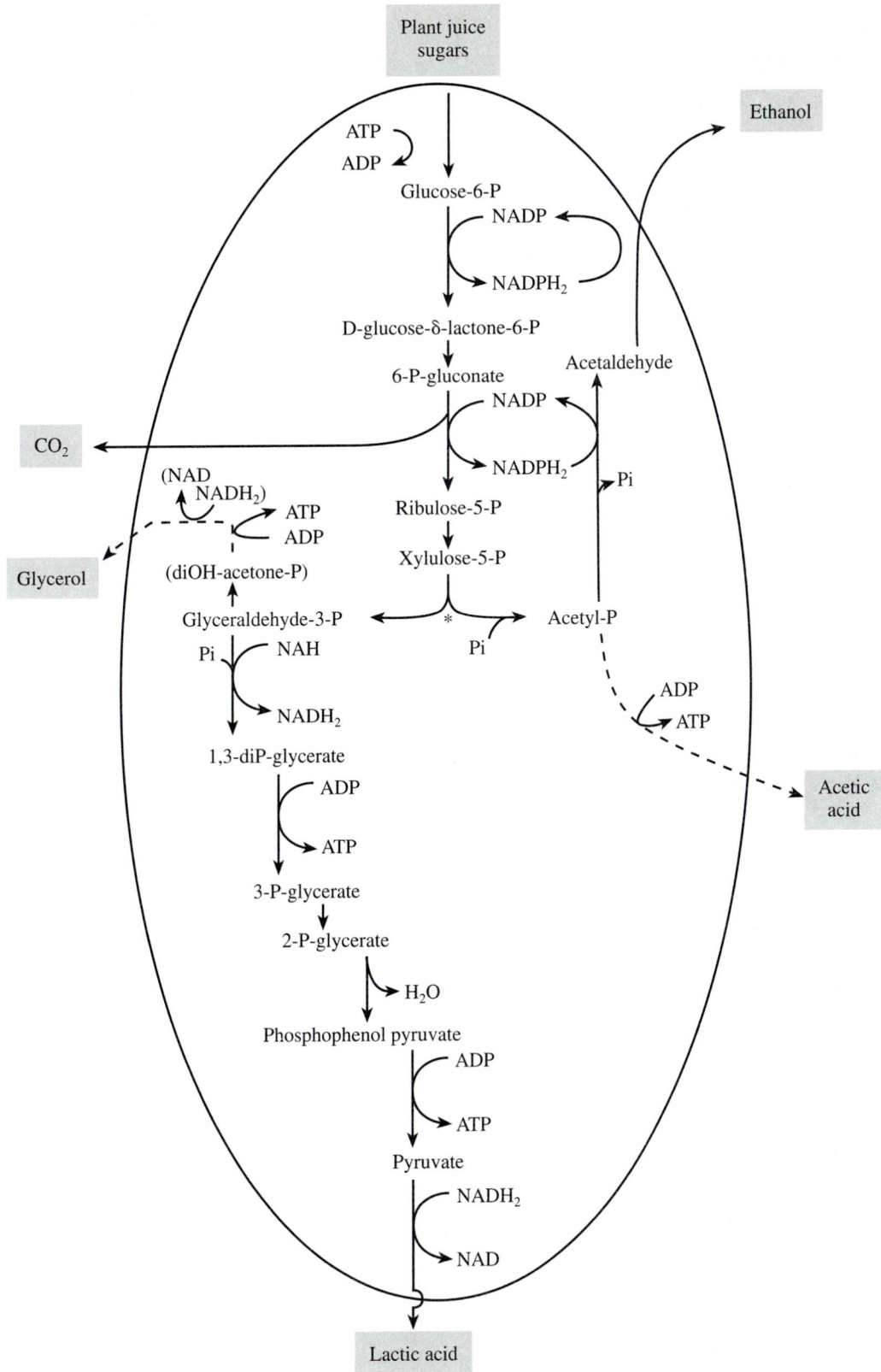

Figure 46–1

The pentose phosphoketolase (PPK) pathway. The sequence of reactions shown is used by the heterolactic bacteria to ferment plant sugars to a mixture of almost equal amounts of carbon dioxide, ethanol, and lactic acid. *Leuconostoc* species are examples of heterolactic bacteria. Explanation of symbols CO_2, carbon dioxide; P, phosphate; OH, hydroxy. (Refer to your text for additional details of *metabolism*.)

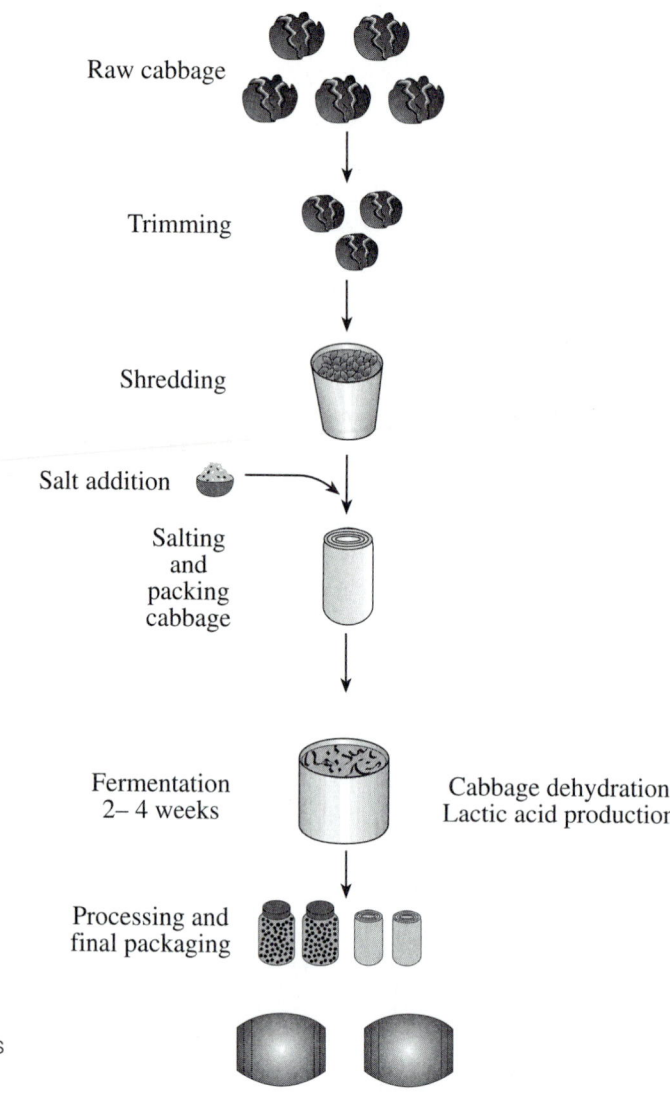

Figure 46–2

Sauerkraut production. The various steps used in the commercial process are shown. The laboratory exercise follows along the general sequence shown.

Procedure

This procedure is to be performed by students in groups of 4. Steps 1 through 11 are to be carried out during the first period of the exercise.

☐ 1. Wash cabbage and remove all outer leaves. (Save the outer leaves for step 8.)

☐ 2. Using a kitchen knife and on a clean surface, halve and core the cabbage head.

☐ 3. Shred the entire cabbage head finely. The core may be shredded and added to the rest of the cabbage. Place the shreds on the wax paper and weigh them.

☐ 4. Divide the cabbage into 100-gram amounts, and weigh out 3 grams of salt for each cabbage portion.

☐ 5. Thoroughly mix the salt with the shredded cabbage. Use a wooden applicator or plastic spoon.

☐ 6. Pack the salted cabbage in the container provided by successively adding the cabbage in 1-inch layers. Press or tamp the layers with the plastic spoon or wooden applicator. (Tamping begins the extraction of the plant juices, and eliminates pockets of air in the packed salted cabbage.)

☐ 7. Continue to pack the vessel until it is about three-quarters full.

☐ 8. Place one or more of the saved outer cabbage leaves to cover the last layer of the salted cabbage. Lightly salt these leaves. (Refrigerate extra cabbage leaves as possible replacements during the first stage of the process.)

Materials

The following materials should be provided per group of 4 students performing the exercise:

❏ 1. One fresh cabbage head
❏ 2. Commercial noniodized salt (3 grams of salt for each 100 grams of cabbage used)
❏ 3. One 2-liter to 4-liter polyethylene pail or transparent polycarbonate jar
❏ 4. One cover for the fermentation container or sufficient clean cheesecloth
❏ 5. A wooden board cover to fit into the container being used
❏ 6. A heavy weight
❏ 7. One sharp kitchen knife or vegetable shredder
❏ 8. Wooden tongue depressors or plastic spoons
❏ 9. Methylene blue stain
❏ 10. Glass slides and cover slips
❏ 11. pH indicator paper or pH meter (enough for class use)
❏ 12. A scale and appropriate weights
❏ 13. A roll of wax paper or aluminum wrap
❏ 14. Disposable 5-mL pipettes

❏ 9. Place the board cover over the top layer of the fermentation mixture and compress until a layer of juice is squeezed from the cabbage. Remove a few drops of this fluid with the aid of a pipette, determine its pH, and record your findings in the Results and Observations section.
❏ 10. Place the heavy weight on top of the board to hold it down.
❏ 11. Cover the preparation with wax paper or aluminum wrap to protect the surface from contamination with dirt or insects. Incubate it at 25°C or at a temperature specified by your instructor.

The following portions of the exercise are to be carried out after 2 days, 1 week, 2 weeks, and 4 weeks of incubation. Record your findings in the Results and Observations section.

❏ 12. Observe, smell, and taste the fermenting mixture. Pay attention to its color and consistency (softness or crispness).
❏ 13. With a pipette, remove 3 mL of the fermenting fluid. Use a portion of the sample to determine the pH, and prepare a temporary wet mount (refer to Procedure Diagram 1). Use a drop each of the methylene blue solution and the fermentation fluid for the wet mount.
❏ 14. Dispose of the pipette as directed.
❏ 15. Examine the temporary wet mount and sketch your findings in the space provided.

Results and Observations

1. Record your results in the table on the following page.
2. Sketch the temporary wet mount of the fermenting mixtures after the table.

Table 46–1

Sauerkraut Production

Properties of Fermenting Mixture	Sampling Periods			
	Day 2	Week 1	Week 2	Week 4
Color				
Odor				
Flavor				
Consistency				
pH				

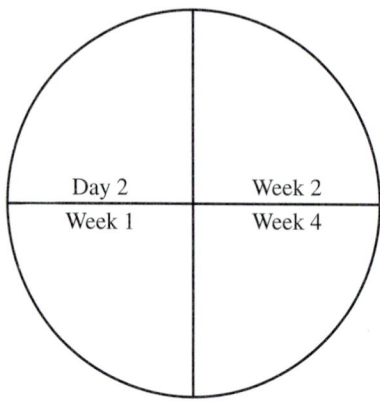

3. Did the sauerkraut preparation show any signs of spoilage? If so, describe them. _____

_____

_____

Laboratory Review 46   The Preparation of a Fermented Food: Sauerkraut

1. Does a succession of microorganisms occur in sauerkraut production? Explain your answer. _____

_____

_____

2. How do fermentation temperature and pH affect the flavor and smell of sauerkraut? _____

_____

_____

3. What other fermented foods are subject to microbial spoilage? _____

_____

_____

4. What is a fermented food? _____

_____

_____

5. Complete Table 46–2 by listing 3 examples of fermented foods, the starting material used in their production, and the types of microorganisms involved. (Use your text as a reference.)

Table 46–2

Examples of Fermented Foods

Fermented Food	Starting Material	Microorganism(s) Involved

6. a. Briefly distinguish between homolactic acid and heterolactic bacteria. _____

_____

_____

   b. What type(s) of bacteria are involved with cabbage fermentation? _____

_____

Key Terms

fermentation (fur-men-TĀ-shun): the enzymatic breakdown of complex organic compounds under anaerobic conditions in which the final hydrogen acceptor is an organic compound

heterofermentation: fermentation of glucose or other sugar to a mixture of products

homofermentation: fermentation of glucose or other sugar ending in the production of a single product, lactic acid

sauerkraut: a product resulting from the lactic acid fermentation of shredded cabbage

Wine: A Product of Alcoholic Fermentation

After completing this exercise: you should be able to:

1. Explain wine production in terms of carbohydrate metabolism.
2. Estimate sugar content and alcohol values by specific gravity determinations.
3. Compare and contrast the processes involved in wine and beer production.

The term *wine* refers to the fermented juice of grapes. Only certain varieties of grapes make good wine (Figure 47–1a). The sugar content of grapes ranges from 12 to 30%, which will produce a wine containing between 6 and 15% alcohol (Figure 47–1c). Although many varieties of wine exist, two basic types, classified by color, are red (Figure 47–1d) and white. Special varieties outside these categories include champagne, rose, and several wines made from fruits other than grapes. Examples of other fruits used in wine making are apples, cherries, and pears. These fruits are used for dessert wines, which are sweet in contrast to white table wines, which are dry (nonsweet). In the case of sparkling wine such as champagne, carbon dioxide (CO_2) is incorporated either naturally or artificially during the bottling process.

(a)   (b)

(c)   (d)

Figure 47–1

Selected aspects of wine making. (a) One example of the many grape (*Vitis*) varieties used in producing red wines. (b) The appearance of *must*, a combination of grape juice and crushed grape pulp. (c) Testing wine for alcohol content. (d) The final products.

The history of wine making goes back as far as that of beer making. The former is a much simpler process, because it starts with fermentable sugars rather than starch. Wine is generally made from *must,* a combination of grape juice and crushed pulp. Essentially, all wines result from the conversion of sugars such as fructose and glucose into ethyl alcohol by a species of yeasts cells called *Saccharomyces cerevisiae* variety *ellipsoideus.* Table 47–1 lists several alcoholic beverages produced by *Saccharomyces* species. The taste of the final product is due to the grape variety, the sugar content, and the yeast strain used. The alcohol concentration of a wine is influenced in part by the concentration of sugar in the *must,* and the alcohol tolerance of the yeast strain. Generally, yeast cells continue to produce alcohol until the concentration interferes with the growth. Most yeast strains can tolerate up to 14% alcohol. After completion of the fermentation, wine is aged in wooden casks for time periods ranging from several months to years. After such periods, wine is bottled and stored for further aging. During the aging process, nonmicrobial changes occur, which give wines their characteristic aroma (bouquet) and flavors. Figure 47–2 provides a general outline of the wine making process.

Table 47–1

Selected Alcoholic Beverages Produced by Saccharomyces Species

Beverage	Starting Material	Procedure
Beer	Germinated grain (malt)	Natural fermentation
Table wine	Fruit juice	Natural fermentation
Sake	Rice	Amylase from mold (*Aspergillus oryzae*) converts starch to sugar, which is naturally fermented
Fortified wine (sherry, port)	Fruit juice	Natural fermentation plus addition of brandy to increase alcohol content to about 20%
Brandy Whiskey Rum Vodka	Fruit juice Grain mash Molasses Potatoes	Natural fermentation followed by distillation to increase alcohol content to about 40 to 50%

In this exercise, various aspects of the alcoholic fermentation process will be demonstrated, including the relationship between the amount of fermentable sugar consumed and the amount of alcohol produced. An instrument known as a *hydrometer* (Figure 47–3) is used to measure the specific gravity before and after the fermentation process. *Specific gravity* is a measure of the relative density of a substance compared to the density of water. The specific gravity of water is 1.000. As sugar or other soluble solids (solutes) are added to water, the specific gravity increases. Wine hydrometers usually contain calibrated scales with which to measure the *must* (see Figure 47–1b) and/or *percent of potential alcohol.* When placed in a liquid whose specific gravity is to be determined, a reading is taken from the floating hydrometer's calibration scale at the *meniscus* (liquid line). Hydrometers with calibration scales ranging from 1.000 to 2.000 or higher are used for liquids whose specific gravities are higher than water.

To begin the process of wine production, enough of a *must* sample (Figure 47–1*b*) is introduced into a clear glass container. Next, a hydrometer is placed in the preparation and spun to dislodge any air bubbles, and the figure on the calibration scale is noted. The calibration scale number should indicate the sugar content in the *must* and the potential alcohol that could result. The amount of sugar needed for a specific type of wine can then be added. Examples of starting specific gravities for three general types of wine are as follows:

Dry wine: 1.085–1.100
Sweet wine (medium): 1.120–1.140
Sweet wine: 1.140–1.160

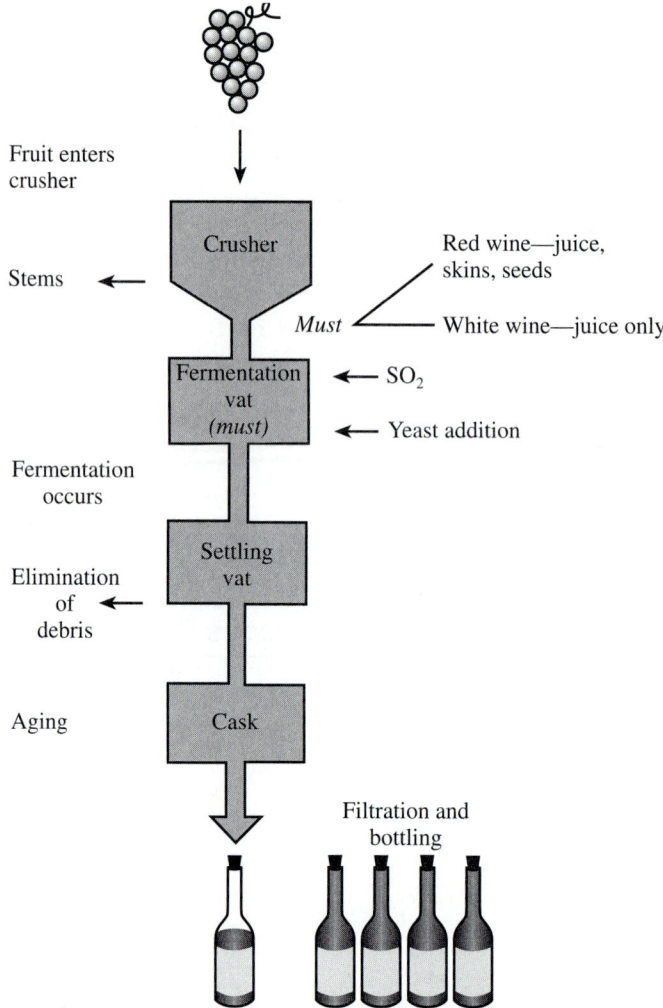

Figure 47–2

The general steps involved with the commercial wine production. The steps used in the exercise are similar in several aspects. Sulfur dioxide (SO_2) is often added to *must* to inhibit the growth of the natural microbial production of the grapes used in wine making. Specifically selected strains of *S. cerevisiae* are added after the SO_2 treatment.

Fruit enters crusher

Crusher

Stems

Red wine—juice, skins, seeds

Must White wine—juice only

Fermentation vat *(must)*

← SO_2

← Yeast addition

Fermentation occurs

Settling vat

Elimination of debris

Aging

Cask

Filtration and bottling

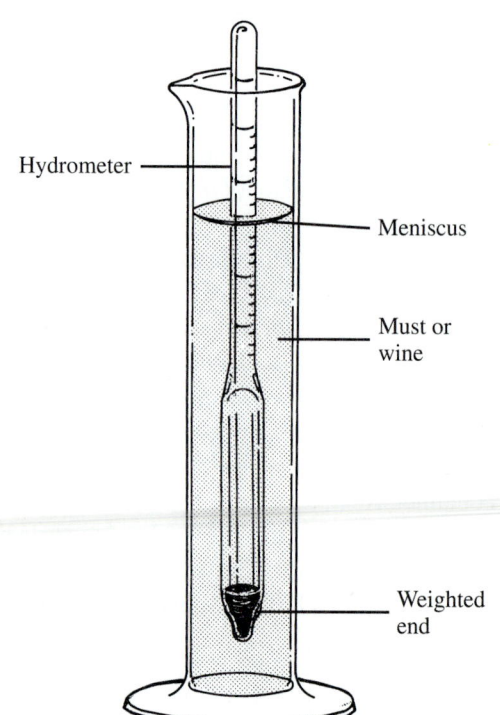

Hydrometer

Meniscus

Must or wine

Weighted end

Figure 47–3

The wine hydrometer. Note that this device must float freely in the solution to be tested for an accurate specific gravity reading.

Table 47–2 shows the alcohol yield that can be expected in relation to the specific gravity of *must* preparations. Such tables are useful if hydrometers are equipped with potential alcohol scale calibrations. To determine the alcohol of a wine, two specific readings are necessary: one before fermentation begins, and the second after fermentation stops. Subtracting the second figure from the first reading should show the alcohol content percentage by volume of the wine. Here is an example of the calculation.

First reading % percentage alcohol	16
Less: Second (or ending) reading % percentage alcohol	−4
% alcohol produced	12

Gas production and yeast growth, both of which are related to the fermentation process, also will be demonstrated in this exercise.

Table 47–2

The Relationship of Specific Gravity to the Estimated Percentage of Sugar Content and Alcohol

Specific Gravity	Percent Sugar	Alcohol Value	Specific Gravity	Percent Sugar	Alcohol Value
1.000	0	0	1.090	22.5	12
1.010	3	1	1.100	24.5	13
1.020	5	2	1.110	26.5	14.5
1.030	8	4.5	1.120	28.5	15.5
1.040	10.5	5	1.130	31	17
1.050	12.5	6	1.140	33	18.5
1.060	15	7.5	1.150	35	20
1.070	17.5	9	1.160	37	21.5
1.080	20	10.5			

A. Laboratory Session 1

Materials

The following items are to be provided per 4 students:

❑ 1. Ten mL of a 24-hour grape juice culture of wine yeast

❑ 2. Two hundred mL of unsweetened grape juice

❑ 3. One 250-mL Erlenmeyer flask equipped with a 1-hole stopper (white rubber or cork), with 2-inch glass tubing to fit and about 12 inches of rubber or plastic tubing, and 1 test tube for making a gas trap, as demonstrated by the instructor (see Figure 47–4)

❑ 4. One 6-inch test tube

❑ 5. One 500-mL beaker

❑ 6. One 250-mL graduated cylinder for use with a hydrometer or wine saccharometer

The following items are to be provided for general class use:

❑ 1. Hydrometers (wine saccharometers)

❑ 2. Sucrose (table sugar)

❑ 3. Weighing papers and spatulas

❑ 4. Mixing rods

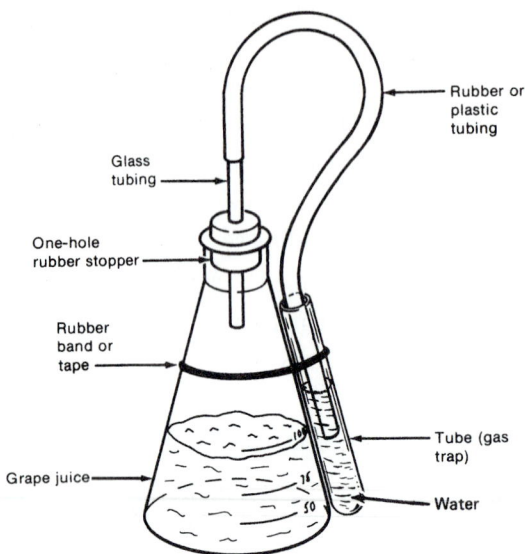

Figure 47–4
The fermentation flask and its parts.

Procedure

This procedure is to be performed by students in pairs.

❏ 1. Determine the specific gravity of your grape juice preparation as demonstrated by the instructor.

❏ 2. Record the specific gravity in the Results and Observations section and estimate the percentage of sugar content using Table 47–2.

❏ 3. Your instructor will assign a sugar value or specific gravity value for your wine. Dissolve the appropriate amount of needed sugar slowly into your juice by mixing the two in the beaker until you obtain the desired value. Record this value and estimate the percentages of sugar and alcohol in Table 47–3.

❏ 4. Calculate the amount of sugar in the juice (e.g., 1% = 1 g sugar per 100 mL juice).

❏ 5. Inoculate the adjusted grape juice with the wine yeast culture and pour into an Erlenmeyer flask. Be sure to leave enough space at the top of the flask to allow for the foaming of the fermenting juice during the early stages.

❏ 6. Stopper and connect the gas trap with a test tube approximately three-quarters full of water.

❏ 7. Label the flask with your name(s) and the initial specific gravity.

❏ 8. Incubate the preparation at room temperature in a location designated by your instructor. Gas production will usually begin in a few hours.

B. Laboratory Session 2
This session should take place 1 week after Laboratory Session 1.

Materials

The following items are to be provided per 4 students:

❏ 1. One 250-mL Erlenmeyer flask

❏ 2. One 250-mL graduated cylinder for use with hydrometers

The following items should be provided for general class use:

❏ 1. Three hydrometers

Procedure

This procedure is to be performed by students in pairs.

❏ 1. Carefully pour the fermenting juice into a graduated cylinder. Avoid carrying over sediment, if possible. This step is called *decanting*.

❏ 2. Determine the specific gravity of the preparation.

❏ 3. Record the specific gravity in the Results and Observations section and estimate the percentage of sugar content remaining.

❏ 4. Place the fermenting juice into a clean Erlenmeyer flask and connect the gas trap.

C. Laboratory Session 3

This session should take place 3 weeks after Laboratory Session 2.

Materials

❏ 1. Hydrometers (3 for class use)

❏ 2. 250-mL graduated cylinders (1 per 4 students)

❏ 3. Two-ounce paper drinking cups (enough for class use)

Procedure

❏ 1. Decant the wine into the graduated cylinder and determine the specific gravity.

❏ 2. Record the specific gravity in the Results and Observations section and estimate the percentage of sugar content and the alcohol value.

❏ 3. Calculate the amount of sugar consumed.

❏ 4. Calculate the alcohol content by subtracting the second alcohol value from the first. This value represents the percentage of alcohol resulting from the fermentation.

❏ 5. Your instructor will set up a wine-tasting exercise so that you can compare the different batches produced based upon the initial sugar contents used.

Results and Observations

1. Enter your findings in the following table.

Table 47–3

Wine Production Measurements

Characteristics	Starting Grape Juice	After 1 Week	Wine After 3 Weeks
Specific gravity			
Percent sugar			
Alcohol value			

2. Complete the following:

    a. amount of sugar in starting grape juice = _____ g

    b. amount of sugar in the wine = _____ g; amount of sugar consumed = _____ g

    c. alcohol content in wine = _____ g

3. In comparing the different batches of wine made by your class, did the differences in initial specific gravity (sugar content) of the different juice preparations affect the following qualities? If so, describe the effects.

    a. alcohol content _____

    _____

    b. relative sweetness _____

    _____

    c. general character (aroma and body) _____

    _____

Laboratory Review 47   Wine: A Product of Alcoholic Fermentation

1. What substances besides alcohol are produced during the fermentation process? _____

    _____

2. How does beer making differ from wine making? (Refer to your text.) _____

    _____

    _____

Key Terms

hydrometer (hī-DROM-e-ter): a sealed, weighted, glass cylinder used to measure the specific gravity of a liquid

must (moost): the juice and crushed pulp of grapes

specific gravity: the density (weight) of a substance relative to the density of water

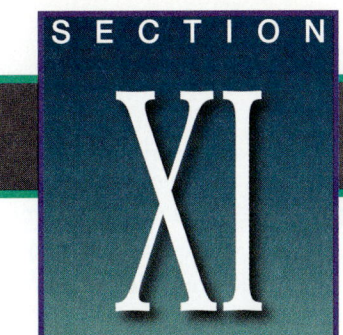

Science advances through tentative answers to a series of more and more subtle questions which reach deeper and deeper into the essence of natural phenomena.

—Louis Pasteur

Originally, the term *immunology* was used in connection with the protective responses of the body resulting from (1) a spontaneous or experimentally induced infection or (2) the injection of various substances into laboratory animals. Today, this word has acquired a much broader meaning. The modern concept of immunology embraces several areas of biological importance, including consideration of the disease-producing capabilities of infectious microorganisms. As several of the aforementioned situations have diagnostic or epidemiologic significance, a major immunologic subdivision, *serology,* was developed to deal adequately with certain antigen[1] and antibody *in vitro* activities.

Antigens have both the capacity to stimulate and provoke the formation of corresponding of homologous antibodies and the ability to react with such antibodies. The term *antibody* refers to the protein substances that are formed in response to an antigenic stimulus. All antibodies belong to a special group of serum proteins now called *immunoglobulins (IG).* Another subdivision of immunology, known as *immunochemistry,* involves laboratory investigation into the chemical and biological properties of the participating immunologic reactants—antigens and antibodies—and the mechanisms involved in their interactions.

Several states of immunity, or *resistance,* are known to exist. For example, *innate,* or *native, immunity,* which is the consequence of hereditarily controlled biochemical and biophysical factors, provides a defensive baseline against harmful foreign agents in the body. These host defense mechanisms can substantially alter the properties of disease agents. One of these mechanisms is phagocytosis. Another resistant state, *acquired immunity,* has received considerably more attention and is caused by the administration into the body of antigens, usually via parenteral routes (other than the alimentary tract). Most antigens are protein in nature, although some are carbohydrates, or a combination of the two types of substances. The response to an antigenic stimulus is the formation of antibodies, which are modified blood globulins (immunoglobulins) that can combine with the antigen responsible for their formation. All immunoglobulins have a similar structural organization. However, they do differ chemically.

The exercises in this section also will demonstrate immunodiagnostic techniques used for the detection and monitoring of infectious diseases and following the recovery from disease. Several of the techniques described have applications not only in the areas of infectious and immunological diseases, but also to the entire spectrum of clinical medicine. Specific procedures are used for measuring levels of certain drugs, hormones, serum proteins, tumor and transplantation antigens, and determining blood group incompatibilities. Forensic medicine, another area of great potential for the application of immunodiagnostic procedures, includes identification of narcotic and hallucinogenic agents, as well as sources of blood and other tissue.

Several *in vitro* techniques can be used to show the presence of antibodies. Historically, the names of the antibodies involved in such tests generally demonstrate how they affect or interact with an antigen. Examples of such procedures include *agglutination, precipitin,* and *complement fixation.* As refinements in testing develop, names are given to immunodiagnostic techniques that reflect the nature of the procedure. Examples of these techniques include *gel diffusion, radial immunodiffusion, latex particle agglutination,* and *flourescent antibody* techniques.

Agglutination is the term that describes the clumping, or aggregation, of particulate antigens such as bacterial and other cells in suspension when mixed with antisera prepared against them (homologous antibodies). In this process, the antibody is referred to as the *agglutinin* and the particulate antigen as the *agglutinogen.* In general, agglutination techniques are relatively simple to perform. They are widely used for the rapid identification of several bacteria, fungi, types of erythrocytes (blood typing), and other antigens, as well as for the detection of homologous antibodies that may be indicative of a disease state. Various factors are important to the performance of diagnostic tests based on the agglutination

---

[1]The term *immunogen* is used for the macromolecules that can stimulate an immune response (antibody production). In contrast to this usage, the term *antigen* would be applied to macromolecules that can react with antibodies but not necessarily cause their production. In this and other sections, *antigen will be used for both types of responses.*

phenomenon: temperature, pH, and electrolyte concentration. The *latex agglutination* and *coagglutination* procedures are variations of the classic agglutination test. In latex agglutination, antibodies to bacterial antigens or other proteins are attached to latex particles. This reagent is then used to detect homologous antigen in body fluids by the clumping of the latex particles. With coagglutination, the Cowan strain of *Staphylococcus aureus,* which is rich in protein A, is coated with antibodies specific for certain bacterial serological groups. Agglutination of the protein A—S, *aureus*-antibody complex, by homologous antigen is used for the immunological identification of bacteria and to detect the presence of antigens in body fluids. Variations of the latex agglutination principle can also be used to determine previous exposure to specific antigens.

Identification of certain viral pathogens can be accomplished by an inhibition of a virus-caused agglutination reaction. An example of the procedure utilizing this approach is the *hemagglutination inhibition (HI) test.* Several viruses, such as those of influenza, mumps, and Newcastle disease, are capable of agglutinating a variety of red blood cells. In these viral infections, HI-identifying antibodies may develop. These antibodies are of great diagnostic importance in identifying infecting viruses and in monitoring the recovery process of patients. In the HI test for the detection of the antibody to a particular virus, the known virus is mixed and subsequently incubated with the serum specimen under investigation. If homologous HI antibodies are present, the hemagglutination activity of the virus will be neutralized. The addition of red blood cells to this virus-antibody mixture will not result in hemagglutination. If HI antibodies had been absent from the serum specimen, hemagglutination would have occurred.

An *immunoprecipitation reaction* is an immunological phenomenon characterized by the formation of a cloudy precipitate when soluble antigen and antiserum—containing homologous antibody in solution—are in the correct quantitative proportion to one another. This state is described as an *optimal antigen-antibody ratio.* Visible reactions do not occur when the reacting substances are not present in optimal concentrations. Immunoprecipitation techniques permit both qualitative and quantitative characterization of antigens with the use of a known antiserum. In addition, the identification and potency of the antibody in preparations can be determined when testing is done with a known antigen. Although several applications of immunoprecipitation are recognized, only three techniques will be described in this section: the *ring test,* the *double-diffusion procedure,* and *radial immunodiffusion.* In the ring test, a narrow tube is partially filled with an antibody solution having a specific gravity comparable to that of the antigen-containing solution, which is carefully layered on top of it. Great care must be exercised to preserve a sharp interface between the two preparations. After a specified period of incubation, antigen-antibody complexes form at or near the interface as viable precipitin rings. In the double-diffusion, or Ouchterlony, technique, antigen- and antibody-containing solutions are placed into separate wells cut in an agar gel plate. These substances diffuse freely toward each other and form immune precipitates that are visible as bands or lines between the respective antigen- and antibody-containing wells (Figure XI–1). The applications of the double-diffusion principle provide valuable methods for the identification of unknown antigens by comparing them with known antigens and homologous antisera. The technique of radial immunodiffusion is a quantitative procedure used to determine unknown protein concentrations in patient or research samples. In this test, antigen placed in a well diffuses into an agarose (special agar) plate containing an antibody preparation. As the two entities meet, an antigen-antibody reaction occurs that produces a visible opaque precipitin ring around the well. The diameter of the ring is proportional to the protein concentration in the test specimen.

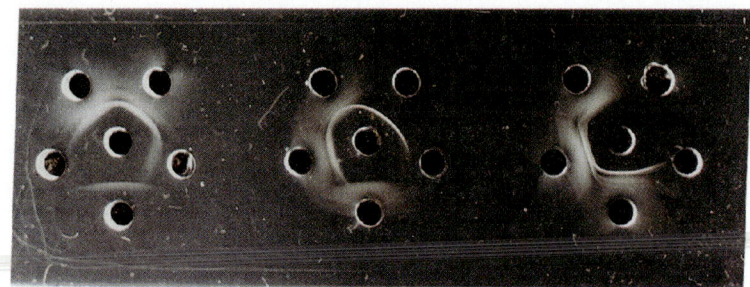

Figure XI–1
Three examples of double-diffusion systems. The visible precipitin lines show reactions of identity between the surrounding wells (antibody containing), and the respective central wells (antigen containing).

Even though the complement fixation test is considered to be classic, it is still very much in use today. *Complement* is a normal protein component of the serum of humans and other vertebrates and is designated by the letter C′. Most antigen-antibody complexes bind or fix complement. *Lysis,* or destruction of cellular antigens, can occur as a result of the combination between such antigens and homologous antibodies in the presence of C´. The reaction is known as *complement fixation*. Complement has long been used as an indicator of the development or existence of immune complexes. Red blood cells sensitized (combined) with antibody are utilized and constitute an indicator system. Diagnostic tests based on the complement fixation phenomenon are very sensitive and can be used to measure minute quantities of either antibody or antigen *(microtiter tests)*.

The enzyme-linked immunosorbent assay (ELISA) has become one of the more prominent procedures in serodiagnosis. This procedure has gained additional importance because of its application to acquired immune deficiency syndrome (AIDS). It is considered in Exercise 53.

New concepts in the delivery of health care are stimulating the development of innovative diagnostic technologies. This is certainly true in the area of diagnostic immunology. The exercises that follow will serve to acquaint you with the application of selected immunological concepts and techniques.

After completing this exercise, you should be able to:

1. Perform microbial slide agglutination tests and interpret typical test results.
2. Recognize the value of agglutination tests in the diagnosis of certain microbial diseases.

The classic serologic reaction involving the aggregation of particulate antigens by homologous antibodies, known as the *agglutination phenomenon,* was described by Gruber and Durham in 1896. Determination of the activity of antisera to agglutinate antigenic material (called *titration*) is accomplished by the addition of constant aliquots of antigen to serial serum dilutions (Figure 48–1).

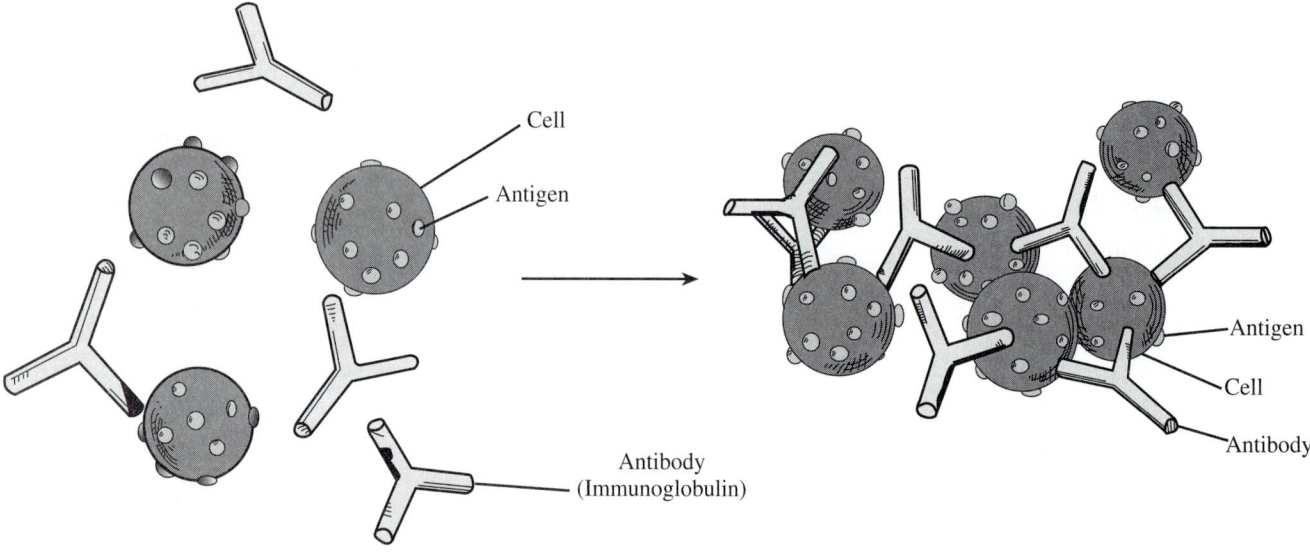

Figure 48–1

A diagrammatic view of agglutination. Agglutination occurs when antibodies bind the antigens on cells together to form large clumps. This reaction is useful in the laboratory for a variety of diagnostic tests.

The agglutination test has become widely utilized for the rapid diagnosis of several infectious diseases. It is especially effective in the case of certain bacterial pathogens in which unknown organisms can be identified on the basis of reactions occurring when suspensions are mixed separately with different antisera.

The agglutination test can be performed in several ways. Examples include macroscopic slide agglutination and tube agglutination techniques (Figure 48–2). In the slide procedure, a loopful of a heavy bacterial suspension is mixed with separate low dilutions of antisera. A control with saline instead of serum is also incorporated. The individual combinations are mixed by a gentle rocking of the slide for 1 to 3 minutes. The appearance of clumping (flocculation or granulation) in the test mixtures on the slide usually constitutes a positive test. This technique is not only rapid, but is also particularly useful in the identification of gram-negative intestinal bacteria. The tube procedure involves mixing a constant aliquot of a bacterial suspension with serial dilutions of serum. After mixing and a short period of incubation, the results are read. Again, the appearance of clumping is usually considered a positive test. In addition to aiding the diagnosis of infectious disease, this test gives a more accurate determination of the antibody level (titer).

(b)

(a)

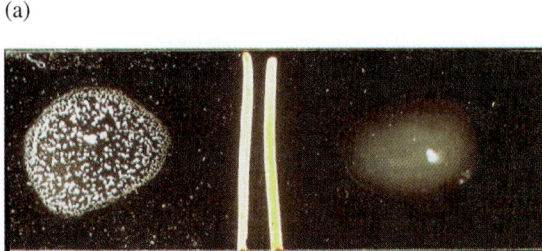

Figure 48–2
Bacterial agglutination (clumping) reactions. (a) The results of a slide agglutination. (b) The appearance of an agglutination reaction in tubes. The positive clumping reaction is obvious in both cases.

The union between the components of the test mixtures can be modified by temperature, electrolytes, and other factors. Therefore, a negative test may not necessarily be conclusive. The importance of these factors is discussed more fully in several of the references cited in the introduction to this section.

In this exercise, the slide agglutination test used in the identification of one or more bacterial pathogens will be considered. Such tests are available for the serological identification of several organisms, including enteropathogenic *Escherichia coli* (EPEC), the causative agent of acute gastroenteritis in children and adults; *Francisella tularensis,* the causative agent of tularemia (deerfly fever): and *Salmonella typhi,* the causative agent of typhoid fever. Depending on the organism under consideration, serological identification can be quite extensive and may require the detection of somatic **(O),** surface **(K),** and flagellar **(H)** antigens. This exercise will also present a variation of the agglutination phenomenon, the *latex agglutination test* (Figure 48–3).

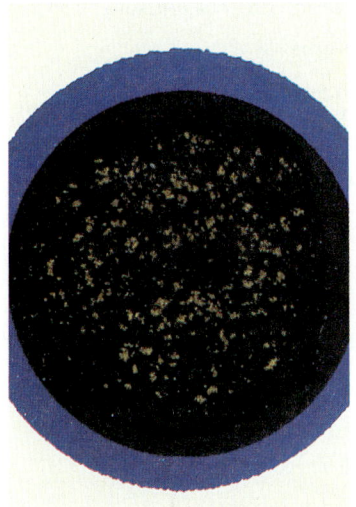

Figure 48–3
The results of a Bacto-Staph Latex Test. A positive yellow reaction (agglutination) shows the presence of *Staphylococcus aureus* clumping factor and protein A.
(Courtesy of Difco Labs., Detroit, MI.)

Approximately 75 to 85% of cases of bacterial meningitis are caused by infections with *Haemophilus influenzae, Neisseria meningitidis,* or *Streptococcus pneumoniae,* with the most common causative agent in pediatric meningitis being *H. influenzae,* type b. Because of the severity of the disease and the need for an early diagnosis in order to provide appropriate chemotherapy, the rapid identification of the causative agent is of great value. "The Directigen *H. influenzae,* type" is an antibody-coated latex test. It is used for the detection of the polyribose-phosphate capsular antigen of *H. influenzae* in specimens such as blood serum, cerebrospinal fluid, and urine. When a specimen containing the capsular antigen is mixed with the *Haemophilus* antibody-latex reagent, agglutination occurs. Appropriate controls are always included in this procedure to verify the test results. The Directigen procedure can be performed on a qualitative or quantitative basis.

Additional variations of the agglutination reaction are presented in other exercises.

A. The Slide Agglutination Test

Materials

❑ 1. Typhoid H antigen and *Salmonella* H antiserum (polyvalent)
❑ 2. Tubes of sterile 0.85% saline (1 per student)
❑ 3. Separate killed suspensions of *Escherichia coli* and *Proteus vulgaris* in sterile saline (1 mL of each). Formaldehyde should be added to give a final concentration of 0.5% (suspensions should be at least 48 hours old)
❑ 4. Two glass slides
❑ 5. Four Pasteur pipettes with rubber bulbs for use with the reagents provided. Be certain not to interchange any of these pipettes
❑ 6. Inoculating loops (1 per student)

Procedure

This procedure is to be performed by students individually.

❑ 1. Using a Pasteur pipette, place 1 drop of *Salmonella* H antiserum on 4 separate areas of a glass slide.
❑ 2. To the first and second drops of the antiserum, add a loopful of the killed suspensions of *E. coli* and *P. vulgaris*. Flame the loop for each addition.
❑ 3. Add a loopful of the typhoid H antigen to the third drop, and a loopful of saline to the drop of antisera.
❑ 4. Rock the slide back and forth gently for 2 minutes to ensure mixing. (Refer to Figure 48–2.)
❑ 5. Record your findings in Table 48–1 in the Results and Observations section. Indicate the presence of agglutination (clumping) by "+" and the absence of the phenomenon by "−."

B. Staph-Latex Test Demonstration

Materials

The following items should be provided for the instructor's demonstration.

❑ 1. One 24-hour trypticase agar divided plate containing isolated colonies of each of the following combinations:
  ❑ a. *Staphylococcus aureus* (ATCC 25923) [coagulase positive] and *S. epidermidis* (ATCC 12228) [coagulase negative]
  ❑ b. Unknown A and unknown B
❑ 2. The following reagents in appropriate dispensers:
  ❑ a. Staph-Latex Reagent
  ❑ b. Staph Positive Control
  ❑ c. Staph Negative Control
  ❑ d. Normal Saline Reagent
❑ 3. Disposable Test Cards (slides)
❑ 4. One inoculating needle
❑ 5. One wax marking pencil (yellow or orange)

Procedure

❏ 1. Obtain 1 Disposable Test Slide and label 1 circle for each of the following materials using the letters shown:
   - ❏ a. Positive control
   - ❏ b. Negative control
   - ❏ c. *S. aureus*
   - ❏ d. *S. epidermidis*
   - ❏ e. Unknown A
   - ❏ f. Unknown B

❏ 2. Place 1 drop of the Staph Positive Control into circle *a* on the test slide.

❏ 3. Place 1 drop of the Staph Negative Control into circle *b*.

❏ 4. Add 1 drop of the Staph-Latex Reagent to each of the above circles containing controls and to circles *c* through *f*.

❏ 5. Flame the inoculating needle and quickly transfer three isolated *S. aureus* colonies to its specified circle, *(c)*, and mix this growth thoroughly with the latex reagent.

❏ 6. Repeat step 5 with each of the cultures provided, using their appropriately marked circles.

❏ 7. Rotate the slide by hand in a circular manner for 45 seconds and read the reactions quickly. Look for the formation of yellow clumps. (Refer to Figure 48–3.)

❏ 8. Indicate the presence of agglutination by a "+" and its absence by a "−" in Table 48–2 of the Results and Observations section.

C. The Directigen Latex Agglutination Test

Materials

❏ 1. One mL each of separate heat-killed suspensions of *Haemophilus influenzae*, type b, and *Escherichia coli* in sterile saline (0.85% NaCl)

❏ 2. One mL of an unknown culture

❏ 3. One mL of sterile 0.85% saline

❏ 4. Four glass slides or serological slides

❏ 5. Eight Pasteur pipettes with rubber bulbs for use with the test materials and test reagents. (Be certain not to interchange any of these pipettes.)

❏ 6. Bactogen *Haemophilus influenzae*, type b reagents including:
   - ❏ a. *Haemophilus* antibody-latex preparation
   - ❏ b. Control latex suspension
   - ❏ c. *Haemophilus* polyribose-phosphate (PRP) positive control
   - ❏ d. *Haemophilus* (PRP) negative control

❏ 7. One serological slide shaker or rotator

❏ 8. One glass marking pen or pencil

❏ 9. One oblique light source

❏ 10. One black paper square 10 × 10 cm or 4 × 4 inches for reading of test reactions

Procedure

This test will be demonstrated by the instructor unless other directions are given.

❏ 1. If serological glass slides are not available, then with a marking pencil mark four separate sections on four individual slides. Allow enough distance between the sections so that reagents used during the procedure will not run into one another.

❏ 2. Label 1 section of each slide for the following: test culture, latex control, positive PRP control, and negative PRP control.

❏ 3. On 1 side of each slide, indicate the culture material to be tested.

❏   4. Carefully shake each of the Bactogen reagent containers and test materials.

❏   5. With a Pasteur pipette, put 1 drop of the *Haemophilus*-antibody-latex preparation on the sections for test culture, positive PRP control, and negative PRP control.

❏   6. With a separate pipette, add 1 drop of one of the test cultures to the appropriate slide section with the *Haemophilus*-antibody-latex preparation.

❏   7. Using another pipette, add 1 drop of the positive PRP control reagent to the sections for latex control and positive PRP control.

❏   8. With separate pipettes, add 1 drop each of the latex control and the negative PRP control reagents to their respective sections on the slide.

❏   9. Rock the slide gently by tilting it at an angle of approximately 45° in order to mix the reagents and test culture.

❏ 10. Continue to rock the slide for 5 minutes, either manually or by using the serological shaker, if provided.

❏ 11. Look for agglutination in the various slide sections by placing the slide on the surface of a square of black paper. Hold the slide at various angles to see the reaction.

❏ 12. Record your findings in Table 48–3 in the Results and Observations section. Indicate the presence of latex agglutination by a "+" and the absence of the reaction by a "−."

❏ 13. Repeat this procedure with each of the test culture materials provided.

Results and Observations

The Slide Agglutination Test

1. Complete the following table.

Table 48–1

Slide Agglutination Results

Test Material	*Salmonella* H Antiserum
E. coli	
P. vulgaris	
Typhoid H antigen	
Saline	

Staph-Latex Test Demonstration

1. Enter your findings in the following table.

2. On the basis of your findings, did either of the unknowns produce a positive test? If so, which one(s)?

    _____

    _____

3. What does a positive reaction mean in this text? _____

    _____

Table 48–2

Staph-Latex Test Results

Material for Testing	Reaction
S. aureus	
S. epidermidis	
Positive control	
Negative control	
Unknown A	
Unknown B	

The Directigen Latex Agglutination Test

1. Complete the following table.

Table 48–3

Directigen Results

Test Material	Test Reagents			
	Haemophilus-Antibody-Latex	Control Latex	Positive Standard Reference	Negative Standard Reference
H. influenzae, type b				
E. coli				
Unknown culture				
Saline (0.85%)				

Laboratory Review 48   Microbial Agglutination Reactions

1. Name two infectious diseases that can be diagnosed with the aid of an agglutination test.

   a. _____   b. _____

2. Briefly describe the principle of the following immunoserological tests:

   a. fluorescent antibody procedure _____

   _____

   b. latex agglutination _____

   _____

3. Complete the following table by indicating applications of the tests listed.

Table 48–4

Microbial Agglutination Applications

Test	Application
Agglutination	1.
	2.
Latex agglutination	1.
	2.

4. What is a heterophile antibody? (Refer to your text.) _____
    _____

5. Briefly define a procedure for the preparation of a bacterial vaccine. (Refer to your text.)
    _____
    _____
    _____

6. List six components of a bacterial cell that could serve as antigenic (immunogenic) components. (Refer to your text.)

    a. _____   d. _____

    b. _____   e. _____

    c. _____   f. _____

Key Terms

agglutination (a-gloo-ti-NĀ-shun): the visible clumping of cells or particles resulting from the mixing of specific antigens with their corresponding immunoglobulins (antibodies)

agglutinin (a-GLOO-ti-nin): an immunoglobulin (antibody) that causes the agglutination or clumping of a specific antigen

agglutinogen (a-gloo-TIN-ō-jen): the specific substance that stimulates the formation of a corresponding agglutinin (antibody)

Aliquot (AL-i-kwot): a measured portion of an original quantity of a substance

gastroenteritis (gas-trō-en-ter-Ī-tis): inflammation of the stomach and intestinal tract

hemolysis (hē-MOL-i-sis): disruption of red blood cells

meningitis (men-in-JĪ-tis): inflammation of the coverings of the brain and spinal cord

somatic (sō-MAT-ik) antigen: refers to antigens located at the cell surface and distinguishable from those of flagella or capsules

titer (TĪ-ter): the highest dilution of an active substance, such as antibody, that still causes a visible or recognizable reaction

After completing this exercise, you should be able to:

1. Recognize the value and limitations of rapid diagnostic tests.

2. Perform and interpret selected rapid diagnostic tests.

3. Understand the immunological bases of certain rapid diagnostic tests.

4. Recognize the wide applications of rapid diagnostic tests in cases of infectious diseases and non-disease situations.

New concepts in the delivery of health care are stimulating the development of innovative diagnostic technologies. Diagnostic testing is increasingly shifting from centralized laboratories to alternative sites, such as physicians' offices and clinics and, eventually perhaps to the home to provide screening and the early detection of medical conditions and diseases. Although the greatest amount of *in vitro* diagnostic testing in the United States is performed in centralized clinical laboratories, the emergence of accurate and cost-effective raid diagnostic assays have made alternate-site testing a reasonable option (Figure 49–1).

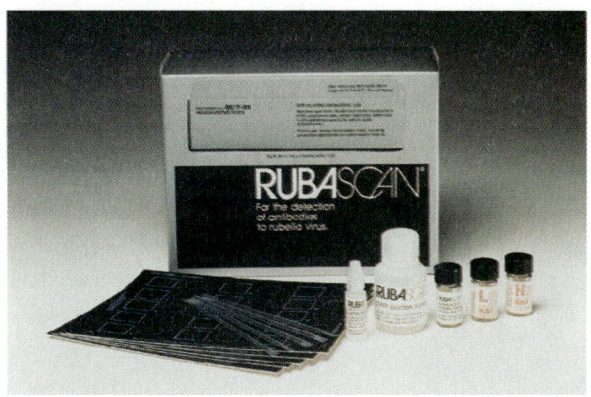

Figure 49–1

An example of one of many available rapid diagnostic tests. The various components of a standard kit are shown. These include test cards, buffer, reagents, and positive and negative controls.
(Photo courtesy of Becton Dickinson Microbiology Systems.)

The rapid assays in order to be usuable must perform well under a wide variety of conditions. Many of these systems offer advantages in terms of accuracy, ease of use, and rapidity of results. Examples of such assays include the *Particle Immuno Filtration Assay* (PIFA), latex agglutination tests (Figure 49–2*b*), immunoenzymatic assays using monoclonal antibodies, and the Single Use Diagnostic System (SUDS) test (Figure 49–2*c*).

The PIFA system is based on the selective filtration of microparticles in response to antibody/antigen binding. The test or assay involves combining a test sample (blood serum, urine, or saliva) with a reagent, which consists of microparticles coated with antigen or antibody. During a 1-minute incubation, the reagent is allowed to react with the test sample. Positive test samples containing corresponding antigens or antibodies cause the particles to form a matrix. Negative test samples without these substances leave the particles dispersed. Once the reagent has reacted with the sample, the mixture is introduced into a membrane device designed to separate matrixed from nonmatrixed particles. This membrane device is a composite of several membranes laminated together, combining controlled pore and liquid flow dynamic properties.

Thus, a positive sample produces a matrix of microparticles that are filtered by the controlled pore membranes. No particles, and hence no color, are able to migrate into successive membrane layers. In this case, color is only visible in the first viewing window of the membrane device. Conversely, a negative sample does not produce a matrix of particles; the dyed particles pass through the controlled pore membrane and into successive membrane layers, where they become visible through a second viewing window in the device.

(b)

(a)

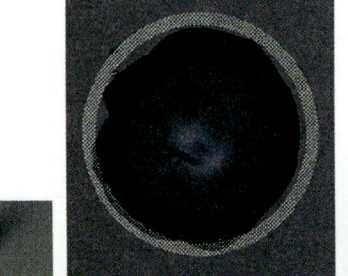

(c)

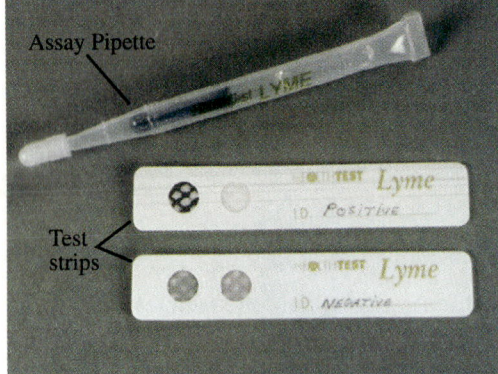

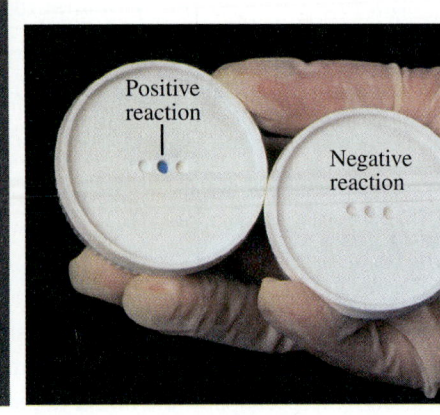

Figure 49–2

(a) The particle Immuno Filtration Assay (PIFA) components. Both positive and negative test results are shown.[1] (b) The results of a latex rubella test. A positive result is indicated by a clumping reaction. Results occur within 10 minutes.[2] (c) A Single Use Diagnostic System (SUDS). Positive and negative results are shown.[3]

[1](Courtesy of Akers Laboratories, Inc.)

[2](Photo courtesy of Becton Dickinson Microbiology Systems.)

[3](Courtesy of Murex Corporation.)

The *Staphaurex test* is a rapid slide agglutination procedure for differentiating staphylococci, which possess coagulase (clumping factor), and/or protein **A** from staphylococci, which possess neither of these factors. The principle involved is based on the fact that staphylococci having the clumping factor and protein **A** can be identified using human plasma-coated latex particles, which agglutinate. The reagent used in this procedure consists of polystyrene latex particles that have been coated with fibrinogen and immunoglobulin G (IgG). When mixed with a suspension of organisms that possess coagulase (the clumping factor) and/or or protein **A,** a strong agglutination of the latex particles results (Figure 49–3).

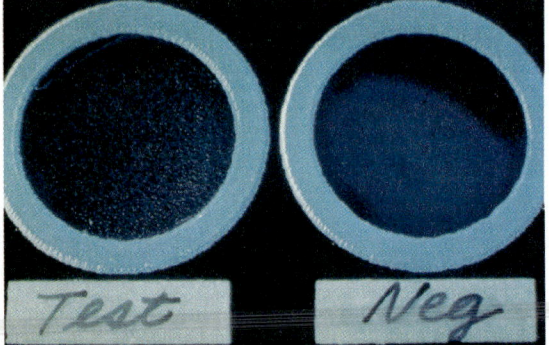

Figure 49–3

The results of the Staphaurex test. A positive reaction is indicated by clumping.
(Courtesy of Burroughs Wellcome.)

Another example of a rapid diagnostic test based on immunological principles is the *Pregnospia test,* an immunoenzymatic assay for pregnancy determination. It uses a revolutionary type of diagnostic tool, the *monoclonal immunoglobulin* (antibody). The production of pure or monoclonal (single-type) antibodies is achieved by an *in vitro* hybridization process in which an antibody-producing cell is fused with a myeloma (particular cancer) cell under special experimental conditions. In this process, the genetic materials of both cells are combined to create a hybrid cell that contains desirable characteristics from both of the parent cells.

The Pregnosia test is an *agglutination immunoassay*. The monoclonal antibodies mentioned earlier are coated onto special gold particles, which have an intense reddish-purple color in solution. These antibodies are noted for their specificity for special sites on the intact pregnancy hormone (human chorionic gonadotropin (HCG) molecule) and one of its polypeptide subunits. When the HCG antibody preparation reacts with HCG present in a urine specimen, connecting molecular bridges are formed, binding the gold particles together and thereby causing them to clump or agglutinate. The resulting agglutination is evidenced by a distinct color change from the original reddish-purple color of the HCG antibody preparation. If there is no HCG present in the urine sample, the agglutination reaction will not change color. This result is interpreted as a negative reaction.

The *Single Use Diagnostic System* (**SUDS**) HIV test is a visually read, rapid immunoassay for the detection of antibodies to the human immunodeficiency virus-1 in human serum or plasma. The test is based on a microfiltration enzyme immunoassay concept. The *solid phase capture reagent* consists of a mixture of latex particles coated with a specific inactivated HIV antigen (p24) prepared from the virus and a purified synthetic peptide similar to the HIV transmembrane protein.

The SUDS device is a small plastic cartridge designed to filter, concentrate, and absorb all liquid reagents added during the test, including the specimen. This absorbent feature is intended to reduce the potential for contamination of the work space and personnel with infectious materials. The SUDS device does not contain any reagents before starting the test procedure.

This exercise will present several examples of rapid diagnostic tests.

Caution

Working with blood or other body fluids will not pose any danger if the following precautions are observed:

❏ 1. Place any and all blood, or materials containing blood, into the disinfectant containers provided.

❏ 2. Dispose of any blood-containing materials as indicated by the instructor.

❏ 3. In cases of blood spillage, wipe the area with a disinfectant-soaked paper towel, and dispose of all materials as indicated by the instructor.

❏ 4. Always wash your hands after handling blood and associated materials.

❏ 5. Use surgical gloves and protective eyewear in specimen-handling situations.

A. Particle Immuno Filtration Assay (PIFA) Demonstration

Materials

The following items should be provided for the instructor's demonstration:

❏ 1. Health Test Infectious Mononucleosis Assay self-contained pipettes (or other Health Test Assay)

❏ 2. Health Test Infections Mononucleosis Assay test strips (Akers Laboratories)

❏ 3. Known positive control in sterile tube

❏ 4. Known negative control in sterile tube

❏ 5. Unknown serum specimen A in sterile tube

❏ 6. Unknown serum specimen B in sterile tube

❏ 7. Container with disinfectant

❏ 8. Disposable surgical gloves

Procedure

❏ 1. Obtain one Health Test Infectious Mononucleosis Assay pipette and test strip.

❏ 2. Remove the guard covering the pipette and draw up the positive control specimen into the pipette. Replace the guard (cover).

❏ 3. Gently crush the ampule in the pipette and mix the contents of the pipette by pressing on the pipette.

❏ 4. Wait one minute.

❏ 5. Place a drop of the reaction mixture (specimen and pipette reagent) onto each of the viewing parts on the test strip.

❏ 6. If the first viewing port on the strip turns black and the second port remains white, the test is *positive*. If the second port also turns black, the test is *negative*.

❏ 7. Repeat steps 1–5 with the other control and specimens.

❏ 8. Enter your findings in Table 49–1 and answer the questions in the Results and Observations section.

B. Staphaurex Test Demonstration

Materials

The following items should be provided for the instructor's demonstration:

❏ 1. One 24-hour trypticase soy or blood agar plate containing isolated colonies of each of the following combinations:
    ❏ a. *Staphylococcus aureus* (ATCC 25923) [Coagulase positive] and *S. epidermidis* (ATCC 12228) [Coagulase negative]
    ❏ b. Unknown A and unknown B
❏ 2. The following reagents in appropriate dispensers with droppers:

    ❏ a. Staphaurex latex (Becton Dickinson Microbiology Systems)
    ❏ b. Normal saline
❏ 3. Disposable test slides with multiple specimen circles
❏ 4. Mixing sticks
❏ 5. Container with disinfectant
❏ 6. Disposable surgical gloves

Procedure

❏ 1. Obtain one disposable test slide and mark one circle for each of the following materials using the letters indicated:
    ❏ a. P, positive control
    ❏ b. N, negative control
    ❏ c. S, *Staphylococcus aureus*
    ❏ d. E, *S. epidermis*
    ❏ e. A, unknown A
    ❏ f. B, unknown B
    ❏ g. C, saline

❏ 2. Place one drop of *Staphaurex latex* into each specimen circle.

❏ 3. Place one drop of the positive and negative controls and saline into their respective circles.

❏ 4. With a mixing stick, pick at least two *S. aureus* colonies from the plate provided and mix them with the reagent in the correctly labeled circle. Dispose of the stick in the container with disinfectant.

❏ 5. Repeat step 4 with each of the cultures provided.

❏ 6. Gently rock the slide(s) for 20–30 seconds and read the results. Look for clumping.

❏ 7. Enter your findings in the Results and Observations section.

C. Human Chorionic Gonadotropin (HCG) Detection

Materials

The following materials should be provided per 2 students. (All reagents are available in the commercial kit.)

❏ 1. Four tubes of Organon Accusphere reagent (gold solid particles bonded with monoclonal antibodies against HCG)

❏ 2. Buffer (2 mL)

❏ 3. One 5-ml pipette, calibrated in 0.1-ml units, and a rubber bulb

❏ 4. Three 1-ml pipettes, calibrated on 0.1-mL units, and rubber bulbs

❏ 5. One known positive urine control

❏ 6. One known negative urine control

❏ 7. Two unknown specimens (one per student)

❏ 8. One marking pen or similar device

❏ 9. Container for disposing of pipettes

❏ 10. Two pairs of disposable surgical gloves

Procedure

This procedure is to be performed by students in pairs.

❏ 1. Obtain 4 tubes containing the Accusphere reagent.

❏ 2. Number the tubes accordingly: **1** (known positive), **2** (known negative), **3** (unknown), and **4** (unknown).

❏ 3. Remove the rubber stoppers from the 4 tubes.

❏ 4. With the 5-mL pipette, introduce 0.4 mL of the buffer provided to each tube.

❏ 5. Replace the stoppers.

❏ 6. Gently shake each tube for approximately 15–30 seconds until the Accusphere reagent is completely resuspended.

❏ 7. Using separate 1-mL pipettes, add 0.1 mL of the positive, negative, and each of the unknown samples to their respectively marked tubes.

❏ 8. Gently shake the tube for about 30 seconds to achieve thorough mixing.

❏ 9. Place the tubes in a rack or on a flat surface, and wait 5 minutes.

❏ 10. Examine the tubes for a color change. If a change does not occur within this time period, continue to observe the tubes for another 25 minutes.

❏ 11. Compare the unknown tubes with the results shown in the positive and negative samples (tubes 1 and 2).

❏ 12. Interpret your findings, and enter them in Table 49–2 in the Results and Observations section.

❏ 13. Check your results with the instructor.

❏ 14. Answer any questions pertaining to this exercise in the Laboratory Review section.

D. Single Use Diagnostic System (SUDS)

Procedure

❏ 1. Place one sample cup in a SUDS test cartridge.

❏ 2. With a micropipette add two drops of the positive control to the sample cup (Figure 49–4).

❏ 3. Discard the micropipette in the container with disinfectant.

❏ 4. Continue by performing the steps shown in Procedure Diagram 38.

❏ 5. Interpret the reaction and enter your findings in Table 49–4 in the Results and Observations section. (Refer to Figure 49–2.)

Materials

The following items should be provided for the instructor's demonstration:

❏ 1. One dropper vial latex-antigen suspension (latex particles with bound HIV-1 virus antigens)

❏ 2. One dropper vial enzyme-antibody conjugate (goat antihuman immunoglobulin G, and antihuman immunoglobulin M conjugated to alkaline pyhosphatase)

❏ 3. One bottle diluent (buffered protein solution)

❏ 4. One bottle wash reagent

❏ 5. One dropper vial, reaction stop solution (0.3M citrate solution)

❏ 6. One container positive control

❏ 7. One container negative control

❏ 8. Unknown serum specimen A

❏ 9. Unknown serum specimen B

❏ 10. Four micropipettes for specimen delivery

❏ 11. Two calibrated droppers

❏ 12. Four SUDS test cartridges

❏ 13. Four sample cups

❏ 14. Container with disinfectant

❏ 15. Disposable surgical gloves

❏ 6. Repeat steps 1 through 3 with the negative control and the unknown samples.

❏ 7. Answer the questions in the Results and Observations and Laboratory Review sections.

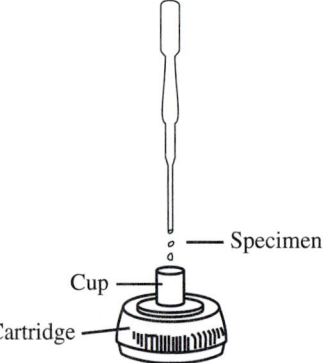

Figure 49–4

Adding a specimen to a sample cup. The cup is contained within a SUDS test cartridge.

Procedure Diagram 38
Single Use Diagnostic System (SUDS)

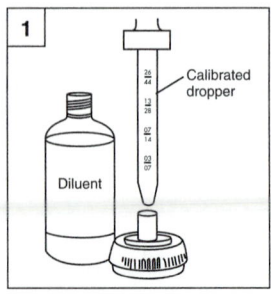

1. Add 0.5 mL of **diluent** to sample cup.

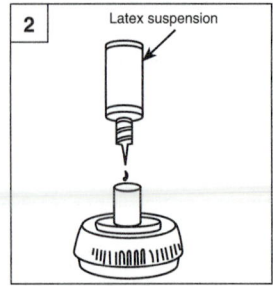

2. Add 1 drop of the **latex-antigen suspension** and wait 3 minutes.

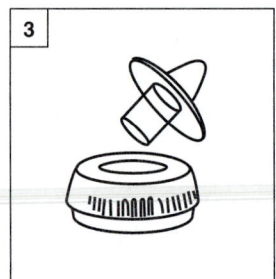

3. Pour complete contents into SUDS cartridge. Discard sample cup into container with disinfectant.

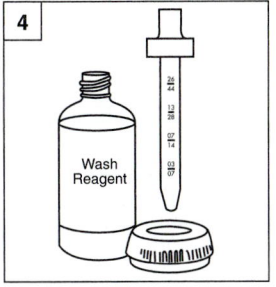

4. With the calibrated dropper, add 0.5 mL of the **wash reagent** and allow the cartridge to drain completely.

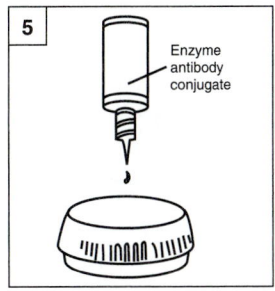

5. Add 1 drop of the **enzyme antibody conjugate**. Wait 3 minutes.

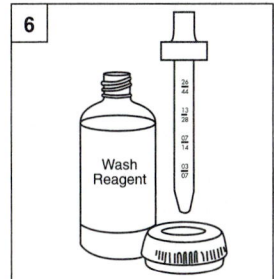

6. With a calibrated dropper add 0.5 mL of **wash reagent**. Allow cartridge to drain completely.

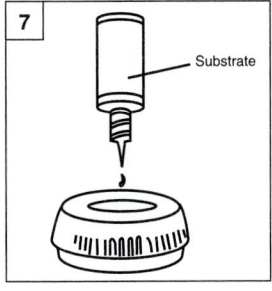

7. Add 1 drop of substrate. Wait for 2 minutes.

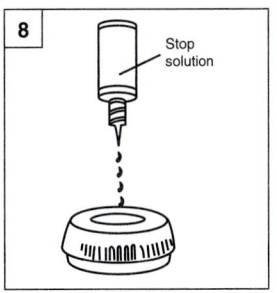

8. Add 4 drops of the **stop solution.** Allow to drain completely.

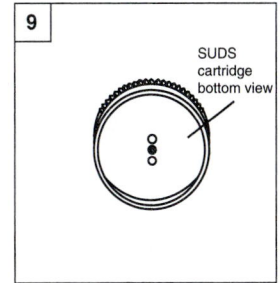

9. Examine cartridge. A positive reaction is indicated by a blue color in the center dot of the **bottom** on the SUDS cartridge.

Results and Observations

Particle Immuno Filtration Assay (PIFA) Demonstration

1. Enter your findings in the following table.

Table 49–1

Particle Immuno Filtration Assay

Material for Testing	Reaction
Positive Control	
Negative Control	
Unknown serum specimen A	
Unknown serum specimen B	

2. Did either of the unknown specimens produce a positive reaction? If so, which one(s)? _____

Staphaurex Test Demonstration

1. Enter your findings in the following table.

Table 49–2

Staphaurex Reactions

Material for Testing	Reaction
S. aureus	
S. epidermidis	
Positive control	
Negative control	
Unknown A	
Unknown B	

2. On the basis of your findings, did either of the unknowns produce a positive test? If so, which one(s)?

_____

_____

3. What does a positive reaction mean in this test? _____

_____

Human Chorionic Gonadotropin (HCG) Detection

1. Enter your findings in the following table.

Table 49–3

HCG Reactions

Specimen	Agglutination		HCG Present	Indicative of Pregnancy (Positive/Negative)
	Positive	Negative		
Known positive				
Known negative				
Unknown 1				
Unknown 2				

2. What is human chorionic gonadotropin? _____

Single Use Diagnostic System (SUDS)

1. Enter your findings in Table 49–4.

2. Which of the unknown specimens gave a positive reaction? _____

3. What does a positive result mean in this test? _____

Table 49–4

Single Use Diagnostic System Reactions

Material for Testing	Reaction
Positive Control	
Negative Control	
Unknown serum specimen A	
Unknown serum specimen B	

Laboratory Review 49 — Survey of Commercially Available Rapid Diagnostic Tests

1. Explain the bases of the positive reactions that occur in the following tests.

   a. *Staphaurex* _____

   _____

   b. Pregnospia test _____

   _____

   c. PIFA _____

   _____

   d. SUDS _____

   _____

2. Complete the following table by indicating the appearance of both positive and negative reactions.

Table 49–5

Rapid Diagnostic Test Results

Test	Positive Reaction	Negative Reaction
PIFA		
Staphaurex		

Key Terms

coagulase (kō-AG-ū-lāz): an enzyme that causes coagulation

fibrinogen (fī-BRIN-ō-jen): protein in blood plasma that can be converted into fibrin; major component of a clot

latex: a form of plastic monomer, similar to rubber latex, which is used to manufacture very small plastic beads, usually of polystyrene

protein A: an antibody-binding protein found in the cell walls of *Staphylococcus aureus* (Cowan I strain)

solid phase capture reagent: in this exercise, the reagent is the component of the test that reacts with HIV antibody if present

Agglutination: Blood Typing and Cross-Matching

After completing this exercise, you should be able to:

1. Understand the basis of blood typing.
2. Determine the blood type of an unknown specimen.
3. Perform a cross-match for blood compatibility.

Blooding typing and antibody detection techniques are important in clinical medicine for: (1) matching donors and recipients of blood transfusions or organ grafts; (2) identifying and protecting against immunizing women exposed to the D antigen of the Rh system, during pregnancy or delivery (also known as *isoimmunization);* (3) predicting, diagnosing, and treating hemolytic disease of the newborn (also known as the Rh baby); and (4) diagnosing and studying red blood cell destruction caused by antibodies produced against the body's own tissues.

In recent years, the medicolegal aspects of blood grouping have assumed great importance. Standard blood-grouping procedures have provided and continue to provide such information as the compatibility of blood donors and recipients, identification of blood stains, detection of rare blood types, potential dangers in pregnancy, and facts that help decide cases of disputed parentage.

At the present time, more than 50 blood factors are known to exist in the human species. Examples of these types include A, B, O, AB, M, N, S, Rh_o, Hr_o, and rh'. The first three major blood factors—A, B, and O—were discovered by Karl Landsteiner in 1900. Landsteiner's findings firmly established the basis for modern-day procedures in determining blood compatibility or incompatibility. As a consequence of mixing serum with erythrocytes from donors, he noted agglutination reactions that demonstrated the existence of the A and B blood groups. In some blood samples, no visible antigen-antibody interactions developed. Such blood was first classified as C, but later this designation was changed to O. The fourth and last of the major blood types, AB, was discovered by von Decastello and Sturli in 1902. Since the time of Landsteiner's findings, several other subgroups of the A antigen have been discovered. The two most common subgroups are A_1 and A_2. In addition, about four-fifths of blood type AB persons have the subgroup A_1 antigen.

In addition to having specific blood group antigens, the human body may possess antibodies capable of dissolving or agglutinating red blood cells from individuals with other blood group antigens (Table 50–1). These immune globulins are referred to as *isohemolysins* and *isohemagglutinins,* respectively. Normally, antibodies found in the blood serum or plasma of an individual are not directed against blood factors present. However, it is possible for an individual to have antibodies against blood factors that are absent from his or her red blood cells. The International System of Nomenclature, proposed by Landsteiner, Von Dungern, and Hirschfeld, listed together with the relationship of the blood groups as

Table 50–1

Blood Group Properties

Blood Groups	ABO System Agglutinogens	Isohemagglutinin
A	A	β
B	B	α
AB	A,B	—
O	O	α,β

to agglutinogens and agglutinins, follows. The specific antibodies anti-A and anti-B are designated by the Greek letters α (alpha) and β (beta), respectively.

Chemically, the BO antigenic factors (*agglutinogens*) are composed of polysaccharide amino acid complexes. A and B antigens are found not only in red blood cells, but in several other types of tissue cells as well, including sperm, liver, and spleen. The O factor, on the other hand, appears to be associated only with erythrocytes. The blood group substances A and B have also been demonstrated in several blood fluids, such as gastric juice, perspiration, saliva, and seminal fluid. Individuals producing blood group substances in this manner are referred to as *secretors*. The secretor characteristic is a genetically controlled property.

AB individuals do not possess antibodies against any of the remaining major blood types. Persons with this blood type are referred to as *universal recipients*. Furthermore, none of the major blood types have antibodies against O blood cells. Consequently, members of this group are referred to as universal donors.

One can determine the blood type of an individual by observing agglutination reactions that incorporate sera from A and B individuals. These sera are mixed separately, either on a glass slide or in a test tube, with unknown blood cell specimens. The possible reactions are shown in Table 50–2 and Figure 50–1. A positive agglutination reaction is represented by + and a negative reaction by −. The anti-A serum used in most blood type determinations will agglutinate both A_1 and A_2 red cells. In routine testing, it is not necessary to distinguish A_1 and A_2 red cells. When necessary, anti-A_1 serum is used to identify blood type A_1 individuals. This exercise will be limited to the routine blood-typing procedure.

Table 50–2

Blood-Typing Reactions

Antigen on Donor's Cells	Agglutinated by Serum Antibodies	
	Anti-A	Anti-B
A	+	−
B	−	+
AB	+	+
O	−	−

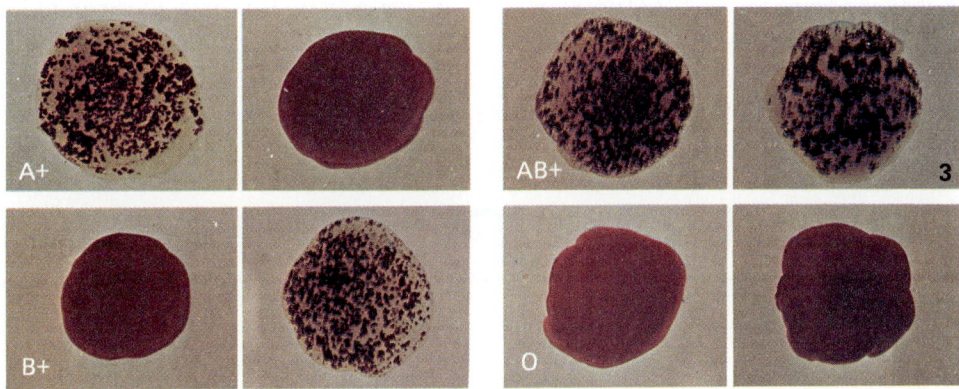

Figure 50–1
Typical blood agglutination reaction. The left-hand side of each slide contains anti-A serum, while the right sides contain anti-B serum. Slide 1 shows that characteristic reaction for blood type A. Slide 2 represents blood type B. Slide 3 shows blood type AB. Slide 4 demonstrates the reaction for blood type O.

Both false-positive and false-negative reactions may be encountered in testing. In order to prevent these reactions, proper controls must always be carried out. Suitable control measures include pretesting of sera for activity, dilution of cells or sera in physiological saline if a reaction has not occurred after a sufficient length of time, and the incorporation of tests with known cells and sera.

Hemolysis of erythrocytes can also occur occasionally. Usually, this reaction results from mixing incompatible blood cells with unheated sera. It is caused by isohemolysins rather than isohemagglutinins. The heating of sera at 56°C for 30 minutes will usually prevent hemolysis from taking place. For blood-grouping purposes, this phenomenon is considered to be equivalent to agglutination.

With the discovery of the ABO system, it was believed that transfusion problems could not develop if the individuals involved belonged to the same blood group. Unfortunately, between 1921 and 1939, hemolytic transfusion reactions were reported even though blood-typing tests showed donor and recipient compatibility. No explanation for these reactions was proposed.

In 1937, Landsteiner and A.S. Wiener uncovered antisera that agglutinated rhesus monkey blood as well as cells of 85% of Caucasians (Figure 50–2). This phenomenon was independent of the reactions obtained with the four major blood types. The factor was designated *Rh,* for *rhesus,* indicating the source of the agglutinogen for the investigation as well as the manner in which the finding was made. The first formal report on the Rh factor was made in 1940. The importance of the factor was realized when the relationship of the antibody response of sensitized Rh-negative individuals to the administration of Rh-positive antigens was demonstrated. This reaction was recognized as the principle cause of *erythroblastosis fetalis,* or the *hemolytic disease of the newborn,* and as the cause of many transfusion reactions.

Figure 50–2

$Rh_O(D)$ reactions. The presence of clumping (agglutination) indicates a positive reaction for the $RH_O(D)$ antigen.

Several other Rh factors were found after the initial discovery by Landsteiner and Wiener. The complexity of the system designated *Rh-Hr* was soon realized. Three principal factors were uncovered and comprise the determinants of the eight principal Rh types known. The small letter *r* associated with certain of these factors designates them to be less antigenic than the factor discovered originally by Landsteiner and Wiener, namely Rh_o. The latter agglutinogen, also designated *D,* is the most important clinically, and is responsible for the majority of cases involving sensitization. The Rh-Hr system designations are shown in Table 50–3 according to the two different nomenclature schemes commonly employed.

Table 50–3

The Rh-Hr System

Wiener Scheme	Fisher-Race Scheme
Rh_O	D
Hr_O	d
rh′	C
hr′	c
rh″	E
hr″	e

Rh antibodies do not occur naturally in humans. They may develop by transfusing an Rh-negative individual with Rh-positive blood, or as a consequence of an Rh-negative woman bearing an Rh-positive fetus.

Persons about to receive or donate blood are tested for the presence or absence of the Rh$_o$ factor. In the event that this factor is absent, Rh-negative blood is administered. It should be noted at this point, especially in the case of transfusions, that even though the Rh$_o$ factor is lacking, the other factors may be present. The safest procedure is to transfuse with Rh-negative blood.

Blood group antibodies are important in cases of transfusion. The procedure known as *compatibility testing* or *cross-matching* is carried out to prevent a transfusion reaction because of blood group-incompatible recipients and donors. Specifically, the cross-matching technique is designed to detect any incompatibility between the recipient's serum and the donor's cells (this procedure is known as the *major cross-match*) or between the recipient's cells and the donor's serum (this procedure is referred to as the *minor cross-match*). The cross-match procedure is done after the respective blood specimens have been typed as to the ABO and Rh factors and any other factor that appears to be indicated. Agglutination or lysis of the red cells from either the donor or the recipient in these tests indicates an incompatible situation. In this exercise, the procedures for blood typing and cross-matching will be performed.

A. Blood Typing

Materials

The following materials should be provided per 4 students:

❏ 1. One container of physiological saline with an eye dropper

❏ 2. Four commercial alcohol preparation pads (Prep Pads)

❏ 3. Four sterile gauze square (5 × 5 cm or 2 × 2 inches)

❏ 4. Four sterile blood lancets

❏ 5. Eight Disposo-type blood-typing slides or glass slides

❏ 6. Eight wooden applicator sticks

❏ 7. Blood-typing antisera, anti-A, anti-B, and anti-D (Rh$_o$), to be dispensed by the instructor

❏ 8. Unknown blood specimens (1 per student)

❏ 9. Container of disinfectant

❏ 10. Four pairs of surgical gloves

Procedure

Caution

Working with blood or other body fluids will not pose any danger if the following precautions are observed:

❏ 1. Place any and all blood or materials containing blood into the disinfectant containers provided.

❏ 2. Dispose of any blood-containing materials as indicated by the instructor.

❏ 3. In cases of blood spillage, wipe the area with a disinfectant-soaked paper towel, and dispose of all materials as indicated by the instructor.

❏ 4. Always wash your hands after handling blood and associated materials.

❏ 5. Use surgical gloves and protective eyewear in specimen-handling situations.

This procedure is to be performed by students individually. Surgical gloves should be worn while working with a blood sample.

❏ 1. Place 1 drop of each antiserum on separate areas of the blood-typing slide.

❏ 2. Carefully wipe the tip of the middle finger with an alcohol prep pad.

❏ 3. Allow the wiped area to dry.

❏ 4. Unwrap the blood lancet and prick the prepared finger with it.

❏ 5. Allow 1 full-sized drop of blood to fall directly into each respective drop of antisera.

❏ 6. Place a fresh, sterile gauze square on the punctured area and hold it tightly in place until the bleeding stops.

❏ 7. Take an applicator stick and mix the components in each area. *Use separate ends or different sticks for each mixture.*

❏ 8. If a reaction does not develop after 2 minutes, warm the slide over a desk lamp or similar device. If this procedure still does not alter the result, add 1 or 2 drops of physiological saline to the mixture. Mix the suspension and observe. This procedure should rule out a false-negative result.

❏ 9. Record your findings in the Results and Observations section. (See Figures 50–1 and 50–2.)

Results and Observations

1. Blood type _____

2. Rh factor _____

B. Unknown Specimen

Procedure

This procedure is to be performed by students individually.

❏ 1. Each student will receive 1 unknown blood specimen.

❏ 2. Determine the major blood type and Rh factor or factors.

❏ 3. Record your results on the form provided (page 441).

C. Cross-Matching

Procedure

This procedure is to be performed by students in groups of 4.

❏ 1. Perform the 4 major cross-matches using the combinations of donor's cells and recipient's serum preparations indicated in Table 50–4 of the Results and Observations section.

❏ 2. Observe the matches for the presence of agglutination.

❏ 3. Record your findings and answer the questions in the Results and Observations section.

❏ 4. Next, perform 4 minor cross-matches using the combinations of recipient's cells and donor's serum preparations indicated in Table 50–5 of the Results and Observations section.

❏ 5. Observe the matches for the presence of agglutination.

❏ 6. Record your findings and answer the questions in the Results and Observations section.

Materials

The following items should be provided for general class use in sufficient amounts. (*Note:* All plasma or serum preparations should be heated at 56°C for 30 minutes prior to use.)

- ❏ 1. Donor's blood cell preparation type A in physiological saline
- ❏ 2. Donor's blood cell preparation type B in physiological saline
- ❏ 3. Plasma or serum from recipient type A
- ❏ 4. Plasma or serum from recipient type O
- ❏ 5. Recipient's blood cell preparation type A in physiological saline
- ❏ 6. Recipient's blood cell preparation type O in physiological saline
- ❏ 7. Two unknown plasma or serum heated preparations from recipients (blood type unknown)
- ❏ 8. Heated plasma or serum donor type A
- ❏ 9. Heated plasma or serum donor type O
- ❏ 10. Two unknown heated plasma or serum preparations from donors (blood type unknown)
- ❏ 11. Glass slides
- ❏ 12. Pasteur pipettes with rubber bulbs
- ❏ 13. Wooden applicator sticks

Results and Observations

1. Major cross-matches

   a. Enter your findings for the major cross-matches performed in Table 50–4. Indicate incompatibility by "+" and compatibility "−."

Table 50–4

Major Cross-Matching Reactions

Donor's Cells (Blood Type)	Recipient's Serum Blood Type			
	A	O	Unknown	Unknown
A				
O				

   b. Which of the combinations for the major cross-matches showed incompatibility? _____

   _____

2. Minor cross-matches

   a. Enter your findings for the minor cross-matches performed in Table 50–5. Indicate incompatibility by "+" and compatibility by "−."

Table 50–5

Minor Cross-Matching Reactions

Recipient's Cells (Blood Type)	Donor's Serum Blood Type			
	A	O	Unknown	Unknown
A				
O				

b. Which of the combinations for the minor cross-matches showed incompatibility? _____

_____

Laboratory Review 50 — Agglutination: Blood Typing and Cross-Matching

1. Define or explain the following terms:

   a. titer _____

   b. hemolysis _____

   c. cross-matching _____

2. List 4 blood group factors other than those of the ABO system.

   a. _____   c. _____

   b. _____   d. _____

3. Distinguish between active and passive forms of immunity. _____

_____

_____

4. a. What is the basis of erythroblastosis fetalis, or the Rh baby? _____

_____

   b. How is the Rh condition corrected in a newborn? _____

_____

5. What is the Coombs' test? (Use your text as a reference.) _____

_____

_____

6. In the screening of blood donors, what general precautions should be taken before blood is considered to be acceptable for transfusions? List at least 2.

   a. _____

   b. _____

Key Terms

cross-match: a procedure used to determine if blood is compatible before a transfusion

isohemolysin (ī-sō-he-MOL-i-sin): an antibody that destroys red blood cells of animals of the same species

isoimmunization: immunization of an individual against the blood of an individual of the same species; also known as *alloimmunization*

Rh (factor) blood group: a blood group agglutinogen discovered on the red blood cells of rhesus monkeys with the designation Rh; the factors are found to a variable degree in human blood cells

secretor: an individual whose body fluids and cells contain A and B blood group antigens

universal donor: an individual belonging to blood group O and whose blood cells may be transfused, generally without danger of life-threatening reactions, into persons belonging to blood groups A, B, and AB

universal recipient: an individual belonging to blood group AB who normally does not have antibodies against the major blood groups A and B

Photograph Quiz 22

In the slide shown, the left-hand side contains anti-A serum, the middle anti-B serum, and the right-hand side anti-Rh_o (D) serum. One drop of an unknown blood sample was added to each of the sera. On the basis of the reactions shown, determine (a) the major blood type and (b) whether the blood specimen contained the D or Rh_o factor.

a. Major blood factor _____

b. Rh_o factor _____

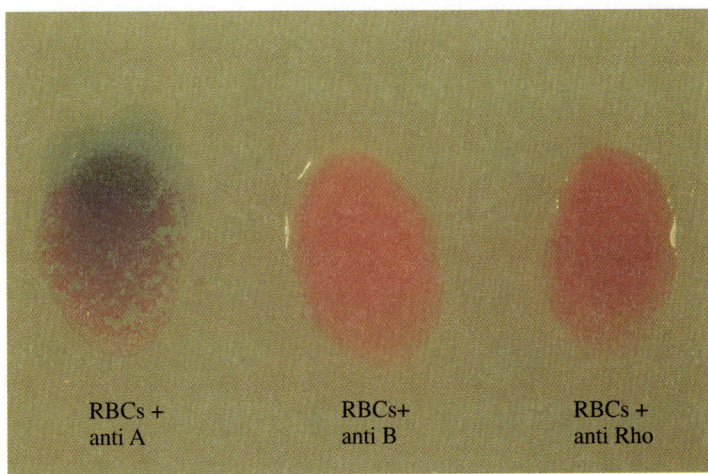

RBCs +
anti A

RBCs+
anti B

RBCs +
anti Rho

Figure 50–3
Blood typing unknown.

Unknown Form for Blood Typing

Student's Name _____

Date _____

Unknown Number _____

Score _____

Laboratory Section _____

Major Blood Type _____

Rh_o Type _____

After completing this exercise, you should be able to:

1. Recognize the major differentiating features between precipitation and agglutination reactions.
2. Perform and interpret a general type of precipitin test.

The precipitin test was first reported in 1897 by Kraus. In the original experiment, when bacteria-free filtrates of the microorganisms *Vibrio cholerae* and *Salmonella typhi* were mixed with their respective homologous antisera, a flocculent precipitate resulted. Similar results did not occur when the antigens and antisera were interchanged, demonstrating the specificity of the reaction. This flocculation test can be inhibited by use of excessive antigen or, in the case of certain antisera, by excessive antibody. The terms *precipitin* and *precipitinogen,* which are often used in conjunction with this type of serological reaction, were introduced by Kraus to designate the type of antibody and antigen involved. The state of dispersion of the antigenic material is the major differentiating feature between precipitation and agglutination reactions. The molecules of precipitinogen are free in solution, while the molecules of agglutinogens are bound in some manner with a surface.

During the normal sequence of events in an infection, precipitating antibody is often produced. This reaction is usually in response to soluble microbial substances released as a consequence of a disintegrative process. The presence of these antibodies can be demonstrated through the *ring, or interfacial, test,* in which a useful form of the precipitin reaction is observed. This procedure, first introduced by Ascoli in 1920, involves the use of antigenic material carefully layered over an equal quantity of antisera in order to form a sharp liquid interface. If the reactants are homologous, a ring or precipitation develops at the interface. In this test, the reactants diffuse into one another until the immunologically optimal proportions of each for precipitation are achieved. Although precipitin tests incorporating agar gel bases are utilized quite extensively for qualitative and semiquantitative studies, the general form of the test is still functional. Uses of the ordinary test include streptococcal extract grouping; the identification of blood, flesh, and milk; and urine precipitin tests, which are of value in the diagnosis of helminthic infections such as trichinosis and schistosomiasis.

Materials

❑ 1. Rabbit antibovine serum albumin (anti-BSA) containing at least 50–100 µg antibody protein/mL (2 mL per student)

❑ 2. Normal rabbit serum (2 mL per student)

❑ 3. Bovine serum albumin (BS) 0.05% (1:2,000 dilution) in physiological saline (2 mL per student)

❑ 4. Durham tubes (5 per student)

❑ 5. Physiological saline (2 mL per student)

❑ 6. Pasteur pipettes with long tips (5 per student)

❑ 7. Rubber bulb (1 per student)

❑ 8. Molding clay

❑ 9. Wax marking pencil

Procedure

This procedure is to be performed by students individually.

❑ 1. Mold the clay into a form to support the Durham tubes for the tests.

❑ 2. Mark tubes 1 through 5 with a wax marking pencil.

❑ 3. With a Pasteur pipette, introduce 0.3 mL of saline into tubes 2, 4, and 5. (Refer to Figure 51–1.)

❑ 4. With a separate pipette, introduce 0.3 mL of BSA into tubes 1 and 3.

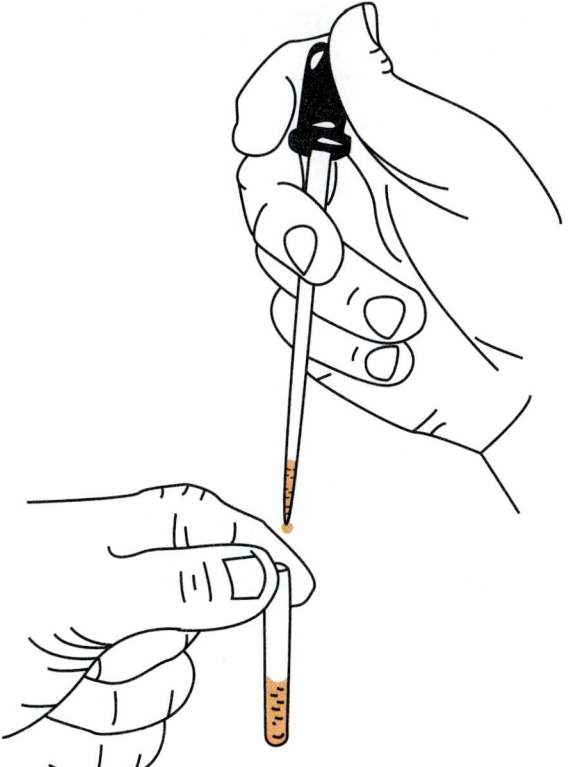

Figure 51–1
The proper position for holding tubes and overlaying antigen solutions onto an antibody-containing solution using a Pasteur pipette in the performance of the ring test.

❏ 5. Carefully overlay 0.3 mL of anti-BSA with a fresh pipette in tubes 1 and 5.

❏ 6. Overlay 0.3 mL of normal serum in tubes 3 and 4.

❏ 7. Overlay 0.3 mL of BSA in tube 2.

❏ 8. Look for a ring of precipitate in the tubes.

❏ 9. Record your findings in the Results and Observations section. Use "+" to designate the presence of a precipitation ring, and "−" to represent a negative finding. (See Figure 51–2.)

(a)   (b)   (c)   (d)   (e)

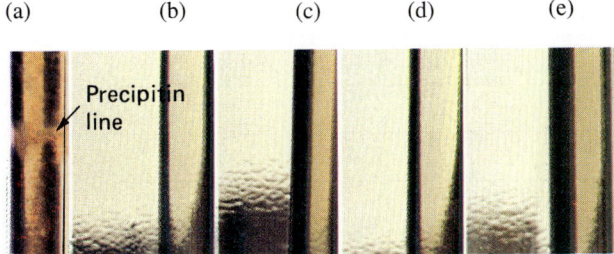

Precipitin line

Figure 51–2
The precipitin reaction. *Tube a,* positive reaction (note ring of precipitation at the interface). *Tubes b* and *c,* antigen controls. *Tubes d* and *e* antibody controls.

Results and Observations

1. Precipitin reactions.

    Tube number: 1   2   3   4   5

    Reaction: _____   _____   _____   _____   _____

2. Test result interpretation. To fully understand the reactions, indicate in the table on the following page the reacting substances involved in each tube. Use "+" to indicate the presence of a reagent and "−" to show its absence.

Table 51–1

Precipitin Reactions

Reagent	1	2	3	4	5
Saline					
BSA					
Anti-BSA					
Normal serum					

Laboratory Review 51   The Precipitin Reaction: The Ring, or Interfacial, Test

1. What is gel diffusion? _____

   _____

   _____

2. Define (use your text as a reference):

   a. VDRL _____

   _____

   b. Reagin _____

   _____

   c. IgE _____

   _____

   d. Soluble antigen _____

   _____

   e. Immunogen _____

   _____

Key Terms

flocculent (FLOK-ū-lent): refers to floating white tufts

helminthic (hel-MINTH-ik): pertains to worms

interface: the area or zone where two phases of a substance or two different substances contact each other

precipitin (prē-SIP-i-tin): the antibody formed in response to a soluble antigen

precipitinogen (prē-sip-i-TIN-ō-jen): the antigen (usually protein) that stimulates the formation of a specific precipitin

schistosomiasis (shis-tō-sō-MĪ-a-sis): a disease caused by the blood fluke (flat worm) belonging to the genus *Schistosoma*

trichinosis (trik-in-Ō-sis): the parasitic disease caused by the pork roundworm *Trichinella spiralis*

After completing this exercise, you should be able to:

1. Perform a basic gel diffusion procedure.
2. Interpret general types of gel diffusion reaction.
3. Recognize the diagnostic advantages and disadvantages of gel diffusion.
4. Describe the steps involved with the radial immunodiffusion (RID) procedure.
5. Interpret the results obtained with an RID procedure.

Precipitation assays can also be carried out in agar gels where antigen or antigen and antibody are allowed to diffuse into or toward one another. One of these assays is the double-diffusion, or Ouchterlony, technique developed mainly by Ouchterlony in 1953. In this application of the precipitin reaction, antigens and antibodies diffuse toward one another in a thin layer of agar gel contained in a Petri dish or a comparably workable system. Precipitin lines, which are immune precipitates, form where these reactants are in equivalent proportions.

The double-diffusion technique has several uses, such as for research studies and diagnosis. As a research tool, it permits identification of unknown antigens by comparing them with known reference antigens and homologous antisera. In the clinical laboratory, the technique can be used to detect the presence of hepatitis-associated antigen (HBsAg) or antibody to this antigen and complement-fixing antibodies to the fungus diseases, coccidioidomycosis, and histoplasmosis.

To perform the double-diffusion tests, wells or depressions are generally cut into the surface of an agar gel plate. The exact number, position, and shape of these wells are a matter of choice. However, it should be noted that square or rectangular wells have the advantage of forming sharp angles of merging precipitin bands or lines in comparative systems, thus making identification easier. The system, containing a series of wells, is filled with appropriate solutions of antigens or antisera, covered in some fashion to prevent evaporation of these solutions, placed into an environment that will facilitate the diffusion of the antigens and immunoglobulins in the antisera, and observed periodically for the formation of precipitin bands. As the diffusion progresses, visible lines (precipitates) will occur at points where equivalent proportions of both antigen and antibody are attained. If concentrated reagents and/or microconcentrations are used in the procedure, precipitin bands may appear in a few hours. If two wells for antigen are placed in opposition to a well containing antisera, three principal types of reactions may be detected. These reactions are shown in Figure 52–1. In this exercise, students will see typical precipitin bands that can occur in the double-diffusion procedure.

(a)   (b)   (c)

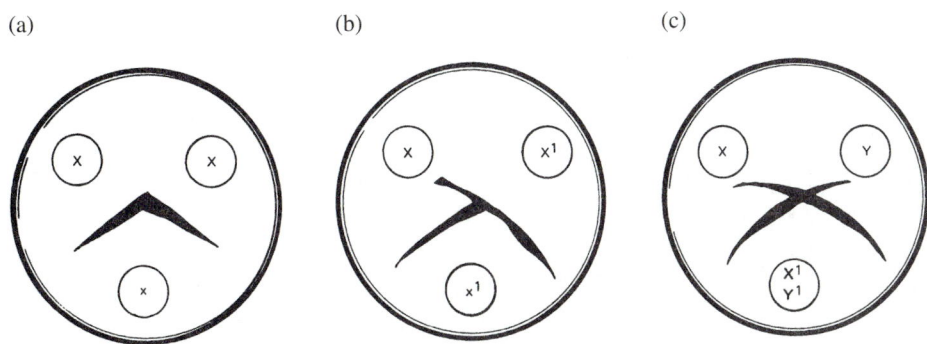

Figure 52–1

Diagrams of three general types of double-diffusion reactions. X = antigen X; Y = antigen Y; X^1 antigen = X plus additional antigenic determinants; x = anti-X; X^3 = anti-X1. (a) A reaction of identity. (b) A reaction of partial identity. (c) A reaction of nonidentity.

In recent years, radial immunodiffusion (RID) has become widely used for the quantitative measurement of a variety of proteins in serum and other body fluids. Such proteins include immunoglobulins, albumin, and certain components of complement. RID procedures are simple to perform and provide rapid, highly accurate, and specific results. In this technique, a specific antihuman serum equal to or greater than 1% is incorporated into an agarose gel support medium (Figure 52–2a). The support medium contains several standard-sized wells (holes) into which various control antigens and test samples are placed. Antigen immunoglobulin reactions take place in the agarose gel and appear as opaque precipitin rings (Figure 52–2a). The diameter squared of an individual precipitin ring (Figure 52–2b) is directly proportional to the concentration of the test antigen. A reference curve can be constructed on graph paper by plotting (1) the precipitin ring diameter's squared value against (2) several samples with known antigen concentrations (Figure 52–3). The antigen concentration of an unknown test sample can be determined by applying the squared diameter of the precipitin ring to the reference curve. This quantitative immunological technique will be demonstrated in this exercise.

(a)   (b)

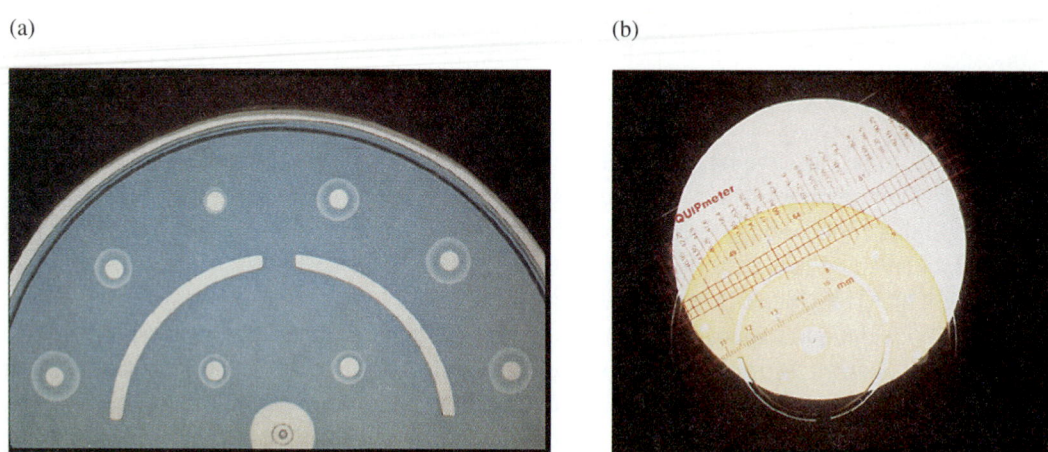

Figure 52–2
Radial immunodiffusion (RID). (a) The appearance of an agarose gel support medium with precipitin rings having varying diameters. (b) A close-up of a precipitin reaction being measured with a Helena Quipmeter. (Courtesy of Helena Laboratories, Beaumont, TX.)

A. Gel Diffusion

Materials

The following materials should be provided per 2 students:

- ❏ 1. One gel diffusion agar deep
- ❏ 2. One 60 × 15-mm sterile Petri plate
- ❏ 3. 0.5-mL quantities of the following:
  - ❏ a. Bovine serum albumin
  - ❏ b. Goat serum
  - ❏ c. Horse serum
- ❏ d. Human serum
- ❏ e. Unknown antibody
- ❏ 4. Five sterile Pasteur pipettes and 1 rubber bulb
- ❏ 5. Filter paper strip
- ❏ 6. One 15 × 100-mm sterile Petri plate

The following items should be provided for general class use:

- ❏ 1. One quarter-inch, or size 2, cork borer
- ❏ 2. A boiling water bath
- ❏ 3. A suction or vacuum device equipped with a safety flask and appropriate tubing
- ❏ 4. A wax marking pencil or felt-tip marker

Procedure 1: Plate Preparation

If diffusion plates are not already prepared, carry out steps 1 through 3 before performing the gel diffusion procedure. This procedure is to be performed by students in pairs.

❑ 1. Melt the agar deep provided in a boiling water bath and pour it into a sterile Petri plate. Allow the agar to harden.

❑ 2. With the aid of a quarter-inch, or size 2, cork borer, make five holes or wells in the hardened agar using the pattern shown in the Results and Observations section. You can place the Petri plate directly over the pattern and use it as a guide.

❑ 3. Use the suction apparatus provided to remove the excess agar from the wells.

Procedure 2: Test

❑ 1. With the wax marking pencil, indicate on the back of the plate the name of the preparation that each is to contain. (*Note:* The unknown antiserum is to be placed into the center well.)

❑ 2. Using a sterile Pasteur pipette, fill the center well with the unknown antiserum. Do not overfill this well or any of the others.

❑ 3. Using a separate Pasteur pipette each time, fill each of the remaining wells with 1 of the 4 antigens provided. Dispose of the used pipettes as indicated by the instructor.

❑ 4. Place the diffusion plate into a larger Petri plate containing a well-moistened filter paper. Cover the plate and refrigerate.

❑ 5. Observe the gel diffusion plate for 7 days and look for the formation of precipitin lines. Add more water to the filter paper if it appears to be drying out.

❑ 6. Draw any and all precipitation lines in the diagram in the Results and Observations section. If the experiment is done well, you should be able to identify the nature of the antibody in the center well.

B. Radial Immunodiffusion (RID)

Materials

The following items should be provided for the demonstration of RID:

❑ 1. One 19-well (specimen chamber) hexagon-shaped, IgG test plate (Helena Laboratories)

❑ 2. One set of immunoglobulin reference standards (usually 3, representing low [422 milligrams/deciliter], normal [845 mg/dL], and high [1,690 mg/dL] concentrations of the specific immunoglobulin)

❑ 3. One mL of a human serum sample

❑ 4. Three tubes of 0.85% saline, 0.9 mL each, for sample dilutions

❑ 5. Three cotton-plugged, 1-mL disposable pipettes graduated in 0.1 mL and bulbs

❑ 6. One Quipmeter (Helena Laboratories), or a ruler scale for the reading and measurement of precipitin ring diameter

❑ 7. Seven individual microsyringes or a Helena Ziptrol delivery device and 7 100-μL capillary tubes

❑ 8. One plastic Petri dish top (100 × 15 mm)

❑ 9. One circular filter paper sheet (90 mm in diameter)

❑ 10. One plastic squeeze water bottle with distilled water

❑ 11. One marking pen or pencil

❑ 12. One test tube rack

❑ 13. One container for pipette disposal

Procedure 1: Specimen Preparation

The instructor will demonstrate the RID materials and technique unless other assignments are made. Dispose of used pipettes in the container provided.

❏ 1. Place the tube of human serum and the 3 tubes of saline into a test tube rack. Label the saline tubes 10^{-1}, 10^{-2}, and 10^{-3}, respectively. These labeled tubes will represent 1:10, 1:100, and 1:1,000 dilutions, respectively, of the serum sample provided.

❏ 2. With a 1-mL pipette, obtain 0.1 mL of the human serum sample and transfer it to the saline tube marked 10^{-1}.

❏ 3. Gently mix the contents of the tube by shaking and then, with another 1-mL pipette, transfer 0.1 mL of the 10^{-1} serum dilution to the saline tube marked 10^{-2}.

❏ 4. Mix the contents of the 10^{-2} dilution, take another 1-mL pipette, and transfer 0.1 mL of 10^{-2} serum dilution to the saline tube marked 10^{-3}. Gently shake the contents of the tube and, with the pipette just used, remove 0.1 mL and dispose of this sample and the pipette.

Procedure 2: Sample Application and Test Interpretation

❏ 1. Assemble microsyringe or Helena Ziptrol delivery device for each reference standard and serum dilution.

❏ 2. Place the filter paper sheet into the top of the Petri dish and saturate it with distilled water. This combination will serve as a humidity chamber.

❏ 3. Carefully remove the cover of the 19-well test plate and note the number of each well.

❏ 4. Using individual delivery devices, apply 2 μL of the low, normal, and high reference standards to wells 1, 2, and 3, respectively (see Figure 52–3).

❏ 5. With four delivery devices, apply 2 μL of the undiluted serum samples and the 10^{-1}, 10^{-2}, and 10^{-3} serum dilutions to wells, 4, 5, 6, and 7, respectively.

Figure 52–3
The technique for delivery of reference standards and samples.

❏ 6. Replace the cover on the Quiplate and put the system into the top of the Petri dish (humidity chamber). Incubate the system at room temperature in a drawer for 18 to 24 hours.

❏ 7. After incubation, remove the cover and, with the aid of the Quipmeter, determine the diameters of the individual precipitin rings. For best results, gradually slide the Quipmeter over the precipitin ring until the 2 diverging lines of the ruler just touch the outer borders of the ring (see Figure 52–2b).

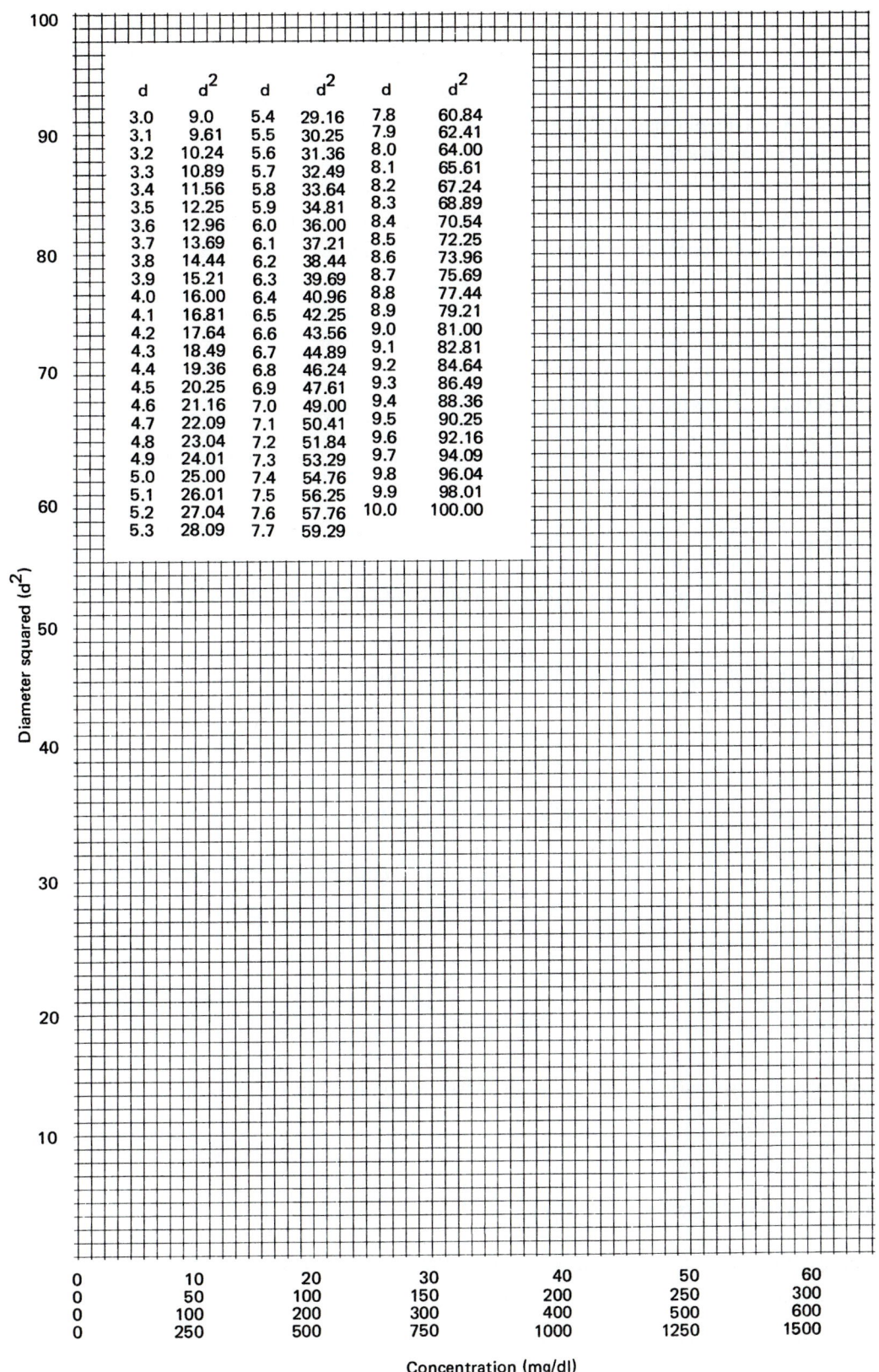

Figure 52–4

A Quiplate System reference curve for immunoglobulin determinations.

❑   8. Enter the measurements for each reference standard and dilution sample in Table 52–1.

❑   9. Square the diameter of each precipitin ring. The graph in Figure 52–4 contains a table from which diameter squared values may be obtained. Enter the squared values in Table 52–1.

❑ 10. On the graph provided in the Results and Observations section (Figure 52–4), plot the squared ring diameters of the standards against the corresponding immunoglobulin concentrations given in milligrams/deciliter. With a ruler, connect the points plotted. The resulting line forms a *reference line* or *curve*.

❑ 11. Mark the squared ring diameter values of the individual serum dilutions on the reference curve. Determine the immunoglobulin concentrations from the graph and enter the values in Table 52–1. Answer the questions in the Results and Observations section.

Results and Observations

Gel Diffusion

1.   a. Indicate on the diffusion plate diagram the respective positions of the different antigens and unknown antibody preparation used.

    b. Show the location of any and all precipitin bands that formed.

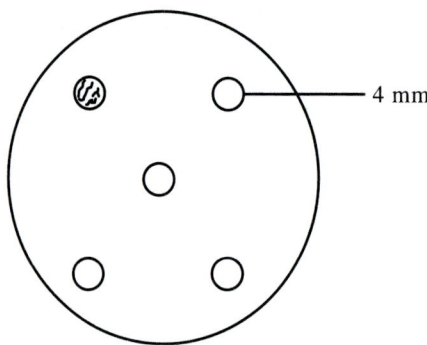

2.   On the basis of your observations, identify the unknown antibody in the center well. _____

_____

Radial Immunodiffusion (RID)

1.   Enter all measurements in the appropriate columns of Table 52–1. Indicate the immunoglobulin concentrations for each reference standard.

2.   Which of the serum wells had the largest precipitin zone diameter? _____

_____

3.   What does this zone represent? _____

_____

Table 52-1

Gel Diffusion Reactions

Sample	Measurements		
	Precipitin Zone Diameter	Precipitin Zone Diameter Squared	Immunoglobulin Concentration (From Graph)
Low reference standard _____ mg/dl			
Normal reference standard _____ mg/dl			
High reference standard _____ mg/dl			
Undiluted serum sample			
10^{-1} serum sample			
10^{-2} serum sample			
10^{-3} serum sample			

Laboratory Review 52   Gel Diffusion and Radial Immunodiffusion

1. What other techniques are used to demonstrate precipitation *in vitro?* _____
   _____
   _____

2. Why didn't microbial contamination of the diffusion plate occur during the incubation period? _____
   _____
   _____

3. List two applications of radial immunodiffusion.

   a. _____

   b. _____

4. Define or explain the following terms:

   a. immunoglobulin _____
   _____

   b. antigen _____
   _____

   c. electrophoresis _____
   _____

   d. alpha globulins _____
   _____

  e. albumin _____

_____

5. Distinguish between plasma and serum. _____

_____

6. How does RID differ from double diffusion? _____

_____

_____

Key Terms

coccidioidomycosis (kok-sid-i-oyd-ō-mī-KŌ-sis): a deep-seated (systemic) fungus infection caused by *Coccidioides immitis*

histoplasmosis (HIS-tō-plaz-mō-sis): a deep-seated (systemic) fungus infection caused by *Histoplasma capsulatum*

immunodiffusion: refers to a form of precipitation test in which an agar gel is used to allow antigens and antibodies to move from a given center toward each other

precipitin (prē-SIP-eh-tin) reaction: the basis of an immunological reaction in which antigens and antibodies diffuse toward one another, resulting in a precipitate where these reactants are in equivalent proportions

radial (RĀ-dē-al) immunodiffusion: a form of immunodiffusion test in which antibody only to the antigen to be detected is incorporated into the agar gel; as the antigen diffuses from its well, a ring of precipitate forms at the point where antigen and antibody are in proper proportion

reaction of identity: a gel diffusion result that indicates that antigen in both wells contains the identical antigenic component

reaction of nonidentity: a gel diffusion result that indicates that the antigen in 1 well does not share in common a certain antigenic component with the antigen in the other well

reaction of partial identity: a gel diffusion result that indicates that antigens in both wells contain a common antigenic component

---

Photograph Quiz 23

a. With which of the surrounding wells is a reaction of identity obvious? _____

_____

b. With which of the wells is a reaction of nonidentity apparent? _____

_____

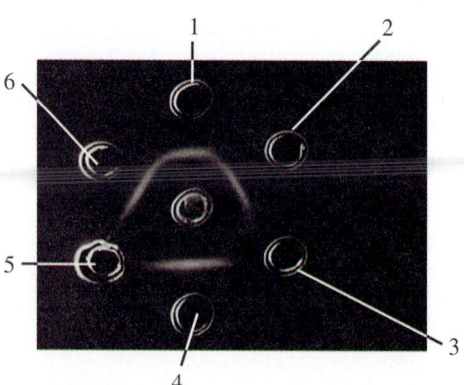

Figure 52–5
Interpretation of double diffusion reactions.
(From Wheat, R. W., *et al.*, *Inf. Immun* 21:585–593, [1978].)

An Enzyme Immunoassay: The Enzyme-Linked Immunosorbent Assay (ELISA)

After completing this experimental exercise, you should be able to:

1. Outline the ELISA procedure.
2. Recognize the diagnostic value of an enzyme immunoassay.
3. Perform the ELISA procedure.
4. Distinguish between the direct and indirect forms of ELISA.
5. Interpret the results obtained from ELISA.

Technological advances during the past several years have led to more rapid, sensitive, and accurate serodiagnostic tests for infectious diseases. Moreover, the development of automated technology has made such tests easier to perform.

Traditionally, the diagnosis of infectious diseases has been made by the growth, isolation, and identification of the pathogen in pure culture, and through the use of artificial media or living systems such as tissue culture. In certain situations, the cultivation of a suspected pathogen involves an extensive period of time, or special equipment or materials available only in central or reference laboratories. For this reason, great interest has been given to the development of assay systems capable of the direct detection of microbial antigens in body fluids without the need for cultivation. Among the most promising of the procedures is the *enzyme-linked immunosorbent assay* (ELISA) also referred to as the enzyme immunosorbent assay (EIA). With this procedure, almost any antigen can be used to produce a highly sensitive and specific test.

An enzyme immunoassay such as ELISA is based on the ability of antigen or antibody to: (1) adsorb to a solid-phase support while retaining immunological activity and (2) be linked to an enzyme, with the antigen or antibody-enzyme complex exhibiting both immunological and enzymatic activities. Figure 60–1 shows the general features of the ELISA reactions. Both the direct and indirect forms are illustrated. ELISA is of particular value in the detection of various infections, especially in cases of AIDS. It should be noted that the procedure is also of great importance for hormone, drug, and unusual protein assays.

Two ELISA methods are currently in use: the double-antibody sandwich technique, which determines the level of antigen from a patient; and the indirect technique, which measures the patient's antibody level to a specific antigen. In the double-antibody sandwich procedure, a specimen containing the suspected antigen is added to a polystyrene well that has been coated with antibody specific for this antigen. This step is followed by the addition of a conjugate (combination) of enzyme-labeled specific antibody (see Figure 53–1). An enzyme substrate is the last reagent added to the system. In a positive test, hydrolysis of the enzyme substrate occurs and will be proportional to the amount of antigen bound. In the indirect method, the polystyrene well is coated with a specific test antigen. This step is then followed by the addition of a patient's serum, and then by the addition of an enzyme-labeled anti-human globulin. An enzyme substrate is added to complete the system. Again, as in the double-antibody sandwich technique, a positive test will result in the hydrolysis of the enzyme substrate. The extent of the reaction will be proportional to the antibody level in the patient's serum.

The end result of an ELISA test may be assessed qualitatively by eye, or it can be accurately measured photometrically. Positive results are indicated by a specific color reaction occurring in the polystyrene wells of microplates or tubes used (Figure 53–2). If a titer determination is required, serial dilutions of the test specimen are made. The last dilution giving a detectable color reaction is reported as the titer.

Photometric readings are necessary for high precision. Such determinations may be made by transferring the reaction produced to a cuvette for spectrophotometric analysis (see Experimental Exercise A in Section XV). An alternate approach uses a rapid microplate reader, which is capable of reading an entire plate without the need for the removal or transfer of the reaction product.

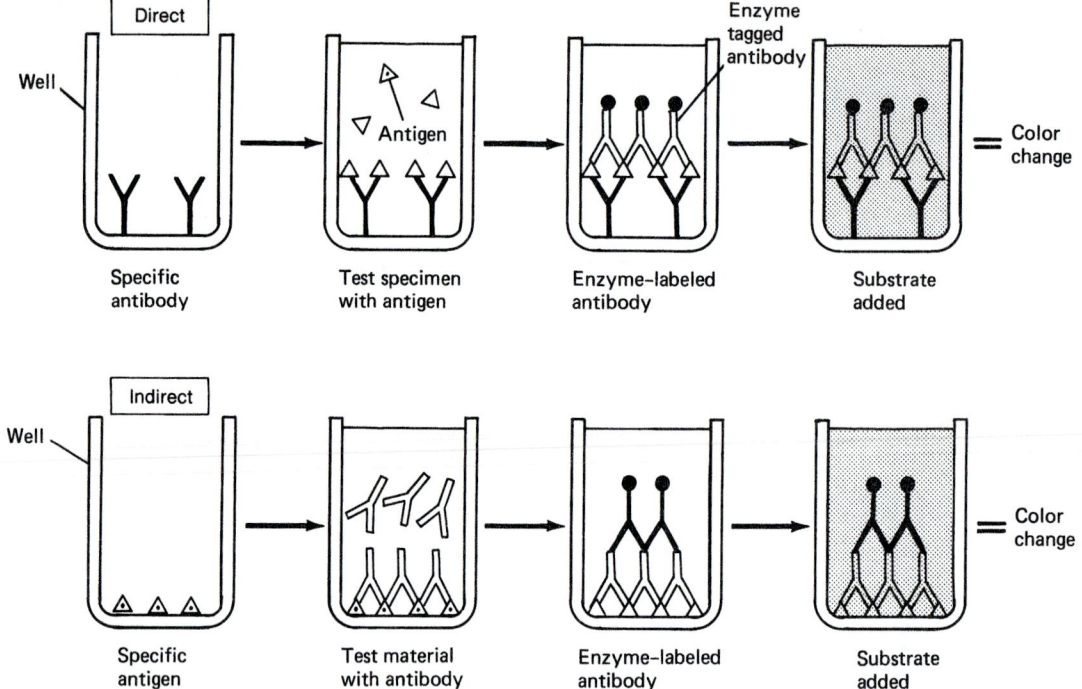

Figure 53–1

Enzyme-linked immunosorbent assay (ELISA). Both the direct, or double sandwich, technique and the indirect method are shown. Antibody to the antigen to be detected (captured antibody) is attached to the well. A specimen possibly containing antigen (triangles) is added and washed. Antibody to the antigen is then added. In the direct method, this antibody is tagged with enzyme so that when substrate is added a color change occurs. In the indirect method, the captured antigen is first treated with heterologous antibody. Then an enzyme-labeled antibody to the heterologous antibody is added next, before the addition of substrate.

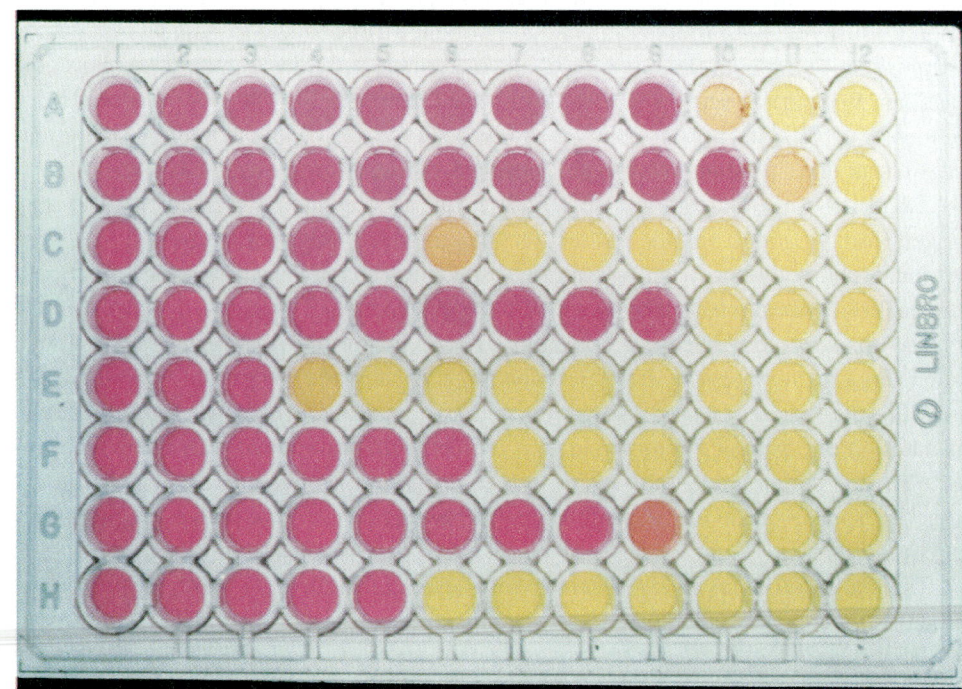

Figure 53–2

The dramatic difference between positive and negative enzyme immunoassay results is shown.

This experimental exercise will present a representative indirect ELISA procedure. Kits are available commercially for the detection of a wide variety of infectious diseases, including: infections caused by herpesvirus, hepatitis B virus, measles virus, mumps virus, chickenpox (varicella), toxoplasmosis, malaria, trichinosis, mycoplasma, and the yeast *Candida albicans*.

A. VARICELISA Test

Materials

The following materials, equipment, and procedures are for 8 tests using the BioWhittaker procedure for the indirect VARICELISA (var-i-sel-I-za) test. If additional tests are to be performed, the brochure accompanying the test materials should be consulted for directions relating to the preparation and dispensing of all materials needed. The first 8 items are used in the test itself. (*Note:* Other test systems and/or sources may be used to demonstrate the features of an ELISA.)

- ❏ 1. Two VARICELISA solid Phase 96 Removawell Trays (1 tray is for examination)
- ❏ 2. Nine mL phosphate-buffered saline (PBS) with Tween
- ❏ 3. One mL reconstituted serum diluent
- ❏ 4. One-tenth (0.1) mL alkaline phosphatase conjugated (rabbit) anti-human immunoglobulin (IgG)
- ❏ 5. Six-tenths (0.6) mL diethanolamine buffer
- ❏ 6. Two mL para-nitrophenyl phosphate (pNPP) enzyme substrate
- ❏ 7. Two and four-tenths (2.4) mL sterile deionized water
- ❏ 8. Four mL normal sodium hydroxide (NaOH)
- ❏ 9. One plastic wash bottle with PBS and Tween added
- ❏ 10. One timer (45-minute range)
- ❏ 11. Micropipettes for the following volumes: 250 microliters (μL), 50 μL, and 10 μL
- ❏ 12. One adjustable multichannel micropipette
- ❏ 13. Disposable micropipette tips
- ❏ 14. Sterile disposable pipettes (assorted sizes)
- ❏ 15. Paper towels or absorbent paper
- ❏ 16. Two graduated cylinders (1-L capacity)
- ❏ 17. A disposable tray or basin containing sodium hypochlorite (household bleach) solution for the disposal of all wash and incubation solutions
- ❏ 18. One liter sodium hypochlorite (household bleach) solution (50 mL household bleach in 950 mL water)
- ❏ 19. Disposable plastic or glass dilution tubes (30-mL capacity)
- ❏ 20. Capped specimen tubes for aliquots of pNPP (5-mL capacity)
- ❏ 21. Two plastic bags for storage of VARICELISA system during incubation steps
- ❏ 22. One micromixer, BioWhittaker products catalog number 18-502, or equivalent shaker system
- ❏ 23. One Dynatech MR 590 enzyme-linked immunosorbent assay or EIA plate reader or equivalent system
- ❏ 24. One marking pen

Procedure 1: The VARICELISA Test

The instructor will demonstrate the VARICELISA test materials and procedure, unless other assignments are made. Dispose of all wash and incubation solutions and used pipettes in the disinfectant container provided.

❏   1. Examine the VARICELISA Removawell tray provided as a demonstration and note the numbering and lettering system on the demonstration tray. The *V* equals wells containing varicella antigen, and the *C* equals wells containing control antigen. One test includes 1 of each type well. (Refer to Figure 53–3.)

❏   2. Remove the tray to be used for the VARICELISA test from its package by cutting one end of the sealed edge.

❏   3. Remove 2 complete vertical rows or columns, and reseal the remaining strips in the original package.

❏   4. Obtain the respective materials to be used in the procedure. The different samples of the specimens should be labeled as follows: unknown serum specimens, #1, #2, and #3; calibrator sera, S1, S2, and S3; and the control sera, CA and CB.

❏   5. Carefully shake each of the labeled tubes to ensure that the contents of each one are well suspended.

❏   6. Dispense the unknown serum specimens, calibrator sera, and control sera according to Procedure Diagram 39, step 1, and Figure 53–4, showing the respective wells for each serum sample. Note that there is 1 V and C well for each sample. Use different pipette tips for each serum.

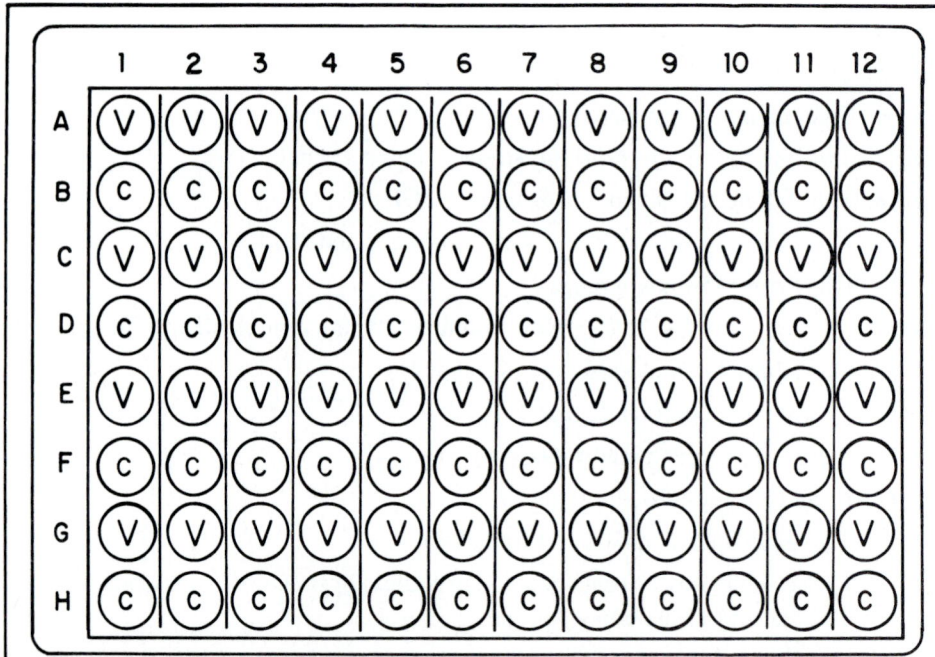

Figure 53–3
A VARICELISA Removawell Tray. It is important to note that the tray wells are coated with inactivated varicella virus. Explanation of symbols: V = wells containing varicella antigen; c = wells containing control antigen.

	1	2	3	4	5
A (V)	#1	S2			
B (C)	#1	S2			
C (V)	#2	S3			
D (C)	#2	S3			
E (V)	#3	CA			
F (C)	#3	CA			
G (V)	S1	CB			
H (C)	S1	CB			

Figure 53–4
The respective wells for unknown serum specimens, calibration sera, and control sera.

❏ 7. Follow the steps shown in Procedure Diagram 39.

❏ 8. Obtain the VARICELISA value of each calibrator sera, and record them in the Results and Observations section. These values are indicated on the respective original vials.

❏ 9. Record all absorbance values (readings) in the Results and Observations section in the spaces provided.

❏ 10. After the test readings have been made, calculate the test results according to the steps described in Procedure 2.

❏ 11. Answer the questions in the Results and Observations section.

Procedure Diagram 39
Enzyme Immunoassay

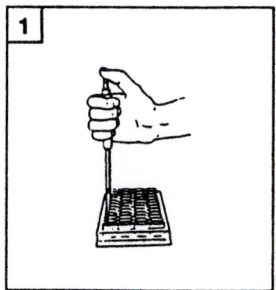

1. Dispense 250 μl serum diluent into each well in the tray.

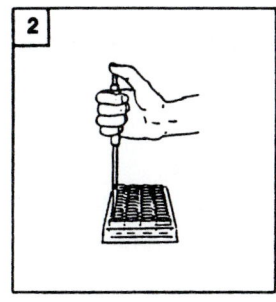

2. Dispense 10 μl unknown specimens, 3 calibrator sera, and the 2 control sera in their respective wells (See Figure 1-3).

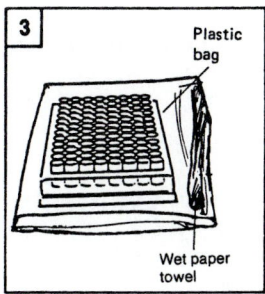

3. Place tray in a plastic bag with a moist paper towel for 45 minutes at room temperature.

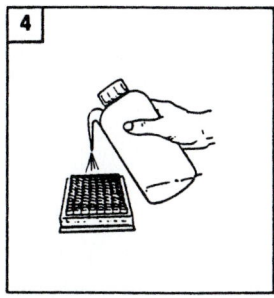

4. Shake out all liquid into disposal tray. Rinse all wells with PBS-Tween 2 times and refill each well with this solution and allow to soak for 3 min.

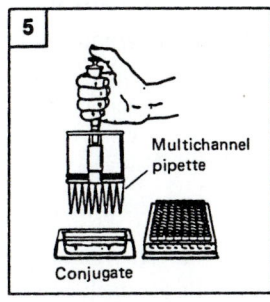

5. Shake out all liquid into disposal tray. Add 250 μl of conjugate preparation to each well and incubate the tray as in step 3 for 45 minutes.

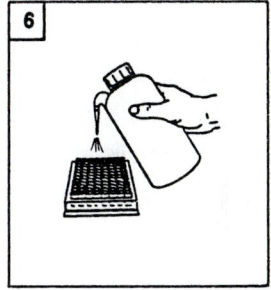

6. After incubation, shake out all liquid into disposal tray. Repeat step 4.

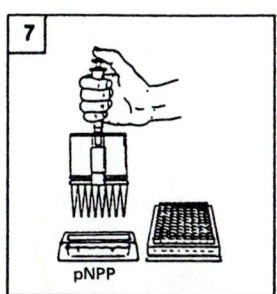

7. Add 250 μl pNPP solution to each well and incubate as in step 3 for 45 minutes.

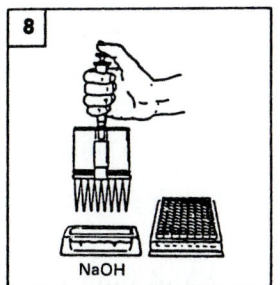

8. Add 50 μl NaOH to each well. This stops the reaction. Mix thoroughly by gentle shaking.

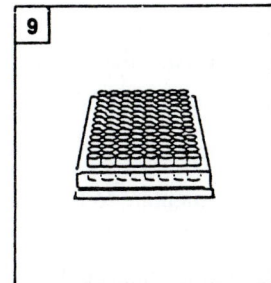

9. Read results at 405nm in a spectrophotometer or as directed. Continue with procedure 2.

Procedure 2: Calculating Test Results

Determining Enzyme-Linked Immunosorbent (EIA) Values of Calibrator Sera, Control Sera, and Unknown Specimens

❏ 1. Subtract the absorbance value of each serum's control (C) antigen reaction from the corresponding absorbance value obtained with the serum's varicella (V) antigen reaction. Use the recorded information in Tables 53–1 and 53–2. The resulting number is the *test EIA value.*

❏ 2. Record the test EIA values in Table 53–3.

❏ 3. Using the graph paper provided, plot the points representing the test EIA values of the calibrator sera on the x-axis versus their respective VARICELISA values on the y-axis.

❏ 4. Draw the best-fitting straight line through these points. This *calibrator line* is to be used to obtain the VARICELISA values for all of the remaining sera used in the test. Use Figure 53–5 for this purpose in the Results and Observations section.

Sample Calculations for the Calibrator Sera and the Construction of the Calibrator Line

Needed Information:

VARICELISA value for Calibrator #1 (listed on the original vial) = 0.27
Absorbance value of varicella antigen well = 0.29
Absorbance value of control antigen well = 0.05
Test EIA Value **= 0.19**

❏ 1. Find VARICELISA value of 0.27 on the y-axis of the graph.

❏ 2. Find the test EIA value of 0.19 on the x-axis of the graph.

❏ 3. Draw a point on the graph where the 2 values meet. (This is the procedure to follow in constructing the *calibrator line.*) Refer to the graph.

❏ 4. Repeat these steps with the remaining calibrator sera.

Determining VARICELISA Values for Controls and Unknown Specimens

❏ 1. Determine the test EIA values by subtracting the absorbance values of the control antigens from the absorbance values of the varicella antigen wells for each control serum and unknown specimen used in the test. (Refer to the earlier section under Procedure 2.)

❏ 2. Record all test EIA values in Table 53–3 in the Results and Observations section.

❏ 3. Find the test EIA values on the x-axis of the graph for each serum used.

❏ 4. Draw a perpendicular line from this point to the *calibrator line* prepared earlier.

❏ 5. Next, with each point on the calibrator line, draw a parallel line from the *calibrator line* to the y-axis. (The y-axis contains the VARICELISA values.)

❏ 6. Obtain the VARICELISA values for each of the sera and enter them in Table 53–3.

Sample Determination of VARICELISA Value Using a Calibrator Line

❏ 1. A perpendicular, dotted line from the test EIA value of 0.50 to a point on the *calibrator line* is shown on the graph.

❏ 2. The point is connected to its corresponding VARICELISA value by a solid line on the y-axis.

❏ 3. These steps should be repeated in the same manner to determine the VARICELISA values of the respective sera used in the test.

Interpretation of Test Results

Your instructor will provide the values representing negative and positive test results for the unknown sera used in the test.

Results and Observations

1. Enter the VARICELISA values and absorbance values of the 3 calibrator sera in Table 53–1.

Table 53–1

Calibrator Sera Values

| Calibrator Serum | VARICELISA Values | Absorbance Values | |
		Control Antigen	Varicella Antigen
S1			
S2			
S3			

2. Enter the absorbance values for the unknown serum specimens and control sera in Table 53–2.

Table 53–2

Absorbance Values of Unknown Serum Specimens and Control Sera

| Specimen | Absorbance Values | |
	Control Antigen	Varicella Antigen
Unknown #1		
Unknown #2		
Unknown #3		
CA		
CB		

3. Record all test EIA values in the appropriate spaces provided in Table 53–3.

Table 53-3

Test EIA Values

Specimen	Test EIA Values	VARICELISA Values
Calibrator serum #1		
Calibrator serum #2		
Calibrator serum #3		
Unknown #1		
Unknown #2		
Unknown #3		
Control A		
Control B		

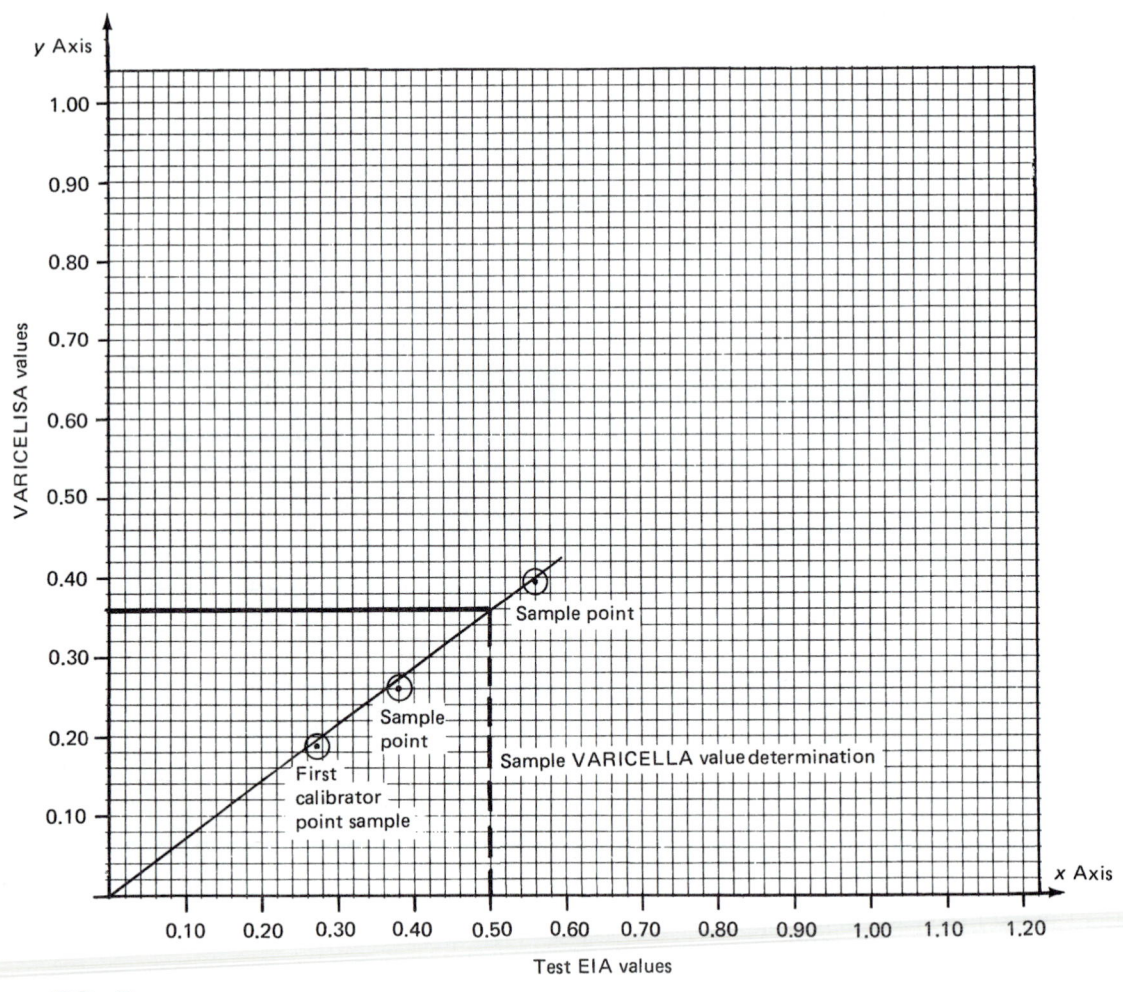

Figure 53–5
Graph for EIA value determinations.

4. Which of the unknown serum samples were positive? _____

_____

5. Which of the unknown serum samples were negative? _____

_____

6. What does a positive test result indicate? _____

_____

7. Plot your findings in the graph shown in Figure 53–5 on the previous page.

Laboratory Review 53   **An Enzyme Immunoassay: The Enzyme-Linked Immunosorbent Assay (ELISA)**

---

1. What is the major advantage of ELISA? _____

_____

2. Is ELISA applicable to the detection of either antigen or antibody? _____

_____

3. What forms of ELISA are in current use? _____

_____

4. How is the end result of an ELISA procedure assessed? _____

_____

5. Give one example of a bacterial, viral, mycotic, protozoan, and helminthic disease for which ELISA is applicable.

_____

_____

Key Terms

AIDS: acquired immune deficiency syndrome: a disease caused by human immunodeficiency virus (HIV), resulting in the destruction of T-helper lymphocytes and other components of the immune system

cuvette (kuv-ET): a small, transparent, glass container that is used for photometric analyses

ELISA (enzyme-linked immunosorbent assay): a highly specific immunologic diagnostic technique that involves the linking of soluble antigens or antibodies to an insoluble solid surface in a way in which the reactivity of the antigen or antibody is still retained

hydrolysis (hi-DROL-is-is): any chemical reaction involving the splitting of a bond by the addition of a water molecule

serial dilutions: a sequence of dilutions in numerical order (e.g., 10^{-1}, 10^{-2}, 10^{-3})

titer (TI-ter): the highest dilution (of serum or antibody) that will cause a specific reaction

SECTION XII

An Introduction to Epidemiology and Related Topics

Infectious diseases result from the interplay of three main factors, the host, the parasite and the environment.

T. Aidan Cockburn

Epidemiology is the study of distribution and causes of diseases and injuries in human populations. It is concerned with the frequencies and types of illnesses and injuries found in groups of people and with the factors that influence their distribution. Knowledge about human health and disease is the sum contributions of a large number of specialty areas or disciplines. The various disciplines however can be grouped according to their methods and underlying concepts. When this is done, three major categories emerge: (1) *basic sciences,* (2) *clinical sciences,* and (3) *public health.* The basic sciences include anatomy, biochemistry, physiology, microbiology, immunology, and pathology. The discipline of clinical medicine focuses largely on the medical care of individuals, while public health is concerned with the study of health and disease in human populations. Epidemiologic knowledge, largely based on observations, is collected from a variety of sources by professionals called *epidemiologists.* Such individuals have the responsibility to develop a comprehensive picture of health problems in a community.

A wide variety of infectious agents, ranging from submicroscopic viruses to complex multicellular organisms such as worms, can produce disease in humans and other forms of life. Many characteristics of an infectious agent are determined by the agent itself *(intrinsic)* and do not depend on any interaction with a host. Such intrinsic properties include ability to produce toxins, range of hosts, resistance to antibiotics and other chemicals, and the ability to survive outside a host in a variety of vehicles *(e.g.,* food, water, and soil). Understanding a specific intrinsic property may be essential to understanding an agent's epidemiology, especially its mode of transmission. It should also be noted that several properties considered to belong to an infectious agent are not actually a part of the agent, but depend on the interaction between the agent and the host.

The various mechanisms by which disease agents reach and infect their hosts form a central aspect of the *transmission of infection.* This involves the agent's exit from a source or reservoir, spread to a susceptible host, and entry into the host. A *reservoir* generally is defined as the living organism or inanimate material in which an infectious disease agent normally lives or reproduces.

Transmission may be *direct* or *indirect.* Direct transmission refers to the immediate transfer of an infectious agent from an infected host or reservoir to the portal of entry of a susceptible host. Indirect transmission may take place by any *vehicleborne, vectorborne,* or *airborne mechanism.* Vehicleborne transmission involves indirect contact with inanimate objects *(fomites)* such as contaminated toys, clothes, or surgical instruments, as well as contaminated food, water, and intravenously administered materials. Vectorborne transmission refers to the arthropod transmission of a disease agent to a susceptible host (Figure XII–1). Airborne transmission may occur by means of dusts or droplet nuclei in the form of *aerosols.* An aerosol is a cloud of very small water droplets or fine solid particles suspended in air. Dusts consist of varying sized particles containing disease agents that have settled onto various surfaces and are resuspended by the wind, or air currents. *Droplet nuclei* are small particles that are the dried remains of droplets released into the environment by sneezing and/or coughing.

The exercises in this section deal with various aspects of disease transmission and control including a demonstration of Koch's postulates, features of specimen handling, the importance of aerosols in laboratory and hospital environments, and approaches to disease prevention in the form of universal precautions.

(a) (b)

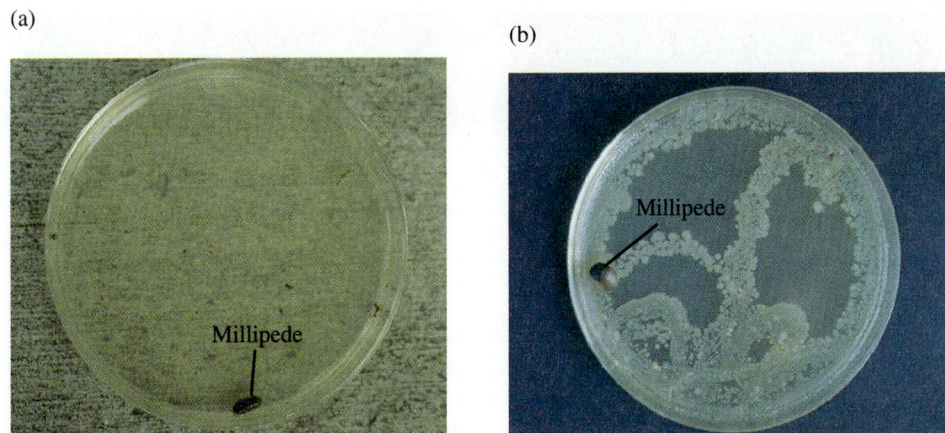

Figure XII-1

The pill millipede, *Glomeris* species, an arthropod source of environmental microorganisms. (a) A Petri plate with the pill millipede found in the environment. The millipede was allowed to move over the surface of the agar medium for about one hour. (b) The appearance of the agar plate after 24 hours of incubation.

After completing this exercise, you should be able to:

1. List the steps of Koch's Postulates.
2. Understand and explain the approach behind Koch's Postulates.
3. Apply Koch's Postulates in the identification of a suspected pathogen.

From the earliest times, diseases were considered to result from some mysterious or even supernatural force. Ancient Greek and Roman physicians suspected that invisible agents caused certain diseases that could be transmitted in some manner from one individual to another. Until the nineteenth century, however, there was no direct proof that microorganisms caused such diseases.

The fact that certain bacteria produce disease was first clearly shown by the German bacteriologist Robert Koch in 1876. He demonstrated that the spore-former, *Bacillus anthracis,* the etiologic or causative agent of anthrax, was then epidemic in cattle, sheep, and other domestic animals and also occurred in humans. Koch followed an experimental approach, which included the isolation and obtaining of pure cultures of the suspected disease agent. These cultures then were used to inoculate experimental animals that died, showing typical symptoms of anthrax. Upon autopsy, the bacteria found in the blood and organs of the infected experimental animals had the same characteristics as those used for inoculation. This approach formed the basis of a clear-cut demonstration of the causal relationship between a suspected microbial disease agent and the disease. Koch's experimental steps, which he later stated in the form of rules, are known as *Koch's Postulates.* They are as follows:

1. Find and isolate the suspected disease organism in all cases of the disease and show its absence in healthy individuals.
2. Obtain the suspected organism in pure culture.
3. Reproduce the same disease in appropriate experimental animals.
4. Reisolate the organism characteristically identical to the suspected agent from the experimentally infected animals.

Koch's Postulates not only provided a logical basis for concluding that a particular microorganism is the etiologic agent of a given disease, but also opened the door for an enormous enthusiastic search for the causes of infectious diseases. Because viruses and their properties were not known during Koch's time, his postulates required certain modifications before they could be applied to these agents of disease. In 1932, Rivers made the necessary changes so that the approach could be extended to most infectious diseases of animals and plants.

In this exercise, Koch's Postulates will be demonstrated with the aid of one animal system represented by the goldfish and one plant system represented by the tomato plant. The approach to be followed resembles the one used by Koch and his colleagues in that cultures of isolated organisms suspected of causing a disease will be used as the inocula. Imagine yourself here as the microbiologist faced with the challenge of establishing that a specific microorganism causes a particular disease.

A. Animal System

Vibrio anguillarium is a known pathogen of various species of fish. If infected, fish generally develop a septicemia (invasion of the bloodstream) and eventually die. In this portion of the exercise, the goldfish (*Carassius* species) will be used for the detection of *V. anguillarium.*

Materials

The following items should be provided per 4 students:

❏ 1. Four 1-mL disposable syringes with 22-gauge needles, each labeled to indicate their respective contents as follows:
    ❏ a. 18-hour culture of *Vibrio anguillarium* (VIB-rē-ō, ANG-il-ar-ē-um), (ATCC strain 14181)
    ❏ b. 18-hour culture of unknown A
    ❏ c. 18-hour culture of unknown B
    ❏ d. Sterile nutrient broth

❏ 2. Four goldfish (*Carassius* species)

❏ 3. Four guppies (*Lebistes* species)

❏ 4. Four 500-mL beakers filled with about 300 mL of chlorine-free or aged water

❏ 5. Four glass plates (to partially cure the beakers in which inoculated fish are placed)

❏ 6. Four nutrient agar plates

❏ 7. Four salt (1.5%)-starch agar plates

❏ 8. Four sterile scalpels and scissors, individually wrapped

❏ 9. One beaker of disinfectant for contaminated instruments

The following items should be provided for general class use:

❏ 1. Paper towels

❏ 2. Sterile gauze pads (2 inches × 2 inches)

❏ 3. Glass slides

❏ 4. Plastic squeeze bottles with disinfectant

❏ 5. Gram stain reagents

❏ 6. Small-sized fish nets

❏ 7. Oxidase reagent

❏ 8. One-gallon container with chlorine-free or aged water

❏ 9. Disposable surgical gloves

❏ 10. Sterile scalpels and forceps

❏ 11. Sterile cotton swabs in individual containers

❏ 12. Plastic sandwich bags

❏ 13. Felt-tip marking pen

Procedure 1: Laboratory Animal Inoculation

(Modified from the technique described in Umbreit, W., and E. Ordal, *ASM News* 38:93–98 [1972].) Procedures 1 and 2 are to be performed by students in groups of 4.

Additional Techniques Required for This Portion of the Exercise:

❏ 1. Streak Plate Technique, Procedure Diagram 10, Exercise 5

❏ 2. The Gram Stain, Procedure Diagram 17, Exercise 14

❑ 1. With a marking pen, divide and label 1 nutrient agar plate into 4 sections, 1 for each of the preparations (prep.) to be used for inoculation.

❑ 2. Repeat step 1 with a salt-starch agar plate.

❑ 3. Take 4 clean glass slides and label 1 for each of the preparations.

❑ 4. Remove the needle guard of the syringe containing *Vibrio anguillarium* and place 1 small drop of the culture onto its respective nutrient agar and salt-starch agar plate section and on its labeled slide. Replace the needle guard and put the syringe down.

❑ 5. Streak each drop over its sector (Figure 54–1) and make a heat-fixed smear with the material on the slide. Store the smear in your laboratory drawer for later staining.

Figure 54–1
The streaking pattern for the inoculation preparations. The pattern is for both the nutrient agar and salt-starch agar plates.

❑ 6. Repeat steps 4 and 5 with each of the remaining preparations. Incubate the agar plates at room temperature for 48 hours.

❑ 7. Observe the fish in their container and locate the various parts indicated in Figure 54–2.

❑ 8. Take the syringe containing *V. anguillarium,* examine the barrel, and note the graduation markings on it. Note the number of lines needed to deliver 0.1 mL. (If in doubt, ask your instructor.)

❑ 9. Next, prepare for the inoculations.

❑ 10. Place 2 or 3 paper towels over the area to be used for the inoculations.

❑ 11. Wet 4 sterile gauze pads with chlorine-free water and squeeze out the excess on the paper towels.

❑ 12. Take a small-sized fish net and dip it into the container of fish to wet it.

❑ 13. Select the fish for the first inoculation and net it.

❑ 14. Take 1 of the moistened gauze pads and use it to remove the fish from the net.

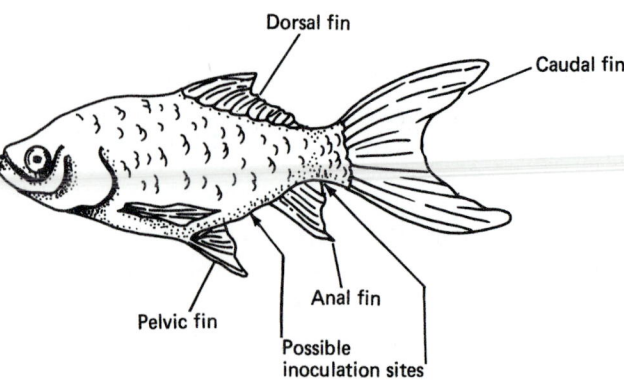

Figure 54–2
A general orientation to the goldfish and to inoculation sites indicated by arrows.

❏ 15. Orient the fish in the gauze so that its belly is toward you and its head is pointed away from you. (See Figure 54–3*a*.)

❏ 16. Remove the needle guard from the *V. anguillarium* syringe, tilt the fish's head slightly downward, and inject 0.1 mL of the *Vibrio* culture in front of the anal fin (Figure 54–3*b*).

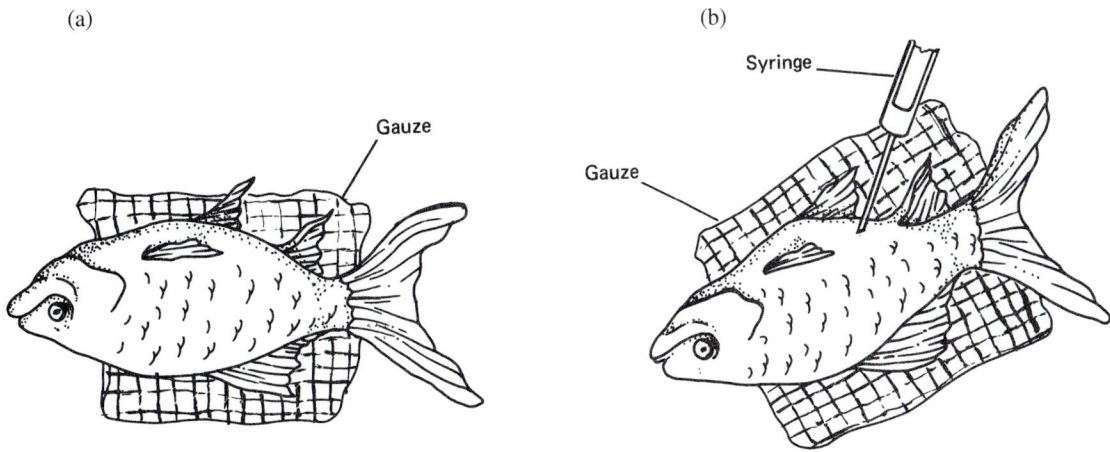

(a)   (b)

Figure 54–3
(a) Orienting the fish. (b) Preparing the fish for inoculation.

❏ 17. Place the inoculated fish into 1 of the beakers with chlorine-free water. Label the beaker with the preparation used. Cover the beaker *partially* with a glass plate to prevent the fish from jumping out.

❏ 18. Repeat steps 7 through 17 with the other materials provided for inoculation.

❏ 19. Gram stain the smears previously prepared and examine them under oil immersion. Record your microscopic findings in Table 54–1 in the Results and Observations section.

❏ 20. After 48 hours of incubation, examine the plates and make a Gram stain from similar colonies growing on the nutrient agar and salt-starch agar plates. Record your microscopic findings in Table 54–1 in the Results and Observations section.

❏ 21. Perform the oxidase test on the colonies picked for the Gram stain from the nutrient agar plates. Add a drop of the oxidase reagent to the colony. The formation of a dark red color is a positive test. (Refer to Exercise 24.) Record your findings in Table 54–1.

❏ 22. Perform the starch hydrolysis test on the colonies picked for the Gram stain from the salt-starch agar plate. Add a few drops of iodine onto the colony. The formation of a yellow zone around the colony is a positive test. (Refer to Exercise 20.) Record your findings in Table 54–1.

❏ 23. Continue on to Procedure 2.

Procedure 2: Demonstration of a Fish Pathogen

❏ 1. Observe the inoculated experimental animals daily for 1 week. Note any unusual behavior, growths, or discharge such as blood or excessive mucus. Record your findings in Table 54–2 in the Results and Observations section.

❏ 2. If a fish dies, remove it with a fish net, wrap it in a moist paper towel, and place it into a small plastic bag. Label the bag with your name and indicate the specimen used for inoculation. Keep the animal in a refrigerator or as indicated by your instructor until you have time to autopsy (dissect) it.

❏ 3. Disinfect the fish net by placing it in disinfectant for at least 10 minutes. Rinse the net thoroughly until all traces of the disinfectant have been removed.

❑ 4. Use a marking pen to label 1 salt-starch plate and 1 nutrient agar plate. Indicate the specimen used for inoculation and your name.

❑ 5. To autopsy the dead animal, place the fish on a paper towel saturated, *but not dripping,* with disinfectant, and perform the 3 steps shown in Procedure Diagram 40.

Procedure Diagram 40
Dissection of Koch's Postulates

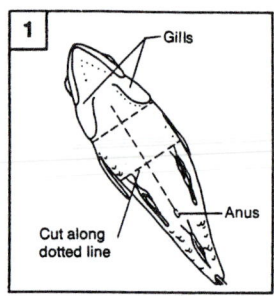

1. With a scapel slit the belly from the anus to the gillaren.

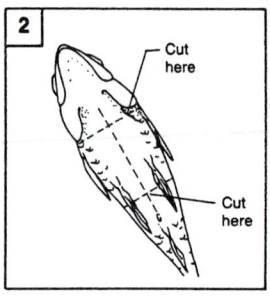

2. Make horizontal cuts. These will form flaps.

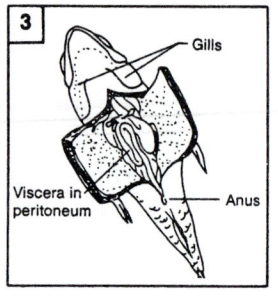

3. Move the flaps to expose the viscera (internal contents).

❑ 6. With the peritoneal cavity open, examine the exposed organs and the cavity. Look for unusual growths, blood, or other features such as blackening of organs indicating an abnormal condition. Record your findings in Table 54–3 of the Results and Observations section.

❑ 7. Obtain specimens from the peritoneal cavity with an inoculating loop and streak the earlier labeled plates for isolated colonies. Incubate the plates at 25°C for 48 hours, or as directed by your instructor.

❑ 8. Prepare 2 Gram stain smears of additional fluid obtained from the peritoneal cavity. Examine both smears under oil immersion and note the Gram reaction morphology and morphological arrangements. Record your findings in Table 54–4 of the Results and Observations section.

❑ 9. Discard the remains of the autopsied animal as directed.

❑ 10. After incubation, select those isolated colonies by circling them on the Petri plate bottom of salt-starch agar medium. Note their characteristics and enter your findings in Table 54–4.

❑ 11. Prepare and examine Gram stains of the selected colonies in step 10. Perform oxidase and starch hydrolysis tests on these colonies as follows:

   ❑ a. Add 1 or 2 drops of the oxidase reagent to each colony. The formation of a dark red color indicates a positive test. (See Exercise 24.)
   ❑ b. Flood the plate with Gram's iodine. Note the reactions of the circled colonies. Starch hydrolysis is indicated by a clear (usually yellow) zone surrounding the colony. (See Exercise 20.)

   Record all findings in Table 54–4 in the Results and Observations section.

❑ 12. Examine the colonies formed on the nutrient agar plate. Select 3 colonies for Gram stains and the oxidase test. Examine the preparations, and record your findings in Table 54–5 and answer the questions in the Results and Observations section.

❑ 13. Repeat steps 2 through 12 with each animal that dies within the experimental period.

❑ 14. Your instructor will indicate how the remaining fish are to be handled.

Results and Observations

Animal Inoculations

1. **Gram Stains, Oxidase, and Starch Hydrolysis Reactions.**

   Indicate positive Gram stain, oxidase, and starch hydrolysis reactions by a "+" and negative reactions by a "−".

Table 54–1

Gram Stains, Oxidase, and Starch Hydrolysis Reactions

| Preparation | Gram Reactions and Morphology | | | Oxidase Test | Starch Hydrolysis |
	Inoculation Preparation	Nutrient Agar Colony	Salt-Agar Colony		
V. anguillarium					
Unknown A					
Unknown B					
Sterile nutrient broth					

2. **Observations.**

   Enter your findings in the following table.

Table 54–2

Fish Observations

| Inoculated Fish | Day | | | | | | |
	1	2	3	4	5	6	7
1.							
2.							
3.							
4.							

3. **Dissection Observations.**

   Enter your autopsy findings in Table 54–3.

Table 54–3

Autopsy Findings

Fish	General Description and/or Findings
1.	
2.	
3.	
4.	

4. **Isolations from Peritoneal Cavity.**

Enter your findings in the following table. Use a "+" for positive reactions and a "−" for negative reactions with the Gram stain, oxidase, and starch hydrolysis tests.

Table 54–4

Fish Pathogen Properties

Fish	Gram Rx and Morphology of Peritoneal Fluid	Colony Description (Salt-Starch Agar)	Gram Rx and Morphology of Colony	Oxidase Test	Starch Test
1.					
2.					
3.					
4.					

5. **Nutrient Agar Isolations.**

Enter your findings in the following table. Use a "+" for positive reactions and a "−" for negative reactions with the Gram stain, oxidase, and starch hydrolysis tests.

Table 54–5

Fish Pathogen Isolation Results

Fish	Colony Descriptions (Nutrient Agar)	Gram Rx and Morphology	Oxidase Test	Starch Test
1.				
2.				
3.				
4.				

6. Which of the cultures caused the death of the inoculated fish? _____

_____

7. Were all of the reactions of the organism isolated similar to those of the original inoculum? (Compare the various

   tables.) If not, explain. _____

   _____

   _____

8. What conclusions can be made from your results? _____

   _____

   _____

B. Plant System

Members of the genus *Agrobacterium* are associated with the formation of tumors known as crown galls on the roots or stems of many different plants. The main species, which is *A. tumefaciens,* harbors the plasmid *pTi*. This plasmid contains the gene sequence associated with starting disease. During the infection process, some or all of *pTi* appears in the cells of the infected plant, while *A. tumefaciens* remains outside of such cells. The newly acquired genetic material causes a hormonal imbalance within the plant, which leads to increased plant cell reproduction and the tumor or gall formation.

Galls may form anywhere on the plant, but they appear most often in the stems near the soil line or on roots. The size of tumors ranges from less than 1 cm to 30 cm or more in diameter. In the following experiment, gall formation will indicate the presence of disease. Figure 54–4 shows the appearance of galls.

Materials

The following items should be provided per 4 students:

❑ 1. Four 1-mL disposable syringes with 22-gauge needles and guards, each labeled to indicate their respective contents as follows:

    ❑ a. 24-hour nutrient broth culture of *Agrobacterium tumefaciens,* (ag-rō-bak-TĒ-rē-um too-mē-FA-shē-enz) (ATCC strain 15955)

    ❑ b. 24-hour nutrient broth culture of unknown A

    ❑ c. 24-hour nutrient broth culture of unknown B

    ❑ d. Sterile nutrient broth (2 mL)

❑ 2. Four tomato plants in individual containers

❑ 3. Two nutrient agar plates

❑ 4. Four tubes of 5 mL sterile water

❑ 5. Four sterile scalpels or single-edge razor blades

❑ 6. Four sterile mortar and pestle combinations

❑ 7. One beaker of disinfectant for contaminated instruments

The following items should be provided for general class use:

❑ 1. White tape or labels

❑ 2. Tap water for periodic watering of plants during the period of the experiment

❑ 3. Felt-tip marking pens (2 different colors)

❑ 4. Gram stain reagents

❑ 5. Sterile Petri plates

❑ 6. Sterile scalpels and forceps

❑ 7. Plastic rulers

Figure 54–4
The appearance of plant galls or tumors.
(From Miller, H.N., *Phytopathology* 65: 850–851, [1975].)

Procedure: Plant Inoculation and Demonstration of Crown Gall Disease

This procedure is to be performed by students in groups of 4.

> **Additional Techniques Required for This Portion of the Exercise:**
>
> ❏ 1. Streak Plate Technique, Procedure Diagram 10, Exercise 5
> ❏ 2. The Gram Stain, Procedure Diagram 17, Exercise 14

❏ 1. With a marking pen, divide and label 1 nutrient agar plate into 4 sections, 1 for each of the preparations to be used. (See Figure 54–1.)

❏ 2. Divide 4 clean glass slides into 2 sections, *a* and *b,* and label 1 for each of the preparations. Section *a* will be used for a Gram stain of each preparation before inoculation and section *b* will be for the Gram stain of a specimen from a lesion resulting from the inoculation with each preparation.

❏ 3. Remove the needle guard of the syringe containing *Agrobacterium tumefaciens,* and place 1 small drop onto its section of the nutrient agar plate and the *a* section of its labeled slide. Replace the needle guard and put the syringe down.

❏ 4. Streak the drop of the preparation on the agar surface, and make heat-fixed smears with the slide specimen. Store the labeled slide in an empty Petri plate for staining later.

❏ 5. Repeat steps 3 and 4 with each of the remaining preparations. Incubate the nutrient agar plate for 48 hours at room temperature.

❏ 6. Examine the colonial growth and prepare Gram stains of a representative colony. Record your colonial description and Gram stain findings in Table 54–7 in the Results and Observations section.

❏ 7. Examine the tomato plant diagram (Figure 54–5) and note the inoculation sites. Locate these sites on the plants provided.

❏ 8. Take the syringe containing *A. tumefaciens,* examine the barrel, and note the graduation markings on it.

❏ 9. Remove the needle guard and with the needle pointing in as near an upward direction as possible, inject 0.05 mL of the preparation into 4 separate injection sites on 1 tomato plant.

❏ 10. Replace the needle guard, and put down the syringe.

❏ 11. Label the container of the inoculated plant with your name, date, and the preparation used for inoculation.

❏ 12. Indicate the specific inoculation sites on 1 of the plant diagrams provided in the Results and Observations section.

❏ 13. Repeat steps 6 through 12 with each of the other preparations provided for plant inoculations.

❏ 14. Observe the inoculated plants on a weekly basis. Look for the appearance of galls. (Refer to Figure 54–4.) Enter your weekly findings in Table 54–6 in the Results and Observations section.

❏ 15. When galls appear on 1 of the plants, measuring about 10 to 20 millimeters (mm), remove them with the aid of a sterile scalpel and forceps. Place them into 1 of the mortars provided. Note your findings in the Results and Observations section.

❏ 16. With a pestle, grind the galls of the plant. Prepare a slush or slurry by adding some sterile water to the ground plant tissue.

❏ 17. Label another nutrient agar plate as described in step 1.

❏ 18. Remove 1 loopful of the gall tissue slurry and streak it on the appropriate agar section.

❏ 19. Remove another loopful of the slurry and make a smear on the *b* section of the previously labeled slide stored in your laboratory drawer.

❏ 20. Repeat steps 15 through 19 with all plants showing galls. Incubate the nutrient agar plate for 48 hours at room temperature.

❏ 21. Examine the colonial growth and prepare a Gram stain of a representative colony. Record your colonial description and Gram stain reactions in Tables 54–7 and 54–8, respectively, in the Results and Observations section.

❏ 22. Gram stain all smears of the starting materials. Examine them under the oil-immersion objective. Record your findings as to Gram stain reaction and the morphological arrangement of the cells seen in Table 54–8 in the Results and Observations section.

❏ 23. Answer all questions in the Results and Observations section.

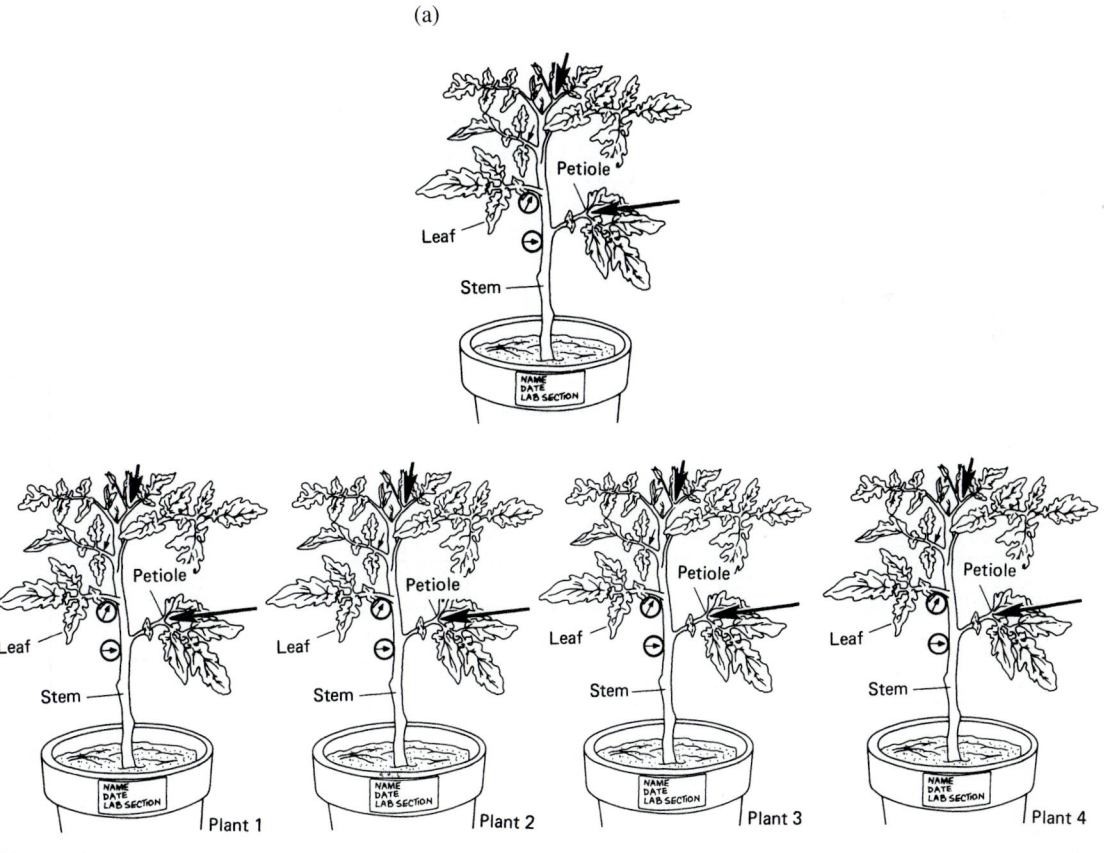

(a)

(b)

Figure 54–5

(a) Plant inoculation sites (arrows). (b) Test plant diagrams.

Results and Observations

Plant Inoculations

1. Complete the following table by indicating with a "+" the first day gall formation was observed and according to the inoculation site.

Table 54–6

Weekly Observations

Inoculation Site Plant #	Day													
	1	2	3	4	5	6	7	8	9	10	11	12	13	14
1.														
2.														
3.														
4.														

2. Did all preparations cause gall formation? If not, which ones did not? _____

_____

3. Gall descriptions: _____

_____

_____

_____

4. Colonial features, Gram reactions, and morphology of inoculation preparations. Enter your findings in Table 54–7.

Table 54–7

Inoculation Preparation Properties

Preparation	Colonial Description	Gram Reaction and Morphology
1.		
2.		
3.		
4.		

5. Gram stain comparisons of starting cultures and gall isolates.

Table 54–8

Starting Culture and Gall Isolate Properties

Preparation	Starting Culture	Gall Isolates
1.		
2.		
3.		
4.		

6. In the cases of gall formation, compare the colonial descriptions, Gram stain reactions, and morphological features (Tables 54–7 through 54–8), and answer the following questions:

a. Were these properties similar in all cases? _____

b. If not, explain why. _____

_____

c. In this portion of the exercise, which preparations served as controls? _____

_____

Laboratory Review 54   Koch's Postulates in Action

1. What value do Koch's Postulates have? _____

_____

_____

2. In the animal inoculation portion of this exercise, give the functions of the following:

a. salt-starch agar _____

b. sterile broth inoculum _____

c. nutrient agar _____

3. List 4 general characteristics of *Vibrio anguillarium.*

    a. _____   c. _____

    b. _____   d. _____

4. a. Briefly indicate how you would disprove the possibility that toxin or virus was the cause of death of the fish in
       this exercise. _____

       _____

    b. Do the same for gall formation. _____

       _____

5. Define:

    a. gall _____

    b. plasmid _____

    c. *pTi* _____

    d. toxin _____

Key Terms

anthrax (AN-thraks): an infectious disease of cattle, goats, horses, and sheep caused by *Bacillus anthracis;*
    humans contract it from diseased animals or their products

gall: plant tumor

Koch's Postulates: an experimental procedure developed by Robert Koch to demonstrate the causal relationship
    between a suspected pathogen and a disease

pathogen: infectious disease agent

plasmid: an extrachromosomal DNA fragment

septicemia (sep-ti-SĒ-mē-a): the presence of pathogenic bacteria in the blood

After completing this exercise, you should be able to:

1. Recognize the importance of using universal precautions to prevent infections.

2. List the main types of precaution factors that should be followed to prevent infection in and out of clinical settings.

3. Give examples of the applications of universal precaution factors.

Because the risk of the unknown is always present in health care, the possibility always exists for exposure to infectious disease agents and/or the development of illness in health care workers or others involved with the care of the ill. To protect health care workers as much as possible, it is generally assumed that every patient is a potential source of disease agents, and appropriate infection control measures must be taken. This is particularly true for pathogens that may be present in blood or other body fluids. Such pathogens may include human immunodeficiency viruses (HIVs); hepatitis virus A, B, C, D, and E; cytomegalovirus; and a number of bacteria and protozoa known to cause gastrointestinal diseases. The spread of hospital-acquired *(nosocomial)* infections can occur in a number of ways including from patient to patient, patient to health care worker, health care worker to other workers, health care worker to patient, and of course from patient to visitors.

The *body substance isolation* or *universal precautions* approach developed by the Centers for Disease Control and Prevention makes the health care workplace safer for both health care workers and their patients. Figure 55–1 shows a typical chart that can be found in health care facilities emphasizing the safety guidelines for health care personnel. Table 55–1 lists a number of the precautions followed by health care workers. Employers are responsible for monitoring the work setting to make certain that protective measures are being observed. If they are not, either more education or disciplinary actions usually are used.

Because most precautions are aimed at preventing punctures by sharp objects and exposure to the splash and splatter of blood and other body fluids, various types of barriers are used. Such barriers generally serve to prevent, restrict, or stop entry and include safe hypodermic needle disposal containers (Figure 55–2a), latex or vinyl gloves (Figure 55–2b), drapes, gowns, face masks, and eye shields.

Historical records indicate the use of protective clothing or apparel dates back to about 1721. At that time, the fear of contagion during the plague epidemics in Europe was a legitimate concern, and physicians wore special garments to avoid infection from coming into contact with fleas and the body fluids of their patients. Today, an extensive selection of protective apparel is available, usually in the form of drapes and gowns. Such apparel is designed to truly keep blood and other potentially infectious material away from health care workers.

Accidental needle-sticks represent a common problem and one that is associated with needle disposal. Several types of containers are in use and are intended to provide a system of safe needle and syringe disposal.

Gloves are worn in a number of patient situations, especially those associated with anticipated contact with blood and other body fluids or secretions, broken skin, and moist body surfaces. The appropriate use of gloves protects not only the health care worker, but the patient as well. A wide selection of different latex and vinyl gloves is available to meet a broad range of needs.

Many of the universal precautions followed by health care workers are also applicable to situations outside of hospitals and medical or dental offices. This would include households as well as public facilities in which infected persons may be present. Certain precautions such as proper hand washing or even the wearing of gloves and face masks may be appropriate to avoid exposure to disease agents.

This exercise will demonstrate the importance of various aspects of universal precautions including handwashing and the use of gloves.

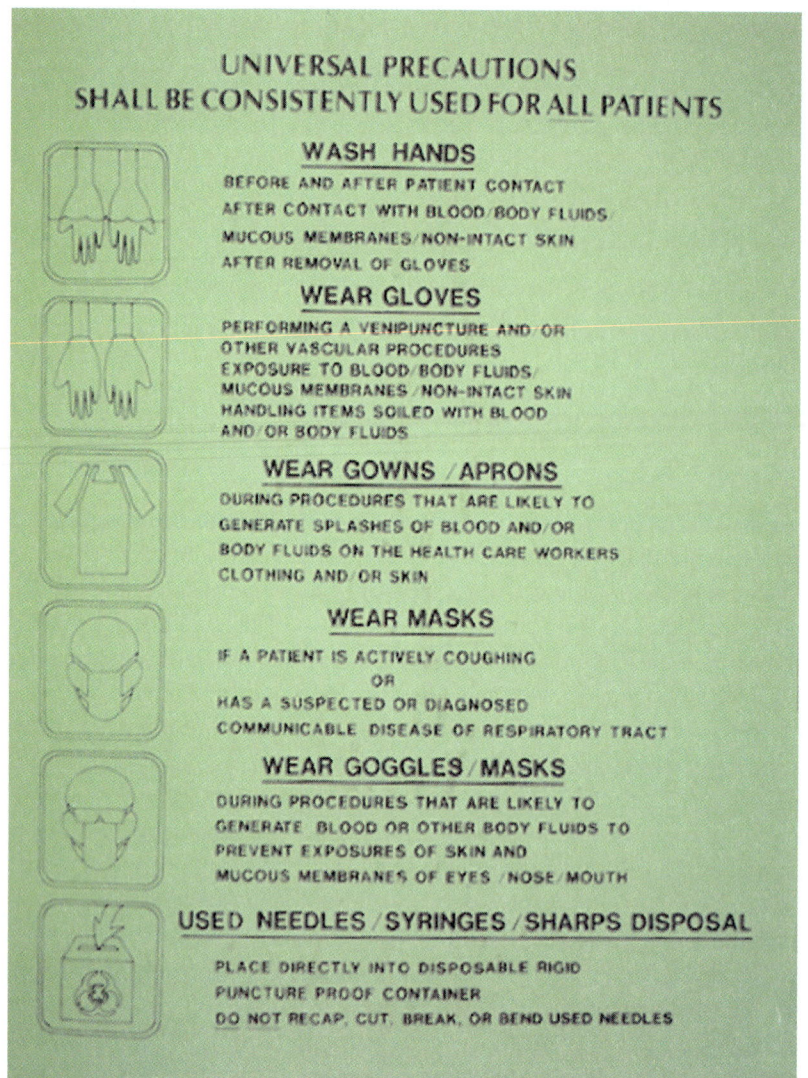

Figure 55–1
An example of a Universal Precautions chart posted in a health care facility.

The chart reads:

**UNIVERSAL PRECAUTIONS
SHALL BE CONSISTENTLY USED FOR ALL PATIENTS**

WASH HANDS
BEFORE AND AFTER PATIENT CONTACT
AFTER CONTACT WITH BLOOD/BODY FLUIDS/
MUCOUS MEMBRANES/NON-INTACT SKIN
AFTER REMOVAL OF GLOVES

WEAR GLOVES
PERFORMING A VENIPUNCTURE AND/OR
OTHER VASCULAR PROCEDURES
EXPOSURE TO BLOOD/BODY FLUIDS/
MUCOUS MEMBRANES/NON-INTACT SKIN
HANDLING ITEMS SOILED WITH BLOOD
AND/OR BODY FLUIDS

WEAR GOWNS/APRONS
DURING PROCEDURES THAT ARE LIKELY TO
GENERATE SPLASHES OF BLOOD AND/OR
BODY FLUIDS ON THE HEALTH CARE WORKERS
CLOTHING AND/OR SKIN

WEAR MASKS
IF A PATIENT IS ACTIVELY COUGHING
OR
HAS A SUSPECTED OR DIAGNOSED
COMMUNICABLE DISEASE OF RESPIRATORY TRACT

WEAR GOGGLES/MASKS
DURING PROCEDURES THAT ARE LIKELY TO
GENERATE BLOOD OR OTHER BODY FLUIDS TO
PREVENT EXPOSURES OF SKIN AND
MUCOUS MEMBRANES OF EYES/NOSE/MOUTH

USED NEEDLES/SYRINGES/SHARPS DISPOSAL
PLACE DIRECTLY INTO DISPOSABLE RIGID
PUNCTURE PROOF CONTAINER
DO NOT RECAP, CUT, BREAK, OR BEND USED NEEDLES

(b)

(a)

Figure 55–2
Examples of items used to carry out Universal Precautions. (a) A needle disposal container.
(Courtesy of Becton Dickinson Vacutainer Systems, Rutherford, New Jersey.)

(b) An example of sterile latex gloves used in surgical situations as well as handling various types of body fluids.
(Courtesy of Thomas Scientific, Swedesboro, New Jersey.)

Table 55–1

Universal Precautions

Precautions Factor	Application
Protective apparel and related barriers	Gloves should be worn to protect hands from coming into direct contact with blood or other possible infectious body fluids or secretions. They should be changed and disposed of after each contact or use. Gowns, face masks, and protective eyewear should be worn, whenever practical, during procedures that generate droplets or other splatter of blood and other body fluids.
Hand-washing	Hands and skin surfaces should be washed and scrubbed immediately with germicidal if contaminated with blood or other body fluids or secretions. Hands also should be washed immediately after gloves, masks, or other barrier devices are removed.
Needle and other sharp object disposal	Care should be taken to prevent injuries by needles or other sharp objects. Needles should not be recapped, bent, or broken by hand. After use, needles should be disposed of in puncture-resistant containers.
Dental hand pieces and/or other related items	All dental equipment should be sterilized between patients. If this is not possible, they should be thoroughly disinfected (peroxide hypochlorite). Blood and saliva should be carefully removed completely from all contaminated dental instruments and intraoral devices prior to sterilization.
Open and/or weeping cuts	Workers with open, weeping skin cuts or sores should not give direct care to patients or handle equipment used in such care situations until the cuts heal. Cuts should be covered with appropriate bandages that repel liquids.
Ventilation devices (e.g., mouthpieces)	Ventilation devices should be available for use during resuscitation and testing procedures. Disposable devices should be used whenever possible.
Pregnancy	Pregnant employees should be especially familiar with and strictly follow the universal precautions.

A. Examples of Barrier and Related Items Associated with Universal Precautions

Materials

The following items should be provided for class demonstration and examination:

❏ 1. One needle disposal container, and hypodermic needle or similar device

❏ 2. One typical protective gown or similar garment

❏ 3. An assortment of latex or vinyl gloves

❏ 4. An assortment of protective devices for the face

Procedure

❏ 1. The instructor will demonstrate the proper use of the items listed in the Materials section. This will include the following:

　❏ a. the proper disposal of a hypodermic needle

　❏ b. the use of latex or vinyl gloves (refer to Figure 55–2)

　❏ c. the use of a face mask or similar device

❏ 2. Examine the various items placed on display.

❏ 3. Answer the questions in the Laboratory Review.

B. The Effectiveness of Hand Washing

Materials

The following items should be provided per 4 students:

❑ 1. Eight individually wrapped surgical or similar type brushes

❑ 2. One dispenser of liquid soap

❑ 3. One unopened roll of paper towels

❑ 4. Container for disposal of used brushes and paper towels

❑ 5. Four nutrient agar plates

❑ 6. Four mannitol-salt agar plates

❑ 7. One marking pen or similar device

The following items should be provided for general class use:

❑ 1. A Quebec Colony Counter

❑ 2. Gram stain sets

Procedure

Additional Technique Required for This Portion of the Exercise:

❑ The Gram Stain, Procedure Diagram 17, Exercise 14

This procedure is to be performed by students in pairs.

❑ 1. With a marking device, divide the bottom of one nutrient agar and one mannitol-salt agar plate into two sectors. Label one sector on each plate "before" and the second "after."

❑ 2. Label one set of plates "right hand" and the other "left hand." Include your initials on all plates.

❑ 3. Using the correct plate, before starting the handwash procedure, gently press the finger tips of your right hand first on the "before" surface of the nutrient agar, and then on the "before" surface of the mannitol-agar plate.

❑ 4. Next follow the steps shown for hand washing in Procedure Diagram 41.

❑ 5. Your laboratory partner should assist you in unwrapping the surgical brushes and providing the paper towels for drying of your hands.

❑ 6. After completing the hand washing procedure, dry your hands thoroughly with the paper towels provided.

❑ 7. Now gently press the finger tips of your right hand on the "after" surface of the nutrient agar, and then on the "after" surface of the mannitol-salt agar plate. Repeat the step with your left hand.

❑ 8. Have your laboratory partner repeat steps 1 through 7.

❑ 9. Incubate all plates at 37°C for 24–48 hours or as indicated by your instructor.

❑ 10. After incubation, count the colonies in each sector on the respective plates. Enter your counts in Table 55–2 of the Results and Observations section.

❑ 11. Select two colonies from each sector and perform the Gram stain.

❑ 12. Enter your findings in the Results and Observations section.

❑ 13. Answer the questions in the Results and Observations and the Laboratory Review sections.

Procedure Diagram 41
Hand Washing

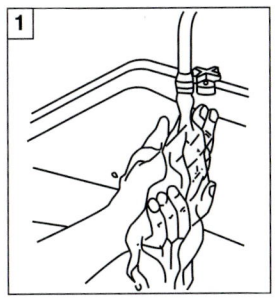

1. Moisten both hands and wrists with warm water.

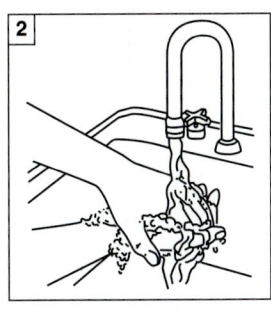

2. Rub and cover hands with a heavy lather of soap for 2 minutes.

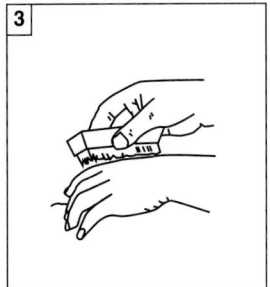

3. Scrub both hands and under the fingernails with the brush provided.

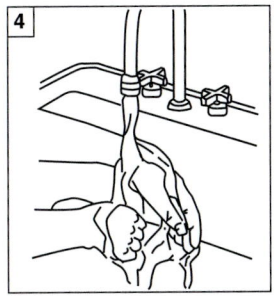

4. Rinse both hands in warm water by allowing water to flow downward from fingertips.

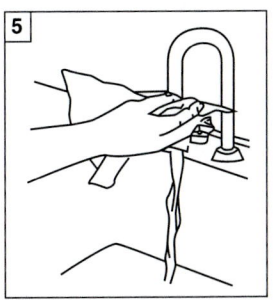

5. Dry hands thoroughly with a clean paper towel.

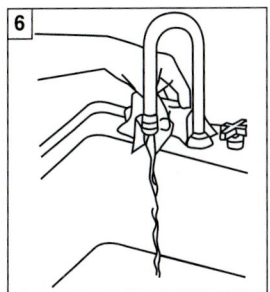

6. Close the faucet with another clean dry paper towel. Continue with the exercise.

Results and Observations

The Effectiveness of Hand Washing

1. Enter your findings in the following table.

Table 55–2

Effectiveness of Handwashing

		Colony Counts			
		Before		After	
Student	Specimen Source	Nutrient Agar	Manitol-Salt Agar	Nutrient Agar	Mannitol-Salt Agar
1	Right Hand				
	Left Hand				
2	Right Hand				
	Left Hand				

2. Were there obvious differences between the before colony counts for the left and right hands? _____

   If so, on which medium? _____

3. Were there any obvious differences between the "after" colony counts on either of the plates for the left and right hands? _____

   _____

   If so, explain. _____

4. Was there a predominant Gram reaction and morphological arrangement found on either plate? _____

   If so, what were the findings? _____

5. On the basis of your findings, would you consider the hand washing procedure to be effective? _____

   If not, why? _____

6. What would you consider to be possible sources of contamination in this procedure? _____

   _____

C. The Use of Surgical Gloves

Materials

The following items should be provided per 4 students:

❏ 1. Four pairs of sterile individually wrapped surgical gloves

❏ 2. Eight individually wrapped surgical or similar type brushes

❏ 3. One dispenser of liquid soap

❏ 4. One unopened roll of paper towels

❏ 5. Container for disposal of used brushes and paper towels

❏ 6. Four nutrient agar plates

❏ 7. Four mannitol-salt agar plates

❏ 8. One marking pen or similar device

❏ 9. One liter of sterile distilled water

The following items should be provided for general class use:

❏ 1. A Quebec Colony Counter

❏ 2. Gram stain sets

Procedure

Additional Technique Required for This Portion of the Exercise:

❏ The Gram Stain, Procedure Diagram 17, Exercise 14

This procedure is to be performed by students in pairs.

❏ 1. With a marking device, divide the bottom of one nutrient agar and one mannitol-salt (MS) agar plates into two sectors. Label one sector on each plate "before" and the second "after." MS plates are of value in the isolation and identification of *Staphylococcus aureus*.

❏ 2. Label one set of plates "right hand" and the other "left hand." Include your initials on all plates.

❏ 3. Follow the steps shown for hand washing in Procedure Diagram 41 in this exercise.

❏ 4. Your laboratory partner should assist you in unwrapping the surgical brushes and providing the paper towels for drying of your hands.

❏ 5. After completing the handwashing procedure, dry your hands thoroughly with the paper towels provided.

❏ 6. Now gently press the finger tips of your right hand on the "before" surface of the nutrient agar, and then on the "before" surface of the mannitol-salt agar plate. Repeat this step with your left hand.

❏ 7. Rinse your hands with the sterile distilled water. Your partner should assist you.

❏ 8. Dry your hands thoroughly.

❏ 9. Have your partner unwrap the packages containing the sterile surgical gloves, and insert your hands into them accordingly.

❏ 10. Wear the gloves for 1–2 hours while performing other duties.

❏ 11. After this time period, have your partner assist you in removing the gloves. DO NOT TOUCH ANY OTHER OBJECTS BEFORE COMPLETING STEP 12.

❏ 12. Gently press the finger tips of your right hand, first on the "after" surface of the nutrient agar, and then on the "after" surface of the mannitol-agar plate.

❏ 13. Have your laboratory partner repeat steps 1 through 12.

❏ 14. Incubate all plates at 37°C for 24–48 hours or as indicated by your instructor.

❏ 15. After incubation, count the colonies in each sector of the respective plates. Enter your counts in Table 55–3 of the Results and Observations section. Compare your findings with Figure 55–3.

❏ 16. Select two colonies from each sector and perform the Gram stain.

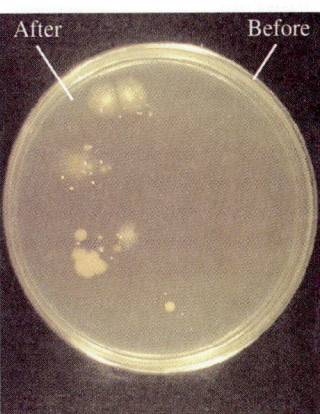

Figure 55–3

The results of a *before* and *after* experiment. The **right side** of the agar plate shows the bacterial outgrowth from fingers washed and before the wearing of a latex glove. The **left side** of the plate shows the bacterial outgrowth from the same washed fingers after the wearing of a latex glove for 2.5 hours. How can this result be explained?

❏ 17. Enter your findings in the Results and Observations section.

❏ 18. Answer the questions in the Results and Observation and the Laboratory Review sections.

Results and Observations

The Use of Surgical Gloves

1. Enter your findings in Table 55–3.

Table 55–3

Surgical Glove Counts

Student	Specimen Source	Colony Counts			
		Before		After	
		Nutrient Agar	Manitol-Salt Agar	Nutrient Agar	Mannitol-Salt Agar
1	Right Hand				
	Left Hand				
2	Right Hand				
	Left Hand				

2. a. Were there obvious differences between the "before" colony counts for the left and right hands? _____

    b. If so, on which medium? _____

    c. How can such differences be explained? _____

    d. Were the results of your laboratory partner different? _____

_____

3. Were there any obvious differences between the "after" colony counts on either of the plates for the left and right

    hands? _____

    If so, explain. _____

4. Was there a predominant Gram reaction and morphological arrangement found on either plate? _____

    If so, what were the findings? _____

5. a. On the basis of your findings, would your hand washing and the use of surgical gloves be effective in

    limiting the spread of microorganisms? _____

_____

_____

    b. How would you test your answer? Propose an experiment. _____

_____

_____

6. What would you consider to be possible sources of contamination in this procedure? _____

_____

Laboratory Review 55

Universal Precautions in Action

1. List 4 examples of barrier and related items that can be used in performing universal precautions.

    a. _____

    b. _____

    c. _____

    d. _____

2. What are universal precautions? _____

    _____

    _____

3. List 6 situations requiring precautions together with the appropriate application of universal precautions (UP).

Situation	**UP Application**
a.	
b.	
c.	
d.	
e.	
f.	

4. What is a nosocomial infection? _____

    _____

Key Terms

nosocomial (nos-ō-KŌ-mē-al) infection: a hospital-acquired infection

universal precautions: a set of safety procedures or steps used to make the workplace safer for both health care workers and patients

After completing this exercise, you should be able to:

1. Obtain a throat swab.

2. Explain the effect of environmental factors on throat-swab specimens prior to their inoculation onto bacteriological media.

3. Explain the effects of refrigeration, incubation, and a typical chemical preservative (thymol) on urine specimens prior to culturing.

4. Explain the effects of refrigeration, freezing, and hemolysis on citrated whole-blood specimens prior to culturing.

5. Recognize the possibility of transmission of a disease associated with collection and handling of specimens.

The proper handling of specimens for microbiological analysis requires: (1) aseptic collection techniques; (2) the use of appropriate containers; (3) suitable means for preservation; and (4) suitable means of transporting specimens to the laboratory. In addition, conditions suitable for analysis or pretreatment must be provided, when necessary, prior to the inoculation of test animals, tissue culture, or laboratory media.

(b)

(a)

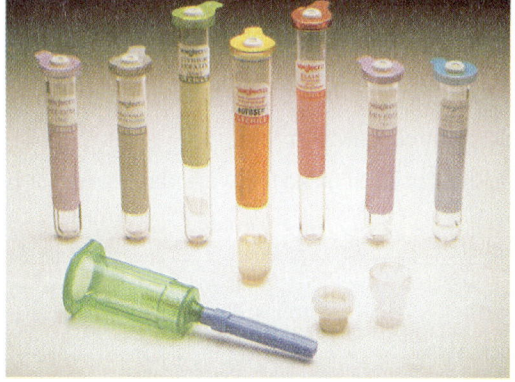
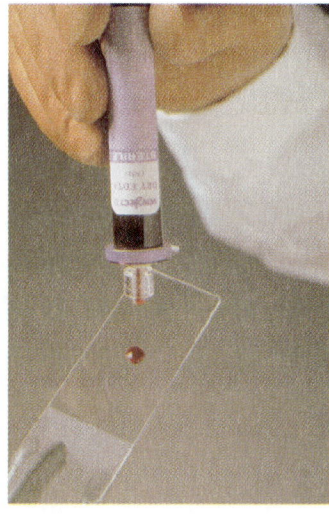

Figure 56–1

Examples of devices used to insure laboratory safety while collecting, transporting, and examining blood specimens. (a) A variety of Venoject plastic blood collection tubes. These tubes are shatter-resistant thereby eliminating the risk of injury and/or exposure to blood caused by breakage. A tube holder and needle guard used in obtaining specimens also are shown. (b) A Venoject piercing sampler. This device permits blood smear preparation without the need to open the tube. This arrangement maintains a closed system and again insures laboratory safety.
(Courtesy of Terumo Medical Corporation.)

All specimens should be handled aseptically and treated as potentially infectious. In cases of spillage or contamination of the outside of a container, some form of disinfection should be carried out immediately. Because the problem of spillage is always potentially present, the use of plastic containers such as sacs and tubes is suggested. In addition, proper hand washing is imperative for the prevention of personnel contamination and the spread of infectious organisms to others.

Many specimens can be refrigerated in the laboratory after collection. Samples of this type include citrated whole blood, urine, stool samples, and specimens of water, food, and milk. However, where serum is required for serological testing, whole blood must be collected without an anticoagulant and kept at room temperature until clot formation has taken place. Then the specimen can be refrigerated until used. Prior to testing, the clot is freed from the sides of the tube (a technique called *ringing*) and centrifuged. Other types of samples that must not be refrigerated and should be kept warm include cough plates for *Bordetella pertussis* and specimens for *Neisseria gonorrhoeae* and *N. meningitidis*. As in the case of *Entamoeba histolytica,* the causative agent of amebic dysentery, trophozoites are extremely susceptible to drying and cold temperatures; therefore, it is imperative that stool specimens be kept at body temperature (37°C) in closed containers and examined within a few hours if trophozoites are to be recovered.

The proper collection and preservation of fecal specimens are important for the isolation and identification of intestinal pathogens. During storage, the pH of feces drops and thereby greatly reduces the possibility of finding disease agents intact. Freshly voided fecal specimens, collected from a freshly voided specimen, are usually placed into a container with a phosphate or related buffer for transport to the laboratory.

Preservatives such as buffered glycerol-saline and ethylene diamine tetraacetic acid (EDTA) can be used in keeping fecal samples that are to be tested for enteric pathogens. The use of preservatives is effective in preventing nonpathogenic organisms from overgrowing the organisms to be examined. However, preservatives used in keeping samples for microbiological analysis must, of necessity, differ from those used for specimens to be subjected to chemical analysis. In the latter case, prevention of microbial contamination is desirable; in the former case, the bactericidal activity of the preservative brings about the need for additional specimens. For example, formalin is commonly employed as a preservative, and some individuals forget that it is germicidal and should not be used to preserve specimens for microbiological analysis.

A widely used multiple chemical system for fecal specimens is the two-vial or polyvinyl alcohol (PVA) fixative and formalin system (Figure 56–2). This system is designed to improve preservation by using two separate and different solutions. Each of these solutions are expected to provide the best preservation of different diagnostic stages of the suspected pathogen. Five to ten percent formalin, one of the solutions, acts as a preservative for helminth (worm) eggs and larval stages, and protozoan cysts. The resulting mixture can be used for direct observation and for concentration preparations. Another portion is added to a vial containing a PVA fixative, which preserves protozoan trophozoites and cysts for permanent staining and slides.

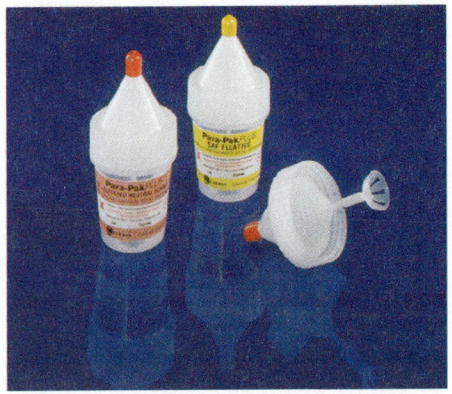

Figure 56–2

The Para-Pak Plus systems. Devices such as these are used for the transport of stool (fecal) specimens. Each device consists of a plastic container, which contains the preservative liquid, and a cone-shaped lid, which contains both a filter and a standard volume collecting scoop. In addition, the vials contain special beads that not only aid in the mixing of the specimen, but also free parasites from fecal debris.
(Courtesy of Meridian Diagnostics, Inc. Cincinnati, OH.)

Several types of specimens do not require the addition of preservatives. For example, swab specimens for throat, wound, or vaginal cultures may be transported in holding media (Stuart's), which will maintain organisms in a viable state.

This exercise has been designed to familiarize you with some of the difficulties encountered in the handling of specimens.

A. Demonstration of Devices Used for the Collection and/or Transport of Specimens

Materials

The following items should be provided for class demonstrations:

❑ 1. A variety of containers used for the collection of clinical specimens including:
    ❑ a. blood   ❑ c. sputum
    ❑ b. urine   ❑ d. feces

❑ 2. Examples of transporting media

Procedure

This procedure is to be performed by students individually.

❑ 1. Examine the various devices and/or containers used for specimen collection.

❑ 2. Note the specific safety features of each one.

B. Throat Swabs

Materials

The following materials should be provided per 2 students:

❑ 1. Three blood agar plates

❑ 2. Three sterile cotton swabs

❑ 3. Three tongue depressors or similar devices to hold the tongue down during specimen taking

❑ 4. One sterile test tube

❑ 5. One screwcap test tube containing Stuart's transport or other comparable medium

❑ 6. One container of germicide

❑ 7. Disposable surgical gloves

Procedure

Additional Technique Required for This Portion of the Exercise:

❑ Streak Plate Technique, Procedure Diagram 10, Exercise 5

Caution

Working with sputum, blood, urine, or other body fluids will not pose any danger if the following precautions are observed:

❑ 1. Place any and all materials containing body fluids into the disinfectant containers provided.

❑ 2. Dispose of any body fluid-containing materials as indicated by the instructor.

❑ 3. In cases of body fluid spillage, wipe the area with a disinfectant-soaked paper towel, and dispose of all used materials as indicated by the instructor.

❑ 4. Always wash your hands after handling body fluids and associated materials.

❑ 5. Use disposable surgical gloves in specimen-handling situations.

This procedure is to be performed by students in pairs.

❑ 1. Obtain a throat specimen from your partner's throat with a sterile swab. (Refer to Figure 56–3.) Place the sterile swab against the back wall of the throat gently and move it up and down.

❑ 2. Put the specimen into a sterile test tube.

❑ 3. Repeat step 1 with a fresh sterile swab and place it in the tube of Stuart's transport medium. Be sure that the swab is pushed into the medium. Break the protruding portion of the swab and replace the cap. Place the broken portion into the germicide.

❑ 4. Repeat step 1 with a fresh swab and inoculate a blood agar plate using the clock-plate manner of streaking (refer to Procedure Diagram 10). Place the swab into the germicide provided.

❑ 5. Store the test tube containing the throat specimen swab and the Stuart's transport medium swab for 48 hours at room temperature. Incubate the inoculated blood agar plate for 48 hours at 37°C.

❑ 6. Record the numbers and types of colonies on the blood agar plate in Table 56–1 of the Results and Observations section. Inoculate each stored swab onto a separate blood agar plate and incubate for 48 hours at 37°C. Dispose of each swab as in step 4.

❑ 7. Record the numbers and types of colonies on the blood agar plate in Table 56–1.

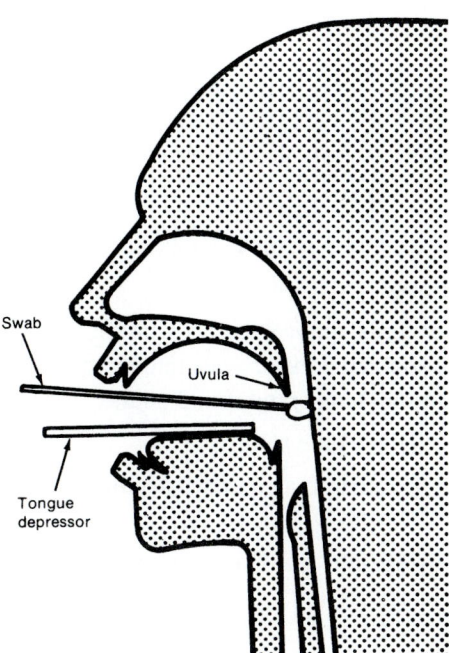

Swab

Uvula

Tongue depressor

Figure 56–3
The proper technique for obtaining a throat specimen. Note the positioning of the swab and tongue depressor in the mouth.

C. Urine Samples

Materials

The following materials should be provided per 4 students:

❑ 1. Five mL of fresh urine in a sterile container

❑ 2. Three sterile test tubes

❑ 3. Four eosin-methylene blue agar plates

❑ 4. One sterile test tube containing thymol crystals

❑ 5. One sterile applicator stick

❑ 6. One test tube with 5 mL of sterile distilled water

❑ 7. One sterile 1-mL pipette and bulb or Pi-Pump

❑ 8. Disposable surgical gloves

Procedure

Additional Technique Required for This Portion of the Exercise:

❏ 1. Streak Plate Technique, Procedure Diagram 10, Exercise 5

❏ 2. The Use of a Pipette Pump, Procedure Diagram 11, Exercise 6

Caution

Working with sputum, blood, urine, or other body fluids will not pose any danger if the following precautions are observed:

❏ 1. Place any and all urine or materials containing urine into the disinfectant containers provided.

❏ 2. Dispose of any urine containing materials as indicated by the instructor.

❏ 3. In cases of urine spillage, wipe the area with a disinfectant-soaked paper towel, and dispose of all materials as indicated by the instructor.

❏ 4. Always wash your hands after handling urine and associated materials.

❏ 5. Use disposable surgical gloves in specimen-handling situations.

This procedure is to be performed by students in groups of 4.

❏ 1. With the aid of a pipette pump, pipette 1 mL of the urine specimen into each of 3 sterile test tubes.

❏ 2. Refrigerate 1 tube and incubate another at room temperature for 2.25 hours.

❏ 3. Place several crystals of thymol into the third test tube. First, moisten a sterile applicator stick with sterile distilled water and use it to pick up the thymol crystals. Then place the stick with the crystals into the tube with urine and rotate the stick to remove the crystals. Store this tube at room temperature for 2 hours.

❏ 4. Prepare a streak plate of the original urine specimen using an eosin-methylene blue plate.

❏ 5. After 2 hours, prepare eosin-methylene blue streak plates from each of the other treated urine specimens. Incubate all plates at 37°C for 48 hours.

❏ 6. Record the relative amounts of growth on each plate in Table 56–2 of the Results and Observations section.

D. Blood Specimens

Materials

The following materials should be provided per student:

❏ 1. Five mL of citrated whole blood

❏ 2. Three sterile test tubes

❏ 3. One test tube with 5 mL of sterile distilled water

❏ 4. One 1-mL sterile pipette and bulb or pipette pump

❏ 5. Disposable surgical gloves

❏ 6. Container with disinfectant

The following item should be provided for general class use:

❏ 1. One refrigerator with a freezer compartment

Procedure

This procedure is to be performed by students individually.

Caution

Working with blood or other body fluids will not pose any danger if the following precautions are observed:

❏ 1. Place any and all blood or materials containing blood into the disinfectant containers provided.

❏ 2. Dispose of any blood containing materials as indicated by the instructor.

❏ 3. In cases of blood spillage, wipe the area with a disinfectant-soaked paper towel, and dispose of all materials as indicated by the instructor.

❏ 4. Always wash your hands after handling blood and associated materials.

❏ 5. Use disposable surgical gloves in specimen-handling situations.

❏ 1. Pipette 1 mL of the citrated whole blood into each of 3 sterile test tubes.

❏ 2. Refrigerate 1 tube and place the second tube in the freezer compartment until the next laboratory period.

❏ 3. Pipette 4 drops of sterile distilled water into the third tube, mix well, and refrigerate until the next laboratory period.

❏ 4. Before making your examination, allow the cellular contents of each tube to settle. Then observe the blood samples for hemolysis, and record the relative amounts of hemolysis in the appropriate section in Table 56–3, and answer the questions in the Results and Observations section.

E. Prevention of Disease Transmission

Materials

The following materials should be provided per 2 students:

❏ 1. One 24-hour nutrient broth culture of *Micrococcus luteus*

❏ 2. One sterile cotton swab

❏ 3. One sterile test tube

❏ 4. One test tube rack

❏ 5. Several sheets of paper towel

❏ 6. Two nutrient agar plates

❏ 7. One container of germicide (disinfectant)

Procedure

This procedure is to be performed by each pair of students.

❏ 1. Moisten the swab in the culture of *M. luteus,* carefully removing excess fluid by pressing the swab against the interior side of the tube as you remove the swab.

❏ 2. Place the sterile tube in a rack on several sheets of paper towel moistened with germicide.

❏ 3. Swab the organism on the outside of the sterile empty test tube. Allow this tube to dry for 2 to 3 minutes. Discard the swab in the container of germicide.

❏ 4. Both students should handle the contaminated tube. Pay particular attention to getting organisms on finger tips.

❏ 5. Next, one student should wash both hands carefully.

❏ 6. Then, both students should select a nutrient agar plate and touch their fingers to the agar surface. Be sure to label each plate appropriately.

❏ 7. Both students should wash their hands carefully after the experiment.

❏ 8. Incubate the agar plates for 48 hours at 37°C.

❏ 9. Examine both plates for the relative numbers of yellow colonies that are characteristic of *M. luteus* and record the findings in Table 56–4 in the Results and Observations section.

❏ 10. Answer the questions in the Results and Observations section.

Results and Observations

Throat Swabs

1. Complete the following table.

Table 56–1

Throat Swab Results

Swabs	Number of Colonies	Numbers of Different Colonies
Fresh		
Room temp. stored		
Transport medium		

Urine Samples

1. Enter your findings in the following table.

Table 56–2

Urine Specimen Results

Urine Specimen	Relative Amount of Growth		
	Slight	Moderate	Heavy
Untreated			
Room temperature			
Refrigerated			
Thymol			

Blood Specimens

1. Enter your findings in Table 56–3.

Table 56–3

Blood Specimen Observations

Specimen	Hemolysis		
	Slight	Moderate	Extreme
Refrigerated blood			
Frozen blood			
Blood with small amount of water			

2. Does there appear to be a significant difference between refrigeration and freezing with respect to the integrity of red blood cells? _____

_____

_____

3. Would hemolysis affect the isolation of bacteria from the blood? Explain. _____

_____

_____

Prevention of Disease Transmission

1. Enter all findings in the following table.

Table 56–4

Hand Washing Results

	Relative Growth of *M. luteus*
Washing	
No washing	

2. What was the observed effect of hand washing on the relative numbers of *M. luteus?* _____

_____

_____

3. Interpret these results in terms of possible disease transmission in a hospital setting. _____

_____

_____

Laboratory Review 56   Specimen Handling and Transport

1. Compare the effect of storing one throat swab (dry), one throat swab in transport medium, and one fresh swab in relation to the survival of microorganisms. _____

    _____

    _____

    _____

2. Which method of handling urine would be more suitable for quantitative bacteriological analysis of urine? For chemical analysis? _____

    _____

    _____

    _____

3. How should urine be collected from a patient to minimize contamination? _____

    _____

    _____

4. What precautions should be taken to ensure the dryness of syringes, test tubes, and other materials that will come into contact with blood? _____

    _____

    _____

    _____

After completing this exercise, you should be able to:

1. Explain how aerosols of infectious microorganisms may be produced by a cough or a sneeze.

2. Explain the hazardous nature of selected microbiological procedures, and how hazards may be minimized.

Airborne pathogenic microorganisms *(aerosols)* constitute a major hazard to many higher forms of life. Viruses, bacteria, and fungi can be easily carried from host to host by air currents caused by thermal shifts, high-pressure and low-pressure areas, or even an unguarded cough or sneeze.

The nature of aerosols and the dangers associated with them are clearly indicated in findings in the National Aeronautics and Space Administration (NASA) and other governmental agencies. NASA reported that a person standing quietly yields a large number of microorganisms from his or her body surfaces. On becoming active, still greater quantities are produced. The increase in activity influences the rate of aerosolization. Another study concerned with skin disinfection showed that the use of soaps with hexachlorophene reduces microbial numbers to a low level but can never bring about sterilization of the skin. The U.S. Army conducted an investigation involving the sampling of air around a laboratory technician who was performing routine microbiological procedures. His aseptic technique appeared to be adequate, but when air samples were taken, representative test bacteria with which he had been working were found several feet in all directions from the working area.

These studies and the knowledge of how easily coughing and sneezing can spread microorganisms emphasize the extreme care that one must take when working with pathogens, whether they are growing under laboratory conditions or in a natural environment (such as the throat of a person with an acute respiratory disease).

The widespread distribution of microorganisms is well known. Usually, wherever one goes, organisms of one type or another are always there. Microorganisms in dust or dried sputum are constantly swept into the environment by air currents made by vehicles, pedestrians, and even microbiology students. The defense mechanisms of the human body normally function well, serving to protect against the constant microbial bombardment. But consider the problems encountered in laboratories, hospitals, clinics, and medical and dental offices: gathered together in one building may be newborn children, patients suffering from a variety of ailments, and individuals undergoing a range of treatments. It is not uncommon for a patient with one type of illness to fall victim to a secondary infection as a result of exposure to a microbial aerosol.

This exercise will demonstrate the ease with which microorganisms are transported through the air in a variety of conditions. As you perform the various parts of the exercise, consider their public health implications.

A. Sneeze Simulation

Materials

❏ 1. Six nutrient agar plates

❏ 2. Ten mL of a 24-hour trypticase soy broth culture of *Micrococcus luteus*

❏ 3. One sterile atomizer

Procedure

This procedure is to be demonstrated by the instructor.

❏ 1. Arrange the 6 nutrient agar plates on a table in a straight line with approximately 12 inches between them. Each plate should be numbered in consecutive order.

❑ 2. Load the atomizer with the broth culture of *M. luteus* and position the atomizer approximately 12 inches in front of plate 1 and approximately 3 to 4 feet above it.

❑ 3. Remove all Petri plate covers and spray the suspension to simulate a sneeze.

❑ 4. Replace the covers after 30–60 minutes and incubate the preparations at room temperature until the next class period.

❑ 5. After incubation, return the plates to their positions on the table.

❑ 6. Students should examine the plates for the relative number of *M. luteus* colonies.

❑ 7. Students should approximate the bacterial load per plate by dotting in the circles provided in the Results and Observations section.

Results and Observations

Record your observations here.

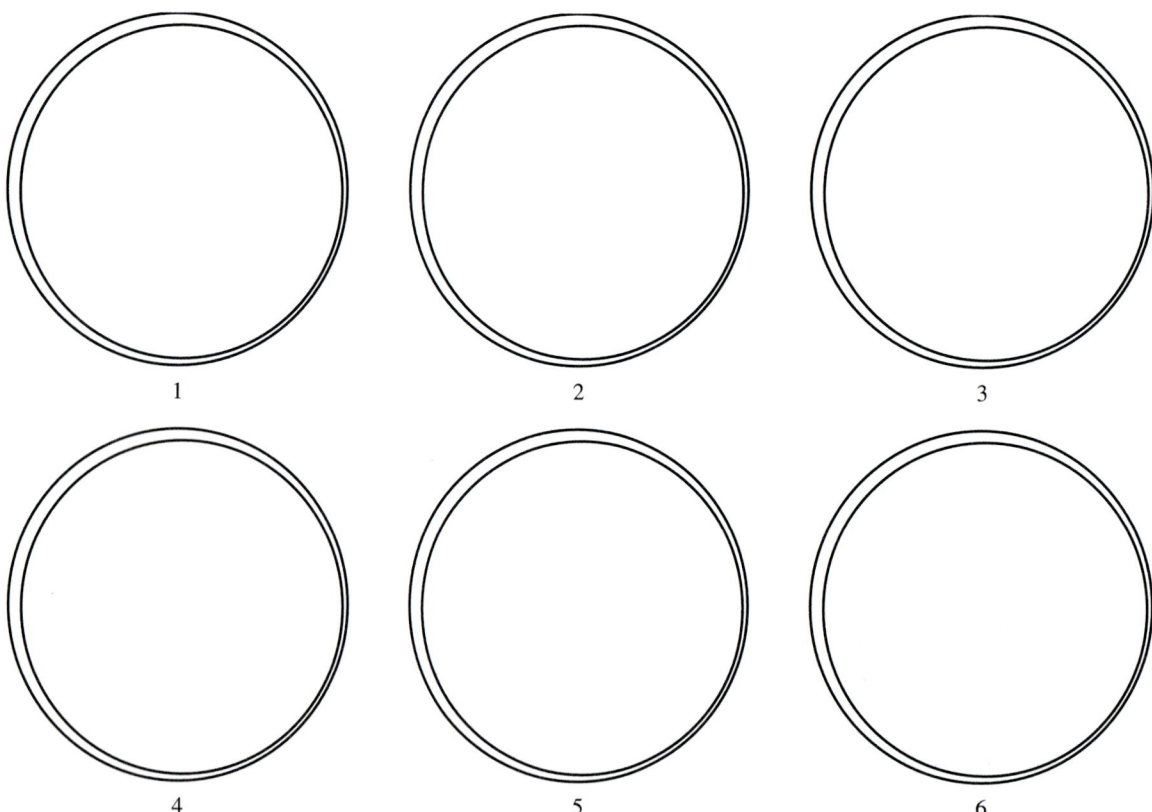

B. Creation of Laboratory Aerosols

Materials

The following items should be provided per pair of students:

❑ 1. One 24-hour trypticase soy broth culture of each of the following in 5-mL amounts:
    ❑ a. *Micrococcus luteus*
    ❑ b. *Bacillus subtilis*

❑ 2. Six sterile 1-mL pipettes and rubber bulbs or pipette pump

❑ 3. Twelve tubes of nutrient broth, dispensed in 9-mL amounts

❑ 4. Six nutrient agar plates

❑ 5. One test tube rack

❑ 6. One container with disinfectant for pipette disposal

Procedure 1: Environmental Sampling

This procedure is to be performed by students in pairs.

General Instructions

❏ 1. Before performing any of the following experiments, place 3 nutrient agar plates in the immediate working area, spaced well apart from each other.

❏ 2. In the Results and Observations section, sketch the relative positions of all plates and label each plate with a number corresponding to its position.

❏ 3. Uncover each labeled Petri plate *before beginning the set of procedures described below.*

❏ 4. Cover each Petri plate upon completion of these procedures.

❏ 5. Incubate all plates at 37°C until the next class period. If the time between classes exceeds 48 hours, the instructor should place the plates in a refrigerator at or before the 48-hour point.

❏ 6. After incubation, return the plates to their respective places and record the numbers of each colonial type observed. Enter the counts in Table 57–1.

❏ 7. The instructor will designate a pair of students to perform 2 of the following 3 procedures with specific cultures. This action will provide a means by which to determine which techniques are probably the most likely to produce an aerosol.

❏ 8. The instructor will record the results obtained with these procedures in order to assess the advantages and disadvantages of the procedures used.

❏ 9. Answer the questions in the Results and Observations section.

Procedure 2: Serial Dilution (Refer to Exercise 6 before beginning this procedure)

Your instructor will review the serial dilution procedure.

❏ 1. Place 6 tubes of nutrient broth in a test tube rack and number each consecutively from 1 to 6. Introduce into the first tube 1 mL of the bacterial culture selected.

❏ 2. Mix well with a fresh pipette by drawing approximately 1 mL of the suspension into the pipette and returning this amount into the tube.

❏ 3. Perform this procedure 10 times.

❏ 4. Remove 1 mL of the resulting suspension and deliver it to the second tube.

❏ 5. Repeat steps 3, 4, and 5 until all 6 tubes have been processed. After the mixing step has been completed for the last tube, do not remove a 1-mL sample.

❏ 6. Dispose of all pipettes in the container with disinfectant.

Procedure 3: Loop Inoculation

❏ 1. Place 6 tubes of nutrient broth in a test tube rack.

❏ 2. Inoculate the first tube with 1 loopful of the culture provided.

❏ 3. Mix the contents of the tube by rotating the tube between the palms of the hands approximately 10 times.

❏ 4. Repeat this procedure with all of the remaining tubes.

Procedure 4: Streak Plate

Additional Technique Required for This Portion of the Exercise:

❏ Streak Plate Technique, Procedure Diagram 10, Exercise 5

❏ 1. Streak each of 3 nutrient agar plates with 1 of the cultures provided.

❏ 2. Dispose of the plates as directed.

Results and Observations

1. Make the sketches for the environment sampling study here.

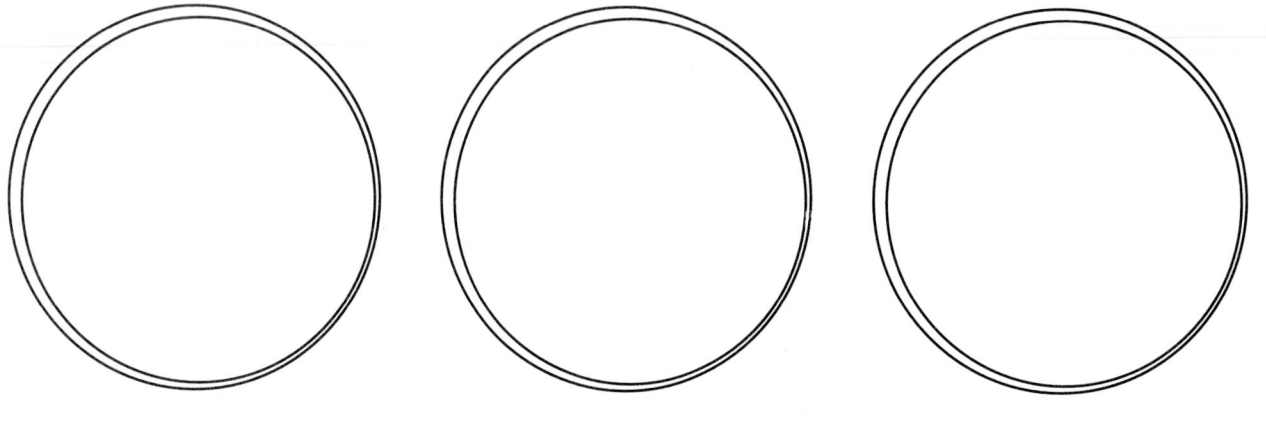

2. List the procedures performed.

   a. _____   b. _____

   _____   _____

3. Colony counts:

   Enter your findings in Table 57–1.

Table 57–1

Organism	Plate 1	Plate 2	Plate 3
Micrococcus luteus			
Bacillus subtilis			
a			
a			

[a]For other species used, if applicable.

4. Which techniques, if any, appeared to be the most potentially hazardous? Why? _____

   _____

   _____

5. How would you minimize the hazard? _____

_____

_____

Laboratory Review 57   **Aerosolization and Its Relationship to Laboratory and Hospital Sepsis**

1. How are organisms aerosolized during a sneeze? _____

_____

_____

2. List 6 infectious diseases that can be spread in this manner.

    a. _____   d. _____

    b. _____   e. _____

    c. _____   f. _____

3. How can people protect themselves from contaminants while performing laboratory procedures? _____

_____

_____

_____

Key Term

aerosol (ĀR-ō-sol): a suspension of fine particles, liquid and/or solid, in a gas

A mighty creature is the germ. Though smaller than the pachyderm. His customary dwelling place is deep within the human race.
His childish pride he often pleases by giving people strange diseases.
Do you, dear reader, feel infirm?
You probably contain a germ.

Ogden Nash

As noted earlier, the relationships between organisms and their respective environments are studied in the branch known as microbial ecology. Many types of interactions are known to occur between different microbial species, as well as between microorganisms and animal and plant life. Several of these relationships involve the living together of two different forms of life, or *symbiosis*. Types of symbiotic relationships include *commensalism, mutualism,* and *parasitism*. Types of interactions that occur between microorganisms include *antagonism* and *synergism*. Additional features of these and related associations are briefly described in the remaining portion of this introduction.

Commensalism is defined as an association between two different species living together in which one form is benefited and the other is neither benefited nor harmed. This situation can be demonstrated clearly where anaerobic organisms are cultured in the presence of aerobic organisms. The aerobic microorganisms will utilize the free oxygen in the microbial environment, causing a reduction in oxygen tension sufficient to enable the anaerobes to flourish.

One of the first and best-known reports demonstrating inhibitory, or *antagonistic,* effects of microorganisms upon one another was published in 1877 by Pasteur and Joubert. They noted that several airborne saprophytic microbial species inhibited the growth of *Bacillus anthracis*. Since then, numerous examples of antagonism between microorganisms have been described. Probably the best-known report is that made by Fleming in 1927, in which he observed that *Penicillium* inhibited the growth of *Staphylococcus aureus*. This finding ultimately resulted in the isolation, characterization, and commercial production of penicillin.

The term *synergism* may be defined as a cooperative action by two or more organisms to produce a reaction neither species could accomplish alone. One example of this type of joint action is the production of gangrene by a mixed infection, an *S. aureus* joining with a microaerophilic, gamma-hemolytic streptococcus. Another example of this phenomenon involves the combination of the metabolic capabilities of two or more species in the production of gas in a carbohydrate medium. This reaction is synergistic, because neither organism could achieve it alone.

The living together of two different organisms for mutual benefit is referred to as *mutualism*. Several examples of two organisms living together for mutual benefit are known. The most familiar is the association of soil bacteria belonging to the genus *Rhizobium* and leguminous plants. The bacteria grow in the nodules (tumorous growths) on the plant roots, where they convert free atmospheric nitrogen into ammonia. These microorganisms furnish nitrogen for the plant to use in the formation of organic compounds, and the microorganisms obtain nutrients from the plant.

Several different forms of parasitism are recognized. *Parasitism* involves the interaction of a parasite and a host, and is defined as the growth of an organism either on or in the body of a host, usually at the host's expense. Many parasites reproduce at a very high rate and consequently cause the death of their hosts. In 1934, Theobald Smith referred to this phenomenon as "bungling" parasitism. In other situations, a balance between parasites and host is achieved, largely as a result of the host's defensive mechanisms and the parasite's limited *virulence* (capability of causing injury). Additional aspects of medical parasitology are covered in Section XIV.

Many microorganisms are incapable of causing disease in a healthy host. However, when the host's defense mechanisms are impaired, severe infections and even death can result as a consequence of microbial action. Organisms that cause disease when defenses are weak are called *opportunists*. Examples of the work of opportunistic microorganisms can often be seen in AIDS victims and in persons who have been given high concentrations of antibiotics for long periods of time. The antibiotics reduce or destroy the populations of microorganisms that constitute the normal flora in the mouth and colon (for example). Since several of these organisms are competitive with or antagonistic to pathogens, their absence provides an opening for the opportunistic microorganisms to cause disease. The disease thrush, caused by the yeast *Candida albicans,* can develop through the effects caused by the excessive use of antibiotics.

This section is concerned with approaches to the demonstration of the normal microbiota in different body regions, the antimicrobial activity of tears, and the properties of various microbial pathogens of the skin, mouth, and respiratory, circulatory, gastrointestinal, and genitourinary systems. Diagnostic techniques for the identification of specific microorganisms are incorporated into several exercises. In addition, skill in identification is tested with the periodic introduction of unknown cultures or specimens. The importance of the proper handling of microorganisms during the application of microbiological procedures is emphasized throughout the section.

Indigenous Flora and the Antimicrobial Activity of Body Fluids

After completing this exercise, you should be able to:

1. Understand the nature of oral flora (microbiota) in terms of total organisms present.

2. Explain the influence of hand washing on the numbers and kinds of organisms comprising the microbiota of the skin.

3. Identify hemolytic reactions on a blood agar medium.

4. Differentiate selected microorganisms associated with the gastrointestinal tract by means of colonial morphology and reactions on MacConkey and Hektoen enteric agars and triple sugar iron agar reactions.

5. Demonstrate the antibacterial activity of tears.

As a result of the remarkable work of many pioneering bacteriologists, including Pasteur, Koch, Semmelweis, and Lister, the relationship of microorganisms to disease states has been demonstrated. Both before and after these discoveries, the transmission of disease agents was widely misunderstood. Little thought was given to the possibility that potential pathogenic microorganisms existed in or on body surfaces; therefore, little attention was given to the role of hospital personnel in harboring and transmitting disease agents to patients. Microorganisms normally exist in many regions of the body. These populations represent what is called *indigenous flora (microbiota)*. But their mere presence should not be interpreted as an indication of a disease. This exercise will present a discussion and demonstration of several aspects of indigenous flora.

Microbial contamination of the body begins during birth and involves surfaces of such anatomical parts as the skin, mouth, throat, esophagus, and intestines. The resulting ecological relationships vary throughout the lifetime of the individual. Although both qualitative and quantitative shifts do occur, the major members of the indigenous flora remain fairly consistent.

The microbiology of the mouth, teeth, gingiva, and saliva is of particular interest because so many different genera exist in these areas, which are indeed all-inclusive ecological niches. Under the limited conditions of this exercise, which involves aerobic incubation on two types of solid media, the majority of such organisms will not grow. But the different morphological types can be seen on direct examination of specimens from the crevices between the teeth. Preparations are observed either in the stained or unstained state with the aid of a dark-field microscope. The organisms that may be present include spirochetes, *Neisseria* spp., streptococci, micrococci, and small bacilli. Table 58–1 lists examples of organisms found in the mouth and related areas.

A number of antibacterial substances have been isolated from animal tissues and body fluids. For example, nasal secretions, saliva, and tears contain the hydrolytic enzyme lysozyme. This enzyme dissolves several gram-positive organisms, including *Bacillus* spp., some staphylococci, and intestinal streptococci. An experiment to test the antibacterial activity of tears is included in this exercise.

The microbiological flora of the skin differ in the various regions of the body surface. The types present are usually determined by the amount of moisture, habits of personal hygiene, and the presence or absence of local infections. The resident flora usually include an aerobic or facultative anaerobic *Staphylococcus* sp. and an anaerobic *Propionibacterium* sp. Transient microorganisms of numerous types are present from time to time. Table 58–2 lists examples of bacteria associated with the skin.

The ecological environment of the intestinal tract includes a wide variety of microbial genera and species. As in saliva, obligate anaerobic organisms are the most numerous types. Generally, one finds enteric organisms that include *Escherichia coli, Proteus* spp., *Bacteroides* spp., *Clostridium* spp., fecal streptococci, lactobacilli, and protozoa. In addition, many viruses, both animal and bacterial, are also found. Here again, one encounters typical resident microbial species as well as transient types. At times, as indicated previously, known pathogens may be included among the residual forms.

Table 58–1

Examples of Throat-Associated Bacteria

Microorganism	Gram Reaction and Morphological Arrangement	Appearance on Blood Agar
Moraxella (Branhamella) catarrhalis	+, diplococci	Small, white or translucent round colonies; no hemolysis
Corynebacterim species	+, club-shaped rods sometimes in picket-fence arrangement	Varies with species; tan to translucent; round colonies; no hemolysis
Diphtheroids	+, solid-staining, club-shaped rods	Variable; some dry, wrinkled tan colonies; no hemolysis
Neisseria species	−, diplococci	Varies with species; tan to yellow in color; some translucent; no hemolysis
Staphylococcus aureus	+, cocci in grapelike clusters	Medium-sized colonies; tan to yellow in color; beta (clear zone) hemolysis[a]
Streptococcus pneumoniae	+, diplococci	Small, translucent, slimy colonies; alpha (green zone) hemolysis[a]
Streptococcus pyogenes and other beta streptococci	+, streptococci	Pin-point to small convex colonies; beta hemolysis[a]
Streptococcus viridans	+, streptococci	Small, translucent colonies; alpha (green zone) hemolysis[a]
Yeast	+, large oval cells, some with buds	Large, round, moist, tan to white colonies; generally no hemolysis[a]

[a]See Figure 58–1.

Table 58–2

Examples of Skin-Associated Bacteria

Microorganism	Gram Reaction and Morphological Arrangement	Appearance on Blood Agar
Bacillus species	+, diplobacilli and/or streptobacilli	Irregular, dull, opaque colonies; beta hemolysis possible[a]
Corynebacterium species	+, club-shaped rods, some arranged in picket-fence formation	Varies with species; tan to translucent round colonies; no hemolysis[a]
Staphylococcus aureus	+, cocci in clusters	Medium, round, glistening, tan to yellow colonies; beta hemolysis[a]
Staphylococcus epidermidis	+, cocci in clusters	Medium, round, white, glistening colonies; generally no hemolysis[a]
Yeasts	+, large oval to round cells, some with buds	Large, round, moist, tan to white colonies; generally no hemolysis[a]

[a]See Figure 58–1.

The resident populations of the various body areas appear to serve several useful functions. The intestinal flora produce various vitamins and serve as a stimulus for maintenance of antibody-producing systems. These organisms also tend to crowd out transient forms, which may be pathogenic. In any ecological situation, a balance between saprophyte and parasite can be achieved without evidence of disease. Factors that upset this relationship may lead to gastrointestinal disturbances or more serious disorders.

In this exercise, representative indigenous bacterial species will be considered for the purpose of differentiation. They include actual specimens from oral and skin flora and typical species found in the large intestine, prepared separately for each student. In addition, several types of media will be used to show specific enzymatic reactions. These include the **differential plating medium,** *blood agar* (Figure 58–1), the **selective and differential media,** *Hektoen enteric medium* (Figure 58–2), and *eosin methylene blue* (Figure 58–3), and the **differential tube medium,** triple sugar iron agar (Figure 58–4). References to these media and associated reactions will be made throughout this and later exercises.

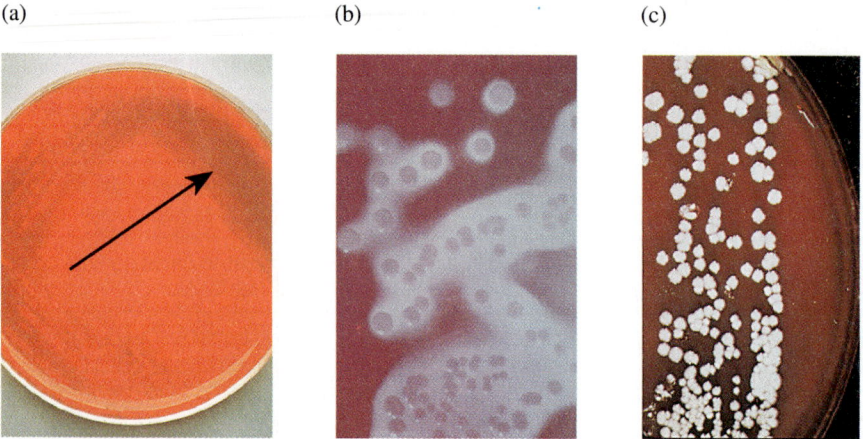

(a)   (b)   (c)

Figure 58–1

Representative hemolytic reactions on blood agar media. (a) Alpha hemolysis. Note the greenish discoloration of the medium surrounding the bacterial colonies.
(Leibovitz, E., A. Kaul, *et al., Cutis* 49:27, [1992].)

(b) Beta hemolysis. Here, clear zones surround the individual bacterial colonies. (c) Gamma hemolysis or nonhemolytic reactions. Discolorations, or clear zones in the medium, are not present.

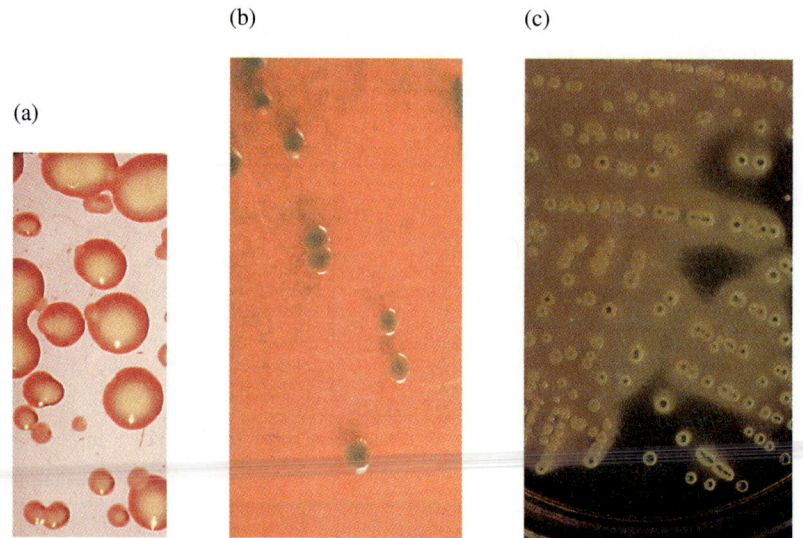

(b)   (c)   (a)

Figure 58–2

Hektoen enteric agar reactions. Three major types of reactions are shown. (a) Rapid lactose fermenters (yellow to salmon-pink colonies). (b) Nonlactose fermenters (green colonies) and (c) Hydrogen sulfide, (H_2S) producers (black centered colonies). *Escherichia coli* is a rapid lactose fermenter, while *Proteus vulgaris* and related species are well-known H_2S producers.

(a)

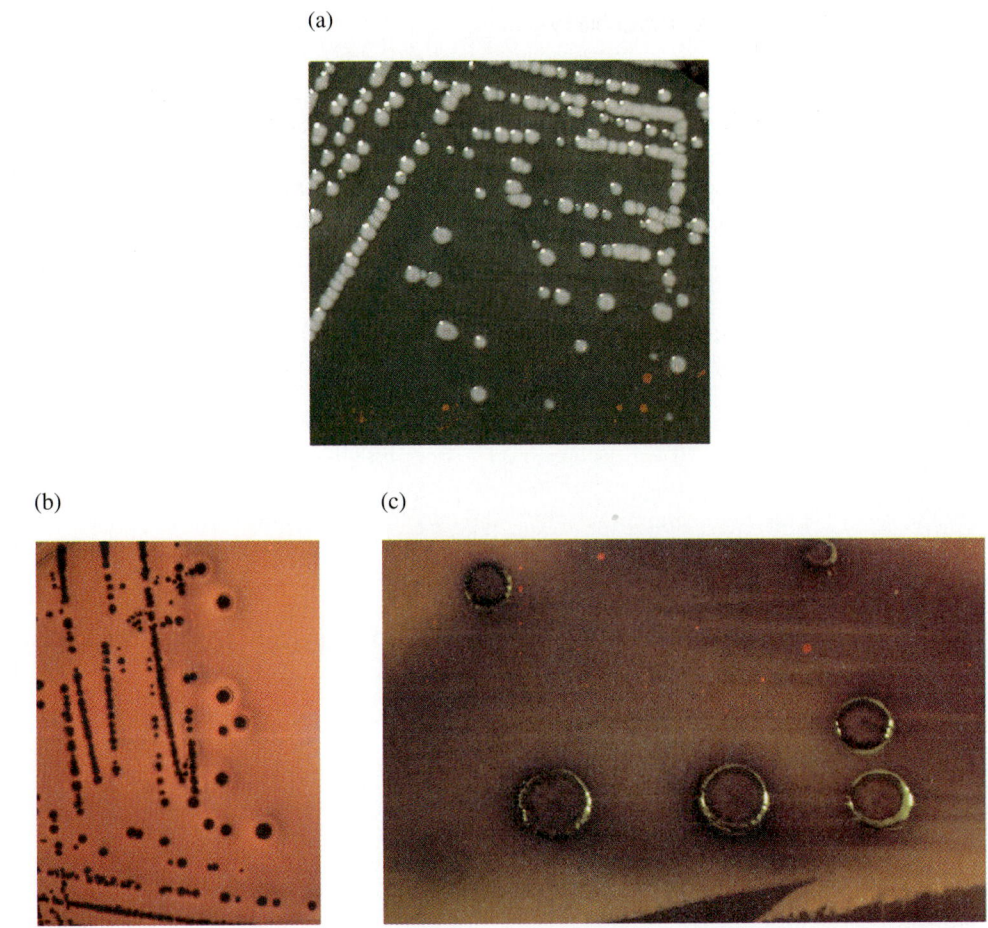

(b)   (c)

Figure 58–3

Eosin-methylene blue agar reactions. (a) Colonies of a lactose nonfermenter. (b) Colonies of a lactose fermenter. (c) A metallic sheen can form on the colony surfaces of some lactose fermenters.

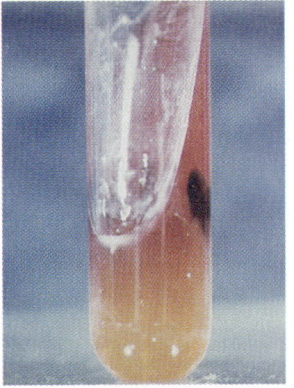

Figure 58–4

Triple sugar iron agar (TSIA) reactions. This medium uses phenol red as the pH indicator. The yellow bottom of the slant indicates an acid reaction from the breakdown of dextrose (glucose). The red slant shows an alkaline reaction and no breakdown of lactose and sucrose, the two other sugars in this medium. The black spots indicated hydrogen sulfide production resulting from the breakdown of amino acids with sulfar. Refer to Exercise 27 for additional details of this medium.

A. Microbial Flora of the Mouth

Materials

The following materials will be supplied per student:

❏ 1. One toothpick

❏ 2. One sterile test tube

❏ 3. Two blood agar plates

The following items will be supplied for general class use:

❏ 1. Glass slides

❏ 2. Gram stain reagent sets

❏ 3. Oxidase reagent

❏ 4. Two nutrient agar deeps

❏ 5. Two sterile Petri plates

❏ 4. Separate containers with disinfectant for used pipettes and toothpicks

Additional Techniques Required for This Portion of the Exercise:

❏ 1. Bacterial Smear Preparation, Procedure Diagram 2, Exercise 2

❏ 2. The Gram Stain, Procedure Diagram 17, Exercise 14

Procedure 1: Staining

This and the following procedure are to be performed by students in pairs.

❏ 1. Remove a small amount of material from the area between your teeth with the aid of a toothpick.

❏ 2. Prepare 2 smears by placing some of the material on each of 2 clean slides. Add 1 drop of distilled water to each slide and spread the material into a smear approximately the size of a dime. Dispose of the used toothpick in the container provided.

❏ 3. Allow the smears to air-dry, and then heat-fix them.

❏ 4. Prepare a simple stain preparation (applying either crystal violet or safranin) with the first smear, and Gram stain the second. (Refer to Procedure Diagrams 2 and 17.)

❏ 5. Examine both preparations under oil. Sketch representative portions of microscope fields in the appropriate areas of the Results and Observations section. (See Figures 2–3 and 14–1.)

❏ 6. Based on your findings, answer the questions in this section.

Results and Observations

1. Sketch your results here.

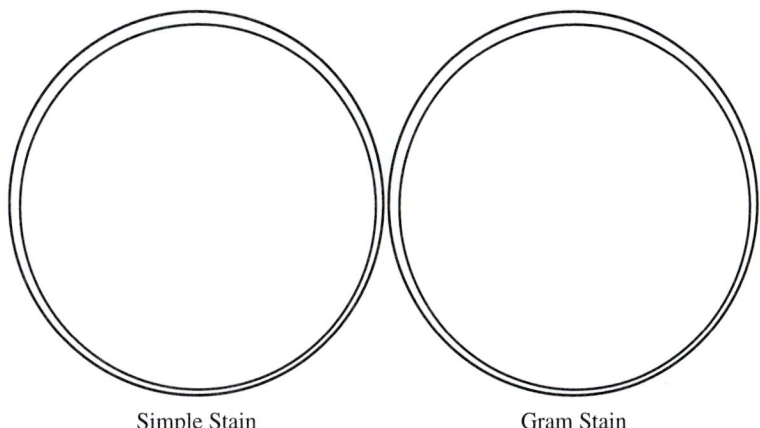

Simple Stain   Gram Stain

2. Is it possible to determine the genus of any spirochetes observed? If so, how? _____

_____

3. What was the predominant morphological type in your preparations?_____

_____

4. What was the predominant Gram stain reaction? _____

5. Was there a significant difference in the two differently stained preparations? Explain. _____

_____

Procedure 2: Isolations

Additional Techniques Required for This Portion of the Exercise:

❏ 1. Streak Plate Technique, Procedure Diagram 10, Exercise 5

❏ 2. The Gram Stain, Procedure Diagram 17, Exercise 14

❏ 1. Melt and pour 1 nutrient agar deep into the sterile Petri dish provided, and allow the agar to harden.

❏ 2. Salivate into the sterile test tube provided.

❏ 3. Remove a loopful of the saliva specimen, and using the clock-plate method of streaking, spread the specimen over the surface of the nutrient agar plate.

❏ 4. Repeat step number 3 with the blood agar plate.

❏ 5. Invert the plates and incubate them at 37°C for 48 hours.

❏ 6. Examine the plates and note the relative number of colonies on each medium.

❏ 7. Select 4 different colonies on the nutrient agar plate that resemble the descriptions of *Moraxella* and *Neisseria* species (refer to Table 58–1). Drip 1 drop of the oxidase reagent on each colony and observe for the formation of a dark red to black color. (See Figure 58–5.) Based on your findings, answer the questions in the Results and Observations section.

(a)   (b)

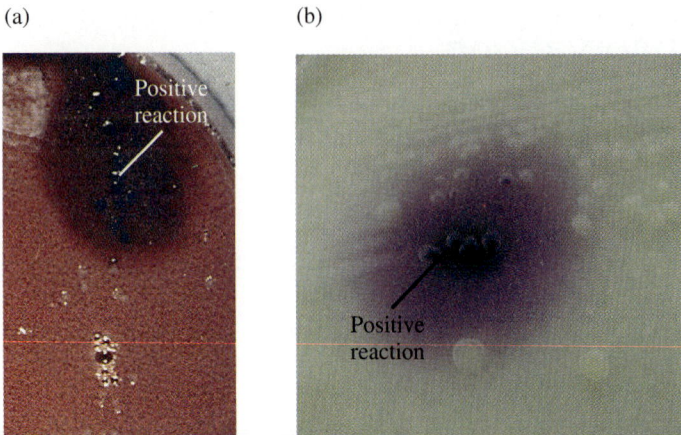

Figure 58–5

The oxidase test. (a) Colonies demonstrating a positive reaction are pink and turn dark red later. Negative reactions are indicated by colorless colonies. (b) Positive oxidase colonies treated with the Difco reagent turn blue.

❏ 8. With the blood agar plate, select 2 distinct colonial types exhibiting different hemolytic reactions. Prepare smears from each colony and Gram stain them accordingly (refer to Procedure Diagrams 5 and 15). Use Figure 58–1 as a guide to hemolytic reactions.

❏ 9. Sketch a representative microscope field in the spaces provided in the Results and Observations section. Color the organisms to correspond with their Gram stain reactions.

❏ 10. Based on your findings, answer the questions in this section.

Results and Observations

1. Sketch your staining results here.

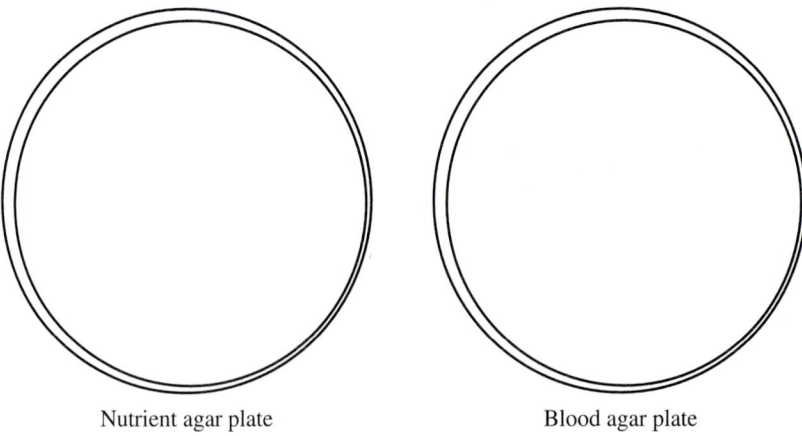

Nutrient agar plate   Blood agar plate

2. Did more colonies appear on 1 type of medium than the other? Explain. _____

_____

_____

3. Were any colonies oxidase-positive? If so, were there large numbers of them compared to non-oxidase-positive

ones? _____

_____

4. Is it possible to culture all the microorganisms in saliva using routine laboratory procedures? Explain.

_____

_____

_____

B. The Antibacterial Activity of Tears

Materials

The following materials will be provided per 4 students:

❏ 1. Two trypticase soy agar deeps

❏ 2. One 24-hour trypticase soy broth culture of each of the following organisms dispensed in 2-mL amounts:
    ❏ a. *Escherichia coli*
    ❏ b. *Staphylococcus aureus*

❏ 3. Two sterile Petri plates

❏ 4. A container of sterile filter paper disks

❏ 5. Two sterile forceps

❏ 6. A mortar and pestle

❏ 7. One-half raw onion or horseradish root

❏ 8. One 50-mL beaker containing disinfectant

Additional Technique Required for This Portion of the Exercise:

❏ Pour Plate Technique, Procedure Diagram 9, Exercise 5

Procedure

This procedure is to be performed by students in groups of 4.

❏ 1. Prepare a pour plate with each of the cultures provided. (Refer to Procedure Diagram 9.)

❏ 2. One student in the group is to crush the cut onion or horseradish in the mortar with the pestle provided.

❏ 3. Another student immediately deeply sniffs the crushed onion or horseradish preparation. If tears do not form, continue sniffing the preparation until obvious tearing occurs.

❏ 4. Using sterile forceps, remove a sterile filter paper disk and carefully touch 1 end of the disk to the formed tear drops.

❏ 5. Place the treated disk on the center of 1 of the pour plate preparations.

❏ 6. Put the forceps into the beaker containing disinfectant.

❏ 7. Repeat steps 3 through 6 with the second pour plate.

❏ 8. Incubate the pour plates at 37°C for 24 hours, or as directed by your instructor.

❏ 9. After incubation, examine the plates for zones of incubation.

❏ 10. Record your findings and answer the questions in the Results and Observations section.

Results and Observations

1. Which of the organisms used in this experiment was (were) sensitive to tears? _____

   _____

2. What is lysozyme? _____

   _____

3. Where is lysozyme found in the body? _____

   _____

C. Microbial Flora of the Skin

Materials

❏ 1. Sterile cotton swabs (2 per student)

❏ 2. Blood agar plates (2 per student)

❏ 3. Gram stain reagent sets (1 per 4 students)

❏ 4. A container with disinfectant for used swabs (enough for class use)

❏ 5. Marking pens or pencils (enough for class use)

Additional Technique Required for This Portion of the Exercise:

❏ The Gram Stain, Procedure Diagram 17, Exercise 14

Procedure

This procedure is to be performed by students in pairs.

❏ 1. Mark one blood agar plate with your name, date, and the word *Before*. Mark the second plate in a similar way, but place the word *After* on it.

❏ 2. Slightly moisten a cotton swab with sterile saline.

❏ 3. Carefully rub a small area of the underside of 1 wrist with the prepared applicator swab.

❏ 4. Inoculate the *Before* blood agar plate with the specimen using the clock-plate method of streaking. Discard the used swab in the container provided.

❏ 5. Wash a general area of the hand, wrist, and forearm with soap and water. Rinse well and dry these regions. Repeat steps 1, 2, and 3, using the same area from which the first specimen was taken.

❏ 6. Inoculate the *After* blood agar plate with the second specimen using the clock-plate method of streaking. Discard the used swab in the container provided.

❏ 7. Incubate both plates at 37°C for 48 hours.

❏ 8. After incubation, compare the colonial types on the 2 plates as to hemolytic reactions, colonial properties, and size. (Use Figure 58–1 as a reference to hemolytic reactions.)

❏ 9. Note the hemolytic reactions and count the number of colonies for each type. Record your findings in the Results and Observations section.

❏ 10. Select 2 different colonial types from each plate and prepare Gram stains. (Refer to Procedure Diagram 17.)

❏ 11. Sketch your findings and answer the questions in the spaces provided in the Results and Observations section.

Results and Observations

1. Number and kinds of hemolytic reactions on blood agar. Enter your findings in the following table.

Table 58–3

Hemolytic Reactions

	Number of Colonies	
Type of Hemolysis	Before	After
alpha		
beta		
gamma		

2. Gram stain reactions of colonial types.

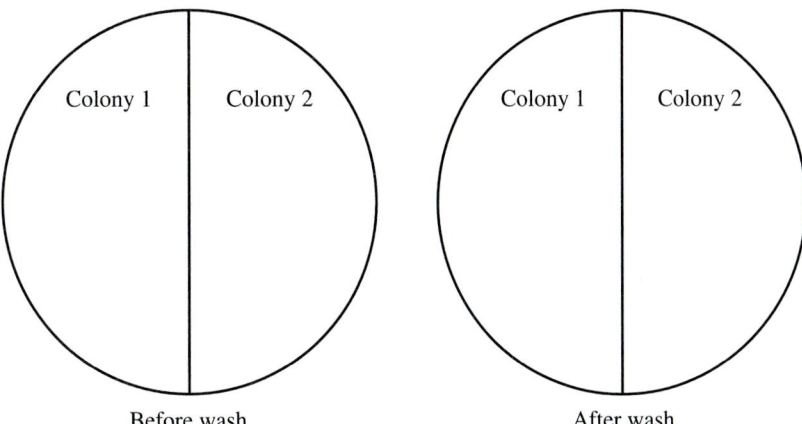

Before wash   After wash

3. Which of the 2 plates represents the resident microbiota? Which represents the transient microbiota? Explain.

_____

_____

4. Name 2 common sources of transient microorganisms.

a. _____   b. _____

D. Microorganisms of the Gastrointestinal Tract

Materials

The following items are to be provided per student:

❑ 1. Mixed broth cultures of 2 organisms indicative of the enteric flora (1 tube)

❑ 2. One eosin-methylene blue (EMB) agar plate

❑ 3. One Hektoen enteric (HE) agar plate

❑ 4. Two tubes of triple sugar iron agar (TSIA) slants. This medium will be used 1 week after primary inoculations

❑ 5. One wax marking pencil or felt-tip marker

Procedure

Additional Techniques Required for This Portion of the Exercise

❏ 1. Streak Plate Technique, Procedure Diagram 10, Exercise 5

❏ 2. The Gram Stain, Procedure Diagram 17, Exercise 14

❏ 3. Triple Sugar Iron Agar (TSIA) Inoculation Technique, Procedure Diagram 24, Exercise 26

This procedure is to be performed by students in pairs.

❏ 1. Streak the eosin-methylene blue and Hektoen enteric agar plates with the mixed broth culture provided, using the clock-plate method of streaking. (Refer to Procedure Diagram 10.)

❏ 2. Incubate both plates at 37°C for 48 hours.

❏ 3. Before discarding the mixed culture, make a smear, Gram stain it, examine it under oil immersion, and record your findings in the appropriate space in the Results and Observations section.

❏ 4. After incubation, examine the plates and determine which of the 2 shows more isolated colonies. Select and circle with a wax marking pencil 2 colonies exhibiting different fermentation reactions. Refer to them as cultures 1 and 2, respectively. (Use Figures 58–2 and 58–3 as guides.) Check the purity of the colonies by preparing Gram stains of each one. Enter your findings in the Results and Observations section.

❏ 5. Use and inoculate 1 tube of triple sugar iron agar medium for each organism.

❏ 6. Incubate the tubes at 37°C for 48 hours.

❏ 7. After incubation, examine the inoculated tubes and record your findings in the Results and Observations section. (Use Figure 58–4 as a reference.)

Results and Observations

1. Gram stain reactions and morphology.

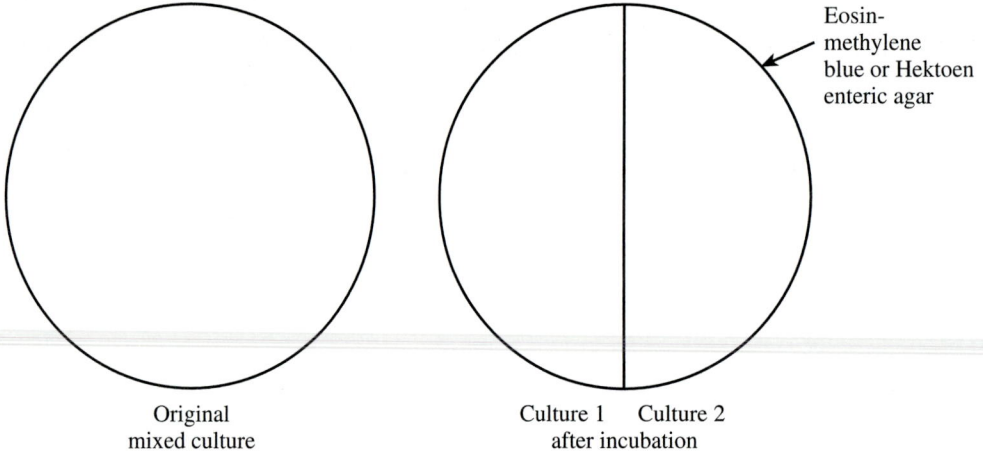

Original
mixed culture

Culture 1   Culture 2
after incubation

Eosin-methylene blue or Hektoen enteric agar

2. Triple sugar iron agar reactions.

    a. Colony Features

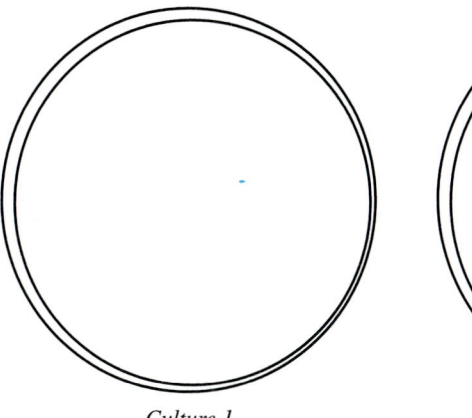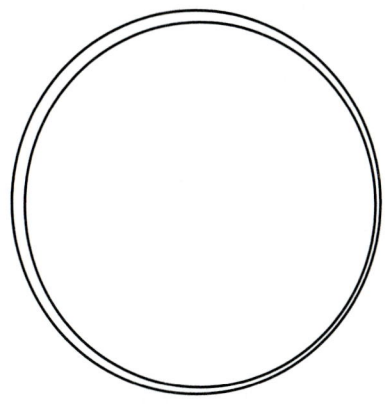

Culture 1
from Eosin-methylene blue
or Hektoen enteric agar

Culture 2
from Eosin-methylene blue
or Hektoen enteric agar

b. TSIA Results

butt _____   butt _____

slant _____   slant _____

H_2S production _____ H_2S production _____

3. By comparing your results with those shown in Figures 58–2 and 58–3, can you identify the organisms isolated? If you can, give the genus of:

    a. culture 1 _____

    b. culture 2 _____

Laboratory Review 58 — Indigenous Flora and the Antimicrobial Activity of Body Fluids

1. List 6 infectious diseases of the skin, together with their causative agents in Table 58–4.

Table 58-4

Infectious Diseases of the Skin

Disease	Genus/Species

2. How would you attempt to sterilize the skin?

   _____

   _____

3. What is the basis for differentiation of enteric microorganisms by the use of eosin-methylene blue agar?

   _____

   _____

4. What is the basis for selection of enteric pathogens by the use of Hektoen enteric agar? _____

   _____

   _____

5. Describe the various reactions and corresponding colors that can be obtained with triple sugar iron agar (TSIA).

   _____

   _____

6. Give the triple sugar iron agar reactions for the organisms listed in the following table. Use these symbols for the following results: A, acid production; Alk, alkaline; and "+" for hydrogen sulfide production.

Table 58-5

Triple Sugar Iron Agar Reactions

Microorganism	Reaction(s)		
	Butt	Slant	H_2S Production
Escherichia coli			
Proteus vulgaris			

Key Terms

indigenous (in-DIJ-en-us) flora: microorganisms native to a region

lysozyme (LĪ-sō-zīm): the enzyme that degrades a peptidoglycan, a bacterial cell wall component

transient: temporary

Photograph Quiz 24

Identify the Hektoen enteric agar plate reactions shown in the following figure.

a. _____ b. _____

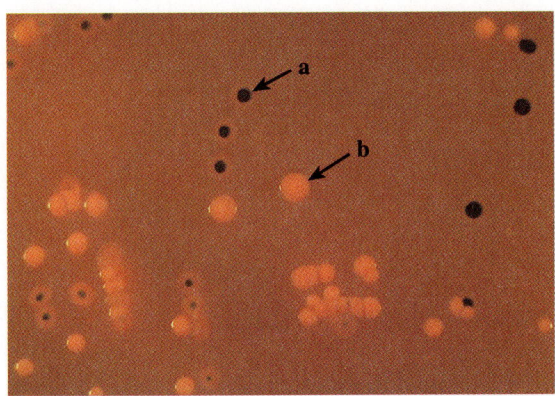

Figure 58–6
Hektoen Enteric agar reactions.

Pathogenic Microorganisms of the Skin

After completing this exercise, you should be able to:

1. List several common microbial infections of the skin.

2. Recognize the significance of the intact skin as a barrier to microbial disease agents.

3. Distinguish between various microbial pathogens of the skin.

4. Interpret specific diagnostic tests for *Staphylococcus aureus,* including Staph-Ident, coagulase, Staph-latex test, and DNAse.

5. Interpret specific tests used for the identification of *Streptococcus* species, including bacitracin sensitivity, CAMP reaction, and L-pyrrolidonyl-β-naphthylamide hydrolysis.

In general, intact skin serves as a natural protective barrier against invasion by most infectious disease agents. But hair follicles and the openings of secreting glands are potential invasion avenues for pathogens. Physiologic factors are also important in providing barriers to skin-invading microorganisms. Such factors include (1) the acidity of the skin, (2) fatty acids on the skin, (3) the presence of indigenous flora, and (4) the temperature of the skin, which is suboptimal for the growth of some disease agents. These factors serve to inhibit or kill invading or transient microorganisms. A simple experiment is included in this exercise to demonstrate the antibacterial activity of the skin against selected typical bacterial contaminants.

Normal human skin is colonized by large numbers of organisms that live harmlessly as commensals on its surface. Of the many different species of microorganisms found in nature, only a few species are found repeatedly on the skin of groups of individuals. This is surprising considering the large variety of organisms found in other areas such as the gastrointestinal tract. These frequently found organisms form the resident *microflora* or *microbiota* of the skin.

For microorganisms to colonize the skin, they must first become attached to the epithelial surface. The capacity is proportional to the ability of the organism to adhere (stick) to the surface. Such adherence involves the attraction of a specific molecular structure on the cell wall of the bacterium, called an *adhesin*, and a specific receptor on the host cell surface. Adhesins are microbial surface antigens, or *lectins,* often in the form of threadlike projections. The adhesin functions as a bridge between the microbe and the host cell. The binding between adhesin and receptor is virtually irreversible. Epithelial cells from different anatomic systems, such as the skin and the gastrointestinal tract, have striking variations in relation to the adherence of bacteria, which helps to explain the differences in resident microflora in different locations on the body. For example, group A streptococci isolated from skin adhere much better to skin surfaces than to the lining of the cheek, whereas streptococci isolated from the mouth bind better to the cheek lining. Pathogenic bacteria have higher potential for adherence to the host, thus making them more virulent. The enhanced potential may be due to several adhesins, which would increase the chance for adherence and give the organism a selective advantage over other organisms.

Normal skin is resistant to colonization and invasion by most bacteria. The presence of bacteria on the skin surface does not make infection inevitable. Many factors prevent colonization and invasion by pathogenic organisms. These include an intact surface *(stratum corneum),* rapid cell turnover of the skin outer layer, and interference and/or antagonistic actions by other bacterial strains of similar species. Participating elements of cellular immunity include inflammation, T lymphocyte activation, and antigen presentation by Langerhans cells. The mechanisms of the elements are vital in preventing skin infection. Individuals with defects in cellular immunity (immunocompromised) are more susceptible to infection, whether they have diseases that cause limited involvement, such as certain yeast infections, or diseases that cause more widespread involvement such as human immunodeficiency virus (HIV) infections (Figure 59–1).

During the course of normal living, breaks of various types occur in the skin that can provide the means for infectious agents to gain entrance into the body. Such injuries usually affect host-parasite relationships in varying degrees. Skin and soft tissue infections (wounds) occur during the life of most individuals. These infections can be rather mild, re-

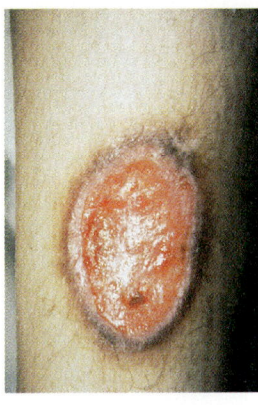

Figure 59–1
An ulcer caused by the varicella-zoster virus (the causative agent of chickenpox and shingles). The patient was infected with HIV.
(From Aditya Kaul, M.D., *et al., Cutis* 49:27, [1992].)

(a)

(b)

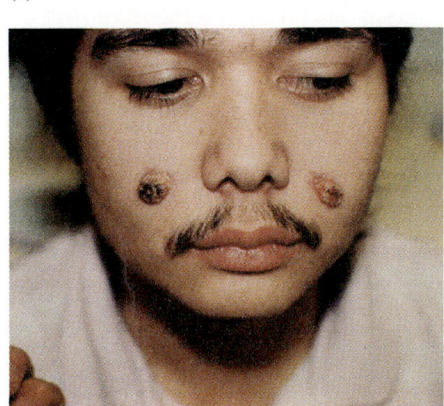

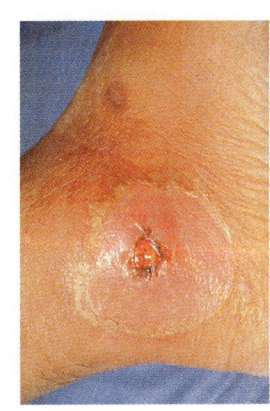

Figure 59–2
Mycotic and bacterial infections of the skin. (a) The appearance of a double fungus skin infection, sporotrichosis and blastomycosis.
[Chuang, Tsu-Yi. *Cutis* 48:1993, [1991].)

(b) An ulcer on the ankle of an individual with cutaneous diphtheria.
(From Höefler, W. *Internat. J. Dermatol.,* 30:845, [1991].)

quiring only local care, or they can be extremely serious with extensive tissue damage (Figure 59–2) and life-threatening complications. Hemorrhage, ischemia, ulceration, or edema are common signs of such infections.

This exercise will be limited to certain pyogenic infections associated with the skin and the superficial fungus diseases known as the *dermatomycoses*.

The bacterial flora of wounds can usually be divided into two major groups—namely, *pyogenic aerobic microorganisms* and the *spore formers* that produce histotoxic and neurotoxic effects and use the skin as their portal of entry. The former include *Staphylococcus aureus,* beta (β)-hemolytic streptococci (group A), and *Proteus* and *Pseudomonas* spp. The latter include *Clostridium perfringens, Clostridium tetani,* and other clostridia (Figure 59–3). Attention in this exercise also will be given to the presumptive grouping of streptococci, with particular attention to group B hemolytic streptococci and other streptococci. The combination of several tests can be used for the identification of these organisms. A specific test used for such purposes is the CAMP (Christie, Atkins, and Munch-Peterson) reaction.

Figure 59–3
A Gram stain of a blood culture from a patient with a case of *Clostridium tertium* bacteremia.
(From LeBar, W.D., *Eur. J. Clin. Microbiol. Inf. Dis.* 8:314, [1989].)

Group B streptococci produce a peptide, known as the **CAMP factor,** that enlarges a portion of the hemolytic zone formed by certain beta-lysin-producing *Staphylococcus aureus* strains. The enlarged zone takes on the appearance of a typical arrowhead- or flame-shaped clear zone (Figure 59–4*a*). Another test that serves to aid in the identification of *Streptococcus pyogenes* and several *Enterococcus* species is the hydrolysis of L-pyrrolidonyl-β-naphthylamide (PYR). A blackening of the medium surrounding bacterial colonies upon the addition of the PYR reagent is a positive reaction (Figure 59–4*b*). Another approach to detecting PYR activity is the *Dry Slide*. This procedure is performed by smearing a loopful of bacteria on the surface of a reaction area on a **Dry Slide PYR.** This step is followed by a two-minute incubation and the addition of the developer reagent. A resulting pink color is a positive reaction (Figure 59–4*d*). Table 59–1 shows the characterization patterns of selected streptococcal groups with these tests.

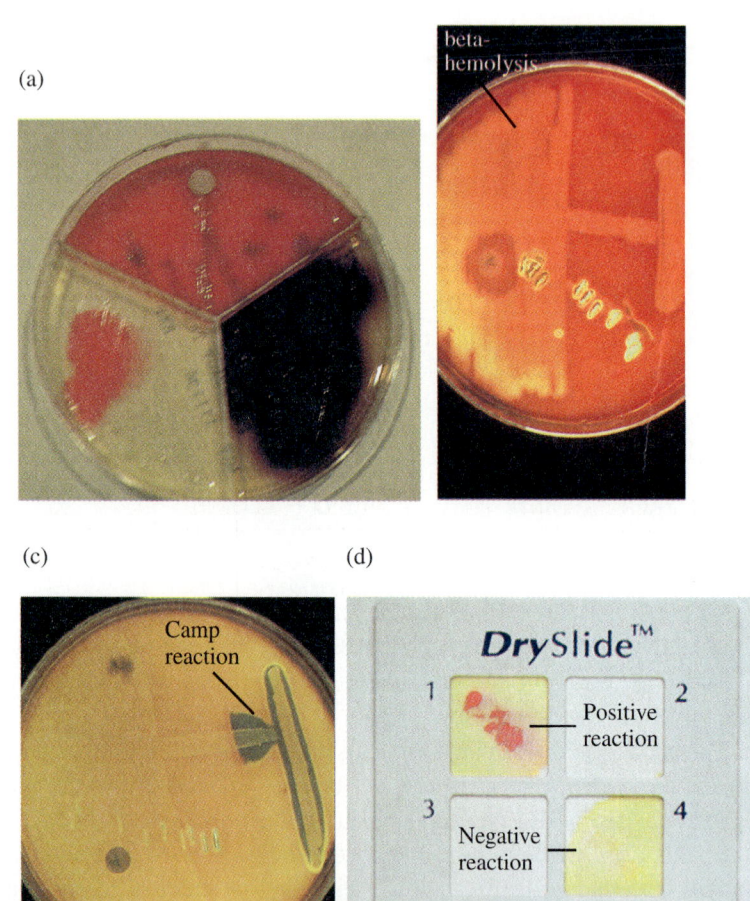

Figure 59–4

Tests used in the identification of streptococci. (a) An agar plate medium showing reactions for the presumptive identification of streptococci. Positive reactions are indicated by the following results: CAMP, formation of a wedge or clear area near CAMP disk; bile-esculin, blackening of the medium; PYR (pyrrolidonyl-β-naphthylamide, a formation of a red color upon addition of PYR reagent. Hemolysis also is determined by the medium. (b) The characteristic growth inhibitory reaction of bacitracin (the disks contain 2 units of bacitracin). Beta hemolysis is quite evident from the agar stabs on the plate. The CAMP test is negative. (c) A negative bacitracin reaction is shown. However, the plate shows a positive CAMP reaction seen as a triangle perpendicular to the streak growth of a beta-lysin producing *Staphylococcus aureus* strain. (Courtesy of Richard R. Facklam, Bureau of Laboratories. Centers for Disease Control, Atlanta.)

(d) The PYR dry slide. (Courtesy of Difco Labs., Detroit, MI.)

Representative infections of the superficial skin layers and hair follicles caused by the pyogenic aerobic agents are impetigo, pustules, septic blisters, furuncles, carbuncles, and folliculitis. More complex infections caused by this group include paronychia (infections of the nail fold), cellulitis (a deep inflammation of the skin and subcutaneous tissues), and occasionally a synergistic form of gangrene. One bacterium, *S. aureus,* is known to be the cause of several of these disease states. In addition to the Gram reaction, the identification of this species requires (1) a differentiation from other staphylococci by biochemical tests, which include mannitol fermentations; (2) demonstration of the enzyme coagulase, which coagulates blood plasma; and (3) demonstration of the heat-stable deoxyribonuclease, an enzyme that breaks down deoxyribonucleic acid (DNA). Mannitol fermentation and the presence of the two enzymes mentioned are frequently found with pathogenic strains of *S. aureus.* A rapid slide agglutination test is available to simultaneously detect coagulase and/or the specific protein A found with *S. aureus.* Staphylococcal colonies containing these factors, when mixed with the yellow reagent of the Bacto Staph Latex Test, form visible clumps within 45 seconds. This test will be demonstrated. (Refer to Exercise 48.)

Table 59–1

Selected Reaction Patterns of Streptococcal Groups

Group	Tests			
	Hemolysis	CAMP	PYR	Bile-Esculin
A	beta	—	+	—
B	beta	+	—	—
C, G, or F	beta	—	—	—
D-enterococcus	alpha, beta, gamma	—	+	+
Viridans	alpha, gamma	—	—	—

The genus *Staphylococcus* contains several species of harmless residents that can inhabit various body regions. Such species can be differentiated by biochemical testing. This exercise will demonstrate methods used to detect various biochemical enzymatic activities of staphylococci. The Staph-Ident system, a rapid, standardized micromethod that uses both miniaturized conventional and color-producing tests for the identification of 13 species of *Staphylococcus*, will be included. Table 59–2 lists the specific tests, together with brief descriptions of expected positive results and interpretations. The method of inoculation is similar to that followed with the miniaturized microbiological system described in Exercise 27.

Table 59–2

Staph-Ident System Tests and Results

Test	Test Symbol	Appearance with Positive Results	Condition(s) Indicated by Positive Results
Phosphatase	PHS	Yellow	Hydrolysis of para-nitrophenylphosphate
Urea utilization	URE	Purple to red	CO_2 and NH_3 formation from urea
β-Glucosidase	GLS	Yellow	Hydrolysis of p-nitrophenyl-β-D-glucopyranoside
Mannose utilization	MNE	Yellow	Utilization of carbohydrate with acid formation
Mannitol utilization	MAN	Yellow	Utilization of carbohydrate with acid formation
Trehalose utilization	TRE	Yellow	Utilization of carbohydrate with acid formation
Salilcin utilization	SAL	Yellow	Utilization of carbohydrate with acid formation
β-Glucuronidase	GLC	Yellow	Hydrolysis of p-nitrophenyl-β-D-glucuronide
Arginine utilization	ARG	Purple to red	Transformation of arginine into alkaline end-products
β-Galactosidase	NGP	Dark purple	Hydrolysis of α-naphthol-β-D-galactopyranoside

Members of the spore-forming group, such as *C. perfringens* type A, *C. novyi*, and *C. histolyticum*, can produce wound infections ranging from localized manifestations (without necessarily involving underlying tissues) to progressive invasion and necrosis of healthy tissue associated with variable gas production and toxemia. A combination of all of these factors is found in a disease known as *gas gangrene (clostridial myonecrosis)*. One *Clostridium species, C. tetani*, does not usually produce extensive local injury. The effects of infection by this bacterium are caused by its neurotoxin.

Diagnosis of infections such as those listed previously involves the use of Gram stains of smears from diseased tissue, as well as identification of microorganisms obtained from culturing clinical specimens. In the case of gas gangrene, stained smears of muscle are usually the only specimens of value in diagnosis.

The dermatomycoses, also commonly referred to as *ringworm,* are caused by fungi that do not invade the deeper tissues. Infections are generally limited to the hair, nails, and skin. The causative agents are members of the class Deuteromycotina (Fungi Imperfecti) and belong to one of the following genera: *Epidermophyton, Microsporum,* or *Trichophyton.* Certain species of these genera may cause the same clinical entity, or one species may cause several different lesions. The following is a list of examples of the more commonly encountered dermatomycoses:

1. Tinea pedis, also called *athlete's foot*
2. Tinea capitis, or ringworm of the scalp
3   Tinea unguium, or ringworm of the nails
4. Tinea cruris, or ringworm of the groin; also called *dhobie itch*

As can be seen from this list, the term *ringworm* is frequently employed. It is used mainly because the infection appears in a circular form. Diagnosis of the dermatomycoses depends on the demonstration of fungi in hair, nail, or skin scrapings and the laboratory culturing of the disease agent.

A. Antibacterial Activity of the Skin

Materials

The following items are to be provided per 2 students:

❏ 1. One 12- to 24-hour mixed broth culture of *Micrococcus luteus* and *Bacillus subtilis*

❏ 2. One nutrient agar plate

❏ 3. Six sterile cotton swabs

❏ 4. Ten mL of sterile saline (0.85%)

❏ 5. One wax marking pencil or felt-tip marker

❏ 6. One 500-mL beaker containing disinfectant

❏ 7. One container of disinfectant for skin disinfection after hand washing

Procedure

This procedure is to be performed by students in pairs.

❏ 1. Using the wax marking pencil, divide the bottom of the nutrient agar plate into 6 sectors. Label these sectors with the following designations: 0, 15, 30, 45, 60, and 75.

❏ 2. Select an area on the back of your laboratory partner's hand. Be certain that this area does not have an obvious cut or abrasion. With the wax marking pencil, mark 5 sectors and number them accordingly.

❏ 3. Moisten 1 sterile swab with the mixed bacterial culture provided, and lightly wipe all 5 marked areas of the skin with it. Then, with the same swab, streak the "0" sector of the agar plate. Discard the swab in the beaker of disinfectant.

❏ 4. After 15 minutes, moisten a fresh sterile swab with saline. Rub the swab over the area marked 1 on the hand, and streak the "15" sector of the agar plate with it. Discard the swab in the beaker of disinfectant.

❏ 5. Repeat the procedure in step 4 with area 2 at "30" minutes, area 3 at "45" minutes, area 4 at "60" minutes, and area 5 at "75" minutes.

❏ 6. The student whose hand has been used should wash both hands with soap and water and rinse with a disinfectant provided for skin disinfection.

❏ 7. Invert the inoculated plate and incubate it in your desk drawer until the next laboratory period.

❏ 8. Examine the plate after incubation and estimate the relative amount of growth of *M. luteus* (yellow colonies) and *B. subtilis* (off-white colonies). Record your findings in Table 59–3 in the Results and Observations sections, using 0 to indicate no growth, 1+ for slight growth, 2+ for medium growth, and 3+ for heavy growth.

B. Cultural Characteristics of Selected Bacterial Pathogens

Materials

The following items should be provided per 4 students:

❑ 1. One trypticase soy broth culture of each of the following microorganisms:
    ❑ a. *Staphylococcus aureus*
    ❑ b. *Staphylococcus albus*
    ❑ c. *Streptococcus pyogenes* (Lancefield's group A hemolytic)
    ❑ d. *Streptococcus agalactiae* (Group B, beta-hemolytic)
    ❑ e. *Streptococcus viridans*
    ❑ f. *Streptococcus faecium*
    ❑ g. *Enterococcus (Streptococcus) faecalis*

❑ 2. One divided streak plate with each of the following microbial combinations:
    ❑ a. *Staphylococcus aureus* and *Streptococcus pyogenes* on a blood agar plate
    ❑ b. The same combination as in (a), but grown on nutrient agar
    ❑ c. *S. aureus* and *Micrococcus luteus* on a blood agar plate
    ❑ d. *S. aureus* and *Aerococcus viridans (Gaffkya tetragena)* on a blood agar plate
    ❑ e. *Streptococcus pyogenes* and *S. viridans* on a blood agar plate
    ❑ f. *S. faecium* and *E. faecalis* on a blood agar plate

❑ 3. Six tubes of each of the following fermentation media:
    ❑ a. Glucose
    ❑ b. Lactose
    ❑ c. Mannitol
    ❑ d. Inulin

❑ 4. Six tubes of 6.5% sodium chloride (NaCl) broth with Brom-cresol purple indicator

❑ 5. Six tubes of bile-esculin medium

❑ 6. One set of Gram-staining reagents and hand lens

Procedure

Additional Techniques Required for This Portion of the Exercise:

❑ 1. Broth Transfer, Procedure Diagram 5, Exercise 4
❑ 2. The Gram Stain, Procedure Diagram 17, Exercise 14

This procedure is to be performed by students in groups of 4.

❑ 1. Using a hand lens, examine all the streak plate cultures provided.

❑ 2. Record the various colonial characteristics (refer to Exercise 7) in Table 59–4 of the Results and Observations section.

❑ 3. Prepare Gram stains of isolated colonies for each microorganism on these streak plates and record the result in Table 59–4.

❑ 4. Inoculate 1 set of the fermentation media and 6.5% NaCl and bile-esculin media with each of the broth cultures provided. The instructor will assign particular organisms to individual students so that all microbial cultures will be utilized.

❑ 5. Incubate the inoculated media for 48 hours or until the next laboratory period.

❏ 6. Discard the other cultures according to the directions of your instructor.

❏ 7. After incubation, record your findings in Table 59–4. Use the following designations in the case of the fermentation reactions: A = acid, Alk = alkaline, G = gas, and NC = no change (refer to Exercise 22); + = growth in 6.5% NaCl medium; and − = blackening of the bile-esculin medium (refer to Figure 59–5).

(a)   (b)

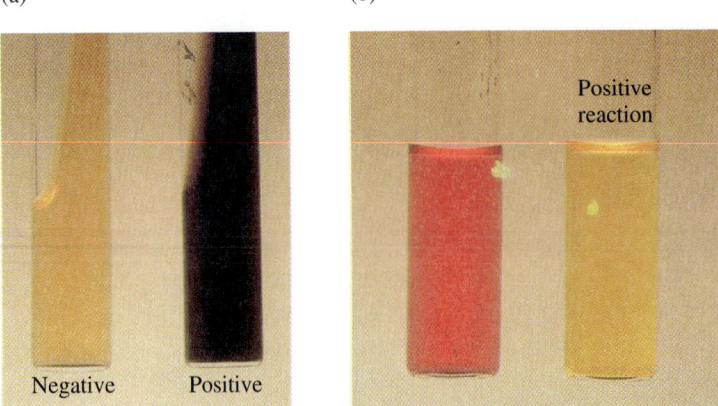

Figure 59–5

Tests for group B streptococci. (a) The bile-esculin test. A positive reaction is a blackened slant. (b) Salt tolerance test. A positive reaction usually is recorded when the indicator brom-cresol purple changes to yellow.

C. Demonstration of Selected Identification Tests

Staph-Ident *(Staphylococcus)* System

Materials

❏ 1. One 24-hour trypticase soy agar plate culture of:
  ❏ a. *Staphylococcus aureus* (ATCC 25923)
  ❏ b. *Staphylococcus simulans* (ATCC 27851)

❏ 2. Two tubes, each with 3 mL of sterile 0.85% saline, pH 7

❏ 3. Two 5-mL Pasteur or other pipettes with rubber bulbs

❏ 4. Two incubation chambers (tray and lid combination)

❏ 5. One marking pen

❏ 6. One 50-mL plastic squeeze bottle containing distilled water

❏ 7. One inoculating loop

Procedures

These procedures will be demonstrated by the instructor, or as directed.

Procedure 1: Preparation of Test Strips

❏ 1. Obtain an incubation tray and lid for the Staph-Ident system.

❏ 2. Mark the elongated flap of the tray with the name of the organism to be tested.

❏ 3. Using the plastic squeeze bottle, introduce about 5 mL of distilled water into the incubation tray to provide a humid atmosphere during incubation. (Refer to Exercise 27.)

❏ 4. Remove one API Staph-Ident strip, containing 10 microcupules (compartments) from the sealed envelope and place it into the incubation tray.

Procedure 2: Preparation of the Inoculum

❑ 1. Label 1 tube containing 0.85% saline for each test organism provided.

❑ 2. With a flamed inoculating loop, remove a bacterial colony from the *Staphylococcus aureus* plate and aseptically introduce the inoculum into its labeled tube of saline.

❑ 3. Remove 2 additional colonies and place them into the same tube of saline to produce a lightly turbid suspension.

❑ 4. Mix the contents by shaking the tube gently.

❑ 5. Repeat steps 2 through 4 for the other *Staphylococcus* species.

Procedure 3: Inoculation of the Staph-Ident Strip

❑ 1. Tilt the incubation tray with the Staph-Ident strip. Using a Pasteur pipette, fill each microcupule with 2 to 3 drops of the bacterial saline suspension. Place the pipette against the side of the cupule while adding the suspension.

❑ 2. After inoculation of the strip, place the plastic lid on the tray.

❑ 3. Repeat steps 1 and 2 with the other *Staphylococcus* species.

❑ 4. Incubate the inoculated systems for 5 hours at 35°C to 37°C.

❑ 5. After incubation, remove the systems and allow them to reach room temperature.

❑ 6. Examine the 2 systems and record the results for all tests not requiring the addition of reagents in the appropriate spaces in Table 59–5. Refer to Table 59–2 for the interpretation of reactions and to Figure 59–6.

❑ 7. Add 2 drops of Staph-Ident reagent to each NGP microcupule on the test strip. Allow 30 seconds to pass and examine. Record your findings in Table 59–4 for each test organism.

❑ 8. Dispose of the entire Staph-Ident strip system as directed.

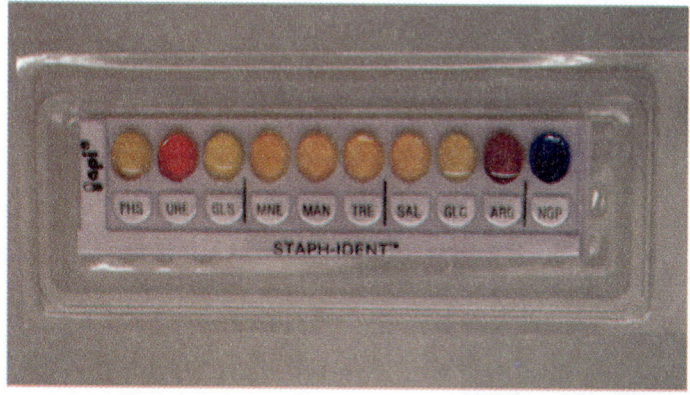

Figure 59–6
The results of the staphylococcal identification system, STAPH-IDENT strip. A battery of 10 tests are used. Symbols: PHS, phosphatase; URE, urease; GLS, beta-glucosidase; MNE, mannose.

Coagulase Test

Procedure 1: Preparation of the Incubation Tray and Test Strip

These procedures will be demonstrated by the instructor.

❑ 1. Obtain an incubation tray and lid for 3 Staphase strips.

❑ 2. Mark the tray at the appropriate locations with the name of each test organism.

❑ 3. Using the plastic squeeze bottle, introduce about 3 mL of distilled water into the incubation tray to provide a humid atmosphere during incubation. (Refer to Procedure Diagram 26.)

❑ 4. Remove 1 test strip from the sealed envelope.

❑ 5. With a sterile Pasteur pipette and bulb, introduce 2 free-falling drops of distilled water into 3 microtubes (chambers). Place the tip of the pipette over, but not touching, the side of each microtube while delivering the water.

Materials

❏ 1. One 24-hour trypticase soy agar divided plate containing isolated colonies of the following:
 ❏ a. *Staphylococcus aureus* (ATCC 25923) [coagulase positive]
 ❏ b. *Staphylococcus epidermidis* (ATCC 12228) [coagulase negative]
 ❏ c. *Micrococcus luteus*

❏ 2. One Staphase strip containing 3 microtubes or chambers

❏ 3. One marking pen

❏ 4. One tube containing 2 mL of sterile distilled water

❏ 5. One Staphase incubation chamber

❏ 6. One 50-mL plastic squeeze bottle containing distilled water

❏ 7. One inoculating loop

❏ 6. Allow 1 minute to pass before inoculating the microtubes. This pause will result in the reconstitution of the plasma in the chamber. Gently squeeze the bottom of the chamber to help the mixing process.

Procedure 2: Inoculation, Incubation, and Reading of the Staphase Strip

❏ 1. Take 1 of the agar plate cultures and aseptically pick 2 colonies with the aid of an inoculating loop.

❏ 2. Transfer the colonies to the reconstituted plasma inside 1 microtubule.

❏ 3. Use the inoculating loop to mix and emulsify thoroughly the colonial material.

❏ 4. Gently squeeze the bottom of the chamber to mix the contents.

❏ 5. Carefully tilt the strip from side to side to ensure that the contents flow freely.

❏ 6. Repeat steps 1 through 5 with each of the cultures provided.

❏ 7. Place the inoculated strip into the incubation tray and cover the lid.

❏ 8. Incubate the inoculated system for no more than 4 days at 35°C to 37°C.

❏ 9. After incubation, carefully remove the strip from the tray.

❏ 10. Tilt the strip and look for a coagulated mass (gel) and a lack of free flow within the chamber. These findings constitute a positive reaction. (Refer to Figure 59–7.)

❏ 11. Examine each chamber and answer the questions in the Results and Observations section.

❏ 12. Dispose of the test systems as directed.

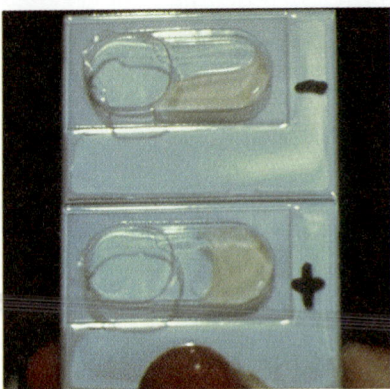

Figure 59–7

A miniaturized coagulase test, STAPHase. This system is easy to inoculate, and results appear in 4 hours. Coagulation is a positive test.
(Courtesy of API, Analytab Products.)

Modifications of this test are in use.

Heat-Stable DNAse Test

Materials

❑ 1. Two 24-hour broth cultures of the following:
   ❑ a. *Staphylococcus aureus* (ATCC 25923) [coagulase positive]
   ❑ b. *Staphylococcus aureus* (ATCC 12228) [coagulase negative]
❑ 2. One toluidine blue DNA agar plate
❑ 3. One boiling water bath
❑ 4. One wax marker or similar marking device

Procedure

The use of DNAse agar will be demonstrated by the instructor.

❑ 1. Place 1 culture into a boiling water bath for 15 minutes and cool. Be certain that these tubes are marked to show their treatment.

❑ 2. Using a wax pencil, mark 4 sectors on the bottom of the DNAse agar plate. Label 1 sector for each culture to be used, 2 heated and 2 unheated.

❑ 3. Place a loopful of each culture into the center of its respective sector.

❑ 4. Incubate the prepared plate at 37° C for 4 hours. (*Note:* Incubation for 8 hours may be necessary to obtain positive results.)

❑ 5. After incubation, examine the plate and look for a change in color on the medium. The formation of a bright pink region around the inoculations is a positive reaction.

❑ 6. Answer the questions in the Results and Observations section.

Bacitracin Sensitivity Test

Materials

❑ 1. Trypticase soy broth cultures of each of the following bacterial species:
   ❑ a. *Staphylococcus aureus* (ATCC 25923) [beta-lysin producer]
   ❑ b. *Streptococcus agalactiae*
   ❑ c. *Enterococcus faecalis*
   ❑ d. *Streptococcus pyogenes*
❑ 2. Four sheep blood agar plates
❑ 3. Taxos A Sensi-Disks (bacitracin differential disks containing 0.04 unit)
❑ 4. Three sterile cotton swabs
❑ 5. One 500-mL beaker with a disinfectant solution

Procedure

❑ 1. The instructor will streak the surfaces of 1 blood agar plate with a cotton swab containing *S. pyogenes,* the second plate with *S. faecalis,* and the third plate with *S. agalactiae.*

❑ 2. Label each plate with the organism streaked.

❑ 3. Place a bacitracin disk on the center of each streaked plate.

❏ 4. Take the remaining plate and at 1 end make a single streak with the *Staphylococcus aureus* strain provided. This step is necessary for the CAMP reaction.

❏ 5. Make individual single streaks with each *Streptococcus* species perpendicular to but not touching the *Staphylococcus aureus* streak line. (Stop these streaks about 1 cm from the *Staphylococcus aureus*.)

❏ 6. Incubate all plates for 48 hours.

❏ 7. Examine each plate for its particular test result. (Refer to Figure 59–4.) Note that the 6.5% NaCl and Bile-Esculin medium tests are also used to differentiate among the streptococci.

❏ 8. Answer the questions in the Results and Observations section.

D. Selected Tests for *Streptococcus* and/or *Enterococcus* Identification

Materials

❏ 1. One trypticase soy agar slant of each of the following microorganisms:
   ❏ a. *Streptococcus pyogenes* (Lancefield's group A hemolytic)
   ❏ b. *Streptococcus agalactiae* (Group B, beta-hemolytic)
   ❏ c. *Streptococcus viridans*
   ❏ d. *Enterococcus faecalis*
   ❏ e. Unknown culture

❏ 2. Five blood agar plates

❏ 3. Five CAMP disks

❏ 4. Five L-pyrrolidonyl-B-naphthylamide (PYR) agar slants

❏ 5. Two PYR DrySlide (Difco) and PYR reagent (Difco)

❏ 6. One container with PYR reagent and dropper

❏ 7. Five bile-esculin agar slants

❏ 8. One marking pen or pencil

❏ 9. Sterile forceps

❏ 10. One pair of scissors

❏ 11. One 500-mL beaker with disinfectant

❏ 12. One tube of sterile distilled water

Procedure 1: CAMP, Hemolysis, and PYR Reactions

❏ 1. Label each plate and other media for each of the cultures provided.

❏ 2. Inoculate each blood agar plate in the following manner with each of the cultures provided:
   ❏ a. For the blood agar plates with a pair of forceps, place a CAMP disk about 3 mm from the edge of the plate. Streak the culture in a straight line about 5 mm from the disk. In addition, stab the blood agar in 2 different areas away from the streak line. This step is done to determine the type of hemolysis.
   ❏ b. Streak the bile-esculin and PYR agars in a regular manner. (See Procedure Diagram 6.)

❏ 3. Invert all plates and incubate all inoculated media at 35°C to 37°C for 48 hours.

❏ 4. After incubation, examine the plates for the following reactions and enter your findings in Table 59–6 in the Results and Observations section.
   ❏ a. A positive CAMP reaction shows a hemolytic half-moon or arrowhead near the CAMP disk.
   ❏ b. The type of hemolysis.
   ❏ c. For the PYR reaction, add 2 drops of the PYR reagent directly to the bacterial growth. The development of cherry-red color within 2 minutes is a positive result.
   ❏ d. A positive reaction for the bile-esculin test is indicated by a blackening of the medium.

Procedure 2: DrySlide PYR Test

❏ 1. With the pair of scissors provided, open the PYR DrySlide pouch and carefully remove the slide. Note that the slide has only 4 compartments.

❏ 2. Aseptically deposit a loopful of distilled water onto the reaction area (center) of each compartment.

❏ 3. Aseptically remove a loopful of growth from the agar slant culture of *Streptococcus pyogenes* and smear it onto the moistened reaction area of the DrySlide.

❏ 4. Incubate the slide at room temperature for 2 minutes.

❏ 5. Place 1 drop of the DrySlide PYR color developer onto the inoculated area.

❏ 6. The appearance of a pink to fuchsia color within 1 minute is a positive reaction. No color change within 1 minute is a negative result.

❏ 7. Repeat steps 2–6 with the other cultures provided and enter your Findings in Table 59–6.

❏ 8. Answer the questions in the Results and Observations and Laboratory Review sections.

E. Staph-Latex Test Demonstration

Materials

The following items should be provided for the instructor's demonstration.

❏ 1. One 24-hour trypticase agar divided plate containing isolated colonies of each of the following combinations:
  ❏ a. *Staphylococcus aureus* (ATCC 25923) [coagulase positive] and *S. epidermidis* (ATCC 12228) [coagulase negative]
  ❏ b. Unknown A and unknown B

❏ 2. The following reagents in appropriate dispensers:
  ❏ a. Bacto Staph-Latex Reagent
  ❏ b. Bacto Staph Positive Control
  ❏ c. Bacto Staph Negative Control
  ❏ d. Bacto Normal Saline Reagent

❏ 3. Bacto Disposable Test Cards (slides)

❏ 4. One inoculating needle

❏ 5. One wax marking pencil (yellow or orange)

Procedure

❏ 1. Obtain 1 Bacto Disposable Test Slide and label 1 circle for each of the following materials using the letters shown:
  ❏ a. Positive control
  ❏ b. Negative control
  ❏ c. *S. aureus*
  ❏ d. *S. epidermidis*
  ❏ e. Unknown A
  ❏ f. Unknown B

❏ 2. Place 1 drop of the Bacto Staph Positive Control into circle *a* on the test slide.

❏ 3. Place 1 drop of the Bacto Staph Negative Control into circle *b*.

❏ 4. Add 1 drop of the Bacto Staph-Latex Reagent to each of the above circles containing controls and to circles *c* through *f*.

❏ 5. Flame the inoculating needle and quickly transfer 3 isolated *S. aureus* colonies to its specified circle, (*c*), and mix this growth thoroughly with the latex reagent.

❏ 6. Repeat step 5 with each of the cultures provided, using their appropriately marked circles.

❏ 7. Rotate the slide by hand in a circular manner for 45 seconds and read the reactions quickly. Look for the formation of yellow clumps. (Refer to Exercise 48.)

❏ 8. Indicate the presence of agglutination by a "+" and its absence by a "−" in Table 59–7 of the Results and Observations section.

F. Demonstration of Microscopic Morphology of Clostridia

Materials

The following slides should be provided for examination by the students:

❏ 1. *Clostridium tetani* spore stain (see Exercise 16)

❏ 2. *Clostridium perfringens* Gram stain

Procedure

This procedure is to be performed by students individually. Make representative sketches of the microorganisms placed on demonstration slides in the Results and Observations section.

Results and Observations

Antibacterial Activity of the Skin

1. Complete the following table.

Table 59–3

Growth Obtained at Various Times of Exposure

Culture	0 Time	15 min.	30 min.	45 min.	60 min.	75 min.
M. luteus						
B. subtilis						

Cultural Characteristics of Selected Bacterial Pathogens

1. Complete Table 59–4.

Table 59–4

Bacterial Pathogen Characteristics

Microorganism	Colonial Characteristics: Pigmentation, Margin, Size, Elevation, and Texture	Gram Reaction	Microscopic Morphology	Type of Hemolysis	Fermentation Pattern				Bile-Esculin Test	6.5% NaCl Test
					Glucose	Lactose	Mannitol	Inulin		

Demonstration of Selected Identification Tests

1. Staph-Ident system. Enter all findings in Table 59–5.

Table 59–5

Staph-Ident Reactions

Microorganism	Staph-Ident Test Results[a]									
	PHS	URE	GLS	MNE	MAN	TRE	SAL	GLC	ARG	NGP
Staphylococcus aureus										
S. simulans										

[a]Refer to Table 66–2 for an explanation of test symbols and the appearance of positive test results.

2. Coagulase test.

    a. Which of the cultures tested produced a positive coagulase reaction? _____

    b. Which cultures produced a negative coagulase reaction? _____

3. DNAse test.

   a. Did all cultures produce a positive reaction? _____

   b. If not, which ones gave a negative reaction? _____

   c. What physical property of the enzyme does this test show? _____

4. Bacitracin sensitivity and CAMP tests.

   a. Which of the *Streptococcus* species tested exhibited:

      i. Bacitracin sensitivity? _____

      ii. A CAMP reaction? _____

   b. Is the CAMP test a functional way to differentiate among certain streptococci? _____

Selected Tests for Streptococcus and/or Enterococcus Identification

1. Enter your findings in the following table. Indicate a positive reaction with a "+" and a negative reaction with a "−" for the CAMP, PYR, and bile-esculin tests. Use alpha, beta, and gamma for the hemolytic reactions observed.

Table 59–6

Streptococcus and/or Enterococcus Identification Tests

| Streptococcus Species | Tests | | | | |
| | Hemolysis | CAMP | PYR | | Bile-Esculin |
			Slant	Dryslide	
S. pyogenes					
S. agalactiae					
S. viridans					
E. faecalis					
Unknown					

2. What is the identity of the unknown culture? _____

_____

Staph-Latex Test Demonstration

1. Enter your findings in Table 59–7.

Table 59–7

Staph-Latex Test Results

Material for Testing	Reaction
S. aureus	
S. epidermidis	
Positive control	
Negative control	
Unknown A	
Unknown B	

2. On the basis of your findings, did either of the unknowns produce a positive test? If so, which one(s)?

_____

_____

3. What does a positive reaction mean in this test? _____

_____

Demonstration of Microscopic Morphology of Clostridia

Sketch your observations in the circles provided.

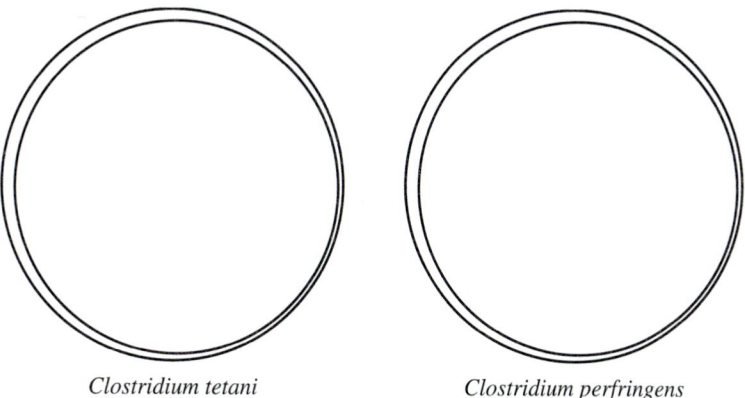

Clostridium tetani   *Clostridium perfringens*

G. Demonstration of Selected Pathogenic Fungi

Materials

❑ 1. Sealed Sabouraud dextrose agar or Georg fungus medium plates of the following fungi:
- ❑ a. *Trichophyton mentagrophytes*
- ❑ b. *Trichophyton rubrum*
- ❑ c. *Microsporum canis*
- ❑ d. *Epidermophyton floccosum*
- ❑ e. *Candida albican*
- ❑ f. *C. albicans* on cornmeal agar

❑ 2. Prepared slides of cultures for the study of microscopic morphological features of the following:
- ❑ a. *T. mentagrophytes*
- ❑ b. *M. canis*
- ❑ c. *E. floccosum*
- ❑ d. *C. albicans* (from cornmeal agar)

Procedure

This procedure is to be performed by students in groups of 4.

❏ 1. Examine the sealed cultures. Note pigmentation patterns, general texture, and so on. Compare the surface of the mycelial growth with the undersurface (backside) of the mycelium. Complete Table 59–8 in the Results and Observations section. Refer to Figures 59–8 to 59–11.

❏ 2. Study the demonstration slides provided. Locate and identify the specific structures shown in Figures 59–12 through 59–15 in the Results and Observations section.

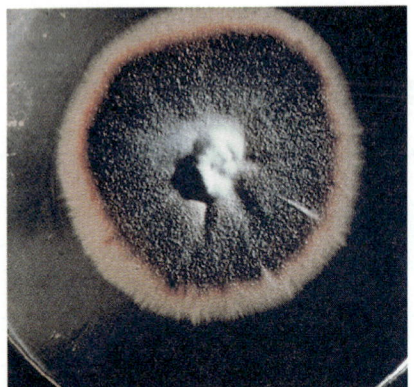

Figure 59–8
Trichophyton rubrum mycelium.

Figure 59–9
Microsporum canis mycelium.

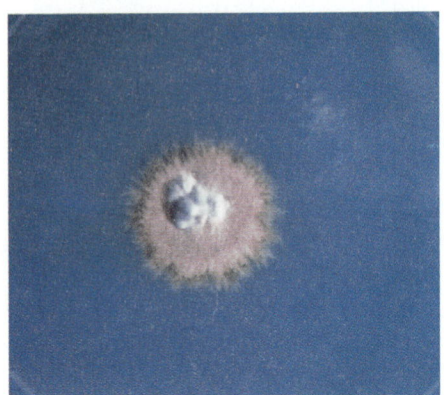

Figure 59–10
Epidermophyton floccosum mycelium.

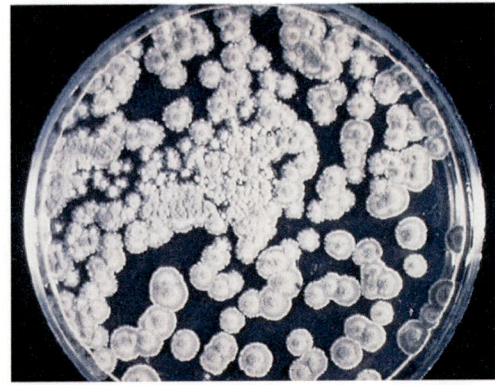

Figure 59–11
Candida albicans.

Results and Observations

Demonstration of Selected Pathogenic Fungi

1. Enter your findings in Table 59–8.

Table 59–8

Properties of Selected Pathogen Fungi

Fungus	Description of Mycelial Surface	Description of Mycelial Undersurface

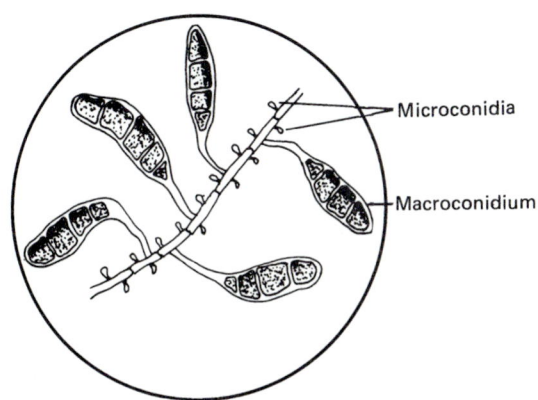

Figure 59–12
Microscopic view of *Trichophyton mentagrophytes* species (a). (b) *Trichophyton* species macroconidium.

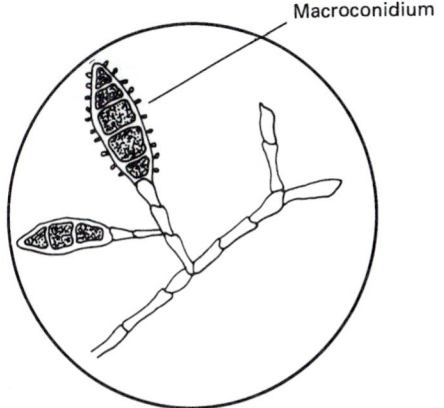

Figure 59–13
Microsporum canis.

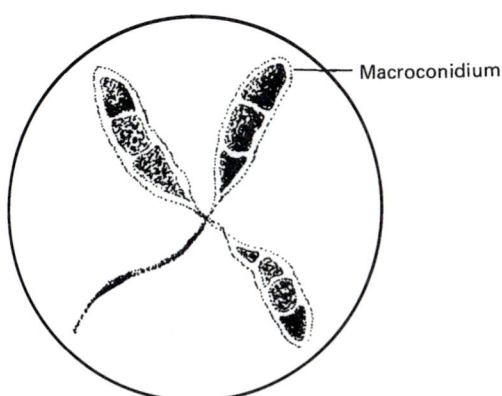

Figure 59–14
Epidermophyton floccosum.

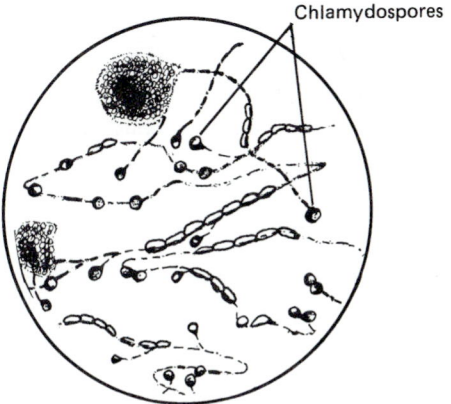

Figure 59–15
Candida albicans (cornmeal agar; different structural forms).

Laboratory Review 59   Pathogenic Microorganisms of the Skin

1. What media would you employ for the cultivation of *Clostridium tetani?* _____

   _____

2. List at least two general methods used for the cultivation of anaerobic bacteria.

   a. _____   b. _____

3. Which of the procedures performed in this exercise are of value in distinguishing between:

   a. *Staphylococcus aureus* and *Streptococcus pyogenes?* _____

   _____

   b. *Staphylococcus aureus* and *Aerococcus viridans?* _____

   _____

   c. *Streptococcus pyogenes* and *Micrococcus luteus?* _____

   _____

   d. *Streptococcus viridans* and *E. faecalis?* _____

   _____

   e. *Staphylococcus aureus* (coagulase positive) and *S. aureus* (coagulase negative)? _____

   _____

   f. Group A and group B streptococci? _____

   _____

   g. Group B and viridans group of streptococci? _____

   _____

4. List the fungus infections that can be detected by the use of a skin test. (Refer to your text.) _____

   _____

   _____

5. Define or explain the following terms:

   a. mycelium _____

   b. hypha _____

   c. pseudomycelium _____

   d. tinea pedia _____

Key Terms

edema (e-DĒ-ma): a localized swelling due to the presence of excessive tissue fluid

histotoxic (his-tō-TOKS-ik): poisonous to tissue

ischemic (is-KĒ-mik): refers to local and temporary deficiency of blood supply caused by a circulatory obstruction

neurotoxic (noo-rō-TOKS-ik): poisonous to nerve tissue

pyogenic (pī-ō-JEN-ik): producing pus

After completing this exercise, you should be able to:

1. List several common infections of the mouth.
2. Recognize general cultural characteristics of selected potential bacterial pathogens.
3. Perform and interpret the Snyder test for dental caries susceptibility.
4. Detect dental plaque on teeth using a color-coating device.

The mouth and associated structures (including the tonsils, oropharynx, and nasopharynx) support a varied and dense population of microbial flora. This condition exists as a consequence of (1) exposure of each newborn infant to the microbial population of the mother's genital tract, (2) exposure to microorganisms in the general environment, and (3) the selective factors present in the infant's oral cavity. The number and variety of microorganisms in the mouth are influenced by several factors, such as the general condition of the teeth and gums, diet, recent use of antibiotics, and the like. Many of the microorganisms constituting the normal flora of this region can readily cause infections of the oral cavity and related anatomical structures. Common bacterial diseases of these areas are gingivitis, periodonitis, periodontosis, dental caries, abscesses associated with teeth, pharyngitis, and tonsillitis. Because the diseases of the mouth and related areas are far too numerous to discuss in any great detail, only a few common infections will be described.

The initiation and progression of diseases affecting the tissues surrounding and supporting teeth are influenced by several factors, including (1) the organisms making up the microbial flora of the mouth, (2) nutritional deficiencies and hormonal imbalances of the host, (3) the presence of calculus, and (4) other factors of inadequate dental hygiene. The protective mechanism, inflammation, appears to be the fundamental response of the host's periodontal tissue to most disease-initiating factors. The inflammatory reaction varies in severity, and several investigators have used response intensity as a factor in classifying periodontal disease. The designations and descriptions of the chronic forms described in this exercise are those of Rosebury. Nomenclatures employed by other investigators, such as Bibby and Fish, are also included.

Dental plaque formation in humans arises from the attachment and reproduction of bacteria on the tooth surface. The bacteria involved are from the commensal flora of the mouth. Thus, plaque is a continuously formed coating of microorganisms and organic matter on tooth enamel. It consists of microorganisms (Figure 60–1), dextran, and proteins from the individual's saliva. Glucose from dietary sources is used to form the polysaccharide dextran. If plaque is not removed thoroughly and regularly, lactobacilli, streptococci, and various other acid-producing bacteria accumulate within it. Plaque formation, although not a disease itself, is the first step in tooth decay and gum disease.

Tooth decay, also known as *dental caries* is the erosion of enamel and the deeper parts of teeth (Figure 60–2). It is one of the most common infectious diseases in industrialized countries where relatively large amounts of refined sugar are in diets.

The *Snyder test* is used to determine caries susceptibility. The procedure involves mixing a measured amount of saliva with Snyder test agar, which contains lactose and the pH indicator bromthymol blue. Bacteria in the saliva fermenting the lactose to lactic acid change the color of the agar from green (neutral pH) to yellow (Figure 60–3). The greater the number of lactose-fermenting bacteria in the saliva, such as the lactobacilli, the faster the agar will turn yellow. If the color changes within 24 to 48 hours of incubation, the individual can be considered to be susceptible to caries formation.

A mild condition involving a painless accumulation of blood (hyperemia) of the gingival margin without gross pus formation is the first type of chronic periodontal disease. This form is referred to as *marginal gingivitis* or *chronic marginal gingivitis*. The second type, known as *ulcerative gingivitis* (as well as *acute gingivitis, Vincent's infection,* and *pyorrhea simplex*), is an acute inflammatory disease involving destruction of exposed gingival tissue. The last type, referred to as *periodontoclasia* (also *periodontitis* and *pyorrhea profunda*), is characterized by an extension of the inflammatory response into the deeper periodontal structures with a protracted destruction of the periodontal membrane. In addition, pus formation, pocket formation, and loss of alveolar bone occurs.

(a)

(b)

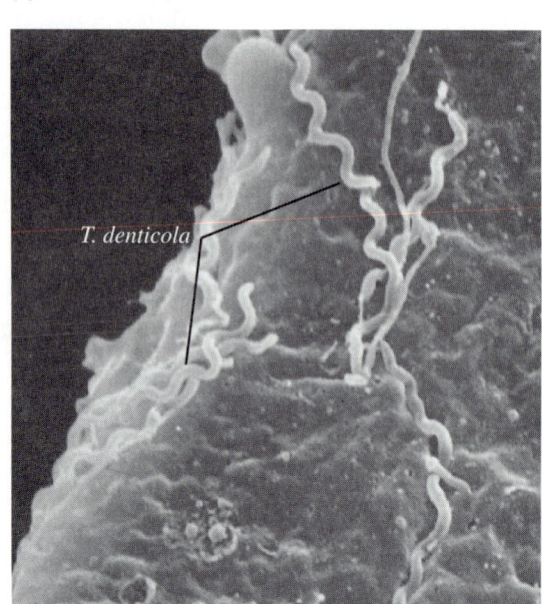

Figure 60–1

Examples of microorganisms associated with dental plaque, caries, and gingivitis. (a) A scanning micrograph of a variety of microorganisms that are found in dental plaques.
(From Kodaka, T., Y. Ohohara, and K. Debari. *Scan Micro.* 6:475–486 [1992].)

(b) A scanning micrograph of *Treponema denticola*.
(From Baehni, P.C. *Inf. Imm.* 60:3360–3368, [1992].)

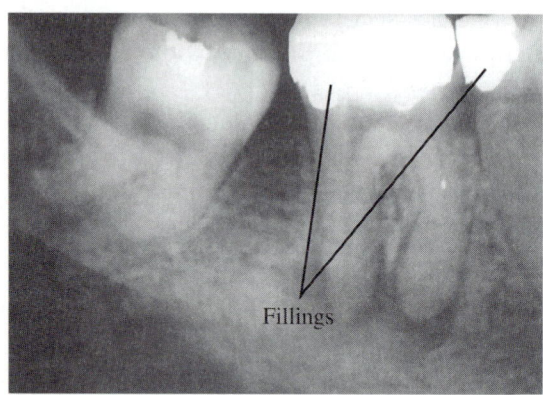

Figure 60–2

An x-ray showing the results of dental caries or tooth decay. Note the parts of the tooth and the fillings (arrows).

Certain yeasts also can cause oral cavity problems for individuals with poorly functioning or defective immune systems. One of these, *Candida albicans,* while present as a member of the normal microbial flora of the digestive tract, can cause **oral thrush.** The condition appears as milky patches of inflammation on the mucous membranes of the mouth. Candidiasis or the older name moniliasis develops in diabetics, persons with AIDS, infants, and persons receiving antibiotics for a long time. Finding oval, budding yeast cells in sputum, and/or lesions is generally sufficient for a diagnosis.

Several specific microorganisms are associated with the causation of dental caries or periodontal disease. Numerous studies have shown the importance of microorganisms in the development of these processes. Organisms involved in periodontal disease states include *Borrelia vincentii, B. bucalis,* fusiform bacilli, spirochetes, and streptococci. The mechanism by which microorganisms exert their various effects involves several of their products (such as toxins and enzymes) that react with host tissues to produce disease.

The indigenous flora of the mouth vary among individuals, but generally include *Moraxella* sp., lactobacilli, staphylococci, streptococci, veillonellae, vibrios, filamentous forms, and fusiforms. Although dental caries are believed to be the result of microbial activities, it has been exceedingly difficult to point to definite species as the causative agents of the disease. Lactobacilli as well as *Actinomyces odontolyticus* and streptococci have been isolated from diseased teeth on several occasions. Investigations have shown that the disease is a consequence of a universal microbial parasitization of teeth that occurs only in the presence of fermentable carbohydrates. Susceptibility to tooth decay varies among individuals. Teeth of certain individuals do not decay in the presence of environmental conditions that normally are associated with dental caries formation.

This exercise will present the characteristics of selected bacterial species found in the mouth (Figure 60–4), plaque detection (Figure 60–5), the Snyder test, and compare the effects of various mouthwashes on microorganisms found in the saliva.

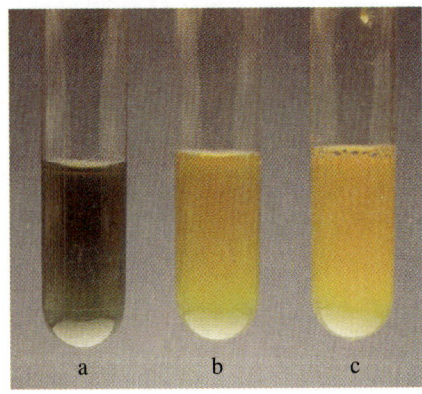

Figure 60–3
The Snyder colorimetric test for caries activity diagnosis. *Tube a,* uninoculated (blue-green). *Tubes b* and *c,* acid production (yellow, indicating caries activity).

(a)

Figure 60–4
Examples of bacteria found in the mouth. (a) *Streptococcus salivarius* on mitis-salivarius medium. (b) *Streptococcus mitis* on mitis-salivarius medium.

A. General Characteristics of Selected Potential Bacterial Pathogens

Materials

The following items should be provided per 4 students:

- ❏ 1. One trypticase soy broth culture of each of the following microorganisms:
    - ❏ a. *Moraxella* catarrhalis
    - ❏ b. *Streptococcus mitis*
    - ❏ c. *Streptococcus salivarius*
    - ❏ d. *S. mutans*
    - ❏ e. *S. sanguis*
- ❏ 2. Five each of the following media:
    - ❏ a. Blood agar plate
    - ❏ b. Mitis-Salivarius agar plate
    - ❏ c. Trypticase soy agar plate
- ❏ 3. One Gram stain reagent set

Procedure

Additional Techniques Required for This Portion of the Exercise:

❏ 1. Streak Plate Technique, Procedure Diagram 10, Exercise 5

❏ 2. The Gram Stain, Procedure Diagram 17, Exercise 14

This procedure is to be performed by students in groups of 4.

❏ 1. Prepare and examine Gram stains of each culture assigned. Enter your results, as well as those of other students, in Table 60–2 of the Results and Observations section.

❏ 2. Using the organisms assigned to you, take 1 each of the 3 plate media and streak with each species. Incubate the inoculated media for 48 hours or until the next laboratory period.

❏ 3. After incubation, record your findings in the Results and Observations section. Compare the colonial morphology, type of hemolysis, and other properties of the organisms in the exercise.

❏ 4. Answer the questions in the Laboratory Review.

B. Dental Plaque Detection and Diagnostic Tests for Dental Caries Susceptibility

Materials

The following items are to be provided per 4 students:

❏ 1. Four Red-Cote pills or a similar preparation for the detection of dental plaque

❏ 2. Four Bacto tomato juice agar plates

❏ 3. Eight Bacto Snyder test agar deeps

❏ 4. Four tubes of trypticase soy broth (4 mL per tube)

❏ 5. Twelve sterile 1-mL pipettes

❏ 6. Four sterile test tubes

❏ 7. Four 1-inch squares of paraffin

The following item should be provided for general class use:

❏ 1. One boiling water bath

Procedure 1: Dental Plaque Detection

This procedure is to be performed by students individually.

❏ 1. Chew, but do not swallow, 1 instant-dissolving tablet provided to detect dental plaque.

❏ 2. Vigorously swish the formed saliva in the mouth for at least 30 seconds.

❏ 3. Collect the stained saliva in a sterile test tube. (*Note:* This material can be used for the following 2 procedures.)

❏ 4. Students should examine one another's mouth for the presence of a *dark red color* remaining on the teeth (usually along the gum lines and beneath the teeth). The presence of the red color may indicate dental plaque accumulations (Figure 60–5).

❏ 5. Answer the questions in the Results and Observations section. The coloring material is water soluble and should readily disappear after use.

❏ 6. Perform the following procedures, which cover diagnostic tests for dental caries susceptibility.

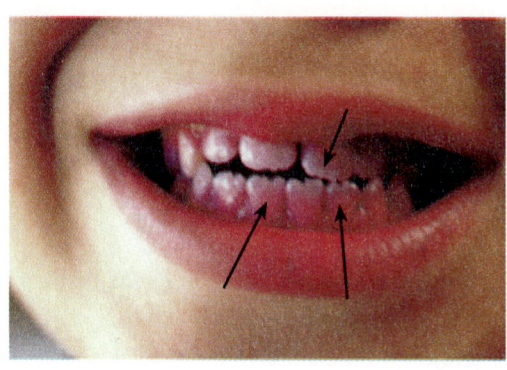

Figure 60–5
Dental plaque detection. Red discolorations on the surfaces of teeth (arrows) may indicate dental plaque accumulation.

Procedure 2: Snyder Test

This procedure is to be performed by students individually.

❏ 1. Chew 1 of the small squares of paraffin for 3 minutes and collect your saliva in a sterile test tube. The best time for specimen collection is usually before lunch or dinner.

❏ 2. Melt 2 Snyder test agar deeps and cool to approximately 45°C.

❏ 3. Shake the saliva specimens vigorously. Remove 0.2 mL of the material with the aid of a sterile pipette and introduce the specimen into 1 tube of Snyder medium. Mix the contents of the tube by gently tapping the tube with your fingers.

❏ 4. Repeat step 3 using the second tube of medium provided.

❏ 5. Allow the inoculated media to solidify and then incubate at 37°C for a total of 72 hours. Examine the tubes daily. Using an inoculated tube as a control, record any color changes that develop. (See Figure 60–3.)

❏ 6. A very general interpretation of the results in terms of caries activity is given in Table 60–1.

Table 60–1

Dental Caries Activity Interpretations

Caries Activity[a]	Hours Incubated		
	24	48	72
Marked	Positive[b]	—	—
Moderate	Negative[c]	Positive	—
Slight	Negative	Negative	Positive
Negative	Negative	Negative	Negative

[a]This table is provided through the courtesy of Difco Laboratories, Inc., Detroit, Michigan.
[b]Positive: This is a reaction in which the color changes, and green is no longer dominant.
[c]Negative: This is a reaction in which no color changes or only a slight one occurs. The medium generally retains a green color throughout.

Procedure 3: Tomato Juice Agar Isolations

Additional Techniques Required for This Portion of the Exercise:

❏ 1. Streak Plate Technique, Procedure Diagram 10, Exercise 5

❏ 2. The Gram Stain, Procedure Diagram 17, Exercise 14

This procedure is to be performed by students individually.

❑ 1. With the aid of a sterile pipette, add 1 mL of the saliva specimen to a tube of trypticase soy broth.

❑ 2. Gently mix the contents by tapping the tube with your fingers.

❑ 3. Remove a loopful of this mixture and spread it over the surface of a tomato juice agar plate using the clock-plate streaking method. (Refer to Procedure Diagram 10.)

❑ 4. Incubate the plate for 48 hours at 37°C.

❑ 5. After the incubation period, examine the plate, note the appearance of the colonies, and determine the number of each type.

❑ 6. Make Gram stains of the different representative types of colonies. (Refer to Procedure Diagram 17.)

❑ 7. Answer the questions in the Results and Observations section.

C. Mouthwashes and Their Effects on Oral Microorganisms

Materials

The following items should be provided per 2 students:

❑ 1. Two nutrient agar deeps

❑ 2. Two sterile Petri plates

❑ 3. Sterile filter paper disks in a sterile Petri plate

❑ 4. One pair of sterile forceps

❑ 5. Two sterile 1-mL pipettes with bulbs or pipette pump device

❑ 6. Two sterile 10–15-mL capacity of disposable tubes

❑ 7. Approximately 2 mL each of three commercial mouthwashes dispensed in tubes marked Brand A, B, and C, respectively

❑ 8. One ruler with a mm scale

Procedure

Additional Techniques Required for this Portion of the Exercise:

❑ 1. Pour Plate Technique, Procedure Diagram 9, Exercise 5

❑ 2. The Use of a Pipette Pump, Procedure Diagram 11, Exercise 6

❑ 3. Filter Paper Disk Technique, Procedure Diagram 29, Exercise 33

This procedure is to be performed by students individually.

❑ 1. Salivate into the sterile test tube provided. (The specimen should be about 2 mL.)

❑ 2. With a sterile 1-mL and pipette helper, aseptically transfer 1 mL of your saliva specimen into a tube of melted nutrient agar.

❑ 3. Discard the pipette in the container with disinfectant or as indicated by your instructor.

❑ 4. Gently mix the contents of the tube and aseptically pour into a sterile Petri plate. (Refer to Procedure Diagram 9.)

❑ 5. Perform the 6 steps shown in Procedure Diagram 29 with each of the mouthwash preparations provided.

❑ 6. Incubate all preparations at room temperature for 48 hours, or as directed.

❑ 7. After incubation, measure the zone of inhibition with the ruler provided and record your findings and your laboratory partner's in Table 60–3 in the Results and Observations section.

❑ 8. Examine the list of ingredients indicated for each preparation used. Your instructor will provide this information.

Results and Observations

General Characteristics of Selected Potential Bacterial Pathogens

1. Complete Table 60–2. Use the following symbols to express the characteristics of the species examined: gram-positive = g+; gram-negative = g−; coccus = c; rod = r; presence of growth = "+"; and absence of growth = "−".

Table 60–2

Selected Pathogen Characteristics

Microorganisms (Genus/Species)	Brief Description of Colonial Morphology on Trypticase Soy Agar	Gram Reaction	Microscopic Morphology	Type of Hemolysis	Presence of Growth		
					Blood Agar	Mitis-Salivarius	Trypticase Soy

Dental Plaque Detection and Diagnostic Tests for Dental Caries Susceptibility

Dental Plaque Detection

1. Did you have any red-colored deposits on your teeth? _____

2. Did your laboratory partner have any such deposits? _____

3. What do your results indicate? _____

_____

Snyder Test

1. Was there marked dental caries activity with your saliva specimen? If not, describe the reaction obtained. _____

_____

_____

2. Can the number of organisms per milliliter in your saliva specimen (as represented by the colony counts) be correlated with the color changes obtained with the Snyder test agar? Explain. _____

_____

_____

Tomato Juice Agar Isolations

1. How many different types of colonies appeared on the tomato juice agar plates? Were the number of colonies similar? _____

_____

_____

2. What type of Gram stain reactions did the isolates produce? _____

_____

_____

Mouthwashes and Their Effects on Oral Microorganisms

1. Enter your findings in Table 60–3.

Table 60–3

Mouthwash Effectiveness

Saliva Specimen(s)	Diameters of Zones Produced (in mm)		
	Brand A	Brand B	Brand C

2. Which of the mouthwash preparations was the most effective? _____

3. Compare the ingredients of the respective mouthwash preparations. What ingredients were shared in common?

_____

_____

4. Were there any ingredients found only in certain preparations? _____ If so, which ones? _____

_____

Laboratory Review 60   Pathogenic Microorganisms of the Mouth and Dental Caries

1. Do yeasts cause infections of the mouth? _____ If so, describe one. _____

    _____

2. Do viruses cause infections of the oral cavity? If so, list and describe a representative type. (Refer to your text.)

    _____

    _____

3. List 2 means of controlling infections of the oral cavity.

    a. _____

    b. _____

4. List 3 diseases of the oral cavity commonly associated with streptococci.

    a. _____   c. _____

    b. _____

5. What antibiotics are in common use for the treatment of oral cavity infections? (Refer to your text.) _____

    _____

    _____

Key Terms

abscess (AB-ses): a localized collection of pus

caries (KĀR-ēz): tooth decay

empyema (em-pī-Ē-ma): put in a body cavity

gingivitis (JIN-ji-vī-tis): inflammation of the gums

periodonitis (per-i-ō-don-TĪ-tis): inflammation and/or degeneration of connective tissue between a tooth and the bone and surrounding gum tissue

periodontoclasia (per-i-ō-don-tō-KLĀ-zē-a): inflammation and degeneration of the pharynx (throat)

tonsillitis (ton-sil-Ī-tis): inflammation of a tonsil

ulcerative (UL-ser-ā-tiv) gingivitis: inflammation of the gums accompanied by open sores

After completing this exercise, you should be able to:

1. List several common infections of the respiratory tract.
2. Recognize general cultural characteristics of selected potential bacterial pathogens.
3. Describe new approaches to the isolation of selected bacterial respiratory system pathogens.
4. Interpret the results of the optochin test for *Streptococcus pneumoniae.*
5. Identify selected fungal pathogens from microscopic and cultural characteristics.

The human respiratory system can be infected by various types of microorganisms. Whether respiratory infections become established depends on host-microbe relationships and the respiratory system and its nonspecific defenses. Respiratory infections are divided into *upper respiratory infections,* which include sore throat *(pharyngitis),* infection of the voice box or larynx *(laryngitis),* infections of the sinus cavities in the head *(sinusitis),* and ear infections *(otitis media),* and the lower respiratory infections (Table 61–1).

Bacterial infections of the upper respiratory tract are common and easily acquired through aerosols. Crowded conditions also favor the spread of such infections. The respiratory tract has several protective substances and processes at its disposal. Protective devices such as mucus secretion, ciliary movement, and phagocytosis function to keep the respiratory passages free of disease agents and materials that may cause substantial injury. Clearing the throat, the cough reflex, and the peristaltic movements of the finer bronchi assist in achieving and maintaining health.

Diseases such as *bronchitis, pneumonia, pulmonary abscess,* and *empyema* are representative of bacterial infections that may occur in this tract. Bronchitis, an inflammation of the tracheobronchial tree, can be produced by chemical, physical, or infectious agents. The bacterial pathogens usually associated with the acute form of this disease include *Bordetella pertussis, Streptococcus pneumoniae, S. pyogenes,* and *Staphylococcus aureus.* In the case of viruses, adeno- and influenza-virus types have usually been implicated. The etiology of chronic bronchitis is not well understood, but these viruses together with *Proteus* spp., *Enterobacter aerogenes,* and *Mycobacterium tuberculosis* may be encountered.

Inflammation of the lungs, or pneumonia, may be the result of infections caused by bacteria, viruses, protozoa, and fungi. The effects of chemical agents may also be involved. Most of the bacterial species listed previously as the etiologic agents of bronchitis, together with *Chlamydia psittaci, Pseudomonas* spp., and coliforms, are capable of producing pneumonia. Viral agents associated with this state include adenoviruses and influenza virus. Serious complications of viral infections can be caused by secondary bacterial invasions, as in the case of influenza-staphylococcal pneumonia. The protozoon, *Pneumocystis carinii,* has caused the death of numerous victims of acquired immune deficiency syndrome (AIDS).

Pneumonia may be classified in several ways. These include grouping by cause or etiology, by anatomic distribution of the inflammation process, or by predisposing factors involved in its development. The most important of these categories is the cause classification because it serves as a guide to treatment. The anatomic classification is used to describe the extent of organ and/or system involvement. For example, if the entire lobe of a lung is involved, the condition is called *lobar pneumonia.* If the inflammatory process involves parts of one lobe or more lobes immediately next to the bronchi, the condition is called *bronchopneumonia.* Classification of pneumonia according to predisposing factors also is commonly done. Categories include *postoperative pneumonia,* which develops in the postsurgical patient who is unable to breathe or cough deeply because of pain and tends to retain secretions, and *aspiration pneumonia,* which results when a foreign body or irritating substance such as food is sucked (aspirated) into the lung. *Obstructive pneumonia* develops in the lung distal to an area where a bronchus is narrowed or obstructed. Blockage of a bronchus by a tumor or foreign body leads to poor aeration and to retention of bronchial secretions in the obstructed portion of the lung.

Empyema, an accumulation of pus in the pleural sacs, is usually a secondary infection. The condition can occur in either an acute or chronic form. In the case of the former, bacteria such as *S. pneumoniae, M. tuberculosis, S. aureus,* hemolytic streptococci, and coliforms have been implicated. Chronic empyema is associated with various species of

Table 61–1

Selected Microbial Diseases of the Lower Respiratory Tract

Bacterial

Disease	Etiologic Agent
Legionnaires' Disease	*Legionella pneumophila*
Mycoplasma pneumonia	*Mycoplasma pneumoniae*
Nocardiosis	*Nocardia asteroides*
Ornithosis	*Chlamydia psittaci*
Q-fever	*Coxiella burnetii*
Pneumonia	*Streptococcus pneumoniae, Chlamydia pneumoniae, Staphylococcus aureus, Klebsiella pneumoniae*
Tuberculosis	*Mycobacterium tuberculosis*
Whooping cough	*Bordetella pertussis*

Mycotic (Fungal)

Blastomycosis	*Blastomyces dermatitidis*
Coccidioidomycosis	*Coccidioides immitis*
Cryptococcosis	*Cryptococcus neoformans*
Histoplasmosis	*Histoplasma capsulatum*

Protozoan

Pneumocystic pneumonia	*Pneumocystis carinii*

Viral

Acute respiratory disease	Adenoviruses
Influenza	Influenza viruses
Respiratory syncytial virus infection	Respiratory syncytial virus

fungi and *M. tuberculosis*. Mention should be made again of the fact that pathogenic microorganisms are important not only because of the infections they are capable of producing, but also because of the complications that may follow.

Transmission of causative agents can occur as a result of direct contact with infected persons or with droplet secretions from the nose or throat of infected persons. Indirect transmission is accomplished through contact with articles recently contaminated by infected individuals.

Humans and many animals harbor *Pneumocystis carinii* (Figure 61–1). The parasite does not affect persons with normally functioning immune systems, but may cause serious pulmonary infections in susceptible individuals. Those at risk include adults whose immune defenses have been impaired by disease, such as by AIDS or with the use of immunsuppressive drugs, and premature infants in whom immune defenses are poorly developed.

Compared to bacteria and viruses, fungi are much less frequent causes of respiratory diseases. Several mycotic (fungal) infections are seen in immunodeficient and debilitated patients. (See Table 61–1 for examples of this group of pathogens.) This exercise is concerned with representative bacteria and fungi that infect the respiratory tract, as well as those that use the tract as a portal of entry. The typical appearance of several organisms on various media (Figures 61–2 and 61–3), together with selected properties of these organisms, will be of significance as a part of this exercise. In addition, the optochin (ethylhydrocupreine) test, which is used in the identification of *Streptococcus pneumoniae,* will be demonstrated. Refer to Figures 61–4 through 61–6 for selected typical reactions of microorganisms on various media. Consideration is also given to a number of fungal respiratory pathogens.

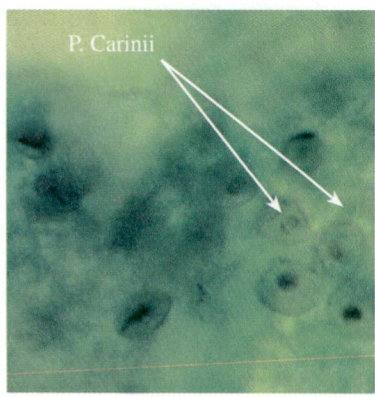

Figure 61–1

Pneumocytis carinii in a bronchial washing specimen.
(Courtesy of Dr. Lesley Alpert, Pathology Department, The Sir Mortimer B. David Jewish General Hospital.)

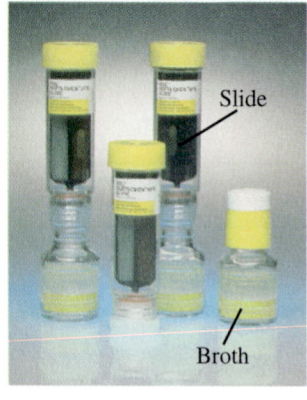

Figure 61–2

The BBL Septi-Chek AFB (acid fast bacteria). This device is a biphasic system (broth and agar) for the isolation of mycobacteria. The system contains Middlebrook 7H9 broth, and a unique slide with modified egg agar on one side and chocolate agar on the other.
(Photos courtesy of Becton Dickinson Microbiology Systems.)

Figure 61–3

The colony of *Coccidioides immitis*.
(Courtesy of the Mycology Section, Laboratory Division, Centers for Disease Control and Prevention, U.S. Public Health Service.)

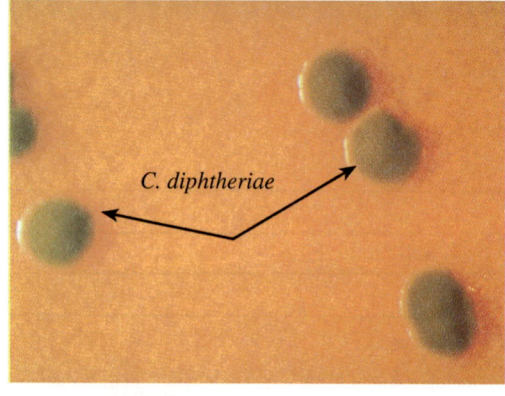

Figure 61–4

The typical black colonies of *Corynebacterium diphtheriae* on a tellurite-containing medium.
(Courtesy of Centers for Disease Control and Prevention.)

Figure 61–5

Beta hemolysis as shown by a blood agar stab.

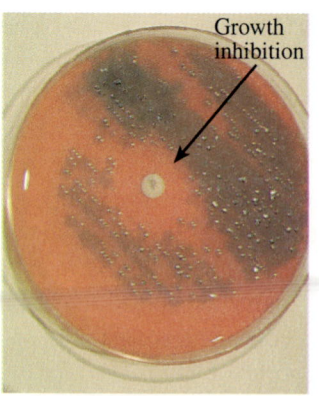

Figure 61–6

A positive optochin (ethylhydrocupriene) reaction with *Streptococcus pneumoniae*. Note the absence of growth around the disk.

A. General Characteristics of Selected Potential Bacterial Pathogens

Materials

The following items should be provided per 4 students:

❏ 1. One trypticase soy broth culture of each of the following microorganisms:
  ❏ a. *Corynebacterium xerosis*   ❏ e. *Serratia marcescens*
  ❏ b. *Escherichia coli*   ❏ f. *Staphylococcus aureus*
  ❏ c. *Pseudomonas aeruginosa*   ❏ g. *Streptococcus pneumoniae*
  ❏ d. *Pseudomonas fluorescens*

❏ 2. Four each of the following media:
  ❏ a. Blood agar plates   ❏ d. Pseudomonas P agar plates
  ❏ b. Trypticase soy agar plates   ❏ e. Mannitol-salt agar plates
  ❏ c. Pseudomonas F agar plates   ❏ f. Mueller-Hinton tellurite medium

❏ 3. Eight phenol red agar deeps, together with (Bacto) carbohydrate differentiation disks of each of the following sugars: dextrose, lactose, and mannitol (disks should be dispensed in sterile test tubes)

❏ 4. Eight sterile Petri plates

❏ 5. One 50-mL beaker containing 20 mL of ethyl alcohol, and pairs of sterile forceps

❏ 6. One Gram stain reagent set

The following item should be provided for general class use:

❏ 1. A boiling water bath

Procedure

Additional Techniques Required for This Portion of the Exercise:

❏ 1. Pour Plate Technique, Procedure Diagram 9, Exercise 5
❏ 2. The Gram Stain, Procedure Diagram 17, Exercise 14

This procedure is to be performed by students in groups of 4. The instructor will assign two specific organisms to each student so that all microorganisms provided will be used in inoculation. Be certain to examine the plates inoculated by other members of the class, so that the significance of the different media and the growth characteristics of all microorganisms used in this exercise can be noted. Refer to Figures 61–4 to 61–7 for selected typical reactions of organisms on various media.

❏ 1. Prepare and examine Gram stains of each culture assigned. Enter your results, as well as those of other students, in Table 61–2 of the Results and Observations section.

❏ 2. Using the organisms assigned to you, take 1 each of the 6 plate media and streak one-half with each species. Incubate the inoculated media for 48 hours or until the next laboratory period.

❏ 3. After incubation, record your findings in the Results and Observations section. Compare the colonial morphology, type of hemolysis, and other properties of the organisms in the exercise.

❏ 4. Prepare pour plates using heavy inocula of the cultures assigned and the phenol red agar deeps (refer to Procedure Diagram 9). After the plates have solidified, place 1 of each carbohydrate differentiation disk onto the surface of the agar by first inserting the forceps into the beaker containing ethyl alcohol and then holding it in the flame of a Bunsen burner. Repeat this step twice. Next, remove 1 of the carbohydrate disks and place it on the agar. Press on the disk lightly to ensure its contact with the medium. Repeat this procedure, flame-sterilizing the forceps each time. The disks should be placed an equal distance from one another.

❏ 5. Again, incubate the prepared plates at 37°C and examine them at 4 and 18 hours. Record your findings in the Results and Observations section. use the following designations to indicate the reactions obtained: A = acid (yellow zone around the disk): Alk = alkaline (no discoloration zone around the disk); and G = gas (splitting of the medium around the disk). Refer to Figure 61–7.

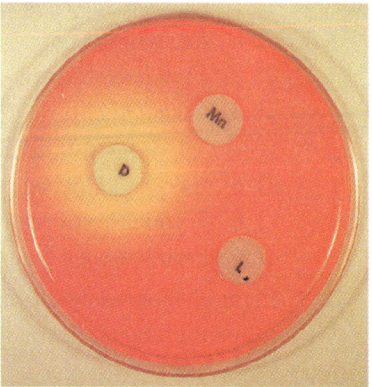

Figure 61–7
Carbohydrate fermentation reactions as demonstrated by fermentation disks on a seeded phenol red pour plate. The area around the dextrose disk (D) shows acid production, while the reactions surrounding the lactose (L) and mannitol (M) disks are negative.

B. BBL Septi-Chek AFB (Mycobacteria Culture/Subculture) System

Materials

The following items should be provided for class demonstration:

❏ 1. One 1-mL broth sample of each of the following bacterial species in 1-mL disposable hypodermic syringes equipped with appropriate-sized needles:
   ❏ a. *Mycobacterium phlei*
   ❏ b. *M. smegmatis*
   ❏ c. *Staphylococcus aureus*
❏ 2. Three Septi-Chek AFB systems
❏ 3. One marking pen or pencil
❏ 4. One container with disinfectant for contaminated material disposal
❏ 5. Acid-fast reagents

Procedure

Additional Technique Required for This Portion of the Exercise:

❏ The Acid-Fast Stain, Procedure Diagram 18, Exercise 15

❏ 1. Examine the Septi-Chek AFB System provided as a demonstration. Note the following parts, and their combination: culture bottle, slide chamber, and the media panels of the slide. (Refer to Figure 61–2.) Your instructor will demonstrate how the different parts are to be joined.
❏ 2. Next, obtain a syringe containing 1 mL of 1 broth culture for inoculation.

❏ 3. Label the culture bottle with the name of the culture, the date, and your initials.

❏ 4. Perform the 6 steps of Procedure Diagram 42.

❏ 5. Repeat steps 2 and 3 with the other cultures provided.

❏ 6. Incubate the system at 37°C for 48 to 72 hours, or as directed.

❏ 7. After incubation, examine the broth culture bottle and the presence of growth and/or colonies.

❏ 8. Note the presence of colonies on the green Middlebrook 7H11 slide medium and the modified egg medium and chocolate agar. Estimate the number of colonies on each medium panel. Record your findings in the Results and Observations section.

❏ 9. Prepare and examine acid-fast stains of representative colonies from each medium. Record your findings in Table 61–3, and answer the questions in the Results and Observations and Laboratory Review sections.

Procedure Diagram 42
Septi-Chek AFB Technique

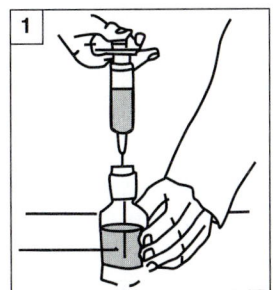

1. Inject the sample into the culture bottle.

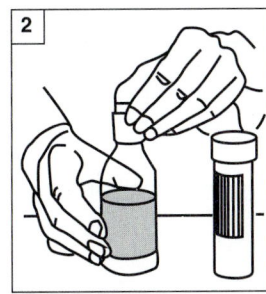

2. Remove the covers of the culture bottle and slide chamber.

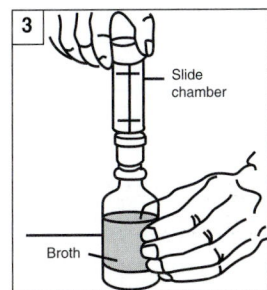

3. Join the threaded end of the slide chamber to the culture bottle.

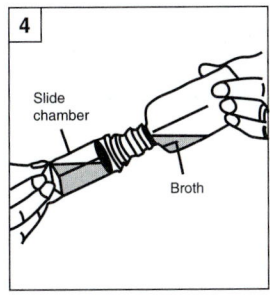

4. Tilt and turn the assembled system so that the broth fills the slide chamber.

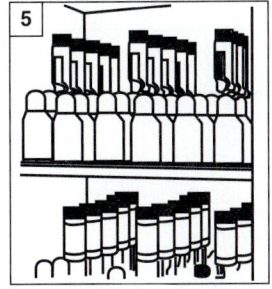

5. Incubate at 37°C or as directed.

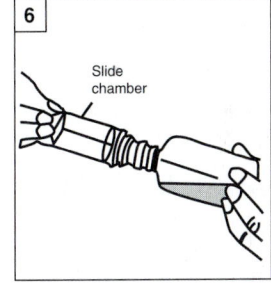

6. Unscrew the top of the slide chamber to obtain specimens for the smears

Results and Observations

General Characteristics of Selected Potential Bacterial Pathogens

Complete Table 61–2. Use the following symbols to express the characteristics of the species examined: gram-positive = g+; gram-negative = g−; coccus = c; rod = r; presence of growth = "+"; and absence of growth = "−."

Table 61–2

Characteristics of Selected Potential Pathogens

Microorganisms (Genus/Species)	Brief Description of Colonial Morphology on Trypticase Soy Agar	Gram Reaction	Microscopic Morphology	Type of Hemolysis	Presence of Growth			
					Pseudomonas F	Pseudomonas P	Mannitol Salt	Mueller-Hinton

BBL Septi-Chek AFB System

1. Did growth occur in all broth culture bottles? _____ If not, which culture(s) did not grow? _____

2. If growth did not occur in any of the broths, offer an explanation for the finding. _____

   _____

3. Complete the following table by entering your findings obtained with the Septi-Chek AFB systems.

Table 61–3

BBL Septi-Chek AFB Results

Medium	Colonies		Morphology
	Present	Absent	
Middlebrook 7H11			
Modified egg			
Chocolate agar			

C. Demonstration of Selected Pathogenic Fungi

Materials

❑ 1. Sealed Sabouraud dextrose agar or Georg fungus medium plates of the following fungi:
  ❑ a. *Aspergillus fumigatus*
  ❑ b. *Coccidioides immitis*
  ❑ c. *Histoplasma capsulatum*
❑ 2. Prepared slides for the study of microscopic features of the following:
  ❑ a. *A. fumigatus*
  ❑ b. *C. immitis*
  ❑ c. *H. capsulatum*
❑ 3. Lens paper

Procedure

This procedure is to be performed by students in groups of 4.

❑ 1. Examine the sealed cultures. Note the pigmentation, general texture, and other distinctive properties. Compare the surface and undersurface (backside) of each mycelium. (Refer to Exercise 11 and Figure 61–3.) Complete Table 61–4 of the Results and Observations section.

❑ 2. Examine and study the demonstration slides provided. Locate and identify the specific structures shown in Figures 61–8 through 61–10 in the Results and Observations section.

Results and Observations

1. Microscopic features of fungi.

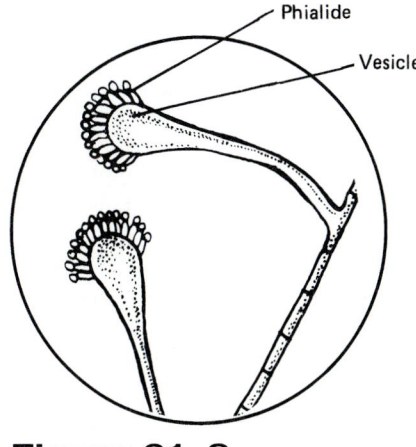

Figure 61–8
Aspergillus fumigatus.

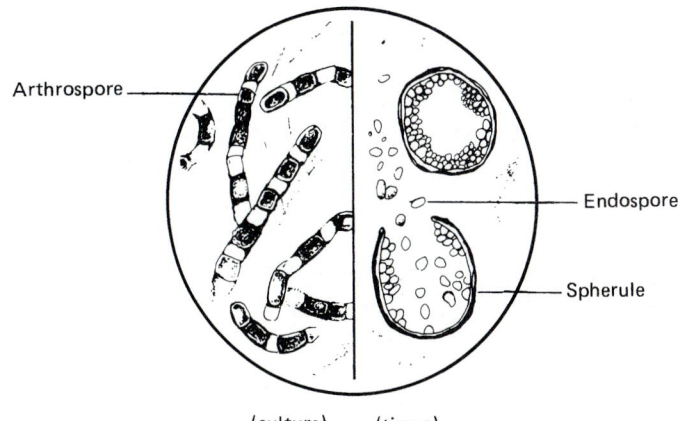

Figure 61–9
Coccidioides immitis.

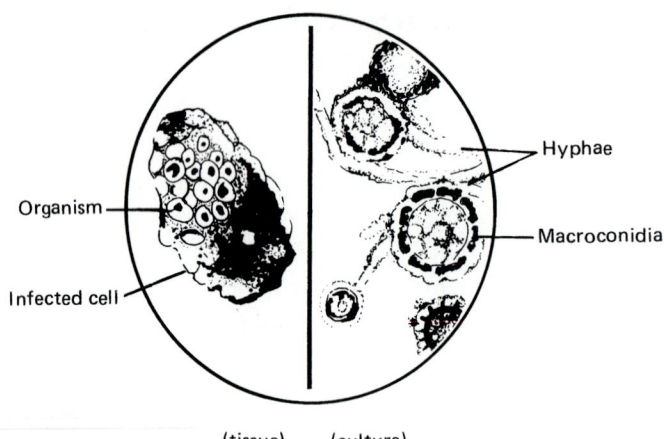

Figure 61–10
Histoplasma capsulatum.   (tissue)   (culture)

2. Mycelial properties.

Table 61–4

Selected Fungal Pathogens

Fungus	Description of Mycelial Surface	Description of Mycelial Undersurface

D. The Optochin Test (A Demonstration)

Materials

The following items are to be provided for the demonstration:

❏ 1. One divided blood agar plate containing *Streptococcus pneumoniae* and an unknown *Streptococcus* species
❏ 2. One blood agar plate
❏ 3. Optochin disks in a dispenser (either 7 or 10 mm in diameter)
❏ 4. One marking pen or similar device
❏ 5. One candle jar or CO_2 incubator

Procedure

Additional Technique Required for This Portion of the Exercise:

❏ Streak Plate Technique, Procedure Diagram 10, Exercise 5

This procedure is to be performed by the instructor, or as directed.

❏ 1. With the marking pen, label one-half of each plate for the species or streptococci to be used.

❏ 2. Remove 3 or 4 colonies of *S. pneumoniae* from the blood agar plate provided and streak one-half of the new plate provided.

❏ 3. Repeat step 2 with the unknown *Streptococcus* species provided, using the second half of the blood agar plate.

❏ 4. Place 1 optochin disk into the center of each streaked area.

❏ 5. Incubate the prepared plate in a candle jar or CO_2 incubator for 48 hours.

❏ 6. Examine the plate and determine which of the cultures was inhibited by the optochin disk. (Refer to Figure 61–6.)

❏ 7. Answer the questions in the Results and Observations section.

Results and Observations

1. Which of the cultures showed a sensitivity to the optochin disk? _____

2. Why was the candle jar or CO_2 incubator necessary in this test procedure? _____
_____

Laboratory Review 61   Pathogens of the Respiratory Tract

1. List at least 2 fungal (mycotic) infections of the respiratory system.

    a. _____   b. _____

2. List 2 viral infections involving the respiratory system.

    a. _____   b. _____

3. List 3 diseases of the respiratory tract associated with streptococci.

    a. _____   c. _____

    b. _____

4. What antibiotics are in common use for the treatment of mycotic respiratory infections? (Refer to your text.) ____
_____

5. Differentiate between superficial and deep-seated mycotic infections. (Refer to your text.) _____
_____
_____

Key Terms

abscess (AB-ses): a localized collection of pus

empyema (em-pī-Ē-ma): pus in a body cavity

tonsillitis (ton-sil-Ī-tis): inflammation of a tonsil

Photograph Quiz 25

Identify the fungus shown. _____

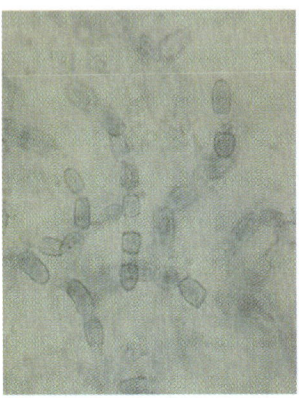

Figure 61–11
Microscopic view of a fungal specimen.

After completing this exercise, you should be able to:

1. List several common microbial infections of the gastrointestinal tract.

2. Perform and interpret standard procedures used in the identification of selected gastrointestinal pathogens.

3. Recognize the relative values of different tests used to distinguish among various gastrointestinal bacterial pathogens.

4. Identify pathogenic protozoa in permanent smears.

The gastrointestinal tracts of humans and other animals generally harbor numerous saprophytic and pathogenic microorganisms. Their distribution and number vary with different parts of the anatomy. Several abdominal structures, such as the liver, biliary system, and peritoneum, normally do not contain microorganisms and are known to exhibit a high resistance to invasion by microbial agents.

A short time after birth, microorganisms from the environment invade and populate the mouth and, in turn, enter the intestinal region to make up a portion of the resident microbial flora. The intestinal flora become varied and abundant as the individual ages, but eventually specific microbial species become established in the intestinal region. The lower portion of the small intestine and the large intestine contain the greatest number and kinds of microorganisms. Representatives of such physiologic flora include *Escherichia coli* (often referred to as *colon bacilli*), aerobic and anaerobic cocci and bacilli, spore formers, and fungi.

The small intestine is a highly efficient organ for nutrient absorption. However, disturbances in nutrient absorption, fluid secretion, and resorption may be caused by microbial and other agents resulting in malabsorption or diarrhea. Malabsorption results from heavy colonization of regions of the small intestine not usually colonized. Such colonization may produce disturbances in vitamin and fat metabolism resulting in reduced absorption. Diarrhea may result from the ingestion of preformed microbial toxins or from the colonization of the intestine by bacteria or viruses. Several examples of such situations will be described.

Antibiotic therapy can significantly alter the normal ecology of the gastrointestinal system of many individuals. This development can be a desirable one, especially in certain cases of intestinal surgery and in situations designed to eliminate enteropathogenic strains of *E. coli* and enterotoxic staphylococci. Unfortunately, antibiotic treatment may encourage the growth of fungi capable of causing disease states themselves, and it may favor development of antibiotic-resistant microorganisms or toxin producers such as *Clostridium difficile*. Several bacterial infections of the intestinal tract are known. A few of the causative agents will be considered here.

Microorganisms entering the digestive system along with food or water may be destroyed by the gastric contents of the stomach. However, if they are not destroyed, some of them can produce severe inflammations of the intestinal tract. Shigellosis, a pyogenic inflammation of the large intestine, is one example. Four species of the genus *Shigella* are known to cause bacterial dysentery in humans: *S. dysenteriae, S. flexneri, S. boydii,* and *S. sonnei.* All these agents possess an extremely irritating endotoxin believed to be the causative element of the disease. *S. shiga,* in addition to possessing endotoxin, produces a soluble exotoxin capable of attacking not only the intestine but the nervous system as well. Diagnosis of the disease is dependent upon (1) isolating the organism, (2) performing differential biochemical tests, and (3) performing agglutination tests with known antisera.

Salmonella infections include typhoid fever, septicemia, and acute gastroenteritis. The best-known of these, typhoid fever, is acquired by infestation of the causative agent, *Salmonella typhi.* The organisms reach the small intestine, where they soon pass through the intestinal walls and gain entrance to various portions of the lymphatic system. Once within the lymphatic system, the organisms multiply and then invade the bloodstream via the thoracic duct to produce a septicemia. Diagnosis of the disease is similar to the procedural approach indicated above for shigellosis.

Cholera is another well-known bacterial disease. It is an acute gastrointestinal tract infection, the etiologic agent being *Vibrio cholerae.* The disease seriously disables the individual by causing profound dehydration and electrolyte imbalance. Diagnosis of the disease is based upon (1) microscopic examination of specimens, (2) isolation of *V. cholerae* on alkaline media, and (3) administration of serological tests.

Helicobacter pylori, a curved or spiral-shaped gram-negative rod has been recognized recently as the major cause of gastritis and duodenal ulcer in humans. This bacterium is motile, microaerophilic, and rapidly hydrolyzes urea in laboratory tests by means of an unusual urease enzyme. There is speculation that *H. pylori*'s urease production may serve as significant virulence factor by neutralizing the acidity in the stomach to allow the pathogen to reproduce more easily in the tissues of the stomach. Isolated *H. pylori* are oxidase and catalase positive. The urea-breath test is an example of a clinical procedure used to identify infected patients. Several serologic tests including the ELISA procedure have been introduced to aid in diagnosis. Newer diagnostic tests are under development.

An inflammatory response to foreign material in the abdominal cavity is peritonitis. The agents responsible for this condition can be either nonbacterial or bacterial. Examples of nonbacterial agents include blood, bile, and pancreatic and gastroduodenal secretions. Examples of bacterial agents include various mixtures of bacterial species such as clostridia, streptococci, and gram-negative bacilli. Peritonitis can also be caused by pneumococci in children with nephrosis. Establishment of the disease state is favored by a continuous source of microorganisms from other loci of infection within the body and a substrate that these organisms can utilize. This inflammatory response can also readily constitute a complication in the majority of intra-abdominal lesions.

Cholecystitis, or gall bladder inflammation, may be either an acute or chronic condition. Although the exact cause of this inflammatory response is not known, three factors have been implicated: (1) chemical irritation, (2) pancreatic reflux, and (3) bacterial infection. The microorganisms most often encountered in cholecystitis include *E. coli, Yersina enterolitica,* and aerobic and anaerobic streptococci.

Obstruction of a bile duct, resulting in an inflammation of this structure, cholangitis, has been known to accompany several types of infections. Several bacterial species have been recovered from cases of this disease, including gram-negative organisms such as *E. coli, Proteus* spp., and *Pseudomonas aeruginosa,* and gram-positive bacteria including *Clostridium perfringens* and *Streptococcus* spp.

Several medically important protozoa are associated with the gastrointestinal tract (Figure 62–1). These include *Balantidum coli* (balantidiasis), *Cryptosporidium* spp. (cryptosporidiasis), *Entamoeba histolytica* (amebic dysentery), *Giarda lamblia* (giardiasis), and *Toxoplasma gondii* (toxoplasmosis). The majority of these intestinal protozoa live in the large intestine, with the exception of *G. lamblia* and *Cryptosporidium* species, which are found in the small intestine. *Cryptosporidium* and *Toxoplasma* infections are found with a certain degree of frequency in AIDS victims. The gastrointestinal tract also harbors several nonhuman or lesser pathogens that may be confused with pathogenic forms. These include *Chilomastix mesnili* and *Dientamoeba fragilis.* In several parasitic species and free-living protozoa found in temporary bodies of water, the normal, motile feeding form, known as the *trophozoite,* often cannot withstand the effects of various chemicals, food deficiencies, temperature or pH changes, and other harsh factors in their environments. To overcome such conditions, many protozoa secrete a thick, resistant covering and develop into a resting stage called a *cyst* (Figure 62–2).

The identification of intestinal protozoa is dependent on the demonstration of their characteristic microscopic features which, in turn, depends on the stool specimen, correct specimen collection, adequate preservation and fixation of the specimen, and appropriate laboratory techniques.

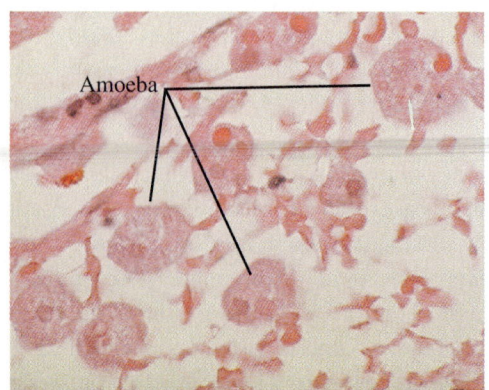

Figure 62–1

A stained preparation of the large intestine showing oval to round amoeba in the tissue.
(Courtesy of Dr. Lesley Alpert, Pathology Department, The Sir Mortimer B. David Jewish General Hospital.)

(a)   (b)

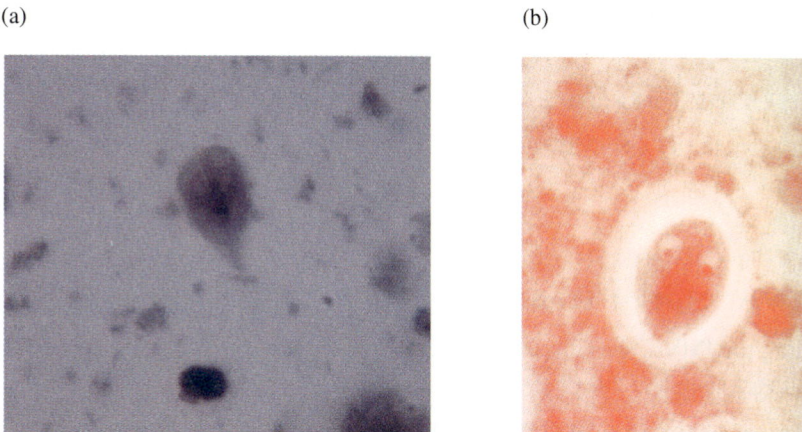

Figure 62–2

Stained preparations of *Giardia lamblia* in fecal smears. (a) A trophozoite. (b) A cyst.

The consistency of stool specimens may provide an indication of the protozoan stage present. When the moisture content of fecal material is decreased during normal passage through the intestinal tract, the trophozoite stages of the protozoa encyst to survive. Trophozoites of intestinal protozoa are usually found in liquid or soft specimens, while cysts are generally found in formed or semiformed stools.

The detection and correct identification of intestinal protozoan pathogens are frequently dependent on the examination of permanently stained smears of fecal specimens. Variations in shape and size, however, occur with protozoa, which may create difficulties in identification. In addition, many artifacts such as vegetable fibers, debris, and human cells may mimic protozoa. Detailed searches of several well-prepared smears should provide a sufficient number of recognizable forms.

Many of the causative agents of intestinal disorders have distinctive properties that can aid in their identification. This exercise will demonstrate certain of these characteristics.

A. Selected Characteristics

Procedures

The tests in this exercise are to be carried out by students in pairs. Photographs of typical reactions can be found in Exercise 22, 23, 25, and 26.

Materials

❑ 1. Trypticase soy broth 24-hour cultures of the following bacterial species (per 2 students):

    ❑ a. *Escherichia coli*   ❑ d. *Pseudomonas aeruginosa*

    ❑ b. *Enterobacter aerogenes*   ❑ e. *Staphylococcus aureus*

    ❑ c. *Proteus vulgaris*   ❑ f. *Enterococcus (Streptococcus) faecalis*

❑ 2. The following media (6 per 2 students, except for item (b), where 12 tubes of media are necessary):

    ❑ a. Tryptone broth   ❑ f. Urea broth

    ❑ b. Methyl red broth   ❑ g. Bacto-SIM media

    ❑ c. Koser's citrate medium   ❑ h. Triple sugar iron agar

    ❑ d. Simmon's citrate medium ❑ i. Endo agar

    ❑ e. Nitrate broth

❑ 3. The following reagents should either be contained in dispenser bottles or accompanied by eye droppers and provided after incubation:

    ❑ a. Kovac's reagent   ❑ d. Dimethyl-alpha-naphthalamine solution

    ❑ b. Methyl red indicator   ❑ e. Powdered zinc

    ❑ c. Alpha (α)-naphthol and potassium   ❑ f. Chloroform

       hydroxide-creatine solutions

continued

Test 1: The IMViC Set of Reactions

Each of the 4 tests forming the IMViC was discussed in Exercise 25. Refer to this exercise for further features of the indole, methyl red, Voges-Proskauer, and citrate tests.

❏ 1. Inoculate each organism into 1 tube of tryptone broth, 2 methyl red broth tubes, and 1 tube each of the Koser's citrate and Simmon's citrate media.

❏ 2. Incubate all tubes at 37°C for 24 hours, and then perform the following procedures.

❏ 3. **Indole test:**
   ❏ a. Add 10 drops of Kovac's reagent to each tube of tryptone broth.
   ❏ b. The development of a red color is a positive test.

❏ 4. **Methyl red test:**
   ❏ a. Add 5 drops of the indicator to 1 set of methyl red broth cultures.
   ❏ b. The development of a red color is a positive test. A yellow color is a negative test.
   ❏ c. Record questionable color reactions as ±.

❏ 5. **Voges-Proskauer test:**
   ❏ a. Add 10 drops of the alpha-naphthol solution to 1 set of methyl red broth cultures.
   ❏ b. Add 5 drops of the potassium hydroxide-creatine solution and shake well for 1 minute.
   ❏ c. A reddish color (which may develop slowly) is a positive test.

❏ 6. **Citrate test:**
   ❏ a. The presence of growth in Koser's citrate medium constitutes a positive reaction. If Simmon's citrate medium is used, the development of a blue color is a positive reaction.

❏ 7. Record your data in Table 62–2 of the Results and Observations section. (Refer to Exercise 25 figures for selected typical reactions.)

Test 2: Nitrogen Metabolism

Two different enzymatic reactions are included here: nitrate reduction and urease activity. Because both of these aspects of nitrogen metabolism have been discussed previously (in Exercise 23), their significance and interpretation will not be considered further here.

❏ 1. Inoculate 1 tube of nitrate broth and 1 tube of urea broth with each bacterial culture provided.

❏ 2. Incubate at 37° C for 24 hours. Refer to Exercise 23 for the remainder of the testing procedure.

❏ 3. Refer to Exercise 24 figures for an interpretation of urease activity.

❏ 4. Enter all findings in Table 62–3.

Test 3: Bacto-SIM Medium and Triple Sugar Iron Agar

Preparations such as SIM and TSIA are examples of differential media that can be used to determine several characteristics of pure cultures at the same time. In general, media of this type can contain (1) carbohydrate, (2) a combination of carbohydrates, or (3) several carbohydrates and substrates rich in sulfur. The last combination can be used to detect the carbohydrate-fermentative capabilities of an organism as well as its ability to produce hydrogen sulfide (H_2S). All these media contain suitable indicators for the detection of end products, such as acids and alkali, and H_2S.

Bacto-SIM medium can be used to determine hydrogen sulfide production (S), indole production (I), and motility (M). The presence of H_2S is indicated by a blackening along the line of inoculation in the medium. Quite often, the dark color diffuses when a motile organism is involved. The presence of indole is detected by the use of Kovac's reagent. Motility is recognized by the appearance of a diffuse growth or turbidity from the inoculation site.

Triple sugar iron agar medium is employed for the detection of dextrose (glucose), saccharose, and lactose fermentation and H_2S production. The color reactions representing the different possible reactions are presented in Table 62–1. A detailed explanation of the bases of these reactions can be found in Exercise 26.

Table 62–1

Triple Sugar Iron Agar Reactions

Appearance of	Positive Glucose Fermentation	Positive Saccharose or Lactose Fermentation	Negative Fermentation	H_2S Production
Slant	Red	Yellow	Red	Blackening of
Butt	Yellow	Yellow	Red	the medium

❑ 1. Inoculate 1 tube of SIM medium and 1 tube of TSIA with each organism provided. Perform the inoculation by first streaking the slant and then stabbing the medium in the butt region.

❑ 2. Incubate the media at 37°C for 24 hours.

❑ 3. Observe the appearance of SIM medium and TSIA for the various test reactions and record your findings in Table 62–4 of the Results and Observations section.

❑ 4. Carry out the indole test using the SIM medium by overlaying 10 drops of chloroform on the surface of the medium. After 2 minutes, add several drops of Kovac's reagent. The formation of a pink or red color is indicative of indole. (Refer to Exercise 26 figures.)

Test 4: Endo Agar

Endo agar is used to differentiate lactose-fermenting and lactose-nonfermenting microorganisms isolated from various types of specimens. The latter type of organisms form clear, colorless colonies, while the former are red and usually color the surrounding medium (Figure 62–3). The typical reactions of this medium are produced by the intermediate product, acetaldehyde, reacting with the indicator system. The latter is a combination of sodium sulfite and basic fuchsin.

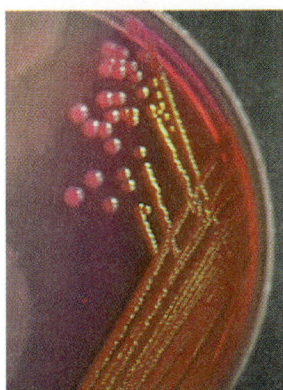

Figure 62–3
Lactose fermentation on an Endo agar plate.

❑ 1. Prepare streak plates of Endo agar with each of the cultures provided.

❑ 2. Incubate the media at 37°C for 24 hours.

❑ 3. Examine the preparations for the presence of growth and the distinguishing features of bacterial colonies.

❑ 4. Record your findings in Table 62–5 in the Results and Observations section.

Results and Observations

The IMViC Set of Reactions

1.   Complete the following table. Indicate a positive test by "+" and a negative test by "−."

Table 62–2

IMViC Results

Microorganism	Test			
	I	M	Vi	C
E. coli				
E. aerogenes				
P. vulgaris				
P. aeruginosa				
S. aureus				
E. faecalis				

Nitrogen Metabolism

1.   Complete the following table.

Table 62–3

Nitrogen Metabolism Results

Organism	Urease	Nitrate Reductase
E. coli		
E. aerogenes		
P. vulgaris		
P. aeruginosa		
S. aureus		
E. faecalis		

Bacto-SIM Medium and Triple Sugar Iron Agar

1.   Complete Table 62–4. Use a "+" to indicate a positive reaction and a "−" for a negative one in the SIM and TSIA tests. Use A for acid, G for gas, and NC for no reaction in the case of carbohydrate fermentations.

Table 62–4

SIM and TSIA Results

Microorganism	SIM			TSIA		
	H₂S S	Indole I	Motility M	Glucose Fermentation	Lactose Fermentation	H₂S
E. coli						
E. aerogenes						
P. vulgaris						
P. aeruginosa						
S. aureus						
S. faecalis						

Endo Agar

1. Complete the following table.

Table 62–5

Endo Agar Reactions

Microorganism	Presence of Growth	Description of Colonial Appearance
E. coli		
E. aerogenes		
P. vulgaris		
P. aeruginosa		
S. aureus		
E. faecalis		

B. Microscopic Examination

Materials

❏ 1. Prepared slides of the following protozoa:
  ❏ a. *Balantidium coli* fecal smears containing both trophozoites and cysts
  ❏ b. *Chilomastix mesnili* fecal smears containing both trophozoites and cysts
  ❏ c. *Dientamoeba fragilis* fecal smears
  ❏ d. *Entamoeba coli* fecal smears containing both trophozoites and cysts
  ❏ e. *E. histolytica* fecal smears containing both trophozoites and cysts
  ❏ f. *Giardia lamblia* fecal smears containing both trophozoites and cysts
  ❏ g. *Toxoplasma gondii* tissue smears

❏ 2. Unknown slides (2 per student)

❏ 3. Immersion oil

❏ 4. Lens paper

Procedure 1: Prepared Slides

This procedure is to be performed by students individually.

- ❑ 1. Examine the slides of protozoa under high-power and oil-immersion objectives. Sketches of several protozoa, showing their respective microscopic characteristics, are included in the Results and Observations section to help identify and compare specific organisms. If 1 slide does not demonstrate the protozoon sufficiently, go to another slide and continue searching for clear details.
- ❑ 2. Using Figures 62–1, 62–2, and 62–4 through 62–7, identify the various trophzoite and cyst stages on the respective slides and answer the questions that follow.
- ❑ 3. After you have seen all the protozoan preparations, clean the used slides thoroughly with xylene and return them as directed.

Procedure 2: Unknowns

This procedure is to be performed by students individually.

- ❑ 1. Each student will be given 2 unknown slides containing 1 of the pathogens studied in this exercise.
- ❑ 2. Identify the unknowns and enter the results on the Unknown Form on page 565.

Results and Observations

Microscopic examination of prepared slides.

1. Compare your observations with the following micrographs and diagrams:

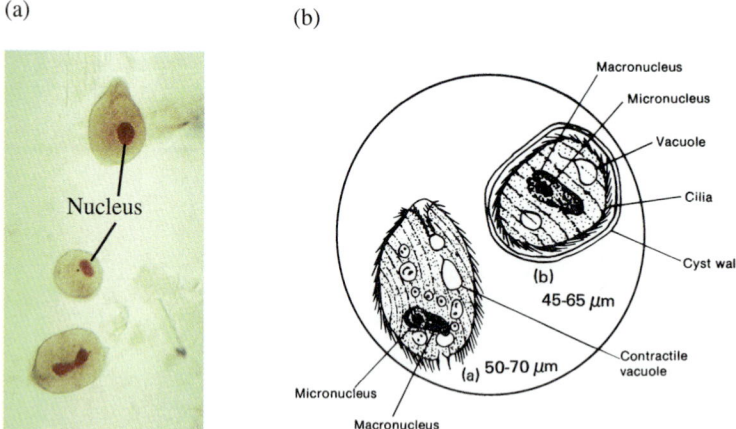

Figure 62–4

The ciliate *Balantidium coli.* This protozoon has both trophozoite and cyst stages. *B. coli* can cause enteritis in humans. (a) Trophozoites from a stool specimen. (b) Diagrammatic views of the trophozoite and cyst.

(a)

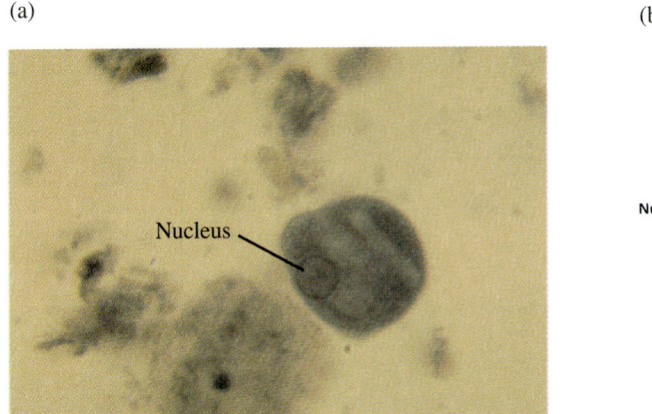

(b)

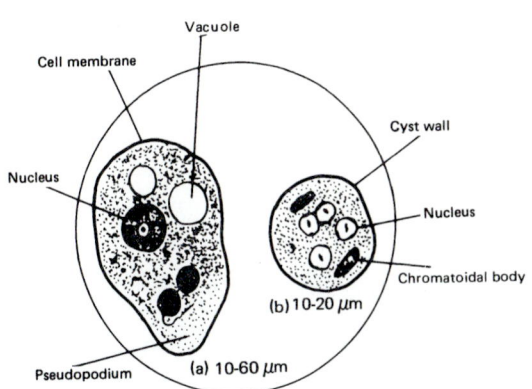

Figure 62–5
Entamoeba histolytica the cause of amebic dysentery. (a) A trophozoite in a stool specimen. (b) Diagrammatic views of the trophozoite and cyst.

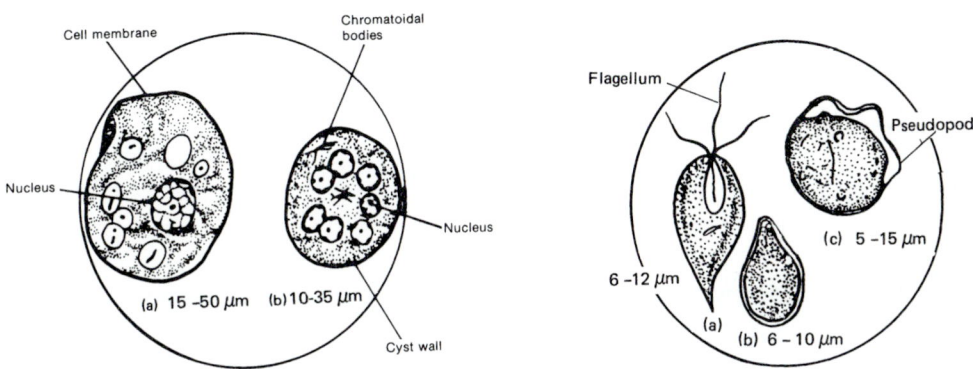

Figure 62–6
(a) *Entamoeba coli.* (b) *Chilomastix mesnili.* (c) *Dientamoeba fragilis.*

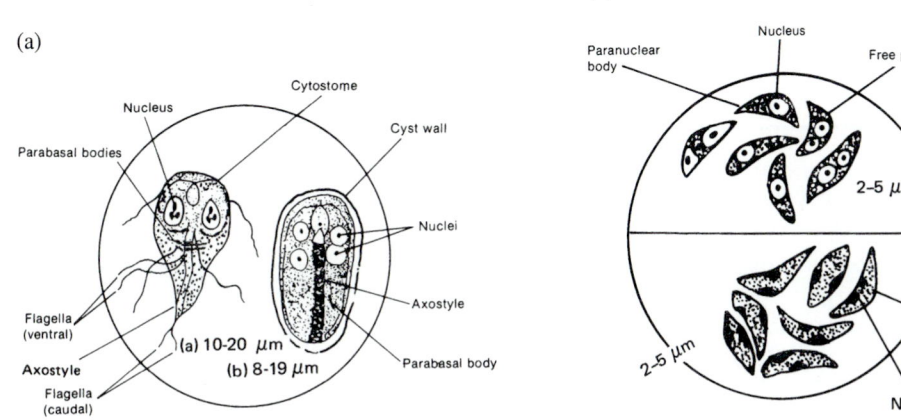

Figure 62–7
(a) *Giardia lamblia.* (b) *Toxoplasma gondii.*

2. What specific morphological features distinguish *Entamoeba histolytica* from *Entamoeba coli?*_____

_____

_____

3. Which intestinal protozoa do not have a cyst stage? _____

_____

Laboratory Review 62 — Pathogens of the Gastrointestinal Tract

1. Name 2 diseases of the digestive system caused by bacteria that are not true infections but poisonings.

   a. _____ b. _____

2. How might the tests described in this exercise be used in the classification of bacteria? _____

   _____

   _____

3. Do viruses cause digestive-system diseases? If so, list 2.

   a. _____ b. _____

4. Complete Table 62–6 for each of the protozoan pathogens listed.

Table 62–6

Pathogenic Protozoa

Protozoan Pathogen	Means of Locomotion	Host	Means of Transmission
E. coli			
E. histolytica			
G. intestinalis			
T. gondii			

5. List 2 distinguishing microscopic features of the protozoa given in Table 62–7.

Table 62–7

Microscopic Features of Protozoa

Microbial Pathogen	Distinguishing Microscopic Feature(s)
D. fragilis	
E. coli	
E. histolytica	
G. intestinalis	

Key Terms

endotoxin (en-dō-TOKS-in): a harmful toxic lipopolysaccharide contained within a gram-negative bacterial cell wall

enteropathic (en-ter-ō-PATH-ik): refers to intestinal disease production

exotoxin (eks-ō-TOKS-in): a secreted harmful protein

gastroenteritis (gas-trō-en-ter-Ī-tis): inflammation of the stomach and intestines

peritonitis (per-i-tō-NĪ-tis): inflammation of the peritoneum (abdominal lining)

septicemia (sep-ti-SĒ-mē-a): the presence of microorganisms in the blood

Photograph Quiz 26

Identify the protozoon shown. _____

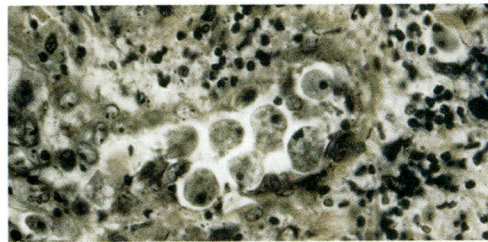

Figure 62–8

A photomicrograph of a laboratory specimen.

Form for Unknown 1

Student's Name _____	Score _____
Date _____	Laboratory Section _____
Unknown Number _____	Organism Identified _____

Form for Unknown 2

Student's Name _____	Score _____
Date _____	Laboratory Section _____
Unknown Number _____	Organism Identified _____

Pathogenic Microorganisms Found in Blood and Selected Microbial Pathogens of the Nervous System

After completing this exercise, you should be able to:

1. List several bacterial pathogens likely to be found in blood cultures.
2. Identify protozoan pathogens found in the blood.
3. Perform a broth blood culture technique.
4. Perform a Septi-Chek inoculation.
5. Recognize the importance of a blood specimen to clinical diagnosis.
6. Associate specific infectious diseases with specific arthropod vectors.
7. List and identify selected microbial pathogens associated with the nervous system.

Microorganisms, worms, and their products are associated with a wide range of diseases. Certain pathogens are able to overcome the local defenses of the host and enter the blood. A septicemia, the presence of living (viable) microorganisms in a patient's blood, almost always is a clear-cut indication of an active and possibly spreading infection in the tissues. The temporary presence of bacteria frequently occurs during the course of several diseases such as bacterial meningitis, generalized *Salmonella* infections, pneumococcal pneumonia, typhoid fever, and urinary tract infections. Infections of the gallbladder, osteomyelitis, peritonitis, and certain wound infections may be accompanied by a bacteremia. Many different bacterial species may be found in cases of pneumonia. The species most commonly isolated from patients include *Bacteroides* spp., *Escherichia coli,* group D streptococci, *Klebsiella* spp., *Enterobacter* spp., *Pseudomonas aeruginosa,* *Staphylococcus aureus,* and *Streptococcus pneumoniae. Bacteremia,* the presence of actively multiplying bacteria and their toxins in the bloodstream, presents a more serious situation. Early isolation is extremely important and may be the only reliable means available for making a diagnosis.

In certain diseases, the ability to find pathogens in the blood depends on the stage of the disease at which a culture is made. Bacteria can be cultured from the blood during the first few days of illness in diseases such as typhoid fever and zoonoses including anthrax, plague, and tularemia. With other disease states such as brucellosis, organisms can be isolated during the attack phase.

With all infectious bacterial processes, appropriate cultures must be obtained. The cornerstone for the laboratory diagnosis of bacteremia is the blood culture (Figure 63–1). The media for such purposes include combinations of broth and an agar slant (known as a *biphasic system*), or liquid media alone such as thioglycollate broth, brain heart infusion, and trypticase soy broth.

Figure 63–1

The ESP blood culture system offers a choice of three different blood culture bottle formats. The blood culture system continuously monitors and signals microbial growth based on the detection of gas production and consumption. The culture medium is specially formulated to ensure detection of a wide range of organisms. The ESP EX View bottle format (center) contains 80 mL of aerobic broth only and accommodates up to 10 mL of blood. The bottle also contains a paddle with a solid agar medium on each side. The paddle does not have to be removed from the bottle to look for growth on the agar.
(Courtesy of Difco Laboratories, Detroit, MI.)

In biphasic bottles, bacterial growth generally appears as various-sized colonies on the agar slant, with cloudiness in the broth (Figure 63–2). Broth preparations usually contain sodium polyanethol-sulfonate, a chemical that has anticomplementary and antiphagocytic activity and inactivates some antibiotics. Thus, this compound in blood culture media enhances the isolation of bacteria that otherwise might be inhibited by the serum in blood specimens. Most culture media for anaerobic bacteria isolation require ingredients not found in standard preparations. The best results for the isolation of anaerobes are provided by the use of prereduced, anaerobically sterilized culture media. These media are prepared under oxygen-free gas, usually nitrogen, reducing exposure to air or oxygen. Culture media exposed to air and oxygen during preparation may form toxic peroxides that will inhibit the growth of sensitive bacteria. Some containers are kept under vacuum with carbon dioxide to ensure an optimal growth environment. A blood specimen is injected directly through a stopper in the top of a blood culture container (Figure 63–2). The Septi-Chek system, which will be demonstrated in this exercise, combines the techniques needed for the isolation of pathogens from blood specimens and their subsequent identification. Figure 63–3 shows a complete assembly of this most modern system.

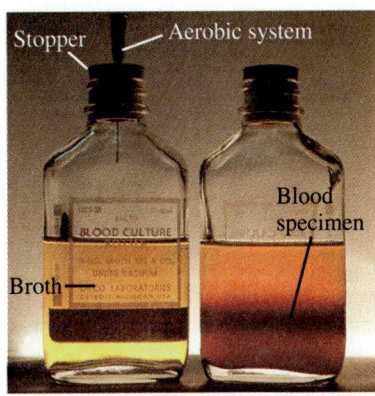

Figure 63–2
Blood culture bottles. a. Uninoculated system. b. Blood culture showing growth and some hemolysis.

Figure 63–3
The appearance of the slide chamber components of the Septi-Chek system. The media shown are chocolate (left), MacConkey, and malt agars (center and right).
(Courtesy of Becton Dickinson Microbiology Systems.)

Many different organisms can infect the nervous system, including bacteria, viruses, and protozoa (Table 63–1). An infection that mainly involves the coverings or meninges surrounding the brain and spinal cord is called a *meningitis*, while an infection of brain tissue is known as an *encephalitis*. When both the brain and the meninges are involved, the term *meningoencephalitis* is often used.

Several medically important protozoa are associated with the blood, nervous system, and related tissues. These include *Babesia* (Texas cattle fever), *Leishmania donovani* (kala azar), *Plasmodium falciparum, P. malariae, P. ovale,* and *P. vivax* (malaria), *Trypanosoma brucei gambiense* and *T. rhodesiense* (African sleeping sickness), and *Trypanosoma cruzi* (Chagas' disease). While most human protozoan pathogens are found largely in tropical areas, relocation of individuals from such areas has caused changes in the distribution and occurrence of these blood parasites. The identification of protozoa is dependent on morphological features, which, in turn, are dependent on correct specimen collection, adequate fixation, and appropriate laboratory techniques. Diagnosis is based on the demonstration of the causative protozoon

Table 63–1

Examples of Microbial Diseases of the Nervous System

Bacterial

Brain and Meninges	Causative Agent(s)
Bacterial meningitis	*Haemophilus influenzae, Neisseria meningitidis*
Listeriosis	*Listeria monocytogenes*
Brain abscesses	Various anaerobes

Nerves, Central Nervous System	Causative Agent
Leprosy (Hansen's Disease)	*Mycobacterium leprae*
Tetanus	*Clistridium tetani*
Botulism	*Clostridium botulinum*

Protozoan

Central Nervous System	Causative Agent
African sleeping sickness	*Trypanosoma gambiense, T. rhodesiense*
Chagas' disease	*Trypanosoma cruzi*

Viral

Central Nervous System	Causative Agent
Rabies	Rabies virus
Encephalitis	Several encephalitis viruses (Arbor viruses, human immunodeficiency virus, HIV, Enteroviruses)
Herpes meningoencephalitis	Herpesvirus

in blood or related tissue. Temporary wet mounts and/or stained blood smears are most often prepared. Blood smears are either thick or thin. Thick film allows the examination of a larger amount of blood and is used for screening. Thin film permits exact identification (speciation) of the protozoan pathogen. Other types of specimens used for diagnosis include cerebrospinal fluid, and biopsy material for appropriate target organs such as the liver and spleen. This exercise will present the distinguishing properties of several protozoa associated with blood and the nervous system.

The phylum Arthropoda is the largest in the animal kingdom. Many arthropods are medically important because of their ability to transmit causative agents of disease and to produce trauma and allergic reactions through the introduction of their toxins or venoms into the tissues of humans. Arthropods transmit disease organisms such as viruses, bacteria (which include the rickettsia), protozoa, and helminths either by mechanical or biological means. In the case of mechanical vectors, the disease agent is carried on body parts such as feet, body hairs, or the proboscis and then deposited on the surfaces of food, water, or other substances. Typhoid fever and infectious hepatitis are spread in this manner. In the case of biological vectors, the disease agent undergoes a period of incubation or development in the host. The transmission of African sleeping sickness, filariasis, Lyme disease, yellow fever, and malaria incorporates biological vectors. In general, the most important arthropod-associated diseases utilize the latter means of transmission.

Certain arthropods produce injury by causing blood loss, injecting poison, or burrowing into the tissues of the host. This exercise will present several medically important arthropods.

A. Microscopic Examination

Materials

- ❏ 1. Prepared slides of bacterial pathogens:
    - ❏ a. Smears of *Borrelia recurrentis*
    - ❏ b. Gram-stained smears of *Francisella tularensis*
    - ❏ c. Gram-stained smears of *Rickettsia rickettsii*
    - ❏ d. Gram-stained smears of *Yersinia pestis*
- ❏ 2. Prepared slides of protozoan pathogens:
    - ❏ a. Normal stained blood smears
    - ❏ b. Blood smears with *Leishmania donovani*
    - ❏ c. Blood smears with *Plasmodium falciparum, P. malariae,* and *P. vivax*
    - ❏ d. Blood smears with *Trypanosoma brucei* variety *gambiense* and *T. cruzi*
- ❏ 3. Unknown slides (2 per student)
- ❏ 4. Immersion oil
- ❏ 5. Lens paper

Procedure 1: Prepared Slides

This procedure is to be performed by students individually.

- ❏ 1. Examine and compare the prepared slides of bacterial pathogens.
- ❏ 2. Sketch representative fields in the spaces provided in the Results and Observations section.
- ❏ 3. Examine the normal blood smear and the slides of protozoan pathogens under high-power and oil-immersion objectives. Sketches of several showing their respective characteristics are included in the Results and Observations section to help you to identify and compare specific organisms.
- ❏ 4. Using the normal blood smear and Figures 63–4 to 63–7 as guides, identify the various pathogenic protozoa on the respective slides and answer the questions that follow.
- ❏ 5. Pay particular attention to sizes, cellular locations, and differences in the organelles (cellular structures) of the various disease agents. *Look for distinguishing characteristics.* Although most of these cultures are indicated in the sketches, it may be difficult to locate them on a slide. If 1 slide does not demonstrate the protozoon sufficiently, go to another slide and continue searching for clear details.
- ❏ 6. After you have viewed all the pathogens, clean the used slides thoroughly with xylene and return them.

Procedure 2: Unknowns

This procedure is to be performed by students individually.

- ❏ 1. Each student will be given 2 unknown slides containing 1 of the pathogens studied in this exercise.
- ❏ 2. Identify the unknowns and enter the results on the Unknown Form on page 581.

Results and Observations

1. Sketch representative fields of bacterial pathogens.

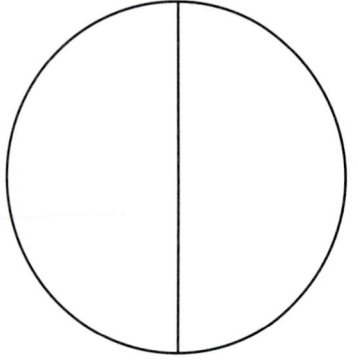

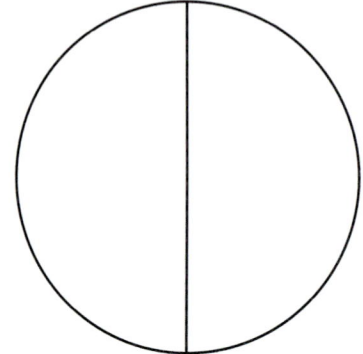

Borrelia recurrentis   *Francisella tularensis*   *Rickettsia rickettsii*   *Yersinia pestis*

2. Compare the following illustrations with your observations of the slides with the following pathogens.
   a. *Leishmania donovani*

(a)   (b)

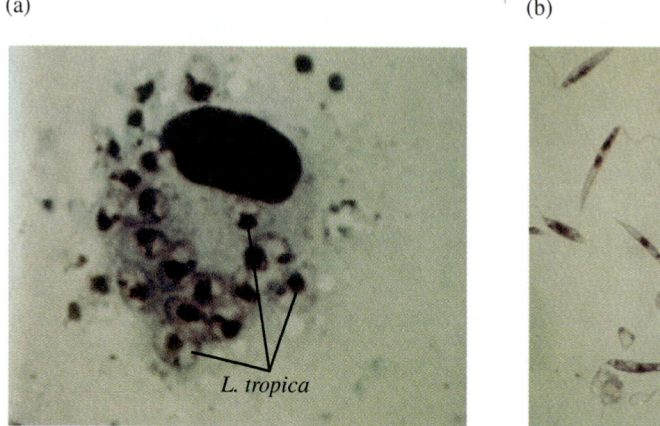

L. tropica

Figure 63–4
(a) The appearance of *Leishmania tropica* isolated from a patient. (b) *L. Tropica* in culture showing the flagellated state known as promastigotes.
(Courtesy of Dr. Muna Al-Taqu.)

b. *Plasmodium species*

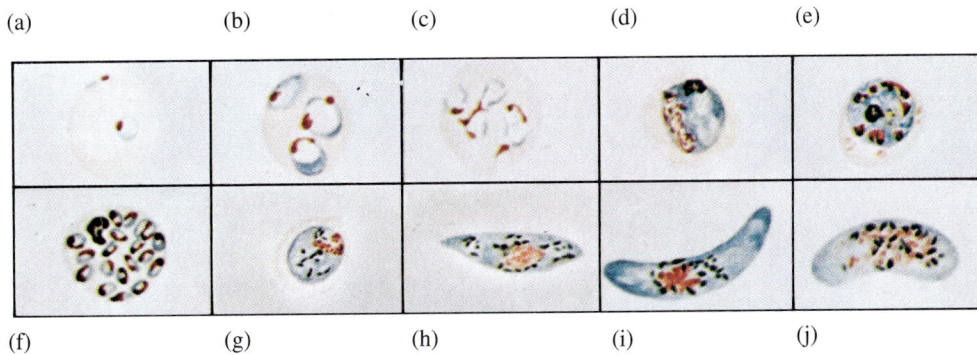

Figure 63–5a

(The descriptions given for the following figures 70–5a, 70–5b and 70–5c are adapted from the 1960 U.S. Department of Health, Education and Welfare publication *Manual for the Microscopical Diagnosis of Malaria in Man.*) *Plasmodium falciparum,* (a) A single erythrocyte showing a double infection with young trophozoites. The parasite close to the center of the red cell is a "signet ring" form, while the other organism located at the periphery is referred to as a "marginal form." (b) One red blood cell with 3 somewhat more developed trophozoites. (c) The parasites shown are called *aestivo-automnal tenue* forms. (d) The parasite is undergoing initial chomatin (red area) division. (e), (f). Mature *schizonts* with *merozoites.* Note the number of small merozoites (g) and (h). These stages are representative of the successive events that take place in gametocyte (sex cell) development. Such forms generally are not found in the peripheral circulation. (i) A mature macrogametocyte (female sex cell). (j) A mature microgametocyte (male sex cell).

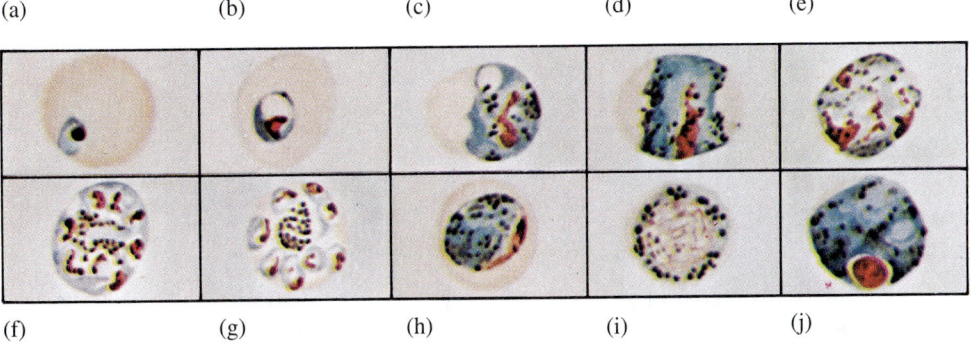

Figure 63–5b

Plasmodium malariae. (a) An erythrocyte with an early ring form (trophozoite) of the parasite. (b) A developing trophozoite. (c) One of several forms that trophozoites may exhibit during their development. (d) A mature trophozoite. The parasite is exhibiting a band form. (e) and (f). Representative stage found during schizont development. (g) A mature schizont. Note the number of merozoites present. (h) An immature microgametocyte. (i) and (j). The appearance of mature micro- and macrogametocytes.

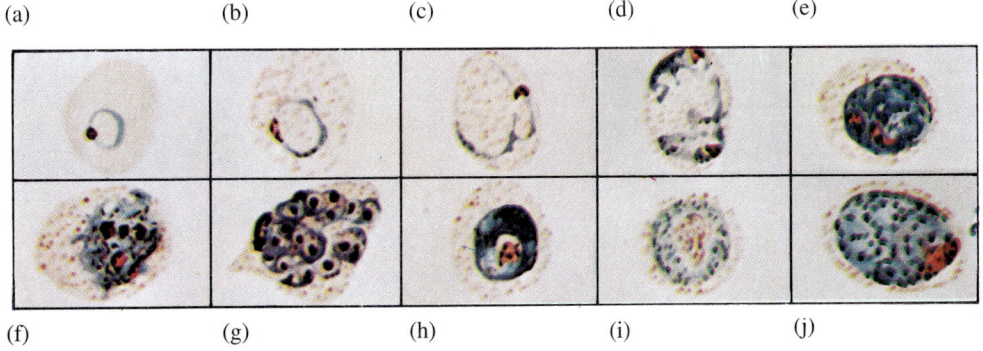

Figure 63–5c

Plasmodium vivax. (a) A typical "signet ring" stage. (b) An enlarged erythrocyte with a ring form of trophozoite. The cell also contains Schuffner's stippling. It should be noted that such stippling may not always be present in infected red blood cells. (c) Another trophozoite form. (d) The presence of two amoeboid-shaped trophozoites. (e) and (f). Erythrocytes showing progressive stages in schizont division. (g) A mature schizont. Note the number of merozoites. (h) A developing gametocyte. (i) and (j). Mature micro- and macrogametocytes.

c. *Trypanosoma brucei* variety *gambiense*

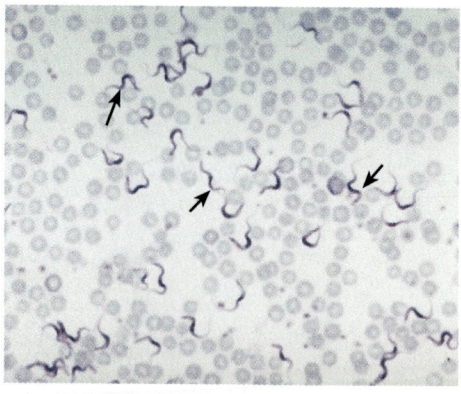

Figure 63–6

A photomicrograph of a blood smear containing the causative flagellate of African sleeping sickness *Trypanosoma brucei variety gambiense* (arrows). The arthropod vectors for this pathogenic protozoon are *Glossina* spp. (tsetse flies).

d. *Trypanosoma cruzi*

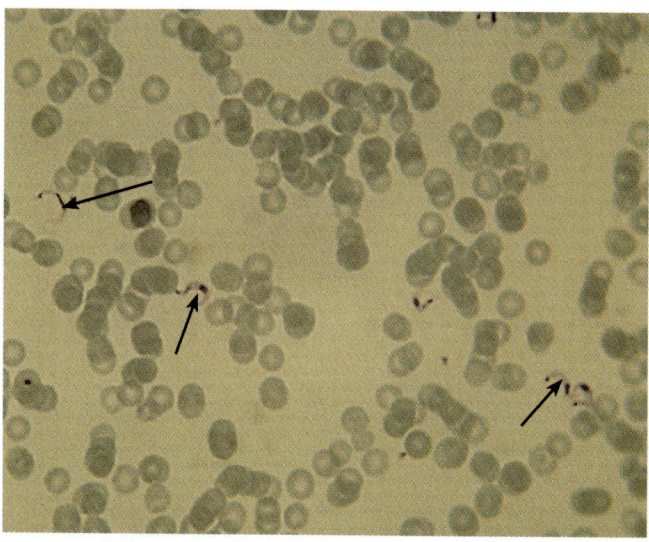

Figure 63–7

Photomicrograph of a blood smear containing *Trypanosoma cruzi*, Cone-nosed bugs of the family Reduviidae are its vectors.

B. Demonstration of Blood Culture Systems

Materials

The following items should be provided for class demonstrations:

❑ 1. Five mL each of 2 blood samples in separate 5-mL disposable hypodermic syringes with 20- or 22-gauge needles and needle guards

❑ 2. Two Bacto or other blood culture bottles containing thiol broth, sodium polyanetholsulfonate and sucrose, carbon dioxide

❑ 3. One sterile air filter (consisting of sterile absorbent cotton in the hub portion of a 20- or 22-gauge syringe needle, as shown in Figure 63–2)

❑ 4. Blood-collecting system for class examination (including sterile tubing with needles attached)

❑ 5. One marking pen or pencil

❑ 6. One bottle of 3.5% iodine solution

❑ 7. Sterile cotton swabs

❑ 8. One container with disinfectant for the disposal of contaminated material

Procedure

The following procedure will be performed by the instructor.

❑ 1. Examine the blood-collecting system used for specimen collection and the inoculation of blood culture systems.

❑ 2. Take 1 blood culture bottle and remove the seal and bottle cap, thereby revealing the underlying rubber seal. Place the cap right side up on a clean laboratory surface.

❑ 3. Wipe the exposed surface of the rubber seal with a cotton swab dipped in 3.5% iodine solution. Allow the stopper to dry while proceeding to the next step.

❑ 4. Label 1 blood culture bottle for each specimen provided.

❑ 5. Remove the needle guard from the syringe containing 1 blood sample, inject the needle directly through the rubber seal, and push the syringe plunger to introduce 5 mL of the inoculum.

❑ 6. Remove the needle from the rubber seal and wipe the surface of the seal with another cotton swab dipped in iodine solution.

❑ 7. Before placing the entire syringe and needle in the disinfectant container, pull the syringe plunger out slightly to draw some disinfectant into the barrel. This is a safety precaution. Allow the syringe and needle to sink into the disinfectant.

❑ 8. Mark the box on the bottle label to specify anaerobic incubation for this system.

❑ 9. Repeat steps 5 through 8 with the syringe containing the second blood sample.

❑ 10. Take the sterile air filter and insert it into the rubber seal of the second blood culture bottle. This step provides aerobic conditions.

❑ 11. Mark the box on the bottle label to specify aerobic incubation for this system.

❑ 12. Incubate both systems, without caps, at 37°C until growth appears (for at least 2 weeks). Look for turbidity (cloudiness), destruction of blood cells (hemolysis), and other signs of activity. (Refer to Figure 63–2.)

❑ 13. Record your findings in Table 63–2 and answer the questions in the Results and Observations section.

❑ 14. Dispose of the bottle caps and blood cultures in the usual manner.

Results and Observations

1. Complete the following table by indicating the day growth was first observed. Use a "+" for growth and "−" for no growth.

Table 63–2

Blood Culture Systems

Day of Incubation	Aerobic System	Anaerobic System
1.		
2.		
3.		
4.		
5.		
6.		
7.		
8.		
9.		
10.		
11.		
12.		
13.		
14.		

2. Did hemolysis occur in either system? If so, which one? _____

_____

3. What is the function of the air filter? _____

_____

C. Septi-Chek System

Materials

The following items should be provided for class demonstration:

❑ 1. Ten mL of a blood sample in a 10-mL or larger disposable hypodermic syringe with a 20- or 22-gauge needle and needle guard

❑ 2. One Septi-Chek System with thioglycollate broth or comparable medium, and a slide chamber

❑ 3. One marking pen or pencil

❑ 4. One container with disinfectant for contaminated material disposal

❑ 5. Gram stain reagents

Procedure

Additional Technique Required for This Portion of the Exercise:

❑ The Gram Stain, Procedure Diagram 17, Exercise 14

❑ 1. Examine the Septi-Chek System provided as a demonstration. Note the following parts, and their combination: blood culture bottle, slide chamber, and the media panels of the slide. (Refer to Figure 63–3.) Your instructor will demonstrate how the different parts are to be joined.

❑ 2. Next, obtain a syringe containing 1 mL of 1 blood specimen for inoculation.

❑ 3. Label the blood culture bottle with the number of the blood specimen, the date, and your initials.

❑ 4. Perform the 6 steps of Procedure Diagram 43.

❑ 5. Incubate the system at 37°C for 18 to 24 hours, or as directed.

❑ 6. After incubation, examine the broth culture bottle and the presence of growth and/or colonies. Note the presence of red colonies (lactose fermenters) and/or pink colonies (lactose nonfermenters) on the MacConkey agar. Estimate the number of colonies on each medium panel. Record your findings in the Results and Observations section.

❑ 7. Prepare and examine Gram stains of representative colonies from each medium. Record your findings in Table 63–3 and answer the questions in the Results and Observations section.

Procedure Diagram 43
Septi-Chek Technique

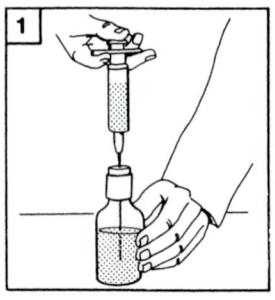

1. Inject the blood sample into the blood culture bottle.

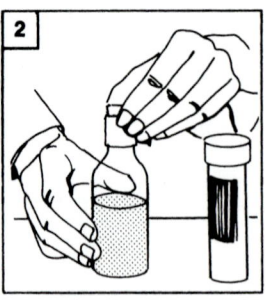

2. Remove the covers of the blood culture bottle and slide chamber.

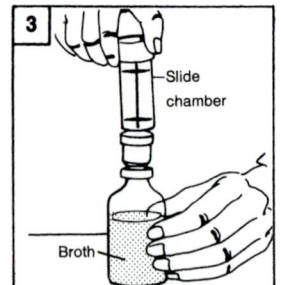

3. Join the threaded end of the slide chamber to the blood culture bottle.

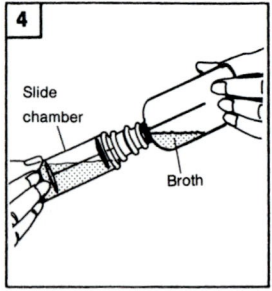

4. Tilt and turn the assembled system so that the broth fills the slide chamber.

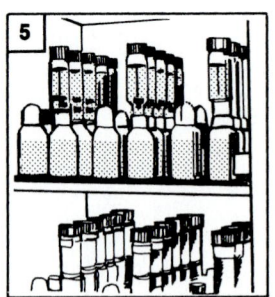

5. Incubate at 37°C or as directed.

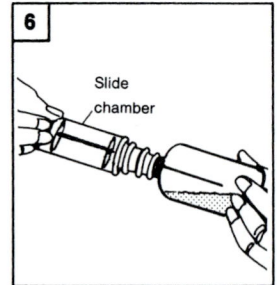

6. Unscrew the top of the slide chamber to obtain specimens for smears.

Results and Observations

1. Was there any growth in the blood culture bottle? _____

2. Enter your findings in the following table.

Table 63–3

Septi-Chek System Results

Medium	Colonies		Number of		Gram Stain and Morphology
	Present	Absent	Lactose Fermenters	Lactose Nonfermenters	
Chocolate agar					
MacConkey agar					
Malt agar					

D. Arthropods and Disease

Materials

❏ 1. Six dissecting microscopes for demonstrations

❏ 2. Magnifying glasses (1 per 4 students)

❏ 3. Charts showing general anatomical features of medically important arthropods

❏ 4. Preserved specimens of the following for demonstrations:
   ❏ a. *Ixodes* spp. (soft tick)
   ❏ b. *Latrodectus mactans* (black widow spider)
   ❏ c. *Dermacentor andersoni* (tick), or other *Dermacentor* species
   ❏ d. *Blatella germanica* (cockroach)
   ❏ e. *Centrurides* spp. (scorpion)
   ❏ f. *Lycosa tarantula* (tarantula)

❏ 5. Prepared slides of the following for demonstrations:
   ❏ a. *Sarcoptes scabiei*
   ❏ b. *Pediculus humanus* var. corporis (human body louse)
   ❏ c. *Culex pipiens* (male and female)
   ❏ d. *Glossina* sp. (African tsetse fly)
   ❏ e. *Stomoxys* sp. (stinging fly)
   ❏ f. *Anopheles quadrimaculatus* (male and female)
   ❏ g. *Trombicula akamushi* (mite)

Procedure

This procedure is to be performed by students individually.

❏ 1. Examine the preserved specimens provided. In the Results and Observations section, make a detailed comparative description of the arthropods examined. Give particular attention to size, coloration, and distinctive markings. Photographs of three representative arthropod vectors are shown in Figures 63–8*a* and 63–8*b*.

❏ 2. Examine and study the prepared slides. Look for distinctive features of the arthropods.

(a)

(b)

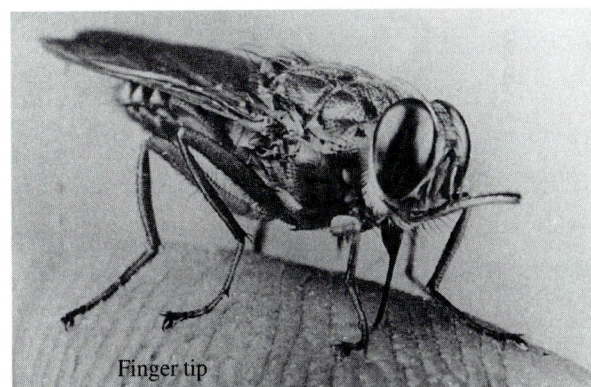

Figure 63–8

a. Ticks. *Ixodes dammini* the vector of Lyme disease and *Dermacentor variabilis* the vector of Rocky Mountain spotted fever. (Courtesy of Univ. Prof. Dr. Franz-Rainer Matuschka, Freie Universitat Berlin.)

b. The tse-tse fly (*Glossina* species) the vector of African sleeping sickness and other infectious diseases. (Courtesy the World Health Organization.)

Results and Observations

Comparative descriptions. Complete the following table.

Table 63–4

Arthropod Vector Properties

Arthropod	General Features		
	Size	Coloration	Other

Laboratory Review 63   Pathogenic Microorganisms Found in Blood and Selected Microbial Pathogens of the Nervous System

1. Give the specific disease caused by each of the following microorganisms:

   a. *Yersenia pestis* _____

   b. *Francisella tularensis* _____

   c. *Rickettsia rickettsii* _____

   d. *Borrelia recurrentis* _____

   e. *Plasmodium vivax* _____

   f. *Trypanosoma cruzi* _____

2. Complete the following table for each of the protozoan pathogens listed.

Table 63–5

Protozoan Blood Pathogens

Protozoon	Means of Locomotion	Host	Means of Transmission
Leishmania donovani			
Plasmodium vivax			
Trypanosoma brucei variety *gambiense*			
T. cruzi			

3. List 2 distinguishing microscopic features of the following microbial pathogens:

Table 63–6

Distinguishing Properties of Blood Pathogens

Microbial Pathogen	Distinguishing Microscopic Feature(s)
Plasmodium falciparum	
P. vivax	
Trypanosoma brucei variety *gambiense*	
Yersinia pestis	

4. What is the purpose of sodium polyanetholsulfonate in blood culture media? _____

_____

5. List 6 commonly isolated bacterial species from cases of bacteremia.

a. _____   d. _____

b. _____   e. _____

c. _____   f. _____

6. Define or explain the following terms. (Refer to your text.)

a. bacterial endocarditis _____

b. bacteremia _____

c. zoonosis _____

d. sporogony _____

e. schizogony _____

Key Terms

bacteremia (bak-ter-Ē-mē-a): the presence of bacteria in blood

meningitis (men-in-JĪ-tis): inflammation of the coverings (meninges) of the brain and spinal cord

peritonitis (per-i-tō-NĪ-tis): inflammation of the membranous abdominal cavity

septicemia (sep-ti-SĒ-mē-a): the presence and multiplication of pathogenic bacteria in the blood

zoonosis (zō-NŌ-sis): a disease communicable from lower animals to humans under natural conditions

---

Photograph Quiz 27

Identify the protozoa shown.

_____

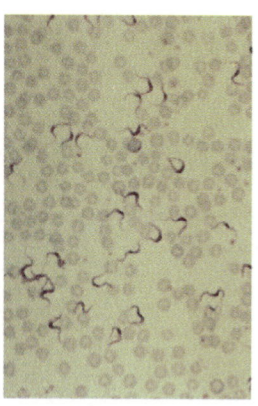

Figure 63–9
Microscopic view of an unknown
blood sample.

Photograph Quiz 28

Identify the arthropod shown.

_____

Figure 63–10
Arthropod identification.

Form for Unknown 1

Student's Name _____   Score _____

Date _____   Laboratory Section _____

Unknown Number _____   Organism Identified _____

Form for Unknown 2

Student's Name _____   Score _____

Date _____   Laboratory Section _____

Unknown Number _____   Organism Identified _____

Pathogens of the Urogenital System and Sexually Transmitted Diseases (STDs)

After completing this exercise, you should be able to:

1. Describe, perform, and interpret basic tests in a routine urinalysis.
2. List common microbial pathogens and diseases of the genitourinary tract.
3. Describe the general features of representative sexually transmitted diseases (STDs).
4. List selected cultural, biochemical, and diagnostic characteristics of certain pathogens of the urogenital system.
5. List selected diagnostic features of selected STDs.

A large number of microorganisms can be found in the external genitalia in both sexes. Potential pathogens are also present in this group and, when introduced into the usually sterile inner portions of the urinary tract, can cause persistent infections. Such conditions can involve one part of the system and easily spread to several others. By definition, a *urinary tract infection* (**UTI**) is the multiplication of microorganisms in the urinary tract. *Urinary tract disease,* on the other hand, refers to any condition in which there is an interruption, stopping, or disorder of the urinary tract function, or its parts. Furthermore, any condition that allows the presence and the multiplication of organisms within the tract is abnormal.

UTIs are among the most common of all infections seen in clinical situations. Second only to sexually transmitted diseases and respiratory infections, they account for over 3 million office visits per year in the United States. Urinary tract infections include *urethritis* (u-rē-THRĪ-tis), or inflammation of the urethra, and *cystitis* (sis-TĪ-tis), or inflammation of the urinary bladder. Because infection spreads easily from the urethra to the bladder, most infections are generally referred to as *urethrocystitis* (ū-rē-thrō-sis-TĪ-tis). Such infections are much more common in women than in men because pathogens reach the bladder more easily through the short female urethra (which measures 4 cm) than through the longer male urethra, (which measures 20 cm). In males, an inflammation of the prostate gland or *prostatitis* (pros-ta-TĪ-tis) often is found with UTIs. UTIs originating in one area of the urogenital system often spread throughout the entire urinary tract by *ascending* (moving upward) or *descending* (moving downward). Infections usually begin in the lower urethra and can ascend to cause inflammation of the kidney, or *pyelonephritis* (pī-e-lō-ne-FRĪ-tis).

Among the various hypotheses offered to account for the sources and routes of urinary tract infection, two currently seem most readily acceptable. The first emphasizes the spread of infection by hematogenous means, while the second describes an ascending or retrograde spread of infection. An example of the latter would be the spread of cystitis to the pelvis or kidney via the ureters. In addition to the presence of uropathogenic bacteria, several factors have been implicated in the development of urinary infections. These factors include congenital defects, biochemically induced lesions, physiologic abnormalities, and traumatic injuries. Prevention of urine outflow by obstructions, large or small, is a well-known cause of persistent infection, as the condition provides an ideal environment for the growth of microorganisms. A major cause of UTIs is incomplete emptying of the bladder during urination. Retained urine provides a medium for microbial growth, thereby encouraging infection. Any factor that interferes with the flow of urine and the complete emptying of the bladder can predispose an individual to a UTI.

Sources of etiologic agents include various respiratory, intestinal, and ear infections together with skin lesions. Seasonal attacks are generally associated with bacterial infections such as summer diarrhea and respiratory conditions including sinusitis, tonsillitis, bronchitis, and pneumonia.

While the pathogen associated with a UTI may be a virus, fungus, bacterium, protozoon, or helminth, the great majority of infections are caused by a limited number of bacterial species. The microorganisms most commonly encountered in nontuberculosis urinary tract infections are *Chlamydia trachomatis, Escherichia coli, Pseudomonas aeruginosa, Klebsiella pneumoniae, Ureaplasma urealyticum, Mycoplasma hominis, Proteus vulgaris,* and *Staphylococcus aureus.*

Chlamydia or *Ureaplasma* infections are usually sexually transmitted and result in nongonococcal urethritis. Renal tuberculosis, an infection that usually affects both kidneys, is also believed to play a significant role in urogenital system infections in various parts of the world.

Several viruses have been associated with UTIs. These include cytomegalovirus and the herpes viruses.

Thousands of individuals throughout the world suffer from urogenital tract diseases caused by worms (helminths) and protozoa. The helminths causing such diseases include *Schistosoma haematobium* and *S. mansomi* (which cause urinary schistosomiasis) and *Wuchereria bancrofti* (which causes elephantitis). Protozoa that cause disease include *Trichomonas* vaginalis (causative agent of some cases of prostatitis) and *Entamoeba histolytica* (the causative agent of amoebic dysentery). The helminths will be described in more detail in Exercise 66.

Various fungi, such as *Blastomyces dermatitidis* and *Coccidioides immitis,* have also been found to cause infections. These agents also are associated with rare infections of the prostate gland.

Urine and blood are analyzed more frequently than any other body fluid. As with blood, the urine composition is a functional indicator of the body's state of health at any given time. Urine is normally sterile, as produced by kidneys and stored in the urinary bladder. When it is voided, urine passes over other parts of the urogenital system and becomes contaminated by the normal flora on these surfaces. Thus, the presence of bacteria in urine is not necessarily an indication of urinary tract infection. However, the presence of large numbers of certain species can be significant.

Normal urine is generally sterile or contains 1,000 or fewer bacteria per milliliter, these few being the microorganisms making up the normal urethral flora. The presence of a microbial infection in any portion of the urinary system usually produces significant changes in urine specimens. The presence of large numbers and kinds of bacteria in urine, or *bacteriuria,* can be determined by specific bacteriological techniques and the use of commercially produced dipsticks or related devices (Figure 64–1). Some of these tests are automated.

In patients with signs of a UTI, urine is collected for screening and culture at the onset of discomfort, and may be repeated 48 to 72 hours later. Urine may be collected by *clean-catch, catheterization,* or *suprapubic aspiration.* Urine is best collected in wide-mouth, sterile, plastic containers with screwcapped lids (Figure 64–2). Upon receipt in the laboratory, urine is processed as soon as possible, but may be refrigerated for up to 24 hours. Such specimens can be rapidly used in screening for a UTI.

If delays in processing are expected, urine may be collected in sterile vacutainer tubes (Figure 64–2) containing boric acid-formate transportation media. Most bacteria present will be kept viable for up to 48 hours at room temperature.

The presence of abnormal constituents such as hemoglobin, erythrocytes, leukocytes, or pus usually is associated with conditions involving severe tissue destruction. The detection of various abnormal constituents is possible by tests that are based on color changes caused by enzymes or chemical reagents. For most of these tests, the color-reacting substance is impregnated on individual plastic strips so that a urine sample can be easily applied. An example of such a di-

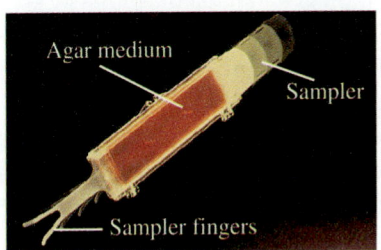

Figure 64–1

An example of a urine culturing device, the *Diaslide.* This device contains two types of media to isolate and identify organisms. The two agar media face one another, separated by a plastic sampler. The plastic sampler has two bent "fingers" for dipping into urine specimens and media inoculations.
(Courtesy of Diatech Diagnostics Inc.)

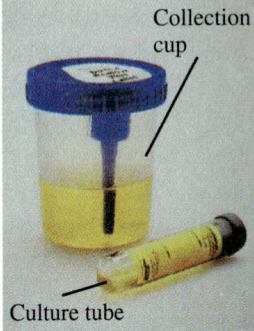

Figure 64–2

An example of a urine collection cup with a screw cup lid. A VACUTAINER Brand Culture tube also is shown.
(Courtesy Becton Dickinson VACUTAINER Systems.)

(a)

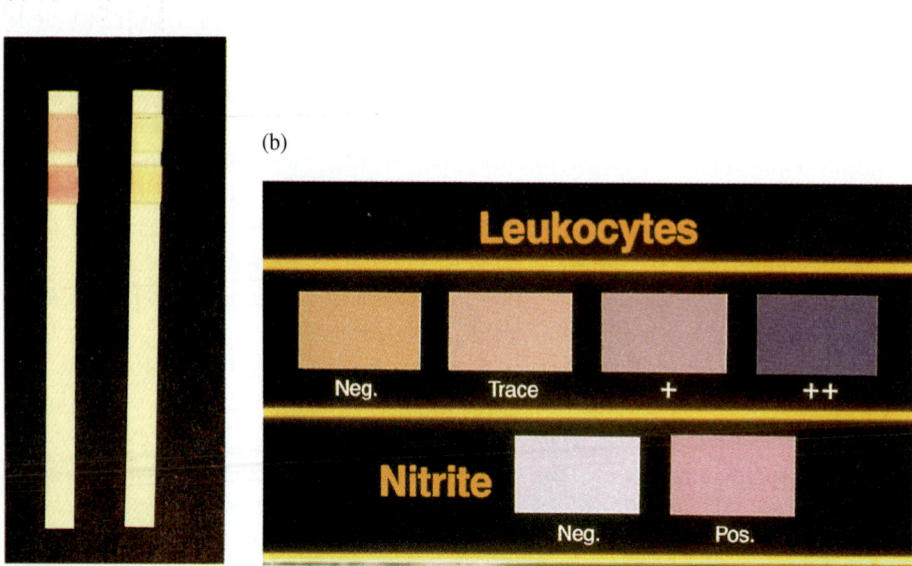

Figure 64–3

(a) An example of dipsticks used in screening tests for urinary tract infections. The Chemstrip LN shows a positive (pink) reaction for nitrite. A positive (blue) reaction would indicate leukocyte esterase. A negative strip (colorless) is provided for purposes of comparison. (b) The various color reactions that can occur with the Chemstrip LN dipstick. (Courtesy Bio-Dynamics.)

agnostic urinary test strip is the Chemstrip LN, used for the detection of leukocytes and determining nitrite in urine (Figure 64–3). The leukocyte-nitrite combination is useful in detecting possible urinary tract infections. The leukocyte reaction is based on a series of reactions involving the esterases of granulocytic leukocytes and the substrate in the test strip. A purple color on the upper patch is a positive result. The nitrite test is based on its production by infecting bacteria. A red-to-violet color develops on the lower patch if sufficient nitrite is present. This or related type of dipstick will be used in this exercise as a representative of the commercially available diagnostic urine screening tests.

Urine is an excellent culture medium for common microbial urinary tract pathogens. When bacteria are present in urine, they multiply freely often exceeding one million per milliliter (10^6 mL). Urine specimens, either clean voided or catheterized, are often contaminated on collection. Therefore, as indicated previously, the recovery of microorganisms, even pathogens, does not necessarily establish the diagnosis of a urinary tract infection. In general, the diagnosis of genitourinary infections requires: (1) establishment of the presence and location of the infection, (2) isolation and identification of the etiologic agent or agents, and (3) determination of the infection's origin. The device known as the Diaslide will be used to demonstrate an approach to identifying bacterial causes of UTIs.

A routine analysis of a urine sample involves a description of physical properties such as color, odor, pH, specific gravity, and turbidity, and determining the presence or absence of abnormal constituents. Microscopic examination of a urinary tract sediment and tests for the detection and identification of pathogenic bacteria, yeasts, and protozoa may also be required. Several of these aspects of urinalysis are considered in this exercise. In addition, this exercise will consider the properties of selected bacterial agents of genitourinary tract infections. The use of commercially available screening and diagnostic systems will also be demonstrated.

The sexually transmitted diseases (STDs), also known as venereal diseases, represent specific infections of the reproductive system. Several STDs cause not only discomfort and disability, but may result in death. These diseases have become a major worldwide public health problem. Most cases are usually spread during sexual contact, although there are other means of transmission. for example, cases of syphilis have been acquired congenitally and by kissing. Because of the public health significance of syphilis, several states require premarital examinations and serological of expectant mothers.

Currently there are more than 25 different STDs (Figures 64–4, 64–5, and 64–6). In addition to syphilis, the bacterial venereal diseases include gonorrhea (*Neisseria gonorrhoeae*), chancroid (*Haemophilus ducreyi*), granuloma inguinale (*Calymmatobacterium granulomatis*), and *Chlamydia trachomatis* infections. Control of these diseases involves early diagnosis, prompt treatment, and proper handling of contaminated articles.

Another possible consequence following sexual activity is *pelvic inflammatory disease* (PID). The condition is caused by pathogenic microorganisms moving from the lower genital tract through the cervix into the uterus, the uterine

(a)   (b)

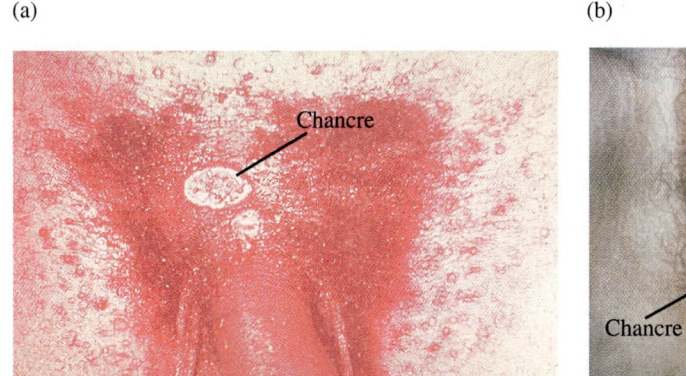
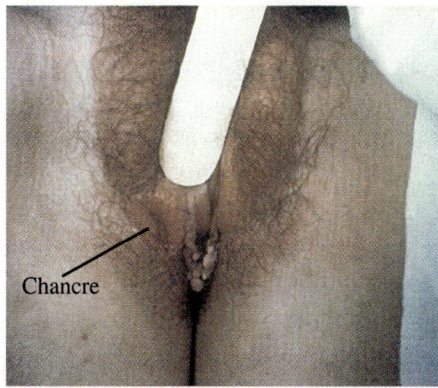

Figure 64–4
Syphilis caused by the bacterium *Treponema pallidum.* The typical lesion known as the chancre in the male (a) and in the female
(b). Microscopic examinations of specimens from chancres show the presence of the causative agent.

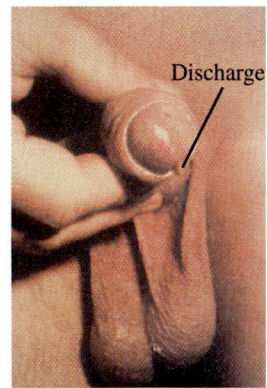

Figure 64–5
A case of gonorrhea, caused by the bacterium *Neisseria gonorrhoeae.* There is a typical discharge
associated with this infection.

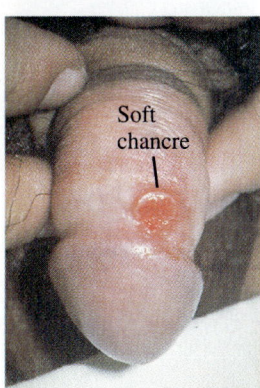

Figure 64–6
A case of the bacterial infection, chancroid caused by *Haemaphilus ducreyi.*

tubes, and the neighboring structures of the reproductive system. A small percentage of cases develops following certain surgical procedures, such as *dilation and curettage* (scrapping the walls of the uterus), and the insertion of an intrauterine device (IUD), a form of birth control. The agents responsible for PID come from the large number of bacteria found in the individual's vagina or cervix. It is important to remember that the urinary and genital (reproductive) systems are closely associated. Infections of one system can easily spread to the other. A number of other pathogenic agents are known to cause STDs. These include the viral agents of infections such as genital herpes (Figure 64–7), genital warts, hepatitis B, and acquired immune deficiency syndrome (AIDS). The yeast *Candida albicans,* the protozoon *Trichomonas vaginalis,* and the itch mite *Sarcoptes* (Figure 64–8) also are STD agents.

Screening and/or culturing specimens for selected STD disease agents frequently may involve collecting specimens with sterile and/or special types of swabs. Specimens are either used for microscopic examinations and/or the inoculation of special media (Figure 64–9). This exercise will only present selected features of bacterial, yeast, and protozoon STD agents.

(b)

(a)

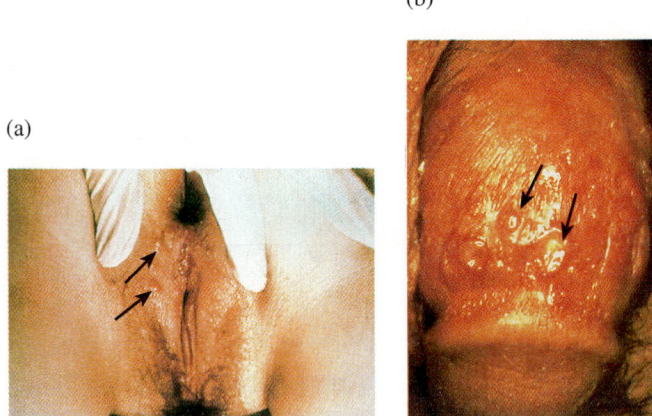

Figure 64–7

The appearance of genital herpes caused by Herpes simplex virus (*Herpesvirus* type 2). (a) In the female. (b) In the male. Note the blister-like lesions.

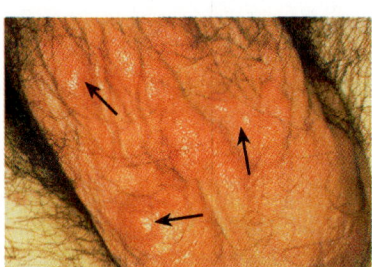

Figure 64–8

The *Scabies* mite and its effects on the scrotum.

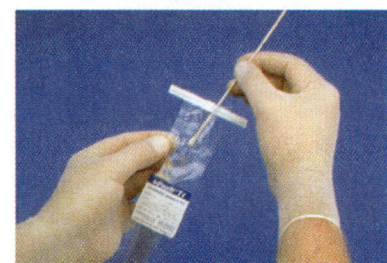

Figure 64–9

Inoculating a specimen on a cotton-tipped swab into *The In Pouch TV* culture system. This medium is used in the diagnosis of the protozoan STD, human trichomoniasis.
(Courtesy of BioMed Diagnostics, Inc.)

A. Physical Properties of Urine

Materials

The following materials should be provided for general class use:

❏ 1. Urinometer (cylinder and hydrometer float device)

❏ 2. pH indicator paper strips (pHydrion or nitrazine test papers)

❏ 3. Two freshly voided urine samples (*Note:* The same specimens should be used in both of the following procedures.)

❏ 4. Thermometers

❏ 5. 250-mL beakers

❏ 6. Filter paper

❏ 7. Liquid soap and brushes for washing of containers

❏ 8. Containers with disinfectant

❏ 9. Disposable surgical gloves

❏ 10. Containers for disposal of urine and used materials.

Caution

Working with urine or other body fluids will not pose any danger if the following precautions are observed:

- ❏ 1. Place any and all urine or materials containing urine into the disinfectant containers provided.
- ❏ 2. Dispose of any urine-containing materials as indicated by the instructor.
- ❏ 3. In cases of urine spillage, wipe the area with a disinfectant-soaked paper towel, and dispose of all materials as indicated by the instructor.
- ❏ 4. Always wash your hands after handling urine and associated materials.
- ❏ 5. Use disposable surgical gloves in specimen-handling situations.

Procedure 1: Physical Properties

This procedure is to be performed by students in pairs.

- ❏ 1. Put on surgical gloves before handling any sample.
- ❏ 2. Pour about 30 mL of one of the samples into a beaker. Carefully pour the urine from the beaker into a clean urinometer cylinder until it is three-fourths full. Obtain more of the specimen, if necessary. If foam forms on the surface, remove it with a piece of filter paper. Discard the paper as directed by your instructor.
- ❏ 3. Note the color, odor, and turbidity of the sample. Refer to Table 64–1. Record your findings in the appropriate specimen column of the Results and Observations section.
- ❏ 4. Check the reading on a thermometer. If it is not at 37°C, shake it down to that level. Insert the thermometer in the urinometer and read the temperature after 3 minutes. The temperature of the sample is used in the determination of specific gravity in the next step. Clean the thermometer as directed by your instructor.
- ❏ 5. Float the hydrometer in the cylinder. Make certain that it is not touching the bottom or the sides of the cylinder. Adjust the value of hydrometer reading according to temperature reading; for each 3° below 25°C, subtract 0.001. Record your findings in the appropriate column of the Results and Observation section. Remove the hydrometer and clean it as directed by your instructor.
- ❏ 6. Dip a pH indicator test strip in the urinometer cylinder 3 consecutive times. Remove excess urine by touching the paper to the inside surface of the cylinder. After 1 minute, compare the color of the strip with the color guide on the strip container or other available guide. Compare your findings with the values in Table 64–1. Record your findings in the Results and Observations section.
- ❏ 7. Clean the urinometer as directed by your instructor.
- ❏ 8. Repeat steps 1 through 7 with the second urine sample.

B. Urinary Tract Infection (UTI) Detection

Materials

The following materials should be provided for class use:

- ❏ 1. Chemistrip LN reagent strips and container (Bio-Dynamics)
- ❏ 2. *Uriscreen* tubes (one per student)
- ❏ 3. *Uriscreen* 10% hydrogen peroxide reagent
- ❏ 4. Urine specimen #1 (with a low bacterial count)
- ❏ 5. Urine specimen #2 (with a high bacterial count)
- ❏ 6. Sterile, small (250-mL) beakers
- ❏ 7. Sterile (5mL) pipettes with rubber bulbs or other pipetting aid
- ❏ 8. Urine disposal container with disinfectant
- ❏ 9. Containers for *Uriscreen* tube pipette, and dropper disposal
- ❏ 10. Disposable eye droppers
- ❏ 11. Sterile pipettes and pipetting aids
- ❏ 12. Disposable surgical gloves

Procedure 1: Chemstrip LN Reagent Strip

This procedure is to be performed by students individually.

❏ 1. Pour about 125 mL of one of the urine samples into a beaker.

❏ 2. Remove one Chemstrip LN strip from its container. Be careful not to handle the specific test squares. Examine the strip and the identification or test-interpretation portion of the label on the container. Note that this strip contains material for the detection of leukocytes and nitrites.

❏ 3. Dip the test strip into the urine sample and tap the excess liquid off on the inside top surface of the beaker. Wait 15 to 30 seconds and compare the color reactions with the test interpretation label. (Refer to Figure 64–3.) Record your findings in Table 64–2 in the Results and Observations section. Discard the test strip as directed by your instructor.

❏ 4. Repeat steps 1 through 3 with the second urine sample.

Procedure: The *Uriscreen*

This procedure is to be performed by students in pairs.

❏ 1. Obtain a small amount of one of the urine specimens provided and one *Uriscreen* tube.

❏ 2. With the aid of a sterile pipette and rubber bulb, carefully transfer 2 mL of the specimen into the *Uriscreen* tube.

❏ 3. Add 4–5 drops of the *Uriscreen* reagent.

❏ 4. Gently shake the mixture and observe the tube for a reaction. (Reactions occur within 2 minutes.)

❏ 5. Bubbling and foam formation indicate urinary tract infection (Figure 64–10).

❏ 6. Record your findings in Table 64–1 in the Results and Observations section.

❏ 7. Dispose of the *Uriscreen* tube, used beaker, pipette, and eyedropper as indicated by your instructor.

❏ 8. Have your laboratory partner repeat steps with the other urine specimen provided.

Figure 64–10
The results of the Uriscreen test. A negative result is indicated by no observable change. The presence of foam or bubbles indicates the presence of a urinary tract infection *(UTI)*.
(Courtesy of Diatech Diagnostics Inc.)

C. Microbial Identifications From Urine Specimens

Materials

The following materials should be provided for class use:

❏ 1. Diaslide (one per student)

❏ 2. Urine specimen with a low bacterial count

❏ 3. Urine specimen with a high bacterial content

❏ 4. Sterile, small (250-mL) beakers

❏ 5. Urine disposal container with disinfectant

❏ 6. Incubator set at 37°C

❏ 7. Container for disposal of used Diaslide

❏ 8. Marking pens or wax marking pencils

❏ 9. Disposable surgical gloves

Procedure

❏ 1. Obtain a Diaslide and examine it. Compare the device to Figure 64–1.

❏ 2. Pour approximately 25 mL of the low bacterial count urine specimen into a sterile beaker.

❏ 3. Peel back the wrapper and remove the Diaslide, being careful not to touch the sampler fingers (Figure 64–11a).

❏ 4. Dip the sampler "fingers" into the specimen. Be certain that the fingers are immersed up to the point where they meet (Figure 64–11b).

❏ 5. With one hand, hold the Diaslide vertically as shown in Figure 64–11c. Use the other hand to pull the sampler through the casing with straight upward motion.

❏ 6. Discard the sampler as directed by your instructor.

❏ 7. Label the Diaslide with your initials and indicate the specimen used (Figure 64–11d).

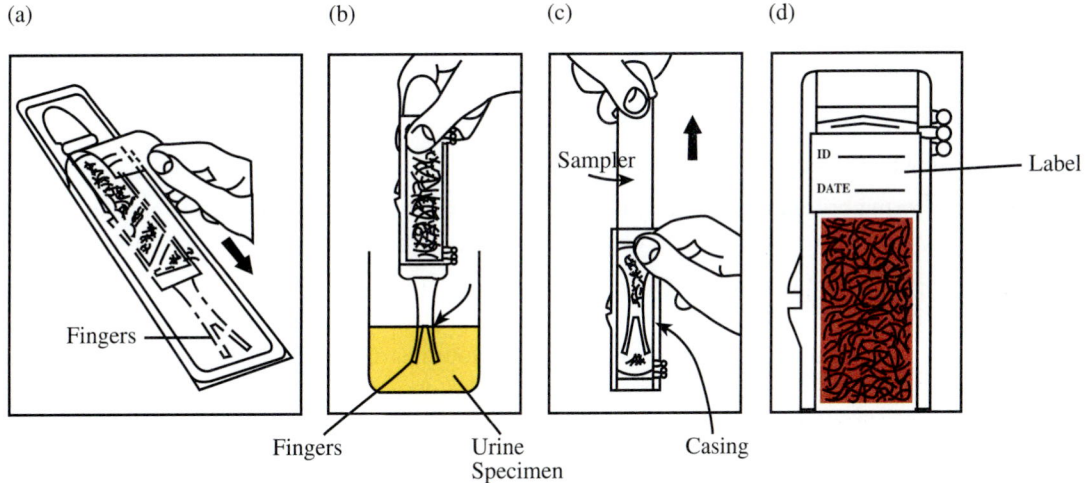

Figure 64–11

Steps in the use of a *Diaslide*. (a) Removing the wrapper. (b) Dipping the sampler fingers into the specimen. (c) Pulling the sampler through the casting. (d) Labeling the device with initials and the type of specimen.

❏ 8. Incubate the device as indicated by your instructor.

❏ 9. Dispose of the urine specimen as directed by your instructor.

❏10. Have your laboratory partner repeat steps 1–9 with the other urine specimen provided.

❏11. After incubation, count the number of colonies (small dotlike or larger solid circular growth) on the green (CLED) medium and on the pink (MacConkey) agar medium. Eighty or more colonies on the green-colored medium and 50 or more on the pink medium indicate a urinary tract infection. These numbers correspond to about 100,000 bacteria per 1 mL of specimen.

❏12. The color of the colonies formed also can be used to identify the type of bacteria causing the infection. Examples of common causes of urinary tract infection are listed in Table 64–1 on the following page. Identify the organism(s) on your Diaslide.

❏13. Record your findings and answer the questions in the Results and Observation section.

D. Unknown Specimens

Materials

The following materials should be provided for class use:

❏ 1. All materials listed for the procedures in Parts A, B, and C

❏ 2. Unknown urine specimen (one per student)

Table 64–1

Examples of Colored Colonies Formed by Bacteria Causing UTIs[a]

Bacterial Cause	Medium	
	CLED	MacConkey
Escherichia coli	Yellow, with dark center	Pink, red
Klebsiella pneumoniae	Slimy and yellow	Pink, red
Pseudomonas aeruginosa	Greenish blue	Clear

[a]Note: This is only a partial listing.

Procedure

This procedure is to be performed by students individually.

❏ 1. Obtain an unknown specimen from your instructor.

❏ 2. Perform the following procedures and tests on the specimens, and time yourself while carrying out the entire analysis. Enter your findings and indicate your time in the Results and Observations section.
    ❏ a. Description of physical properties (color, odor, turbidity, pH, and specific gravity)
    ❏ b. Urinary tract infection determination, and microbial identification

❏ 3. Discard the specimens and materials used as directed by your instructor. Clean all glassware.

Results and Observations

Physical Properties of Urine

1. Complete Table 64–2 on the following page.

2. Did either of the urine samples show abnormalities? If so, which specimen and what abnormalities? _____

_____

Urinary Tract Infection (UTI) Detection

1. Which of the urine specimens showed signs of an infection?

   a. Specimen 1 _____

   b. Specimen 2 _____

2. Was this procedure simple to perform? _____

_____

Microbial Identifications from Urine Specimens

1. Which of the urine specimens showed signs of an infection?

   a. Specimen 1 _____

   b. Specimen 2 _____

Table 64-2

Urine Specimen Analysis

Property	Normal Feature(s)	Abnormal Feature(s)	Specimen 1	Specimen 2
Color	Generally light yellow	Dark red or brown		
Turbidity (cloudiness)	Aromatic	Fishy, ammonialike		
Specific gravity at (25°C)	Clear	Hazy, may contain obvious shreds of mucus		
pH (hydrogen ion concentration)	1.002–1.030	Any major change from normal range		
Volume	6.0 average, but ranges from 4.8–7.5	Any major variation from normal range		
Leukocyte strip	Negative (Cream)	Positive (Pink-Purple)		
Nitrite strip	Negative (White)	Positive (Pink)		
Uriscreen	No foaming	Foaming		

2. Describe the colonies as to color that formed on the surface of the Diaslide.

CLED   MAC

   a. Specimen 1 _____   _____

   b. Specimen 2 _____   _____

3. Was this procedure simple to perform? _____ If not, why? _____
    _____

Unknown Specimens

1. Enter your findings in Table 64–3.

2. Were any of the results obtained with the unknown specimen abnormally high? If so, indicate which ones. _____
    _____
    _____

3. How long did it take you to perform the urinalysis? _____
    _____

4. How did your time compare with others in your laboratory?
    _____

Table 64–3

Unknown Specimen Results

Procedure or Test	Results for Unknown Specimen # _____
Color	
Odor	
Turbidity	
Specific Gravity	
pH[a]	
Leukocyte strip[a]	
Nitrite Strip[a]	
Uriscreen	
Bacterial Colony Count (CLEO)[b]	
Bacterial Colony Count (MAC)[b]	

[a]These determinations are to be made with reagent strips.
[b]This result can be determined only after 24 hours. The count of more than 100,000 indicates a urinary tract infection. Refer to the section describing the culturing of bacteria for additional details.

E. Characteristics of Selected Potential Bacterial Pathogens

Materials

The following materials should be provided per 4 students:

❑ 1. Two 24-hour trypticase soy broth cultures of the following:

    ❑ a. *Moraxella (Branhamella) catarrhalis*   ❑ d. *Pseudomonas aeruginosa*
    ❑ b. *Escherichia coli*   ❑ e. *Staphylococcus aureus*
    ❑ c. *Proteus vulgaris*   ❑ f. *Enterococcus (Streptococcus) faecalis*

❑ 2. One divided streak plate with the following microbial combinations:

    ❑ a. *Proteus vulgaris* and *Moraxella (Branhamella) catarrhalis* on (Bacto) Pseudomonas P agar
    ❑ b. *Escherichia coli* and *Pseudomonas aeruginosa* on (Bacto) Pseudomonas P agar
    ❑ c. *P. aeruginosa* and *E. coli* on eosin-methylene blue agar
    ❑ d. *Staphylococcus aureus* and *Enterococcus faecalis* on blood agar
    ❑ e. *S. aureus* and *E. faecalis* on chocolate agar

❑ 3. One 20-mL container of 3% hydrogen peroxide

❑ 4. One 20-mL container of dimethyl-p-phenylenediamine (oxidase reagent)

❑ 5. Six trypticase soy agar plates

❑ 6. Six sets of antibiotic sensitivity disks including penicillin and several broad-spectrum preparations

❑ 7. Six sterile cotton swabs

❑ 8. Two Gram-staining reagent sets

❑ 9. Four hand lenses

❑ 10. Pasteur pipettes with rubber bulbs to dispense reagents (enough for class use)

Procedure 1: General Characteristics

This and the following procedures are to be performed by students in groups of 4.

❏ 1. With the aid of a hand lens, examine all the streak plate preparations provided. Note any and all distinguishing characteristics. Record these findings in the appropriate portions of Table 64–4 in the Results and Observations section. (Refer to Exercise 7.)

❏ 2. Prepare Gram stains of isolated colonies for each organism and record your results in the Results and Observations section.

Procedure 2: The Oxidase Reaction

❏ 1. Place 1 drop of oxidase reagent on an isolated colony of each organism cultured on chocolate agar.

❏ 2. If the test is positive, organisms should first turn red and then blue-black or blue depending on the source of the reagent used. (Refer to Exercise 24.)

Procedure 3: The Catalase Reaction

❏ 1. Add 2 or 3 drops of the 3% hydrogen peroxide solution to 1 of each broth culture.

❏ 2. Mix by gently shaking the contents of the tubes.

❏ 3. A positive test is indicated by the abundant production of gas bubbles. (Refer to Exercise 24.)

Procedure 4: Antibiotic Sensitivities

❏ **Additional Technique Required for This Portion of the Exercise:**

Commercial-Type Multiple Antibiotic Disk Dispenser, Procedure Diagram 32, Exercise 37

❏ 1. Prepare streak plates of each organism using the trypticase soy agar plates, the cotton swabs, and the second tubes of trypticase soy broth cultures.

❏ 2. Place the antibiotic sensitivity disks onto the agar surface as indicated in Procedure Diagram 32.

❏ 3. Incubate the plates at 37°C for 24 hours. Record your findings in Table 64–4 of the Results and Observations section. (Refer to Exercise 37.)

Results and Observations

1. Complete Table 64–4 on the following page.

2. Which of the cultures tested were:

   a. oxidase positive? _____

   b. catalase positive? _____

   c. pigmented or produced pigment (indicate color)? _____

Table 64–4

Characteristics of Selected Bacterial Pathogens

Microorganisms	Colonial Characteristics[a]	Gram Reaction	Oxidase Reaction[b]	Catalase Reaction[b]	Antibiotic Sensitivities[c]						
					P	S	E	T			

[a]Pay particular attention to pigmentation, elevation, size, margin, and texture characteristics. Refer to Exercise 7.
[b]Use "+" for a positive reaction and "−" for a negative reaction. Refer to Exercise 24, and the color photographs.
[c]These letters represent common antibiotics: P = penicillin; S = dihydrostreptomycin; E = erythromycin; and T = tetracycline. Use R for a resistant reaction (no effect by antibiotic) and S to indicate sensitivity to an antibiotic.

F. Microscopic Examination of Sexually Transmitted Disease Agents

Materials

❏ 1. Prepared slides of the following microbial pathogens:
    ❏ a. *Candida albicans* (yeast)
    ❏ b. Gram-stained *Neisseria gonorrhoeae* (bacterium)
    ❏ c. Gram-stained *Staphylococcus aureus* (bacterium)
    ❏ d. *Treponema pallidum* (bacterium)
    ❏ e. *Trichomonas vaginalis* (protozoon)
❏ 2. Unknown slide (1 per student)
❏ 3. Immersion oil
❏ 4. Lens paper

Procedure 1: Prepared Slides

This procedure is to be performed by students individually.

❏ 1. Examine and compare the prepared slides of the microbial causative agents of sexually transmitted diseases (STDs). (Refer to Figures 64–4b, 64–12, and 64–13.)

❏ 2. Sketches showing the microscopic characteristics of these microorganisms are included in the Results and Observations section to help you identify and compare specific organisms. If one slide does not demonstrate the microorganisms sufficiently, go to another slide and continue searching for clear details.

❏ 3. Using Figures 64–12 and 64–13 as guides, answer the questions in the Results and Observations section.

❏ 4. After observing all preparations, clean the used slides thoroughly and return them as directed.

(a)   (b)   (c)

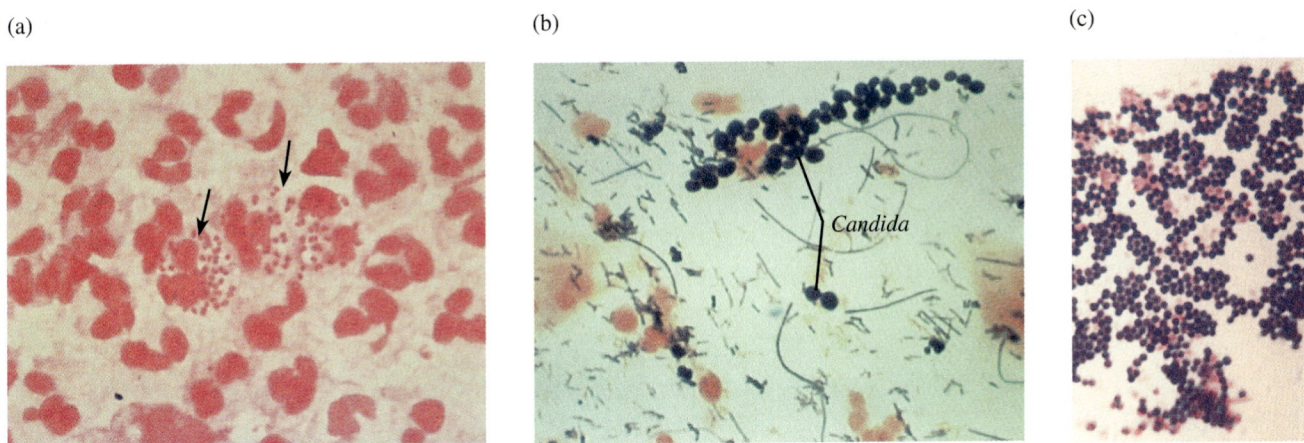

Figure 64–12

Microscopic views of selected STD agents. (a) *Neisseria gonorrhoeae.* Note the diplococcus arrangement of the bacterium within white blood cells. (b) *Candida albicans* (large purple oval cells) with numerous bacteria and other cells. (c) *Staphylococcus aureus.*

(a)   (b)

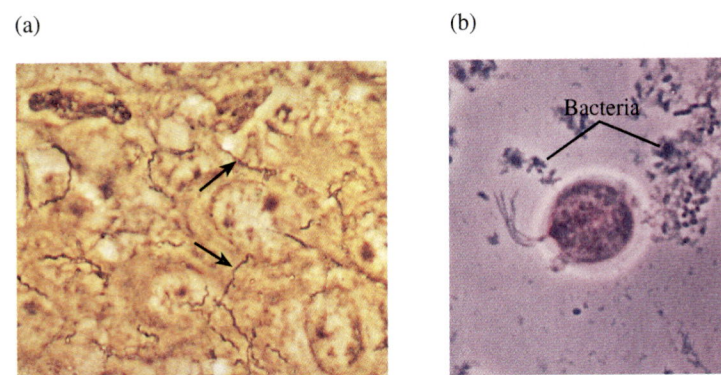

Figure 64–13

Examples of other STD agents. (a) *Treponema pallidum.* (b) *Trichomonas vaginalis.*

Procedure 2: Unknowns

This procedure is to be performed by students individually.

❑ 1. Each student will be given an unknown slide containing one of the pathogens studied in this exercise.

❑ 2. Identify the unknowns and enter the results on the Unknown Form on page 598.

Results and Observations

Microscopic examination of prepared slides

1. What microscopic properties distinguish N. gonorrhoeae from S. aureus? (Refer to Figures 64–12 and 64–13.)

_____

2. List the STD agents in this portion of the exercise in order of increasing size. _____

_____

G. Demonstration of the VDRL Slide Flocculation Test

Several diagnostic tests for syphilis are in use. Of these, probably the Wasserman test, the Venereal Disease Research Laboratory (VDRL) test, the rapid plasma reagin (RPR) card test, and the Treponema pallidum immobilization test are the most familiar. This exercise will cover the VDRL test.

In this test, the VDRL antigen is mixed with heat-inactivated patient's serum on a glass slide. The mixture is rotated either manually or mechanically for approximately 4 minutes and then examined for the presence of clumping. Large aggregates of the antigen emulsion are considered to be a strongly positive test, while small clumps are indicative of a weak reaction. A negative result is characterized by unchanged antigen.

Materials

❑ 1. Commercially prepared VDRL antigen
❑ 2. Heat-inactivated reactive serum (4+)
❑ 3. Heat-inactivated nonreactive serum
❑ 4. Buffered saline solution
❑ 5. Four 0.2-mL pipettes graduated in 0.01-mL increments or suitably calibrated hypodermic needles without sharp points, attached to clear eye droppers or Luer-type syringes (1 to 2 mL)
❑ 6. One flat, clear glass slide with several paraffin or ceramic rings, approximately 14 mm in diameter

Procedure

This procedure is to be demonstrated by the instructor.

❑ 1. Number 5 rings on the glass slide.
❑ 2. Add 0.05 of the 4+ serum to rings 1 and 4 and 0.05 mL of the nonreactive serum to rings 2 and 5.
❑ 3. Add 0.05 mL of buffered saline to rings 3, 4, and 5.
❑ 4. Place 1 drop (1/60 mL) of VDRL antigen into rings 1, 2, and 3.
❑ 5. Rotate the slides of a flat surface, making a circle with an approximate diameter of 2 inches. Regulate this movement so that the slide is rotated 120 times per minute. Hand rotation should be continued for 4 minutes.
❑ 6. Examine the slide immediately under a microscope set at 100× magnification.
❑ 7. Remember: (a) No clumps indicate a nonreactive reaction; (b) small clumps indicate a weakly reactive situation; and (c) large clumps indicate a strong reactive result.
❑ 8. Answer the questions in the Results and Observations section.

Results and Observations

1. Which of the mixtures were controls? _____

   _____

2. What would the result be if the sera were not heat-inactivated before use? _____

   _____

Laboratory Review 64   Pathogens of the Urogenital System and Sexually Transmitted Diseases

1. List 4 tests used in the diagnosis of syphilis. (Refer to your text.)

    a. _____   c. _____

    b. _____   d. _____

2. What tests are used in the diagnosis of the following infections? (Refer to your text.)

    a. gonorrhea _____

    b. lymphogranuloma venereum _____

    c. *Trichomonas* infection _____

    d. herpesvirus type 2 _____

    e. candidiasis _____

3. List 4 specific infectious diseases of the urinary tract.

    a. _____   c. _____

    b. _____   d. _____

4. List 6 sexually transmitted diseases other than those listed in question #2 above.

    a. _____   d. _____

    b. _____   e. _____

    c. _____   f. _____

5. What is the purpose of the Chemstrip LN? _____

    _____

Key Terms

bacteriuria (bak-tē-rē-Ū-rē-a): the presence of bacteria in urine

catheterization (kath-e-ter-i-ZĀ-shun): in this exercise, the term refers to the use or passage of a catheter (a tube) through the urethra into the bladder for urine removal

chancre (SHANG-ker): a hard primary stage ulcer found in syphilis

clean-catch method: a general procedure used to obtain a urine specimen for culture that will have minimal chance for contamination

cystitis (sis-TĪ-tis): urinary bladder infection

pyuria (pī-Ū-rē-a): pus in the urine

suprapubic aspiration: in this exercise, this term refers to drawing out of urine from the urinary bladder above the pubic arch

urinary tract disease: a condition in which urinary tract parts or functions are affected abnormally

urinary tract infection (UTI): a condition in which microbial reproduction occurs in the urinary tract

Form for Unknown

Student's Name _____   Score _____

Date _____   Laboratory Section _____

Unknown Number _____   Organism Identified _____

EXERCISE 65   Methods for Identifying an Unknown Bacterial Specimen

After completing this exercise, you should be able to:

1. Identify an unknown bacterial culture using the series of media and tests listed in the keys provided.

2. Outline and explain an orderly approach to the identification of an unknown culture.

3. Interpret correctly the reactions obtained with selective and/or differential media, multiple-test media systems, and other tests indicated in the exercise.

In the case of many infectious diseases, chemotherapy procedures can be selected on the basis of symptoms and/or preliminary examination of clinical specimens. Specimen examination does not necessarily involve isolation and identification of the etiological agent. For example, the antibiotic sensitivity of a suspected pathogen can be determined through tests performed on the clinical specimen while the organism is being identified. Eventual classification of the organism is important not only for epidemiological purposes but also for verification of the pathogen.

The classic approach to bacterial identification involves preliminary microscopic examination of a portion of the specimen by either the Gram stain or the acid-fast procedure, depending on the symptoms. Meanwhile, the specimen is usually inoculated into or onto a variety of media that often include differential or selective types. When the suspected pathogen is obtained in pure culture, its biochemical properties can be studied. Its genus and species can often be determined on the basis of sugar-fermentation patterns and the production of various metabolic by-products such as hydrogen sulfide, acetylmethylcarbinol, and indole.

One way to facilitate microbial identification without exhaustive biochemical testing is through serological testing. Specific antibodies for many pathogens, such as *Salmonella, Shigella,* and *Neisseria,* are commercially available. The specificity of bacterial viruses (bacteriophages) has led to their utilization for typing some bacterial strains. This procedure has been of particular use in the identification of *Mycobacterium* and *Staphylococcus* species. In this technique, the organism is spread onto the surface of a standard growth medium, and then several known virus strains are spotted onto specifically designated areas. If a particular virus attacks and lyses the test organism, it will produce a clear zone in the bacterial growth. The application of this typing method (phage typing) in the case of *Staphylococcus aureus* has been particularly advantageous in epidemiological studies concerned with the spread of a specific pathogenic strain in hospitals and among carriers.

Although this exercise is primarily designed to use classic fermentation methods for biochemical testing, changes are occurring in the realm of clinical technology. For example, the Difco Company of Detroit, Michigan, uses disposable slides containing chips impregnated with specific substrates with specific substrates. The product is known as *DrySlide.* After smearing a test organism onto the surface of the slide, and a short incubation of a few minutes, a specific test reagent is applied to detect the reaction produced by the test organism. This type of spot test provides an opportunity for multiple testing of different organisms on a single slide, and a convenient, cost-effective approach to the identification of a number of bacterial unknown cultures (Figure 65–1). Other approaches, such as the miniaturized systems including the API and Enterotube systems discussed in Exercise 27, are valuable in the identification of a wide variety of bacterial species.

In this exercise, each student will be provided with two bacterial unknowns. These organisms will exhibit either the same or different Gram stain reactions. Known cultures of *Staphylococcus aureus* and *Escherichia coli* will be provided for Gram reaction controls, while *S. aureus* and *Mycobacterium smegmatis* will constitute similar controls for the acid-fast procedure. Be certain and careful in your approach to the problem. Identification of organisms requires careful application of the techniques previously studied. To avoid problems of limited time and supplies of microorganisms, complex media requirements, and danger to the experimenter, a relatively simple, yet thorough, exercise has been constructed. However simple it may be, carrying it out should give you some understanding of the problems encountered in clinical laboratory work.

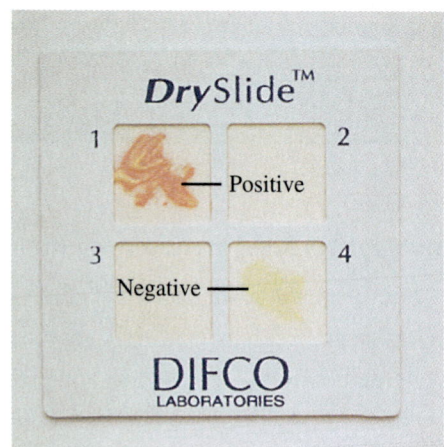

Figure 65–1
The DrySlide for indol. A positive reaction is indicated by a red color.
(Courtesy of Difco Labs., Detroit, MI.)

Hundreds of media and tests have been developed for the identification of unknown organisms. From a practical point of view, however, it is important that the number of tests used for identification be well chosen and kept to a minimum. A **test pattern,** or **key,** based on reactions with known organisms, can serve as a rapid and efficient approach to the identification of an unknown organism. Keys of this type are provided in this exercise.

A. Bacterial Unknowns

Materials

❏ 1. The following organisms will be provided as controls for the Gram and acid-fasting staining reactions (1 per 4 students):
  ❏ a. One each of 24-hour trypticase soy agar slant cultures of *Escherichia coli* and *Staphylococcus aureus*
  ❏ b. One 48-hour trypticase soy agar slant of *Mycobacterium smegmatis*
❏ 2. Trypticase soy agar slants (2 per student)
❏ 3. The following media and reagents will be provided for this exercise only upon authorization by the instructor:

Media

❏ a. Trypticase soy agar plates
❏ b. Blood agar plates
❏ c. Tomato juice agar plates
❏ d. Mueller-Hinton tellurite agar plates
❏ e. 6.5% NaCl in brain-heart transfusion broth
❏ f. Mitis-Salivarius agar plates
❏ g. Eosin-methylene blue agar plates
❏ h. Triple sugar iron agar

❏ i. Pseudomonas P agar plates
❏ j. Bacto-SIM medium
❏ k. Koser's citrate medium
❏ l. Tryptone broth
❏ m. Methyl red broth
❏ n. Urea broth
❏ o. Carbohydrate fermentation media, including glucose, maltose, sucrose, and mannitol

Reagents

❏ a. Gram stain reaction
❏ b. Acid-fast stain reaction
❏ c. Coagulase: 3 mL citrated plasma
❏ d. Oxidase: dimethyl-*p*-phenylene-diamine
❏ e. Catalase: 3% hydrogen peroxide

❏ f. Indole: Kovac's solution
❏ g. Methyl red: methyl red indicator
❏ h. Voges-Proskauer alpha-naphthol and potassium hydroxide-creatine solutions

❏ 4. Enterotube II Code Manual (Computer Identification System)

Procedure

This procedure is to be performed by students individually.

Additional Techniques Required for This Portion of the Exercise:

❑ 1. Bacterial Smear Preparation, Procedure Diagram 2, Exercise 2

❑ 2. Broth Transfer, Procedure Diagram 5, Exercise 4

❑ 3. Agar Slants as Sources of Inoculum, Procedure Diagram 6, Exercise 4

❑ 4. Colony Selection, Procedure Diagram 8, Exercise 4

❑ 5. Streak Plate Technique, Procedure Diagram 10, Exercise 5

❑ 6. The Gram Stain, Procedure Diagram 17, Exercise 14

❑ 7. The Acid-Fast Stain, Procedure Diagram 18, Exercise 15

❑ 8. The Schaeffer-Fulton Spore Stain, Procedure Diagram 20, Exercise 16

❑ 9. Triple Sugar Iron Agar (TSIA) Inoculation Technique, Procedure Diagram 25, Exercise 26

❑ 1. Each student will receive 2 different bacterial species streaked on separate halves of a trypticase soy agar plate.

❑ 2. Observe the colonial morphology of well-isolated colonies on each half of the trypticase soy agar. Give each unknown a number for identification. Record all your observations in the appropriate portions of the Results and Observations section.

❑ 3. Select 1 of the colonies of each culture. Remove a small inoculum and inoculate separate trypticase soy agar slants. Label each slant with the correct unknown number. These stock cultures, after incubation, will serve as sources of organisms for subsequent inoculations.

❑ 4. Perform a Gram stain on the remaining portions of the colonies used for the stock culture preparations. Record the Gram stain reaction and morphology on page 607.

❑ 5. On the basis of the Gram stain reaction and cellular morphology of the 2 different bacterial cultures, select an identification scheme from those presented following this section. Representative reactions of certain pathogens have been included for comparative purposes.

❑ 6. Record the necessary steps for identification of the suspected organism *in pencil* on the form provided on page 607.

❑ 7. Obtain your instructor's approval prior to checking out the necessary materials and performing the tests.

❑ 8. Successful identification of an unknown culture depends on common sense and the correct application of basic procedures and identification keys. If you have difficulty, perform the following *unknown check:*

    a. Check the purity of your stock culture.
    b. Go over the identification key if you performed the correct tests.
    c. Make certain that tests have been made carefully and interpreted correctly. Use the appropriate figures in Exercises 20 through 27.

Identification Keys

In the **Identification schemes** that follow, approaches are given for the identification of important bacteria in general, dairy, food, industrial, and medical microbiology. These diagnostic keys are separated into the following categories:

1. Gram-positive rods

2. Gram-positive cocci

3.   Gram-negative rods

4.   Gram-negative cocci

5.   Enterotube identification system for selected gram-negative rods

The symbols used in the various approaches (unless otherwise indicated) are as follows:

M, medium	X, late or irregularly positive
R, result	O, no growth
T, test	A, acid
+, positive reaction	Alk, alkaline
−, negative reaction	NC, no change
±, slightly or weakly positive	

Identification Scheme for
Selected Gram-Positive Bacteria

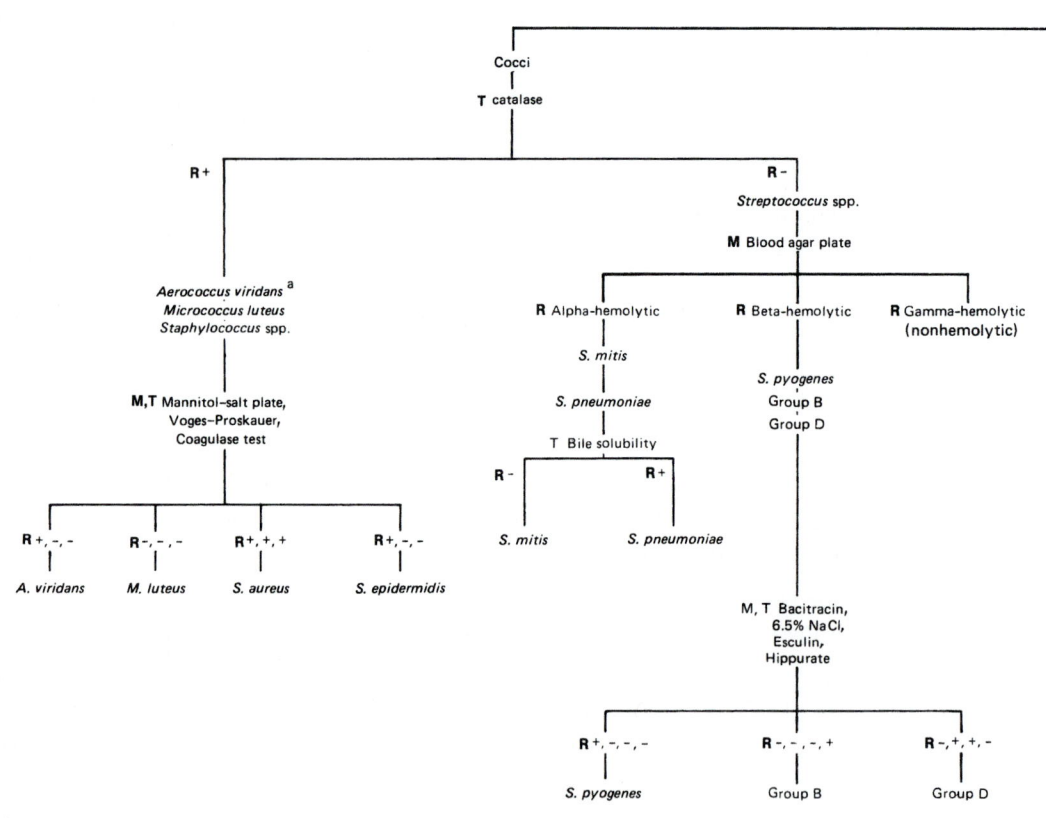

^{a} Colonial and microscopic morphology should differentiate these organisms from one another.

^{b} Refer to the index for location of these tests.

Identification Scheme for Selected Gram-Positive Bacteria

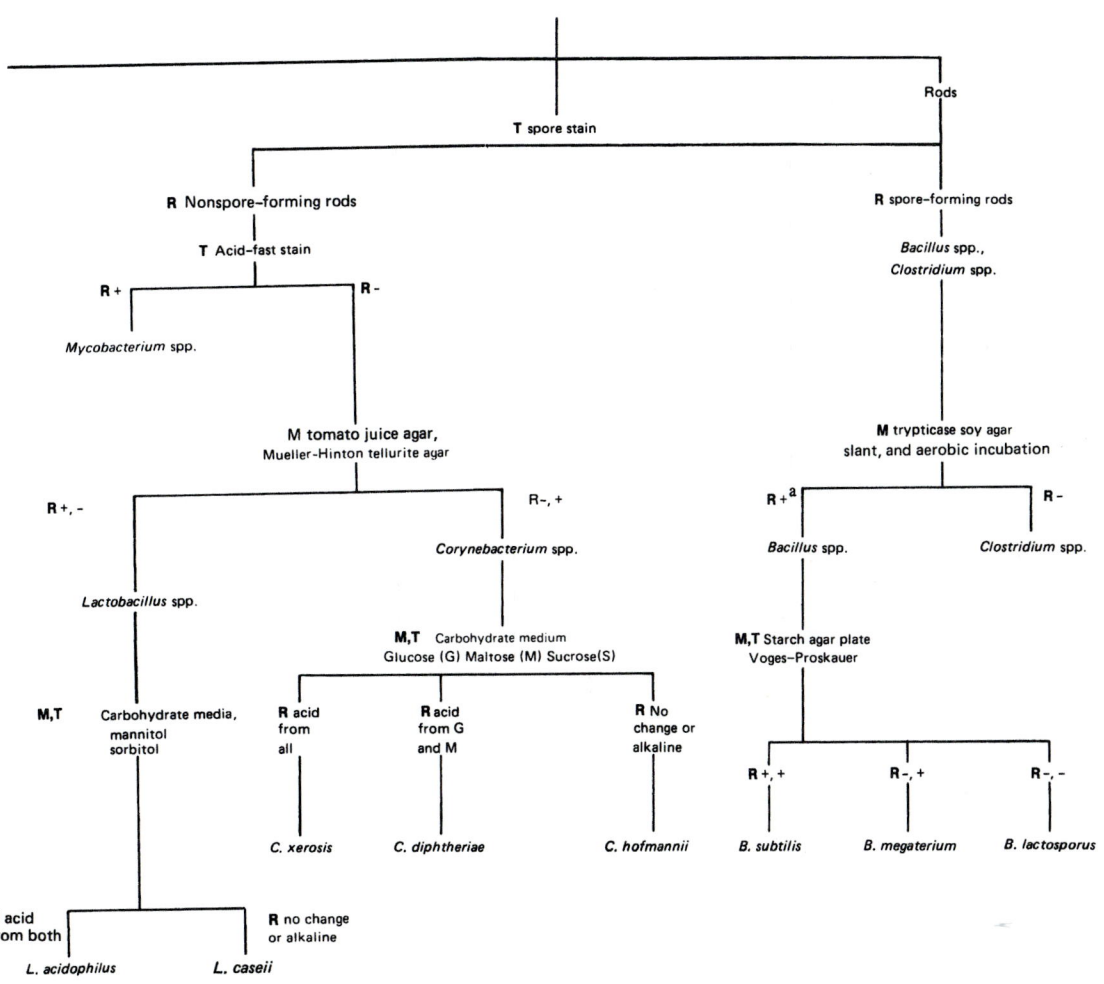

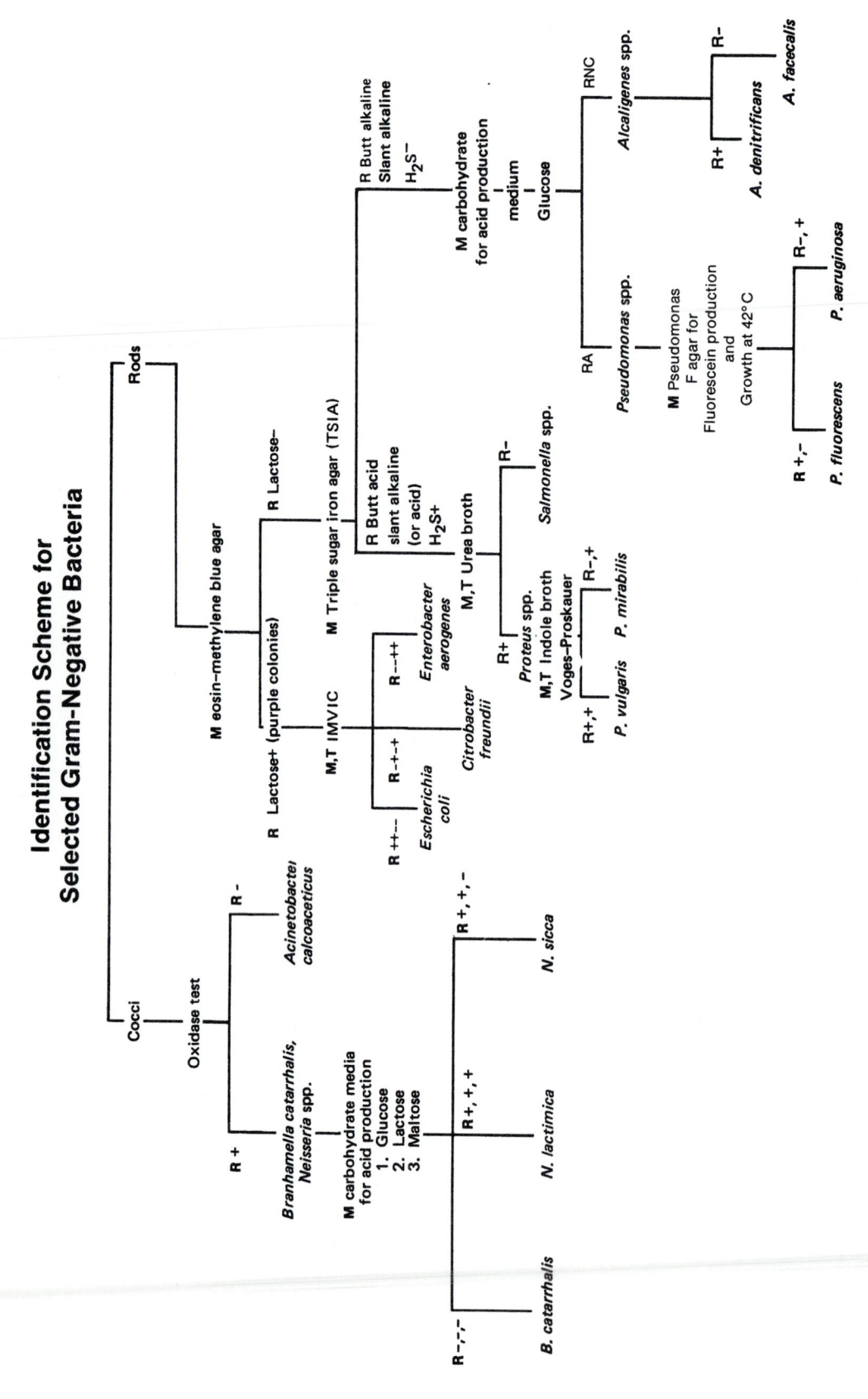

Identification Scheme for Selected Gram-Negative Bacteria

B. Unknown Bacterium Identification with Enterotube

Procedure

This procedure is to be performed only with gram-negative bacterial rods. Refer to Exercise 22 and Procedure Diagram 27 for additional details of Enterotube usage. The approach followed here is a modified version of the one used with Enterotubes in a clinical setting.

Inoculation

❑   1. Remove the caps from each end of an Enterotube system. *Do not flame the needle.*

❑   2. Pick a well-isolated colony directly with the tip of the Enterotube II inoculating wire.

❑   3. Inoculate the Enterotube by pulling the wire through all compartments using a twisting and turning motion.

❑   4. Reinsert the inoculating wire into the Enterotube until the tip of the wire is in the H₂S/indole compartment. A *notch* should now be obvious on the wire.

❑   5. Break the inoculating wire at the notch by bending.

❑   6. Discard the handle end of the wire as directed, and replace the caps on the Enterotube.

❑   7. Punch holes with the broken-off part of the inoculating rod through the thin tapelike material on the bottom of the Enterotube covering the last 8 compartments (adonitol, lactose, arabinose, sorbitol, Voges-Proskauer, dulcitol/PA, urease and citrate). The holes will provide for aerobic growth in these comparements.

❑   8. Label each Enterotube with your unknown, and incubate at 35°C to 37°C for 18 to 24 hours or as indicated by your instructor.

❑   9. After incubation, add the necessary reagents as described in Procedure Diagram 27 in Exercise 27. However, interpret and record all reactions with the exception of the indole and Voges-Proskauer tests. These tests must be done last because they may affect the other results obtained with the Enterotube system.

❑ 10. A number of Enterotube II test systems are shown in Figure 65–2.

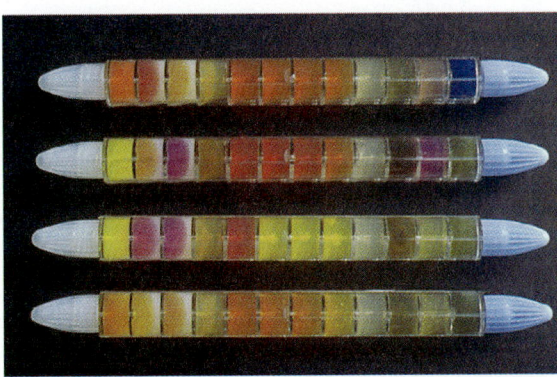

Figure 65–2
A number of Enterotube II systems showing a variety of possible reactions. The system at the bottom is uninoculated.

❑ 11. The reactions of the Enterotube test system are arranged in specific groups to yield a 5-digit code number from which an organism's identity can be determined. This arrangement is shown in Procedure Diagram 27 of Exercise 27. Table 65–1 lists several examples of the 5-digit code.

❑ 12. Using positive and negative reactions as the basis for an organism's identification, convert the test results obtained with your unknown to digital form using the diagram provided. An example of this step is shown in Procedure 27 in Exercise 27.

Determining an Unknown's Identity with the Enterotube II

❑   1. Place the incubated Enterotube system in line with figure or the unknown form so that the appropriate compartments are next to it.

❑   2. Circle the number appearing below the positive test compartment in the figure.

❑ 3. Add the numbers in the bracketed sections and enter the totals in the spaces provided below each respective arrow.

❑ 4. Locate the 5-digit number either in Table 65–1 or in an appropriate code book if one is available. Find the best or closest answer.

❑ 5. Enter the 5-digit number and the identity of the organism it represents.

❑ 6. Check your Unknown Form for completeness and turn it in to the instructor for grading.

Table 65–1

Enterotube II ID Values[a]

Enterotube II ID Value	Bacterial Species	Enterotube II ID Value	Bacterial Species
#40673	*Klebsiella pneumoniae*	#56001	*Arizona sp.*
#41447	*Providencia stuartii*	#56150	*Salmonella enteriditis*
#42226	*Proteus mirabilis*	#60050	*Shigella flexneri*
#45162	*Yersinia enterocolitica*	#60371	*Enterobacter cloacae*
#50040	*Salmonella typhi*	#62006; 36007; 16007	*Proteus vulgaris*
#50063	*Serratia marcescens*	#62302	*Citrobacter freundii*
#55241; 55650; 55601	*Escherichia coli*	#70020	*Hafnia alvei*
#55760	*Enterobacter aerogenes*		

[a]Refer to Enterotube Code Book for additional values.

Results and Observations

Bacterial Unknowns

1. Enter your observations and test results in the following report form. Be certain that all descriptions and results are indicated together with the correct spelling of the identified bacterial species.

Bacterial Unknown Report Form

Unknown Culture 1 Code No. _____	Unknown Culture 2 Code No. _____
Description of colony: 1. Pigment _____ 2. Margin _____ 3. Colonial growth _____ 4. Elevation _____ 5. Odor _____ 6. Other _____	Description of colony: 1. Pigment _____ 2. Margin _____ 3. Colonial growth _____ 4. Elevation _____ 5. Odor _____ 6. Other _____
Gram reaction and morphology _____	Gram reaction and morphology _____

Unknown Culture 1

Tests (approved by _____)

	Result:
1.	1.
2.	2.
3.	3.
4.	4.
5.	5.
6.	6.
7.	7.
8.	8.
9.	9.
10.	10.

This bacterium is probably

_____   _____
Genus   species

Student's name _____

Score _____

Remarks (for instructor only)

_____

Unknown Culture 2

Tests (approved by _____)

	Result:
1.	1.
2.	2.
3.	3.
4.	4.
5.	5.
6.	6.
7.	7.
8.	8.
9.	9.
10.	10.

This bacterium is probably

_____   _____
Genus   species

Student's name _____

Score _____

Remarks (for instructor only)

_____

Unknown Bacterial Identification with Enterotube

1. Enter your observations and test results in the following report form.

Enterotube Unknown Report Form

Unknown Culture Number _____

Description of Colony:

1. Pigment _____   4. Elevation _____

2. Margin _____   5. Odor _____

3. Colonial growth _____   6. Other _____

Enterotube II: Results by Compartment

| GLU. | GAS | LYS. | ORN. | H₂S | IND. | ADON. | LAC. | ARAB. | SORB. | V.–P | DUL. | P.A. | UREA | CIT. |

4 + 2 + 1   4 + 2 + 1   4 + 2 + 1   4 + 2 + 1   4 + 2 + 1

Id Value

_____

Final ID   Bacterial

Value _____   Identification _____
                                      Genus   Species

Student's name: _____   Score _____

Remarks (for instructor only) _____

(a)   (b)

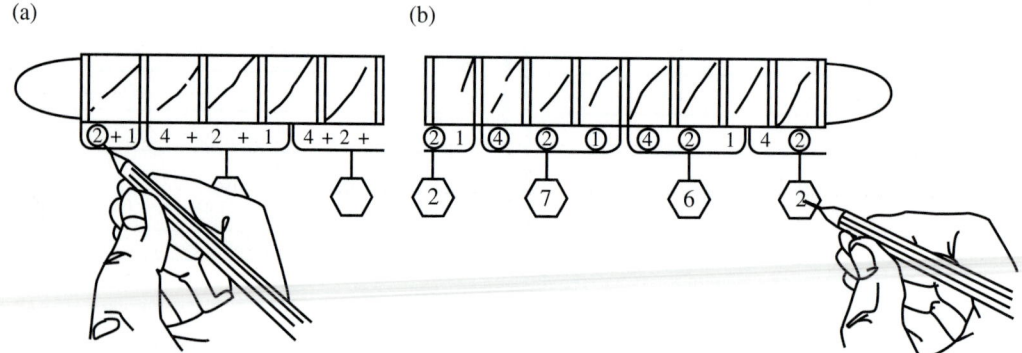

Figure 65–3

Recording test results. (a) Indicating the position reaction of a compartment. (b) Adding circled numbers of Enterotube sections and entering the total into the ID space. The order of the ID values specifies a particular bacterial species.

Laboratory Review 65 Methods for Identifying an Unknown Bacterial Specimen

1. Describe an important characteristic that differentiates each set of organisms:

   a. *Proteus* sp. versus *Salmonella typhi* ______________________

   b. *Bacillus* sp. versus *Clostridium* sp. ______________________

   c. *Escherichia* sp. versus *Klebsiella* sp. ______________________

   d. *Proteus vulgaris* versus *P. mirabilis* ______________________

   e. *Escherichia coli* versus *Citrobacter freundii* ______________________

   f. *Streptococcus pneumoniae* and *Staphylococcus aureus* ______________________

2. List several selective and/or differential media from this exercise and describe the means by which selection or differentiation is made possible.

Table 65–2

Selective and/or Differential Media

Medium	Basis for Selection or Differentiation

So, naturalists observe, a flea Hath smaller fleas that on him prey.
And These have smaller still to bite'em;
And so proceed ad infinitum.

Jonathan Swift

Parasitology deals with organisms that live either in or on another organism. The latter is referred to as the *host*. The close association of these organisms is known as *symbiosis*. If the relationship is one of mutual benefit to the individuals involved, it is called *mutalism*. However, if only one member of the pair benefits, while the other is neither benefited nor harmed, the relationship is referred to as *commensalism*. Any of these symbiotic relationships can be developed satisfactorily by a parasite, which usually does not settle into its environment as a result of chance but is forced into its form of existence as a consequence of complete dependence on the host for survival. Some parasites multiply in tremendous numbers and eventually kill their host. This is an example of *unbalanced parasitism,* because the death of the host eliminates the only immediate means of continued survival for the parasite.

The alternative to the predator existence mentioned is one of *normal parasitism,* in which the parasite has a low level of virulence and the host can control the parasite's location and ability to reproduce. Thus, a balanced relationship can be established between the two members of the association.

Animals and plants have competed among themselves for millions of years for food and space. Over this time, parasites have invaded almost every type of living host, exhibiting various degrees of dependence. While some organisms are capable of either a parasitic or free-living existence, more kinds and numbers of animal parasites than free-living animals exist. The major groups of animal-like or animal parasites are found among the Protista (**protozoa**); the **helminths,** which include the flatworms (Figures XIV–1*a* and XIV–1*b*) and roundworms (Figure XIV–1*c*); and the **arthropods.** The host and its parasites represent a community of organisms living in close contact and exerting a significant effect upon each other.

(a)

(b)

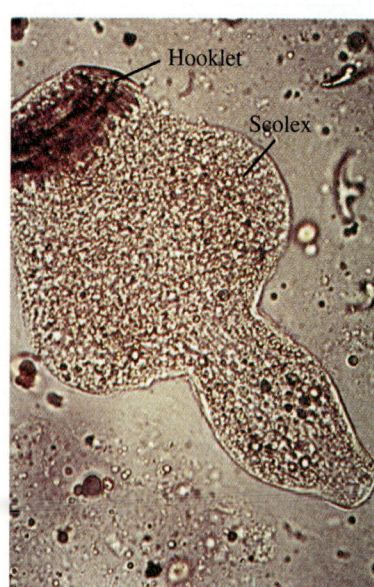

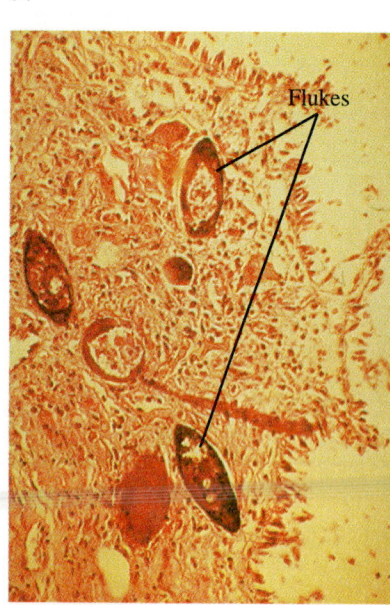

(c)

Figure XIV–1

The helminths. (a) A microscopic view of the sheep tapeworm. Note the hooklets on the head or scolex. (b) A number of flukes (trematodes) in a tissue specimen. (c) An adult *Ascaris,* a roundworm, that is visible to the eye.

The term **infection** is used in microbiology to indicate the relationship of the parasite to its host. The usage is also applicable to animal species that are *endoparasites*. However, for *ectoparasites,* which attach to the skin or temporarily invade the superficial layers of the skin, the term *infestation* is employed. An example of an infestation is scabies (Figure XIV–2), which is caused by the mite, *Sarcoptes scabei.* Thus, a state of *parasitosis* is caused by either an infection or an infestation with an animal parasite.

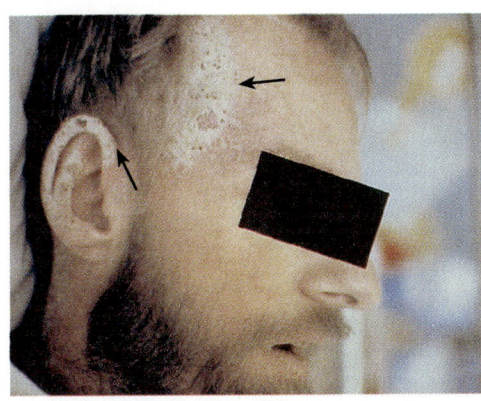

Figure XIV–2
The destructive effects of arthropods. An AIDS patient with the mite-caused *infestation* known as scabies.

Helminths are different from all other infectious agents. Parasitic worms tend to have much longer generation times than pathogenic bacteria, fungi, protozoa, and viruses. In addition, direct multiplication within the host is either absent or occurs at a low rate. Because most worm infections cause relatively limited immune responses, they tend to persist.

Helminthic infections in compromised hosts, individuals in whom normal immune systems are impaired, absent, or bypassed, are becoming more common. Such situations are especially true in health care and related facilities, and present a growing problem in terms of diagnosis and subsequent treatment.

The sources of parasites are many. These include (1) domestic or wild animals in which parasites can live (usually referred to as *domestic* and *sylvatic reservoirs,* respectively); (2) blood-sucking insects, of which mosquitoes, lice, and ticks are good examples; (3) various foods containing immature, infective parasites: (4) contaminated soil or water; and (5) humans and any portion of their environment that has been contaminated. It should be noted that the individual harboring a parasite can cause his or her own reexposure with the same species of parasite. This is called *autoinfection.* A person with a parasitic infection that is transmissible to others, and yet not exhibiting any related signs or symptoms, is referred to as a *carrier.*

The exercise in this section is concerned primarily with the life cycles and distinguishing characteristics of selected parasitic helminths of medical importance.

After completing this exercise, you should be able to:

1. Distinguish among the basic parts of adult tapeworms, roundworms, and flukes.
2. Identify specific helminth species on the basis of the appearance of characteristic ova.
3. Identify selected arthropods associated with infestations.

The term *helminth* is derived from the Greek word meaning worm, and is used in relation to parasitic as well as free-living species. Worms of medical importance include those belonging to the phyla Nematoidea (roundworms, or nematodes), Platyhelminthes (flukes, or trematodes, and tapeworms, or cestodes), and Annelida (segmented worms), a group that includes the blood-sucking aquatic and terrestrial leeches. All these phyla are classified under the subkingdom Metazoa.

Adaptation of a helminth to a parasitic existence is in large part determined by the development of certain structural and metabolic modifications. Many intestinal worms (and some others) have an especially hardened integument that enables them to resist being digested by the host (Figure 66–1a). Other modifications include the possession of hooks (Figure 66–1b), spines, cutting plates, various enzyme secretions, and other weapons for purposes of attachment or penetration and the development of elaborate reproductive systems. The latter feature is represented by hermaphroditism in tapeworms and a large number of flukes. It should be noted that most worm species do not multiply within the human definitive (final) host.

The nematodes, or true roundworms, are the largest group of helminths that parasitize humans. Some of these worms also have other mammalian hosts. Nematodes are unsegmented worms. They are usually long, cylindrical, and vary from a few millimeters (Figure 66–2) to over a meter long. The nematodes have a fully functioning digestive tract, and generally have separate sexes.

Species included in this group are *Ascaris lumbricoides,* the etiological agent of ascariasis; *Trichenella spiralis,* the pork roundworm (Figure 66–3a): *Wuchereria bancrofti,* the causative agent of elephantiasis of filarial origin (Figures

(a)

(b)

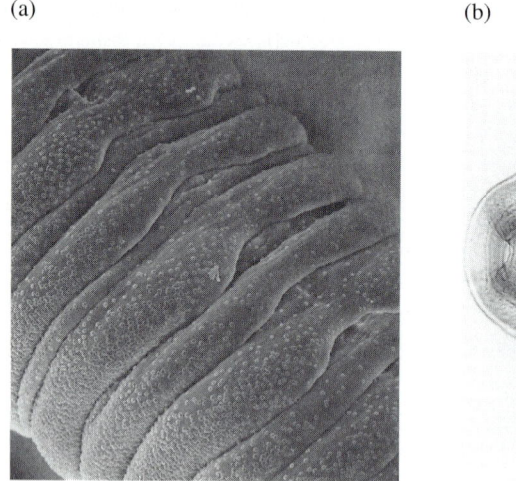 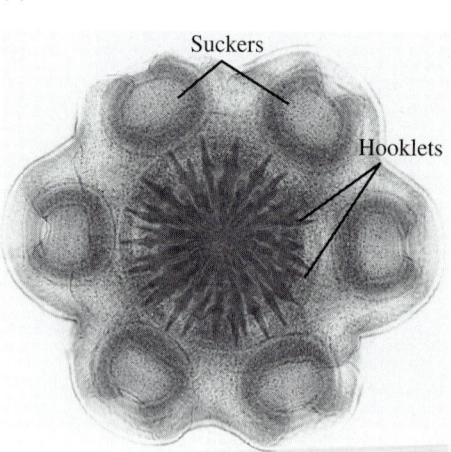

Figure 66–1

Parasitic worms. (a) A scanning micrograph showing the surface of a tapeworm's strobila, or body. The outer covering of this worm's body is covered with small structures that probably function in the uptake of nutrients.
[Reproduced by permission of the National Council of Canada, from Boyce, N.P., *Can. J. Zoo.* 54:610–613 (1976).]

(b) A view of the scolex (head) of the tapeworm *Taenia crassiceps.* The photo clearly shows the devices used for attachment, namely hooklets and suckers.
(From Schiller, E.L., *J. Parasit.* 59:122–129 [1973].)

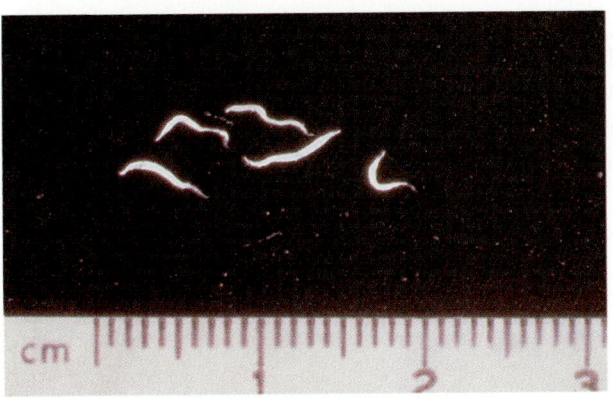

Figure 66–2
Pinworms recovered from the perianal region of a child.
(Courtesy Dr. Lesley Alpert, Pathology Department, The Sir Mortimer B. David Jewish General Hospital.)

66–3*b* and 66–3*c*); and *Necator americanus,* one of the hookworm species. All roundworms are characterized by five successive life-cycle stages, four larval and one adult. Ova or eggs and larvae living in the host are ejected in a variety of ways, depending upon where the parasite is located. Eggs of some pathogenic worms are discharged in sputum, feces, or urine. Others are removed from blood or tissues by blood-sucking insects. Eggs of *Enterobius vermicularis* (the pinworm) are released in and near the perianal area. Scratching the area may help to transfer eggs of the parasite and spread the infection. Wearing contaminated clothes may also initiate an infection or reinfection. The diagnosis of pinworm is normally achieved by applying cellulose tape (Scotch tape), sticky surface down to the skin in the anal or perianal area. The tape is transferred to a glass slide and examined under the microscope for the presence of pinworm eggs (Figure 66–3*c*).

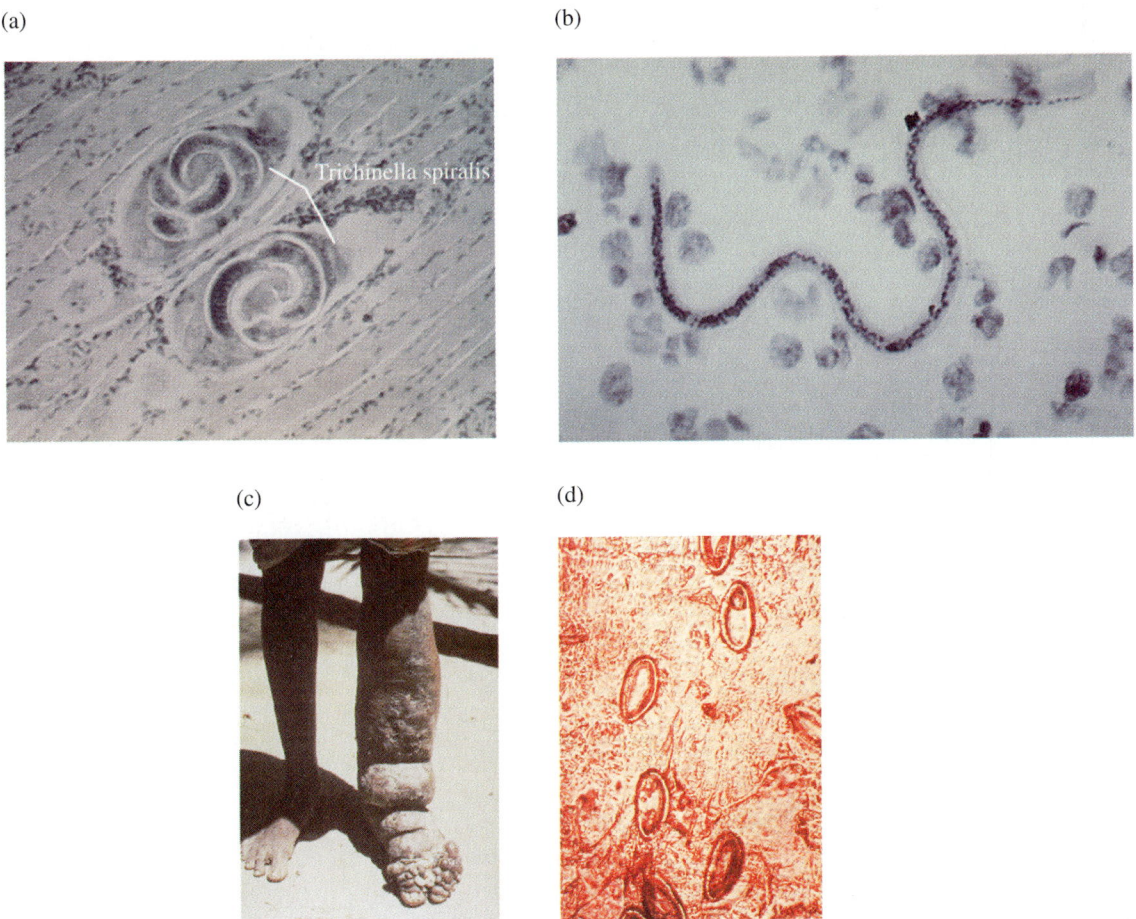

Figure 66–3
Examples of nematodes. (a) A muscle section showing *Trichinella spiralis* larvae. (b) *Wuchereria bancrofti,* a cause of elephantiasis in a stained tissue smear. (c) An actual case of elephantiasis. (d) The ova of *Enterobius vermicularis* (the cause of pinworm) on a cellulose tape.

Eggs of some nematodes undergo a period of maturation in soil before becoming infective. Certain of these agents require the aid of blood-sucking insects for development. Still others are ejected from a host in an advanced state and can become infective very quickly. Upon entering a host, some nematode larvae proceed directly to the bowel, where they develop into adult worms. Other species migrate through the tissues of the host and mature during the journey.

The platyhelminths, which include flukes (Figure 66–4a) and tapeworms, exhibit complicated life cycles (Figure 66–4). Flukes (*Clonorchis sinensis*, the chinese liver fluke, and *Schistosoma japonicum*, the blood fluke) have life cycles that can be divided into three generations. Depending on the fluke species, after the ovum is discharged from the definitive host, it hatches and successively forms the developing stages: *miracidium*, *sporocyst*, *redia*, and *cercaria* (Figure 66–4). Some flukes also form another stage, the metacercaria. Upon completing two generations in the bodies of snails, they spend the third generation in the body of the definitive host. Tapeworms, such as *Taenia solium*, the pork tapeworm;

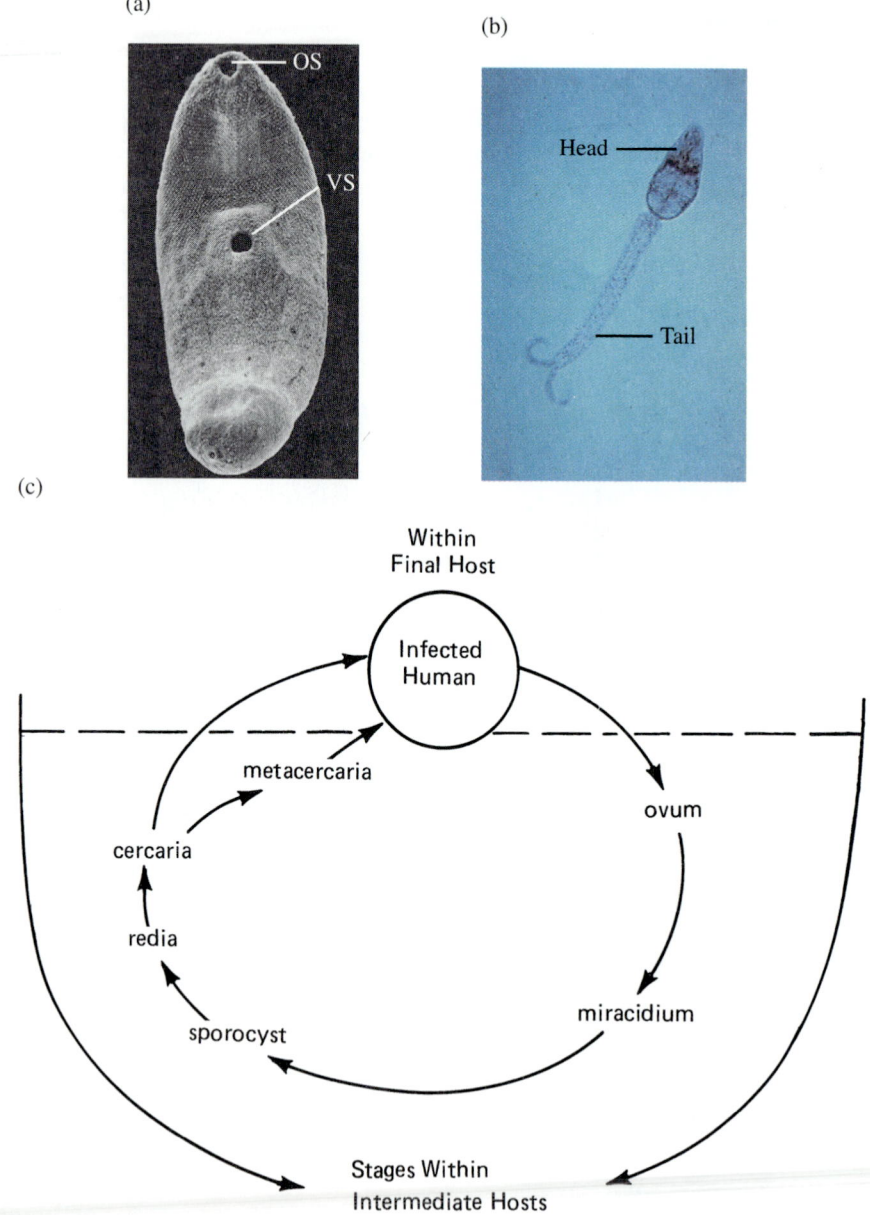

Figure 66–4

The trematodes or flukes. (a) A scanning micrograph of the fluke *Fasciola hepatica*, showing its oral sucker (OS), ventral sucker (VS), and an outer covering of spines.
(From Bennett, C.E., *J. Parasit.* 61:886–891 [1975].)

(b) A cercaria (larval form) of the blood fluke *Schistosoma* species. (c) Features of trematode development within a general life cycle.

Echinococcus granulosus, the causative agent of hydatid disease; and *Diphyllobothrium latum,* the fish tapeworm may live out their lives in a single obligatory host, or they may have two or more hosts during their lives.

Adult flukes are usually dorsoventrally flattened and unsegmented. Some are barely visible to the naked eye, while others measure 3 cm of more in length. Most flukes possess a sucker surrounding their mouths (Figure 66–4a).

Typical adult tapeworms are characterized by the following anatomical portions: (1) a head or attachment organ called the *scolex,* and (2) a chain of individual body segments, referred to as *proglottids,* which produce eggs (Figure 66–5). The entire body is designated as the *strobila.* The egg-producing segments, containing well-developed male and female reproductive organs, are located in the region behind the scolex and neck region.

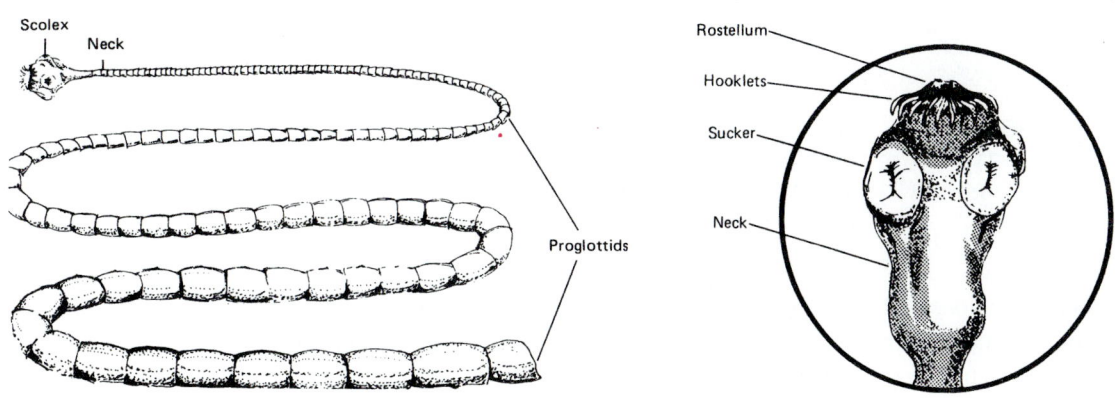

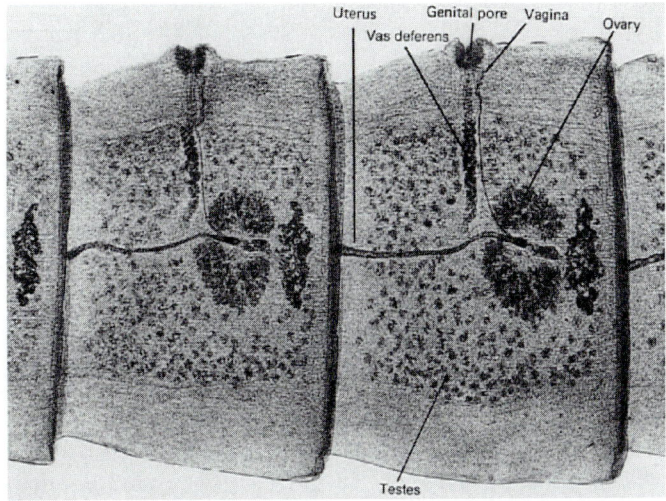

Figure 66–5

The pork tapeworm *Taenia solium.* (a) A composite sketch. (b) The armed scolex. (c) Proglottids of *Taenia* species.

Specific diagnosis of helminth infections may include (1) examination of feces, urine, sputum, and other specimens for eggs and proglottids, in the case of tapeworms, and (2) examination of biopsied material for eggs and adult specimens.

Representative adult parasites, as well as ova and larvae, are provided for examination in this exercise. Students should note the distinctive features of the parasites. Sketches of several parasites are included on the following pages.

Ectoparasites of significance include lice, *Pediculus* species (head, body, and pubic), and mites, such as *Scabies* (Figure 66–6). Head and pubic lice and the scabies mite are fairly common within many parts of the world, including the United States. While the infestations caused by these arthropods are generally not serious problems in child day-care centers and in some situations, health care facilities, they have the potential to cause significant outbreaks.

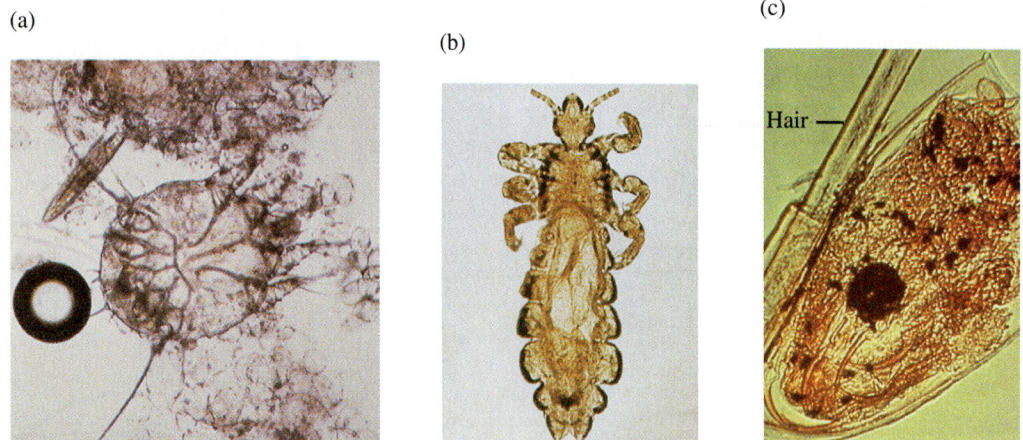

(a)

(b)

(c)

Hair —

Figure 66–6

Examples of arthropods known to cause infestations. (a) The mite *Sarcoptes* species. (b) The human body louse, *Pediculus humanus.* (c) A louse egg on a hair shaft.

Helminthic infections and arthropod infestations in compromised individuals, those persons in whom normal immune systems are impaired, absent, or bypassed, are becoming more common. Such situations are especially true in health care facilities and present a growing problem in terms of diagnosis and appropriate follow-up treatment.

A. Preserved Specimens

Materials

❑ 1. *Nematodes*
  ❑ a. Adult *Ascaris lumbricoides* (AS-kar-ris LUM-brē-koy-dēz), preserved in formalin
  ❑ b. Adult *Necator americanus* (nē-KĀ-tor a-mer-ē-CAN-us), the American hookworm, preserved in formalin
  ❑ c. Adult *Trichuris trichiura* (trī-KŪ-ris trik-ē-ŪR-a), the whipworm
  ❑ d. *Onchocerca volvolus* (ong-kō-SER-ka vol-VU-lus), contained in tumor (plastic-mounted specimens showing worms *in situ*)
❑ 2. *Platyhelminths*
  ❑ a. Adult *Taenia saginata* (TĒ-nē-a saj-Ē-na-ta), preserved in formalin
  ❑ b. Mature segments of *Diphyllobothrium latum* (di-fil-ō-BOTH-rē-um LĀ-tum) and *Taenia* species
  ❑ c. Plastic mount of adult *Dipylidium caninum* (dip-i-LID-ē-um kā-NĪ-num), the dog tapeworm
  ❑ d. Plastic mounts of *Fasciola hepatica* (fa-SĒ-ō-la ha-pat-Ē-ka) and *Clonorchis sinensis* (klō-NOR-kis si-NEN-sis)

Procedure

This procedure is to be performed by students individually.

❑ 1. Examine the various specimens provided.

❑ 2. Pay particular attention to relative sizes, shapes, and other obvious characteristics. Note structures that are recognizable with the unaided eye. Can you see the scolex, proglottids, and sex organs within the proglottids? (Use Figures 66–5a, 66–5b, and 66–5c as guides.)

B. Demonstration Slides

Materials

❑ 1. Nematodes
  ❑ a. Ova of *Ascaris* spp.
  ❑ b. Ova of *Enterobius vermicularis* (en-ter-Ō-bē-us ver-mik-CŪ-lār-is)
  ❑ c. Ova of *Necator americanus*
  ❑ d. Ova of *Trichuris trichiura*
  ❑ e. Tissue sections (t.s.) of *Trichinella spiralis* (trik-i-NEL-la spir-AL-is)
  ❑ f. Adult worms of *Wuchereria bancrofti* (voo-ker-Ē-rē-a ban-krof-tē) in blood

❑ 2. Platyhelminths (cestodes, or tapeworms)
  ❑ a. Ova of *Diphyllobothrium latum*
  ❑ b. Ova of *Taenia saginata* and *T. solium*
  ❑ c. Scolex and proglottids of *T. solium*
  ❑ d. Adult *Echinococcus granulosus* (e-ki-nō-KOK-us gran-Ū-lō-sus)

❑ 3. Platyhelminth (trematodes or flukes)
  ❑ a. Ova of *Clonorchis sinensis*
  ❑ b. Ova of *Fasciola hepatica*
  ❑ c. Ova of *Schistosoma haemotobium* (shis-tō-SŌ-ma hē-ma-TŌ-bē-um), *S. japonicum* (ja-PON-ē-kum), and *S. mansoni* (man-SŌ-nē)
  ❑ d. Miracidia, rediae, and cercaria of *F. hepatica*
  ❑ e. Cercaria of *S. japonicum*
  ❑ f. Adult female and male *Schistosoma japonicum*
  ❑ g. Adult *Clonorchis sinensis*

❑ 4. Arthropods associated with infestations. Prepared slides of the following specimens:
  ❑ a. Adult *Pediculus humanus capitis* (pe-DIK-ū-lus hū-MAN-us KAP-i-tus)
  ❑ b. *P. humanus* eggs on hair shaft
  ❑ c. The itch mite *Sarcoptes* sp. (sar-KOP-tēz)

Procedure 1: Examining Specimens

This and the following procedure are to be performed by students individually.

❑ 1. Composite sketches of a fluke and various helminth ova are given in the Results and Observations section. Use these drawings and Figures 66–3, 66–4, 66–5, and 66–7 as guides for the examination of the parasite specimens.

❑ 2. Complete Table 66–1 in the Results and Observations section.

(b)

(a)

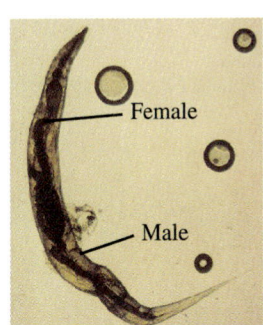

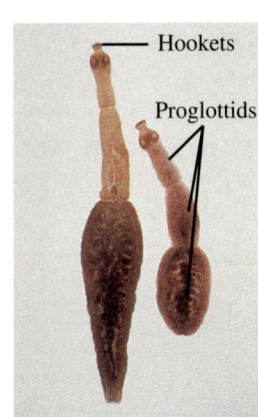

Figure 66–7

Two examples of helminths. (a) The blood flukes *Schistosoma* species. (b) The sheep tapeworm, *Echinococcus graculosus.*

Procedure 2: Identifying Unknowns

❏ 1. Each student will be given 2 unknown slides containing 1 of the parasites studied in this exercise.

❏ 2. The unknowns are to be identified, and the results entered on the Unknown Form on page 622.

Results and Observations

1. Use these diagrams in examining the demonstration slides.

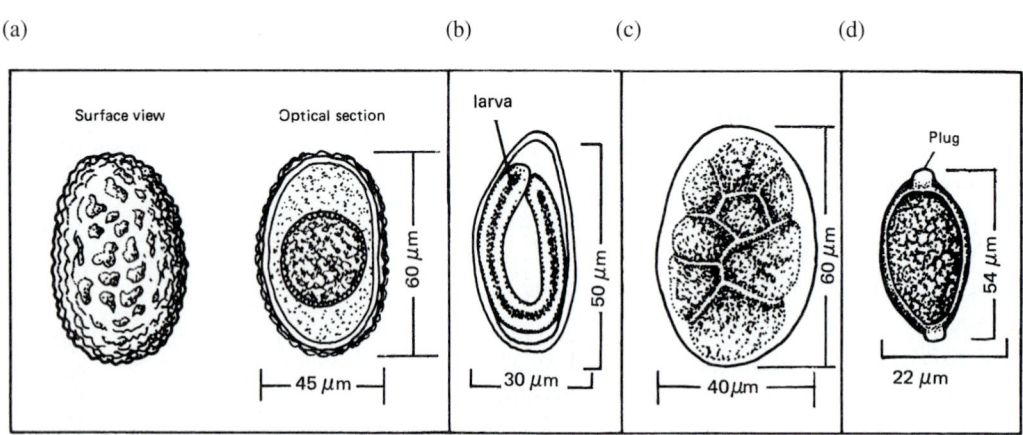

Figure 66–8
Diagrams of selected nematode ova. (a) *Ascaris* spp. (b) *Enterobius vermicularis.* (c) *Necator americanus.* (d) *Trichuris trichiura.*

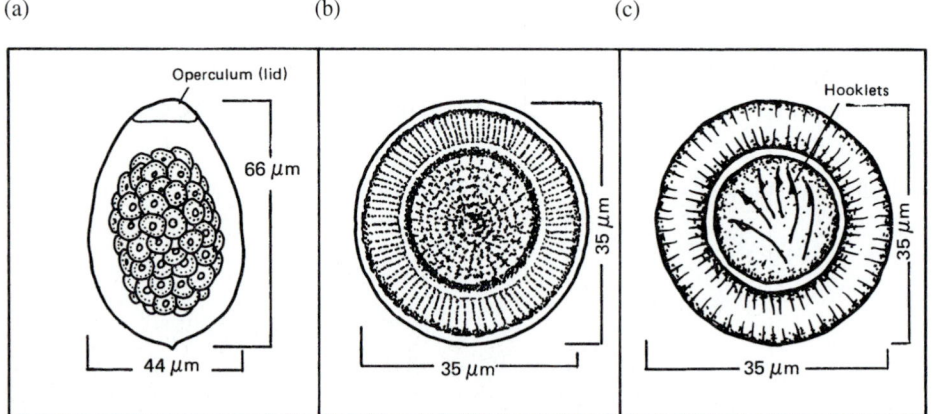

Figure 66–9
Diagrams of selected cestode ova. (a) *Diphyllobothrium latum.* (b) *Taenia saginata.* (c) *T. Solium.*

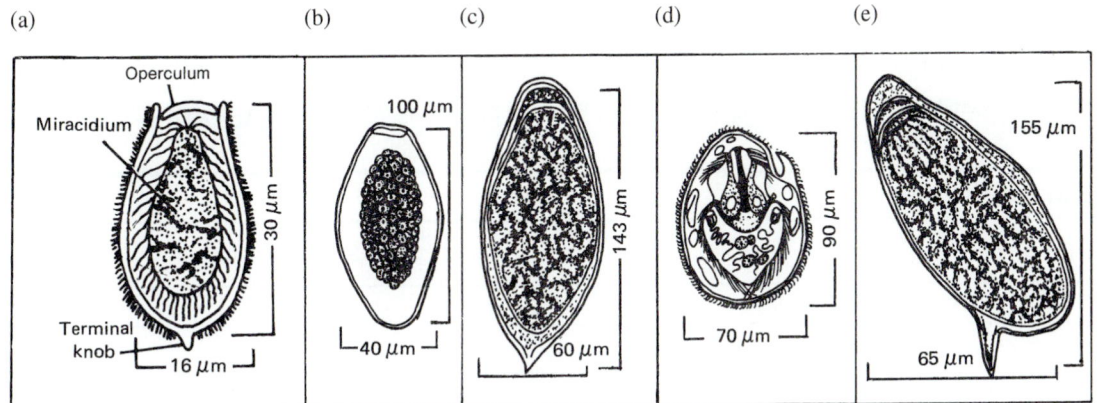

Figure 66–10

Diagrams of selected trematode ova. (a) *Clonorchis sinensis.* (b) *Fasciola hepatica.* (c) *Schistosoma haematobium.* (d) *S. japonicum.* (e) *S. mansoni.*

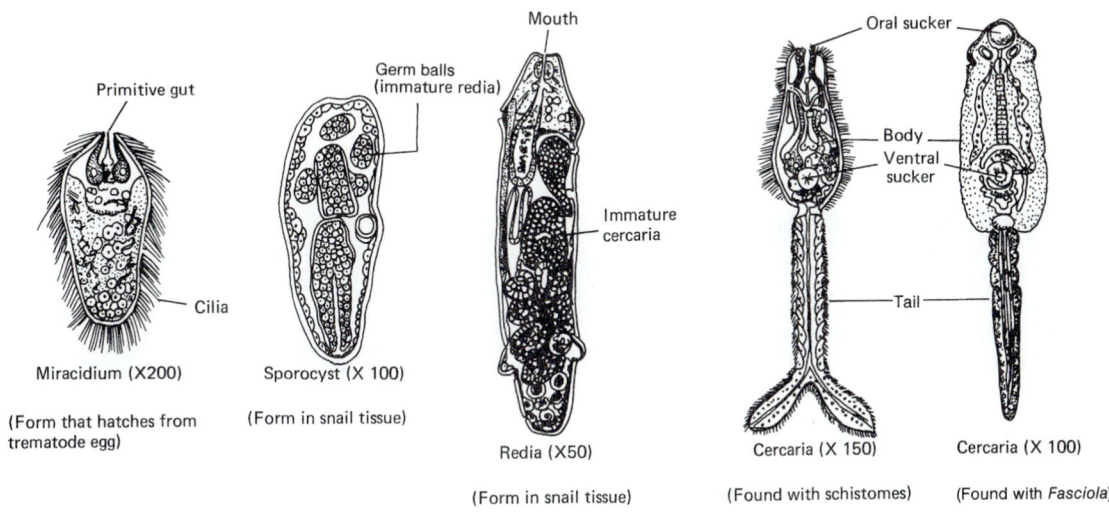

Figure 66–11

Developing or larval stages found among various species of flukes.
(Modified after U.S. Naval Medical School Laboratory Manual.)

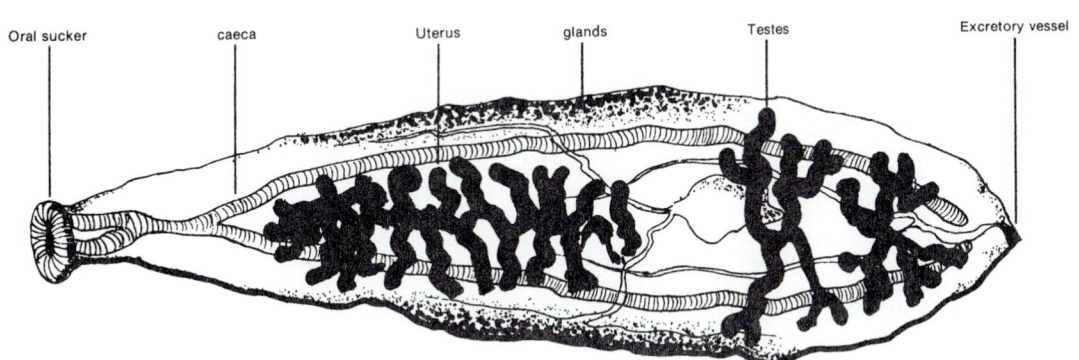

Figure 66–12

A composite sketch of the liver fluke *Clonorchis sinensis.*

2. What specific features differentiate the following helminthic ova from one another? Complete Table 66–1.

Table 66–1

Helminth Features

Helminth	Specific Feature(s)
Trichuris trichiura and Enterobius vermicularis	
Schistosoma haematobium and S. japonicum	
Fasciola hepatica and Taenia solium	
Clonorchis sinensis and Diphyllobothrium latum	

3. What specific features differentiate the cercariae of *Schistosoma japonicum* and *Fasciola hepatica* from one another?_____

_____

_____

Laboratory Review 66   Helminthology and Selected Medically Important Arthropods

1. Complete the following table. (Use your text and/or other references.)

Table 66–2

Helminth Diseases

Organism	Disease	Mode of Transmission	Tissues Involved	Clinical Specimen of Choice
A. lumbricoides				
D. latum				
F. hepatica				
S. mansoni				
T. solium				
T. spiralis				
T. trichiura				
W. bancrofti				

2. Define and/or explain:

   a. scolex _____

   b. proglottid _____

   c. ovum _____

   d. definitive host _____

   e. cercaris _____

   f. hermaphroditism _____

Key Terms

arthropod (ar-thrō-pod): an invertebrate animal generally having a hard jointed outer covering (exoskeleton), segmental body, and jointed paired appendages; examples include lice, fleas, ticks, and mites

carrier an individual with a parasitic infection, transmissible to others, yet not showing any signs or symptoms

cercaria (ser-KĀ-rē-a): a free-living, tail-bearing larval stage of a fluke; it is the infective stage

cestode (SES-tōd): a tapeworm

definitive (de-FIN-i-tiv): the final host

helminth (hel-MINTH): any worm or wormlike animal

hermaphroditism (her-MAF-rō-dit-izm): the presence of both female and male organisms in the same individual

intermediate (in-ter-MĒ-dē-et): a form of life used for development by larval stages of helminths

larva (LAR-va): developing form of an insect or worm

miracidium (me-ra-SID-ē-um): a ciliated free-swimming developmental form of a fluke that hatches from a helminth egg

nematode (NEM-a-tōd): a roundworm

ovum (Ō-vum): the female reproductive cell; egg

proglottid (prō-GLOT-tid): a segment of a tapeworm, containing both female and male reproductive organs

scolex (SKŌ-leks): the head of a tapeworm

sporocyst (SPŌR-ō-sist): a developmental stage of a fluke formed from a miracidium containing reproductive cells, and found in snail (intermediate host) tissues

strobila (strō-BIL-la): the adult form of a tapeworm

trematode (TREM-a-tōd): a parasitic flatworm belonging to the class Trematoda; a fluke

Photograph Quiz 29

Identify the ova shown on the following page.

a. _____   b. _____

c. _____

(a)

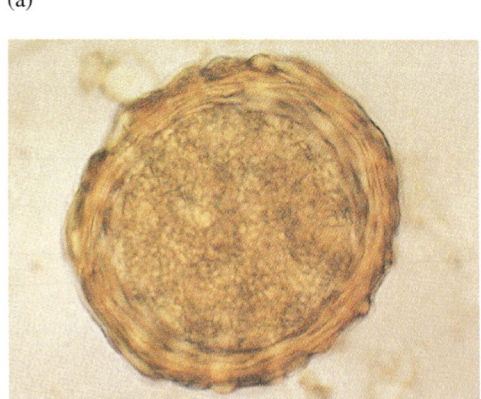

(b)

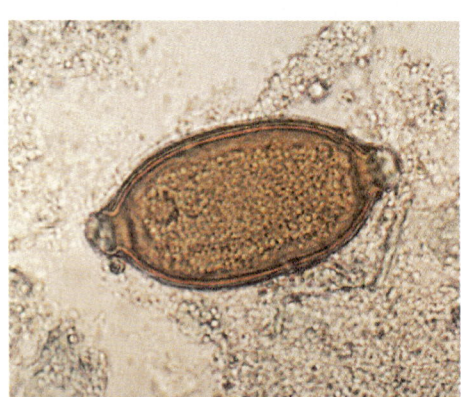

(c)

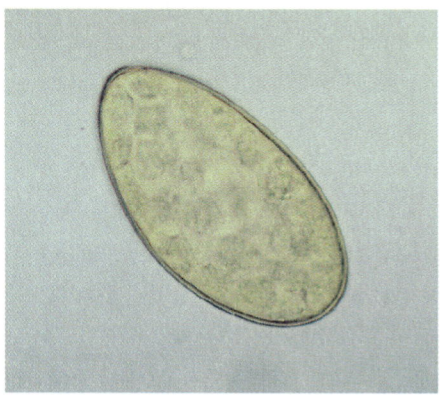

Figure 66–13
Helminth ova.

Unknown Form for Parasitic Helminths and Selected Medically Important Arthropods

Student's Name _____   Score _____

Date _____   Laboratory Section _____

Unknown Number _____   Organism Identified _____

Unknown Form for Parasitic Helminths and Selected Medically Important Arthropods

Student's Name _____   Score _____

Date _____   Laboratory Section _____

Unknown Number _____   Organism Identified _____

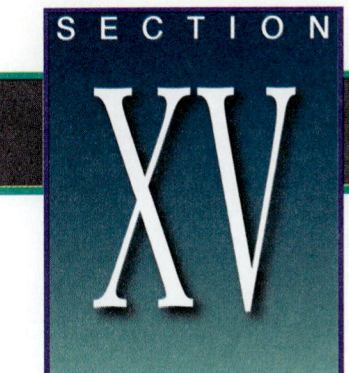

[What is written here] . . . is intended only as a small fragile seed dropped in the wind. A possible beginning. It will either grow or die depending on where it falls and how it is nurtured.

L.F. Buscaglia, 1978

Sophisticated new technology has led to exciting research and advances in microbiology and related areas of the biological sciences. The findings of such research have produced much new knowledge and additions to an ever-expanding collection of microbiological concepts and fundamentals.

This section was developed and designed to introduce examples of techniques used in research studies. Brief descriptions of the principles involved with electrophoresis, and DNA restriction analysis and the polymerase chain reaction also are included in this portion of the manual. Several of these and other techniques presented allow researchers to isolate and identify microorganisms from various sources; to make precise identifications, diagnoses, and predictions; and to design more effective drugs with which to treat disease.

After completing this experimental exercise, you should be able to:

1. Recognize and demonstrate the four phases of a typical bacterial growth curve.

2. Standardize and use a spectrophotometer.

3. Estimate the bacterial growth turbidometrically.

4. Determine viable bacterial numbers by means of a modified pour plate technique.

5. Plot and interpret bacterial growth data.

Bacteria as a group live and grow under a wide range of environmental conditions. Because these microorganisms exert most of their effects through growth, knowledge of the processes involved with microbial growth is essential to the understanding, study, and control of fundamental microbial activities. Bacteria can be studied at all stages of growth—microscopically, physically, and chemically. The information acquired through such studies can be used to correlate the chemicals formed during growth with the appearance of cellular structures. In this exercise, the measurement of bacterial growth and the influence of incubation temperature will be considered.

The term *growth* as applied to microorganisms usually refers to an increase in the number of microorganisms or mass beyond that present in the original inoculum. A single bacterial cell continually increases in size until it is about double its original dimensions. It then divides and gives rise to two cells approximately the size of the original parent cell (Figure A–1). All structural parts of the cell double during this cell growth division cycle.

The time required for the formation of two new cells from one is called the *generation time.* During this time, the population of cells doubles. Thus, starting with one bacterium, the increase in population follows a progressive doubling as shown:

	1 cell →	2 cells →	4 cells →	8 cells →	16 cells →	32 cells →	etc.
Algebraic shorthand	(1)	(2)	(2^2)	(2^3)	(2^4)	(2^5)	(2^n)

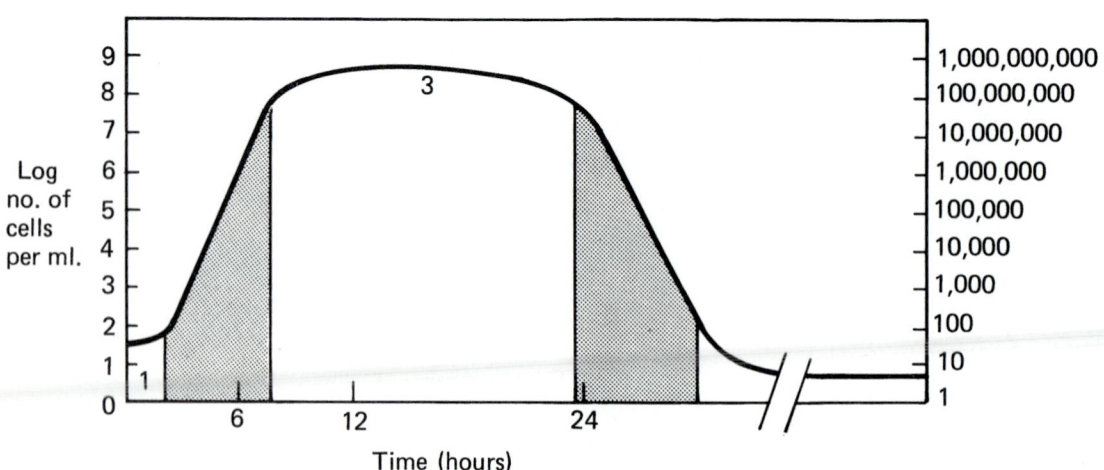

Figure A–1

An idealized growth curve obtained by counting the bacteria in a culture at intervals after inoculation. This curve shows total (living plus dead) cells.

Because cell populations can reach levels into the millions, the logarithmic (algebraic) notation approach that incorporates exponents of 10 is used to handle large numbers. For example, one hundred (100) is expressed as 10^2; one million (1,000,000) is expressed as 10^6; and so on.

To express a cell population of 8,000,000, the figure is written as the unit integer (8), multiplied by the proper power of 10 (10^6). Thus, $8,000,000 = 8 \times 10^6$. The increase in cell numbers is generally plotted (expressed) in terms of the logarithms of cell numbers.

Bacteria and other unicellular microorganisms exhibit characteristic growth cycles or patterns, which can be divided into the four distinct growth phases (Figure A–2): *lag phase* (1); *exponential* or *logarithmic phase* (2); *stationary phase* (3); and *death phase* (4). Their general characteristics are as follows:

1. *Lag Phase:* During this phase, cells in a new inoculum adjust to the medium. Increases occur in enzyme production and cell size.

2. *Exponential* or *Logarithmic Phase:* During this phase, cells and cell mass double at a constant rate. Metabolic activities proceed at a constant rate. Environmental conditions, which include pH, temperature, medium properties, etc., influence this phase. Because the doubling of the population (generation time) occurs at regular time intervals, this phase also is referred to as one of *balanced growth.*

3. *Stationary Phase:* During this phase, toxic products accumulate and/or the availability of nutrients decreases and the number of viable cells reaches a plateau (Figure A–1), resulting from some cells dying and others still growing and dividing. There is no net increase or decrease in cell number.

4. *Death Phase:* If toxic substances accumulate and/or cell starvation occurs, the cells of the population enter this phase. The rate of decline becomes exponential within time.

The measurement of bacterial growth can be determined by several methods. These include microscopic count, plate count, membrane filter count, turbidometric measurement, nitrogen determination, weight (cellular) determination, and biochemical activity measurement. The generation time of bacterial culture can be calculated with the use of a standardized inoculum (known number of bacteria), and optimum incubation conditions. Periodic sampling of the bacterial population is performed to obtain experimental data to calculate the generation time according to the following formula:

$$\text{Generation time (GT)} = \frac{t}{3.3 \log (b/B)}$$

where   t = time interval between the measurement of cell numbers in the population at one point in the log phase (B)
            and then at a later point in time (b)
       B = initial population
       b = population after time t
     log = $\log_{10}$
     3.3 = $\log_{10}$ to $\log_2$ conversion factor

In this exercise, the standard plate count and turbidometric determinations will be used to measure the growth of the same bacterial culture. Although the two approaches yield somewhat similar results, there are distinct differences in technique, in the materials used, and in the information related to living and dead organisms. General descriptions of the methods and the principles involved are presented in the following sections.

A. Standard Plate Count Technique

The standard plate count (SPC) is used to determine the number of living (viable) organisms in various substances such as water, milk, and foods, and during the various stages of bacterial growth curves. The technique is relatively simple to perform and produces excellent reproducible results. It is based on the assumption that each viable cell will form a colony. Thus, the number of resulting colonies on a plate is an indication of the number of viable cells in the original sample under study. The SPC technique consists of diluting a sample and then plating the dilutions (Figure A–5). A sample must be diluted because the number of organisms present may be too numerous to count accurately. For purposes of accuracy and reliability, after incubation only those plates with 30 to 300 colonies are used. Plates with higher counts are reported as too numerous to count (TNTC). The number of organisms in the original sample per milliliter is determined by multiplying the number of colonies formed by the dilution factor for the particular plate being used. For example, if

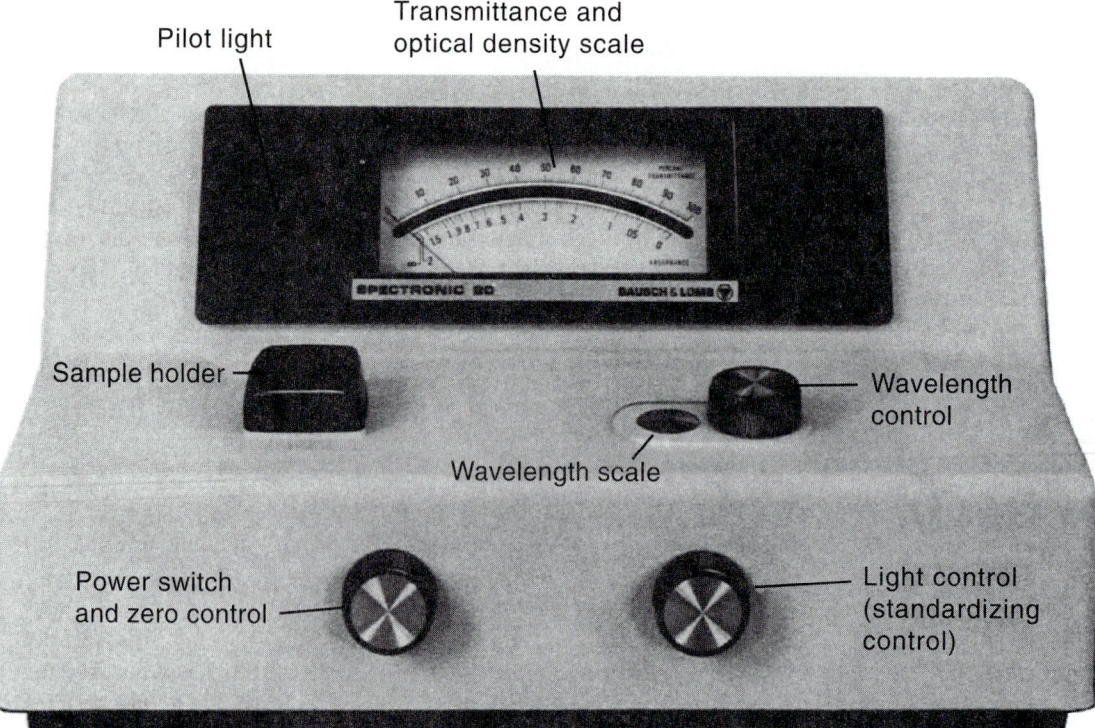

Figure A–2
Spectrophotometer 20.

200 colonies were counted on the 10^{-3} (1:1000) dilution plate, the number of organisms in the original sample would be calculated as follows:

$$\text{Number of viable bacteria/mL} = \frac{200 \text{ (number of colonies)}}{10^{-3} \text{ (dilution used)}}$$

$$\text{Number of viable bacteria/mL} = 200{,}000$$

In this experimental exercise, the growth cycle of the bacterium *Vibrio natriegens* will be followed. This microorganism has a generation time of less than 15 minutes. A complete growth cycle (Figure A–1) can be plotted in approximately 3 hours.

B. Spectrophotometric Measurements

Spectrophotometers (Figure A–2) are instruments that electronically quantify the kinds and amount of light that are absorbed by molecules in solution. Spectrophotometric measurements such as those described in this exercise are based on the Beer-Bouger Law, which shows the relationship between a solution's concentration of suspended particles and its transmission of light. This relationship provides the basis for the quantitative determinations of the concentration of bacterial cells in a solution by measurement of the light transmitted by the solution.

Normal white light is a mixture of many different wavelengths (colors) between 380 and 750 nanometers *(nm)*. The eye and brain perceive these different wavelengths as different colors. The growth measurements require that the light entering solutions (incident light) being analyzed be monochromatic (composed of a single wavelength). A spectrophotometer is equipped with a device to separate white light into its component wavelengths.

The terms *percent transmittance* and *absorbance* are used with optical methods of analysis. Transmittance (T) is the ratio of the transmitted light of the sample to the incident light on the sample. The value obtained is multiplied by 100 to derive the percent transmittance.

Absorbance (A), also called optical density (OD), is directly proportional to concentration (particles suspended) when a solution behaves according to the Beer-Lambert Law (a modification of the Beer-Bouger Law). Absorbance specifically is the logarithm to the base 10 of the reciprocal of the transmittance. In this exercise, absorbance values will

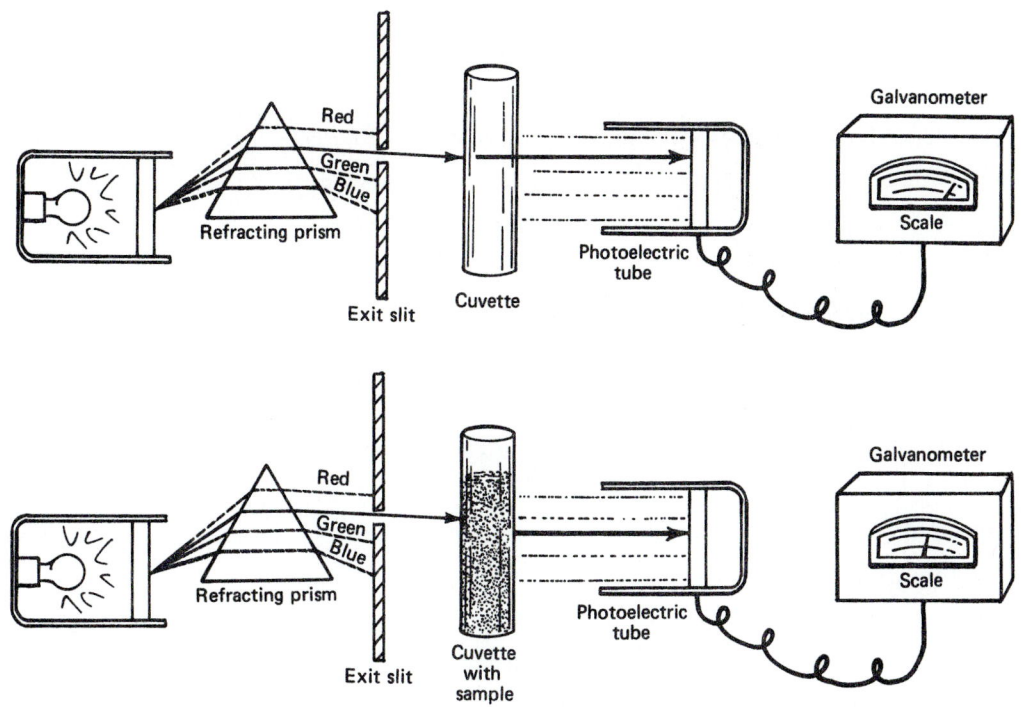

Figure A–3
A representation of the spectrophotometry principle. Spectrophotometry is based upon the physical principle that the degree of absorption of light is directly proportional to the concentration of the absorbing substance.

be used. On spectrophotometers, the absorbance scale normally is present along with the transmittance scale (Figure A–3).

The spectrophotometer measures the intensity of light (at various wavelengths) before and after the light has passed through a solution. If the solution contains particles such as bacteria that absorb light, the concentration of the bacteria can be determined spectrophotometrically by measuring the absorption of light.

The procedure for spectrophotometric measurements is relatively simple to perform. In general, it involves placing a light-absorbing particle solution into a cuvette (a special, small test tube), selecting the appropriate wavelength, and inserting the cuvette in the light pathway of a spectrophotometer so that the wavelength passes through the cuvette and the solution it contains. Any light transmitted through the cuvette is directed onto a photosensitive device, which converts the radiant energy produced into electrical energy. The electrical current generated is then measured by the spectrophotometer's meter (Figure A–3).

The absorbance, or percent transmission, is determined by comparing the value produced by the particle-containing solution to that produced by a *blank* solution, which does not contain the light-absorbing material. Note that standard plate counts and spectrophotometric measurements will be performed simultaneously to follow the bacterial growth cycle and to obtain data with which to plot the corresponding growth curve.

C. Standard Plate Count

Materials

The following items should be provided for students in groups of 4:

❑ 1. Five mL of a 24-hour brain heart infusion (BHI) broth (with 2% NaCl) culture of *Vibrio natriegens* (ATCL 14043)

❑ 2. One hundred and fifty mL of BHI broth (with 2% NaCl) in a 250-mL Erlenmeyer flask

❑ 3. Two tubes of sterile BHI broth (with 2% NaCl) for spectrophotometric standardization (6 mL/tube)

❑ 4. Forty-six melted BHI (with 2% NaCl) agar deeps

continued

Materials (continued)

- ❑ 5. Forty-six sterile Petri plates
- ❑ 6. Thirteen sterile 1.1-mL pipettes with rubber bulbs or other appropriate delivery device
- ❑ 7. Fifteen sterile 5-mL pipettes, and one Pi-Pump or similar delivery device
- ❑ 8. Thirty-one 99-mL sterile water blanks with 2% NaCl
- ❑ 9. Two different colored pencils or pens for data plotting
- ❑ 10. One container with disinfectant for pipette disposal

The following items should be provided for general class use:

- ❑ 1. One Quebec Colony Counter
- ❑ 2. One shaker water bath set at 37°C with brackets for 250-mL flasks
- ❑ 3. One water bath set at 50°C for melted BHI agar deeps
- ❑ 4. One Spectronic 20 (Bausch and Lomb) or other comparable spectrophotometer
- ❑ 5. Photocolorimetric cuvettes
- ❑ 6. Disinfectant solution in plastic wash bottles to decontaminate cuvettes
- ❑ 7. Plastic water bottles with distilled water for the rinsing of cuvettes
- ❑ 8. Test tube brushes and soap solution
- ❑ 9. Test tube racks
- ❑ 10. Kimwipes or other disposable lint-free tissues
- ❑ 11. Labeling tape or felt-tip marking pens

Procedure 1: Operation of the Bausch and Lomb Spectronic 20

(*Note:* Procedures 2 and 3 should be performed simultaneously by students working in pairs. Your instructor will indicate the inoculation volume to be used.)

This procedure is to be performed by students individually.

- ❑ 1. Examine the spectrophotometer and cuvettes provided.
- ❑ 2. With the aid of Figure A–2, locate and learn the functions of each spectrophotometer component shown.
- ❑ 3. Rotate the amplifier control (left-hand) knob clockwise to turn on the instrument. Allow the spectrophotometer to warm up for at least 15 minutes.
- ❑ 4. Close the lid of the sample holder.
- ❑ 5. After the instrument has warmed up, turn the amplifier knob so that the meter needle will read "0" on the percentage transmission (%T) scale. With this step, the instrument has been "zeroed."
- ❑ 6. Obtain a cuvette. Be careful not to handle its bottom portion. This region will be the one through which the light beam will pass.
- ❑ 7. Rinse the cuvette several times with distilled water.
- ❑ 8. Pour about one-half of the sterile BHI broth from one tube into the cuvette. Empty the cuvette and rinse it with the remainder of the broth.
- ❑ 9. Fill the cuvette with 6 mL of the BHI broth and wipe it with a Kimwipe to remove any liquid drops, fingerprints, or smudges on the lower half. The BHI solution is known as the "blank."
- ❑ 10. Insert the cuvette into the sample holder with the index line or a frosted triangle on the tube facing toward the front of the instrument.
- ❑ 11. Close the lid.
- ❑ 12. Set the wavelength for 660 nm. This wavelength is used for yellow to brown solutions.

❏ 13. Rotate the light control (right-hand) knob until the meter needle reads 100 %T. The spectrophotometer has now been standardized, and is ready for measurements.

❏ 14. Open the sample holder lid and remove the cuvette.

❏ 15. If measurements are not to be made or if the instrument is no longer needed, turn off the machine with the amplifier control knob.

Procedure 2: Standard Plate Count

Additional Technique Required for This Portion of the Exercise:

❏ The Use of a Pipette Pump, Procedure Diagram 11, Exercise 6.

This procedure should be performed by students in pairs and in duplicate. Your instructor will demonstrate the use of pipettes.

❏ 1. Label the BHI medium-containing flask with your initials. This preparation will be the source of samples for both colony counts and spectrophotometric measurements.

❏ 2. Obtain 2 99-mL water blanks with 2% NaCl and label them 10^{-2} and 10^{24}, respectively.

❏ 3. Aseptically introduce 5 mL of the *V. natriegens* culture into the labeled flask. Shake the flask to disperse the bacterial cells in the preparation. This inoculation starts the experiment and is the zero (0) time. (See Table A–1.)

❏ 4. Aseptically, with a 1-mL pipette, transfer 1 mL of the inoculated broth into the 10^{-2} water blank. (See Figure A–4.) Place the pipette in the disposal container provided.

❏ 5. Tighten the cap and vigorously shake the water blank about 25 times. This action will break up bacterial clumps and ensure a uniform mixing of the sample.

❏ 6. Repeat steps 4 and 5 with the 10^{-4} water blank.

❏ 7. Obtain 2 sterile Petri plates and 2 melted BHI agar deeps.

❏ 8. Label the plates 0 time and 10^{-4} and 10^{-5}, respectively.

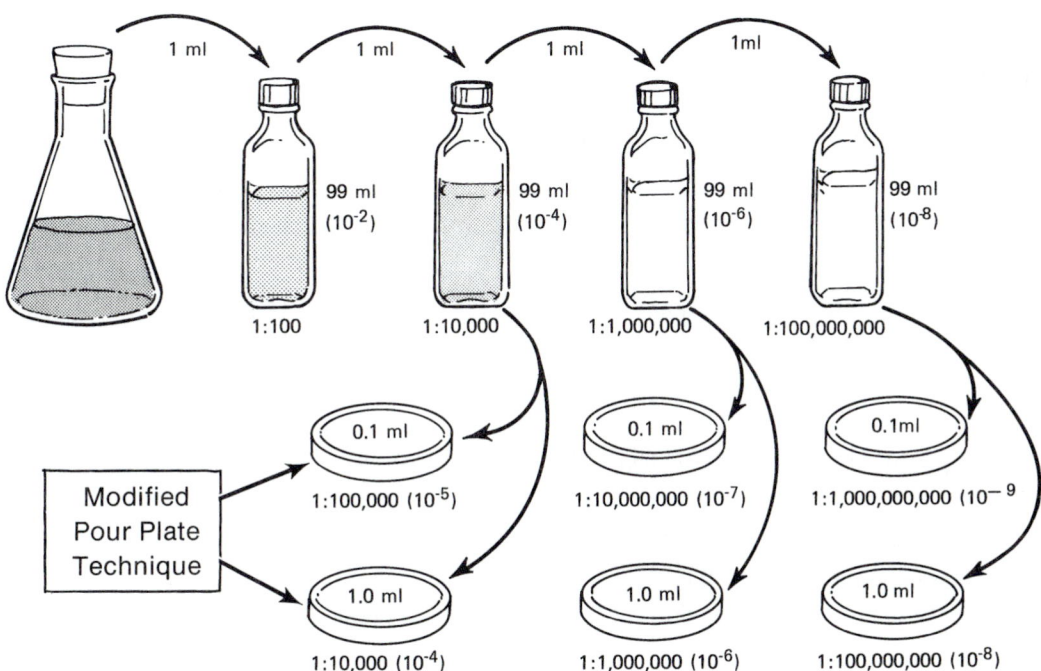

Figure A–4
Standard plate count procedure showing both dilution and plating steps.

❏ 9. With a 1.1-mL pipette, transfer 0.1 mL of the inoculated broth to the 10^{-5} plate and 1.0 mL to the 10^{-4} plate. (See Figure A–4.)

❏ 10. Prepare a modified pour plate by aseptically adding one tube of BHI melted agar to each plate. Rotate the plates on a flat surface to completely mix and fill the Petri dish bottoms.

❏ 11. Allow the agar preparation to harden. Invert and incubate the plates for 24 hours at 37°C.

❏ 12. Repeat steps 4 through 11 20 minutes later.

❏ 13. Repeat the general plate count procedure at the 20-minute intervals specified in Table A–1 of the Results and Observations section. Note the following additional directions.

   a. Make 1 additional serial dilution for sampling times 40 and 60 minutes (dilutions 10^{-5} and 10^{-6}, respectively).

   b. Make 2 additional serial dilutions for sampling times 80, 100, and 120 minutes (dilutions 10^{-5}, 10^{-6}, and 10^{-7}, respectively).

   c. Make 3 additional serial dilutions for sampling time 140 minutes (dilutions 10^{-8} and 10^{-9}).

   d. Incubate all plates for 24 hours at 37°C.

❏ 14. After incubation, count the colonies and enter the numbers in Table A–1 of the Results and Observations section. Indicate plates with less than 30 colonies as TFTC (too few to count) and plates with more than 300 colonies as TNTC (too numerous to count). Use the colony counter for your counts.

❏ 15. Calculate the viable cell count per mL of the original sample according to the formula given earlier in this exercise. Enter the values in Table A–1.

❏ 16. Plot the viable cell count per mL versus time on the graph provided in the Results and Observation section. (Figure A–5.) Use a colored pencil.

Procedure 3: Spectrophotometer Measurements

All measurements are to be made immediately following the sample-taking for the standard plate count procedure.

❏ 1. Allow the spectrophotometer to warm up for at least 10 minutes and carry out procedure 1. (Standardize the instrument to 100% transmission with sterile broth.)

❏ 2. Immediately after inoculating the flask with BHI medium, obtain a cuvette. Be careful not to handle the lower half of the tube.

❏ 3. With a 5-mL pipette delivery device, introduce 5 mL of the inoculated broth culture into the cuvette. Place the pipette into disposal container 1.

❏ 4. Immediately return the culture flask to the shaker.

❏ 5. Wipe the cuvette to remove any fingerprints, liquid, etc., and insert it into the sample holder. Be certain that the index line or frosted triangle on the tube faces the front of the instrument.

❏ 6. Close the lid.

❏ 7. Determine the optical density (OD) or absorbance of the culture and record the reading in Table A–2 in the Results and Observations section. Note this reading is for 0 time.

❏ 8. Pour the contents of the cuvette into the container of disinfectant and rinse it several times with distilled water.

❏ 9. Repeat steps 3 through 8 to take OD readings at 15-minute intervals until the absorbance no longer increases. Enter your readings for the time periods in Table A–2 in the Results and Observations section.

❏ 10. After completing all measurements, turn off the machine.

❏ 11. Using the graph provided in the Results and Observation Section, plot both the absorbance readings and the viable counts per mL versus time. Use a different colored pencil for each type of information.

❏ 12. Answer the questions in the Results and Observations section.

Results and Observations

Standard Plate Counts

1. Enter your findings in Table A–1.

2. Plot the average colony counts (Table A–1) on the graph section provided. (Figure A–5.) Does the plot of the results resemble the growth curve in Figure A–1? _____ If not, explain why. _____

_____

_____

3. Which plate dilution(s) provided the best colony count? _____

_____

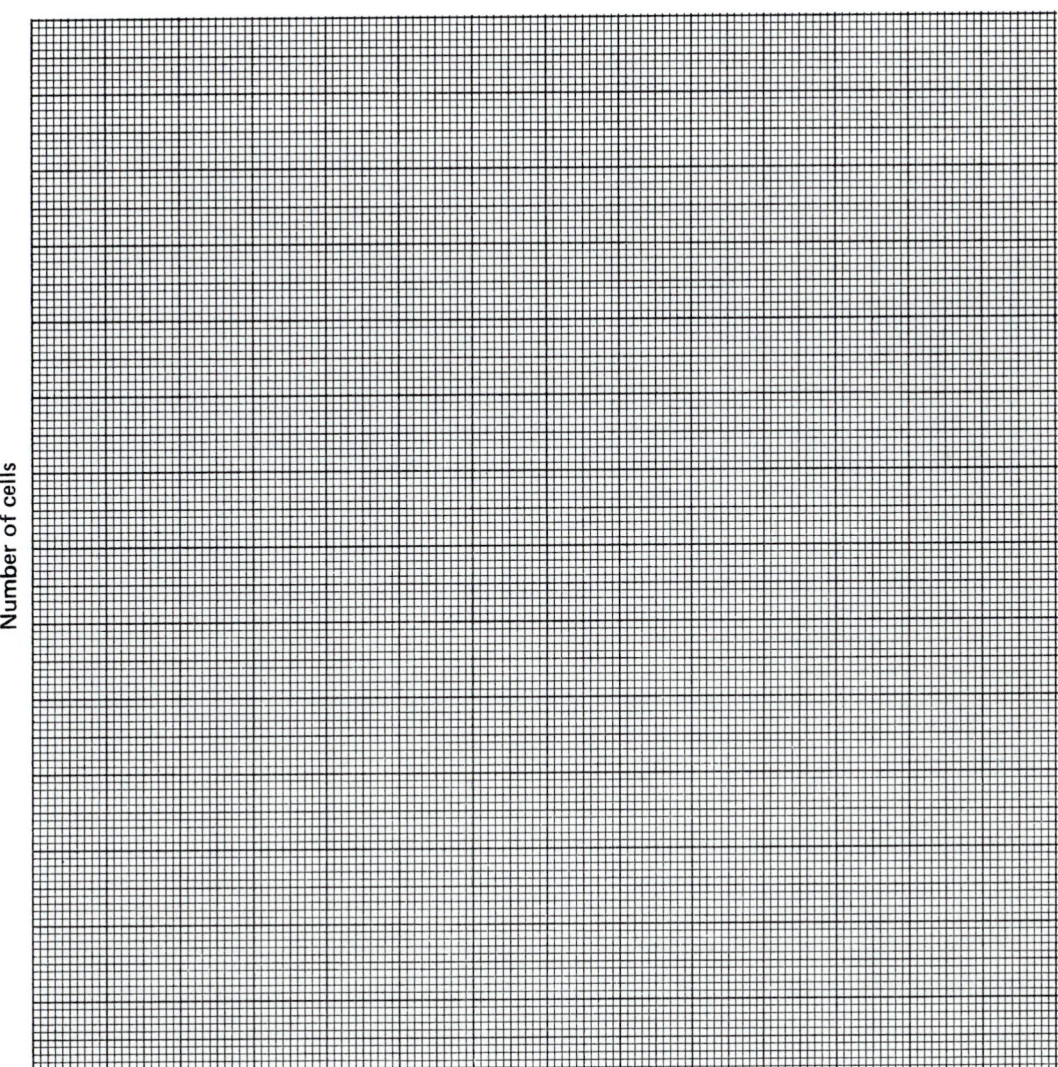

Figure A–5
Cell numbers plot against time.

Table A–1

Standard Plate Counts

Approximate Sampling Time (min.)[a]	Sample Dilution Used (e.g., 10^{-6}, 10^{-7}, etc.)	Colony Counts			Cell Count per mL
		Plate 1	Plate 2	Average	
0	10^{-4}				
	10^{-5}				
20	10^{-4}				
	10^{-5}				
40	10^{-5}				
	10^{-6}				
60	10^{-5}				
	10^{-6}				
80	10^{-5}				
	10^{-6}				
	10^{-7}				
100	10^{-5}				
	10^{-6}				
	10^{-7}				
120	10^{-5}				
	10^{-6}				
	10^{-7}				
140	10^{-8}				
	10^{-9}				
160	10^{-8}				
	10^{-9}				
180	10^{-8}				
	10^{-9}				

[a]Note: These times reflect the age of the culture.

Spectrophotometric Measurements

1. Enter your OD (absorbance readings) in Table A–2.

2. After which reading did the absorbance value of the culture not increase? _____

3. Plot your readings on the graph provided. (Figure A–6.) Can you identify the lag, exponential, and stationary phases? _____ If so, label them accordingly.

4. Do the plots of absorbance and viable cell counts differ? Explain. _____

_____

Table A–2

Spectrophotometric Measurements

Time	OD	Time	OD
0		105	
30		120	
45		135	
60		150	
75		165	
90		180	

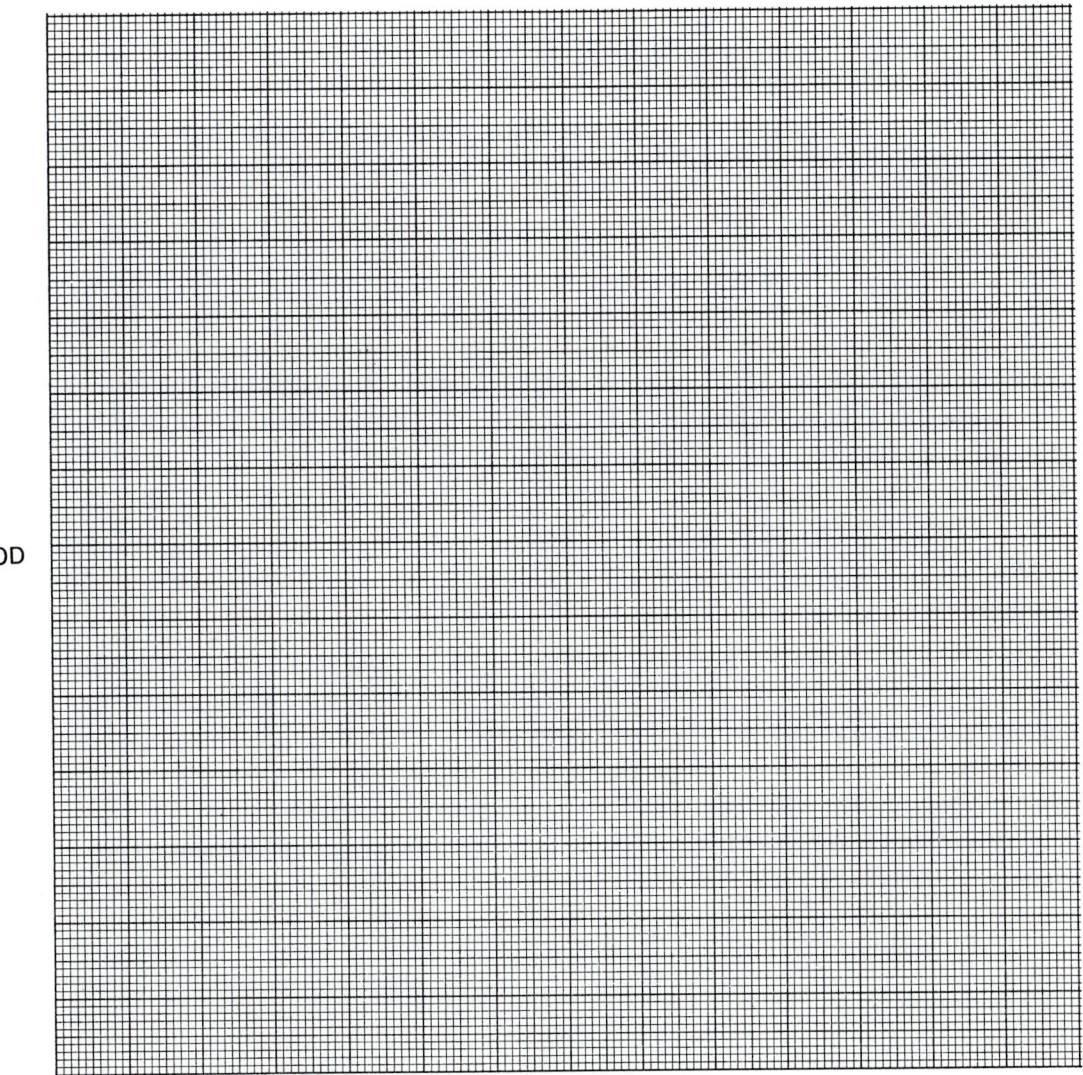

Time

OD

Figure A–6

Absorbance readings (OD) plot against time.

Laboratory Review A   The Bacterial Growth Curve

1. List the four phases of bacterial growth.

   a. _____   c. _____

   b. _____   d. _____

2. During which phase of growth do cells divide at a constant rate? _____

3. During which phase of growth is there an accumulation of poisonous substances? _____

4. During which phase of growth do cells increase in size and undergo changes in chemical composition? _____

5. What is generation time? _____

   _____

6. List 4 methods to measure bacterial growth.

   a. _____   c. _____

   b. _____   d. _____

7. Calculate the number of viable bacteria/mL with the following combinations:

   a. colonies $= 150$, dilution used 10^{-2}, _____

   b. colonies $= 300$, dilutions used 10^{-3}, _____

   c. colonies $= 45$, dilutions used 10^{-2}, _____

Key Terms

absorption spectroscopy: measurement of amount of light of a given wavelength that is absorbed by a substance when the light passes through it and analysis of measurement to determine concentration of the substance

blank: a solution used to adjust the spectrophotometer to read 0 absorbance or 100% transmission without the presence of the light-absorbing substance to be tested

cuvette (KOOV-et): a transparent sample holder

generation time: the time required for the formation of two new cells from one

microbial growth: an increase in the number of microorganisms or mass beyond that present in the original inoculum

serial dilution: preparation of a series of smaller dilutions from an original sample. The total dilution is the product of each dilution in the series. For example: 1 mL diluted with 9 mL, followed by 1 mL of this preparation being diluted with 9 mL would result in a 1:100 or 10^{-2} dilution

spectrophotometer (spek-trō-fō-TOM-et-er): an electronic device used to measure the percent transmission of light through a solution

After completing this experimental exercise, you should be able to:

1. Perform an immunoelectrophoresis.

2. Identify the major plasma or serum fractions in an electrophoretic pattern.

3. Compare plasma and/or serum fractions from different sources.

Electrophoresis is an analytical technique in which charged molecules in solution, mainly proteins, protein-related compounds, such as peptides, and nucleic acids, migrate in response to an electrical field (Figure B–1). The rate of movement of the molecules depends on several factors, including the net electrical charge, the size and shape of the molecules, the strength of the electrical field, and the ionic, strength, thickness, and temperature of the *medium* (material substance) through which the molecules are moving. Electrophoresis is a simple, rapid, and highly sensitive technique used to identify and study the properties of a single charged molecule and to separate a mixture of charged molecules. Proteins or unusual molecules found in serum are the most frequently studied substances.

Electrophoresis may be carried out in a supporting medium or matrix, such as paper, cellulose acetate, or gels formed in tubes, slabs, or flat beds. In most electrophoretic units (Figure B–2), the gel is mounted between two buffer chambers containing separate electrodes (see top left of Figure B–3). With this arrangement, the only electrical connection between the chambers is through the gel.

After the electrophoresis of a protein or nucleic acid sample is completed, the gel or other supporting matrix can be analyzed by one of several methods. The most common analytical procedure is staining. Protein gels are frequently stained by Coomassie blue or Ponceau S. Nucleic acids are usually stained with ethidium bromide, a fluorescent dye that produces an orange glow when exposed to ultraviolet light. This exercise considers the use of electrophoresis with serum proteins.

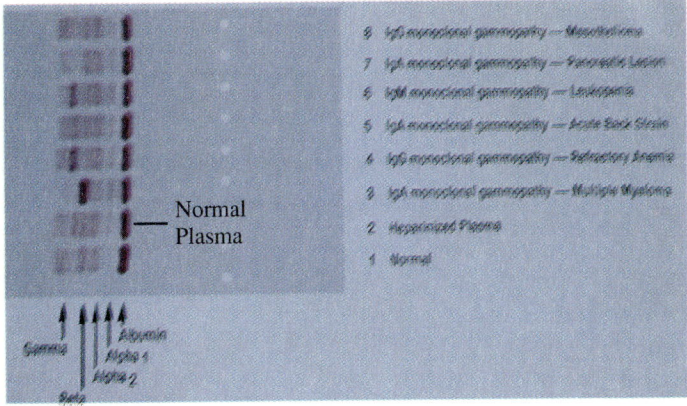

Figure B–1

An example of an immunoelectrophoretic pattern. The various fractions of normal plasma (2nd from bottom), normal serum (bottom), and a number of disease states are shown.

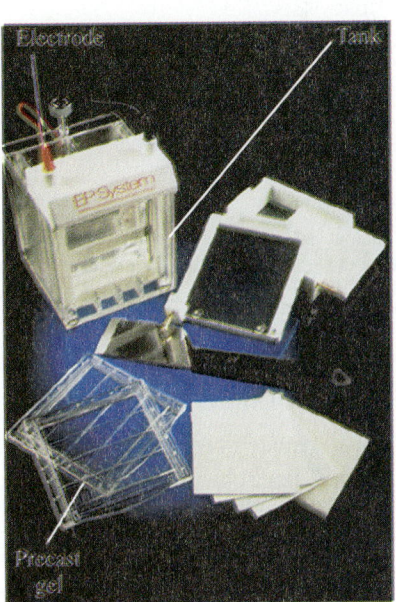

Figure B–2

Complete S&S EP Mini-Electrophoresis System, showing tank, electrodes, and precast gels.
(Courtesy of Schleicher & Schuell, Inc., Keene, New Hampshire.)

Plasma Protein Determinations

Materials

The following items should be provided for general class use:

- ❑ 1. Isophore Universal Gel Electrophoresis system or other electrophoresis unit
- ❑ 2. Silicon grease or similar lubricant
- ❑ 3. Electrophoresis basic buffer (Isophore®)
- ❑ 4. Electrophoresis acidic buffer (Isophore®)
- ❑ 5. Stacking buffer (Isophore®)
- ❑ 6. Application buffer (Isophore®)
- ❑ 7. Fixative Ponceau S dye solution
- ❑ 8. Acetic acid rinse solution
- ❑ 9. Cleaning solution
- ❑ 10. Phenol red pH indicator (0.1%)
- ❑ 11. Power source constant current
- ❑ 12. Separate trays for butter, dye, rinse, and cleaning solutions

- ❑ 13. Glass vials or tubes (small capacity)
- ❑ 14. Test tube racks
- ❑ 15. Disposable surgical gloves
- ❑ 16. 1-mL pipettes, graduated in 0.1 mL, with rubber bulbs
- ❑ 17. Pasteur pipettes with rubber bulbs
- ❑ 18. Single-edge razor blades or scalpels
- ❑ 19. Transparent tape (1-inch wide)
- ❑ 20. Spatulas
- ❑ 21. Flat, blunt-end forceps
- ❑ 22. Filter paper
- ❑ 23. Marking pencil
- ❑ 24. Container with disinfectant for pipette disposal

The following should be provided for students in groups of four, or as indicated by the instructor:

- ❑ 1. Normal plasma samples from the following sources in 0.5-mL aliquots
    - ❑ a. human
    - ❑ b. cow (bovine)
    - ❑ c. chicken
- ❑ 2. Three unknown plasma samples (in 0.5-mL aliquots)
- ❑ 3. Nine 5-lambda pipettes or microsyringes, or one 1-20 µL Eppendorf digital micropipettor with disposable tips
- ❑ 4. One gel cassette (Isofocus survey® gel, pH 3-10)
- ❑ 5. One sample spacer comb (Isophore®)
- ❑ 6. One set of drying weights (to be used during drying of gel after electrophoresis run)

Procedure 1: Electrophoresis Unit and Accessories

This procedure is to be performed by students in groups of four.

- ❑ 1. Examine the Isophore® unit (or other electrophoresis system provided), power unit, and accessories.
- ❑ 2. Refer to Figure B–3 and identify the following components of the Isophore® unit:
    - ❑ a. Anode in the upper chamber
    - ❑ b. Upper chamber support containing the cathode
    - ❑ c. Safety-interlocking lid (prevents contact with the buffers and platinum electrodes)
    - ❑ d. Gel cassettes that form the side walls of the upper chamber
- ❑ 3. Examine the power unit and accessories provided.

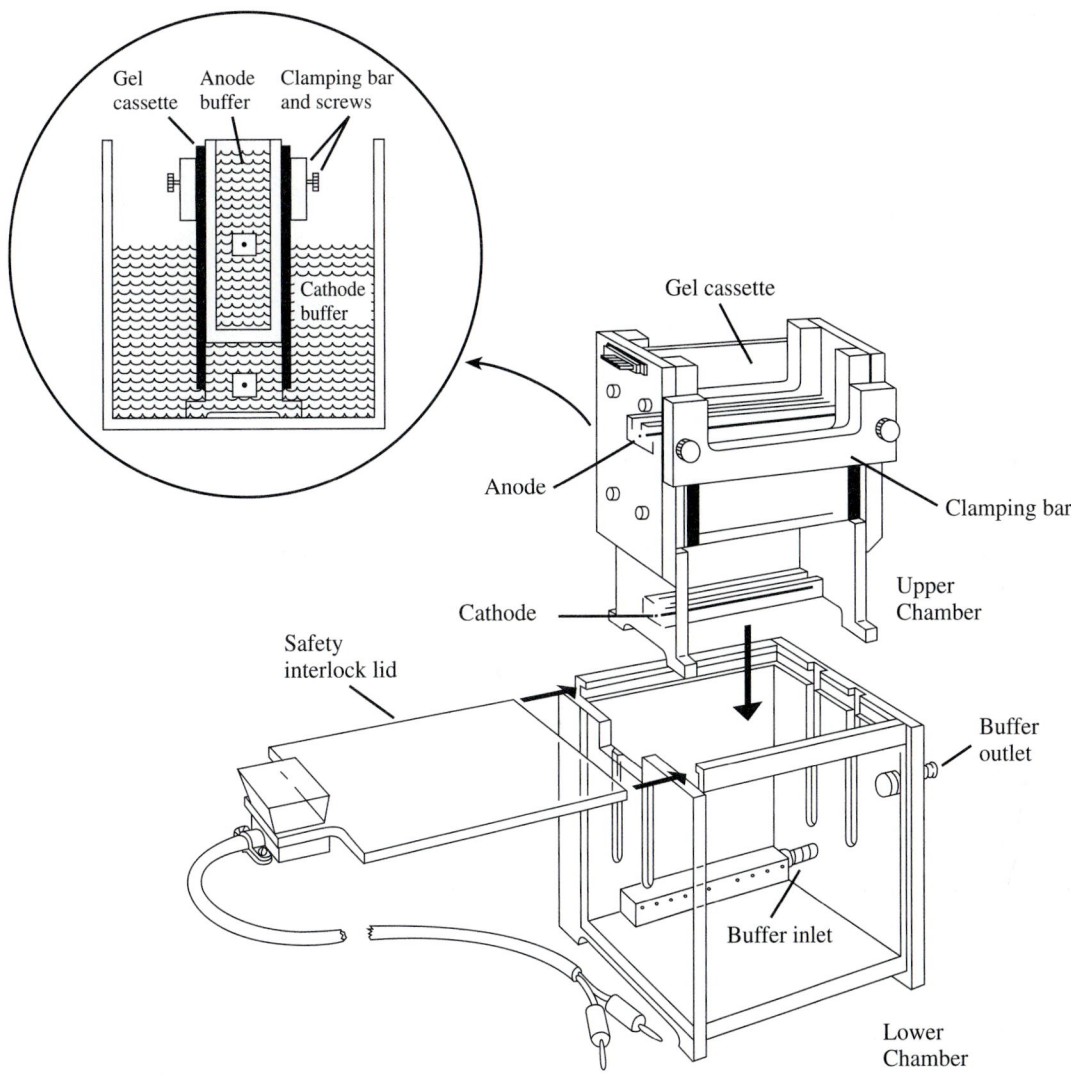

Figure B–3

The Isophore module and its components. This device is quite compact and straightforward in operation. Note the two chambers and their respective parts. The insert at the upper left-hand corner shows a side view of two gel cassettes held in place by clamping bars and screws.
(After Isolab, Inc.; RESOLVE, Isophore, Isofocus, and Isolytes are registered trademarks of Isolab, Inc.)

Procedure 2: Gel Preparation

This procedure is to be performed by students in groups of four.

❑   1. Remove one Isofocus® gel from its package. Shake any excess liquid from the surface. (Do not separate the glass or plastic plate from the gel before performing the electrophoresis run.) Note that the gel is held between a glass plate and a notched plastic plate. A notched line, which extends along the lower edge of the plastic plate, prevents the gel from sliding out of its position.

❑   2. Examine the gel to be sure that the plastic spacers on both sides of the gel are in direct contact with it. If they are not, make a small slit in the tape, and gently push the spacer with a spatula until it touches the gel. (Figure B–4.)

❑   3. Use a Pasteur pipette to fill the space above the gel with a stacking buffer. (Figure B–4.)

❑   4. Place the sample spacer comb on top of the gel so that the tips of the comb enter the gel surface 1 to 2 mm (Figure B–4). When the sample spacer is pressed into the gel, 9 separated sample wells are formed.

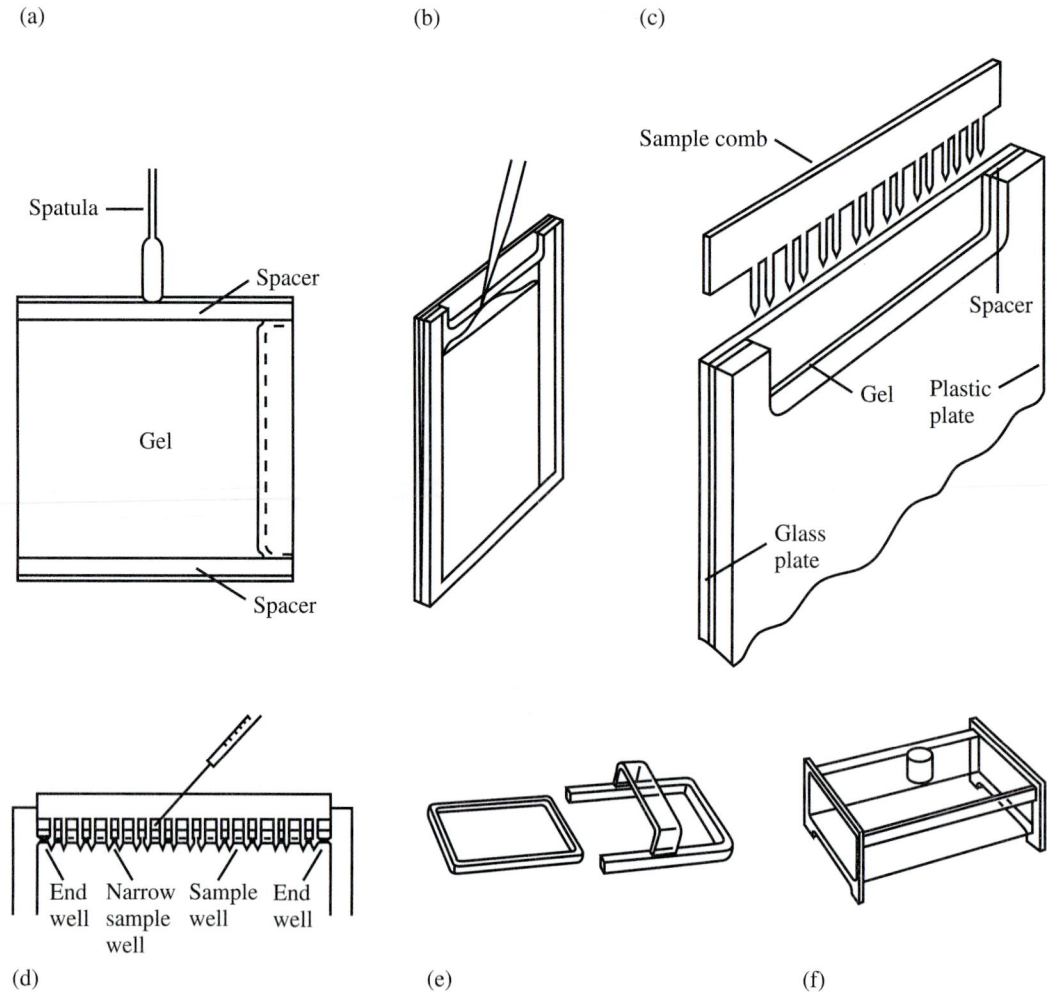

Figure B–4

Steps in sample application and an electrophoresis run; (a) correcting the spacers; (b) applying the stacking buffer; (c) the Isofocus gel and sample-containing comb; (d) applying a sample with a microsyringe—each end well and each narrow sample well receives application buffer only; (e) transferring gel to transporter and tray; (f) fixation and staining.

- ☐ 5. Place the upper chamber of the electrophoresis system on a convenient working surface. Loosen the screws on each side of the machine so that there is space between the gaskets. (Refer to Figure B–3.)
- ☐ 6. Coat the gaskets with silicon grease.
- ☐ 7. Place one gel cassette between each pair of gaskets. If only one gel is to be run, remove one clamping bar, and replace it with a blanking plug. The notched plastic plate should be placed so that it faces toward the center of the upper chamber.
- ☐ 8. Tighten the screws in the clamping bar until each cassette is held snugly in place. Use a light touch when clamping—too much pressure will crack the glass plate.
- ☐ 9. Fill the lower chamber with the cathode (basic) buffer. The buffer level should be about 8.0 cm from the top of the chamber. About 1,500 mL of buffer will be needed.
- ☐ 10. Carefully slide the upper chamber into the lower chamber.
- ☐ 11. Check the top and bottom of the gel cassette for air bubbles. Remove any bubbles by stirring the buffer or flushing with buffer (a syringe works well for this). The buffer must be in direct contact with the entire bottom edge of the gel. It is important that all air bubbles be eliminated.
- ☐ 12. Pour the anode (acidic) buffer into the upper chamber.

Procedure 3: Sample Preparation

This procedure is to be performed by students in groups of four.

- ❑ 1. Obtain 6 small vials, and label them "N" (for normal human plasma), "B" (for normal bovine plasma), "C" (for normal chicken plasma), and "1," "2," and "3" (for each of the unknown samples).

- ❑ 2. With a 1-mL pipette and bulb, introduce 0.2 mL of the application buffer to each vial.

- ❑ 3. With the micropipettor or other delivery device, add 40 μL of each protein sample to its labeled tube. Use a new tip for each sample. (Refer to Exercise 40 for the use of a digital micropipettor.)

- ❑ 4. Dispose of the tip after use as directed by your instructor.

- ❑ 5. With a new micropipettor tip, add 5 μL of application buffer to each *narrow well* and to both *end wells* of the sample comb. (Figure B–4.)

- ❑ 6. With a new micropipettor tip, introduce the protein sample into the sample wells, as follows: To introduce a sample, insert the tip between the teeth of the sample comb so that it is immediately above the gel. Gently express the sample onto the gel.
  - ❑ a. Add 5 μL of Sample N to the first sample well on the left. (See Figure B–4.) The end well and first narrow well have been filled with application buffer.
  - ❑ b. Add 5 μL of Sample B to the next sample well (second from left).
  - ❑ c. Add 5 μL of Sample C to the third sample well from left.
  - ❑ d. Add 5 μL unknown samples 1, 2, and 3 to the fourth, fifth, and sixth sample wells from the left, respectively

- ❑ 7. With a new tip, apply 1 drop of the 0.1% phenol-red dye solution to the sample well containing sample N (the normal plasma sample). **This chemical is a tracking dye, to show the movement of the system.**

- ❑ 8. The samples are now ready for electrophoresis.

Procedure 4: Electrophoretic Run

This procedure is to be performed by students in groups of four.

- ❑ 1. Slide the safety interlock lid over the Isophore® unit, enclosing both the chambers (see Figure B–3). Connect the power cables (generally, **black** = negative, cathode, and lower chamber; **red** = positive, anode, and upper chamber).

- ❑ 2. Turn on the power, and adjust the voltage to 200 volts on the power setting. The milliamperes (mA) should read 3.5 to 5.8 mA.

- ❑ 3. With no more than eight samples, constant voltage is applied for 60 minutes.

- ❑ 4. Continue the electrophoretic run until separation is complete. (*Run times* may generally be shortened by increasing the voltage.)

- ❑ 5. Upon completion of the run, turn off the power supply, remove the safety lid, and pull out the upper chamber.

- ❑ 6. Pour out the anode buffer, and remove the gel for fixation and staining.

Procedure 5: Fixation and Staining

This procedure is to be performed by students in groups of four.

- ❑ 1. After electrophoresis, open the cassette by slitting the tape on one side with a single-edge razor blade.

- ❑ 2. With the aid of a spatula, pry apart the two plates. Remove the plastic plate, leaving the gel on the glass plate.

- ❑ 3. Remove the sample spacer comb and the side spacers. To remove the comb, loosen the top screws on each clamp. Then pull up gently on the comb while rocking it back and forth. Air bubbles should be visible underneath the comb teeth.

- ❑ 4. Float the gel off the glass plate onto the transporter. (Figure B–4e.) Place the holding screen over the gel to prevent it from floating away during the staining and destaining steps.

❏   5. Place the gel in the tray containing the fixative Ponceau S dye solution for 7 to 10 minutes (Figure B–4*f*).

❏   6. Lift the transporter from the stain tray. Allow excess stain to drain back into the tray.

❏   7. Transfer the gel to the clearing solution, and gently agitate the gel for about 1 minute.

❏   8. Remove the gel and place it on a glass plate. Squeegee the gel gently to remove excess solution, and allow it to air dry. Place a drying weight on each edge of the gel, to prevent curling.

❏   9. Place a clean glass plate over the dried gel, making a sandwich.

❏ 10. Place tape over the edges of the plates.

❏ 11. Label the tape to identify the different specimens used.

Procedure 6: Examination

This procedure is to be performed by students in groups of four.

❏ 1. Examine the stained gel, and identify in each sample the plasma proteins, albumin, and the alpha, beta, and gamma globulins. (Refer to Figure B–1.)

❏ 2. Sketch and label a representative electrophoretic pattern for each plasma protein sample (N,B, and C.) in the space provided in the Results and Observations section.

❏ 3. Answer the questions in the Results and Observations and Laboratory Review sections.

Results and Observations

1. Sketch the electrophoretic patterns for each of the known plasma protein samples in the spaces provided.

Human   Bovine   Chicken

2. Were the different plasma protein electrophoretic patterns similar in appearance? _____

_____

3. a. Were any fractions (alpha globulins, albumin, etc.) missing in the various plasma samples? _____

_____

   b. If fractions were missing, indicate which ones and the plasma sample lacking them. _____

_____

4. Which of the plasma protein fractions was the most abundant in each of the specimens? _____

    _____

5. On the basis of your observations what was the source of each of the unknown samples?

    a. Sample 1: _____

    b. Sample 2: _____

    c. Sample 3: _____

Laboratory Review B   Protein Electrophoresis

---

1. What is electrophoresis? _____

    _____

2. What types of substances can be analyzed by electrophoresis? _____

    _____

3. Define the following (refer to your text):

    a. gamma globulin _____

    b. albumin _____

    c. plasma _____

    d. serum _____

    e. alpha globulin _____

    f. anode _____

    g. cathode _____

Key Terms

albumin (al-BŪ-min): one group of simple proteins widely distributed in plant and animal tissues

anode (AN-ōd): the positive pole of an electrical source

cathode (KATH-ōd): the negative electrode of an electrical source

gamma globulin: a protein fraction of blood; contains immunoglobulins or antibodies

plasma: the liquid part of blood and lymph; it is a medium for the circulation of blood cells

serum: the liquid portion of blood found after coagulation

After completing this exercise, you should be able to:

1. Perform a DNA restriction analysis procedure.

2. Understand the role of restriction enzymes in DNA restriction analysis.

3. Manipulate a micropipettor

4. Perform gel electrophoresis.

5. Define or explain the terms: *endonuclease, restriction site,* and *restriction fragments.*

DNA restriction analysis is a method that uses bacterial enzymes, known as *restriction endonucleases* or *restriction enzymes* to reveal exact nucleotide sequences. The first of the specific restriction endonucleases (*Hind*II) that cleave DNA at specific sites into fragments was discovered in 1970 by Hamilton Smith at Johns Hopkins University. The enzyme was isolated from the bacterium *Haemophilus influenzae*. Since the original discovery, restrictive enzymes that cut specific sequences have been isolated from several hundred bacterial strains. These enzymes recognize specific and different *nucleotide* or *base-pair sequences*. More than 70 types are commercially available. To avoid confusion, restriction endonucleases are named according to the following system:

1. The first letter of the enzyme is taken from the genus of the microbial source.

2. The second and third letter of the organism's species name.

3. If a fourth letter is used, it indicates a specific strain of the microbial source.

4. Roman numerals, such as I, II, or III are usually added to indicate the order of discovery. Table C–1 lists examples of restriction endonucleases and their sources.

The various fragments produced when a specific viral DNA is cut by a restriction enzyme can be separated by the use of agarose gel electrophoresis. Prior to electrophoresis, digested DNA samples are mixed with a loading solution containing sucrose and one or more dyes. The sucrose causes the DNA to sink into a well (depression in the agarose gel). This procedure will be performed in this exercise. The rate at which the fragments move through the gel is related to the size of their respective lengths. Smaller fragments move much faster than larger ones. Staining of agarose gels with dyes that bind to DNA generates a series of bands (Figure C–1). Each band corresponds to a restriction fragment, the molecular weight of which can be established by calibration with DNA molecules of known weights. Different restriction enzymes produce different restriction fragments for the same viral DNA molecule.

Table C–1

Examples of Restriction Enzymes

Microbial Source	Enzyme Abbreviation
Bacillus amyloliquefaciens	*Bam*HI
Escherichia coli	*Eco*RI
Haemophilus aegyptius	*Hae*III
Haemophilus influenzae	*Hind*II
Serratia marcescens S_b	*Sma*

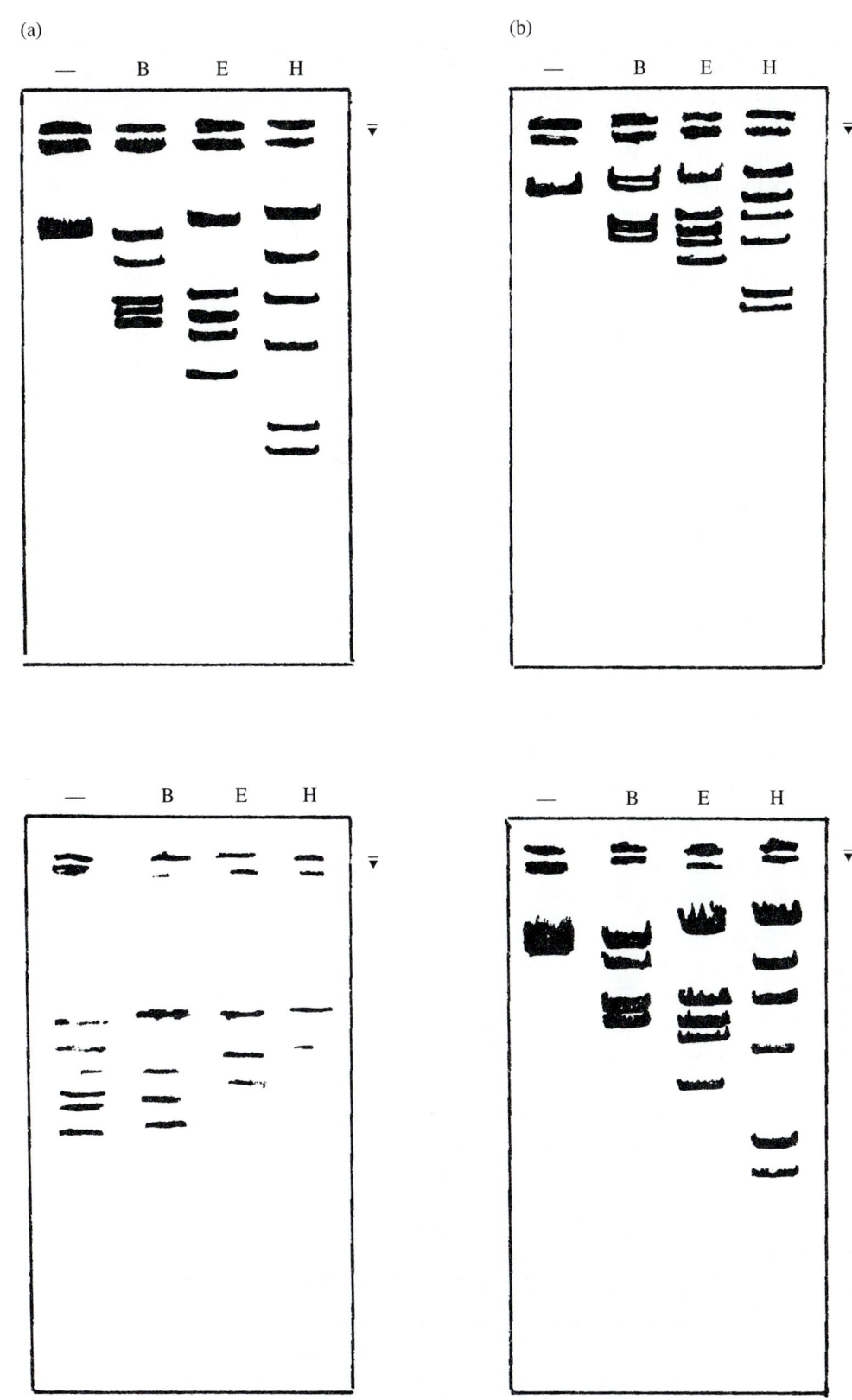

Figure C–1

Examples of various situations affecting electrophoresis. Note that the direction of band movement is indicated by the arrow. (a) An ideal gel. The bands are spread in a nice pattern. (b) A short run. Here the bands are compressed due to the short electrophoresing time. (c) An underloaded gel. The bands are faint because too little DNA was present in the digests. (d) An overloaded gel. The smearing of the bands is caused by too much DNA in the digests.

The information obtained through the use of restriction enzymes to cut DNA molecules, such as circular viral DNA's into specific fragments can be used to produce a *restriction map*. By following the pattern of fragments produced as the digestion process proceeds to completion, the location of sites on the DNA molecule that are attacked by the restriction enzyme (restriction map) can be determined. Repeating the process with other enzymes, produces a more detailed map with several different restriction sites. With such maps it becomes possible to locate the site at which viral DNA replication is initiated, identify the region that specifies the messenger RNA (mRNA) molecules of viral proteins during viral replication, and to locate other genetically significant DNA regions which are of great importance to recombinant DNA technology and DNA sequencing techniques.

This experimental exercise presents the techniques used in the analysis of DNA with the aid of restriction endonucleases and gel electrophoresis. The three restriction endonucleases *Bam*HI, *Eco*RI, and *Hind*II, will be used to produce restriction fragments of different nucleotide or base-pair lengths. Performing DNA restriction analyses requires the use of various types of equipment and supply items. Figures C–2 through C–5 show typical examples.

(a)

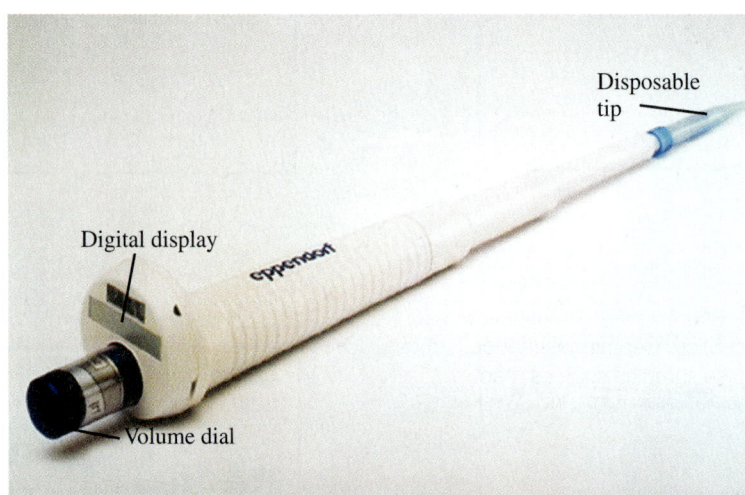

(b)

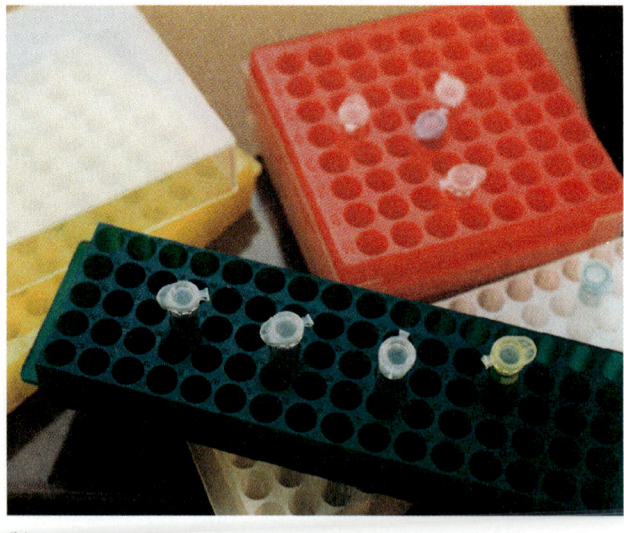

Figure C–2

Micropipette and microtubes. (a) An Eppendorf positive displacement pipette. Note the disposable tip, top-view digital display, and the volume setting dial.
(Courtesy of Brinkmann Instruments, Inc., Westbury, N.Y.)

(b) Microtubes and rack.
(Courtesy of Interlab Products, Inc. Westboro, MA.)

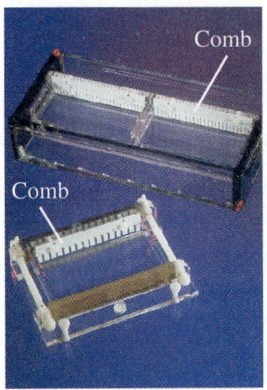

Figure C–3
Examples of electrophoresis systems. Note the presence of the plastic white comb.
(Courtesy of Jordan Scientific Co.)

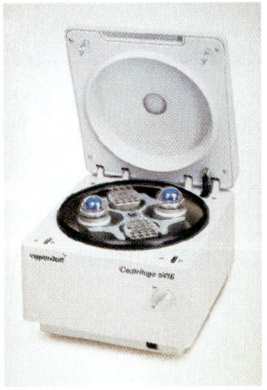

Figure C–4
Two examples of Micro Centrifuges that can be used to spin-down or pulse samples such as reaction mixtures.
(Courtesy of Brinkmann Instruments, Inc., Westbury, N.Y.)

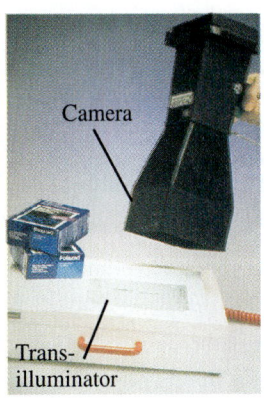

Figure C–5
Transilluminator and camera set up.
(Courtesy of Brinkmann Instruments, Inc., Westbury, N.Y.)

A. Manipulation of Micropipettor

Materials

The following items should be provided for pairs of students.

❏ 1. One *Eppendorf* or related micropipettor with a volume range of 0.5–10 μL or 1–20 μL

❏ 2. Eight disposable micropipettor tips

❏ 3. Six 1.5 mL disposable microtubes

❏ 4. Four test solutions with the volumes as specified:
    ❏ a. solution 1–30 μL   ❏ c. solution 3–4 μL
    ❏ b. solution 2–36 μL   ❏ d. solution 4–4 μL

❏ 5. One microtube rack

❏ 6. Container for micropipettor and microtube disposal.

The following items should be provided for general class use:

❏ 1. Microfuge

❏ 2. Permanent markers

Procedure

Table C-2

The DO NOTS in Using a Micropipettor

1. **DO NOT** rotate the volume adjusting knob past the upper or lower ranges specified by the instrument's manufacturer.

2. **DO NOT** force the volume adjusting knob. (If there is a problem rotating the volume knob tell your instructor.)

3. **DO NOT** use a micropipettor without a tip in place. (The precision piston that measures the volume of fluid can be ruined.)

4. **DO NOT** lay the pipettor down with a filled tip. (Fluid can run back into the precision piston.)

5. **DO NOT** allow the micropipettor plunger to snap back after fluid is either withdrawn or ejected.

6. **DO NOT** immerse the micropipettor barrel into fluid.

7. **DO NOT** flame the micropipettor tip.

The following procedure should be performed by students individually with a permanent marker.

❏ 1. Label three 1.5 mL microtubes A, B, and C, respectively.

❏ 2. Firmly seat (place) a fresh micropipettor tip on the end of the micropipettor tip. (See Figure C–6a.)

❏ 3. Set the micropipettor volume dial to 4 μL and withdraw 4 μL from the test solution tube marked *1*. Withdraw the fluid in the following way:

   ❏ a. Depress the micropipettor with the thumb plunger (see Figure C–6b) to the first stop and hold the pipettor in this position.
   ❏ b. Dip the micropipettor tip into Solution 1.
   ❏ c. Gradually release the plunger to allow the fluid to be drawn into the tip. (Refer to Figure C–6.)
   ❏ d. Carefully slide the pipette tip along the side of the tube to remove any unwanted droplets of fluid sticking to the tip's surface.

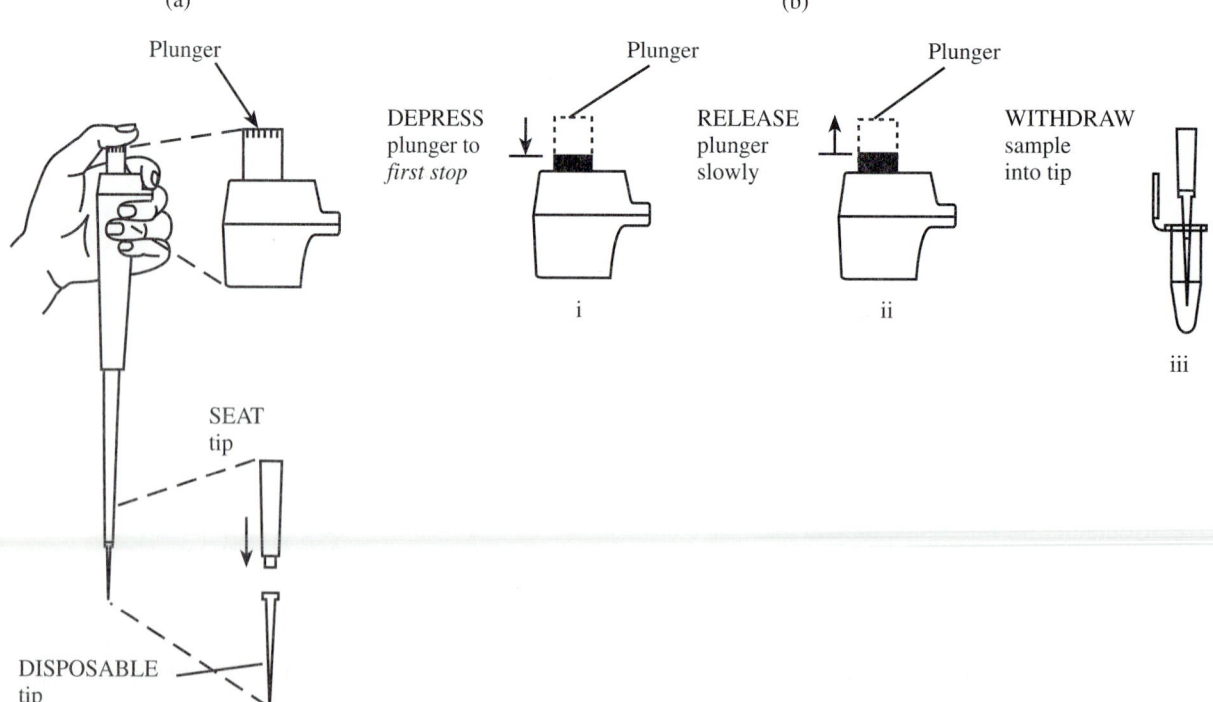

Figure C–6

The use of a micropipettor. (a) The correct way to hold and manipulate a micropipettor. Note the position of the thumb. (b) The different steps used to withdraw fluid. i. Depressing plunger to first stop, ii. Plunger release, iii. Withdrawal of fluid sample into the micropipettor tip.

❑ 4. Expel the fluid into the microtube in the following way:

    ❑ a. Touch the pipette tip to the inside surface of microtube 1. (See Figure C–7a.)

    ❑ b. Slowly depress the micropipettor plunger to the first stop and hold to expel the fluid. (See Figure C–7b.)

    ❑ c. Depress the plunger to the next or second stop and hold to blow out any remaining fluid. (See Figure C–7c.)

    ❑ d. Continue to hold the plunger in the depressed position as you remove the micropipettor from the microtube.

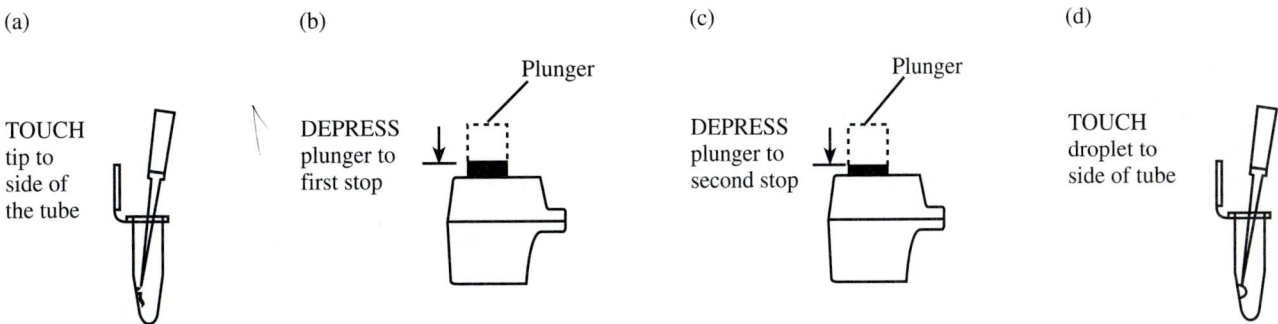

Figure C–7

The different steps used to expel fluid from a micropipettor. (a) Touching tip to inner surface of microtube. (b) Depressing plunger to first stop. (c) Depressing plunger to second stop. (d) Touching droplet to inner surface of microtube.

❑ 5. Repeat steps 3 and 4 with tubes 2 and 3.

❑ 6. After the solution 1 fluid samples have been introduced into microtubes 1, 2, and 3, eject the micropipettor tip into the container provided in the following way:

    ❑ a. Depress the micropipettor plunger beyond the second stop or depress a separate tip-ejection button if it is present on the pipettor. (See Figure C–8.)

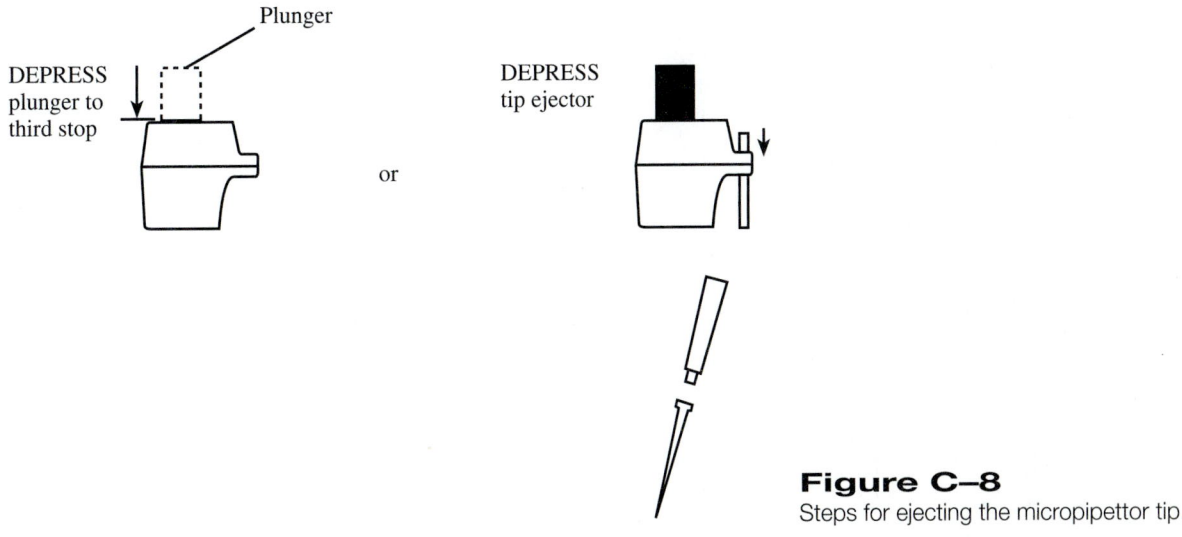

Figure C–8

Steps for ejecting the micropipettor tip.

❑ 7. Obtain a fresh tip to add Solution 2 to tubes #1, 2, and 3 respectively, according to the volumes indicated in Table C–3

Table C–3

Micropipettor Volumes

Tube #	Solution 1	Solution 2	Solution 3	Solution 4
1	4 μL	5 μL	1 μL	—
2	4 μL	5 μL	—	1 μL
3	4 μL	4 μL	1 μL	1 μL

❏ 8. Repeat step 7 with Solution 3 and 4.

❏ 9. After all solutions have been added to the respective microtubes, mix the solutions by one of the following methods:

    ❏ a. Sharply tap tube bottom on bench top. Be certain that the individual drops have pooled into one drop at the bottom of the tube.

    ❏ b. Place the microtubes into a microfuge and apply a short 10 second pulse. Be certain tubes are placed in a balanced configuration in the microfuge rotor. (See Figure C–9.) (Spinning tubes in an unbalanced position will damage the microfuge motor.)

❏ 10. Obtain a fresh micropipettor tip and set the micropipettor volume dial to 10 μL.

❏ 11. Carefully withdraw the fluid from microtube 1. Check to see if the tip is just filled. This finding would indicate that the volumes were correctly added to the microtube.

❏ 12. Check to see if any fluid remained in the microtube.

❏ 13. Repeat steps 11 and 12 with microtubes 2 and 3.

❏ 14. Enter all findings and answer the questions in the Results and Observations section and Laboratory Review.

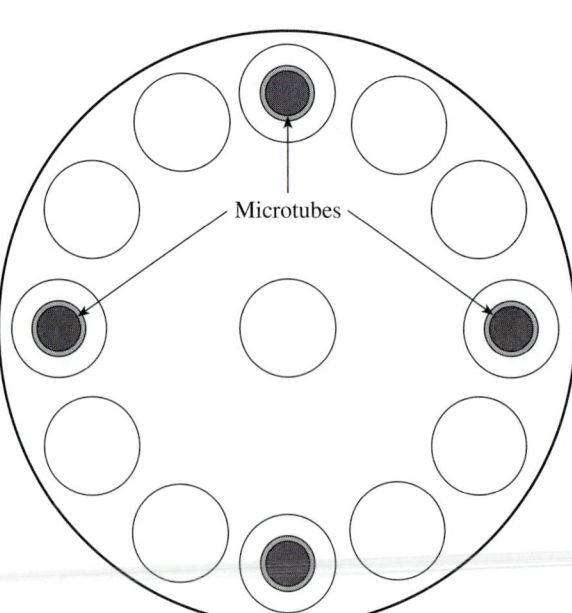

Microtubes

Figure C–9

A top view of a microfuge showing the correct positioning of micro-tubes to balance samples.

B. DNA Restriction Analysis

All procedures in this part of the exercise are to be performed by students in pairs.

Materials

The following items should be provided for pairs of students:

Restriction Digest (Procedure 1)

❑ 1. Sixteen μL of lambda (λ) DNA (0.4 μg/mL)

❑ 2. Twenty μL of restriction enzymes
  ❑ a. *Bam* HI
  ❑ b. *Eco*RI
  ❑ c. *Hind*III

❑ 3. Two μL of distilled water

❑ 4. Four 1.5-mL tubes

❑ 5. One 0.5–10-μL micropipettor and appropriate tips

❑ 6. Container for used micropipettor tips

❑ 7. One container with disinfectant

Agarose Gel Preparation (Procedure 2)

❑ 1. Fifty mL of 0.8 percent agarose solution in flask

Gel Loading and Electrophoresis (Procedures)

❑ 1. Five hundred mL of Tris/Borate/EDTA (TBE buffer)

❑ 2. Six μL of loading dye

Methylene Blue or Ethidium Bromide Staining (Procedures 4 and 5)

❑ 1. One hundred mL of methylene blue solution (0.025 percent)

❑ 2. One hundred mL of ethidium bromide solution (1 μg/mL)

The following items should be provided for general use.

❑ 1. Masking tape (¾ inch)

❑ 2. Microfuge

❑ 3. Transilluminator

❑ 4. Water bath set at 37°C

❑ 5. Permanent markers or pens

❑ 6. Water bath set at 60°C (for agarose gel)

❑ 7. Rubber gloves

❑ 8. Test tube racks for 1.5-mL tubes

Procedure 1: Restriction Digest Preparation

The following procedures should be performed by students in pairs.

❑ 1. Label one 1.5-mL tube with a permanent marker for each of the restriction reactions and using the following abbreviations to represent the respective reaction mixtures:
  ❑ − = no enzyme
  ❑ B = *Bam*III
  ❑ E = *Eco*RI
  ❑ H = *Hind*III

❑ 2. Add the reagents to each of the respective tubes as shown in Table C–4 with a micropipettor. *Use a fresh micropipettor tip for each reagent.*

Table C–4

Reaction Mixtures

			Reagents			
Tube	λ DNA	Buffer	*Bam*HI	*Eco*RI	*Hind*III	H$_2$O
—	4 µL	5 µL	—	—	—	1 µL
B	4 µL	5 µL	1 µL	—	—	—
E	4 µL	5 µL	—	1 µL	—	—
H	4 µL	5 µL	—	—	1 µL	—

❏ 3. Place all tubes in the order shown in Table C–4:

  ❏ a. Add 4 µL of λDNA to each reaction tube. Touch the micropipettor tip to the reaction tube side surface as near the bottom as possible.

  ❏ b. With a fresh tip add 5 µL of restriction buffer to a clean inner surface of each tube.

  ❏ c. Use fresh tips to add 1 µL of the restriction enzyme *Bam*HI, *Eco*RI, and *Hind*III to the appropriate tubes as indicated in Table C–4.

  ❏ d. Use a fresh tip to add 1 µL of distilled water to the control tube labeled "−."

❏ 4. Close all tube tops firmly.

❏ 5. Mix the reagents in the respective reaction tubes by sharply tapping the bottoms of the tubes on a lab table or bench, or by pulsing in a microfuge.

❏ 6. Place the reaction tubes in a 37°C water bath for a minimum of 30 minutes.

❏ 7. Following incubation at 37°C, freeze the reaction mixtures at −20°C until the electrophoresis procedure is to be performed (Procedure 3).

Procedure 2: Making Agarose Gels

❏ 1. Use masking tape strips to seal the two ends of the gel-casting tray. (Refer to Figure C–10.)

❏ 2. Insert the well-forming comb, and place the gel-casting tray in an area on the laboratory bench where it will not be disturbed after the agarose solution is poured into it.

❏ 3. Carefully pour enough agarose solution into the casting tray to reach only about one-third the level or height of the teeth of the well-forming comb. (*If large bubbles should form, carefully use a pipette tip to move them to the sides or end of the tray.*)

❏ 4. Allow at least 10–15 minutes for the agarose solution to harden. (As the gel forms, the agarose will become cloudy.) The corner of the agarose away from the comb can be touched to determine if the gel has fully solidified.

❏ 5. After the agarose solution has fully set or solidified, carefully remove the masking tape strips from the ends of the casting tray.

❏ 6. Place the tray into the electrophoresis box so that the comb is at the negative (**black**) electrode position.

Figure C–10

The gel-casting tray with its two ends sealed and the well-forming comb in position.

❑ 7. Fill the box with sufficient TBE electrophoresis buffer so that the entire surface of the gel is covered.

❑ 8. Gently rock the comb slightly and remove it from gel without damaging the sample wells that have been formed.

❑ 9. Check to see that the sample wells are completely submerged. If depressions or dry areas are noticed around the wells, simply add additional buffer until they disappear.

Procedure 3: Loading the Gel and Electrophoresis

❑ 1. Carefully add 1 μL of loading dye to each reaction tube.

❑ 2. Close the reaction tube tops and mix the contents by tapping the tube bottom on the lab bench, or pulsing in a microfuge. (Make certain the tubes in the microfuge are arranged in positions facing one another [balanced configuration] before pulsing.)

❑ 3. If the lab bench is not black, place a sheet of black paper under the agarose gel.

❑ 4. With the aid of the micropipettor and fresh tips for each reaction mixture load the contents of each tube into separate gel wells in the following manner. Figures C–11*a* and C–11*b* show the hand positions for loading and the appearance of loading a reaction mixture into wells, respectively.

  ❑ a. Begin with the control mixture.
  ❑ b. Draw up the entire mixture.
  ❑ c. Check the micropipettor tip for an air bubble. If air is present, slightly depress the plunger of the micropipettor to move the mixture to the end of the tip.
  ❑ d. Steady the micropipettor with two hands over the well to be used. (Refer to Figure C–11*a*.)
  ❑ e. Dip the pipettor tip below the surface of the buffer, and center it over the well.
  ❑ f. Depress the micropipettor's plunger slowly to expel the mixture into the well. (Sucrose in the loading dye will cause the mixture to sink to the well bottom.) (Refer to Figure C–11*b*.)

❑ 5. Repeat steps 4b through 4f with each of the other reaction mixtures.

❑ 6. After all the reaction mixtures have been loaded, close the electrophoresis box and connect leads to the same channel of a power supply. *Note:* The anode should be connected to anode (red-red), and cathode to cathode (black-black).

❑ 7. Turn the power supply on and set to 100–150 volts. The ampmeter should register 50–100 milliampere.

❑ 8. Electrophorese for 40–60 minutes. An indication of good separation is when the bromophenol blue of the loading dye (bands) has moved 4–8 cm from the wells. (*Do not* electrophorese until the bands run off the end of the gel.)

(a)

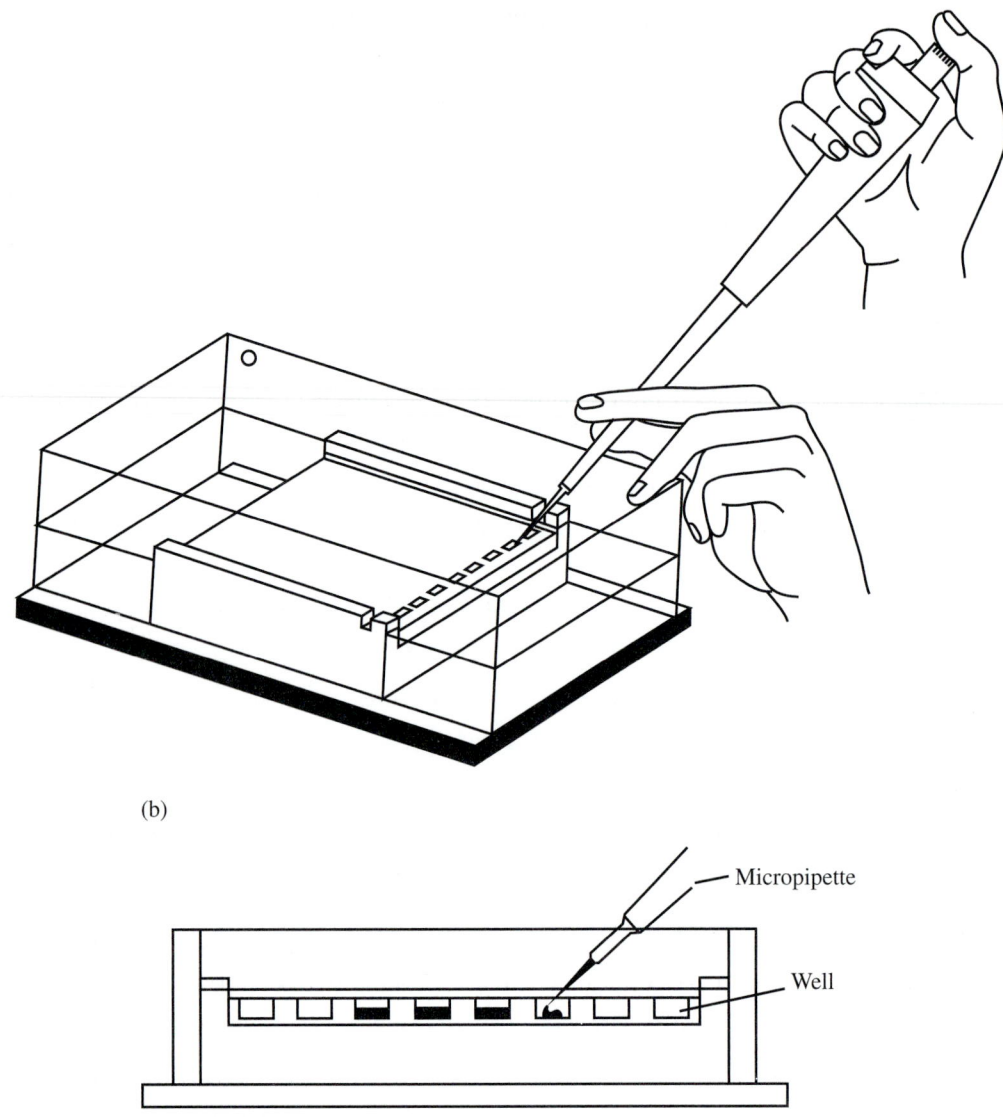

(b)

Micropipette

Well

Figure C–11

Loading the reaction mixture. (a) The hand positions used in loading an agarose gel. Note that one hand is used to deliver the mixture while the other is used to guide the micropipettor tip into proper well. (b) The appearance of the reaction mixture being added to the well. The sucrose in the mixture causes it to sink to the well bottom.

❑   9. After the electrophoresis, turn off the power supply, disconnect the leads, and remove the electrophoresis box top.

❑ 10. Carefully remove the casting tray from the electrophoresis box and slide the agarose gel into a shallow tray.

❑ 11. Label the tray with your initials and continue with one of the two following staining procedures.

Procedure 4: Methylene Blue Gel Staining and Viewing

Rubber gloves should be used for this procedure.

❑   1. Cover the agarose gel with 0.25 percent methylene blue solution.

❑   2. Allow the stain to remain for about 30 minutes.

❑   3. Place your hand gently over the gel to hold it in place and carefully pour the methylene blue solution into a beaker.

❏ 4. Gently rinse the gel in the staining tray with running tap water and allow it to soak for 5 minutes.

❏ 5. Change the water at least 4 times with a soaking period of 5 minutes each time. Distinct DNA bands will begin to appear as the gel destains.

Procedure 5: Ethidium Bromide Gel Viewing (Alternate Method)

(*Note:* Rubber gloves should be worn when staining with ethidium bromide, viewing the agarose gel, and during clean up.)

❏ 1. Cover the agarose gel in the shallow tray with ethidium bromide solution and stain for 10 minutes.

❏ 2. With the aid of a funnel or small beaker, decant as much of the stain as possible from the shallow tray. Return the stain to a storage container for ethidium bromide.

❏ 3. Rinse the stained gel and tray under running tap water or allow the gel to destain in tap water for 5 minutes.

❏ 4. View the gel under an ultraviolet transilluminator. (See Figure C–5.)

❏ 5. Polaroid photos may be taken at this time if appropriate equipment is available.

❏ 6. Compare the results obtained with Figure C–1, sketch a representative view of your gel in the Results and Observations Section.

❏ 7. Answer the questions in the Results and Observations and Laboratory Review sections.

❏ 8. Wash your hands before leaving.

Results and Observations

Manipulation of Micropipettor

1. Were all micropipettor tips filled accurately thus indicating a total volume of $10\,\mu L$? _____

    a. If not which ones showed an inaccurate volume(s)? _____

    b. Explain the finding(s) in 1a. _____

2. List two possible sources of error that might occur with the use of a micropipettor.

    a. _____

    b. _____

DNA Restriction Analysis

1. Sketch a representative view of your agarose gel in the space provided.

2. Was your gel similar to the ideal gel shown in Figure C–1a? _____

   If not, explain why it was not. _____

   _____

Laboratory Review C   DNA Restriction Analysis

---

1. Explain or define the following:

   a. μL = _____

   b. μg = _____

   c. mL = _____

   d. base pair _____

   e. restriction enzyme _____

   _____

2. What is electrophoresis? _____

   _____

3. Of what value are restriction endonucleases? _____

   _____

   _____

Key Terms

anode (ANōd): the positive pole of an electrical source

bacteriophage: a virus thus infects bacteria. Particular forms such as lambda are used as vectors for cloning DNA

Bam HI: the endonuclease obtained from *Bacillus amyliolique faciens*

bp (base pair): a pair of complementary nitrogenous bases in a DNA molecule; also the unit of measurement for DNA sequences

cathode: the negative electrode from which electrons are emitted; opposite of the anode

EcoRI: one type of endonuclease obtained from *Escherichia coli*

endonuclease: a class of enzymes that degrades DNA and/or RNA molecules that cleave bonds linking adjacent nucleotides

gel electrophoresis: the technique used to separate charged molecules in a matrix (gel) to which an electric current is applied

Hind II: an endonuclease obtained from *Haemophilus influenza*

EXPERIMENTAL EXERCISE D

Demonstration of the Polymerase Chain Reaction (PCR)

After completing this exercise, you should be able to:

1. Understand and outline the PCR procedure.

2. Manipulate a micropipettor.

3. Perform gel electrophoresis.

4. Perform both a modified form of PCR and a DNA fingerprint.

5. Define or explain the terms *PCR, polymerase, denaturation, annealing,* and *Taq.*

The development of recombinant DNA technologies has had an enormous impact on molecular biology and biotechnology. One of the more recent breakthroughs, the *polymerase chain reaction (PCR),* devised in the mid-1980s by Kary Mullis, has revolutionized molecular genetics by enabling researchers to produce enormous numbers of copies in a relatively short period of time starting with a single DNA molecule—one billion copies can be made in one hour. The PCR technique makes possible an entire new and highly specific approach to the study and analysis of genes. Its applications include: detection of hidden defective genes, early diagnosis of human immunodeficiency virus (HIV) infection, identifying individuals in cases of questionable paternity and crimes, and research studies involving the identification of DNA in ancient specimens such as mummies, and preserved insects and other forms of life that existed thousands of years ago.

The copying of *amplification* mechanism involved with the PCR technique is shown in Figure D–1 and briefly described in the following section. This exercise will demonstrate the principle of PCR.

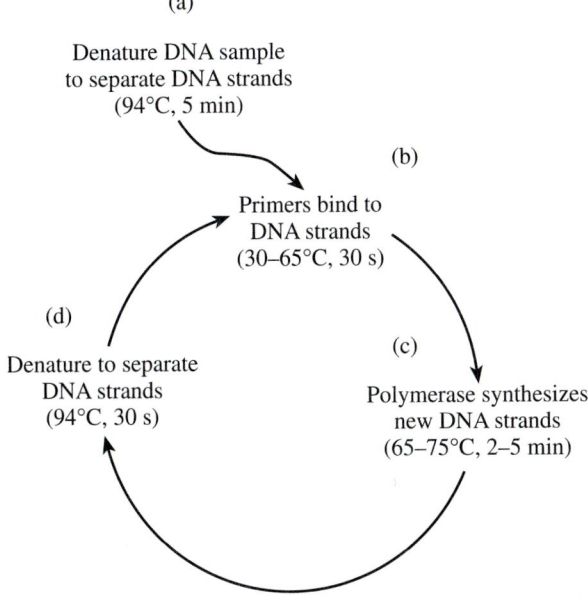

(a)
Denature DNA sample
to separate DNA strands
(94°C, 5 min)

(b)
Primers bind to
DNA strands
(30–65°C, 30 s)

(c)
Polymerase synthesizes
new DNA strands
(65–75°C, 2–5 min)

(d)
Denature to separate
DNA strands
(94°C, 30 s)

Figure D–1

Overall view of the polymerase chain reaction (PCR) cycle. (a) The DNA sample is heated to separate the DNA strands (initial denaturation). The resulting reaction mixture then goes through repeated cycles of primer annealing. (b) Polymerase synthesis of DNA (c), and denaturation (d). The target DNA sequence doubles in concentration for each turn of the cycle. After 32 cycles 1,073,741,824 double-stranded target molecules will form.

PCR is an extremely powerful technique for finding the nucleic acid equivalent of the proverbial needle in the haystack. The copying process begins with the isolation of DNA from a cell or other source and heating it to approximately 100°C. This heat treatment causes the two strands of DNA to separate as hydrogen bonds between adenine (A) and thymine (T), and guanine (G) and cytosine (C), break. (Figure D–2, step 1.)

In the second step of the process, the temperature is lowered and two short pieces of DNA, called **primers,** are added. The primers are complementary to the two ends of the opposite strands of the DNA that are to be copied, termed

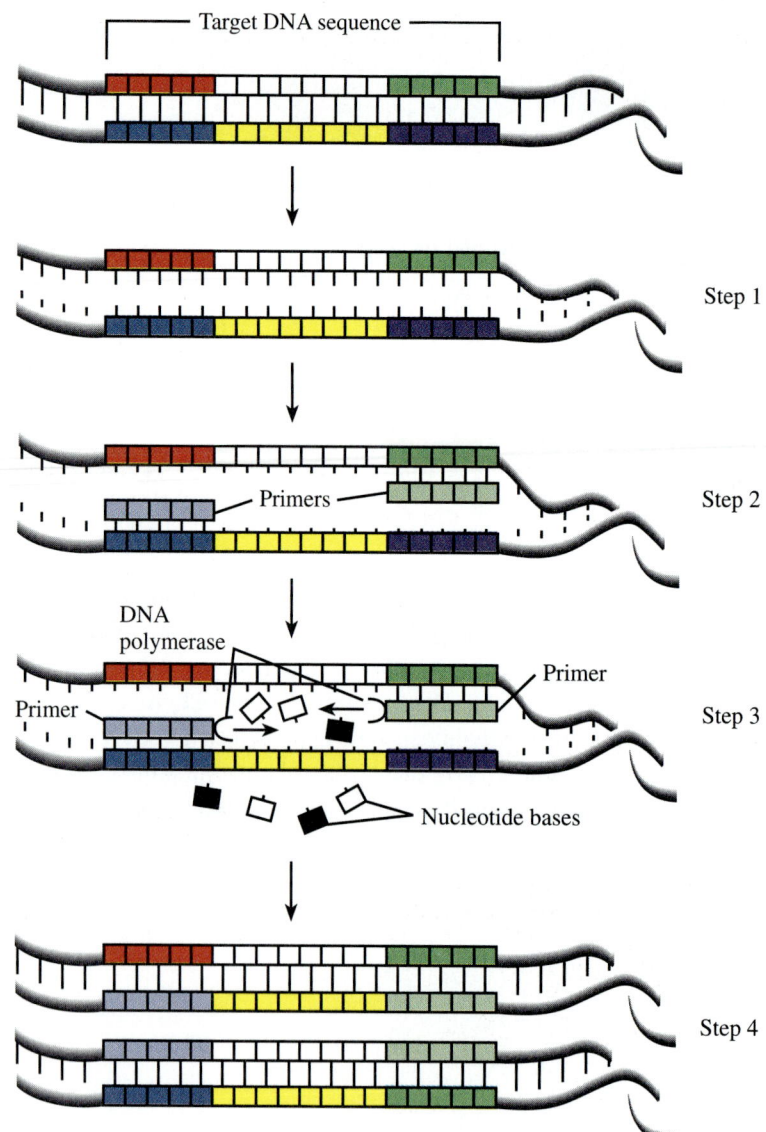

Figure D–2

The molecular events taking place during one round of the polymerase chain reaction. *Step 1:* Separation of the two strands of the target DNA sequence. *Step 2a:* Addition of primers that are complementary to the ends of the target sequence. *Step 2b:* Annealing. The primers bind to the original strands during a cooling process. *Step 3:* Extension of primers by the addition of nucleotides and the action of a heat-resistant polymerase. *Step 4:* Cycle ends, with two copies of the target DNA sequence. The cycle is repeated and results in the production of four copies. Each time the cycle is repeated, the number of copies doubles.

the **target DNA** sequence. The two primers bind by hydrogen bonding to the original, complementary nucleotide sequences when the reaction mixture is cooled (annealing). (See Figure D–2, step 2.)

In the third step, the formation or synthesis of a **complementary strand of new DNA** takes place. Each new strand is an extension of each annealed primer and consists of a string of nucleotides complementary to the opposite strand of DNA, to which the primer is complementary. (See Figure D–2, step 3.) This synthesis requires the use of a thermostable DNA polymerase, which is not destroyed at high temperatures. The *Taq* polymerase, originally isolated from the thermophilic bacterium *Thermus aquaticus,* which grows in hot springs at temperatures of 70–75°C, is commonly used.

After the first amplification cycle, two new strands of DNA will be produced for every two original DNA strands, or a total of four. (Figure D–2, step 4.) The number of strands doubles with each completion of the cycle.

Repeating of the cycle is made easier with the use of a programmable thermal cycler. (Figure D–3.) The thermal cycler, essentially a programmable heating block, is capable of conducting successive heating and cooling cycles unattended, and it eliminates the tedious task of transferring reaction tubes between water baths or heating blocks set at the required temperatures.

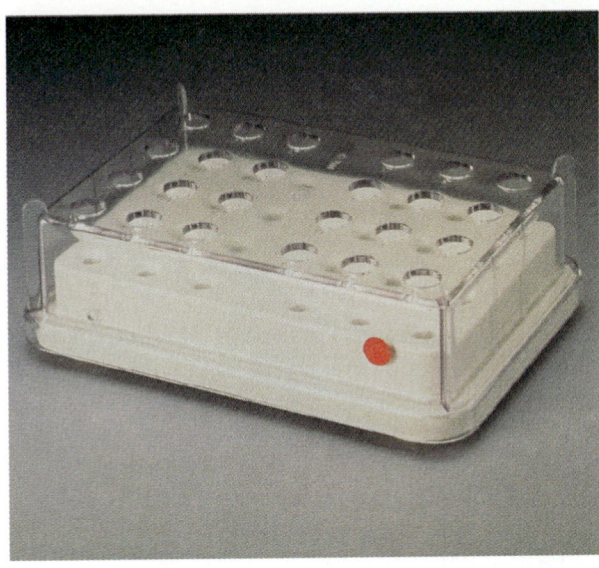

Figure D–3
A thermocycler.
(Courtesy of Brinkmann Instruments, Inc., Westbury, N.Y.)

This exercise will be used to demonstrate the PCR technique, and the influence of the number of cycles on DNA amplification. To perform a PCR reaction, a small quantity of the target DNA will be added to a test tube with a buffered solution containing DNA polymerase, oligonucleotide primers, the four deoxynucleotide (dNTPs) building blocks of DNA, and the cofactor $MgCl_2$.

Following amplification cycles, the PCR products are usually loaded into wells of an agarose gel and electrophoresed. Because PCR amplifications can generate microgram (μg) quantities of product, amplified fragments can be visualized easily following staining with chemical stains such as ethidium bromide or methylene blue (Figure D–4).

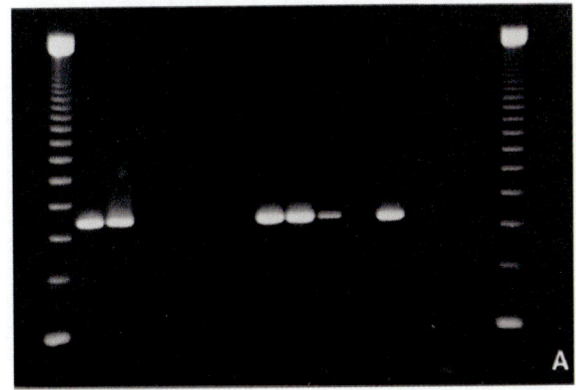

Figure D–4
The results of PCR reactions as shown by an ethidium bromide-stained gel.
(Courtesy P.H. Gumerlock, Y.J. Tang, J.B. Weiss, and J. Silva, Jr., U.C. Davis Medical Center.)

A. Polymerase Chain Reaction (PCR) Amplification

Materials

The following items are to be provided per group of 4 students:

❏ 1. One microcentrifuge tube rack

❏ 2. One micropipettor (0.5–10 μL)

❏ 3. One micropipettor (10–100 μL)

❏ 4. One box micropipettor tips

❏ 5. Five empty 0.5-mL microcentrifuge tubes

❏ 6. Five empty 1.5-mL microcentrifuge tubes

❏ 7. One marking pen

❏ 8. One stopwatch or other timer device

❏ 9. One container for used micropipettor tips and microcentrifuge tubes

❏ 10. One container for disposal of solutions

continued

Materials (continued)

The following items should be provided per groups of 4 students in 0.5-mL microcentrifuge tubes:

❏ 1. 180 μL of 0.05 nanograms/μL Lambda DNA

❏ 2. 100 μL of 25 mM magnesium chloride ($MgCl_2$)

❏ 3. 180 μL of PCR reagent mixture containing:
   ❏ a. 44 μL of 10X PCR buffer (magnesium chloride, potassium chloride, KCl, Trishydrochloride, and gelatin)
   ❏ b. 72 μL of deoxynucleotide triphosphates (dNTPs) 1.25 mM of each of dATP, dCTP, dGTP, and dTTP
   ❏ c. 61.6 μL of PCR primers (4 picograms/μL of each primer)
   ❏ d. 2.4 μL *Thermus aquaticus* (*Taq*) polymerase

The following items are to be provided for general class use:

❏ 1. One water bath with microcentrifuge tube racks set at each of the following temperatures:
   ❏ a. 55°C
   ❏ b. 72°C
   ❏ c. 98°C

❏ 2. Individual wire strands, about 15 cm (6 inches), for microcentrifuge tube handles

❏ 3. Tabletop microcentrifuge

❏ 4. Microcentrifuge tubes

Procedure 1

Additional Technique Required for This Portion of the Exercise:

❏ The Use of a Digital Pipette, Procedure Diagram 37, Exercise 41

This procedure is to be performed by students in groups of four. Your instructor will demonstrate the use of the digital pipettor.

❏ 1. Obtain five 0.5-mL microcentrifuge tubes and the materials to be used in this procedure.

❏ 2. Label the lid of one tube for each PCR reaction system as follows: #1, 0 cycles; #2, 5 cycles; #3, 10 cycles; #4, 15 cycles; and #5, 30 cycles (standard). Place the tubes in a rack.

❏ 3. With the aid of the micropipettor, add the materials indicated in the following matrix to each labeled tube. Use a fresh micropipettor tip for each reagent.

Tube #	Lambda DNA	PCR Reaction Mix	$MgCl_2$
1	40 μL	40 μL	20 μL
2	40 μL	40 μL	20 μL
3	40 μL	40 μL	20 μL
4	40 μL	40 μL	20 μL
5[a]	0	0	0

[a]100 μL of a standard 30 cycle PCR reaction system will be provided for Part C.

❏ 4. Carefully add one drop of mineral oil to PCR reaction tubes #2 through #4. *Do not touch any part of the PCR tube with the dropper or micropipettor tip.*

❏ 5. Close the tops of each tube and mix the reagents by pulsing (centrifuging) in a microfuge, or by sharply tapping the tube buttons on the lab table. The PCR tube can be centrifuged by first placing it inside an empty 1.5-mL reaction tube and then into the microfuge.

❑ 6. Obtain one wire strand for each of the PCR reaction systems to be amplified, and twist one end to fit under the lid as shown in Figure D–5.

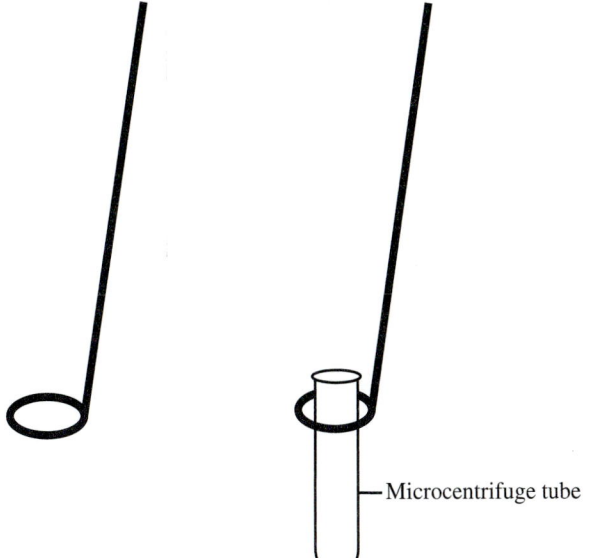

Microcentrifuge tube

Figure D–5
The microcentrifuge holding wire.

❑ 7. Follow the heating cycles indicated in Table D–1 for the tubes specified. Tube #2, 5 cycles; Tube #3, 10 cycles; and Tube #4, 15 cycles. Tubes #1 and #5 are not to be heated. (*Caution:* Do not leave any reaction system in the 98°C for longer than 20 seconds. Prolonged exposure will destroy *Taq* enzyme activity.)

❑ 8. After performing the heating cycles, refrigerate the PCR reaction systems until they are needed for the next procedure.

Table D–1

The Factor of the Number of Cycles

	Group 1			Group 2			Group 3	
Cycle Number	Time (in Seconds)	Temp. (°C)	Cycle Number	Time (in Seconds)	Temp. (°C)	Cycle Number	Time (in Seconds)	Temp. (°C)
1	0–0:20	98 ____	1	0–0:20	98 ____	1	0–0:20	98 ____
	0:20–1:30	55 ____		0:20–1:30	55 ____		0:20–1:30	55 ____
	1:30–3:00	72 ____		1:30–3:00	72 ____		1:30–3:00	72 ____
2	3:00–3:20	98 ____	2	3:00–3:20	98 ____	2	3:00–3:20	98 ____
	3:20–4:30	55 ____		3:20–4:30	55 ____		3:20–4:30	55 ____
	4:30–6:00	72 ____		4:30–6:00	72 ____		4:30–6:00	72 ____
3	6:00–6:20	98 ____	3	6:00–6:20	98 ____	3	6:00–6:20	98 ____
	6:20–7:30	55 ____		6:20–7:30	55 ____		6:20–7:30	55 ____
	7:30–9:00	72 ____		7:30–9:00	72 ____		7:30–9:00	72 ____
4	9:00–9:20	98 ____	4	9:00–9:20	98 ____	4	9:00–9:20	98 ____
	9:20–10:30	55 ____		9:20–10:30	55 ____		9:20–10:30	55 ____
	10:30–12:00	72 ____		10:30–12:00	72 ____		10:30–12:00	72 ____

continued

Table D–1 (continued)

Group 1			Group 2			Group 3		
Cycle Number	Time (in Seconds)	Temp. (°C)	Cycle Number	Time (in Seconds)	Temp. (°C)	Cycle Number	Time (in Seconds)	Temp. (°C)
5	12:00–12:20	98 ⎯⎯	5	12:00–12:20	98 ⎯⎯	5	12:00–12:20	98 ⎯⎯
	12:20–13:30	55 ⎯⎯		12:20–13:30	55 ⎯⎯		12:20–13:30	55 ⎯⎯
	13:30–15:00	72 ⎯⎯		13:30–15:00	72 ⎯⎯		13:30–15:00	72 ⎯⎯
	No further cycles		6	15:00–15:20	98 ⎯⎯	6	15:00–15:20	98 ⎯⎯
				15:20–16:30	55 ⎯⎯		15:20–16:30	55 ⎯⎯
				16:30–18:00	72 ⎯⎯		16:30–18:00	72 ⎯⎯
			7	18:00–18:20	98 ⎯⎯	7	18:00–18:20	98 ⎯⎯
				18:20–19:30	55 ⎯⎯		18:20–19:30	55 ⎯⎯
				19:30–21:00	72 ⎯⎯		19:30–21:00	72 ⎯⎯
			8	21:00–21:20	98 ⎯⎯	8	21:00–21:20	98 ⎯⎯
				21:20–22:30	55 ⎯⎯		21:20–22:30	55 ⎯⎯
				22:30–24:00	72 ⎯⎯		22:30–24:00	72 ⎯⎯
			9	24:00–24:20	98 ⎯⎯	9	24:00–24:20	98 ⎯⎯
				24:20–25:30	55 ⎯⎯		24:20–25:30	55 ⎯⎯
				25:30–27:00	72 ⎯⎯		25:30–27:00	72 ⎯⎯
			10	27:00–27:20	98 ⎯⎯	10	27:00–27:20	98 ⎯⎯
				27:20–28:30	55 ⎯⎯		27:20–28:30	55 ⎯⎯
				28:30–30:00	72 ⎯⎯		28:30–30:00	72 ⎯⎯
				No further cycles		11	30:00–30:20	98 ⎯⎯
							30:20–30:30	55 ⎯⎯
							30:30–31:00	72 ⎯⎯
						12	31:00–31:20	98 ⎯⎯
							31:20–31:30	55 ⎯⎯
							31:30–32:00	72 ⎯⎯
						13	32:00–32:20	98 ⎯⎯
							32:20–32:30	55 ⎯⎯
							32:30–33:00	72 ⎯⎯
						14	33:00–33:20	98 ⎯⎯
							33:20–33:30	55 ⎯⎯
							33:30–34:00	72 ⎯⎯
						15	34:00–34:20	98 ⎯⎯
							34:20–34:30	55 ⎯⎯
							34:30–35:00	72 ⎯⎯

B. Phenol/Chloroform Extraction

Materials

The following items are to be provided per group of 4 students:

- ❏ 1. Student-prepared PCR samples
- ❏ 2. One microcentrifuge tube rack
- ❏ 3. One micropipettor (1–20 μL)
- ❏ 4. One box micropipettor tips
- ❏ 5. Eight empty 1.5-mL microcentrifuge tubes
- ❏ 6. ❏ a. 450 μL chloroform
  - ❏ b. 450 μL phenol/chloroform

- ❏ 7. 450 μL of TE buffer
- ❏ 8. Forty μL 3M sodium acetate (pH 5.2)
- ❏ 9. 1,600 μL ice cold ethanol
- ❏ 10. One marking pen or pencil
- ❏ 11. One container for used micropipettor tips and microcentrifuge tubes
- ❏ 12. One container for disposal of solutions

The following items are to be provided for class use:

- ❏ 1. Rubber gloves
- ❏ 2. Safety glasses
- ❏ 3. Safety hood
- ❏ 4. Safety eye wash
- ❏ 5. Hair dryers

- ❏ 6. Kimwipes
- ❏ 7. Tabletop microcentrifuge
- ❏ 8. Ice in a bucket or other container
- ❏ 9. Paper towels
- ❏ 10. 1.5-mL microcentrifuge tables

Procedure

This procedure is to be performed by students in groups of 4.

- ❏ 1. Use a marking pen to label one 1.5-mL reaction tube for each student-prepared PCR sample (#1, 0 cycles; #2, 5 cycles; #3, 10 cycles; and #4, 15 cycles).

- ❏ 2. With a fresh micropipettor tip, transfer as much PCR sample as possible (80–90 μL) to the respective 1.5-mL tubes. Do not transfer any of the mineral oil in the PCR tube. After withdrawing the aqueous sample, use a Kimwipe to remove any mineral oil clinging to the outside of the pipette tip. Discard the PCR reaction tubes containing the remaining mineral oil as directed by your instructor.

- ❏ 3. Use a fresh tip to add 100-μL TE buffer to each of the PCR tubes. Mix by tapping the tubes with your finger. (The additional volume provided by the TE buffer makes the extraction easier to perform.)

- ❏ 4. Use a fresh tip to add an equal volume (about 100 μL) of phenol/chloroform to the PCR sample. (*Note:* Phenol is corrosive and can cause severe burns. Wear gloves, safety glasses, and protective clothing when working with phenol. Always handle phenol solutions in a chemical hood. If phenol *contacts* skin, rinse with plenty of water and wash with soap and water. Do not use ethanol.)

- ❏ 5. Close each tube cap and mix the respective tube contents by inverting several times to form a homogenous emulsion.

- ❏ 6. Place the tubes together with a balance tube in microcentrifuge and spin for 30 seconds to separate the phases of the samples.

- ❏ 7. Use a fresh tip to remove the organic (bottom) phase to an appropriate waste container. Insert the pipette tip through the water or aqueous (top) phase and position it at the very bottom of one tube. Continue to withdraw the organic phase until a small amount of the aqueous phase enters the pipette tip. Dispose of the tip.

- ❏ 8. Repeat step 7 with each sample.

- ❏ 9. Use a fresh tip to add 100 μL of chloroform to the remaining aqueous phase in each tube.

- ❏ 10. Close tube caps and mix the tube contents by inverting several times to form a homogenous emulsion.

❑ 11. Place the sample tubes, together with a balance tube, in microcentrifuge and spin for 30 seconds to separate the phases.

❑ 12. Use individual fresh tips to remove the chloroform (bottom phase) from each tube and place into an appropriate waste container as directed by your instructor.

❑ 13. Use individual fresh tips to add 10 μL of 3M sodium acetate to each PCR sample. Close the tube caps and mix the respective contents by tapping with a finger.

❑ 14. Use individual fresh tips to add 400 μL (about two volumes) of ice-cold ethanol to each PCR sample. Close the tube caps and mix by rapidly inverting each tube several times.

❑ 15. Incubate the sample tubes on ice for 10 minutes.

❑ 16. Place the sample tubes together with a balance tube in the microcentrifuge and spin for 10 minutes to recover the DNA. Align tubes in centrifuge so that the cap hinges point outward. (Following centrifugation, the nucleic acid residue, visible or not, will collect near the bottom of the tube wall under the hinge.)

❑ 17. Pour off supernatant from each tube. Be careful not to disturb nucleic acid pellet. Invert the tubes, and gently tap them on the surface of a clean paper towel to drain thoroughly.

❑ 18. Dry the nucleic acid pellets by one of the following methods.

    ❑ a. Direct a stream of warm air from a hair dryer into open end of each tube for about three minutes. Be careful not to blow pellet out of tube!

    ❑ b. Close the caps, and pulse the tubes in the microcentrifuge to pool remaining ethanol. Carefully remove drops of ethanol using a 1–10-μL micropipettor. Allow the pellets to air dry at room temperature for 10 minutes.

    All ethanol must be evaporated before proceeding to the next step. Hold each tube up to the light to check that no ethanol droplets remain. If ethanol is still evaporating, an alcohol odor can be detected by smelling the mouth of the individual tubes.

❑ 19. With a fresh tip, add 50 μL of TE buffer to a sample tube. Resuspend the pellet by crushing it with the pipette tip and pipetting in and out vigorously. Rinse down the side of the tube several times, concentrating on the area where the pellet should have formed during centrifugation (beneath cap hinge). Check that all DNA is dissolved and that no particles remain in tip or on side of tube. Dispose of the tip as directed.

❑ 20. Repeat step 19 with the other samples.

❑ 21. Label one 1.5-mL tube for each sample, using the numbering system indicated in step 1.

❑ 22. Use individual fresh tips to transfer 10 μL from each now "extracted" PCR sample to its respectively labeled tube.

C. Electrophoresis of PCR Samples

Materials

The following items should be provided per students in groups of 4:

❑ 1. ❑ a. Ten-μL unamplified DNA sample (control)
    ❑ b. Ten-μL 30 cycled DNA (standard)

❑ 2. Six μL of loading dye

❑ 3. Extracted student-prepared PCR samples

❑ 4. One agarose gel with comb and in casting tray

❑ 5. One electrophoresis box

❑ 6. One 0.5–10 μL micropipettor and appropriate tips

❑ 7. One container for used micropipettor tips

❑ 8. One container with disinfectant

The following items should be provided for general class use:

❑ 1. Electrophoresis power sources

❑ 2. Tris/Borate/EDTA (TBE buffer)

❑ 3. Methylene blue (0.025%) or ethidium bromide (1 μg/mL) solution

❑ 4. Rubber gloves

❑ 5. Safety glasses

❑ 6. Transilluminator

❑ 7. Polaroid camera and film (optional)

Procedure 1: Preparing the Electrophoresis System

Your instructor will demonstrate this procedure.

❏ 1. After the agarose solution has fully set or solidified in its casting tray, carefully remove the masking tape strips from the end of the casting tray.

❏ 2. Place the tray into the electrophoresis box so that the comb is at the negative (*black*) electrode position.

❏ 3. Fill the box with sufficient TBE electrophoresis buffer so that the entire surface of the gel is covered.

❏ 4. Gently rock the comb slightly and remove it from gel without damaging the sample wells that have been formed.

❏ 5. Check to see that the sample wells are completely submerged. If depressions or dry areas are noticed around the wells, simply add additional buffer until they disappear.

Procedure 2: Loading the Gel and Electrophoresis

Your instructor will demonstrate the gel loading procedure.

❏ 1. With a micropipettor, carefully add 1 μL of loading dye to each PCR sample tube.

❏ 2. Close the tube tops and mix the contents by tapping the tube bottom on the lab bench, or pulsing in a microfuge. Make certain the tubes in the microfuge are arranged in positions facing one another (balanced configuration) before pulsing.

❏ 3. If the lab bench is not black, place a sheet of black paper under the agarose gel.

❏ 4. With the aid of the micropipettor and fresh tips for each reaction mixture, load the contents of each tube into separate gel wells in the following manner. Your instructor will show the hand positions for loading and the appearance of loading a reaction mixture into wells respectively.

  ❏ a. Begin with the control mixture.
  ❏ b. Draw up the entire mixture.
  ❏ c. Check the micropipettor tip for an air bubble. If air is present, slightly depress the plunger of the micropipettor to move the mixture to the end of the tip.
  ❏ d. Steady the micropipettor with two hands over the well to be used.
  ❏ e. Dip the pipettor tip below the surface of the buffer, and center it over the well.
  ❏ f. Depress the micropipettor's plunger slowly to expel the mixture into the well. (Sucrose in the loading dye will cause the mixture to sink to the well bottom.)

❏ 5. Repeat steps 4b through 4f with each of the other reaction mixtures.

❏ 6. After all reaction mixtures have been loaded, close the electrophoresis box and connect leads to the same channel of a power supply. *Note:* Anode should be connected to anode (red-red) and cathode to cathode (black-black).

❏ 7. Turn the power supply on and set to 100–150 volts. The ampmeter should register 50–100 milliampere.

❏ 8. Electrophorese for 40–60 minutes. An indication of good separation is when the bromophenol blue of the loading dye (bands) has moved 4–8 cm from the wells. (*Do not* electrophorese until the bands run off the end of the gel.)

❏ 9. After the electrophoresis, turn off the power supply, disconnect the leads, and remove the electrophoresis box top.

❏ 10. Carefully remove the casting tray from the electrophoresis box and slide the agarose gel into a shallow tray.

❏ 11. Label the tray with your initials and continue with one of the two following staining procedures.

Procedure 3: Methylene Blue Gel Staining and Viewing

Rubber gloves should be used for this procedure.

- ❏ 1. Cover the agarose gel with 0.25% methylene blue solution.
- ❏ 2. Allow the stain to remain for about 30 minutes.
- ❏ 3. Place your hand gently over the gel to hold it in place and carefully pour the methylene blue solution into a beaker.
- ❏ 4. Gently rinse the gel in the staining tray with running tap water and allow it to soak for 5 minutes.
- ❏ 5. Change the water at least 4 times with a soaking period of 5 minutes each time. Distinct DNA bands will begin to appear as the gel destains.

Procedure 4: Ethidium Bromide Gel Viewing (Alternate Method)

(*Note:* Rubber gloves should be worn when staining with ethidium bromide, viewing the agarose gel, and during clean up.)

- ❏ 1. Cover the agarose gel in the shallow tray with ethidium bromide solution and stain for 10 minutes.
- ❏ 2. With the aid of a funnel or small beaker, decant as much of the stain as possible from the shallow tray. Return the stain to a storage container for ethidium bromide.
- ❏ 3. Rinse the stained gel and tray under running tap water or allow the gel to destain in tap water for 5 minutes.
- ❏ 4. View the gel under an ultraviolet transilluminator.
- ❏ 5. Polaroid photos may be taken at this time if appropriate equipment is available.
- ❏ 6. Sketch a representative view of your gel in the Results and Observations section.
- ❏ 7. Answer the questions in the Results and Observations and Laboratory Review sections.
- ❏ 8. Wash your hands before leaving.

Results and Observations

1. a. Sketch a representative view of your agarose gel in the space provided.

   b. Indicate the contents of each lane in the space provided.

   Lane 1 ———————
   Lane 2 ———————
   Lane 3 ———————
   Lane 4 ———————
   Lane 5 (control) ———————
   Lane 6 (standard) ———————

2. a. Did the PCR technique work with each assigned number of cycles? ————————————————

   b. If not, why? ————————————————————————————————————

3. What can you state with certainty about the influence of the number of cycles on PCR results? _____

_____

Laboratory Review D

Demonstration of the Polymerase Chain Reaction (PCR)

1. What is the value of PCR? _____

    _____

2. List the main steps in a full cycle of the PCR.

    a. _____

    b. _____

    c. _____

    d. _____

3. List 3 specific applications of PCR.

    a. _____

    b. _____

    c. _____

4. Define or explain:

    a. *Taq* _____

    b. polynucleotide _____

    _____

    c. primer _____

    d. target DNA _____

Key Terms

amplification (am-pli-fi-KĀ-shun): the process by which the number of copies of a DNA sequence are increased

annealing (an-NĒL-ing): the process by which complementary DNA or RNA sequences are paired by means of hydrogen binding to form a double-stranded polynucleotide

nucleotide: a building block of DNA or RNA, consisting of a nitrogenous base, a five-carbon sugar, and a phosphate group

polymerase (pol-IM-er-ās): an enzyme that catalyzes the addition of multiple subunits to a substrate molecule

polymerase chain reaction (PCR): a procedure to enzymatically increase (amplify) any specific DNA fragment of up to approximately 6,000 base pairs through repeated replication by a DNA polymerase

polynucleotide: several nucleotides linked together

primer: short pieces of DNA complementary to the ends of the DNA strands that are to be copied

Taq polymerase: a heat-stable DNA polymerase isolated from the bacterium *Thermus aquaticus*

target DNA: the DNA strands to be copied

Appendix 1   *Answers to the Photograph Quizzes*

Photograph Quiz Number	Page	Answers
1	30	Approximately 1.5 μm
2	76	a. Serrate b. Entire c. Entire
3	76	Flocculent
4	101	a. Rod b. Spiral
5	111	The protozoon, *Vorticella*
6	128–129	a. *Penicillium* b. 1. Sterigma; 2. Condidiophore; 3. Conidium
7	128–129	a. 1. Cap; 2. lamellae; 3. annulus; 4. Stipe or stalk b. Basidiospores c. The mycelium is in the soil
8	136	a. Bacterial colony b. Fungal mycelium c. Fungal mycelium
9	149	a. Control culture was too old (that is, more than 24 hours old) b. Overdecolorization c. Too much exposure to the counterstain
10	150	Gram-positive, staphylococci
11	159	Small acid-fast rods
12	169	Spores are present. The differential spore stain should be used for confirmation.
13	169–170	a. Endospore b. Cortex c. Cell wall d. Bacterial nucleoid
14	176	a. Capsule b. Bacterial cell, rods
15	185	a. Monotrichous b. Peritrichous
16	190	a. Cell wall b. Nucleoid c. Cytoplasmic (cell) membrane d. Mesosome (an artefact formed during specimen preparation) e. Cytoplasm and ribosomes
17	202	Lactose nonfermentation and hydrogen sulfide positive

18	212	The tube on the left (yellow)
19	229	I,+; M,+; VP,−; C (Kosers'), −; C (Simmon's), −
20	240	a. Acid, gas, and litmus reduction b. Litmus reduction, coagulation, and peptonization (whey formation)
21	299	No oligodynamic activity is indicated.
22	441	a. A b. Rh_o [D] positive
23	452	a. 1, 2, 4, and 6 b. None of the wells exhibit this reaction
24	515	a. H_2S producer b. Lactose fermenter
25	554	Arthrospores of *Coccidioides immitis*
26	565	*Entamoeba histolytica*
27	580	*Trypanosoma gambiense*
28	580	*Pediculus humanus* (the body louse)
29	621–622	a. *Ascaris* species (nematode) b. *Trichuris* (nematode) c. *Fasciola hepatica* (trematode)

Index